Dynamics: Theory and Application of Kane's Method

This book is ideal for teaching students in engineering or physics the skills necessary to analyze motions of complex mechanical systems such as spacecraft, robotic manipulators, and articulated scientific instruments. Kane's method, which emerged recently, reduces the labor needed to derive equations of motion and leads to equations that are simpler and more readily solved by computer, in comparison to earlier, classical approaches. Moreover, the method is highly systematic and thus easy to teach. This book is a revision of *Dynamics: Theory and Applications* by T. R. Kane and D. A. Levinson and presents the method for forming equations of motion by constructing generalized active forces and generalized inertia forces. Important additional topics include approaches for dealing with finite rotation, an updated treatment of constraint forces and constraint torques, an extension of Kane's method to deal with a broader class of nonholonomic constraint equations, and other recent advances.

Carlos M. Roithmayr is a senior aerospace engineer in the Systems Analysis and Concepts Directorate at the NASA Langley Research Center in Hampton, Virginia. He earned a Bachelor of Aerospace Engineering degree at the Georgia Institute of Technology, both an M.S. and a Degree of Engineer in Aeronautics and Astronautics from Stanford University, and a Ph.D. in Aerospace Engineering from the Georgia Institute of Technology. He began his career with NASA at the Johnson Space Center in Houston, Texas. His research interests include dynamics of multibody mechanical systems, spacecraft attitude dynamics and control, and orbital mechanics, and he has contributed to a wide variety of Agency projects and missions. He is author or coauthor of numerous refereed journal papers. Dr. Roithmayr is a senior member of the American Institute of Aeronautics and Astronautics.

Dewey H. Hodges is a professor of aerospace engineering at the Georgia Institute of Technology. He holds a B.S. in Aerospace Engineering from the University of Tennessee at Knoxville and both M.S. and Ph.D. degrees in Aeronautics and Astronautics from Stanford University. His research interests include aeroelasticity, structural mechanics, rotorcraft dynamics, finite element analysis, and computational optimal control. He has authored or coauthored five books and more than 200 technical papers in refereed journals. Professor Hodges is a Fellow of the American Helicopter Society, the American Institute of Aeronautics and Astronautics, the American Society of Mechanical Engineers, and the American Academy of Mechanics. He serves on the editorial boards of the *Journal of Fluids and Structures*, the *Journal of Mechanics of Materials and Structures*, and *Nonlinear Dynamics*.

Dynamics: Theory and Application of Kane's Method

Carlos M. Roithmayr
NASA Langley Research Center

Dewey H. Hodges
Georgia Institute of Technology

CAMBRIDGE
UNIVERSITY PRESS

CAMBRIDGE
UNIVERSITY PRESS

University Printing House, Cambridge CB2 8BS, United Kingdom

One Liberty Plaza, 20th Floor, New York, NY 10006, USA

477 Williamstown Road, Port Melbourne, VIC 3207, Australia

4843/24, 2nd Floor, Ansari Road, Daryaganj, Delhi - 110002, India

79 Anson Road, #06-04/06, Singapore 079906

Cambridge University Press is part of the University of Cambridge.

It furthers the University's mission by disseminating knowledge in the pursuit of education, learning and research at the highest international levels of excellence.

www.cambridge.org
Information on this title: www.cambridge.org/9781107005693

First published 2016

A catalogue record for this publication is available from the British Library

Library of Congress Cataloging in Publication data
Names: Roithmayr, Carlos M., author. | Hodges, Dewey H., author.
Title: Dynamics : theory and application of Kane's method / Carlos M. Roithmayr (NASA Langley Research Center), Dewey H. Hodges (Georgia Institute of Technology).
Description: New York, NY : Cambridge University Press, [2015] | 2015 | Includes bibliographical references and index.
Identifiers: LCCN 2014026343 | ISBN 9781107005693 (hardback : alk. paper)
Subjects: LCSH: Motion–Mathematical models. | Constraints (Physics) | Nonholonomic dynamical systems. | Structural dynamics. | Structural optimization.
Classification: LCC QC133 . R65 2015 | DDC 531/.11–dc23
LC record available at http://lccn.loc.gov/2014026343

ISBN 978-1-107-00569-3 Hardback

Contents

PREFACE

The authors of the earlier version of this book succeeded in accomplishing the goals stated in their preface. Since it was written, *Dynamics: Theory and Applications* has served as a textbook for teaching graduate students a method of formulating dynamical equations of motion for mechanical systems. The method has proved especially useful for dealing with the complex multibody mechanical systems that in the twentieth and twenty-first centuries have challenged engineers in industry, government, and universities: the *Galileo* spacecraft sent to Jupiter, the International Space Station, and the robotic manipulator arms aiding astronauts on the Space Shuttle and International Space Station are but a few examples. Kane's method is systematic and easily taught, in a way that enables the student to be conversant with colleagues trained to apply traditional approaches found in the classical literature.

Although the fundamental aspects of the method have not changed during the past three decades, advances and refinements have been made in a number of areas. In certain cases the newer developments facilitate exposition of the topic at hand and lend themselves well to integration with material in the original textbook. The primary purpose of this text, then, is to make the benefits of this progress available for current courses in dynamics.

The preface to the earlier version (which immediately follows this Preface) includes a discussion of the organization of the original book and supporting rationale. Here, we give an overview of the modest alterations made to the earlier structure.

The initial chapter now begins with three brief sections that put the student into position to give a mathematical description of the orientation of a rigid body with respect to a reference frame, when the rigid body has been subjected to successive rotations. Inclusion of these sections provides a formal presentation of topics that typically were covered in classroom discussion. The final section of the first chapter is concerned with differentiation of a scalar function of vectors, which subsequently comes into play in Chapter 6. The original second chapter is divided in two; Chapter 2 deals solely with kinematics, and Chapter 3 is devoted to constraints. The separation focuses attention on the subject of constraints, where there are important distinctions to be made between Kane's method and the classical approaches. The treatment of motion constraints has been broadened. Satisfaction of a constraint entails application of certain forces and torques that are the center of attention in Chapter 6. The practice of expressing constraint equations in terms of vectors, as illustrated in Chapter 3, makes it possible to identify, by inspection, the direction of each constraint force and the point at which it must be applied, as well as the direction of the torque of each constraint force couple, together with the body on which the couple acts. Constraint forces and constraint torques can be identified in this manner if they are of interest in a particular analysis. If, however, they are immaterial, they need not enter the picture at all; indeed, this is a central feature of Kane's method. Thus, Chapter 6 concludes with a discussion of noncontributing forces in two sections that have been relocated from Chapter 5. Extraction of information from equations of motion, formerly covered in the final chapter, is now taken up in Chapter 9. The checking function, introduced in an additional section, can be constructed even when an energy integral does not exist, and is used for the same purpose; namely, to test the results of numerical integrations of equations of motion. The section dealing with momentum integrals has been revised to demonstrate that they can be regarded as nonholonomic constraint equations. Finally, the orientation of a rigid body in a reference frame, addressed at a basic level at the beginning of Chapter 1, receives advanced treatment in Chapter 10. With the exception of two sections dealing with Wiener-Milenković parameters, the material in this chapter is drawn largely from the book *Spacecraft Dynamics*. An understanding of this chapter is especially helpful to the dynamicist who is tackling a problem involving a rigid body (for example, an aircraft or a spacecraft) that is not mechanically attached to the reference frame in question. Nevertheless, the preceding chapters can be mastered without referring to the last one.

A small number of problems have been revised to be consistent with the revisions made to the text. Likewise, problems have been created to cover the newly added material. The significance of a star in connection with a problem, and the importance of solving all unstarred problems, remains unchanged. As before, and for the same pedagogical reasons, results are supplied for all problems.

It is our sincere hope that this updated book will serve as the basis for continued graduate instruction in dynamics so that Kane's method can be applied to the challenging problems that face us now and in the future. We are indebted to the authors of the earlier version for instructing us in their classrooms, and for their generosity in allowing us to make use of their material here.

Carlos M. Roithmayr
Dewey H. Hodges

PREFACE TO *DYNAMICS: THEORY AND APPLICATIONS*

Dissatisfaction with available graduate-level textbooks on the subject of dynamics has been widespread throughout the engineering and physics communities for some years among teachers, students, and employers of university graduates; furthermore, this dissatisfaction is growing at the present time. A major reason for this is that engineering graduates entering industry with advanced degrees, when asked to solve dynamics problems arising in fields such as multibody spacecraft attitude control, robotics, and design of complex mechanical devices, find that their education in dynamics, based on the textbooks currently in print, has not equipped them adequately to perform the tasks confronting them. Similarly, physics graduates often discover that, in their education, so much emphasis was placed on preparation for the study of quantum mechanics, and the subject of rigid body dynamics was slighted to such an extent, that they are handicapped, both in industry and in academic research, by their inability to design certain types of experimental equipment, such as a particle detector that is to be mounted on a planetary satellite. In this connection, the ability to analyze the effects of detector scanning motions on the attitude motion of the satellite is just as important as knowledge of the physics of the detection process itself. Moreover, the graduates in question often are totally unaware of the deficiencies in their dynamics education. How did this state of affairs come into being, and is there a remedy?

For the most part, traditional dynamics texts deal with the exposition of eighteenth-century methods and their application to physically simple systems, such as the spinning top with a fixed point, the double pendulum, and so forth. The reason for this is that, prior to the advent of computers, one was justified in demanding no more of students than the ability to formulate equations of motion for such simple systems, for one could not hope to extract useful information from the equations governing the motions of more complex systems. Indeed, considerable ingenuity and a rather extensive knowledge of mathematics were required to analyze even simple systems. Not surprisingly, therefore, ever more attention came to be focused on analytical intricacies of the *mathematics* of dynamics, while the process of formulating equations of motion came to be regarded as a rather routine matter. Now that computers enable one to extract highly valuable information from large sets of complicated equations of motion, all this has changed. In fact, the inability to *formulate* equations of motion effectively can be as great a hindrance at present as the inability to *solve* equations was formerly. It follows that the subject of formulation of equations of motion demands careful reconsideration. Or, to say it another way, a major goal of a modern dynamics course must be to produce students who are proficient in the use of the best available methodology for formulating equations of motion. How can this goal be attained?

In the 1970s, when extensive dynamical studies of multibody spacecraft, robotic devices, and complex scientific equipment were first undertaken, it became apparent that straightforward use of classical methods, such as those of Newton, Lagrange, and Hamilton, could entail the expenditure of very large, and at times even prohibitive, amounts of analysts' labor, and could lead to equations of motion so unwieldy as to render computer solutions unacceptably slow for technical and/or economic reasons. Now, while it may be impossible to overcome this difficulty entirely, which is to say that it is unlikely that a way will be found to reduce formulating equations of motion for complex systems to a truly simple task, there does exist a method that is superior to the classical ones in that its use leads to major savings in labor, as well as to simpler equations. Moreover, being highly systematic, this method is easy to teach. Focusing attention on motions, rather than on configurations, it affords the analyst maximum physical insight. Not involving variations, such as those encountered in connection with virtual work, it can be presented at a relatively elementary mathematical level. Furthermore, it enables one to deal directly with nonholonomic systems without having to introduce and subsequently eliminate Lagrange multipliers. It follows that the resolution of the dilemma before us is to instruct students in the use of this method (which is often referred to as Kane's method). This book is intended as the basis for such instruction.

Textbooks can differ from each other not only in content but also in organization, and the sequence in which topics are presented can have a significant effect on the relative ease of teaching and learning the subject. The rationale underlying the organization of the present book is the following. We view dynamics as a deductive discipline, knowledge of which enables one to describe in quantitative and qualitative terms how mechanical systems move when acted upon by given forces, or to determine what forces must be applied to a system in order to cause it to move in a specified manner. The solution of a dynamics problem is carried out in two major steps, the first being the formulation of equations of motion, and the second the extraction of information from these equations. Since the second step cannot be taken fruitfully until the first has been completed, it is imperative that the distinction between the two be kept clearly in mind. In this book, the extraction of information from equations of motion is deferred formally to the last chapter, while the preceding chapters deal with the material one needs to master in order to be able to arrive at valid equations of motion.

Diverse concepts come into play in the process of constructing equations of motion. Here again it is important to separate ideas from each other distinctly. Major attention must be devoted to kinematics, mass distribution considerations, and force concepts. Accordingly, we treat each of these topics in its own right. First, however, since differentiation of vectors plays a key role in dynamics, we devote the initial chapter of the book to this topic. Here we stress the fact that differentiation of a vector with respect to a scalar variable requires specification of a reference frame, in which connection we dispense with the use of limits because such use tends to confuse rather than clarify matters; but we draw directly on students' knowledge of scalar calculus. Thereafter, we devote one chapter each to the topics of kinematics, mass distribution, and generalized forces, before discussing energy functions, in Chapter 5, and the formulation of equations of motion, in Chapter 6. Finally, the extraction of information from equations of motion is considered in Chapter 7. The material in these seven chapters has formed the basis for a one-year course for first-year graduate students at Stanford University for more than 20 years.

Dynamics is a discipline that cannot be mastered without extensive practice. Accordingly, the book contains 14 sets of problems intended to be solved by users of the book. To learn the material presented in the text, the reader should solve *all* of the *unstarred* problems, each of which covers some material not covered by any other. In their totality, the unstarred problems provide complete coverage of the theory set forth in the book. By solving also the starred problems, which are not necessarily more difficult than the unstarred ones, one can gain additional insights. Results are given for *all* problems, so that the correcting of problem solutions needs to be undertaken only when a student is unable to reach a given result. It is important, however, that both students and instructors expend whatever effort is required to make certain that students know what the *point* of each problem is, not only how to solve it. Classroom discussion of selected problems is most helpful in this regard.

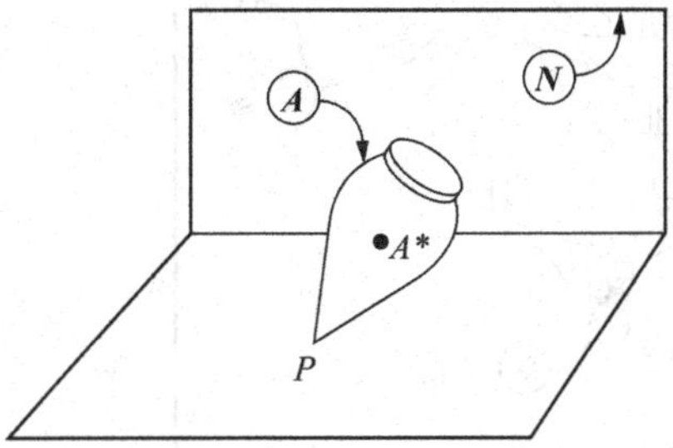

Figure i

Finally, a few words about notation will be helpful. Suppose that one is dealing with a simple system, such as the top A, shown in Fig. *i*, the top terminating in a point P that is fixed in a Newtonian reference frame N. The notation needed here certainly can be simple. For instance, one can let $\boldsymbol{\omega}$ denote the angular velocity of A in N, and let $\mathbf{v}$ stand for the velocity in N of point $A^{\star}$, the mass center of A. Indeed, notations more elaborate than these can be regarded as objectionable because they burden the analyst with unnecessary writing. But suppose that one must undertake the analysis of motions of a complex system, such as the Galileo spacecraft, modeled as consisting of eight rigid bodies $A, B, \ldots, H$, coupled to each other as indicated in Fig. *ii*. Here, unless one employs notations more elaborate than $\boldsymbol{\omega}$ and $\mathbf{v}$, one cannot distinguish from each other such quantities as, say, the angular velocity of A in a Newtonian reference frame N, the angular velocity of B in N, and the angular velocity of B in A, all of which may enter the analysis. Or, if $A^{\star}$ and $B^{\star}$ are points of interest fixed on A and B, perhaps the respective mass centers, one needs a notation that permits one to distinguish from each other, say, the velocity of $A^{\star}$ in N, the velocity of $B^{\star}$ in N, and the velocity of $B^{\star}$ in A. Therefore, we establish, and use consistently throughout this book, a few notational practices that work well in such situations. In particular, when a vector denoting an angular velocity or an angular acceleration of a rigid body in a certain reference frame has two superscripts, the right superscript stands for the rigid body, whereas the left superscript refers to the reference frame. Incidentally, we use the terms "reference frame" and "rigid body" interchangeably. That is, every rigid body can serve as a reference frame, and every reference frame can be regarded as a massless rigid body. Thus, for example, the three angular velocities mentioned in connection with the system depicted in Fig. *ii*, namely, the angular velocity of A in N, the angular velocity of B in N, and the angular velocity of B in A, are denoted by ${}^N\boldsymbol{\omega}^A$, ${}^N\boldsymbol{\omega}^B$, and ${}^A\boldsymbol{\omega}^B$, respectively. Similarly, the right superscript on a vector denoting a velocity or acceleration of a point in a reference frame is the name of the point, whereas the left superscript identifies the reference frame. Thus, for example, the aforementioned velocity of $A^\star$ in N is written ${}^N\mathbf{v}^{A^\star}$, and ${}^A\mathbf{v}^{B^\star}$ represents the velocity of $B^\star$ in A. Similar conventions are established in connection with angular momenta, kinetic energies, and so forth.

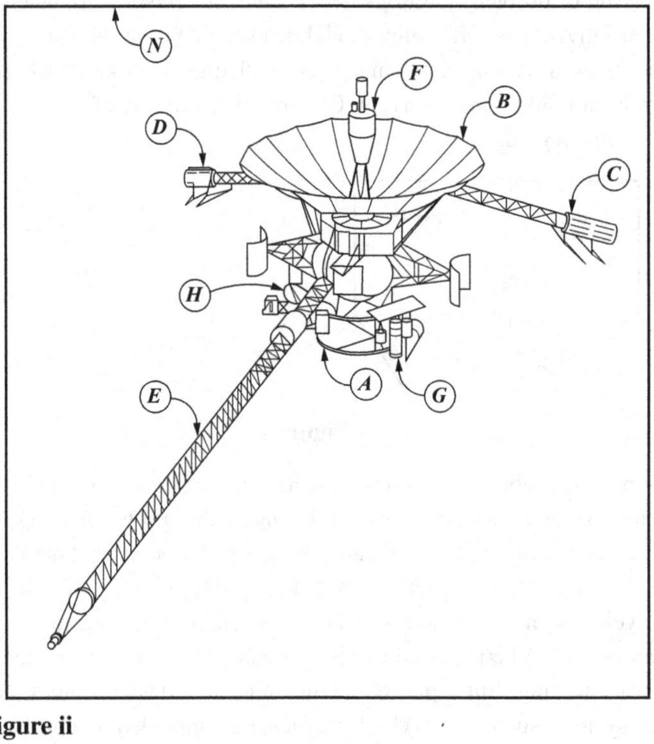

Figure ii

While there are distinct differences between our approach to dynamics, on the one hand, and traditional approaches, on the other hand, there is no fundamental conflict between the new and the old. On the contrary, the material in this book is entirely compatible with the classical literature. Thus, it is the purpose of this book not only to equip students with the skills they need to deal effectively with present-day dynamics problems, but also to bring them into position to interact smoothly with those trained more conventionally.

Thomas R. Kane
David A. Levinson

TO THE READER

Each of the ten chapters of this book is divided into sections. A section is identified by two numbers separated by a decimal point, the first number referring to the chapter in which the section appears, and the second identifying the section within the chapter. Thus, the identifier 3.6 refers to the sixth section of the third chapter. A section identifier appears at the *top of each page*.

Equations are numbered serially within sections. For example, the equations in Secs. 3.6 and 3.9 are numbered (1)–(31) and (1)–(50), respectively. References to an equation may be made both within the section in which the equation appears and in other sections. In the first case, the equation number is cited as a single number; in the second case, the section number is included as part of a three-number designation. Thus, within Sec. 3.6, Eq. (2) of Sec. 3.6 is referred to as Eq. (2); in Sec. 3.9, the same equation is referred to as Eq. (3.6.2). To locate an equation cited in this manner, one may make use of the section identifiers appearing at the tops of pages.

Figures appearing in the chapters are numbered so as to identify the sections in which the figures appear. For example, the two figures in Sec. 5.7 are designated Fig. 5.7.1 and Fig. 5.7.2. To avoid confusing these figures with those in the problem sets and in Appendix III, the figure number is preceded by the letter P in the case of problem set figures, and by the letter A in the case of Appendix III figures. The double number following the letter P refers to the problem statement in which the figure is introduced. For example, Fig. P13.3 is introduced in Problem 13.3. Similarly, Table 4.4.1 is the designation for a table in Sec. 4.4, and Table P13.19(*b*) is associated with Problem 13.19.

1 DIFFERENTIATION OF VECTORS

The discipline of dynamics deals with changes of various kinds, such as changes in the position of a particle in a reference frame and changes in the configuration of a mechanical system. To characterize the manner in which some of these changes take place, one employs the differential calculus of vectors, a subject that can be regarded as an extension of material usually taught under the heading of the differential calculus of scalar functions. The extension consists primarily of provisions made to accommodate the fact that *reference frames* play a central role in connection with many of the vectors of interest in dynamics. A reference frame can be regarded as a massless rigid body, and a rigid body can serve as a reference frame. (A reference frame should not be confused with a coordinate system. Many coordinate systems can be embedded in a given reference frame.) The importance of reference frames in connection with change in a vector can be illustrated by considering the following example. Let A and B be reference frames moving relative to each other, but having one point O in common at all times, and let P be a point fixed in A, distinct from O and thus moving in B. Then the velocity of P in A is equal to zero, whereas the velocity of P in B differs from zero. Now, each of these velocities is a time derivative of the same vector, $\mathbf{r}^{OP}$, the position vector from O to P. Hence, it is meaningless to speak simply of *the* time derivative of $\mathbf{r}^{OP}$. Clearly, therefore, the calculus used to differentiate vectors must permit one to distinguish between differentiation with respect to a scalar variable in a reference frame A and differentiation with respect to the same variable in a reference frame B.

When working with elementary principles of dynamics, such as Newton's second law or the angular momentum principle, one needs only the ordinary differential calculus of vectors, that is, a theory involving differentiations of vectors with respect to a single scalar variable, generally the time. Consideration of advanced principles of dynamics, such as those presented in later chapters of this book, necessitates, in addition, partial differentiation of vectors with respect to several scalar variables, such as generalized coordinates and motion variables. Accordingly, the present chapter is devoted to the exposition of definitions, and consequences of these definitions, needed in the chapters that follow.

1.1 SIMPLE ROTATION

Let $\hat{\mathbf{a}}_1$, $\hat{\mathbf{a}}_2$, and $\hat{\mathbf{a}}_3$ be a set of right-handed, mutually perpendicular unit vectors fixed in a reference frame A, and let $\hat{\mathbf{b}}_1$, $\hat{\mathbf{b}}_2$, and $\hat{\mathbf{b}}_3$ be a similar set of unit vectors fixed in a reference frame B. Suppose that each unit vector $\hat{\mathbf{b}}_i$ initially has the same direction as $\hat{\mathbf{a}}_i$ $(i = 1,2,3)$. B is said to undergo a *simple rotation* relative to A when B is rotated about a line whose orientation relative to A and to B does not change as a result of the rotation. After a particular unit vector parallel to the line of rotation is selected, the angle of rotation q is regarded as positive when a right-handed screw fixed in B, with its axis parallel to the line, advances in the direction of the selected unit vector. Figure 1.1.1 illustrates simple rotation of B in A about $\hat{\mathbf{b}}_1$.

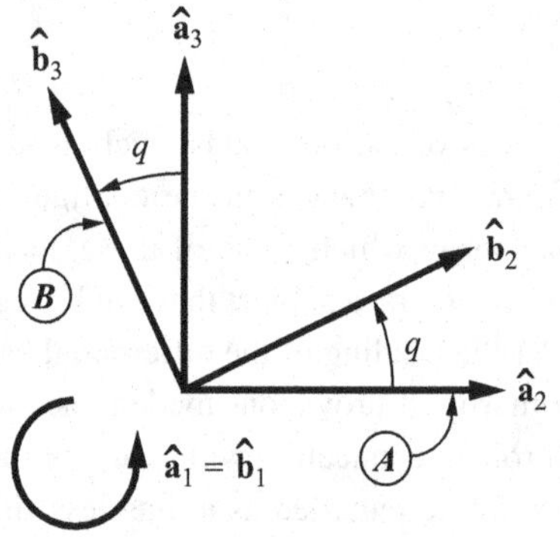

Figure 1.1.1

A simple rotation of B in A can also be performed about $\hat{\mathbf{b}}_2$, or about $\hat{\mathbf{b}}_3$. In fact, simple rotation can be performed about a unit vector whose direction in B is arbitrary, as discussed in Sec. 10.1.

1.2 DIRECTION COSINE MATRIX

Any vector $\mathbf{v}$ can be expressed in terms of the set of unit vectors $\hat{\mathbf{a}}_1$, $\hat{\mathbf{a}}_2$, $\hat{\mathbf{a}}_3$ introduced in Sec. 1.1, or in terms of the set $\hat{\mathbf{b}}_1$, $\hat{\mathbf{b}}_2$, $\hat{\mathbf{b}}_3$. (Expressions for $\mathbf{v}$ need not be limited to those involving only one set of unit vectors; at times it can be advantageous to work with unit vectors fixed in two or more reference frames.) As will soon become apparent, the study of dynamics frequently involves transforming an expression for $\mathbf{v}$ in terms of $\hat{\mathbf{a}}_1$, $\hat{\mathbf{a}}_2$, $\hat{\mathbf{a}}_3$ to one involving $\hat{\mathbf{b}}_1$, $\hat{\mathbf{b}}_2$, $\hat{\mathbf{b}}_3$, or vice versa, and this is greatly facilitated when one has in hand the nine *direction cosines* $\hat{\mathbf{a}}_i \cdot \hat{\mathbf{b}}_j$ $(i, j = 1,2,3)$.

Direction cosines for unit vectors $\hat{\mathbf{a}}_1$, $\hat{\mathbf{a}}_2$, $\hat{\mathbf{a}}_3$ and $\hat{\mathbf{b}}_1$, $\hat{\mathbf{b}}_2$, $\hat{\mathbf{b}}_3$ can be arranged advantageously in a table. For example, Table 1.2.1 contains the direction cosines obtained by examining Fig. 1.1.1,

Table 1.2.1

	$\hat{\mathbf{b}}_1$	$\hat{\mathbf{b}}_2$	$\hat{\mathbf{b}}_3$
$\hat{\mathbf{a}}_1$	1	0	0
$\hat{\mathbf{a}}_2$	0	$\cos q$	$-\sin q$
$\hat{\mathbf{a}}_3$	0	$\sin q$	$\cos q$

where, for instance, the entry in the row containing $\hat{\mathbf{a}}_2$ and in the column containing $\hat{\mathbf{b}}_3$ signifies that

$$\hat{\mathbf{a}}_2 \cdot \hat{\mathbf{b}}_3 = -\sin q \tag{1}$$

Moreover, one may refer to, say, the row containing $\hat{\mathbf{a}}_2$ and write

$$\hat{\mathbf{a}}_2 = \cos q \hat{\mathbf{b}}_2 - \sin q \hat{\mathbf{b}}_3 \tag{2}$$

or, for example, to the last column of the table and state

$$\hat{\mathbf{b}}_3 = -\sin q \hat{\mathbf{a}}_2 + \cos q \hat{\mathbf{a}}_3 \tag{3}$$

A square matrix containing the direction cosines for $\hat{\mathbf{a}}_1$, $\hat{\mathbf{a}}_2$, $\hat{\mathbf{a}}_3$ and $\hat{\mathbf{b}}_1$, $\hat{\mathbf{b}}_2$, $\hat{\mathbf{b}}_3$ is referred to as a *direction cosine matrix*, ${}^AC^B$, whose elements are defined as

$${}^AC^B{}_{ij} \triangleq \hat{\mathbf{a}}_i \cdot \hat{\mathbf{b}}_j \qquad (i,j = 1,2,3) \tag{4}$$

Additional material regarding direction cosines and the direction cosine matrix is provided in Sec. 10.2.

1.3 SUCCESSIVE ROTATIONS

When an analysis involves two sets of unit vectors $\hat{\mathbf{a}}_1$, $\hat{\mathbf{a}}_2$, $\hat{\mathbf{a}}_3$ and $\hat{\mathbf{b}}_1$, $\hat{\mathbf{b}}_2$, $\hat{\mathbf{b}}_3$ such as those introduced in Sec. 1.1, as well as a third similar set $\hat{\mathbf{c}}_1$, $\hat{\mathbf{c}}_2$, $\hat{\mathbf{c}}_3$ fixed in a reference frame C, the direction cosine matrix ${}^AC^C$ whose elements are $\hat{\mathbf{a}}_i \cdot \hat{\mathbf{c}}_j$ $(i,j = 1,2,3)$ can be expressed as the matrix product

$${}^AC^C = {}^AC^B\,{}^BC^C \tag{1}$$

where ${}^AC^B$ is the direction cosine matrix whose elements are $\hat{\mathbf{a}}_i \cdot \hat{\mathbf{b}}_j$, and where ${}^BC^C$ is the direction cosine matrix whose elements are $\hat{\mathbf{b}}_i \cdot \hat{\mathbf{c}}_j$ $(i,j = 1,2,3)$.

Derivation The element of the direction cosine matrix ${}^AC^C$ in the ith row and j^{th} column is

$${}^AC^C{}_{ij} \triangleq \hat{\mathbf{a}}_i \cdot \hat{\mathbf{c}}_j \qquad (i,j = 1,2,3) \tag{2}$$

Now, $\hat{\mathbf{a}}_i$ can be expressed in terms of $\hat{\mathbf{b}}_1, \hat{\mathbf{b}}_2, \hat{\mathbf{b}}_3$ as[†]

$$\begin{aligned}\hat{\mathbf{a}}_i &= (\hat{\mathbf{a}}_i \cdot \hat{\mathbf{b}}_1)\hat{\mathbf{b}}_1 + (\hat{\mathbf{a}}_i \cdot \hat{\mathbf{b}}_2)\hat{\mathbf{b}}_2 + (\hat{\mathbf{a}}_i \cdot \hat{\mathbf{b}}_3)\hat{\mathbf{b}}_3 \\ &\underset{(1.2.4)}{=} {}^{A}C^{B}{}_{i1}\,\hat{\mathbf{b}}_1 + {}^{A}C^{B}{}_{i2}\,\hat{\mathbf{b}}_2 + {}^{A}C^{B}{}_{i3}\,\hat{\mathbf{b}}_3 \qquad (i = 1,2,3) \qquad (3)\end{aligned}$$

so that the direction cosine of interest is given by

$$\begin{aligned}{}^{A}C^{C}{}_{ij} &\underset{(2)}{=} \hat{\mathbf{a}}_i \cdot \hat{\mathbf{c}}_j \\ &\underset{(3)}{=} {}^{A}C^{B}{}_{i1}\,\hat{\mathbf{b}}_1 \cdot \hat{\mathbf{c}}_j + {}^{A}C^{B}{}_{i2}\,\hat{\mathbf{b}}_2 \cdot \hat{\mathbf{c}}_j + {}^{A}C^{B}{}_{i3}\,\hat{\mathbf{b}}_3 \cdot \hat{\mathbf{c}}_j \\ &\underset{(1.2.4)}{=} {}^{A}C^{B}{}_{i1}\,{}^{B}C^{C}{}_{1j} + {}^{A}C^{B}{}_{i2}\,{}^{B}C^{C}{}_{2j} + {}^{A}C^{B}{}_{i3}\,{}^{B}C^{C}{}_{3j} \qquad (i,j = 1,2,3) \qquad (4)\end{aligned}$$

which is recognized as the matrix product of the ith row of ${}^{A}C^{B}$ with the jth column of ${}^{B}C^{C}$. Because Eqs. (4) are applicable to every element of ${}^{A}C^{C}$, they establish the validity of Eq. (1).

Example Four rectangular parallelepipeds, A, B, C, and D, are arranged as shown

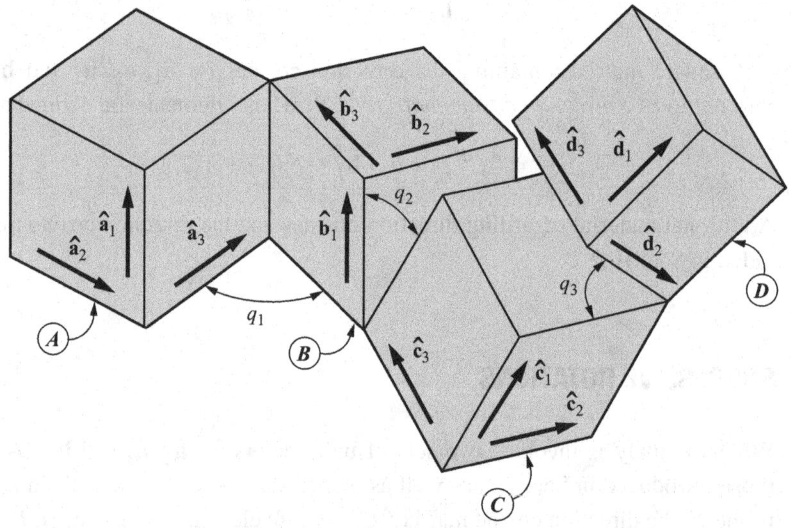

Figure 1.3.1

in Fig. 1.3.1, where $\hat{\mathbf{a}}_1, \hat{\mathbf{a}}_2, \hat{\mathbf{a}}_3$ are unit vectors parallel to edges of A; $\hat{\mathbf{b}}_1, \hat{\mathbf{b}}_2, \hat{\mathbf{b}}_3$ are unit vectors parallel to edges of B; and so forth; while q_1, q_2, and q_3 are the radian measures of angles that determine the relative orientations of $A, \ldots, D$. The configuration shown is one for which q_1, q_2, and q_3 are positive.

[†] Numbers beneath signs of equality or beneath other symbols refer to equations numbered correspondingly. For example, (2) refers to Eq. (2) in the present section. When it is necessary to refer to an equation from an earlier section, the section number is cited together with the equation number. For example, (1.2.4) refers to Eq. (4) in Sec. 1.2.

The direction cosine matrix ${}^{A}C^{D}$ whose elements are $\hat{\mathbf{a}}_i \cdot \hat{\mathbf{d}}_j$ $(i,j = 1,2,3)$ is to be obtained by appealing twice to Eq. (1), after forming three direction cosine matrices that, respectively, relate unit vectors $\hat{\mathbf{a}}_1$, $\hat{\mathbf{a}}_2$, $\hat{\mathbf{a}}_3$ to $\hat{\mathbf{b}}_1$, $\hat{\mathbf{b}}_2$, $\hat{\mathbf{b}}_3$; $\hat{\mathbf{b}}_1$, $\hat{\mathbf{b}}_2$, $\hat{\mathbf{b}}_3$ to $\hat{\mathbf{c}}_1$, $\hat{\mathbf{c}}_2$, $\hat{\mathbf{c}}_3$; and $\hat{\mathbf{c}}_1$, $\hat{\mathbf{c}}_2$, $\hat{\mathbf{c}}_3$ to $\hat{\mathbf{d}}_1$, $\hat{\mathbf{d}}_2$, $\hat{\mathbf{d}}_3$.

A simple rotation about $\hat{\mathbf{b}}_1$ through an angle q_1 separates unit vectors $\hat{\mathbf{b}}_1$, $\hat{\mathbf{b}}_2$, $\hat{\mathbf{b}}_3$ from $\hat{\mathbf{a}}_1$, $\hat{\mathbf{a}}_2$, $\hat{\mathbf{a}}_3$; therefore, one may construct Table 1.3.1, similar to Table 1.2.1, where c_1 and s_1 are abbreviations for $\cos q_1$ and $\sin q_1$, respectively.

Table 1.3.1

	$\hat{\mathbf{b}}_1$	$\hat{\mathbf{b}}_2$	$\hat{\mathbf{b}}_3$
$\hat{\mathbf{a}}_1$	1	0	0
$\hat{\mathbf{a}}_2$	0	c_1	$-\mathrm{s}_1$
$\hat{\mathbf{a}}_3$	0	s_1	c_1

Similarly, two more tables may be constructed, Table 1.3.2 and Table 1.3.3.

Table 1.3.2

	$\hat{\mathbf{c}}_1$	$\hat{\mathbf{c}}_2$	$\hat{\mathbf{c}}_3$
$\hat{\mathbf{b}}_1$	c_2	0	s_2
$\hat{\mathbf{b}}_2$	0	1	0
$\hat{\mathbf{b}}_3$	$-\mathrm{s}_2$	0	c_2

Table 1.3.3

	$\hat{\mathbf{d}}_1$	$\hat{\mathbf{d}}_2$	$\hat{\mathbf{d}}_3$
$\hat{\mathbf{c}}_1$	c_3	$-\mathrm{s}_3$	0
$\hat{\mathbf{c}}_2$	s_3	c_3	0
$\hat{\mathbf{c}}_3$	0	0	1

Equation (1) can be used first to obtain

$$
{}^{A}C^{C} \underset{(1)}{=} {}^{A}C^{B}\,{}^{B}C^{C} = \begin{bmatrix} 1 & 0 & 0 \\ 0 & \mathrm{c}_1 & -\mathrm{s}_1 \\ 0 & \mathrm{s}_1 & \mathrm{c}_1 \end{bmatrix} \begin{bmatrix} \mathrm{c}_2 & 0 & \mathrm{s}_2 \\ 0 & 1 & 0 \\ -\mathrm{s}_2 & 0 & \mathrm{c}_2 \end{bmatrix}
= \begin{bmatrix} \mathrm{c}_2 & 0 & \mathrm{s}_2 \\ \mathrm{s}_1\mathrm{s}_2 & \mathrm{c}_1 & -\mathrm{s}_1\mathrm{c}_2 \\ -\mathrm{c}_1\mathrm{s}_2 & \mathrm{s}_1 & \mathrm{c}_1\mathrm{c}_2 \end{bmatrix} \tag{5}
$$

A second appeal to Eq. (1) now can be made, first letting D play the role of C, and then allowing C to play the role of B,

$$
{}^{A}C^{D} \underset{(1)}{=} {}^{A}C^{C}\,{}^{C}C^{D} \underset{(5)}{=} \begin{bmatrix} \mathrm{c}_2 & 0 & \mathrm{s}_2 \\ \mathrm{s}_1\mathrm{s}_2 & \mathrm{c}_1 & -\mathrm{s}_1\mathrm{c}_2 \\ -\mathrm{c}_1\mathrm{s}_2 & \mathrm{s}_1 & \mathrm{c}_1\mathrm{c}_2 \end{bmatrix} \begin{bmatrix} \mathrm{c}_3 & -\mathrm{s}_3 & 0 \\ \mathrm{s}_3 & \mathrm{c}_3 & 0 \\ 0 & 0 & 1 \end{bmatrix}
= \begin{bmatrix} \mathrm{c}_2\mathrm{c}_3 & -\mathrm{c}_2\mathrm{s}_3 & \mathrm{s}_2 \\ \mathrm{s}_1\mathrm{s}_2\mathrm{c}_3 + \mathrm{s}_3\mathrm{c}_1 & -\mathrm{s}_1\mathrm{s}_2\mathrm{s}_3 + \mathrm{c}_3\mathrm{c}_1 & -\mathrm{s}_1\mathrm{c}_2 \\ -\mathrm{c}_1\mathrm{s}_2\mathrm{c}_3 + \mathrm{s}_3\mathrm{s}_1 & \mathrm{c}_1\mathrm{s}_2\mathrm{s}_3 + \mathrm{c}_3\mathrm{s}_1 & \mathrm{c}_1\mathrm{c}_2 \end{bmatrix} \tag{6}
$$

With this result in hand, one can immediately write, for example,

$$^{A}C^{D}{}_{21} = \hat{\mathbf{a}}_2 \cdot \hat{\mathbf{d}}_1 = s_1 s_2 c_3 + s_3 c_1 \tag{7}$$

Before one leaves this example, it is worth noting that q_1, q_2, and q_3 are known as *orientation angles*, and more will be said regarding such angles in Sec. 10.3. Here, the rotations through angles q_1, q_2, and q_3 are performed, respectively, about unit vectors $\hat{\mathbf{b}}_1$, $\hat{\mathbf{c}}_2$, and $\hat{\mathbf{d}}_3$. Other sequences can be employed; for example, the rotations through angles q_1, q_2, and q_3 can be performed first about $\hat{\mathbf{b}}_3$, then about $\hat{\mathbf{c}}_1$, and finally about $\hat{\mathbf{d}}_2$. Direction cosines for six *body-three* sequences are tabulated in Appendix I, beginning with those appearing in Eq. (6). Six other *body-two* sequences exist, and corresponding tables are also included in Appendix I; for instance, rotations through angles q_1, q_2, and q_3 can be performed first about $\hat{\mathbf{b}}_3$, then about $\hat{\mathbf{c}}_1$, and finally about $\hat{\mathbf{d}}_3$.

1.4 VECTOR FUNCTIONS

When either the magnitude of a vector $\mathbf{v}$ and/or the direction of $\mathbf{v}$ in a reference frame A depends on a scalar variable q, $\mathbf{v}$ is called a *vector function of q in A*. Otherwise, $\mathbf{v}$ is said to be *independent of q in A*.

Example In Fig. 1.4.1, P represents a point moving on the surface of a rigid sphere

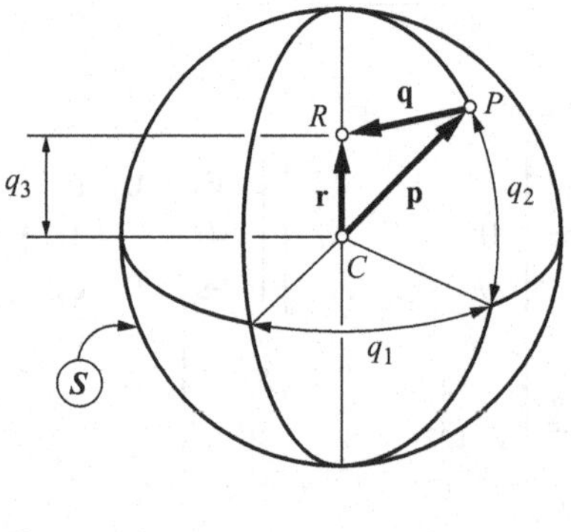

Figure 1.4.1

S, which may be regarded as a reference frame. If $\mathbf{p}$ is the position vector from the center C of S to point P, and if q_1 and q_2 are the angles shown, then $\mathbf{p}$ is a vector function of q_1 and q_2 in S because the direction of $\mathbf{p}$ in S depends on q_1 and q_2, but $\mathbf{p}$ is independent of q_3 in S, where q_3 is the distance from C to a point R situated as shown in Fig. 1.4.1. The position vector $\mathbf{r}$ from C to R is a vector function of q_3 in S but is independent of q_1 and q_2 in S, and the position vector $\mathbf{q}$ from P to R is a vector function of q_1, q_2, and q_3 in S.

1.5 SEVERAL REFERENCE FRAMES

A vector **v** may be a function of a variable q in one reference frame but be independent of q in another reference frame.

Example The outer gimbal ring A, inner gimbal ring B, and rotor C of the gyroscope depicted in Fig. 1.5.1 each can be regarded as a reference frame. If **p** is the position vector from point O to a point P of C, then **p** is a function of q_1 both in A and in B, but is independent of q_1 in C; **p** is a function of q_2 in A, but is independent of q_2 both in B and in C; and **p** is independent of q_3 in each of A, B, and C, but is a function of q_3 in reference frame D.

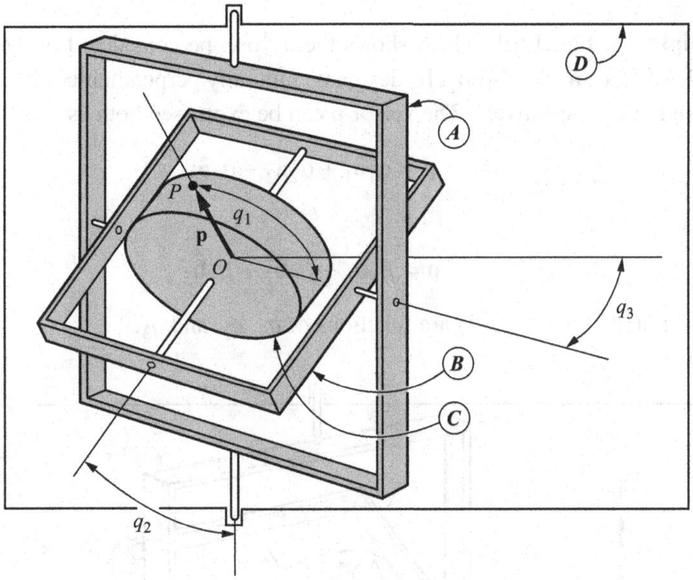

Figure 1.5.1

1.6 SCALAR FUNCTIONS

Given a reference frame A and a vector function **v** of n scalar variables $q_1, \ldots, q_n$ in A, let $\hat{\mathbf{a}}_1$, $\hat{\mathbf{a}}_2$, $\hat{\mathbf{a}}_3$ be a set of nonparallel, noncoplanar (but not necessarily mutually perpendicular) unit vectors fixed in A. Then there exist three unique scalar functions v_1, v_2, v_3 of $q_1, \ldots, q_n$ such that

$$\mathbf{v} = v_1\hat{\mathbf{a}}_1 + v_2\hat{\mathbf{a}}_2 + v_3\hat{\mathbf{a}}_3 \tag{1}$$

This equation may be regarded as a bridge connecting scalar to vector analysis; it provides a convenient means for extending to vector analysis various important concepts familiar from scalar analysis, such as continuity and differentiability. The vector $v_i\hat{\mathbf{a}}_i$ is called the $\hat{\mathbf{a}}_i$ *component* of $\mathbf{v}$, and v_i is known as the $\hat{\mathbf{a}}_i$ *measure number* of $\mathbf{v}$ $(i = 1,2,3)$.

When $\hat{\mathbf{a}}_1, \hat{\mathbf{a}}_2, \hat{\mathbf{a}}_3$ are *mutually perpendicular* unit vectors, then it follows from Eq. (1) that the $\hat{\mathbf{a}}_i$ measure number of $\mathbf{v}$ is given by

$$v_i = \mathbf{v} \cdot \hat{\mathbf{a}}_i \qquad (i = 1,2,3) \tag{2}$$

and that Eq. (1) may, therefore, be rewritten as

$$\mathbf{v} = \mathbf{v} \cdot \hat{\mathbf{a}}_1\hat{\mathbf{a}}_1 + \mathbf{v} \cdot \hat{\mathbf{a}}_2\hat{\mathbf{a}}_2 + \mathbf{v} \cdot \hat{\mathbf{a}}_3\hat{\mathbf{a}}_3 \tag{3}$$

Conversely, if $\hat{\mathbf{a}}_1$, $\hat{\mathbf{a}}_2$, and $\hat{\mathbf{a}}_3$ are mutually perpendicular unit vectors and Eqs. (2) are regarded as definitions of v_i $(i = 1,2,3)$, then it follows from Eq. (3) that $\mathbf{v}$ can be expressed as in Eq. (1).

Example In Fig. 1.6.1, which shows the gyroscope considered in the example in Sec. 1.5, $\hat{\mathbf{a}}_1, \hat{\mathbf{a}}_2, \hat{\mathbf{a}}_3$ and $\hat{\mathbf{b}}_1, \hat{\mathbf{b}}_2, \hat{\mathbf{b}}_3$ designate mutually perpendicular unit vectors fixed in A and in B, respectively. The vector $\mathbf{p}$ can be expressed both as

$$\mathbf{p} = \alpha_1\hat{\mathbf{a}}_1 + \alpha_2\hat{\mathbf{a}}_2 + \alpha_3\hat{\mathbf{a}}_3 \tag{4}$$

and as

$$\mathbf{p} = \beta_1\hat{\mathbf{b}}_1 + \beta_2\hat{\mathbf{b}}_2 + \beta_3\hat{\mathbf{b}}_3 \tag{5}$$

where α_i and β_i $(i = 1,2,3)$ are functions of q_1, q_2, and q_3.

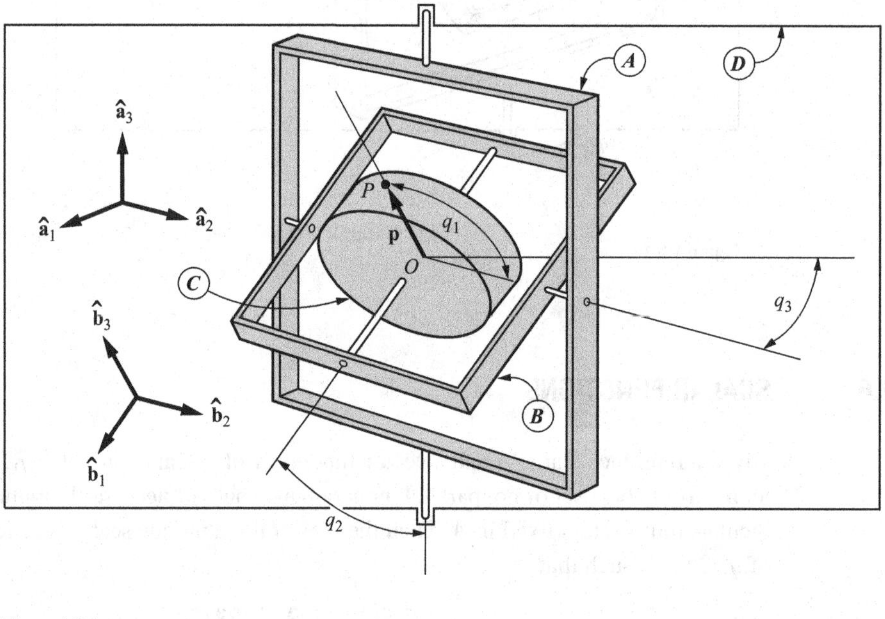

Figure 1.6.1

To determine these functions, note that, if C has a radius R, one can proceed from O

to P by moving through the distances $R\cos q_1$ and $R\sin q_1$ in the directions of $\hat{\mathbf{b}}_2$ and $\hat{\mathbf{b}}_3$ (see Fig. 1.6.2), respectively, which means that

$$\mathbf{p} = R(c_1\hat{\mathbf{b}}_2 + s_1\hat{\mathbf{b}}_3) \tag{6}$$

where c_1 and s_1 are abbreviations for $\cos q_1$ and $\sin q_1$, respectively. Comparing Eqs. (5) and (6), one thus finds that

$$\beta_1 = 0 \qquad \beta_2 = Rc_1 \qquad \beta_3 = Rs_1 \tag{7}$$

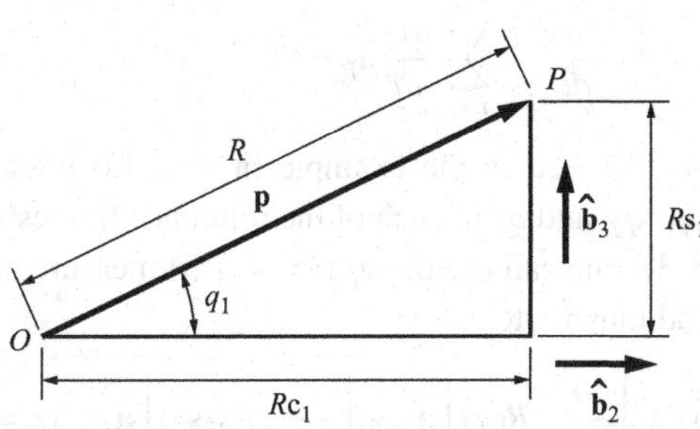

Figure 1.6.2

Moreover, in view of Eq. (4), one can write

$$\alpha_1 \underset{(2)}{=} \mathbf{p}\cdot\hat{\mathbf{a}}_1 \underset{(6)}{=} R(c_1\hat{\mathbf{b}}_2\cdot\hat{\mathbf{a}}_1 + s_1\hat{\mathbf{b}}_3\cdot\hat{\mathbf{a}}_1) \tag{8}$$

$$\alpha_2 \underset{(2)}{=} \mathbf{p}\cdot\hat{\mathbf{a}}_2 \underset{(6)}{=} R(c_1\hat{\mathbf{b}}_2\cdot\hat{\mathbf{a}}_2 + s_1\hat{\mathbf{b}}_3\cdot\hat{\mathbf{a}}_2) \tag{9}$$

$$\alpha_3 \underset{(2)}{=} \mathbf{p}\cdot\hat{\mathbf{a}}_3 \underset{(6)}{=} R(c_1\hat{\mathbf{b}}_2\cdot\hat{\mathbf{a}}_3 + s_1\hat{\mathbf{b}}_3\cdot\hat{\mathbf{a}}_3) \tag{10}$$

From Fig. 1.6.1,

$$\hat{\mathbf{b}}_2\cdot\hat{\mathbf{a}}_1 = 0 \qquad \hat{\mathbf{b}}_2\cdot\hat{\mathbf{a}}_2 = 1 \qquad \hat{\mathbf{b}}_2\cdot\hat{\mathbf{a}}_3 = 0 \tag{11}$$

$$\hat{\mathbf{b}}_3\cdot\hat{\mathbf{a}}_1 = c_2 \qquad \hat{\mathbf{b}}_3\cdot\hat{\mathbf{a}}_2 = 0 \qquad \hat{\mathbf{b}}_3\cdot\hat{\mathbf{a}}_3 = s_2 \tag{12}$$

Hence, the $\hat{\mathbf{a}}_1, \hat{\mathbf{a}}_2, \hat{\mathbf{a}}_3$ measure numbers of $\mathbf{p}$ are

$$\alpha_1 \underset{(8)\ (11,\ 12)}{=} Rs_1c_2 \qquad \alpha_2 \underset{(9)\ (11,\ 12)}{=} Rc_1 \qquad \alpha_3 \underset{(10)\ (11,\ 12)}{=} Rs_1s_2 \tag{13}$$

respectively.

1.7 FIRST DERIVATIVES

If $\mathbf{v}$ is a vector function of n scalar variables $q_1, \ldots, q_n$ in a reference frame A (see Sec. 1.4), then n vectors, called *first partial derivatives* of $\mathbf{v}$ *in* A and denoted by the symbols

$$\frac{{}^A\partial\mathbf{v}}{\partial q_r} \quad \text{or} \quad \frac{{}^A\partial}{\partial q_r}(\mathbf{v}) \quad \text{or} \quad {}^A\partial\mathbf{v}/\partial q_r \qquad (r = 1, \ldots, n)$$

are defined as follows: Let $\hat{\mathbf{a}}_1, \hat{\mathbf{a}}_2, \hat{\mathbf{a}}_3$ be any nonparallel, noncoplanar unit vectors fixed in A, and let v_i be the $\hat{\mathbf{a}}_i$ measure number of $\mathbf{v}$ (see Sec. 1.6). Then

$$\frac{{}^A\partial \mathbf{v}}{\partial q_r} \triangleq \sum_{i=1}^{3} \frac{\partial v_i}{\partial q_r}\hat{\mathbf{a}}_i \qquad (r = 1,\ldots,n) \tag{1}$$

When $\mathbf{v}$ is regarded as a vector function of only a single scalar variable in A —for instance the time t— then this definition reduces to that of the *ordinary derivative of* $\mathbf{v}$ *with respect to t in A*, that is, to†

$$\frac{{}^A d\mathbf{v}}{dt} \triangleq \sum_{i=1}^{3} \frac{dv_i}{dt}\hat{\mathbf{a}}_i \tag{2}$$

Example The vector $\mathbf{p}$ considered in the example in Sec. 1.6 possesses partial derivatives with respect to q_1, q_2, and q_3 in each of the reference frames A, B, and C. To form ${}^A\partial\mathbf{p}/\partial q_r$ $(r = 1,2,3)$, one can use the $\hat{\mathbf{a}}_i$ $(i = 1,2,3)$ measure numbers of $\mathbf{p}$ available in Eqs. (1.6.13) and thus write

$$\frac{{}^A\partial \mathbf{p}}{\partial q_r} \underset{(1)}{=} \left[\frac{\partial}{\partial q_r}(R\mathrm{s}_1\mathrm{c}_2)\right]\hat{\mathbf{a}}_1 + \left[\frac{\partial}{\partial q_r}(R\mathrm{c}_1)\right]\hat{\mathbf{a}}_2 + \left[\frac{\partial}{\partial q_r}(R\mathrm{s}_1\mathrm{s}_2)\right]\hat{\mathbf{a}}_3 \quad (r = 1,2,3) \tag{3}$$

Consequently,

$$\frac{{}^A\partial \mathbf{p}}{\partial q_1} \underset{(3)}{=} R(\mathrm{c}_1\mathrm{c}_2\hat{\mathbf{a}}_1 - \mathrm{s}_1\hat{\mathbf{a}}_2 + \mathrm{c}_1\mathrm{s}_2\hat{\mathbf{a}}_3) \tag{4}$$

$$\frac{{}^A\partial \mathbf{p}}{\partial q_2} \underset{(3)}{=} R\mathrm{s}_1(-\mathrm{s}_2\hat{\mathbf{a}}_1 + \mathrm{c}_2\hat{\mathbf{a}}_3) \tag{5}$$

$$\frac{{}^A\partial \mathbf{p}}{\partial q_3} \underset{(3)}{=} \mathbf{0} \tag{6}$$

The last result agrees with the statement in the example in Sec. 1.5 that $\mathbf{p}$ is independent of q_3 in A.

Proceeding similarly to determine ${}^B\partial\mathbf{p}/\partial q_r$ $(r = 1,2,3)$, one obtains with the aid of Eqs. (1.6.7),

$$\frac{{}^B\partial \mathbf{p}}{\partial q_1} = R(-\mathrm{s}_1\hat{\mathbf{b}}_2 + \mathrm{c}_1\hat{\mathbf{b}}_3) \qquad \frac{{}^B\partial \mathbf{p}}{\partial q_2} = \mathbf{0} \qquad \frac{{}^B\partial \mathbf{p}}{\partial q_3} = \mathbf{0} \tag{7}$$

Finally, since $\mathbf{p}$ is independent of q_r $(r = 1,2,3)$ in C,

$$\frac{{}^C\partial \mathbf{p}}{\partial q_r} = \mathbf{0} \qquad (r = 1,2,3) \tag{8}$$

Suppose now that q_1, q_2, and q_3 are specified as explicit functions of time t, namely,

$$q_1 = t \qquad q_2 = 2t \qquad q_3 = 3t \tag{9}$$

† The importance of reference frames in connection with time differentiation of vectors is obscured by defining the ordinary derivative of a vector $\mathbf{v}$ with respect to time t as the limit of $\Delta\mathbf{v}/\Delta t$ as Δt approaches zero, for this fails to bring any reference frame into evidence.

Then α_i $(i = 1,2,3)$ of Eq. (1.6.4) can be expressed as [see Eqs. (1.6.13)]

$$\alpha_1 = R \sin t \cos 2t \qquad \alpha_2 = R \cos t \qquad \alpha_3 = R \sin t \sin 2t \tag{10}$$

and the ordinary derivative of $\mathbf{p}$ with respect to t in A is seen to be given by

$$\begin{aligned} \frac{{}^A d\mathbf{p}}{dt} &\underset{(2)}{=} \frac{d\alpha_1}{dt}\hat{\mathbf{a}}_1 + \frac{d\alpha_2}{dt}\hat{\mathbf{a}}_2 + \frac{d\alpha_3}{dt}\hat{\mathbf{a}}_3 \\ &\underset{(10)}{=} R[(\cos t \cos 2t - 2\sin t \sin 2t)\hat{\mathbf{a}}_1 - \sin t\,\hat{\mathbf{a}}_2 \\ &\quad + (\cos t \sin 2t + 2 \sin t \cos 2t)\hat{\mathbf{a}}_3] \end{aligned} \tag{11}$$

while the ordinary derivative of $\mathbf{p}$ with respect to t in B is [see Eqs. (1.6.7)]

$$\frac{{}^B d\mathbf{p}}{dt} \underset{(2)}{=} R(-\sin t\,\hat{\mathbf{b}}_2 + \cos t\,\hat{\mathbf{b}}_3) \tag{12}$$

Finally,

$$\frac{{}^C d\mathbf{p}}{dt} = \mathbf{0} \tag{13}$$

because, when $\mathbf{p}$ is expressed as

$$\mathbf{p} = \gamma_1\hat{\mathbf{c}}_1 + \gamma_2\hat{\mathbf{c}}_2 + \gamma_3\hat{\mathbf{c}}_3 \tag{14}$$

where $\hat{\mathbf{c}}_1$, $\hat{\mathbf{c}}_2$, $\hat{\mathbf{c}}_3$ are unit vectors fixed in C, then γ_1, γ_2, γ_3 are necessarily constants, $\mathbf{p}$ being fixed in C.

1.8 REPRESENTATIONS OF DERIVATIVES

When a partial or ordinary derivative of a vector $\mathbf{v}$ in a reference frame A is formed by carrying out the operations indicated in Eqs. (1.7.1) and (1.7.2), the resulting expression involves the unit vectors $\hat{\mathbf{a}}_1$, $\hat{\mathbf{a}}_2$, $\hat{\mathbf{a}}_3$, that is, unit vectors fixed in A. By expressing each of these unit vectors in terms of unit vectors $\hat{\mathbf{b}}_1$, $\hat{\mathbf{b}}_2$, $\hat{\mathbf{b}}_3$ fixed in a reference frame B, one arrives at a new representation of the derivative under consideration, namely, one involving $\hat{\mathbf{b}}_1$, $\hat{\mathbf{b}}_2$, $\hat{\mathbf{b}}_3$, but one is still dealing with derivatives of $\mathbf{v}$ in A, not in B, unless these two derivatives happen to be equal to each other.

Example Referring to Eq. (1.7.4) and noting (see Fig. 1.6.1) that

$$\hat{\mathbf{a}}_1 = s_2\hat{\mathbf{b}}_1 + c_2\hat{\mathbf{b}}_3 \qquad \hat{\mathbf{a}}_2 = \hat{\mathbf{b}}_2 \qquad \hat{\mathbf{a}}_3 = -c_2\hat{\mathbf{b}}_1 + s_2\hat{\mathbf{b}}_3 \tag{1}$$

one can write

$$\begin{aligned} \frac{{}^A \partial\mathbf{p}}{\partial q_1} &= R[c_1c_2(s_2\hat{\mathbf{b}}_1 + c_2\hat{\mathbf{b}}_3) - s_1\hat{\mathbf{b}}_2 + c_1s_2(-c_2\hat{\mathbf{b}}_1 + s_2\hat{\mathbf{b}}_3)] \\ &= R(-s_1\hat{\mathbf{b}}_2 + c_1\hat{\mathbf{b}}_3) \end{aligned} \tag{2}$$

The right-hand sides of this equation and of the first of Eqs. (1.7.7) are identical. Hence, it appears that we have produced ${}^B\partial\mathbf{p}/\partial q_1$ by expressing ${}^A\partial\mathbf{p}/\partial q_1$ in terms of $\hat{\mathbf{b}}_1, \hat{\mathbf{b}}_2, \hat{\mathbf{b}}_3$. It was possible to do this because ${}^A\partial\mathbf{p}/\partial q_1$ and ${}^B\partial\mathbf{p}/\partial q_1$ happen to be equal to each other. To see that the same procedure does not lead to ${}^B\partial\mathbf{p}/\partial q_2$ when one starts with ${}^A\partial\mathbf{p}/\partial q_2$, refer to Eq. (1.7.5) to write, with the aid of Eqs. (1),

$$\begin{aligned}\frac{{}^A\partial\mathbf{p}}{\partial q_2} &= R\mathrm{s}_1[-\mathrm{s}_2(\mathrm{s}_2\hat{\mathbf{b}}_1 + \mathrm{c}_2\hat{\mathbf{b}}_3) + \mathrm{c}_2(-\mathrm{c}_2\hat{\mathbf{b}}_1 + \mathrm{s}_2\hat{\mathbf{b}}_3)] \\ &= -R\mathrm{s}_1\hat{\mathbf{b}}_1 \end{aligned} \tag{3}$$

and compare this with the second of Eqs. (1.7.7)

1.9 NOTATION FOR DERIVATIVES

In general, both partial and ordinary derivatives of a vector $\mathbf{v}$ in a reference frame A differ from corresponding derivatives in any other reference frame B. It follows that notations such as $\partial\mathbf{v}/\partial q_r$ or $d\mathbf{v}/dt$ —that is, ones that involve no mention of any reference frame— are meaningful only either when the context in which they appear clearly implies a particular reference frame or when it does not matter which reference frame is used. In the sequel, it is to be understood that, whenever no reference frame is mentioned explicitly, any reference frame may be used, but all partial or ordinary differentiations indicated in any one equation are meant to be performed in the same reference frame.

Example The equations

$$\mathbf{p}\cdot\frac{\partial\mathbf{p}}{\partial q_r} = 0 \qquad (r = 1,2,3) \tag{1}$$

and

$$\mathbf{p}\cdot\frac{d\mathbf{p}}{dt} = 0 \tag{2}$$

are valid for the vector $\mathbf{p}$ and the quantities q_1, q_2, q_3 introduced in the example in Sec. 1.5, regardless of the reference frame in which $\mathbf{p}$ is differentiated. We shall shortly be in a position to prove this. To verify it for a few specific cases, refer to Eqs. (1.6.6) and (1.7.7), which yield

$$\mathbf{p}\cdot\frac{{}^B\partial\mathbf{p}}{\partial q_1} = R^2(\mathrm{c}_1\hat{\mathbf{b}}_2 + \mathrm{s}_1\hat{\mathbf{b}}_3)\cdot(-\mathrm{s}_1\hat{\mathbf{b}}_2 + \mathrm{c}_1\hat{\mathbf{b}}_3) = 0 \tag{3}$$

or use Eqs. (1.6.4), (1.6.13), and (1.7.5) to write

$$\mathbf{p}\cdot\frac{{}^A\partial\mathbf{p}}{\partial q_2} = R^2\mathrm{s}_1(\mathrm{s}_1\mathrm{c}_2\hat{\mathbf{a}}_1 + \mathrm{c}_1\hat{\mathbf{a}}_2 + \mathrm{s}_1\mathrm{s}_2\hat{\mathbf{a}}_3)\cdot(-\mathrm{s}_2\hat{\mathbf{a}}_1 + \mathrm{c}_2\hat{\mathbf{a}}_3) = 0 \tag{4}$$

Finally, note that Eqs. (1.6.4), (1.7.10), (1.7.11), and (1.7.13) lead to

$$\mathbf{p}\cdot\frac{{}^A d\mathbf{p}}{dt} = \mathbf{p}\cdot\frac{{}^C d\mathbf{p}}{dt} = 0 \tag{5}$$

1.10 DIFFERENTIATION OF SUMS AND PRODUCTS

As an immediate consequence of the definition given in Eqs. (1.7.1), the following rules govern the differentiation of sums and products involving vector functions.

If $\mathbf{v}_1, \ldots, \mathbf{v}_N$ are vector functions of the scalar variables $q_1, \ldots, q_n$ in some reference frame, then

$$\frac{\partial}{\partial q_r} \sum_{i=1}^{N} \mathbf{v}_i = \sum_{i=1}^{N} \frac{\partial \mathbf{v}_i}{\partial q_r} \qquad (r = 1, \ldots, n) \tag{1}$$

If s is a scalar function of $q_1, \ldots, q_n$, and $\mathbf{v}$ and $\mathbf{w}$ are vector functions of these variables in some reference frame, then

$$\frac{\partial}{\partial q_r}(s\mathbf{v}) = \frac{\partial s}{\partial q_r}\mathbf{v} + s\frac{\partial \mathbf{v}}{\partial q_r} \qquad (r = 1, \ldots, n) \tag{2}$$

$$\frac{\partial}{\partial q_r}(\mathbf{v} \cdot \mathbf{w}) = \frac{\partial \mathbf{v}}{\partial q_r} \cdot \mathbf{w} + \mathbf{v} \cdot \frac{\partial \mathbf{w}}{\partial q_r} \qquad (r = 1, \ldots, n) \tag{3}$$

$$\frac{\partial}{\partial q_r}(\mathbf{v} \times \mathbf{w}) = \frac{\partial \mathbf{v}}{\partial q_r} \times \mathbf{w} + \mathbf{v} \times \frac{\partial \mathbf{w}}{\partial q_r} \qquad (r = 1, \ldots, n) \tag{4}$$

More generally, if P is the product of N scalar and/or vector functions F_i $(i = 1, \ldots, N)$, that is, if

$$P = F_1 F_2 \cdots F_N \tag{5}$$

then, if all symbols of operation, such as dots, crosses, and parentheses, are kept in place,

$$\frac{\partial P}{\partial q_r} = \frac{\partial F_1}{\partial q_r} F_2 \cdots F_N + F_1 \frac{\partial F_2}{\partial q_r} \cdots F_N + \cdots + F_1 F_2 \cdots F_{N-1} \frac{\partial F_N}{\partial q_r} \qquad (r = 1, \ldots, n) \tag{6}$$

Relationships analogous to Eqs. (1)–(6) govern the ordinary differentiation [see Eq. (1.7.2)] of vector and/or scalar functions of a single scalar variable.

Example By definition, the square of a vector $\mathbf{v}$, written $\mathbf{v}^2$, is the scalar quantity obtained by dot-multiplying $\mathbf{v}$ with $\mathbf{v}$. Hence, if s is a scalar function of $q_1, \ldots, q_n$, then

$$\begin{aligned} \frac{\partial}{\partial q_r}(s\mathbf{v}^2) &= \frac{\partial}{\partial q_r}(s\mathbf{v} \cdot \mathbf{v}) \\ &\underset{(6)}{=} \frac{\partial s}{\partial q_r}\mathbf{v} \cdot \mathbf{v} + s\frac{\partial \mathbf{v}}{\partial q_r} \cdot \mathbf{v} + s\mathbf{v} \cdot \frac{\partial \mathbf{v}}{\partial q_r} \qquad (r = 1, \ldots, n) \end{aligned} \tag{7}$$

or, since the last two terms are equal to each other,

$$\frac{\partial}{\partial q_r}(s\mathbf{v}^2) = \frac{\partial s}{\partial q_r}\mathbf{v}^2 + 2s\mathbf{v} \cdot \frac{\partial \mathbf{v}}{\partial q_r} \qquad (r = 1, \ldots, n) \tag{8}$$

This result can be used to establish the validity of Eqs. (1.9.1) by taking $s = 1$, writing $\mathbf{p}$ in place of $\mathbf{v}$, and letting $n = 3$, which yields

$$\frac{\partial}{\partial q_r}(\mathbf{p}^2) \underset{(8)}{=} 2\mathbf{p} \cdot \frac{\partial \mathbf{p}}{\partial q_r} \qquad (r = 1, 2, 3) \tag{9}$$

and then noting that, for the vector $\mathbf{p}$ in Eqs. (1.9.1), $\mathbf{p}^2$ is a constant, so that

$$\frac{\partial}{\partial q_r}(\mathbf{p}^2) = 0 \qquad (r = 1,2,3) \tag{10}$$

Since $2 \neq 0$, Eqs. (9) and (10) imply that

$$\mathbf{p} \cdot \frac{\partial \mathbf{p}}{\partial q_r} = 0 \qquad (r = 1,2,3) \tag{11}$$

in agreement with Eqs. (1.9.1).

1.11 SECOND DERIVATIVES

In general, ${}^A\partial\mathbf{v}/\partial q_r$ (see Sec. 1.7) is a vector function of $q_1, \ldots, q_n$ both in A and in any other reference frame B and can, therefore, be differentiated with respect to any one of $q_1, \ldots, q_n$ both in A and in B. The result of such a differentiation is called a *second partial derivative*. Similarly, the ordinary derivative ${}^A d\mathbf{v}/dt$ (see Sec. 1.7) can be differentiated with respect to t both in A and in any other reference frame B.

The order in which successive differentiations are performed can affect the results. For example, in general,

$$\frac{{}^B\partial}{\partial q_s}\left(\frac{{}^A\partial\mathbf{v}}{\partial q_r}\right) \neq \frac{{}^A\partial}{\partial q_r}\left(\frac{{}^B\partial\mathbf{v}}{\partial q_s}\right) \qquad (r,s = 1,\ldots,n) \tag{1}$$

and

$$\frac{{}^B d}{dt}\left(\frac{{}^A d\mathbf{v}}{dt}\right) \neq \frac{{}^A d}{dt}\left(\frac{{}^B d\mathbf{v}}{dt}\right) \tag{2}$$

However, if successive partial differentiations with respect to various variables are performed in the same reference frame, then the order is immaterial; that is,

$$\frac{\partial}{\partial q_s}\left(\frac{\partial\mathbf{v}}{\partial q_r}\right) = \frac{\partial}{\partial q_r}\left(\frac{\partial\mathbf{v}}{\partial q_s}\right) \qquad (r,s = 1,\ldots,n) \tag{3}$$

and

$$\frac{\partial}{\partial t}\left(\frac{\partial\mathbf{v}}{\partial q_r}\right) = \frac{\partial}{\partial q_r}\left(\frac{\partial\mathbf{v}}{\partial t}\right) \qquad (r = 1,\ldots,n) \tag{4}$$

Example Referring to the example in Sec. 1.6, suppose that a vector $\mathbf{v}$ is given by

$$\mathbf{v} = t\hat{\mathbf{a}}_1 \tag{5}$$

and that q_2 is a specified function of t. Then

$$\frac{{}^A d\mathbf{v}}{dt} \underset{(5)}{=} \hat{\mathbf{a}}_1 \underset{(1.8.1)}{=} \mathrm{s}_2\hat{\mathbf{b}}_1 + \mathrm{c}_2\hat{\mathbf{b}}_3 \tag{6}$$

and

$$\frac{{}^B d}{dt}\left(\frac{{}^A d\mathbf{v}}{dt}\right) \underset{(6)}{=} \dot{q}_2(\mathrm{c}_2\hat{\mathbf{b}}_1 - \mathrm{s}_2\hat{\mathbf{b}}_3) \underset{(1.8.1)}{=} -\dot{q}_2\hat{\mathbf{a}}_3 \tag{7}$$

where $\dot{q}_2$ denotes the first derivative of q_2 with respect to t. Also,

$$\mathbf{v} \underset{(5)}{=} t\,(\underset{(1.8.1)}{s_2\hat{\mathbf{b}}_1 + c_2\hat{\mathbf{b}}_3}) \tag{8}$$

so that

$$\begin{aligned}\frac{^B d\mathbf{v}}{dt} &\underset{(8)}{=} \frac{dt}{dt}(s_2\hat{\mathbf{b}}_1 + c_2\hat{\mathbf{b}}_3) + t\,\frac{^B d}{dt}(s_2\hat{\mathbf{b}}_1 + c_2\hat{\mathbf{b}}_3) \\ &= s_2\hat{\mathbf{b}}_1 + c_2\hat{\mathbf{b}}_3 + t\dot{q}_2(c_2\hat{\mathbf{b}}_1 - s_2\hat{\mathbf{b}}_3) \\ &\underset{(1.8.1)}{=} \hat{\mathbf{a}}_1 - t\dot{q}_2\hat{\mathbf{a}}_3 \end{aligned} \tag{9}$$

and

$$\frac{^A d}{dt}\left(\frac{^B d\mathbf{v}}{dt}\right) \underset{(9)}{=} -(\dot{q}_2 + t\,\ddot{q}_2)\hat{\mathbf{a}}_3 \tag{10}$$

Comparing Eqs. (7) and (10), one sees that, in general, one must expect the result of successive differentiations in various reference frames to depend on the order in which the differentiations are performed.

1.12 TOTAL AND PARTIAL DERIVATIVES

If $q_1, \ldots, q_n$ are scalar functions of a single variable t, it is sometimes convenient to regard a vector $\mathbf{v}$ as a vector function of the $n + 1$ independent variables $q_1, \ldots, q_n$ and t in a reference frame A. The ordinary derivative of $\mathbf{v}$ with respect to t in A (see Sec. 1.7), called a *total derivative* under these circumstances, then can be expressed in terms of partial derivatives as

$$\frac{^A d\mathbf{v}}{dt} = \sum_{r=1}^{n} \frac{^A \partial\mathbf{v}}{\partial q_r}\dot{q}_r + \frac{^A \partial\mathbf{v}}{\partial t} \tag{1}$$

where $\dot{q}_r$ denotes the first derivative of q_r with respect to t. Moreover, if $\mathbf{v}$ is differentiated both totally with respect to t and partially with respect to q_r, then the order in which the differentiations are performed is immaterial; that is,

$$\frac{d}{dt}\frac{\partial\mathbf{v}}{\partial q_r} = \frac{\partial}{\partial q_r}\frac{d\mathbf{v}}{dt} \qquad (r = 1, \ldots, n) \tag{2}$$

Derivations Let $\hat{\mathbf{a}}_i$ $(i = 1, 2, 3)$ be nonparallel, noncoplanar unit vectors fixed in A, and regard v_i, the $\hat{\mathbf{a}}_i$ measure number of $\mathbf{v}$ (see Sec. 1.6), as a function of $q_1, \ldots, q_n$, and t. From scalar calculus,

$$\frac{dv_i}{dt} = \sum_{r=1}^{n} \frac{\partial v_i}{\partial q_r}\dot{q}_r + \frac{\partial v_i}{\partial t} \qquad (i = 1, 2, 3) \tag{3}$$

and, if m and q_m are defined as

$$m \triangleq n + 1 \tag{4}$$

and

$$q_m \stackrel{\Delta}{=} t \tag{5}$$

so that

$$\dot{q}_m \underset{(5)}{=} 1 \tag{6}$$

then Eqs. (3) can be rewritten as

$$\frac{dv_i}{dt} = \sum_{r=1}^{n} \frac{\partial v_i}{\partial q_r}\dot{q}_r + \frac{\partial v_i}{\partial q_m}\underset{(5)}{}\underset{(6)}{\dot{q}_m} = \underset{(4)}{\sum_{r=1}^{m} \frac{\partial v_i}{\partial q_r}\dot{q}_r} \qquad (i = 1,2,3) \tag{7}$$

and substitution into Eq. (1.7.2) yields

$$\begin{aligned}\frac{{}^A d\mathbf{v}}{dt} &= \sum_{i=1}^{3}\left(\sum_{r=1}^{m} \frac{\partial v_i}{\partial q_r}\dot{q}_r\right)\hat{\mathbf{a}}_i = \sum_{r=1}^{m}\left(\sum_{i=1}^{3} \frac{\partial v_i}{\partial q_r}\hat{\mathbf{a}}_i\right)\dot{q}_r \\ &= \sum_{r=1}^{m} \underset{(1.7.1)}{\frac{{}^A\partial\mathbf{v}}{\partial q_r}}\dot{q}_r = \sum_{r=1}^{n} \frac{{}^A\partial\mathbf{v}}{\partial q_r}\dot{q}_r + \frac{{}^A\partial\mathbf{v}}{\partial q_m}\dot{q}_m\end{aligned} \tag{8}$$

which, in view of Eqs. (5) and (6), establishes the validity of Eq. (1). Furthermore, replacing $\mathbf{v}$ in Eq. (1) with $\partial\mathbf{v}/\partial q_s$ produces

$$\begin{aligned}\frac{d}{dt}\frac{\partial\mathbf{v}}{\partial q_s} &= \sum_{r=1}^{n}\left[\frac{\partial}{\partial q_r}\left(\frac{\partial\mathbf{v}}{\partial q_s}\right)\right]\dot{q}_r + \frac{\partial}{\partial t}\left(\frac{\partial\mathbf{v}}{\partial q_s}\right) \\ &= \sum_{r=1}^{n}\underset{(1.11.3)}{\left[\frac{\partial}{\partial q_s}\left(\frac{\partial\mathbf{v}}{\partial q_r}\right)\right]}\dot{q}_r + \underset{(1.11.4)}{\frac{\partial}{\partial q_s}\left(\frac{\partial\mathbf{v}}{\partial t}\right)} \\ &\underset{(1.10.1)}{=} \frac{\partial}{\partial q_s}\left(\sum_{r=1}^{n}\frac{\partial\mathbf{v}}{\partial q_r}\dot{q}_r + \frac{\partial\mathbf{v}}{\partial t}\right) \\ &\underset{(1)}{=} \frac{\partial}{\partial q_s}\frac{d\mathbf{v}}{dt} \qquad (s = 1,\ldots,n)\end{aligned} \tag{9}$$

in agreement with Eq. (2).

Example To see that one can proceed in a variety of ways to find the ordinary derivative of a vector function in a reference frame, consider once more the vector **p** introduced in the example in Sec. 1.6, and again let q_1, q_2, and q_3 be given by Eqs. (1.7.9). Then the ordinary time-derivative of **p** in A, previously found by using Eq. (1.7.2), is given by Eq. (1.7.11). Now refer to Eqs. (1.6.4) and (1.6.13) to express **p** as

$$\mathbf{p} = R(s_1c_2\hat{\mathbf{a}}_1 + c_1\hat{\mathbf{a}}_2 + s_1s_2\hat{\mathbf{a}}_3) \tag{10}$$

and use Eqs. (1.7.9) to rewrite the $\hat{\mathbf{a}}_3$ measure number of **p** as an explicit function of t, that is, to replace Eq. (10) with

$$\mathbf{p} = R(s_1c_2\hat{\mathbf{a}}_1 + c_1\hat{\mathbf{a}}_2 + \sin t \sin 2t\ \hat{\mathbf{a}}_3) \tag{11}$$

Furthermore, regard **p** as a function of the independent variables q_1, q_2, and q_3 and t, and then appeal to Eq. (1) to write

$$\begin{aligned}\frac{{}^A d\mathbf{p}}{dt} &= \frac{{}^A\partial\mathbf{p}}{\partial q_1}\dot{q}_1 + \frac{{}^A\partial\mathbf{p}}{\partial q_2}\dot{q}_2 + \frac{{}^A\partial\mathbf{p}}{\partial q_3}\dot{q}_3 + \frac{{}^A\partial\mathbf{p}}{\partial t} \\ &\underset{(11)}{=} R[\mathrm{c}_1\mathrm{c}_2\hat{\mathbf{a}}_1 - \mathrm{s}_1\hat{\mathbf{a}}_2)\dot{q}_1 + (-\mathrm{s}_1\mathrm{s}_2\hat{\mathbf{a}}_2)\dot{q}_2 \\ &\quad + (\cos t \sin 2t + 2\sin t\cos 2t)\hat{\mathbf{a}}_3]\end{aligned} \tag{12}$$

Finally, make the substitutions [see Eqs. (1.7.9)]

$$\dot{q}_1 = 1 \qquad \mathrm{s}_1 = \sin t \qquad \mathrm{c}_1 = \cos t \tag{13}$$

$$\dot{q}_2 = 2 \qquad \mathrm{s}_2 = \sin 2t \qquad \mathrm{c}_2 = \cos 2t \tag{14}$$

and verify that the resulting equation is precisely Eq. (1.7.11). The point here is not that use of Eq. (1) facilitates the evaluation of ordinary derivatives; indeed, it may complicate matters. What is important is to realize that one may treat the same vector in a variety of ways, that the formalism one uses to construct the ordinary derivative of the vector in a given reference frame depends on the functional character one attributes to the vector, but that the result one obtains is independent of the approach taken. In the sequel, Eq. (1) will be used primarily in the course of certain derivations rather than for the actual evaluation of ordinary derivatives.

1.13 SCALAR FUNCTIONS OF VECTORS

A scalar variable can sometimes be regarded as a function of vector variables. For example, the scalar $s = \mathbf{v}\cdot\mathbf{w}$ is a function of the vectors **v** and **w**. Similarly, the scalar $r = \mathbf{a}\cdot\mathbf{b}\times\mathbf{c}$ is a function of vectors **a**, **b**, and **c**. When a scalar s is a function of a single scalar variable t, and n vector variables $\mathbf{v}_1,\ldots,\mathbf{v}_n$, each of which is regarded as a function of t, the ordinary derivative of s with respect to t can be expressed as

$$\frac{ds}{dt} = \sum_{i=1}^{n}\frac{\partial s}{\partial \mathbf{v}_i}\cdot\frac{d\mathbf{v}_i}{dt} + \frac{\partial s}{\partial t} \tag{1}$$

In accordance with Sec. 1.9, all ordinary derivatives indicated in the right hand member of this equation are to be performed in the same reference frame, which can be any reference frame whatsoever. In practice, the vectors $\partial s/\partial \mathbf{v}_i$ can be formed simply by performing differentiation of products of vectors according to instructions given in Sec. 1.10.

Example Consider the scalar function

$$s = \mathbf{a}\cdot\mathbf{b}\times\mathbf{c} - 5t^2 \tag{2}$$

The ordinary derivative of s with respect to t is given by

$$\begin{aligned}\frac{ds}{dt} &= \frac{d\mathbf{a}}{dt}\cdot(\mathbf{b}\times\mathbf{c}) + \mathbf{a}\cdot\left(\frac{d\mathbf{b}}{dt}\times\mathbf{c}\right) + \mathbf{a}\cdot\left(\mathbf{b}\times\frac{d\mathbf{c}}{dt}\right) - 10t \\ &= (\mathbf{b}\times\mathbf{c})\cdot\frac{d\mathbf{a}}{dt} + (\mathbf{c}\times\mathbf{a})\cdot\frac{d\mathbf{b}}{dt} + (\mathbf{a}\times\mathbf{b})\cdot\frac{d\mathbf{c}}{dt} - 10t \qquad (3)\end{aligned}$$

whereupon one can readily identify the partial derivatives $\partial s/\partial\mathbf{a} = \mathbf{b}\times\mathbf{c}$, $\partial s/\partial\mathbf{b} = \mathbf{c}\times\mathbf{a}$, and $\partial s/\partial\mathbf{c} = \mathbf{a}\times\mathbf{b}$.

2 KINEMATICS

Considerations of kinematics play a central role in dynamics. Indeed, one's effectiveness in formulating equations of motion depends primarily on one's ability to construct correct mathematical expressions for kinematical quantities such as angular velocities of rigid bodies, velocities of points, and so forth. Therefore, mastery of the material in this chapter is essential.

Sections 2.1–2.5 are concerned with *rotational motion of a rigid body*. The principal kinematical quantity introduced here is the angular velocity of a rigid body in a reference frame. Next, *translational motion of a point* is treated in Secs. 2.6–2.8, where four theorems frequently used in practice are derived from definitions of the velocity and acceleration of a point in a reference frame.

The reason for discussing translational motion after rotational motion is that the theorems on translational motion in Secs. 2.6–2.8 involve angular velocities and angular accelerations of rigid bodies, whereas the material on rotational motion in Secs. 2.1–2.5 does not involve velocities or accelerations of points. It is important to keep in mind that a point, which can be regarded as a massless particle, can possess a position, a velocity, and an acceleration in a reference frame, but a point does not have an orientation, an angular velocity, or an angular acceleration in a reference frame. Likewise, although a rigid body can possess orientation, angular velocity, and angular acceleration, it is in general only meaningful to speak of position, velocity, and acceleration of the individual particles belonging to a rigid body; the rigid body itself does not possess these quantities.

2.1 ANGULAR VELOCITY

The use of angular velocities greatly facilitates the analysis of motions of systems containing rigid bodies. We begin our discussion of this topic with a formal definition of angular velocity; while it is abstract, this definition provides a sound basis for the derivation of theorems [see, for example, Eq. (2)] used to solve physical problems.†

Let $\hat{\mathbf{b}}_1$, $\hat{\mathbf{b}}_2$, $\hat{\mathbf{b}}_3$ form a right-handed set of mutually perpendicular unit vectors fixed in a rigid body B moving in a reference frame A. The *angular velocity* of B in A, denoted

† The frequently employed definition of angular velocity as the limit of $\Delta\theta/\Delta t$ as Δt approaches zero is deficient in this regard.

by ${}^A\boldsymbol{\omega}^B$, is defined as

$$ {}^A\boldsymbol{\omega}^B \triangleq \hat{\mathbf{b}}_1 \frac{{}^A d\hat{\mathbf{b}}_2}{dt} \cdot \hat{\mathbf{b}}_3 + \hat{\mathbf{b}}_2 \frac{{}^A d\hat{\mathbf{b}}_3}{dt} \cdot \hat{\mathbf{b}}_1 + \hat{\mathbf{b}}_3 \frac{{}^A d\hat{\mathbf{b}}_1}{dt} \cdot \hat{\mathbf{b}}_2 \tag{1} $$

One task facilitated by the use of angular velocity vectors is the time differentiation of vectors fixed in a rigid body, for it enables one to obtain the first time-derivative of such a vector by performing a cross multiplication. Specifically, if $\boldsymbol{\beta}$ is any vector fixed in B, then

$$ \frac{{}^A d\boldsymbol{\beta}}{dt} = {}^A\boldsymbol{\omega}^B \times \boldsymbol{\beta} \tag{2} $$

Derivation Using dots to denote time differentiation in A, one can rewrite Eq. (1) as

$$ {}^A\boldsymbol{\omega}^B \triangleq \hat{\mathbf{b}}_1 \dot{\hat{\mathbf{b}}}_2 \cdot \hat{\mathbf{b}}_3 + \hat{\mathbf{b}}_2 \dot{\hat{\mathbf{b}}}_3 \cdot \hat{\mathbf{b}}_1 + \hat{\mathbf{b}}_3 \dot{\hat{\mathbf{b}}}_1 \cdot \hat{\mathbf{b}}_2 \tag{3} $$

and cross multiplication of Eq. (3) with $\hat{\mathbf{b}}_1$ gives

$$ {}^A\boldsymbol{\omega}^B \times \hat{\mathbf{b}}_1 \underset{(3)}{=} \hat{\mathbf{b}}_2 \times \hat{\mathbf{b}}_1 \dot{\hat{\mathbf{b}}}_3 \cdot \hat{\mathbf{b}}_1 + \hat{\mathbf{b}}_3 \times \hat{\mathbf{b}}_1 \dot{\hat{\mathbf{b}}}_1 \cdot \hat{\mathbf{b}}_2 \tag{4} $$

Now, since $\hat{\mathbf{b}}_1$, $\hat{\mathbf{b}}_2$, $\hat{\mathbf{b}}_3$ form a right-handed set of mutually perpendicular unit vectors, each can be expressed as a cross product involving the remaining two. For example,

$$ \hat{\mathbf{b}}_2 = \hat{\mathbf{b}}_3 \times \hat{\mathbf{b}}_1 \qquad \hat{\mathbf{b}}_3 = \hat{\mathbf{b}}_1 \times \hat{\mathbf{b}}_2 \tag{5} $$

and substitution into Eq. (4) yields

$$ {}^A\boldsymbol{\omega}^B \times \hat{\mathbf{b}}_1 \underset{(4)}{=} -\underset{(5)}{\hat{\mathbf{b}}_3}\, \dot{\hat{\mathbf{b}}}_3 \cdot \hat{\mathbf{b}}_1 + \underset{(5)}{\hat{\mathbf{b}}_2}\, \dot{\hat{\mathbf{b}}}_1 \cdot \hat{\mathbf{b}}_2 \tag{6} $$

Moreover, time differentiation of the equations $\hat{\mathbf{b}}_1 \cdot \hat{\mathbf{b}}_1 = 1$ and $\hat{\mathbf{b}}_3 \cdot \hat{\mathbf{b}}_1 = 0$ produces

$$ \dot{\hat{\mathbf{b}}}_1 \cdot \hat{\mathbf{b}}_1 = 0 \qquad \dot{\hat{\mathbf{b}}}_3 \cdot \hat{\mathbf{b}}_1 = -\dot{\hat{\mathbf{b}}}_1 \cdot \hat{\mathbf{b}}_3 \tag{7} $$

and with the aid of these one can rewrite Eq. (6) as

$$ {}^A\boldsymbol{\omega}^B \times \hat{\mathbf{b}}_1 \underset{(6)}{=} \hat{\mathbf{b}}_1 \underset{(7)}{\dot{\hat{\mathbf{b}}}_1 \cdot \hat{\mathbf{b}}_1} + \hat{\mathbf{b}}_2 \dot{\hat{\mathbf{b}}}_1 \cdot \hat{\mathbf{b}}_2 + \hat{\mathbf{b}}_3 \underset{(7)}{\dot{\hat{\mathbf{b}}}_1 \cdot \hat{\mathbf{b}}_3} \tag{8} $$

But the right-hand member of this equation is simply a way of writing $\dot{\hat{\mathbf{b}}}_1$ [see Eq. (1.6.3)]. Consequently,

$$ {}^A\boldsymbol{\omega}^B \times \hat{\mathbf{b}}_1 \underset{(8)}{=} \dot{\hat{\mathbf{b}}}_1 \tag{9} $$

Similarly,

$$ {}^A\boldsymbol{\omega}^B \times \hat{\mathbf{b}}_2 = \dot{\hat{\mathbf{b}}}_2 \qquad {}^A\boldsymbol{\omega}^B \times \hat{\mathbf{b}}_3 = \dot{\hat{\mathbf{b}}}_3 \tag{10} $$

and, after expressing any vector $\boldsymbol{\beta}$ fixed in B as

$$ \boldsymbol{\beta} = \beta_1 \hat{\mathbf{b}}_1 + \beta_2 \hat{\mathbf{b}}_2 + \beta_3 \hat{\mathbf{b}}_3 \tag{11} $$

where β_1, β_2, β_3 are constants, so that

$$\dot{\boldsymbol{\beta}} \underset{(11)}{=} \beta_1 \dot{\hat{\mathbf{b}}}_1 + \beta_2 \dot{\hat{\mathbf{b}}}_2 + \beta_3 \dot{\hat{\mathbf{b}}}_3 \tag{12}$$

one arrives at

$$\begin{aligned}\dot{\boldsymbol{\beta}} &\underset{(12)}{=} \beta_1 \underset{(9)}{{}^A\boldsymbol{\omega}^B} \times \hat{\mathbf{b}}_1 + \beta_2 \underset{(10)}{{}^A\boldsymbol{\omega}^B} \times \hat{\mathbf{b}}_2 + \beta_3 \underset{(10)}{{}^A\boldsymbol{\omega}^B} \times \hat{\mathbf{b}}_3 \\ &= {}^A\boldsymbol{\omega}^B \times (\beta_1 \hat{\mathbf{b}}_1 + \beta_2 \hat{\mathbf{b}}_2 + \beta_3 \hat{\mathbf{b}}_3) \underset{(11)}{=} {}^A\boldsymbol{\omega}^B \times \boldsymbol{\beta}\end{aligned} \tag{13}$$

Examples Figure 2.1.1 shows a rigid satellite B in orbit about the Earth A. A dextral set of mutually perpendicular unit vectors $\hat{\mathbf{b}}_1$, $\hat{\mathbf{b}}_2$, $\hat{\mathbf{b}}_3$ is fixed in B, and a similar such set, $\hat{\mathbf{a}}_1$, $\hat{\mathbf{a}}_2$, $\hat{\mathbf{a}}_3$, is fixed in A. Measurements are made to determine the time histories of α_i, β_i, γ_i, defined as

$$\alpha_i \triangleq \hat{\mathbf{b}}_1 \cdot \hat{\mathbf{a}}_i \qquad \beta_i \triangleq \hat{\mathbf{b}}_2 \cdot \hat{\mathbf{a}}_i \qquad \gamma_i \triangleq \hat{\mathbf{b}}_3 \cdot \hat{\mathbf{a}}_i \qquad (i = 1,2,3) \tag{14}$$

as well as the time histories of $\dot{\alpha}_i$, $\dot{\beta}_i$, $\dot{\gamma}_i$, defined as the time derivatives of α_i, β_i, γ_i, respectively. At a certain time t^* these quantities have the values recorded in Tables 2.1.1 and 2.1.2. The angular velocity of B in A at time t^* is to be determined.

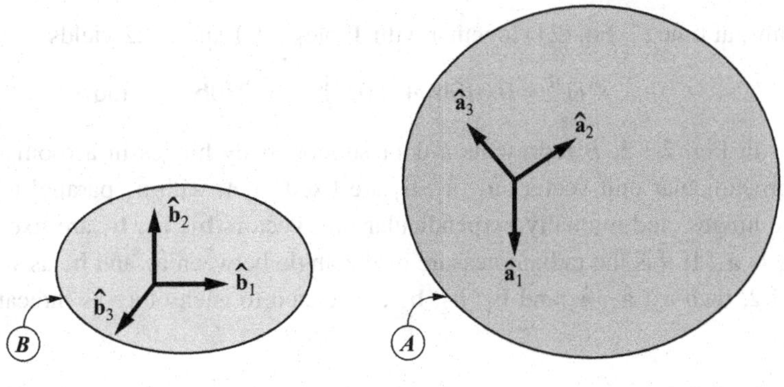

Figure 2.1.1

It follows from Eqs. (14) that

$$\hat{\mathbf{b}}_1 = \alpha_1 \hat{\mathbf{a}}_1 + \alpha_2 \hat{\mathbf{a}}_2 + \alpha_3 \hat{\mathbf{a}}_3 \tag{15}$$

$$\hat{\mathbf{b}}_2 = \beta_1 \hat{\mathbf{a}}_1 + \beta_2 \hat{\mathbf{a}}_2 + \beta_3 \hat{\mathbf{a}}_3 \tag{16}$$

and

$$\hat{\mathbf{b}}_3 = \gamma_1 \hat{\mathbf{a}}_1 + \gamma_2 \hat{\mathbf{a}}_2 + \gamma_3 \hat{\mathbf{a}}_3 \tag{17}$$

Consequently, for all values of the time t,

$$\frac{{}^A d\hat{\mathbf{b}}_1}{dt} \underset{(15)}{=} \dot{\alpha}_1 \hat{\mathbf{a}}_1 + \dot{\alpha}_2 \hat{\mathbf{a}}_2 + \dot{\alpha}_3 \hat{\mathbf{a}}_3 \tag{18}$$

Table 2.1.1

i	α_i	β_i	γ_i
1	0.9363	−0.2896	0.1987
2	0.3130	0.9447	−0.0981
3	−0.1593	0.1540	0.9751

Table 2.1.2

i	$\dot{\alpha}_i$ (rad/s)	$\dot{\beta}_i$ (rad/s)	$\dot{\gamma}_i$ (rad/s)
1	−0.0127	−0.0261	0.0216
2	0.0303	−0.0103	−0.0032
3	−0.0148	0.0145	−0.0047

$$\frac{{}^A d\hat{\mathbf{b}}_2}{dt} \underset{(16)}{=} \dot{\beta}_1 \hat{\mathbf{a}}_1 + \dot{\beta}_2 \hat{\mathbf{a}}_2 + \dot{\beta}_3 \hat{\mathbf{a}}_3 \tag{19}$$

$$\frac{{}^A d\hat{\mathbf{b}}_3}{dt} \underset{(17)}{=} \dot{\gamma}_1 \hat{\mathbf{a}}_1 + \dot{\gamma}_2 \hat{\mathbf{a}}_2 + \dot{\gamma}_3 \hat{\mathbf{a}}_3 \tag{20}$$

and

$$\begin{aligned} {}^A\boldsymbol{\omega}^B \underset{(1)}{=} \; & \hat{\mathbf{b}}_1 \underset{(19,\,17)}{(\dot{\beta}_1\gamma_1 + \dot{\beta}_2\gamma_2 + \dot{\beta}_3\gamma_3)} + \hat{\mathbf{b}}_2 \underset{(20,\,15)}{(\dot{\gamma}_1\alpha_1 + \dot{\gamma}_2\alpha_2 + \dot{\gamma}_3\alpha_3)} \\ & + \hat{\mathbf{b}}_3 \underset{(18,\,16)}{(\dot{\alpha}_1\beta_1 + \dot{\alpha}_2\beta_2 + \dot{\alpha}_3\beta_3)} \end{aligned} \tag{21}$$

Thus, at time t^*, Eq. (21) together with Tables 2.1.1 and 2.1.2 yields

$${}^A\boldsymbol{\omega}^B = 0.010\hat{\mathbf{b}}_1 + 0.020\hat{\mathbf{b}}_2 + 0.030\hat{\mathbf{b}}_3 \qquad \text{rad/s} \tag{22}$$

In Fig. 2.1.2, B represents a door supported by hinges in a room A. Mutually perpendicular unit vectors $\hat{\mathbf{a}}_1$, $\hat{\mathbf{a}}_2$, $\hat{\mathbf{a}}_3$ are fixed in A, with $\hat{\mathbf{a}}_3$ parallel to the axis of the hinges, and mutually perpendicular unit vectors $\hat{\mathbf{b}}_1$, $\hat{\mathbf{b}}_2$, $\hat{\mathbf{b}}_3$ are fixed in B, with $\hat{\mathbf{b}}_3 = \hat{\mathbf{a}}_3$. If θ is the radian measure of the angle between $\hat{\mathbf{a}}_1$ and $\hat{\mathbf{b}}_1$ as shown in Fig. 2.1.2, then $\hat{\mathbf{a}}_1$, $\hat{\mathbf{a}}_2$, $\hat{\mathbf{a}}_3$ and $\hat{\mathbf{b}}_1$, $\hat{\mathbf{b}}_2$, $\hat{\mathbf{b}}_3$ are related to each other as indicated in Table 2.1.3.

Table 2.1.3

	$\hat{\mathbf{b}}_1$	$\hat{\mathbf{b}}_2$	$\hat{\mathbf{b}}_3$
$\hat{\mathbf{a}}_1$	$\cos\theta$	$-\sin\theta$	0
$\hat{\mathbf{a}}_2$	$\sin\theta$	$\cos\theta$	0
$\hat{\mathbf{a}}_3$	0	0	1

The angular velocity of B in A, ${}^A\boldsymbol{\omega}^B$, found with the aid of Eq. (1) and Table 2.1.3, is given by†

$${}^A\boldsymbol{\omega}^B = \dot{\theta}\hat{\mathbf{b}}_3 \tag{23}$$

and the utility of Eq. (2) becomes apparent when one seeks to find, for example, the

† Use of the theorem stated in Sec. 2.2 allows one to write Eq. (23) by inspection.

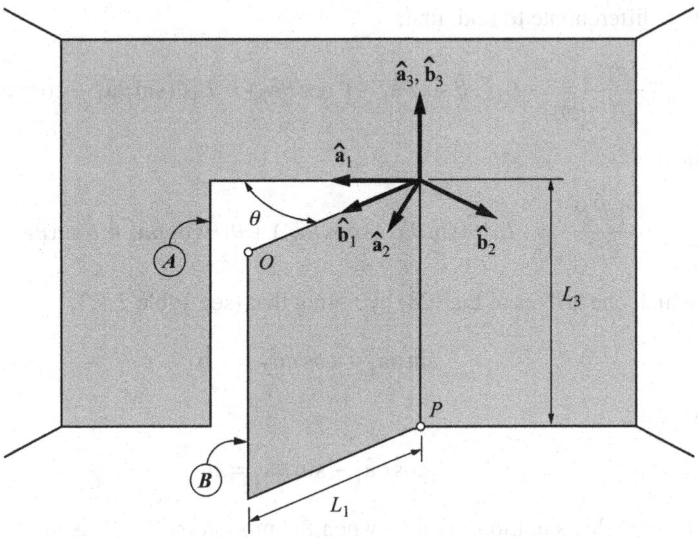

Figure 2.1.2

second time-derivative in A of the position vector from the point O shown in Fig. 2.1.2 to the point P, that is, of the vector $\boldsymbol{\beta}$ given by

$$\boldsymbol{\beta} = -L_1\hat{\mathbf{b}}_1 - L_3\hat{\mathbf{b}}_3 \tag{24}$$

For, using Eq. (2), one immediately has

$$\frac{{}^A d\boldsymbol{\beta}}{dt} \underset{(2)}{=} \underset{(23)}{\dot{\theta}\hat{\mathbf{b}}_3} \times \underset{(24)}{(-L_1\hat{\mathbf{b}}_1 - L_3\hat{\mathbf{b}}_3)} = -L_1\dot{\theta}\hat{\mathbf{b}}_2 \tag{25}$$

so that one can write

$$\frac{{}^A d^2\boldsymbol{\beta}}{dt^2} = \frac{{}^A d}{dt}\left(\frac{{}^A d\boldsymbol{\beta}}{dt}\right) \underset{(25)}{=} -L_1\ddot{\theta}\hat{\mathbf{b}}_2 - L_1\dot{\theta}\frac{{}^A d\hat{\mathbf{b}}_2}{dt} \tag{26}$$

Since $\hat{\mathbf{b}}_2$ is a vector fixed in B, its time derivative in A can be found with the aid of Eq. (2); that is,

$$\frac{{}^A d\hat{\mathbf{b}}_2}{dt} \underset{(2)}{=} {}^A\boldsymbol{\omega}^B \times \hat{\mathbf{b}}_2 \underset{(23)}{=} \dot{\theta}\hat{\mathbf{b}}_3 \times \hat{\mathbf{b}}_2 = -\dot{\theta}\hat{\mathbf{b}}_1 \tag{27}$$

Consequently,

$$\frac{{}^A d^2\boldsymbol{\beta}}{dt^2} \underset{(26,\,27)}{=} L_1(\dot{\theta}^2\hat{\mathbf{b}}_1 - \ddot{\theta}\hat{\mathbf{b}}_2) \tag{28}$$

To obtain the same result without the use of Eq. (2), one must write (see Table 2.1.3)

$$\boldsymbol{\beta} \underset{(24)}{=} -L_1(\cos\theta\hat{\mathbf{a}}_1 + \sin\theta\hat{\mathbf{a}}_2) - L_3\hat{\mathbf{a}}_3 \tag{29}$$

and then differentiate to find, first,

$$\frac{{}^A d\boldsymbol{\beta}}{dt} \underset{(29)}{=} -L_1(-\dot{\theta}\sin\theta\hat{\mathbf{a}}_1 + \dot{\theta}\cos\theta\hat{\mathbf{a}}_2) = L_1\dot{\theta}(\sin\theta\hat{\mathbf{a}}_1 - \cos\theta\hat{\mathbf{a}}_2) \tag{30}$$

and, next,

$$\frac{{}^A d^2\boldsymbol{\beta}}{dt^2} \underset{(30)}{=} L_1[\ddot{\theta}(\sin\theta\hat{\mathbf{a}}_1 - \cos\theta\hat{\mathbf{a}}_2) + \dot{\theta}(\dot{\theta}\cos\theta\hat{\mathbf{a}}_1 + \dot{\theta}\sin\theta\hat{\mathbf{a}}_2)] \tag{31}$$

after which one arrives at Eq. (28) by noting that (see Table 2.1.3)

$$\sin\theta\hat{\mathbf{a}}_1 - \cos\theta\hat{\mathbf{a}}_2 = -\hat{\mathbf{b}}_2 \tag{32}$$

while

$$\cos\theta\hat{\mathbf{a}}_1 + \sin\theta\hat{\mathbf{a}}_2 = \hat{\mathbf{b}}_1 \tag{33}$$

In more complex situations, that is, when the motion of B in A is more complicated than that of a door B in a room A, the use of the angular velocity vector as an "operator," which, through cross multiplication, produces time derivatives, is all the more advantageous.

2.2 SIMPLE ANGULAR VELOCITY

When a rigid body B moves in a reference frame A in such a way that there exists throughout some time interval a unit vector $\hat{\mathbf{k}}$ whose orientation in both A and B is independent of the time t, then B is said to have a *simple angular velocity* in A throughout this time interval, and this angular velocity can be expressed as

$${}^A\boldsymbol{\omega}^B = \omega\hat{\mathbf{k}} \tag{1}$$

with ω defined as

$$\omega \triangleq \dot{\theta} \tag{2}$$

where θ is the radian measure of the angle between a line L_A whose orientation is fixed in A and a line L_B similarly fixed in B (see Fig. 2.2.1), both lines are perpendicular to $\hat{\mathbf{k}}$, and θ is regarded as positive when the angle can be generated by a rotation of B relative to A during which a right-handed screw rigidly attached to B and parallel to $\hat{\mathbf{k}}$ advances in the direction of $\hat{\mathbf{k}}$. The scalar quantity ω is called an *angular speed* of B in A. [The indefinite article "an" is used here because, if $\hat{\mathbf{k}}'$ and ω' are defined as $\hat{\mathbf{k}}' \triangleq -\hat{\mathbf{k}}$ and $\omega' \triangleq -\omega$, then Eq. (1) can be written ${}^A\boldsymbol{\omega}^B = \omega'\hat{\mathbf{k}}'$ so that ω' is no less "the" angular speed than is ω.]

Derivation Let $\hat{\mathbf{a}}_1$, $\hat{\mathbf{a}}_2$, $\hat{\mathbf{a}}_3$ be a right-handed set of mutually perpendicular unit vectors fixed in A, with $\hat{\mathbf{a}}_1$ parallel to a line L_A and $\hat{\mathbf{a}}_3 = \hat{\mathbf{k}}$, and let $\hat{\mathbf{b}}_1$, $\hat{\mathbf{b}}_2$, $\hat{\mathbf{b}}_3$ be a similar set of

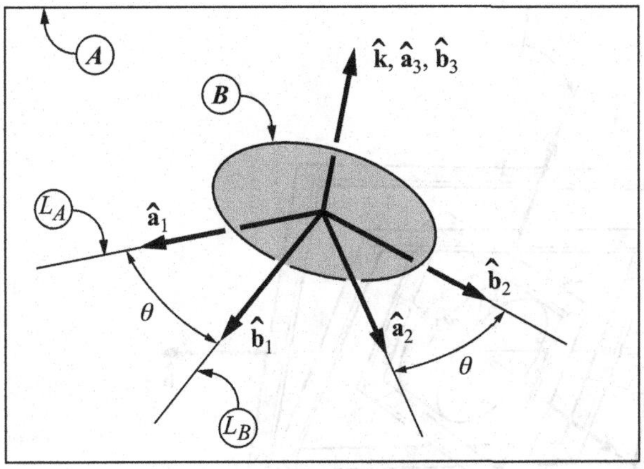

Figure 2.2.1

unit vectors fixed in B, with $\hat{\mathbf{b}}_1$ parallel to L_B and $\hat{\mathbf{b}}_3 = \hat{\mathbf{k}}$. Then

$$\hat{\mathbf{b}}_1 = \cos\theta\hat{\mathbf{a}}_1 + \sin\theta\hat{\mathbf{a}}_2 \tag{3}$$

$$\hat{\mathbf{b}}_2 = -\sin\theta\hat{\mathbf{a}}_1 + \cos\theta\hat{\mathbf{a}}_2 \tag{4}$$

$$\hat{\mathbf{b}}_3 = \hat{\mathbf{a}}_3 \tag{5}$$

$$\frac{^A d\hat{\mathbf{b}}_1}{dt} \underset{(3)}{=} \dot{\theta}(-\sin\theta\hat{\mathbf{a}}_1 + \cos\theta\hat{\mathbf{a}}_2) \underset{(4)}{=} \dot{\theta}\hat{\mathbf{b}}_2 \tag{6}$$

$$\frac{^A d\hat{\mathbf{b}}_2}{dt} \underset{(4)}{=} \dot{\theta}(-\cos\theta\hat{\mathbf{a}}_1 - \sin\theta\hat{\mathbf{a}}_2) \underset{(3)}{=} -\dot{\theta}\hat{\mathbf{b}}_1 \tag{7}$$

$$\frac{^A d\hat{\mathbf{b}}_3}{dt} \underset{(5)}{=} \mathbf{0} \tag{8}$$

and substitution into Eq. (2.1.1) leads directly to

$$^A\boldsymbol{\omega}^B = \dot{\theta}\hat{\mathbf{b}}_3 = \dot{\theta}\hat{\mathbf{k}} \tag{9}$$

Example Simple angular velocities are encountered most frequently in connection with bodies that are meant to rotate relative to each other about axes fixed in the bodies, such as the rotor and the gimbal rings of a gyroscope. This is illustrated in Fig. 2.2.2, where A designates an aircraft that carries a gyroscope consisting of an outer gimbal B, an inner gimbal C, and a rotor D. Here, the angular velocities of B in A, C in B, and D in C all are simple angular velocities, and they can be expressed as

$$^A\boldsymbol{\omega}^B = \dot{q}_1\hat{\mathbf{b}} \qquad ^B\boldsymbol{\omega}^C = \dot{q}_2\hat{\mathbf{c}} \qquad ^C\boldsymbol{\omega}^D = -\dot{q}_3\hat{\mathbf{d}} \tag{10}$$

where q_1, q_2, q_3 measure angles as indicated in Fig. 2.2.2 and $\hat{\mathbf{b}}$, $\hat{\mathbf{c}}$, $\hat{\mathbf{d}}$ are unit vectors directed as shown. (Note that, for example, $^A\boldsymbol{\omega}^C$ and $^B\boldsymbol{\omega}^D$ are *not* simple angular velocities.)

A rigid body B need not be mounted in bearings fixed in a reference frame A in order to have a simple angular velocity in A. Indeed, it is possible for B to have

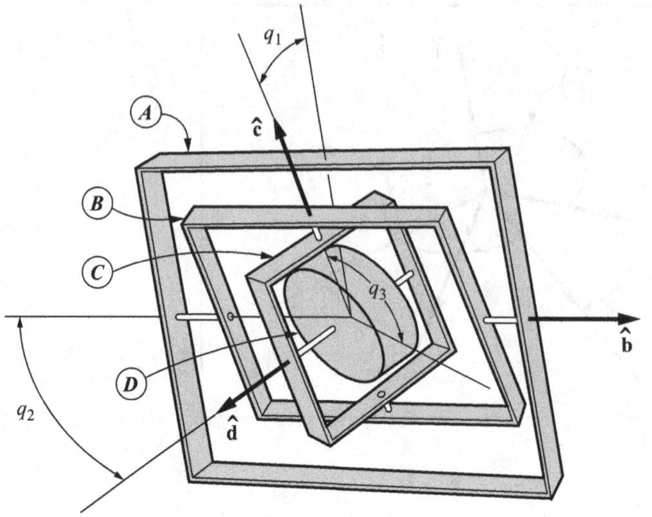

Figure 2.2.2

a simple angular velocity in A when B moves in such a way that *no* point of B remains fixed in A. For example, suppose that A is an aircraft in level flight above a hilly roadway that lies in a vertical plane, as depicted in Fig. 2.2.3, and that B is an automobile wheel traversing the roadway. No point of B is fixed in A, but ${}^A\boldsymbol{\omega}^B$ is a simple angular velocity, the role of $\hat{\mathbf{k}}$ being played by any unit vector that is perpendicular to the middle plane of the wheel.

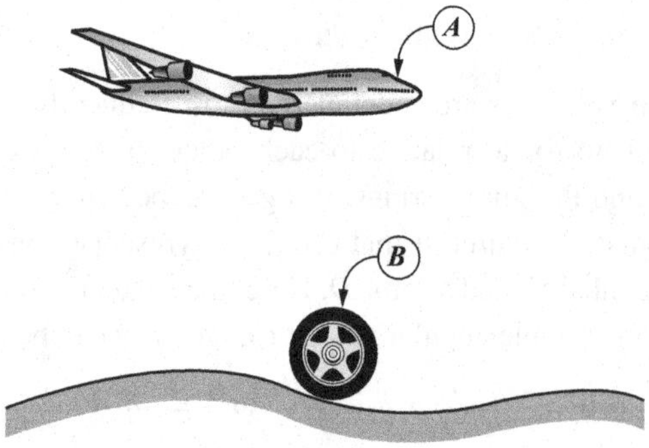

Figure 2.2.3

2.3 DIFFERENTIATION IN TWO REFERENCE FRAMES

If A and B are any two reference frames, the first time-derivatives of any vector $\mathbf{v}$ in A and in B are related to each other as follows:

$$\frac{^A d\mathbf{v}}{dt} = \frac{^B d\mathbf{v}}{dt} + {^A\boldsymbol{\omega}^B} \times \mathbf{v} \tag{1}$$

where ${^A\boldsymbol{\omega}^B}$ is the angular velocity of B in A (see Sec. 2.1).

Derivation With $\hat{\mathbf{b}}_1, \hat{\mathbf{b}}_2, \hat{\mathbf{b}}_3$ as in Sec. 2.1, let

$$v_i \triangleq \mathbf{v} \cdot \hat{\mathbf{b}}_i \qquad (i = 1, 2, 3) \tag{2}$$

so that [see Eqs. (1.6.1) and (1.6.2)]

$$\mathbf{v} = \sum_{i=1}^{3} v_i \hat{\mathbf{b}}_i \tag{3}$$

Then

$$\begin{aligned} \frac{^A d\mathbf{v}}{dt} &\underset{(3)}{=} \sum_{i=1}^{3} \frac{dv_i}{dt}\hat{\mathbf{b}}_i + \sum_{i=1}^{3} v_i \frac{^A d\hat{\mathbf{b}}_i}{dt} \\ &= \underset{(1.7.2)}{\frac{^B d\mathbf{v}}{dt}} + \sum_{i=1}^{3} v_i \underset{(2.1.2)}{{^A\boldsymbol{\omega}^B} \times \hat{\mathbf{b}}_i} \\ &= \frac{^B d\mathbf{v}}{dt} + {^A\boldsymbol{\omega}^B} \times \sum_{i=1}^{3} v_i \hat{\mathbf{b}}_i \\ &= \frac{^B d\mathbf{v}}{dt} + {^A\boldsymbol{\omega}^B} \times \underset{(3)}{\mathbf{v}} \end{aligned} \tag{4}$$

Equation (1) enables one to find the time derivative of $\mathbf{v}$ in A without having to resolve $\mathbf{v}$ into components parallel to unit vectors fixed in A.

Example A vector $\mathbf{H}$, called the *central angular momentum* of a rigid body B in a reference frame A, can be expressed as

$$\mathbf{H} = I_1\omega_1\hat{\mathbf{b}}_1 + I_2\omega_2\hat{\mathbf{b}}_2 + I_3\omega_3\hat{\mathbf{b}}_3 \tag{5}$$

where $\hat{\mathbf{b}}_1, \hat{\mathbf{b}}_2, \hat{\mathbf{b}}_3$ form a certain set of mutually perpendicular unit vectors fixed in B, ω_j is defined as

$$\omega_j \triangleq {^A\boldsymbol{\omega}^B} \cdot \hat{\mathbf{b}}_j \qquad (j = 1, 2, 3) \tag{6}$$

and I_1, I_2, I_3 are constants, called central principal moments of inertia of B. With the aid of Eq. (1), one can find M_1, M_2, M_3 such that the first time-derivative of $\mathbf{H}$ in A is given by

$$\frac{^A d\mathbf{H}}{dt} = M_1\hat{\mathbf{b}}_1 + M_2\hat{\mathbf{b}}_2 + M_3\hat{\mathbf{b}}_3 \tag{7}$$

Specifically,

$$\frac{{}^{A}d\mathbf{H}}{dt} \underset{(1)}{=} \frac{{}^{B}d\mathbf{H}}{dt} + {}^{A}\boldsymbol{\omega}^{B} \times \mathbf{H} \tag{8}$$

$$\frac{{}^{B}d\mathbf{H}}{dt} \underset{(5)}{=} I_1\dot{\omega}_1\hat{\mathbf{b}}_1 + I_2\dot{\omega}_2\hat{\mathbf{b}}_2 + I_3\dot{\omega}_3\hat{\mathbf{b}}_3 \tag{9}$$

$${}^{A}\boldsymbol{\omega}^{B} \underset{(6)}{=} \omega_1\hat{\mathbf{b}}_1 + \omega_2\hat{\mathbf{b}}_2 + \omega_3\hat{\mathbf{b}}_3 \tag{10}$$

$${}^{A}\boldsymbol{\omega}^{B} \times \mathbf{H} \underset{(10,\,5)}{=} (\omega_2 I_3\omega_3 - \omega_3 I_2\omega_2)\hat{\mathbf{b}}_1 + \cdots \tag{11}$$

$$\frac{{}^{A}d\mathbf{H}}{dt} \underset{(8)}{=} [\underset{(9)}{I_1\dot{\omega}_1} - \underset{(11)}{(I_2 - I_3)\omega_2\omega_3}]\hat{\mathbf{b}}_1 + \cdots \tag{12}$$

It follows from Eqs. (7) and (12) that

$$M_1 = I_1\dot{\omega}_1 - (I_2 - I_3)\omega_2\omega_3 \tag{13}$$

$$M_2 = I_2\dot{\omega}_2 - (I_3 - I_1)\omega_3\omega_1 \tag{14}$$

$$M_3 = I_3\dot{\omega}_3 - (I_1 - I_2)\omega_1\omega_2 \tag{15}$$

2.4 AUXILIARY REFERENCE FRAMES

The angular velocity of a rigid body B in a reference frame A (see Sec. 2.1) can be expressed in the following form involving n auxiliary reference frames $A_1, \ldots, A_n$:

$${}^{A}\boldsymbol{\omega}^{B} = {}^{A}\boldsymbol{\omega}^{A_1} + {}^{A_1}\boldsymbol{\omega}^{A_2} + \cdots + {}^{A_{n-1}}\boldsymbol{\omega}^{A_n} + {}^{A_n}\boldsymbol{\omega}^{B} \tag{1}$$

This relationship, the *addition theorem for angular velocities*, is particularly useful when each term in the right-hand member represents a simple angular velocity (see Sec. 2.2) and can, therefore, be expressed as in Eq. (2.2.1). However, Eq. (1) applies even when one or more of ${}^{A}\boldsymbol{\omega}^{A_1}, \ldots, {}^{A_n}\boldsymbol{\omega}^{B}$ are not simple angular velocities.

The reference frames $A_1, \ldots, A_n$ may or may not correspond to actual rigid bodies. Frequently, such reference frames are introduced as aids in analysis but have no physical counterparts.

When the angular velocity of B in A is resolved into components (that is, when ${}^{A}\boldsymbol{\omega}^{B}$ is expressed as the sum of a number of vectors), these components may always be regarded as angular velocities of certain bodies in certain reference frames. Indeed, Eq. (1) represents precisely such a resolution of ${}^{A}\boldsymbol{\omega}^{B}$ into components. In no case, however, are these components themselves angular velocities of B in A, for there exists at any one instant only one angular velocity of B in A. In other words, B cannot possess simultaneously[†] several angular velocities in A.

† In the literature, one encounters not infrequently the equation $\boldsymbol{\omega} = \boldsymbol{\omega}_1 + \boldsymbol{\omega}_2$, accompanied by a discussion of "simultaneous angular velocities of a rigid body" and/or "the vector character of angular velocity." Moreover, $\boldsymbol{\omega}_1$ and $\boldsymbol{\omega}_2$ often are called angular velocities of B about certain axes. This leads one to wonder how many such axes exist in a given case, how one can locate them, and so forth. Since the notion of "angular velocity about an axis" serves no useful purpose, it is best simply to dispense with it.

Derivation For any vector $\boldsymbol{\beta}$ fixed in B,

$$\frac{^{A}d\boldsymbol{\beta}}{dt} \underset{(2.1.2)}{=} {}^{A}\boldsymbol{\omega}^{B} \times \boldsymbol{\beta} \tag{2}$$

$$\frac{^{A_1}d\boldsymbol{\beta}}{dt} \underset{(2.1.2)}{=} {}^{A_1}\boldsymbol{\omega}^{B} \times \boldsymbol{\beta} \tag{3}$$

and

$$\frac{^{A}d\boldsymbol{\beta}}{dt} \underset{(2.3.1)}{=} \frac{^{A_1}d\boldsymbol{\beta}}{dt} + {}^{A}\boldsymbol{\omega}^{A_1} \times \boldsymbol{\beta} \tag{4}$$

Hence,

$$\underset{(2)}{{}^{A}\boldsymbol{\omega}^{B} \times \boldsymbol{\beta}} \underset{(4)}{=} \underset{(3)}{{}^{A_1}\boldsymbol{\omega}^{B} \times \boldsymbol{\beta}} + {}^{A}\boldsymbol{\omega}^{A_1} \times \boldsymbol{\beta} \tag{5}$$

Since this equation is satisfied by *every* $\boldsymbol{\beta}$ fixed in B, it implies that

$$^{A}\boldsymbol{\omega}^{B} = {}^{A}\boldsymbol{\omega}^{A_1} + {}^{A_1}\boldsymbol{\omega}^{B} \tag{6}$$

which shows that Eq. (1) is valid for $n = 1$. Proceeding similarly, one can verify that

$$^{A_1}\boldsymbol{\omega}^{B} = {}^{A_1}\boldsymbol{\omega}^{A_2} + {}^{A_2}\boldsymbol{\omega}^{B} \tag{7}$$

and substitution into Eq. (6) then yields

$$^{A}\boldsymbol{\omega}^{B} = {}^{A}\boldsymbol{\omega}^{A_1} + {}^{A_1}\boldsymbol{\omega}^{A_2} + {}^{A_2}\boldsymbol{\omega}^{B} \tag{8}$$

which is Eq. (1) with $n = 2$. The validity of Eq. (1) for any value of n thus can be established by applying this procedure a sufficient number of times.

Example In Fig. 2.4.1, q_1, q_2, and q_3 denote the radian measures of angles characterizing the orientation of a rigid cone B in a reference frame A. These angles are formed by lines described as follows: L_1 and L_2 are perpendicular to each other and fixed in A; L_3 is the axis of symmetry of B; L_4 is perpendicular to L_2 and intersects L_2 and L_3; L_5 is perpendicular to L_3 and intersects L_2 and L_3; L_6 is perpendicular to L_3 and is fixed in B; L_7 is perpendicular to L_2 and L_4. To find an expression for the angular velocity of B in A, one can designate as A_1, a reference frame in which L_2, L_4, and L_7 are fixed, and as A_2 a reference frame in which L_3, L_5, and L_7 are fixed, observing that L_2 then is fixed in both A and A_1, L_7 is fixed in both A_1 and A_2, and L_3 is fixed in both A_2 and B, so that, in accordance with Eqs. (2.2.1) and (2.2.2), one can write

$$^{A}\boldsymbol{\omega}^{A_1} = \dot{q}_1\hat{\mathbf{k}}_2 \qquad {}^{A_1}\boldsymbol{\omega}^{A_2} = \dot{q}_2\hat{\mathbf{k}}_7 \qquad {}^{A_2}\boldsymbol{\omega}^{B} = \dot{q}_3\hat{\mathbf{k}}_3 \tag{9}$$

where $\hat{\mathbf{k}}_2$, $\hat{\mathbf{k}}_7$, and $\hat{\mathbf{k}}_3$ are unit vectors directed as shown in Fig. 2.4.1. Substituting from Eq. (9) into Eq. (1) with $n = 2$, one arrives at

$$^{A}\boldsymbol{\omega}^{B} = \dot{q}_1\hat{\mathbf{k}}_2 + \dot{q}_2\hat{\mathbf{k}}_7 + \dot{q}_3\hat{\mathbf{k}}_3 \tag{10}$$

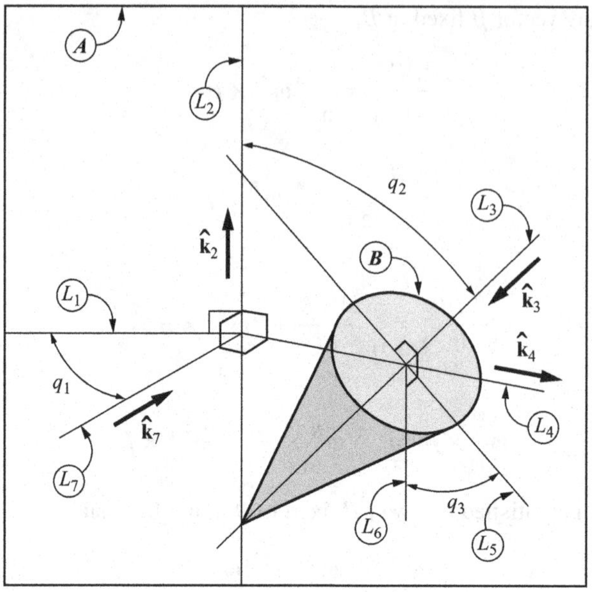

Figure 2.4.1

2.5 ANGULAR ACCELERATION

The *angular acceleration* ${}^A\boldsymbol{\alpha}^B$ of a rigid body B in a reference frame A is defined as the first time-derivative in A of the angular velocity of B in A (see Sec. 2.1):

$$ {}^A\boldsymbol{\alpha}^B \triangleq \frac{{}^A d\,{}^A\boldsymbol{\omega}^B}{dt} \tag{1} $$

Since the first time-derivatives of ${}^A\boldsymbol{\omega}^B$ in A and in B are equal to each other, as becomes evident when one replaces $\mathbf{v}$ in Eq. (2.3.1) with ${}^A\boldsymbol{\omega}^B$, Eq. (1) implies that

$$ {}^A\boldsymbol{\alpha}^B = \frac{{}^B d\,{}^A\boldsymbol{\omega}^B}{dt} \tag{2} $$

which furnishes a convenient way to find ${}^A\boldsymbol{\alpha}^B$ when ${}^A\boldsymbol{\omega}^B$ has been expressed in terms of components parallel to unit vectors fixed in B.

If $A_1, \ldots, A_n$ are n auxiliary reference frames, ${}^A\boldsymbol{\alpha}^B$ is *not*, in general, equal to the sum ${}^A\boldsymbol{\alpha}^{A_1} + {}^{A_1}\boldsymbol{\alpha}^{A_2} + \cdots + {}^{A_n}\boldsymbol{\alpha}^B$. Thus, Eq. (2.4.1) does not, in general, have an angular acceleration counterpart.

The angular velocity of B in A can always be expressed as ${}^A\boldsymbol{\omega}^B = \omega\hat{\mathbf{k}}_\omega$, where $\hat{\mathbf{k}}_\omega$ is a unit vector parallel to ${}^A\boldsymbol{\omega}^B$; similarly, ${}^A\boldsymbol{\alpha}^B$ can always be expressed as ${}^A\boldsymbol{\alpha}^B = \alpha\hat{\mathbf{k}}_\alpha$, where $\hat{\mathbf{k}}_\alpha$ is a unit vector parallel to ${}^A\boldsymbol{\alpha}^B$. In general, $\hat{\mathbf{k}}_\omega$ differs from $\hat{\mathbf{k}}_\alpha$, and $\alpha \neq d\omega/dt$. But when B has a simple angular velocity in A (see Sec. 2.2), and ${}^A\boldsymbol{\omega}^B$ is expressed as in Eq. (2.2.1), then

$$ {}^A\boldsymbol{\alpha}^B = \alpha\hat{\mathbf{k}} \tag{3} $$

where α, called a *scalar angular acceleration*, is given by

$$\alpha = \frac{d\omega}{dt} \tag{4}$$

Example Referring to the example in Sec. 2.4 and to Fig. 2.4.1, one can find an expression for the angular acceleration of the cone B in reference frame A as follows:

$$\begin{aligned} {}^A\boldsymbol{\alpha}^B &\underset{(1)}{=} \frac{{}^Ad}{dt}(\underset{(2.4.10)}{\dot{q}_1\hat{\mathbf{k}}_2 + \dot{q}_2\hat{\mathbf{k}}_7 + \dot{q}_3\hat{\mathbf{k}}_3}) \\ &= \ddot{q}_1\hat{\mathbf{k}}_2 + \dot{q}_1\frac{{}^Ad\hat{\mathbf{k}}_2}{dt} + \ddot{q}_2\hat{\mathbf{k}}_7 + \dot{q}_2\frac{{}^Ad\hat{\mathbf{k}}_7}{dt} + \ddot{q}_3\hat{\mathbf{k}}_3 + \dot{q}_3\frac{{}^Ad\hat{\mathbf{k}}_3}{dt} \end{aligned} \tag{5}$$

Since $\hat{\mathbf{k}}_2$ is fixed in A,

$$\frac{{}^Ad\hat{\mathbf{k}}_2}{dt} = \mathbf{0} \tag{6}$$

The unit vector $\hat{\mathbf{k}}_7$ is fixed in a reference frame previously called A_1 and having an angular velocity in A given by

$${}^A\boldsymbol{\omega}^{A_1} \underset{(2.4.9)}{=} \dot{q}_1\hat{\mathbf{k}}_2 \tag{7}$$

Hence,

$$\frac{{}^Ad\hat{\mathbf{k}}_7}{dt} \underset{(2.1.2)}{=} {}^A\boldsymbol{\omega}^{A_1} \times \hat{\mathbf{k}}_7 = \underset{(7)}{\dot{q}_1\,\hat{\mathbf{k}}_2} \times \hat{\mathbf{k}}_7 \tag{8}$$

Similarly, since $\hat{\mathbf{k}}_3$ is fixed in B,

$$\frac{{}^Ad\hat{\mathbf{k}}_3}{dt} = {}^A\boldsymbol{\omega}^B \times \hat{\mathbf{k}}_3 \underset{(2.4.10)}{=} \dot{q}_1\hat{\mathbf{k}}_2 \times \hat{\mathbf{k}}_3 + \dot{q}_2\hat{\mathbf{k}}_7 \times \hat{\mathbf{k}}_3 \tag{9}$$

Consequently,

$$\begin{aligned} {}^A\boldsymbol{\alpha}^B &\underset{(5)}{=} \ddot{q}_1\hat{\mathbf{k}}_2 + \underset{(6)}{\mathbf{0}} + \ddot{q}_2\hat{\mathbf{k}}_7 + \dot{q}_2\underset{(8)}{\dot{q}_1\,\hat{\mathbf{k}}_2 \times \hat{\mathbf{k}}_7} \\ &\quad + \ddot{q}_3\hat{\mathbf{k}}_3 + \dot{q}_3\,(\underset{(9)}{\dot{q}_1\hat{\mathbf{k}}_2 \times \hat{\mathbf{k}}_3 + \dot{q}_2\hat{\mathbf{k}}_7 \times \hat{\mathbf{k}}_3}) \end{aligned} \tag{10}$$

The angular accelerations of A_1 in A, A_2 in A_1, and B in A_2 are

$${}^A\boldsymbol{\alpha}^{A_1} \underset{(1)}{=} \frac{{}^Ad\,{}^A\boldsymbol{\omega}^{A_1}}{dt} \underset{(7)}{=} \ddot{q}_1\hat{\mathbf{k}}_2 \tag{11}$$

$${}^{A_1}\boldsymbol{\alpha}^{A_2} \underset{(1)}{=} \frac{{}^{A_1}d\,{}^{A_1}\boldsymbol{\omega}^{A_2}}{dt} \underset{(2.4.9)}{=} \ddot{q}_2\hat{\mathbf{k}}_7 \tag{12}$$

$${}^{A_2}\boldsymbol{\alpha}^B \underset{(1)}{=} \frac{{}^{A_2}d\,{}^{A_2}\boldsymbol{\omega}^B}{dt} \underset{(2.4.9)}{=} \ddot{q}_3\hat{\mathbf{k}}_3 \tag{13}$$

Hence,

$${}^A\boldsymbol{\alpha}^{A_1} + {}^{A_1}\boldsymbol{\alpha}^{A_2} + {}^{A_2}\boldsymbol{\alpha}^B = \underset{(11)}{\ddot{q}_1\hat{\mathbf{k}}_2} + \underset{(12)}{\ddot{q}_2\hat{\mathbf{k}}_7} + \underset{(13)}{\ddot{q}_3\hat{\mathbf{k}}_3} \underset{(10)}{\neq} {}^A\boldsymbol{\alpha}^B \tag{14}$$

2.6 VELOCITY AND ACCELERATION

The solution of nearly every problem in dynamics requires the formulation of expressions for velocities and accelerations of points of a system under consideration. At times, the most convenient way to generate the needed expressions is to use the definitions given in Eqs. (1) and (2). Frequently, however, much labor can be saved by appealing to the theorems† stated in Secs. 2.7 and 2.8.

Let **p** denote the position vector from any point O fixed in a reference frame A to a point P moving in A. The velocity of P in A and the acceleration of P in A, denoted by ${}^A\mathbf{v}^P$ and ${}^A\mathbf{a}^P$, respectively, are defined as

$$ {}^A\mathbf{v}^P \triangleq \frac{{}^A d\mathbf{p}}{dt} \tag{1} $$

and

$$ {}^A\mathbf{a}^P \triangleq \frac{{}^A d\,{}^A\mathbf{v}^P}{dt} \tag{2} $$

Example In Fig. 2.6.1, P_1 and P_2 designate two points connected by a line of length L and free to move in a plane B that is rotating at a constant rate ω about a line Y fixed both in B and in a reference frame A. The velocities ${}^A\mathbf{v}^{P_1}$ and ${}^A\mathbf{v}^{P_2}$ of P_1 and P_2 in A are to be expressed in terms of the quantities q_1, q_2, q_3, their time derivatives $\dot{q}_1, \dot{q}_2, \dot{q}_3$, and the mutually perpendicular unit vectors $\hat{\mathbf{e}}_x, \hat{\mathbf{e}}_y, \hat{\mathbf{e}}_z$ shown in Fig. 2.6.1.

The point O shown in Fig. 2.6.1 is a point fixed in A, and the position vector $\mathbf{p}_1$ from O to P_1 can be written

$$ \mathbf{p}_1 = q_1\hat{\mathbf{b}}_x + q_2\hat{\mathbf{b}}_y \tag{3} $$

where $\hat{\mathbf{b}}_x$ and $\hat{\mathbf{b}}_y$ are unit vectors directed as shown in Fig. 2.6.1. It follows that

$$ {}^A\mathbf{v}^{P_1} \underset{(1)}{=} \frac{{}^A d\mathbf{p}_1}{dt} \underset{(2.3.1)}{=} \frac{{}^B d\mathbf{p}_1}{dt} + {}^A\boldsymbol{\omega}^B \times \mathbf{p}_1 \tag{4} $$

where

$$ \frac{{}^B d\mathbf{p}_1}{dt} \underset{(3)}{=} \dot{q}_1\hat{\mathbf{b}}_x + \dot{q}_2\hat{\mathbf{b}}_y \tag{5} $$

and

$$ {}^A\boldsymbol{\omega}^B \times \mathbf{p}_1 = \underset{(2.2.1)}{\omega\hat{\mathbf{b}}_y} \times \underset{(3)}{(q_1\hat{\mathbf{b}}_x + q_2\hat{\mathbf{b}}_y)} = -\omega q_1\hat{\mathbf{b}}_z \tag{6} $$

so that

$$ {}^A\mathbf{v}^{P_1} \underset{(4)}{=} \underset{(5)}{\dot{q}_1\hat{\mathbf{b}}_x + \dot{q}_2\hat{\mathbf{b}}_y} - \underset{(6)}{\omega q_1\hat{\mathbf{b}}_z} \tag{7} $$

† The discussion of velocities and accelerations in Secs. 2.7 and 2.8 involves the concept of angular velocity. Hence, to come into position to present this material without a break in continuity, one must deal with angular velocity before taking up velocity and acceleration. Conversely, as Secs. 2.1–2.4 show, angular velocity can be discussed without any reference to velocity or acceleration. Therefore, it is both natural and advantageous to treat these topics in the order used here, that is, angular velocity before velocity and acceleration, rather than in the reverse order.

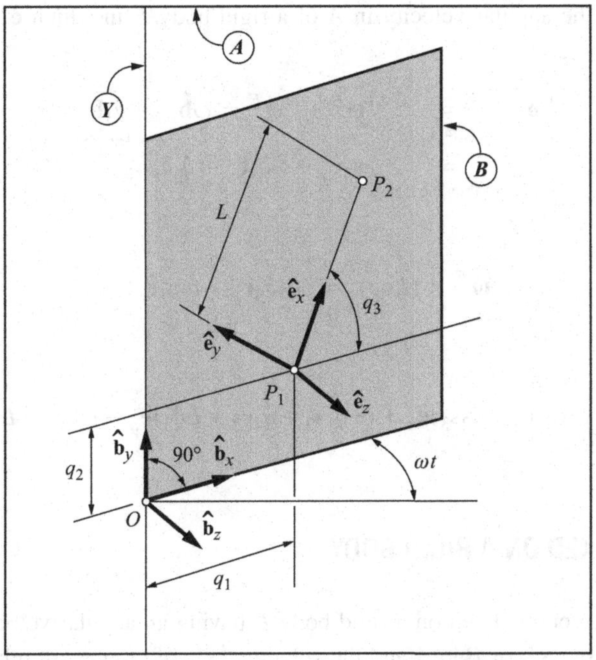

Figure 2.6.1

Since the unit vectors $\hat{\mathbf{b}}_x$, $\hat{\mathbf{b}}_y$, $\hat{\mathbf{b}}_z$ are related to the unit vectors $\hat{\mathbf{e}}_x$, $\hat{\mathbf{e}}_y$, $\hat{\mathbf{e}}_z$ as in Table 2.6.1, where s_3 and c_3 stand for $\sin q_3$ and $\cos q_3$, respectively, Eq. (7) is equivalent to

$$^A\mathbf{v}^{P_1} = (\dot{q}_1 c_3 + \dot{q}_2 s_3)\hat{\mathbf{e}}_x + (-\dot{q}_1 s_3 + \dot{q}_2 c_3)\hat{\mathbf{e}}_y - \omega q_1 \hat{\mathbf{e}}_z \tag{8}$$

which is the desired expression for the velocity of P_1 in A.

Table 2.6.1

	$\hat{\mathbf{e}}_x$	$\hat{\mathbf{e}}_y$	$\hat{\mathbf{e}}_z$
$\hat{\mathbf{b}}_x$	c_3	$-s_3$	0
$\hat{\mathbf{b}}_y$	s_3	c_3	0
$\hat{\mathbf{b}}_z$	0	0	1

The position vector $\mathbf{p}_2$ from O to P_2 can be written

$$\mathbf{p}_2 = \mathbf{p}_1 + L\hat{\mathbf{e}}_x \tag{9}$$

Hence,

$$^A\mathbf{v}^{P_2} \underset{(1)}{=} \frac{^A d\mathbf{p}_1}{dt} + \frac{^A d}{dt}(L\hat{\mathbf{e}}_x) \underset{(4)}{=} {}^A\mathbf{v}^{P_1} + {}^A\boldsymbol{\omega}^E \underset{(2.1.2)}{\times} (L\hat{\mathbf{e}}_x) \tag{10}$$

where ${}^A\boldsymbol{\omega}^E$ is the angular velocity in A of a rigid body E in which $\hat{\mathbf{e}}_x$, $\hat{\mathbf{e}}_y$, $\hat{\mathbf{e}}_z$ are fixed; that is,

$$
\begin{aligned}
{}^A\boldsymbol{\omega}^E &\underset{(2.4.1)}{=} {}^A\boldsymbol{\omega}^B + {}^B\boldsymbol{\omega}^E = \underset{(2.2.1)}{\omega\hat{\mathbf{b}}_y} + \underset{(2.2.1)}{\dot{q}_3\hat{\mathbf{b}}_z} \\
&\underset{\text{(Table 2.6.1)}}{=} \omega s_3\hat{\mathbf{e}}_x + \omega c_3\hat{\mathbf{e}}_y + \dot{q}_3\hat{\mathbf{e}}_z
\end{aligned} \tag{11}
$$

so that

$$
{}^A\boldsymbol{\omega}^E \times (L\hat{\mathbf{e}}_x) \underset{(11)}{=} L(\dot{q}_3\hat{\mathbf{e}}_y - \omega c_3\hat{\mathbf{e}}_z) \tag{12}
$$

and

$$
{}^A\mathbf{v}^{P_2} \underset{(10,\ 8,\ 12)}{=} (\dot{q}_1 c_3 + \dot{q}_2 s_3)\hat{\mathbf{e}}_x + (-\dot{q}_1 s_3 + \dot{q}_2 c_3 + L\dot{q}_3)\hat{\mathbf{e}}_y - \omega(q_1 + Lc_3)\hat{\mathbf{e}}_z \tag{13}
$$

2.7 TWO POINTS FIXED ON A RIGID BODY

If P and Q are two points fixed on a rigid body B having an angular velocity ${}^A\boldsymbol{\omega}^B$ in A, then the velocity ${}^A\mathbf{v}^P$ of P in A and the velocity ${}^A\mathbf{v}^Q$ of Q in A are related to each other as follows:

$$
{}^A\mathbf{v}^P = {}^A\mathbf{v}^Q + {}^A\boldsymbol{\omega}^B \times \mathbf{r} \tag{1}
$$

where $\mathbf{r}$ is the position vector from Q to P. The relationship between the acceleration ${}^A\mathbf{a}^P$ of P in A and the acceleration ${}^A\mathbf{a}^Q$ of Q in A involves the angular acceleration ${}^A\boldsymbol{\alpha}^B$ of B in A and is given by

$$
{}^A\mathbf{a}^P = {}^A\mathbf{a}^Q + {}^A\boldsymbol{\omega}^B \times ({}^A\boldsymbol{\omega}^B \times \mathbf{r}) + {}^A\boldsymbol{\alpha}^B \times \mathbf{r} \tag{2}
$$

Derivation Let O be a point fixed in A, $\mathbf{p}$ the position vector from O to P, and $\mathbf{q}$ the position vector from O to Q. Then

$$
\begin{aligned}
{}^A\mathbf{v}^P &\underset{(2.6.1)}{=} \frac{{}^Ad\mathbf{p}}{dt} = \frac{{}^Ad}{dt}(\mathbf{q}+\mathbf{r}) = \frac{{}^Ad\mathbf{q}}{dt} + \frac{{}^Ad\mathbf{r}}{dt} \\
&= \underset{(2.6.1)}{{}^A\mathbf{v}^Q} + \underset{(2.1.2)}{{}^A\boldsymbol{\omega}^B \times \mathbf{r}}
\end{aligned} \tag{3}
$$

and

$$
\begin{aligned}
{}^A\mathbf{a}^P &\underset{(2.6.2)}{=} \frac{{}^Ad\,{}^A\mathbf{v}^P}{dt} \underset{(3)}{=} \frac{{}^Ad\,{}^A\mathbf{v}^Q}{dt} + \frac{{}^Ad\,{}^A\boldsymbol{\omega}^B}{dt} \times \mathbf{r} + {}^A\boldsymbol{\omega}^B \times \frac{{}^Ad\mathbf{r}}{dt} \\
&= \underset{(2.6.2)}{{}^A\mathbf{a}^Q} + \underset{(2.5.1)}{{}^A\boldsymbol{\alpha}^B} \times \mathbf{r} + \underset{(2.1.2)}{{}^A\boldsymbol{\omega}^B \times ({}^A\boldsymbol{\omega}^B \times \mathbf{r})}
\end{aligned} \tag{4}
$$

Example Since an expression for the velocity of P_1 in A in the example in Sec. 2.6 is available in Eq. (2.6.7), the acceleration of P_1 in A can be found most directly by differentiating this expression, which yields

$$
\begin{aligned}
{}^A\mathbf{a}^{P_1} &\underset{(2.6.2)}{=} \frac{{}^A d\,{}^A\mathbf{v}^{P_1}}{dt} \underset{(2.3.1)}{=} \frac{{}^B d\,{}^A\mathbf{v}^{P_1}}{dt} + {}^A\boldsymbol{\omega}^B \times {}^A\mathbf{v}^{P_1} \\
&= \ddot{q}_1\hat{\mathbf{b}}_x + \ddot{q}_2\hat{\mathbf{b}}_y - \omega\dot{q}_1\hat{\mathbf{b}}_z + \omega\hat{\mathbf{b}}_y \times {}^A\mathbf{v}^{P_1} \quad \text{(2.6.7) (2.2.1)} \\
&\underset{(2.6.7)}{=} (\ddot{q}_1 - \omega^2 q_1)\hat{\mathbf{b}}_x + \ddot{q}_2\hat{\mathbf{b}}_y - 2\omega\dot{q}_1\hat{\mathbf{b}}_z
\end{aligned} \tag{5}
$$

In the case of P_2, it is more convenient to use Eq. (2) together with the result just obtained than it is to differentiate the expression for ${}^A\mathbf{v}^{P_2}$ available in Eq. (2.6.13). Letting P_1 and P_2 play the parts of Q and P, respectively, in Eq. (2), and replacing B with E, since P_1 and P_2 are fixed on E, not on B, one can write

$$
{}^A\mathbf{a}^{P_2} \underset{(2)}{=} {}^A\mathbf{a}^{P_1} + {}^A\boldsymbol{\omega}^E \times [{}^A\boldsymbol{\omega}^E \times (L\hat{\mathbf{e}}_x)] + {}^A\boldsymbol{\alpha}^E \times (L\hat{\mathbf{e}}_x) \tag{6}
$$

Now,

$$
{}^A\boldsymbol{\omega}^E \times [{}^A\boldsymbol{\omega}^E \times (L\hat{\mathbf{e}}_x)] \underset{(2.6.11,\ 2.6.12)}{=} L[-(\omega^2 c_3{}^2 + \dot{q}_3{}^2)\hat{\mathbf{e}}_x + \omega^2 s_3 c_3\hat{\mathbf{e}}_y + \omega\dot{q}_3 s_3\hat{\mathbf{e}}_z] \tag{7}
$$

and

$$
{}^A\boldsymbol{\alpha}^E \underset{(2.5.2)}{=} \frac{{}^E d\,{}^A\boldsymbol{\omega}^E}{dt} \underset{(2.6.11)}{=} \omega\dot{q}_3 c_3\hat{\mathbf{e}}_x - \omega\dot{q}_3 s_3\hat{\mathbf{e}}_y + \ddot{q}_3\hat{\mathbf{e}}_z \tag{8}
$$

so that

$$
{}^A\boldsymbol{\alpha}^E \times (L\hat{\mathbf{e}}_x) \underset{(8)}{=} L(\ddot{q}_3\hat{\mathbf{e}}_y + \omega\dot{q}_3 s_3\hat{\mathbf{e}}_z) \tag{9}
$$

Substitution from Eqs. (5), (7), and (9) into Eq. (6) yields

$$
\begin{aligned}
{}^A\mathbf{a}^{P_2} = (\ddot{q}_1 - \omega^2 q_1)\hat{\mathbf{b}}_x + \ddot{q}_2\hat{\mathbf{b}}_y - 2\omega(\dot{q}_1 - L\dot{q}_3 s_3)\hat{\mathbf{b}}_z \\
+ L[-(\omega^2 c_3{}^2 + \dot{q}_3{}^2)\hat{\mathbf{e}}_x + (\omega^2 s_3 c_3 + \ddot{q}_3)\hat{\mathbf{e}}_y]
\end{aligned} \tag{10}
$$

If one wishes to express this vector solely in terms of $\hat{\mathbf{b}}_x$, $\hat{\mathbf{b}}_y$, $\hat{\mathbf{b}}_z$, one can refer to Table 2.6.1 to obtain

$$
\begin{aligned}
{}^A\mathbf{a}^{P_2} &\underset{(10)}{=} [\ddot{q}_1 - \omega^2 q_1 - L(\ddot{q}_3 s_3 + \dot{q}_3{}^2 c_3 + \omega^2 c_3)]\hat{\mathbf{b}}_x \\
&\quad + [\ddot{q}_2 + L(\ddot{q}_3 c_3 - \dot{q}_3{}^2 s_3)]\hat{\mathbf{b}}_y - 2\omega(\dot{q}_1 - L\dot{q}_3 s_3)\hat{\mathbf{b}}_z
\end{aligned} \tag{11}
$$

2.8 ONE POINT MOVING ON A RIGID BODY

If a point P is moving on a rigid body B while B is moving in a reference frame A, the velocity ${}^A\mathbf{v}^P$ of P in A is related to the velocity ${}^B\mathbf{v}^P$ of P in B as follows:

$$
{}^A\mathbf{v}^P = {}^A\mathbf{v}^{\overline{B}} + {}^B\mathbf{v}^P \tag{1}
$$

where ${}^A\mathbf{v}^{\overline{B}}$ denotes the velocity in A of the point $\overline{B}$ of B that coincides with P at the instant under consideration. The acceleration ${}^A\mathbf{a}^P$ of P in A is given by

$$
{}^A\mathbf{a}^P = {}^A\mathbf{a}^{\overline{B}} + {}^B\mathbf{a}^P + 2\,{}^A\boldsymbol{\omega}^B \times {}^B\mathbf{v}^P \tag{2}
$$

where ${}^A\mathbf{a}^{\overline{B}}$ is the acceleration of $\overline{B}$ in A, ${}^B\mathbf{a}^P$ is the acceleration of P in B, and ${}^A\boldsymbol{\omega}^B$ is the angular velocity of B in A. The term $2{}^A\boldsymbol{\omega}^B \times {}^B\mathbf{v}^P$ is referred to as "Coriolis acceleration."

Derivation Let $\tilde{A}$ be a point fixed in A, $\tilde{B}$ a point fixed in B, $\mathbf{p}$ the position vector from $\tilde{A}$ to P, $\mathbf{q}$ the position vector from $\tilde{B}$ to P, and $\mathbf{r}$ the position vector from $\tilde{A}$ to $\tilde{B}$, as shown in Fig. 2.8.1. Then, in accordance with Eq. (2.6.1), the velocities ${}^A\mathbf{v}^P$, ${}^B\mathbf{v}^P$, and ${}^A\mathbf{v}^{\tilde{B}}$ are given by

$$ {}^A\mathbf{v}^P = \frac{{}^A d\mathbf{p}}{dt} \tag{3} $$

$$ {}^B\mathbf{v}^P = \frac{{}^B d\mathbf{q}}{dt} \tag{4} $$

$$ {}^A\mathbf{v}^{\tilde{B}} = \frac{{}^A d\mathbf{r}}{dt} \tag{5} $$

As can be seen in Fig. 2.8.1,

$$ \mathbf{p} = \mathbf{r} + \mathbf{q} \tag{6} $$

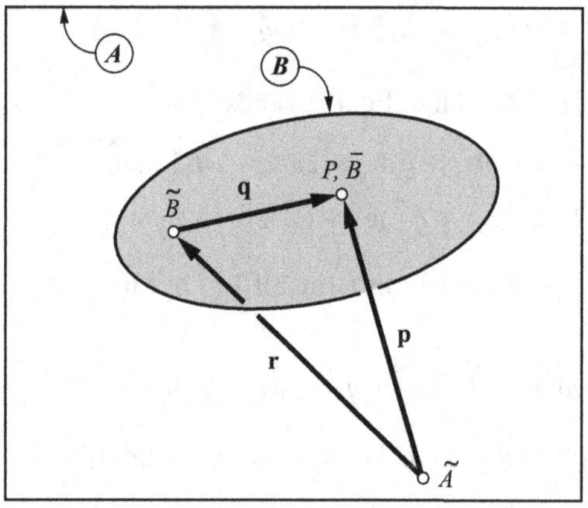

Figure 2.8.1

Hence,

$$ \begin{aligned} {}^A\mathbf{v}^P &\underset{(3,\,6)}{=} \frac{{}^A d\mathbf{r}}{dt} + \frac{{}^A d\mathbf{q}}{dt} \\ &= \underset{(5)}{{}^A\mathbf{v}^{\tilde{B}}} + \frac{{}^B d\mathbf{q}}{dt} \underset{(2.3.1)}{+} {}^A\boldsymbol{\omega}^B \times \mathbf{q} \\ &= {}^A\mathbf{v}^{\tilde{B}} + \underset{(4)}{{}^B\mathbf{v}^P} + {}^A\boldsymbol{\omega}^B \times \mathbf{q} \end{aligned} \tag{7} $$

$\widetilde{B}$ may always be taken as $\overline{B}$, that is, as the point of B that coincides with P at the instant under consideration. In that event,

$$ {}^{A}\mathbf{v}^{\widetilde{B}} = {}^{A}\mathbf{v}^{\overline{B}} \qquad \mathbf{q} = \mathbf{0} \tag{8}$$

and substitution from Eqs. (8) into Eq. (7) leads to Eq. (1).

Referring to Eq. (2.6.2), one can write

$$ {}^{A}\mathbf{a}^{P} = \frac{{}^{A}d\,{}^{A}\mathbf{v}^{P}}{dt} \tag{9}$$

$$ {}^{B}\mathbf{a}^{P} = \frac{{}^{B}d\,{}^{B}\mathbf{v}^{P}}{dt} \tag{10}$$

$$ {}^{A}\mathbf{a}^{\widetilde{B}} = \frac{{}^{A}d\,{}^{A}\mathbf{v}^{\widetilde{B}}}{dt} \tag{11}$$

Substitution from Eq. (7) into Eq. (9) yields

$$\begin{aligned} {}^{A}\mathbf{a}^{P} &= \frac{{}^{A}d\,{}^{A}\mathbf{v}^{\widetilde{B}}}{dt} + \frac{{}^{A}d\,{}^{B}\mathbf{v}^{P}}{dt} + \frac{{}^{A}d\,{}^{A}\boldsymbol{\omega}^{B}}{dt} \times \mathbf{q} + {}^{A}\boldsymbol{\omega}^{B} \times \frac{{}^{A}d\mathbf{q}}{dt} \\ &= \underset{(11)}{{}^{A}\mathbf{a}^{\widetilde{B}}} + \frac{{}^{B}d\,{}^{B}\mathbf{v}^{P}}{dt} \underset{(2.3.1)}{+ {}^{A}\boldsymbol{\omega}^{B} \times {}^{B}\mathbf{v}^{P}} + \underset{(2.5.1)}{{}^{A}\boldsymbol{\alpha}^{B}} \times \mathbf{q} \\ &\quad + {}^{A}\boldsymbol{\omega}^{B} \times \left(\frac{{}^{B}d\mathbf{q}}{dt} \underset{(2.3.1)}{+ {}^{A}\boldsymbol{\omega}^{B} \times \mathbf{q}} \right) \end{aligned} \tag{12}$$

and, if $\widetilde{B}$ is once again taken as $\overline{B}$, this reduces to

$$ {}^{A}\mathbf{a}^{P} \underset{(12)}{=} {}^{A}\mathbf{a}^{\overline{B}} + \underset{(10)}{{}^{B}\mathbf{a}^{P}} + {}^{A}\boldsymbol{\omega}^{B} \times {}^{B}\mathbf{v}^{P} + {}^{A}\boldsymbol{\omega}^{B} \times \underset{(4)}{{}^{B}\mathbf{v}^{P}} \tag{13}$$

in agreement with Eq. (2).

Example In the examples in Secs. 2.6 and 2.7, expressions for the velocity and the acceleration of P_1 in A were found by appealing to the definitions of velocity and acceleration, respectively. Alternatively, one can proceed as follows.

If $\overline{B}$ is the point of B that coincides with P_1 (see Fig. 2.6.1), then $\overline{B}$ moves on a circle of radius q_1, and

$$ {}^{A}\mathbf{v}^{\overline{B}} = -\omega q_1 \hat{\mathbf{b}}_z \tag{14}$$

while (remember that ω is a constant)

$$ {}^{A}\mathbf{a}^{\overline{B}} = -\omega^2 q_1 \hat{\mathbf{b}}_x \tag{15}$$

The velocity and acceleration of P_1 in B are

$$ {}^{B}\mathbf{v}^{P_1} = \dot{q}_1 \hat{\mathbf{b}}_x + \dot{q}_2 \hat{\mathbf{b}}_y \tag{16}$$

and

$$ {}^{B}\mathbf{a}^{P_1} = \ddot{q}_1 \hat{\mathbf{b}}_x + \ddot{q}_2 \hat{\mathbf{b}}_y \tag{17}$$

respectively, and the angular velocity of B in A is given by

$$ {}^{A}\boldsymbol{\omega}^{B} \underset{(2.2.1)}{=} \omega\hat{\mathbf{b}}_y \tag{18}$$

Thus,

$$ {}^{A}\mathbf{v}^{P_1} \underset{(1)}{=} \underset{(14)}{-\omega q_1\hat{\mathbf{b}}_z + \dot{q}_1\hat{\mathbf{b}}_x} \underset{(16)}{+ \dot{q}_2\hat{\mathbf{b}}_y} \tag{19}$$

in agreement with Eq. (2.6.7), and

$$ {}^{A}\mathbf{a}^{P_1} \underset{(2)}{=} \underset{(15)}{-\omega^2 q_1\hat{\mathbf{b}}_x} + \underset{(17)}{\ddot{q}_1\hat{\mathbf{b}}_x + \ddot{q}_2\hat{\mathbf{b}}_y} \underset{(18,\ 16)}{- 2\omega\dot{q}_1\hat{\mathbf{b}}_z} \tag{20}$$

which is the result previously recorded as Eq. (2.7.5).

3 CONSTRAINTS

A mechanical system is frequently constrained in the way it can be configured and/or in the way it can move. This chapter deals with mathematical treatments of both types of constraints. A *configuration constraint* imposes restrictions on the positions of certain particles in a system. Such constraints, and the use of *generalized coordinates* to describe the configuration of a system, are discussed in Secs. 3.1–3.3. Motion of a system can be characterized succinctly with the aid of *motion variables*, which are introduced in Sec. 3.4. *Motion constraints*, which entail restrictions on the angular velocities of rigid bodies and the velocities of particles in a system, are the subject of Secs. 3.5–3.8. In Sec. 3.6 vectors referred to as *partial angular velocities* of a rigid body and *partial velocities* of a point are introduced; other vectors, known as *partial angular accelerations* of a rigid body and *partial accelerations* of a point, are presented in Sec. 3.8. It is these quantities that ultimately enable one to form in a straightforward manner the terms that make up dynamical equations of motion. In later chapters we will have occasion to make use of certain relationships involving velocity, acceleration, and partial velocities of a particle set forth in Sec. 3.9.

3.1 CONFIGURATION CONSTRAINTS

The *configuration* of a set S of ν particles $P_1, \ldots, P_\nu$ in a reference frame A is known whenever the position vector of each particle relative to a point fixed in A is known. Thus, ν vector quantities, or, equivalently, 3ν scalar quantities, are required for the specification of the configuration of S in A.

If the motion of S is affected by the presence of bodies that come into contact with one or more of $P_1, \ldots, P_\nu$, restrictions are imposed on the positions that the affected particles may occupy, and S is said to be subject to *configuration constraints*; an equation expressing such a restriction is called a *holonomic constraint equation*. If $\mathbf{p}_i$ is the position vector from a point O fixed in A to P_i, then a holonomic constraint equation can be expressed as

$$g(\mathbf{p}_1, \mathbf{p}_2, \ldots, \mathbf{p}_\nu, t) = 0 \tag{1}$$

where t is the time. Alternatively, if $\hat{\mathbf{a}}_x$, $\hat{\mathbf{a}}_y$, $\hat{\mathbf{a}}_z$ are mutually perpendicular unit vectors fixed in A, and x_i, y_i, z_i, called *Cartesian coordinates* of P_i in A, are defined as

$$x_i \triangleq \mathbf{p}_i \cdot \hat{\mathbf{a}}_x \qquad y_i \triangleq \mathbf{p}_i \cdot \hat{\mathbf{a}}_y \qquad z_i \triangleq \mathbf{p}_i \cdot \hat{\mathbf{a}}_z \qquad (i = 1, \ldots, \nu) \tag{2}$$

then a holonomic constraint equation can have the form

$$f(x_1, y_1, z_1, \ldots, x_\nu, y_\nu, z_\nu, t) = 0 \tag{3}$$

Holonomic constraint equations are classified as *rheonomic* or *scleronomic*, according to whether the function f does or does not contain t explicitly.

Example Figure 3.1.1 shows two small blocks, P_1 and P_2, connected by a thin rod R of length L, and constrained to remain between two parallel panes of glass that are attached to each other, forming a rigid body B. This body is made to rotate at a *constant* rate ω about a line Y fixed both in B and in a reference frame A. Blocks P_1 and P_2 are treated as a set S of two particles, and $\mathbf{p}_1$ and $\mathbf{p}_2$ denote their position vectors relative to the point O shown in Fig. 3.1.1.

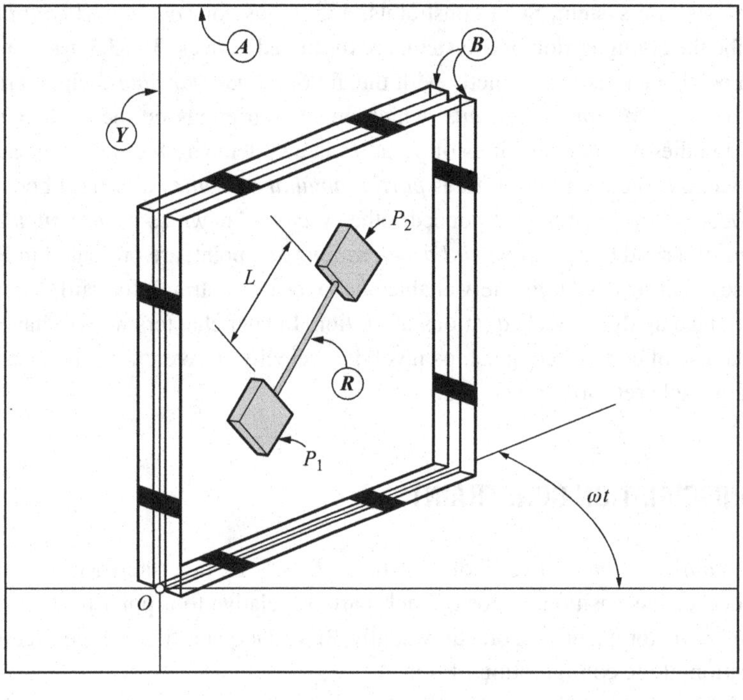

Figure 3.1.1

The requirement that P_1 and P_2 remain at all times between the two panes of glass is fulfilled if, and only if,

$$\mathbf{p}_i \cdot \hat{\mathbf{b}}_z = 0 \qquad (i = 1, 2) \tag{4}$$

where $\hat{\mathbf{b}}_z$ is a unit vector normal to the plane determined by the panes of glass, as indicated in Fig. 3.1.2. The fact that P_1 and P_2 are connected by a rod of length L constitutes one more configuration constraint, for this implies, and is implied by,

$$(\mathbf{p}_1 - \mathbf{p}_2) \cdot (\mathbf{p}_1 - \mathbf{p}_2) - L^2 = 0 \tag{5}$$

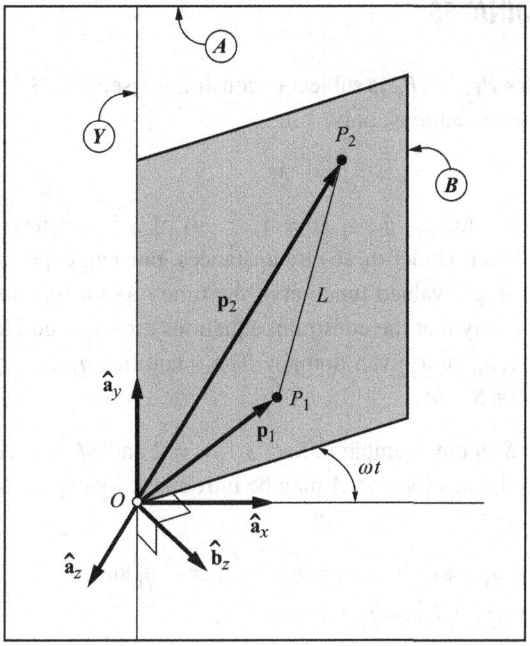

Figure 3.1.2

Equations (4) and (5) thus furnish a set of three holonomic constraint equations cast in the form of Eq. (1).

Now, one can express $\mathbf{p}_1$ and $\mathbf{p}_2$ as

$$\mathbf{p}_i = x_i\hat{\mathbf{a}}_x + y_i\hat{\mathbf{a}}_y + z_i\hat{\mathbf{a}}_z \qquad (i = 1,2) \tag{6}$$

where $\hat{\mathbf{a}}_x$, $\hat{\mathbf{a}}_y$, $\hat{\mathbf{a}}_z$ are mutually perpendicular unit vectors fixed in A, as shown in Fig. 3.1.2. Furthermore,

$$\hat{\mathbf{b}}_z = \cos\omega t\hat{\mathbf{a}}_z + \sin\omega t\hat{\mathbf{a}}_x \tag{7}$$

Hence,

$$\mathbf{p}_i \cdot \hat{\mathbf{b}}_z \underset{(6,\,7)}{=} z_i\cos\omega t + x_i\sin\omega t \qquad (i = 1,2) \tag{8}$$

and substitution into Eqs. (4) leads to the rheonomic holonomic constraint equations

$$z_i\cos\omega t + x_i\sin\omega t = 0 \qquad (i = 1,2) \tag{9}$$

Substitution from Eqs. (6) into Eq. (5) yields

$$(x_1 - x_2)^2 + (y_1 - y_2)^2 + (z_1 - z_2)^2 - L^2 = 0 \tag{10}$$

Since in this equation, in contrast with Eqs. (9), t does not appear explicitly, Eq. (10) is a scleronomic holonomic constraint equation. Equations (9) and (10) have the form of Eq. (3); written entirely in terms of scalars, they are counterparts of Eqs. (4) and (5), respectively.

3.2 GENERALIZED COORDINATES

When a set S of ν particles $P_1, \ldots, P_\nu$ is subject to constraints (see Sec. 3.1) represented by M holonomic constraint equations, only

$$n \triangleq 3\nu - M \tag{1}$$

of the 3ν Cartesian coordinates x_i, y_i, z_i $(i = 1, \ldots, \nu)$ of S in a reference frame A are independent of each other. Under these circumstances, one can express each of x_i, y_i, z_i $(i = 1, \ldots, \nu)$ as a single-valued function of the time t and n functions of t, say, $q_1(t), \ldots, q_n(t)$, in such a way that the constraint equations are satisfied identically for all values of t and $q_1, \ldots, q_n$ in a given domain. The quantities $q_1, \ldots, q_n$ are called *generalized coordinates* for S in A.

Example For the set S in the example in Sec. 3.1, $\nu = 2$ and $M = 3$. Hence $n = 3$. Three generalized coordinates for S in A may be introduced by expressing x_i, y_i, z_i $(i = 1, 2)$ as

$$x_1 = q_1 \cos \omega t \qquad y_1 = q_2 \qquad z_1 = -q_1 \sin \omega t \tag{2}$$

$$x_2 = (q_1 + L \cos q_3) \cos \omega t \tag{3}$$

$$y_2 = q_2 + L \sin q_3 \tag{4}$$

$$z_2 = -(q_1 + L \cos q_3) \sin \omega t \tag{5}$$

That q_1, q_2, q_3 are, indeed, generalized coordinates of S in A may be verified by substituting from Eqs. (2)–(5) into the left-hand members of Eqs. (3.1.9) and (3.1.10). For example,

$$z_1 \cos \omega t + x_1 \sin \omega t \underset{(2)}{=} -q_1 \sin \omega t \cos \omega t + q_1 \cos \omega t \sin \omega t \equiv 0 \tag{6}$$

so that Eq. (3.1.9) is seen to be satisfied identically for $i = 1$.

The geometric significance of q_1, q_2, q_3 and, hence, the rationale underlying the introduction of generalized coordinates as in Eqs. (2)–(5), will be discussed presently. First, however, it is important to point out that other choices of generalized coordinates are possible. Suppose, for example, that x_i, y_i, z_i $(i = 1, 2)$ are expressed as

$$x_1 = q_1 \cos q_2 \cos \omega t \tag{7}$$

$$y_1 = q_1 \sin q_2 \tag{8}$$

$$z_1 = -q_1 \cos q_2 \sin \omega t \tag{9}$$

$$x_2 = [(q_1 + L \cos q_3) \cos q_2 - L \sin q_2 \sin q_3] \cos \omega t \tag{10}$$

$$y_2 = (q_1 + L \cos q_3) \sin q_2 + L \cos q_2 \sin q_3 \tag{11}$$

$$z_2 = -[(q_1 + L \cos q_3) \cos q_2 - L \sin q_2 \sin q_3] \sin \omega t \tag{12}$$

Then Eqs. (3.1.9) and (3.1.10) are satisfied identically, which means that, once again, q_1, q_2, q_3 are generalized coordinates of S in A.

The generalized coordinates introduced in Eqs. (2)–(5) may be regarded as measures of two distances and an angle as indicated in Fig. 3.2.1. This may be seen as

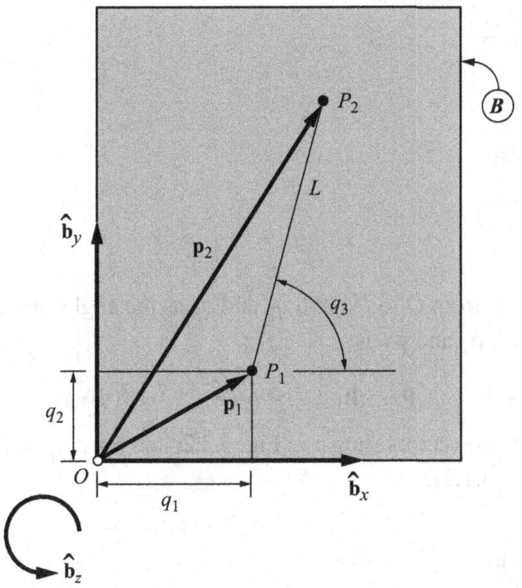

Figure 3.2.1

follows. If $\hat{\mathbf{b}}_x$, $\hat{\mathbf{b}}_y$, $\hat{\mathbf{b}}_z$ are mutually perpendicular unit vectors directed as in Fig. 3.2.1, then $\mathbf{p}_1$ and $\mathbf{p}_2$ can be written

$$\mathbf{p}_1 = q_1\hat{\mathbf{b}}_x + q_2\hat{\mathbf{b}}_y \qquad \mathbf{p}_2 = \mathbf{p}_1 + L\cos q_3\hat{\mathbf{b}}_x + L\sin q_3\hat{\mathbf{b}}_y \tag{13}$$

and

$$x_1 \underset{(3.1.2)}{\triangleq} \mathbf{p}_1 \cdot \hat{\mathbf{a}}_x \underset{(13)}{=} q_1\hat{\mathbf{b}}_x \cdot \hat{\mathbf{a}}_x + q_2\hat{\mathbf{b}}_y \cdot \hat{\mathbf{a}}_x \tag{14}$$

$$y_1 \underset{(3.1.2)}{\triangleq} \mathbf{p}_1 \cdot \hat{\mathbf{a}}_y \underset{(13)}{=} q_1\hat{\mathbf{b}}_x \cdot \hat{\mathbf{a}}_y + q_2\hat{\mathbf{b}}_y \cdot \hat{\mathbf{a}}_y \tag{15}$$

$$z_1 \underset{(3.1.2)}{\triangleq} \mathbf{p}_1 \cdot \hat{\mathbf{a}}_z \underset{(13)}{=} q_1\hat{\mathbf{b}}_x \cdot \hat{\mathbf{a}}_z + q_2\hat{\mathbf{b}}_y \cdot \hat{\mathbf{a}}_z \tag{16}$$

The dot products appearing in these equations are evaluated most conveniently by referring to Table 3.2.1, which is a concise way of stating the six equations that relate $\hat{\mathbf{a}}_x$, $\hat{\mathbf{a}}_y$, $\hat{\mathbf{a}}_z$ to $\hat{\mathbf{b}}_x$, $\hat{\mathbf{b}}_y$, $\hat{\mathbf{b}}_z$. Thus one finds that Eqs. (14)–(16) give way to precisely Eqs. (2). Similarly,

$$\begin{aligned} x_2 \underset{(3.1.2)}{\triangleq} \mathbf{p}_2 \cdot \hat{\mathbf{a}}_x &\underset{(13)}{=} \mathbf{p}_1 \cdot \hat{\mathbf{a}}_x + L\cos q_3\hat{\mathbf{b}}_x \cdot \hat{\mathbf{a}}_x + L\sin q_3\hat{\mathbf{b}}_y \cdot \hat{\mathbf{a}}_x \\ &\underset{(14,\ \text{Table 3.2.1})}{=} q_1\cos\omega t + L\cos q_3\cos\omega t \end{aligned} \tag{17}$$

which is the same as Eq. (3), and when y_2 and z_2 are formed correspondingly, Eqs. (4) and (5) are recovered.

In the case of the generalized coordinates appearing in Eqs. (7)–(12), q_1 may be

Table 3.2.1

	$\hat{\mathbf{b}}_x$	$\hat{\mathbf{b}}_y$	$\hat{\mathbf{b}}_z$
$\hat{\mathbf{a}}_x$	$\cos\omega t$	0	$\sin\omega t$
$\hat{\mathbf{a}}_y$	0	1	0
$\hat{\mathbf{a}}_z$	$-\sin\omega t$	0	$\cos\omega t$

interpreted as the distance from O to P_1, and q_2 and q_3 as the angles indicated in Fig. 3.2.2. To see this, express $\mathbf{p}_1$ and $\mathbf{p}_2$ as

$$\mathbf{p}_1 = q_1\hat{\mathbf{c}}_x \qquad \mathbf{p}_2 = \mathbf{p}_1 + L\cos q_3\hat{\mathbf{c}}_x + L\sin q_3\hat{\mathbf{c}}_y \tag{18}$$

where $\hat{\mathbf{c}}_x$ and $\hat{\mathbf{c}}_y$ are the unit vectors shown in Fig. 3.2.2, and form x_i, y_i, z_i $(i = 1,2)$ in accordance with Eqs. (3.1.2).

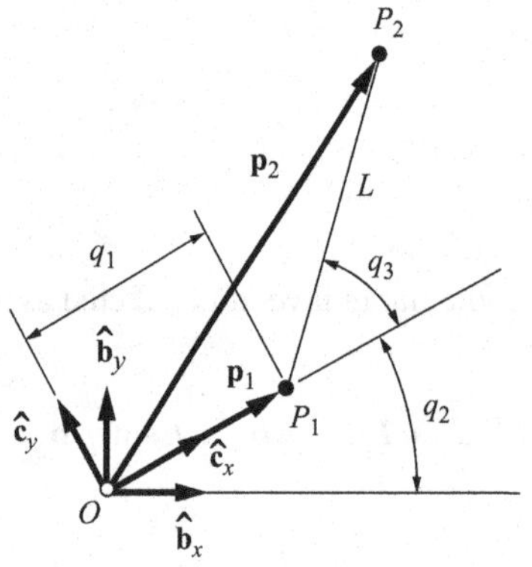

Figure 3.2.2

3.3 NUMBER OF GENERALIZED COORDINATES

The number n of generalized coordinates of a set S of ν particles in a reference frame A (see Sec. 3.2) is the smallest number of scalar quantities such that to every assignment of values to these quantities and the time t (within a domain of interest) there corresponds a definite admissible configuration of S in A. Frequently, one can find n by inspection rather than by determining the number M of holonomic constraint equations (see Sec. 3.1) and then subtracting M from 3ν. For example, suppose that S consists of ν particles $P_1,\ldots,P_\nu$ forming a rigid body B that is free to move in A. Then there corresponds a definite admissible configuration of S in A to every assignment of values to three Cartesian coordinates of one particle of B and three angles that characterize the orientation of B in A. Hence, $n = 6$. The same conclusion is obtained formally by letting $\mathbf{p}_1,\ldots,\mathbf{p}_\nu$ be the position vectors from a point fixed in A to $P_1,\ldots,P_\nu$, respectively, and noting that rigidity can be ensured by letting P_1, P_2, and P_3 be noncollinear particles and requiring, first, that the distances between P_1 and P_2, P_2 and P_3, and P_3 and P_1 remain constant, so that

$$(\mathbf{p}_1 - \mathbf{p}_2)^2 = c_1 \tag{1}$$
$$(\mathbf{p}_2 - \mathbf{p}_3)^2 = c_2 \tag{2}$$
$$(\mathbf{p}_3 - \mathbf{p}_1)^2 = c_3 \tag{3}$$

where c_1, c_2, c_3 are constants, and, second, that the distances between each of the remaining $\nu - 3$ particles and each of P_1, P_2, and P_3 remain constant, that is,

$$(\mathbf{p}_i - \mathbf{p}_1)^2 = c_{i1} \qquad (i = 4,\ldots,\nu) \tag{4}$$
$$(\mathbf{p}_i - \mathbf{p}_2)^2 = c_{i2} \qquad (i = 4,\ldots,\nu) \tag{5}$$
$$(\mathbf{p}_i - \mathbf{p}_3)^2 = c_{i3} \qquad (i = 4,\ldots,\nu) \tag{6}$$

where c_{ij} $(i = 4,\ldots,\nu; j = 1,2,3)$ are constants. The number M of holonomic constraint equations is thus given by

$$M = 3 + 3(\nu - 3) = 3\nu - 6 \tag{7}$$

and it follows that

$$n \underset{(3.2.1)}{=} 3\nu - M = 3\nu - \underset{(7)}{(3\nu - 6)} = 6 \tag{8}$$

3.4 MOTION VARIABLES

As will be seen presently, expressions for angular velocities of rigid bodies and velocities of points of a system S whose configuration in a reference frame A is characterized by n generalized coordinates $q_1,\ldots,q_n$ (see Sec. 3.2) can be brought into particularly advantageous forms through the introduction of n quantities $u_1,\ldots,u_n$, called *motion variables* for S in A, these being quantities defined by equations of the form

$$u_r \triangleq \sum_{s=1}^{n} Y_{rs}\,\dot{q}_s + Z_r \qquad (r = 1,\ldots,n) \tag{1}$$

where Y_{rs} and Z_r are functions of $q_1,\ldots,q_n$, and the time t. These functions must be chosen such that Eqs. (1) can be solved uniquely for $\dot{q}_1,\ldots,\dot{q}_n$. Equations (1) are called *kinematical differential equations* for S in A.

Example Letting S be the set of two particles considered in the example in Sec. 3.1, and using as generalized coordinates the quantities q_1, q_2, q_3 indicated in Fig. 3.2.1, one may define three motion variables as

$$u_1 \triangleq \dot{q}_1 \cos\omega t - \omega q_1 \sin\omega t \qquad u_2 \triangleq \dot{q}_2 \qquad u_3 \triangleq \dot{q}_3 \tag{2}$$

In that event, the functions Y_{rs} and Z_r $(r, s = 1, 2, 3)$ of Eqs. (1) are

$$Y_{11} = \cos\omega t \qquad Y_{12} = Y_{13} = 0 \qquad Z_1 = -\omega q_1 \sin\omega t \tag{3}$$

$$Y_{21} = 0 \qquad Y_{22} = 1 \qquad Y_{23} = 0 \qquad Z_2 = 0 \tag{4}$$

$$Y_{31} = 0 \qquad Y_{32} = 0 \qquad Y_{33} = 1 \qquad Z_3 = 0 \tag{5}$$

and, solved for $\dot{q}_1, \dot{q}_2, \dot{q}_3$, Eqs. (2) yield

$$\dot{q}_1 = u_1 \sec\omega t + \omega q_1 \tan\omega t \qquad \dot{q}_2 = u_2 \qquad \dot{q}_3 = u_3 \tag{6}$$

Since $\sec\omega t$ and $\tan\omega t$ become infinite whenever ωt is equal to an odd multiple of $\pi/2$ rad, Eqs. (2) furnish acceptable definitions of u_1, u_2, u_3 except when ωt takes on one of these values.

As an alternative to Eqs. (2), one might let

$$u_1 \triangleq \dot{q}_1 c_3 + \dot{q}_2 s_3 \qquad u_2 \triangleq -\dot{q}_1 s_3 + \dot{q}_2 c_3 \qquad u_3 \triangleq \dot{q}_3 \tag{7}$$

where s_3 and c_3 stand for $\sin q_3$ and $\cos q_3$, respectively. Then $Z_r = 0$ $(r = 1, 2, 3)$, and $\dot{q}_1, \dot{q}_2, \dot{q}_3$ are given by

$$\dot{q}_1 = u_1 c_3 - u_2 s_3 \qquad \dot{q}_2 = u_1 s_3 + u_2 c_3 \qquad \dot{q}_3 \triangleq u_3 \tag{8}$$

Here, no value of ωt needs to be excluded. Finally, suppose that u_1, u_2, u_3 are defined as

$$u_1 \triangleq \dot{q}_1 \qquad u_2 \triangleq \dot{q}_2 \qquad u_3 \triangleq \dot{q}_3 \tag{9}$$

Although these definitions of u_1, u_2, u_3 are simpler than those in Eqs. (2) and Eqs. (7), the latter are preferable in certain contexts, as will presently become apparent.

The motivation for introducing u_1, u_2, and u_3 as in Eqs. (2), (7), and (9) is the following. The velocity of P_1 in A can be expressed in a variety of ways, such as

$$^A\mathbf{v}^{P_1} \underset{(2.6.7)}{=} \dot{q}_1 \hat{\mathbf{b}}_x + \dot{q}_2 \hat{\mathbf{b}}_y - \omega q_1 \hat{\mathbf{b}}_z \tag{10}$$

$$^A\mathbf{v}^{P_1} \underset{(2.6.8)}{=} (\dot{q}_1 c_3 + \dot{q}_2 s_3)\hat{\mathbf{e}}_x + (-\dot{q}_1 s_3 + \dot{q}_2 c_3)\hat{\mathbf{e}}_y - \omega q_1 \hat{\mathbf{e}}_z \tag{11}$$

and (see Fig. 3.1.2 for $\hat{\mathbf{a}}_x, \hat{\mathbf{a}}_y, \hat{\mathbf{a}}_z$)

$$^A\mathbf{v}^{P_1} = (\dot{q}_1 \cos\omega t - \omega q_1 \sin\omega t)\hat{\mathbf{a}}_x + \dot{q}_2 \hat{\mathbf{a}}_y - (\dot{q}_1 \sin\omega t + \omega q_1 \cos\omega t)\hat{\mathbf{a}}_z \tag{12}$$

When Eqs. (9) are used to define u_1, u_2, and u_3, Eq. (10) can be rewritten as

$$^A\mathbf{v}^{P_1} = u_1 \hat{\mathbf{b}}_x + u_2 \hat{\mathbf{b}}_y - \omega q_1 \hat{\mathbf{b}}_z \tag{13}$$

Similarly, the definitions of u_1, u_2, and u_3 in accordance with Eqs. (7) permit one to replace Eq. (11) with a relationship having the same simple form as Eq. (13), namely,

$$^A\mathbf{v}^{P_1} = u_1 \hat{\mathbf{e}}_x + u_2 \hat{\mathbf{e}}_y - \omega q_1 \hat{\mathbf{e}}_z \tag{14}$$

Finally, with the aid of Eqs. (2), (6), and (12), one obtains

$$^A\mathbf{v}^{P_1} = u_1 \hat{\mathbf{a}}_x + u_2 \hat{\mathbf{a}}_y - (u_1 \tan\omega t + \omega q_1 \sec\omega t)\hat{\mathbf{a}}_z \tag{15}$$

Here, the third component is a bit more complicated than in Eqs. (13) and (14), but the introduction of motion variables has led to a noticeable simplification, nevertheless. The guiding idea in writing Eqs. (7) and (2) was thus to enable one to replace Eqs. (11) and (12), respectively, with expressions having, as nearly as possible, the same simple form as Eq. (10). As for Eqs. (9), their use does not lead to any simplifications since Eq. (10) cannot be simplified further, but they were included to show that the concept of motion variables remains applicable even under these circumstances.

The simplification of an angular velocity expression through the use of motion variables can be illustrated by returning to the example in Sec. 2.4. The angular velocity expression recorded in Eq. (2.4.10), though simple in form, is unsuitable for certain purposes because $\hat{\mathbf{k}}_2$, $\hat{\mathbf{k}}_7$, and $\hat{\mathbf{k}}_3$ are not mutually perpendicular. To overcome this difficulty, one can let $\hat{\mathbf{k}}_4$ be a unit vector directed as shown in Fig. 2.4.1 and note that $\hat{\mathbf{k}}_3$ then is given by

$$\hat{\mathbf{k}}_3 = -(\cos q_2 \hat{\mathbf{k}}_2 + \sin q_2 \hat{\mathbf{k}}_4) \tag{16}$$

so that Eq. (2.4.10) may be replaced with

$$\begin{aligned} {}^A\boldsymbol{\omega}^B &= \dot{q}_1 \hat{\mathbf{k}}_2 + \dot{q}_2 \hat{\mathbf{k}}_7 - \dot{q}_3 (\cos q_2 \hat{\mathbf{k}}_2 + \sin q_2 \hat{\mathbf{k}}_4) \\ &= (\dot{q}_1 - \dot{q}_3 \cos q_2) \hat{\mathbf{k}}_2 + \dot{q}_2 \hat{\mathbf{k}}_7 - \dot{q}_3 \sin q_2 \hat{\mathbf{k}}_4 \end{aligned} \tag{17}$$

which reduces to

$$ {}^A\boldsymbol{\omega}^B = u_1 \hat{\mathbf{k}}_2 + u_2 \hat{\mathbf{k}}_7 + u_3 \hat{\mathbf{k}}_4 \tag{18}$$

if motion variables u_1, u_2, and u_3 are defined as

$$u_1 \triangleq \dot{q}_1 - \dot{q}_3 \cos q_2 \qquad u_2 \triangleq \dot{q}_2 \qquad u_3 \triangleq -\dot{q}_3 \sin q_2 \tag{19}$$

While motion variables can be time derivatives of a function of the generalized coordinates and the time t [for example, u_1, u_2, and u_3 as defined in Eqs. (2) are, respectively, the time derivatives of $q_1 \cos \omega t$, q_2, and q_3], this is not always the case. Consider, for instance, u_1 as defined in Eq. (7), and assume the existence of a function f of q_1, q_2, q_3, and t such that

$$\frac{df}{dt} = u_1 \tag{20}$$

for all values of q_1, q_2, q_3, and t in some domain of these variables. Then

$$\begin{aligned} \frac{df}{dt} &\underset{(1.12.3)}{=} \frac{\partial f}{\partial q_1}\dot{q}_1 + \frac{\partial f}{\partial q_2}\dot{q}_2 + \frac{\partial f}{\partial q_3}\dot{q}_3 + \frac{\partial f}{\partial t} \\ &\underset{(7,\ 20)}{=} \dot{q}_1 c_3 + \dot{q}_2 s_3 \end{aligned} \tag{21}$$

which imples that

$$\frac{\partial f}{\partial q_1} = c_3 \qquad \frac{\partial f}{\partial q_2} = s_3 \qquad \frac{\partial f}{\partial q_3} = 0 \qquad \frac{\partial f}{\partial t} = 0 \tag{22}$$

The first and third of these equations are incompatible with each other, for they lead to different expressions for $\partial^2 f / \partial q_1 \partial q_3$. Thus, the hypothesis that f exists such that Eq. (20) is satisfied is untenable.

3.5 MOTION CONSTRAINTS

It can occur that restrictions are placed on ${}^A\mathbf{v}^{P_1},\ldots,{}^A\mathbf{v}^{P_\nu}$, the velocities in a reference frame A of the particles $P_1,\ldots,P_\nu$ belonging to a system S. In general, a relationship describing such a restriction can be written as

$$f({}^A\mathbf{v}^{P_1},\ldots,{}^A\mathbf{v}^{P_\nu},t) = 0 \tag{1}$$

In certain cases a relationship of this form cannot be obtained by taking the time derivative of a function having the form of Eq. (3.1.1), because no such function exists. In that event, S is said to be subject to a *motion constraint*, and Eq. (1) is called a *nonholonomic constraint equation.*

When a system S is not subject to motion constraints, then S is said to be a *holonomic system possessing n degrees of freedom in A*. If S is subject to motion constraints, S is called a *nonholonomic system.*

As a consequence of a nonholonomic constraint equation, the motion variables $u_1,\ldots,u_n$ for S in A (see Sec. 3.4) are not independent of each other. Frequently, the physical nature of a motion constraint gives rise to a nonholonomic constraint equation that is linear in ${}^A\mathbf{v}^{P_1},\ldots,{}^A\mathbf{v}^{P_\nu}$ and, hence, linear in $u_1,\ldots,u_n$. When all nonholonomic constraint equations can be expressed as the m relationships

$$u_r = \sum_{s=1}^{p} A_{rs}u_s + B_r \qquad (r = p+1,\ldots,n) \tag{2}$$

where

$$p \triangleq n - m \tag{3}$$

and where A_{rs} and B_r are functions of $q_1,\ldots,q_n$, and the time t, S is referred to as a *simple nonholonomic system possessing p degrees of freedom* in A. Motion variables that are independent of each other are referred to as *generalized velocities*; thus, $u_1,\ldots,u_n$ are generalized velocities for a holonomic system S in A, whereas $u_1,\ldots,u_p$ are generalized velocities for a simple nonholonomic system S in A.

Alternatively, a relationship having the form of Eq. (1) can be nonlinear in the velocity vectors and, hence, nonlinear in $u_1,\ldots,u_n$. In this case the equation can be regarded as describing intended motion of S, and imposition of the motion constraint requires specialized devices associated with a feedback control system, such as sensors, actuators, and computer equipment. Motion constraints of this kind are considered in Sec. 3.7.

Example The particles P_1 and P_2 considered in the example in Sec. 3.1 form a holonomic system possessing three degrees of freedom in A. Suppose that P_2 is replaced with a small sharp-edged circular disk D whose axis is normal to the rod R and parallel to the plane in which R moves, as indicated in Fig. 3.5.1; further, that D comes into contact with the two panes of glass at the points D_1 and D_2.

The sharp edge permits $D^\star$, the center of D, to move freely in B in a direction parallel to $\hat{\mathbf{e}}_x$ and, at the same time, prevents $D^\star$ from moving in a direction parallel

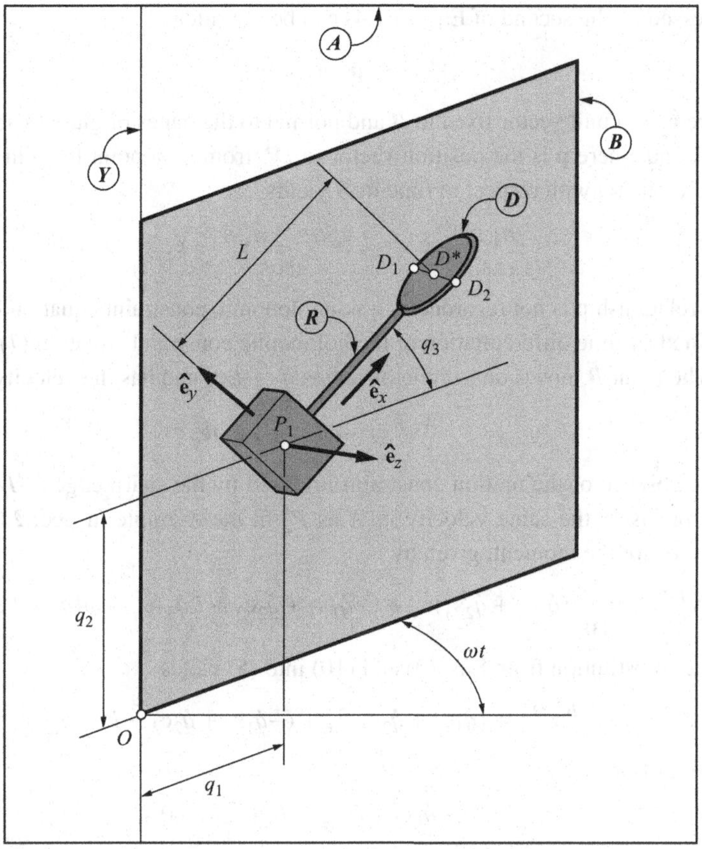

Figure 3.5.1

to $\hat{\mathbf{e}}_y$, where $\hat{\mathbf{e}}_x$ and $\hat{\mathbf{e}}_y$ are unit vectors directed as shown in Fig. 3.5.1. The latter condition may be stated analytically in terms of ${}^B\mathbf{v}^{D^\star}$, the velocity of $D^\star$ in B, as

$$ {}^B\mathbf{v}^{D^\star} \cdot \hat{\mathbf{e}}_y = 0 \tag{4} $$

Now,

$$ {}^B\mathbf{v}^{D^\star} \underset{(2.8.1)}{=} {}^A\mathbf{v}^{D^\star} - {}^A\mathbf{v}^{\overline{B}} \tag{5} $$

where $\overline{B}$ is the point of B with which $D^\star$ coincides. Therefore, Eq. (4) can be restated as

$$ (\underset{(5)}{{}^A\mathbf{v}^{D^\star} - {}^A\mathbf{v}^{\overline{B}}}) \cdot \hat{\mathbf{e}}_y \underset{(4)}{=} 0 \tag{6} $$

With the considerations that $D^\star$ is a particle belonging to S, but that $\overline{B}$ is not such a particle, Eq. (6) is seen to have the form of Eq. (1).

Upon letting $D^\star$ play the role of P_2, the configuration constraint (see Sec. 3.1) expressed by the second of Eqs. (3.1.4) can be rewritten

$$\mathbf{p} \cdot \hat{\mathbf{e}}_z = 0 \tag{7}$$

where $\hat{\mathbf{e}}_z$ is a unit vector fixed in B and normal to the panes of glass as shown in Fig. 3.5.1, and where $\mathbf{p}$ is the position vector to $D^\star$ from O, a point fixed in A and in B. Differentiation with respect to time in B yields

$$\underset{(2.6.1)}{{}^B\mathbf{v}^{D^\star} \cdot \hat{\mathbf{e}}_z =} \underset{(5)}{({}^A\mathbf{v}^{D^\star} - {}^A\mathbf{v}^{\overline{B}})} \cdot \hat{\mathbf{e}}_z \underset{(7)}{=} 0 \tag{8}$$

This relationship is not regarded as a nonholonomic constraint equation because it is obtained by time differentiation of the holonomic constraint equation (7).

The point $\overline{B}$ moves on a circle of radius $q_1 + Lc_3$ and has the velocity

$${}^A\mathbf{v}^{\overline{B}} = -\omega(q_1 + Lc_3)\hat{\mathbf{e}}_z \tag{9}$$

In the absence of the motion constraint imposed by the sharp edge of D, $D^\star$ would have precisely the same velocity in A as P_2 in the example in Sec. 2.6; therefore, ${}^A\mathbf{v}^{D^\star}$ is, for the moment, given by

$${}^A\mathbf{v}^{D^\star} \underset{(2.6.13)}{=} (\dot{q}_1 c_3 + \dot{q}_2 s_3)\hat{\mathbf{e}}_x + (-\dot{q}_1 s_3 + \dot{q}_2 c_3 + L\dot{q}_3)\hat{\mathbf{e}}_y - \omega(q_1 + Lc_3)\hat{\mathbf{e}}_z \tag{10}$$

Hence, substitution from Eqs. (9) and (10) into (5) yields

$${}^B\mathbf{v}^{D^\star} = (\dot{q}_1 c_3 + \dot{q}_2 s_3)\hat{\mathbf{e}}_x + (-\dot{q}_1 s_3 + \dot{q}_2 c_3 + L\dot{q}_3)\hat{\mathbf{e}}_y \tag{11}$$

and Eq. (4) leads to

$$-\dot{q}_1 s_3 + \dot{q}_2 c_3 + L\dot{q}_3 \underset{(4,\ 11)}{=} 0 \tag{12}$$

while Eq. (8) is satisfied identically. Furthermore, if motion variables u_1, u_2, and u_3 are introduced as in Eqs. (3.4.2), so that Eqs. (3.4.6) apply, then Eq. (12) gives rise to the nonholonomic constraint equation

$$u_3 = \frac{1}{L}[(u_1 \sec \omega t + \omega q_1 \tan \omega t)s_3 - u_2 c_3] \tag{13}$$

Thus, $m = 1$, $p = n - m = 3 - 1 = 2$, and P_1 and D are seen to form a simple nonholonomic system possessing two degrees of freedom in A. The generalized velocities for the system are u_1 and u_2. The functions A_{rs} and B_r ($r = 3$; $s = 1,2$) of Eq. (2) are

$$A_{31} = \frac{s_3}{L}\sec \omega t \qquad A_{32} = \frac{-c_3}{L} \qquad B_3 = \frac{s_3}{L}\omega q_1 \tan \omega t \tag{14}$$

If Eqs. (3.4.7), rather than Eqs. (3.4.2), are used to define u_1, u_2, and u_3, then Eq. (13) gives way to the much simpler relationship

$$u_3 \underset{(12)}{=} \frac{-u_2}{L} \tag{15}$$

and if u_1, u_2, and u_3 are defined as in Eqs. (3.4.9), the nonholonomic constraint equation is

$$u_3 \underset{(12)}{=} \frac{1}{L}(u_1 s_3 - u_2 c_3) \tag{16}$$

Before leaving this example, it is worth noting that Eq. (12) is nonintegrable; that is, there exists no function $f(q_1, q_2, q_3)$ which is constant throughout every time interval in which Eq. (12) is satisfied. If such a function existed, then q_1, q_2, and q_3 would not be independent of each other and thus would not be generalized coordinates.

3.6 PARTIAL ANGULAR VELOCITIES, PARTIAL VELOCITIES

If $q_1, \ldots, q_n$ and $u_1, \ldots, u_n$ are, respectively, generalized coordinates (see Sec. 3.2) and motion variables (see Sec. 3.4) for a simple nonholonomic system S possessing p degrees of freedom in a reference frame A, then $\boldsymbol{\omega}$, the angular velocity in A of a rigid body B belonging to S, and $\mathbf{v}$, the velocity in A of a particle P belonging to S, can be expressed *uniquely* as

$$\boldsymbol{\omega} = \sum_{r=1}^{n} \boldsymbol{\omega}_r u_r + \boldsymbol{\omega}_t \tag{1}$$

and

$$\mathbf{v} = \sum_{r=1}^{n} \mathbf{v}_r u_r + \mathbf{v}_t \tag{2}$$

where $\boldsymbol{\omega}_r$, $\mathbf{v}_r$ $(r = 1, \ldots, n)$, $\boldsymbol{\omega}_t$, and $\mathbf{v}_t$ are functions of $q_1, \ldots, q_n$, and the time t. The vector $\boldsymbol{\omega}_r$ is called the rth *holonomic partial angular velocity* of B in A, and $\mathbf{v}_r$ is referred to as the rth *holonomic partial velocity* of P in A.

The vectors $\boldsymbol{\omega}$ and $\mathbf{v}$ can also be expressed *uniquely* as

$$\boldsymbol{\omega} = \sum_{r=1}^{p} \tilde{\boldsymbol{\omega}}_r u_r + \tilde{\boldsymbol{\omega}}_t \tag{3}$$

and

$$\mathbf{v} = \sum_{r=1}^{p} \tilde{\mathbf{v}}_r u_r + \tilde{\mathbf{v}}_t \tag{4}$$

where $\tilde{\boldsymbol{\omega}}_r$, $\tilde{\mathbf{v}}_r$ $(r = 1, \ldots, p)$, $\tilde{\boldsymbol{\omega}}_t$, and $\tilde{\mathbf{v}}_t$ are functions of $q_1, \ldots, q_n$, and t. The vector $\tilde{\boldsymbol{\omega}}_r$ is called the rth *nonholonomic partial angular veclocity* of B in A, while $\tilde{\mathbf{v}}_r$ is known as the rth *nonholonomic partial velocity* of P in A.

When speaking of partial angular velocities and/or partial velocities, one can generally omit the adjectives "holonomic" and "nonholonomic" without loss of clarity, but the tilde notation should be used to distinguish nonholonomic partial angular velocities from holonomic ones, and similarly for partial velocities. When $p = n$, that is, when S is a holonomic system possessing n degrees of freedom in A, then $\tilde{\boldsymbol{\omega}}_r = \boldsymbol{\omega}_r$ and $\tilde{\mathbf{v}}_r = \mathbf{v}_r$ $(r = 1, \ldots, n)$. It is customary not to write any tildes under these circumstances.

Derivation Solution of Eqs. (3.4.1) for $\dot{q}_1, \ldots, \dot{q}_n$ yields

$$\dot{q}_s = \sum_{r=1}^{n} W_{sr} u_r + X_s \qquad (s = 1, \ldots, n) \tag{5}$$

where W_{sr} and X_s are certain functions of $q_1, \ldots, q_n$, and t. Now, if $\hat{\mathbf{b}}_1$, $\hat{\mathbf{b}}_2$, $\hat{\mathbf{b}}_3$ form a right-handed set of mutually perpendicular unit vectors fixed in B, and if $\dot{\hat{\mathbf{b}}}_i$ denotes the first time-derivative of $\hat{\mathbf{b}}_i$ in A, then

$$\dot{\hat{\mathbf{b}}}_i \underset{(1.12.1)}{=} \sum_{s=1}^{n} \frac{\partial \hat{\mathbf{b}}_i}{\partial q_s} \dot{q}_s + \frac{\partial \hat{\mathbf{b}}_i}{\partial t}$$

$$\underset{(5)}{=} \sum_{s=1}^{n} \frac{\partial \hat{\mathbf{b}}_i}{\partial q_s} \left(\sum_{r=1}^{n} W_{sr} u_r + X_s \right) + \frac{\partial \hat{\mathbf{b}}_i}{\partial t}$$

$$= \sum_{r=1}^{n} \sum_{s=1}^{n} \frac{\partial \hat{\mathbf{b}}_i}{\partial q_s} W_{sr} u_r + \sum_{s=1}^{n} \frac{\partial \hat{\mathbf{b}}_i}{\partial q_s} X_s + \frac{\partial \hat{\mathbf{b}}_i}{\partial t} \qquad (i = 1,2,3) \tag{6}$$

where all partial differentiations of $\hat{\mathbf{b}}_i$ are performed in A. Consequently,

$$\boldsymbol{\omega} \underset{(2.1.1)}{=} \hat{\mathbf{b}}_1 \dot{\hat{\mathbf{b}}}_2 \cdot \hat{\mathbf{b}}_3 + \hat{\mathbf{b}}_2 \dot{\hat{\mathbf{b}}}_3 \cdot \hat{\mathbf{b}}_1 + \hat{\mathbf{b}}_3 \dot{\hat{\mathbf{b}}}_1 \cdot \hat{\mathbf{b}}_2$$

$$\underset{(6)}{=} \hat{\mathbf{b}}_1 \left(\sum_{r=1}^{n} \sum_{s=1}^{n} \frac{\partial \hat{\mathbf{b}}_2}{\partial q_s} \cdot \hat{\mathbf{b}}_3 W_{sr} u_r + \sum_{s=1}^{n} \frac{\partial \hat{\mathbf{b}}_2}{\partial q_s} \cdot \hat{\mathbf{b}}_3 X_s + \frac{\partial \hat{\mathbf{b}}_2}{\partial t} \cdot \hat{\mathbf{b}}_3 \right)$$

$$+ \hat{\mathbf{b}}_2 \left(\sum_{r=1}^{n} \sum_{s=1}^{n} \frac{\partial \hat{\mathbf{b}}_3}{\partial q_s} \cdot \hat{\mathbf{b}}_1 W_{sr} u_r + \sum_{s=1}^{n} \frac{\partial \hat{\mathbf{b}}_3}{\partial q_s} \cdot \hat{\mathbf{b}}_1 X_s + \frac{\partial \hat{\mathbf{b}}_3}{\partial t} \cdot \hat{\mathbf{b}}_1 \right)$$

$$+ \hat{\mathbf{b}}_3 \left(\sum_{r=1}^{n} \sum_{s=1}^{n} \frac{\partial \hat{\mathbf{b}}_1}{\partial q_s} \cdot \hat{\mathbf{b}}_2 W_{sr} u_r + \sum_{s=1}^{n} \frac{\partial \hat{\mathbf{b}}_1}{\partial q_s} \cdot \hat{\mathbf{b}}_2 X_s + \frac{\partial \hat{\mathbf{b}}_1}{\partial t} \cdot \hat{\mathbf{b}}_2 \right) \tag{7}$$

and, if $\boldsymbol{\omega}_r$ and $\boldsymbol{\omega}_t$ are defined as

$$\boldsymbol{\omega}_r \triangleq \sum_{s=1}^{n} \left(\hat{\mathbf{b}}_1 \frac{\partial \hat{\mathbf{b}}_2}{\partial q_s} \cdot \hat{\mathbf{b}}_3 + \hat{\mathbf{b}}_2 \frac{\partial \hat{\mathbf{b}}_3}{\partial q_s} \cdot \hat{\mathbf{b}}_1 + \hat{\mathbf{b}}_3 \frac{\partial \hat{\mathbf{b}}_1}{\partial q_s} \cdot \hat{\mathbf{b}}_2 \right) W_{sr} \qquad (r = 1, \ldots, n) \tag{8}$$

and

$$\boldsymbol{\omega}_t \triangleq \sum_{s=1}^{n} \left(\hat{\mathbf{b}}_1 \frac{\partial \hat{\mathbf{b}}_2}{\partial q_s} \cdot \hat{\mathbf{b}}_3 + \hat{\mathbf{b}}_2 \frac{\partial \hat{\mathbf{b}}_3}{\partial q_s} \cdot \hat{\mathbf{b}}_1 + \hat{\mathbf{b}}_3 \frac{\partial \hat{\mathbf{b}}_1}{\partial q_s} \cdot \hat{\mathbf{b}}_2 \right) X_s$$

$$+ \hat{\mathbf{b}}_1 \frac{\partial \hat{\mathbf{b}}_2}{\partial t} \cdot \hat{\mathbf{b}}_3 + \hat{\mathbf{b}}_2 \frac{\partial \hat{\mathbf{b}}_3}{\partial t} \cdot \hat{\mathbf{b}}_1 + \hat{\mathbf{b}}_3 \frac{\partial \hat{\mathbf{b}}_1}{\partial t} \cdot \hat{\mathbf{b}}_2 \tag{9}$$

respectively, then substitution from Eqs. (8) and (9) into Eq. (7) leads directly to Eq. (1).

To establish the validity of Eq. (2), let $\mathbf{p}$ be the position vector from a point fixed in

A to P, and let $\dot{\mathbf{p}}$ denote the first time-derivative of $\mathbf{p}$ in A. Then

$$\begin{aligned}\mathbf{v} \underset{(2.6.1)}{=} \dot{\mathbf{p}} &\underset{(1.12.1)}{=} \sum_{s=1}^{n} \frac{\partial \mathbf{p}}{\partial q_s}\dot{q}_s + \frac{\partial \mathbf{p}}{\partial t} \\ &\underset{(5)}{=} \sum_{s=1}^{n} \frac{\partial \mathbf{p}}{\partial q_s}\left(\sum_{r=1}^{n} W_{sr}u_r + X_s\right) + \frac{\partial \mathbf{p}}{\partial t} \\ &= \sum_{r=1}^{n}\sum_{s=1}^{n} \frac{\partial \mathbf{p}}{\partial q_s} W_{sr}u_r + \sum_{s=1}^{n} \frac{\partial \mathbf{p}}{\partial q_s} X_s + \frac{\partial \mathbf{p}}{\partial t}\end{aligned} \tag{10}$$

and, after defining $\mathbf{v}_r$ and $\mathbf{v}_t$ as

$$\mathbf{v}_r \triangleq \sum_{s=1}^{n} \frac{\partial \mathbf{p}}{\partial q_s} W_{sr} \qquad (r = 1, \ldots, n) \tag{11}$$

and

$$\mathbf{v}_t \triangleq \sum_{s=1}^{n} \frac{\partial \mathbf{p}}{\partial q_s} X_s + \frac{\partial \mathbf{p}}{\partial t} \tag{12}$$

respectively, one obtains Eq. (2) by substituting from Eqs. (11) and (12) into Eq. (10).

Suppose now that S is subject to motion constraints such that $u_1, \ldots, u_n$ are governed by Eqs. (3.5.2). Then, after rewriting Eq. (1) as

$$\boldsymbol{\omega} = \sum_{r=1}^{p} \boldsymbol{\omega}_r u_r + \sum_{r=p+1}^{n} \boldsymbol{\omega}_r u_r + \boldsymbol{\omega}_t \tag{13}$$

one can use Eqs. (3.5.2) to obtain

$$\begin{aligned}\boldsymbol{\omega} &\underset{(13)}{=} \sum_{r=1}^{p} \boldsymbol{\omega}_r u_r + \sum_{r=p+1}^{n} \boldsymbol{\omega}_r \left(\sum_{s=1}^{p} A_{rs}u_s + B_r\right)_{(3.5.2)} + \boldsymbol{\omega}_t \\ &= \sum_{r=1}^{p} \boldsymbol{\omega}_r u_r + \sum_{s=1}^{p}\sum_{r=p+1}^{n} \boldsymbol{\omega}_r A_{rs} u_s + \sum_{r=p+1}^{n} \boldsymbol{\omega}_r B_r + \boldsymbol{\omega}_t \\ &= \sum_{r=1}^{p} \boldsymbol{\omega}_r u_r + \sum_{r=1}^{p}\sum_{s=p+1}^{n} \boldsymbol{\omega}_s A_{sr} u_r + \sum_{r=p+1}^{n} \boldsymbol{\omega}_r B_r + \boldsymbol{\omega}_t \\ &= \sum_{r=1}^{p}\left(\boldsymbol{\omega}_r + \sum_{s=p+1}^{n} \boldsymbol{\omega}_s A_{sr}\right) u_r + \sum_{r=p+1}^{n} \boldsymbol{\omega}_r B_r + \boldsymbol{\omega}_t\end{aligned} \tag{14}$$

and, after defining $\tilde{\boldsymbol{\omega}}_r$ and $\tilde{\boldsymbol{\omega}}_t$ as

$$\tilde{\boldsymbol{\omega}}_r \triangleq \boldsymbol{\omega}_r + \sum_{s=p+1}^{n} \boldsymbol{\omega}_s A_{sr} \qquad (r = 1, \ldots, p) \tag{15}$$

and

$$\tilde{\boldsymbol{\omega}}_t \triangleq \boldsymbol{\omega}_t + \sum_{r=p+1}^{n} \boldsymbol{\omega}_r B_r \tag{16}$$

one arrives at Eq. (3) by substituting from Eqs. (15) and (16) into Eq. (14). A completely analogous derivation leads from Eq. (2) to Eq. (4), provided that $\tilde{\mathbf{v}}_r$ and $\tilde{\mathbf{v}}_t$ are defined as

$$\tilde{\mathbf{v}}_r \triangleq \mathbf{v}_r + \sum_{s=p+1}^{n} \mathbf{v}_s A_{sr} \qquad (r = 1, \ldots, p) \tag{17}$$

and

$$\tilde{\mathbf{v}}_t \triangleq \mathbf{v}_t + \sum_{r=p+1}^{n} \mathbf{v}_r B_r \tag{18}$$

As will become evident later, the use of partial angular velocities and partial velocities greatly facilitates the formulation of equations of motion. Moreover, the constructing of expressions for these quantities is a simple matter involving nothing more than the *inspecting of expressions* for angular velocities of rigid bodies and/or expressions for velocities of particles.

Example In the example in Sec. 3.4, three sets of motion variables were introduced and the corresponding expressions for the velocity of the particle P_1 in reference frame A (see Fig. 3.1.1) were recorded in Eqs. (3.4.13)–(3.4.15). Each of these equations has precisely the same form as Eq. (2), and inspection of the equations thus permits one to identify the associated holonomic partial velocities of P_1 in A as

$$^A\mathbf{v}_1^{P_1} \underset{(3.4.13)}{=} \hat{\mathbf{b}}_x \quad ^A\mathbf{v}_2^{P_1} \underset{(3.4.13)}{=} \hat{\mathbf{b}}_y \quad ^A\mathbf{v}_3^{P_1} \underset{(3.4.13)}{=} \mathbf{0} \tag{19}$$

$$^A\mathbf{v}_1^{P_1} \underset{(3.4.14)}{=} \hat{\mathbf{e}}_x \quad ^A\mathbf{v}_2^{P_1} \underset{(3.4.14)}{=} \hat{\mathbf{e}}_y \quad ^A\mathbf{v}_3^{P_1} \underset{(3.4.14)}{=} \mathbf{0} \tag{20}$$

$$^A\mathbf{v}_1^{P_1} \underset{(3.4.15)}{=} \hat{\mathbf{a}}_x - \tan\omega t \hat{\mathbf{a}}_z \quad ^A\mathbf{v}_2^{P_1} \underset{(3.4.15)}{=} \hat{\mathbf{a}}_y \quad ^A\mathbf{v}_3^{P_1} \underset{(3.4.15)}{=} \mathbf{0} \tag{21}$$

The angular velocity of E in A, introduced in the example in Sec. 2.6, can be written

$$^A\boldsymbol{\omega}^E \underset{(2.6.11)}{=} \omega s_3 \hat{\mathbf{e}}_x + \omega c_3 \hat{\mathbf{e}}_y + u_3 \hat{\mathbf{e}}_z \tag{22}$$

in all three cases because u_3 was defined as $\dot{q}_3$ in Eqs. (3.4.2), (3.4.7), and (3.4.9). Comparing Eq. (22) with Eq. (1), one can write down the holonomic partial angular velocities of E in A,

$$^A\boldsymbol{\omega}_1^E = {}^A\boldsymbol{\omega}_2^E = \mathbf{0} \qquad ^A\boldsymbol{\omega}_3^E = \hat{\mathbf{e}}_z \tag{23}$$

To illustrate the idea of nonholonomic partial angular velocities and nonholonomic partial velocities, we confine our attention to the motion variables of Eqs. (3.4.7), which means that Eqs. (20) and (23) apply, and explore the effect of the motion constraint considered in the example in Sec. 3.5, which was there shown to give rise to the nonholonomic constraint equation

$$u_3 \underset{(3.5.15)}{=} -\frac{u_2}{L} \tag{24}$$

As regards the partial velocities of P_1 in A, Eq. (24) makes no difference whatsoever, for u_3 is absent from Eq. (3.4.14), the relevant expression for ${}^A\mathbf{v}^{P_1}$. In other words, the two nonholonomic partial velocities of P_1 in A (there are two because $n = 3$, $m = 1$, and $p = n - m = 3 - 1 = 2$) are

$$ {}^A\tilde{\mathbf{v}}_1^{P_1} \underset{(3.4.14)}{=} \hat{\mathbf{e}}_x \qquad {}^A\tilde{\mathbf{v}}_2^{P_1} \underset{(3.4.14)}{=} \hat{\mathbf{e}}_y \tag{25} $$

and these are the same as their holonomic counterparts in Eqs. (20). In connection with the partial angular velocities of E in A, however, Eq. (24) matters very much, for substitution from Eq. (24) into Eq. (22) produces

$$ {}^A\boldsymbol{\omega}^E = \omega s_3 \hat{\mathbf{e}}_x + \omega c_3 \hat{\mathbf{e}}_y - \left(\frac{u_2}{L}\right)\hat{\mathbf{e}}_z \tag{26} $$

which has the form of Eq. (3) and permits one to identify the two nonholonomic partial angular velocities of E in A as

$$ {}^A\tilde{\boldsymbol{\omega}}_1^E \underset{(26)}{=} \mathbf{0} \qquad {}^A\tilde{\boldsymbol{\omega}}_2^E \underset{(26)}{=} -\frac{1}{L}\hat{\mathbf{e}}_z \tag{27} $$

The second of these differs noticeably from its counterpart in Eq. (23).

Finally, still working with the motion variables of Eqs. (3.4.7), let us examine the three holonomic and two nonholonomic partial velocities of $D^\star$ in A (see Fig. 3.5.1). To determine these, we refer to Eqs. (2.6.13) and (3.4.7) to express ${}^A\mathbf{v}^{D^\star}$ as

$$ {}^A\mathbf{v}^{D^\star} = u_1\hat{\mathbf{e}}_x + (u_2 + Lu_3)\hat{\mathbf{e}}_y - \omega(q_1 + Lc_3)\hat{\mathbf{e}}_z \tag{28} $$

and note that, when Eq. (24) is taken into account, ${}^A\mathbf{v}^{D^\star}$ is given by

$$ {}^A\mathbf{v}^{D^\star} = u_1\hat{\mathbf{e}}_x - \omega(q_1 + Lc_3)\hat{\mathbf{e}}_z \tag{29} $$

Consequently, the holonomic partial velocities of $D^\star$ in A are

$$ {}^A\mathbf{v}_1^{D^\star} \underset{(28)}{=} \hat{\mathbf{e}}_x \qquad {}^A\mathbf{v}_2^{D^\star} \underset{(28)}{=} \hat{\mathbf{e}}_y \qquad {}^A\mathbf{v}_3^{D^\star} \underset{(28)}{=} L\hat{\mathbf{e}}_y \tag{30} $$

while the nonholonomic partial velocities of $D^\star$ in A are

$$ {}^A\tilde{\mathbf{v}}_1^{D^\star} \underset{(29)}{=} \hat{\mathbf{e}}_x \qquad {}^A\tilde{\mathbf{v}}_2^{D^\star} \underset{(29)}{=} \mathbf{0} \tag{31} $$

3.7 MOTION CONSTRAINTS WITH NONLINEAR EQUATIONS

In some cases a motion constraint (see Sec. 3.5) is described by a mathematical relationship that is inherently nonlinear in the velocities of particles and/or angular velocities of rigid bodies belonging to a system S; in that event it is not possible to express the corresponding nonholonomic constraint equation in the form of Eqs. (3.5.2), which are linear in the motion variables $u_1, \ldots, u_p$ for S in a reference frame A. Nevertheless, differentiation of the nonholonomic constraint equation with respect to time will result in a relationship that is linear in the time derivatives of the motion variables. When all inherently nonlinear nonholonomic constraint equations give rise in this way to the ℓ relationships

$$\dot{u}_r = \sum_{s=1}^{c} \widetilde{A}_{rs}\dot{u}_s + \widetilde{B}_r \qquad (r = c+1,\ldots,p) \tag{1}$$

where

$$c \triangleq p - \ell \tag{2}$$

and where $\widetilde{A}_{rs}$ and $\widetilde{B}_r$ are functions of $q_1,\ldots,q_n, u_1,\ldots,u_p$, and the time t, S is referred to as a *complex nonholonomic system possessing c degrees of freedom* in A.

Example In Fig. 3.7.1, P_1 and P_2 denote two particles moving in a plane fixed in a reference frame A. The configuration in A of the system S composed of P_1 and P_2 is specified by generalized coordinates $q_1,\ldots,q_4$. It is convenient to choose four motion variables for S in A as $u_r = \dot{q}_r$ $(r = 1,2,3,4)$, in which case the velocities in A of P_1 and P_2 are written as

$$^A\mathbf{v}^{P_1} = u_1\hat{\mathbf{a}}_1 + u_2\hat{\mathbf{a}}_2 \qquad ^A\mathbf{v}^{P_2} = u_3\hat{\mathbf{a}}_1 + u_4\hat{\mathbf{a}}_2 \tag{3}$$

where $\hat{\mathbf{a}}_1$, $\hat{\mathbf{a}}_2$, $\hat{\mathbf{a}}_3$ form a dextral set of mutually perpendicular unit vectors directed as shown in Fig. 3.7.1. In this case S is a holonomic system possessing $p = n = 4$ degrees of freedom in A.

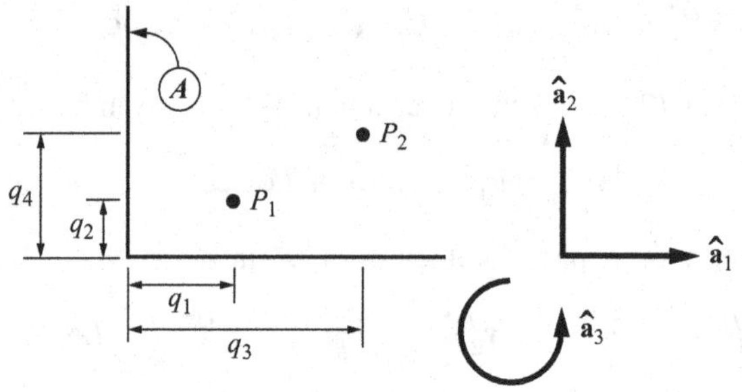

Figure 3.7.1

Now suppose that S is to be subject to a motion constraint, the nature of which is to have $^A\mathbf{v}^{P_1}$ perpendicular to $^A\mathbf{v}^{P_2}$ at every instant of time. The restriction is described by the nonholonomic constraint equation

$$^A\mathbf{v}^{P_2} \cdot {^A\mathbf{v}^{P_1}} = 0 \tag{4}$$

This constraint equation, which has the form of Eqs. (3.5.1), is evidently nonlinear in velocity because more than one velocity vector appears in a dot product. Upon substitution from Eqs. (3) into (4), and evaluation of the dot products, it can be seen that the resulting constraint equation

$$u_1u_3 + u_2u_4 = 0 \tag{5}$$

is nonlinear in the motion variables and, therefore, cannot be cast in the form of Eqs.

(3.5.2). When S is subject to the constraint described by either of the two foregoing equivalent expressions, $\ell = 1$, $c = p - \ell = 4 - 1 = 3$, and S is seen to be a complex nonholonomic system possessing three degrees of freedom in A.

Relationships that result from performing time differentiation of Eqs. (4) and (5) are seen to be, in the first case, linear in acceleration and, in the second case, linear in the time derivatives of the motion variables. First, the left- and right-hand members of Eq. (4) are differentiated with respect to the time t,

$$\frac{d}{dt}({}^A\mathbf{v}^{P_2} \cdot {}^A\mathbf{v}^{P_1}) \underset{(4)}{=} \frac{d}{dt} 0 = 0 \tag{6}$$

Next, bearing in mind the instructions given in Secs. 1.9 and 1.10, A is chosen as the reference frame in which to differentiate ${}^A\mathbf{v}^{P_1}$ and ${}^A\mathbf{v}^{P_2}$, whence

$$\frac{{}^A d\,{}^A\mathbf{v}^{P_2}}{dt} \cdot {}^A\mathbf{v}^{P_1} + {}^A\mathbf{v}^{P_2} \cdot \frac{{}^A d\,{}^A\mathbf{v}^{P_1}}{dt} \underset{(2.6.2)}{=} {}^A\mathbf{a}^{P_2} \cdot {}^A\mathbf{v}^{P_1} + {}^A\mathbf{a}^{P_1} \cdot {}^A\mathbf{v}^{P_2} = 0 \tag{7}$$

This relationship is linear in acceleration because only one acceleration vector appears in each of the two dot products. As for Eq. (5), time differentiation yields a relationship that is equivalent to Eq. (7),

$$\frac{d}{dt}(u_1u_3 + u_2u_4) = u_3\dot{u}_1 + u_4\dot{u}_2 + u_1\dot{u}_3 + u_2\dot{u}_4 \underset{(5)}{=} 0 \tag{8}$$

The result is linear in $\dot{u}_1, \ldots, \dot{u}_4$ and can readily be rearranged as

$$\dot{u}_4 = -\frac{1}{u_2}(u_3\dot{u}_1 + u_4\dot{u}_2 + u_1\dot{u}_3) \tag{9}$$

whereupon the functions $\widetilde{A}_{rs}$ and $\widetilde{B}_r$ $(r = 4; s = 1,2,3)$ of Eqs. (1) are identified to be

$$\widetilde{A}_{41} = -\frac{u_3}{u_2} \qquad \widetilde{A}_{42} = -\frac{u_4}{u_2} \qquad \widetilde{A}_{43} = -\frac{u_1}{u_2} \qquad \widetilde{B}_4 = 0 \tag{10}$$

3.8 PARTIAL ANGULAR ACCELERATIONS, PARTIAL ACCELERATIONS

If $q_1, \ldots, q_n$ and $u_1, \ldots, u_p$ are, respectively, generalized coordinates (see Sec. 3.2) and motion variables (see Sec. 3.4) for a complex nonholonomic system S possessing c degrees of freedom in a reference frame A (see Sec. 3.7), then $\boldsymbol{\alpha}$, the angular acceleration in A of a rigid body B belonging to S, and $\mathbf{a}$, the acceleration in A of a particle P belonging to S, can be expressed *uniquely* as

$$\boldsymbol{\alpha} = \sum_{r=1}^{p} \boldsymbol{\alpha}_r\, \dot{u}_r + \boldsymbol{\alpha}_t \tag{1}$$

and

$$\mathbf{a} = \sum_{r=1}^{p} \mathbf{a}_r\, \dot{u}_r + \mathbf{a}_t \tag{2}$$

where $\boldsymbol{\alpha}_r = \widetilde{\boldsymbol{\omega}}_r$ and $\mathbf{a}_r = \widetilde{\mathbf{v}}_r$ $(r = 1,\ldots,p)$ are functions of $q_1,\ldots,q_n$, and the time t as discussed in Sec. 3.6. The vector $\boldsymbol{\alpha}_r$ is called the rth *partial angular acceleration* of B in A, and $\mathbf{a}_r$ is referred to as the rth *partial acceleration* of P in A. The vectors $\boldsymbol{\alpha}_t$ and $\mathbf{a}_t$ are functions of $q_1,\ldots,q_n, u_1,\ldots,u_p$, and t.

The vectors $\boldsymbol{\alpha}$ and $\mathbf{a}$ can also be expressed *uniquely* as

$$\boldsymbol{\alpha} = \sum_{r=1}^{c} \widetilde{\boldsymbol{\alpha}}_r \, \dot{u}_r + \widetilde{\boldsymbol{\alpha}}_t \tag{3}$$

and

$$\mathbf{a} = \sum_{r=1}^{c} \widetilde{\mathbf{a}}_r \, \dot{u}_r + \widetilde{\mathbf{a}}_t \tag{4}$$

where $\widetilde{\boldsymbol{\alpha}}_r$, $\widetilde{\mathbf{a}}_r$ $(r = 1,\ldots,c)$, $\widetilde{\boldsymbol{\alpha}}_t$, and $\widetilde{\mathbf{a}}_t$, are functions of $q_1,\ldots,q_n, u_1,\ldots,u_p$, and t. The vector $\widetilde{\boldsymbol{\alpha}}_r$ is called the rth *nonholonomic partial angular acceleration* of B in A, while $\widetilde{\mathbf{a}}_r$ is known as the rth *nonholonomic partial acceleration* of P in A.

When $c = p$, that is, when S is a simple nonholonomic system possessing p degrees of freedom in A, then $\widetilde{\boldsymbol{\alpha}}_r = \boldsymbol{\alpha}_r$ and $\widetilde{\mathbf{a}}_r = \mathbf{a}_r$ $(r = 1,\ldots,p)$.

Derivation Differentiation of both sides of Eq. (3.6.3) with respect to t in A yields

$$\boldsymbol{\alpha} \underset{(2.5.1)}{=} \frac{{}^A d\boldsymbol{\omega}}{dt} \underset{(3.6.3)}{=} \sum_{r=1}^{p} \widetilde{\boldsymbol{\omega}}_r \, \dot{u}_r + \sum_{r=1}^{p} \frac{{}^A d\, \widetilde{\boldsymbol{\omega}}_r}{dt} u_r + \frac{{}^A d\, \widetilde{\boldsymbol{\omega}}_t}{dt} \tag{5}$$

After defining $\boldsymbol{\alpha}_r$ and $\boldsymbol{\alpha}_t$ as

$$\boldsymbol{\alpha}_r \triangleq \widetilde{\boldsymbol{\omega}}_r \qquad (r = 1,\ldots,p) \tag{6}$$

and

$$\boldsymbol{\alpha}_t \triangleq \sum_{r=1}^{p} \frac{{}^A d\, \widetilde{\boldsymbol{\omega}}_r}{dt} u_r + \frac{{}^A d\, \widetilde{\boldsymbol{\omega}}_t}{dt} \tag{7}$$

one arrives at Eq. (1) by substituting from Eqs. (6) and (7) into (5). Similarly, Eq. (2) is obtained by differentiating Eq. (3.6.4)

$$\mathbf{a} \underset{(2.6.2)}{=} \frac{{}^A d\mathbf{v}}{dt} \underset{(3.6.4)}{=} \sum_{r=1}^{p} \widetilde{\mathbf{v}}_r \, \dot{u}_r + \sum_{r=1}^{p} \frac{{}^A d\, \widetilde{\mathbf{v}}_r}{dt} u_r + \frac{{}^A d\, \widetilde{\mathbf{v}}_t}{dt} \tag{8}$$

and defining $\mathbf{a}_r$ and $\mathbf{a}_t$, respectively, as

$$\mathbf{a}_r \triangleq \widetilde{\mathbf{v}}_r \qquad (r = 1,\ldots,p) \tag{9}$$

and

$$\mathbf{a}_t \triangleq \sum_{r=1}^{p} \frac{{}^A d\, \widetilde{\mathbf{v}}_r}{dt} u_r + \frac{{}^A d\, \widetilde{\mathbf{v}}_t}{dt} \tag{10}$$

A complex nonholonomic system S is subject to motion constraints such that $\dot{u}_1,\ldots,\dot{u}_p$ are governed by Eqs. (3.7.1). After rewriting Eq. (1) as

$$\boldsymbol{\alpha} = \sum_{r=1}^{c} \boldsymbol{\alpha}_r \, \dot{u}_r + \sum_{r=c+1}^{p} \boldsymbol{\alpha}_r \, \dot{u}_r + \boldsymbol{\alpha}_t \tag{11}$$

one can use Eqs. (3.7.1) to obtain

$$\begin{aligned}
\boldsymbol{\alpha} &\underset{(11)}{=} \sum_{r=1}^{c} \boldsymbol{\alpha}_r \dot{u}_r + \sum_{r=c+1}^{p} \boldsymbol{\alpha}_r \left(\sum_{s=1}^{c} \underset{(3.7.1)}{\widetilde{A}_{rs}\dot{u}_s + \widetilde{B}_r} \right) + \boldsymbol{\alpha}_t \\
&= \sum_{r=1}^{c} \boldsymbol{\alpha}_r \dot{u}_r + \sum_{s=1}^{c} \sum_{r=c+1}^{p} \boldsymbol{\alpha}_r \widetilde{A}_{rs} \dot{u}_s + \sum_{r=c+1}^{p} \boldsymbol{\alpha}_r \widetilde{B}_r + \boldsymbol{\alpha}_t \\
&= \sum_{r=1}^{c} \boldsymbol{\alpha}_r \dot{u}_r + \sum_{r=1}^{c} \sum_{s=c+1}^{p} \boldsymbol{\alpha}_s \widetilde{A}_{sr} \dot{u}_r + \sum_{r=c+1}^{p} \boldsymbol{\alpha}_r \widetilde{B}_r + \boldsymbol{\alpha}_t \\
&= \sum_{r=1}^{c} \left(\boldsymbol{\alpha}_r + \sum_{s=c+1}^{p} \boldsymbol{\alpha}_s \widetilde{A}_{sr} \right) \dot{u}_r + \sum_{r=c+1}^{p} \boldsymbol{\alpha}_r \widetilde{B}_r + \boldsymbol{\alpha}_t
\end{aligned} \tag{12}$$

and, after defining $\widetilde{\boldsymbol{\alpha}}_r$ and $\widetilde{\boldsymbol{\alpha}}_t$ as

$$\widetilde{\boldsymbol{\alpha}}_r \triangleq \boldsymbol{\alpha}_r + \sum_{s=c+1}^{p} \boldsymbol{\alpha}_s \widetilde{A}_{sr} \qquad (r = 1, \ldots, c) \tag{13}$$

and

$$\widetilde{\boldsymbol{\alpha}}_t \triangleq \boldsymbol{\alpha}_t + \sum_{r=c+1}^{p} \boldsymbol{\alpha}_r \widetilde{B}_r \tag{14}$$

one arrives at Eq. (3) by substituting from Eqs. (13) and (14) into Eq. (12). The path from Eq. (2) to Eq. (4) entails an analogous derivation and the following definitions of $\widetilde{\mathbf{a}}_r$ and $\widetilde{\mathbf{a}}_t$:

$$\widetilde{\mathbf{a}}_r \triangleq \mathbf{a}_r + \sum_{s=c+1}^{p} \mathbf{a}_s \widetilde{A}_{sr} \qquad (r = 1, \ldots, c) \tag{15}$$

$$\widetilde{\mathbf{a}}_t \triangleq \mathbf{a}_t + \sum_{r=c+1}^{p} \mathbf{a}_r \widetilde{B}_r \tag{16}$$

It will become apparent in the sequel that nonholonomic partial angular accelerations and nonholonomic partial accelerations are especially useful in forming equations of motion for complex nonholonomic systems. These vectors are obtained easily by inspecting expressions for angular accelerations of rigid bodies and/or expressions for accelerations of particles, just as the partial angular velocities and partial velocities presented in Sec. 3.6 are obtained by inspection.

Example In the example presented in Sec. 3.7, particles P_1 and P_2 form a complex nonholonomic system S when subject to the constraint expressed with either Eq. (3.7.4) or Eq. (3.7.5). The accelerations in A of P_1 and P_2, respectively, are given by

$$^{A}\mathbf{a}^{P_1} \underset{(2.6.2)}{=} \frac{^{A}d\,^{A}\mathbf{v}^{P_1}}{dt} \underset{(3.7.3)}{=} \dot{u}_1 \hat{\mathbf{a}}_1 + \dot{u}_2 \hat{\mathbf{a}}_2 \tag{17}$$

$$^{A}\mathbf{a}^{P_2} \underset{(2.6.2)}{=} \frac{^{A}d\,^{A}\mathbf{v}^{P_2}}{dt} \underset{(3.7.3)}{=} \dot{u}_3 \hat{\mathbf{a}}_1 + \dot{u}_4 \hat{\mathbf{a}}_2 \tag{18}$$

The partial accelerations ${}^A\mathbf{a}_r^{P_1}$ and ${}^A\mathbf{a}_r^{P_2}$ $(r = 1,2,3,4)$ in A of P_1 and P_2 are obtained simply by inspecting the foregoing expressions for the vector coefficients of $\dot{u}_r$ $(r = 1,2,3,4)$.

$$ {}^A\mathbf{a}_1^{P_1} = \hat{\mathbf{a}}_1 \qquad {}^A\mathbf{a}_2^{P_1} = \hat{\mathbf{a}}_2 \qquad {}^A\mathbf{a}_3^{P_1} = \mathbf{0} \qquad {}^A\mathbf{a}_4^{P_1} = \mathbf{0} \tag{19} $$

$$ {}^A\mathbf{a}_1^{P_2} = \mathbf{0} \qquad {}^A\mathbf{a}_2^{P_2} = \mathbf{0} \qquad {}^A\mathbf{a}_3^{P_2} = \hat{\mathbf{a}}_1 \qquad {}^A\mathbf{a}_4^{P_2} = \hat{\mathbf{a}}_2 \tag{20} $$

Inspection likewise reveals that

$$ {}^A\mathbf{a}_t^{P_1} = \mathbf{0} \qquad {}^A\mathbf{a}_t^{P_2} = \mathbf{0} \tag{21} $$

Now, $\dot{u}_4$ can be removed from the picture on the basis of Eq. (3.7.9); substitution into Eqs. (17) and (18) produces

$$ {}^A\mathbf{a}^{P_1} \underset{(17)}{=} \dot{u}_1\hat{\mathbf{a}}_1 + \dot{u}_2\hat{\mathbf{a}}_2 \tag{22} $$

$$ {}^A\mathbf{a}^{P_2} \underset{(18)}{=} \dot{u}_3\hat{\mathbf{a}}_1 - \frac{1}{u_2}\underset{(3.7.9)}{(u_3\dot{u}_1 + u_4\dot{u}_2 + u_1\dot{u}_3)}\,\hat{\mathbf{a}}_2 \tag{23} $$

The nonholonomic partial accelerations ${}^A\tilde{\mathbf{a}}_r^{P_1}$ and ${}^A\tilde{\mathbf{a}}_r^{P_2}$ $(r = 1,2,3)$ in A of P_1 and P_2 are easily identified by inspecting these expressions for the vector coefficients of $\dot{u}_r$ $(r = 1,2,3)$.

$$ {}^A\tilde{\mathbf{a}}_1^{P_1} = \hat{\mathbf{a}}_1 \qquad {}^A\tilde{\mathbf{a}}_2^{P_1} = \hat{\mathbf{a}}_2 \qquad {}^N\tilde{\mathbf{a}}_3^{P_1} = \mathbf{0} \tag{24} $$

$$ {}^A\tilde{\mathbf{a}}_1^{P_2} = -\frac{u_3}{u_2}\hat{\mathbf{a}}_2, \qquad {}^A\tilde{\mathbf{a}}_2^{P_2} = -\frac{u_4}{u_2}\hat{\mathbf{a}}_2, \qquad {}^A\tilde{\mathbf{a}}_3^{P_2} = \hat{\mathbf{a}}_1 - \frac{u_1}{u_2}\hat{\mathbf{a}}_2 \tag{25} $$

Additional inspection of Eqs. (22) and (23) shows that

$$ {}^A\tilde{\mathbf{a}}_t^{P_1} = \mathbf{0} \qquad {}^A\tilde{\mathbf{a}}_t^{P_2} = \mathbf{0} \tag{26} $$

The vectors ${}^A\tilde{\mathbf{a}}_r^{P_1}$, ${}^A\tilde{\mathbf{a}}_r^{P_2}$ $(r = 1,2,3)$, ${}^A\tilde{\mathbf{a}}_t^{P_1}$, and ${}^A\tilde{\mathbf{a}}_t^{P_2}$ can also be obtained by an alternative route: with Eqs. (19), (20), and (3.7.10) in hand, one may then appeal to Eqs. (15) and (16).

A comparison of Eqs. (19) with (24) shows that ${}^A\tilde{\mathbf{a}}_r^{P_1} = {}^A\mathbf{a}_r^{P_1}$ $(r = 1,2,3)$; however, comparison of Eqs. (20) with (25) reveals that ${}^A\tilde{\mathbf{a}}_r^{P_2}$ are quite distinct from ${}^A\mathbf{a}_r^{P_2}$ $(r = 1,2,3)$. This observation provides an illustration of the conclusion that can be reached by examining Eqs. (13) and (15); namely, that nonholonomic partial angular accelerations differ in general from partial angular accelerations, and nonholonomic partial accelerations are not the same as partial accelerations.

3.9 ACCELERATION AND PARTIAL VELOCITIES

When $q_1, \ldots, q_n$ are generalized coordinates characterizing the configuration of a system S in a reference frame A (see Sec. 3.2), then $\mathbf{v}^2$, the square of the velocity in A of a generic particle P of S, may be regarded as a (scalar) function of the $2n + 1$ independent variables $q_1, \ldots, q_n$, $\dot{q}_1, \ldots, \dot{q}_n$, and t, where $\dot{q}_r$ denotes the first time-derivative of q_r $(r = 1, \ldots, n)$. If motion variables (see Sec. 3.4) are defined as

$$u_r \triangleq \dot{q}_r \qquad (r = 1, \ldots, n) \tag{1}$$

and $\mathbf{v}_r$ denotes the rth holonomic partial velocity of P in A (see Sec. 3.6), then $\mathbf{v}_r$, the acceleration $\mathbf{a}$ of P in A, and $\mathbf{v}^2$ are related to each other as follows:

$$\mathbf{v}_r \cdot \mathbf{a} = \frac{1}{2}\left(\frac{d}{dt}\frac{\partial \mathbf{v}^2}{\partial \dot{q}_r} - \frac{\partial \mathbf{v}^2}{\partial q_r}\right) \qquad (r = 1, \ldots, n) \tag{2}$$

If, in accordance with Eqs. (3.4.1), motion variables are defined as

$$u_r \triangleq \sum_{s=1}^{n} Y_{rs}\,\dot{q}_s + Z_r \qquad (r = 1, \ldots, n) \tag{3}$$

where Y_{rs} and Z_r are functions of $q_1, \ldots, q_n$, and the time t, and $\mathbf{v}_r$ denotes the associated r^{th} holonomic partial velocity of P in A (see Sec. 3.6), then

$$\mathbf{v}_r \cdot \mathbf{a} = \frac{1}{2}\sum_{s=1}^{n}\left(\frac{d}{dt}\frac{\partial \mathbf{v}^2}{\partial \dot{q}_s} - \frac{\partial \mathbf{v}^2}{\partial q_s}\right) W_{sr} \qquad (r = 1, \ldots, n) \tag{4}$$

where W_{sr} is a function of $q_1, \ldots, q_n$, and t such that solution of Eqs. (3) for $\dot{q}_1, \ldots, \dot{q}_n$ yields

$$\dot{q}_s = \sum_{r=1}^{n} W_{sr} u_r + X_s \qquad (s = 1, \ldots, n) \tag{5}$$

Finally, when S is a simple nonholonomic system possessing p degrees of freedom in A (see Sec. 3.5), so that there exist m nonholonomic constraint equations of the form

$$u_r = \sum_{s=1}^{p} A_{rs} u_s + B_r \qquad (r = p+1, \ldots, n) \tag{6}$$

then $\tilde{\mathbf{v}}_r \cdot \mathbf{a}$, where $\tilde{\mathbf{v}}_r$ is the rth nonholonomic partial velocity of P in A (see Sec. 3.6), can be expressed in terms of $\mathbf{v}^2$ (still regarded as a function of $q_1, \ldots, q_n$, $\dot{q}_1, \ldots, \dot{q}_n$, and t) as

$$\begin{aligned}\tilde{\mathbf{v}}_r \cdot \mathbf{a} = {} & \frac{1}{2}\left(\frac{d}{dt}\frac{\partial \mathbf{v}^2}{\partial \dot{q}_r} - \frac{\partial \mathbf{v}^2}{\partial q_r}\right) \\ & + \frac{1}{2}\sum_{s=p+1}^{n}\left(\frac{d}{dt}\frac{\partial \mathbf{v}^2}{\partial \dot{q}_s} - \frac{\partial \mathbf{v}^2}{\partial q_s}\right) A_{sr} \qquad (r = 1, \ldots, p)\end{aligned} \tag{7}$$

when u_r is defined as in Eqs. (1), and as

$$\tilde{\mathbf{v}}_r \cdot \mathbf{a} = \frac{1}{2}\sum_{s=1}^{n}\left[\left(\frac{d}{dt}\frac{\partial \mathbf{v}^2}{\partial \dot{q}_s} - \frac{\partial \mathbf{v}^2}{\partial q_s}\right)\left(W_{sr} + \sum_{k=p+1}^{n} W_{sk} A_{kr}\right)\right] \qquad (r = 1, \ldots, p) \tag{8}$$

when u_r is defined as in Eq. (3).

Equations (2), (7), and (8) play essential parts in the derivations of Lagrange equations and Passerello-Huston equations (see Problem 12.14). Additionally, Eqs. (2) can facilitate the determination of accelerations, as will be shown presently.

Derivation When u_r is defined as in Eqs. (1), then W_{sr} in Eqs. (3.6.5) is equal to unity for $s = r$ and vanishes otherwise, while X_s vanishes for $s = 1,\dots,n$. Consequently, Eqs. (3.6.11) and (3.6.12) reduce to

$$\mathbf{v}_r \underset{(3.6.11)}{=} \frac{\partial \mathbf{p}}{\partial q_r} \qquad (r = 1,\dots,n) \tag{9}$$

$$\mathbf{v}_t \underset{(3.6.12)}{=} \frac{\partial \mathbf{p}}{\partial t} \tag{10}$$

respectively. From the first of these it follows that

$$\frac{\partial \mathbf{v}_r}{\partial q_s} \underset{(9)}{=} \frac{\partial}{\partial q_s}\left(\frac{\partial \mathbf{p}}{\partial q_r}\right) \underset{(1.11.3)}{=} \frac{\partial}{\partial q_r}\left(\frac{\partial \mathbf{p}}{\partial q_s}\right) \underset{(9)}{=} \frac{\partial \mathbf{v}_s}{\partial q_r} \qquad (r, s = 1,\dots,n) \tag{11}$$

while the two together lead to

$$\frac{\partial \mathbf{v}_t}{\partial q_r} \underset{(10)}{=} \frac{\partial}{\partial q_r}\left(\frac{\partial \mathbf{p}}{\partial t}\right) \underset{(1.11.4)}{=} \frac{\partial}{\partial t}\left(\frac{\partial \mathbf{p}}{\partial q_r}\right) \underset{(9)}{=} \frac{\partial \mathbf{v}_r}{\partial t} \tag{12}$$

These relationships will be used shortly.

When $\mathbf{v}$ is expressed as in Eq. (3.6.2) and u_r is replaced with $\dot{q}_r$ in accordance with Eqs. (1), one can regard $\mathbf{v}$ as a function of the independent variables $q_1,\dots,q_n$, $\dot{q}_1,\dots,\dot{q}_n$, and t, in which event

$$\frac{\partial \dot{q}_s}{\partial q_r} = 0 \qquad (r, s = 1,\dots,n) \tag{13}$$

and partial differentiation of $\mathbf{v}$ with respect to q_r gives

$$\begin{aligned}
\frac{\partial \mathbf{v}}{\partial q_r} &\underset{(3.6.2)}{=} \frac{\partial}{\partial q_r}\Bigg(\sum_{s=1}^{n} \mathbf{v}_s \underset{(1)}{\dot{q}_s} + \mathbf{v}_t\Bigg) \\
&\underset{(1.10.1,\ 1.10.2)}{=} \sum_{s=1}^{n}\left(\frac{\partial \mathbf{v}_s}{\partial q_r}\dot{q}_s + \mathbf{v}_s\frac{\partial \dot{q}_s}{\partial q_r}\right) + \frac{\partial \mathbf{v}_t}{\partial q_r} \\
&= \underset{(11)}{\sum_{s=1}^{n} \frac{\partial \mathbf{v}_r}{\partial q_s}\dot{q}_s} + \underset{(13)}{0} + \underset{(12)}{\frac{\partial \mathbf{v}_r}{\partial t}} \underset{(1.12.1)}{=} \frac{d\mathbf{v}_r}{dt} \qquad (r = 1,\dots,n)
\end{aligned} \tag{14}$$

while partial differentiation with respect to $\dot{q}_r$ produces

$$\frac{\partial \mathbf{v}}{\partial \dot{q}_r} \underset{(3.6.2)}{=} \frac{\partial}{\partial \dot{q}_r}\left(\sum_{s=1}^{n} \mathbf{v}_s \dot{q}_s + \mathbf{v}_t\right) = \mathbf{v}_r \qquad (r = 1,\dots,n) \tag{15}$$

since $\mathbf{v}_s$ and $\mathbf{v}_t$ are independent of $\dot{q}_r$, and $\partial \dot{q}_s / \partial \dot{q}_r$ vanishes except for $s = r$, in which case it is equal to unity.

To conclude the derivation of Eqs. (2), we note that

$$\frac{d}{dt}(\mathbf{v}_r \cdot \mathbf{v}) = \frac{d\mathbf{v}_r}{dt} \cdot \mathbf{v} + \mathbf{v}_r \cdot \frac{d\mathbf{v}}{dt} = \underset{(14)}{\frac{\partial \mathbf{v}}{\partial q_r}} \cdot \mathbf{v} + \underset{(2.6.2)}{\mathbf{v}_r \cdot \mathbf{a}} \qquad (r = 1,\ldots,n) \tag{16}$$

which, solved for $\mathbf{v}_r \cdot \mathbf{a}$, yields

$$\mathbf{v}_r \cdot \mathbf{a} \underset{(16)}{=} \frac{d}{dt}(\mathbf{v}_r \cdot \mathbf{v}) - \frac{\partial \mathbf{v}}{\partial q_r} \cdot \mathbf{v} \qquad (r = 1,\ldots,n) \tag{17}$$

or

$$\mathbf{v}_r \cdot \mathbf{a} \underset{(17)}{=} \frac{d}{dt}\left(\underset{(15)}{\frac{\partial \mathbf{v}}{\partial \dot{q}_r}} \cdot \mathbf{v}\right) - \frac{\partial \mathbf{v}}{\partial q_r} \cdot \mathbf{v} = \frac{d}{dt}\left(\frac{1}{2}\frac{\partial \mathbf{v}^2}{\partial \dot{q}_r}\right) - \frac{1}{2}\frac{\partial \mathbf{v}^2}{\partial q_r} \qquad (r = 1,\ldots,n) \tag{18}$$

and this is equivalent to Eqs. (2).

To establish the validity of Eqs. (4), we begin by exploring the relationship between the partial velocities associated with motion variables defined as in Eqs. (1), on the one hand, and partial velocities associated with motion variables defined as in Eqs. (3), on the other hand. Denoting the former by $\mathbf{v}_1,\ldots,\mathbf{v}_n$, as heretofore, we have

$$\mathbf{v} \underset{(3.6.2)}{=} \sum_{r=1}^{n} \mathbf{v}_r \underset{(1)}{\dot{q}_r} + \mathbf{v}_t \tag{19}$$

Hence, when $u_1,\ldots,u_n$ are defined as in Eqs. (3), so that Eqs. (5) apply, we can write

$$\mathbf{v} \underset{(19,\,5)}{=} \sum_{r=1}^{n} \mathbf{v}_r \left(\sum_{s=1}^{n} W_{rs} u_s + X_r\right) + \mathbf{v}_t = \sum_{r=1}^{n}\sum_{s=1}^{n} \mathbf{v}_s W_{sr} u_r + \sum_{r=1}^{n} \mathbf{v}_r X_r + \mathbf{v}_t \tag{20}$$

and now we can identify the partial velocities associated with $u_1,\ldots,u_n$, which we denote temporarily by $\overline{\mathbf{v}}_1,\ldots,\overline{\mathbf{v}}_n$, as the coefficients of $u_1,\ldots,u_n$, respectively, in Eq. (20); that is,

$$\overline{\mathbf{v}}_r \triangleq \sum_{s=1}^{n} \mathbf{v}_s W_{sr} \qquad (r = 1,\ldots,n) \tag{21}$$

The derivation of Eqs. (4) then can be completed by dot-multiplying Eqs. (21) with $\mathbf{a}$, using Eqs. (2) to eliminate $\mathbf{v}_s \cdot \mathbf{a}$ $(s = 1,\ldots,n)$, and writing $\mathbf{v}_r$ in place of $\overline{\mathbf{v}}_r$ $(r = 1,\ldots,n)$.

Lastly, to obtain Eqs. (7) and (8), dot-multiply Eqs. (3.6.17) with $\mathbf{a}$, showing that

$$\tilde{\mathbf{v}}_r \cdot \mathbf{a} = \mathbf{v}_r \cdot \mathbf{a} + \sum_{s=p+1}^{n} \mathbf{v}_s \cdot \mathbf{a} A_{sr} \qquad (r = 1,\ldots,p) \tag{22}$$

and then use Eqs. (2) in connection with Eqs. (7), and Eqs. (4) in the case of Eqs. (8), to eliminate $\mathbf{v}_r \cdot \mathbf{a}$ $(r = 1, \ldots, p)$ and $\mathbf{v}_s \cdot \mathbf{a}$ $(s = p+1, \ldots, n)$.

Example Considering a point P moving in a reference frame A, let $\hat{\mathbf{a}}_1$, $\hat{\mathbf{a}}_2$, $\hat{\mathbf{a}}_3$ be mutually perpendicular unit vectors fixed in A, and express $\mathbf{p}$, the position vector from a point fixed in A to point P, as

$$\mathbf{p} = p_1\hat{\mathbf{a}}_1 + p_2\hat{\mathbf{a}}_2 + p_3\hat{\mathbf{a}}_3 \tag{23}$$

where p_r $(r = 1,2,3)$ are single-valued functions of three scalar variables, q_r $(r = 1,2,3)$. Then there corresponds to every set of values of q_r $(r = 1,2,3)$ a unique position of P in A; q_r $(r = 1,2,3)$ are called *curvilinear coordinates* of P in A; and the partial derivatives of $\mathbf{p}$ with respect to q_r $(r = 1,2,3)$ in A can be expressed as

$$\frac{\partial \mathbf{p}}{\partial q_r} = f_r\hat{\mathbf{n}}_r \qquad (r = 1,2,3) \tag{24}$$

where f_r is a function of q_1, q_2, q_3, and $\hat{\mathbf{n}}_r$ is a unit vector. If $\hat{\mathbf{n}}_1$, $\hat{\mathbf{n}}_2$, and $\hat{\mathbf{n}}_3$ are mutually perpendicular, then q_1, q_2, and q_3 are called *orthogonal curvilinear coordinates*. For instance, suppose that q_1, q_2, and q_3 measure two distances and an angle as indicated in Fig. 3.9.1. Then

$$\mathbf{p} = q_1 \cos q_2\hat{\mathbf{a}}_1 + q_1 \sin q_2\hat{\mathbf{a}}_2 + q_3\hat{\mathbf{a}}_3 \tag{25}$$

so that

$$\frac{\partial \mathbf{p}}{\partial q_1} \underset{(25)}{=} \cos q_2\hat{\mathbf{a}}_1 + \sin q_2\hat{\mathbf{a}}_2 \tag{26}$$

$$\frac{\partial \mathbf{p}}{\partial q_2} \underset{(25)}{=} -q_1 \sin q_2\hat{\mathbf{a}}_1 + q_1 \cos q_2\hat{\mathbf{a}}_2 \tag{27}$$

$$\frac{\partial \mathbf{p}}{\partial q_3} \underset{(25)}{=} \hat{\mathbf{a}}_3 \tag{28}$$

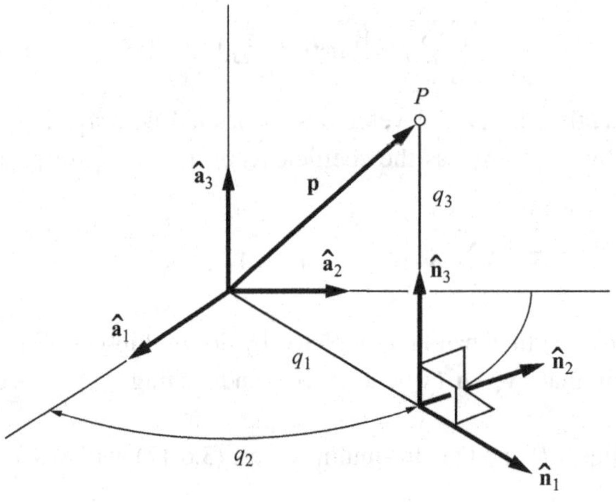

Figure 3.9.1

and the functions f_r and unit vectors $\hat{\mathbf{n}}_r$ $(r = 1,2,3)$ appearing in Eqs. (24) can be identified as

$$f_1 \underset{(26)}{=} 1 \qquad \hat{\mathbf{n}}_1 \underset{(26)}{=} \cos q_2 \hat{\mathbf{a}}_1 + \sin q_2 \hat{\mathbf{a}}_2 \tag{29}$$

$$f_2 \underset{(27)}{=} q_1 \qquad \hat{\mathbf{n}}_2 \underset{(27)}{=} -\sin q_2 \hat{\mathbf{a}}_1 + \cos q_2 \hat{\mathbf{a}}_2 \tag{30}$$

$$f_3 \underset{(28)}{=} 1 \qquad \hat{\mathbf{n}}_3 \underset{(28)}{=} \hat{\mathbf{a}}_3 \tag{31}$$

Moreover, it follows from Eqs. (29)–(31) that

$$\hat{\mathbf{n}}_1 \cdot \hat{\mathbf{n}}_2 = \hat{\mathbf{n}}_2 \cdot \hat{\mathbf{n}}_3 = \hat{\mathbf{n}}_3 \cdot \hat{\mathbf{n}}_1 = 0 \tag{32}$$

which means that $\hat{\mathbf{n}}_1$, $\hat{\mathbf{n}}_2$, and $\hat{\mathbf{n}}_3$ are mutually perpendicular, as indicated in Fig. 3.9.1. Consequently, q_1, q_2, and q_3 are orthogonal curvilinear coordinates of P in A.

When q_r $(r = 1,2,3)$ are orthogonal curvilinear coordinates of P in A, the acceleration $\mathbf{a}$ of P in A can be expressed as

$$\mathbf{a} = a_1 \hat{\mathbf{n}}_1 + a_2 \hat{\mathbf{n}}_2 + a_3 \hat{\mathbf{n}}_3 \tag{33}$$

where $\hat{\mathbf{n}}_r$ $(r = 1,2,3)$ are the unit vectors appearing in Eqs. (24). To find an expression for a_r $(r = 1,2,3)$ in terms of f_1, f_2, f_3 and q_1, q_2, q_3, one can employ Eqs. (2) after noting that the velocity $\mathbf{v}$ of P in A is given by

$$\mathbf{v} \underset{(2.6.1)}{=} \frac{d\mathbf{p}}{dt} \underset{(1.12.1)}{=} \sum_{r=1}^{3} \frac{\partial \mathbf{p}}{\partial q_r} \dot{q}_r \underset{(24)}{=} \sum_{r=1}^{3} f_r \dot{q}_r \hat{\mathbf{n}}_r \tag{34}$$

so that, if u_r is defined as in Eqs. (1), then

$$\mathbf{v} \underset{(34)}{=} \sum_{r=1}^{3} f_r u_r \hat{\mathbf{n}}_r \tag{35}$$

which means that

$$\mathbf{v}_r \underset{(3.6.2)}{=} f_r \hat{\mathbf{n}}_r \qquad (r = 1,2,3) \tag{36}$$

and

$$\mathbf{v}_1 \cdot \mathbf{a} = \underset{(36)}{f_1 \hat{\mathbf{n}}_1} \cdot \underset{(33)}{(a_1 \hat{\mathbf{n}}_1 + a_2 \hat{\mathbf{n}}_2 + a_3 \hat{\mathbf{n}}_3)} = a_1 f_1 \tag{37}$$

But

$$\mathbf{v}_1 \cdot \mathbf{a} \underset{(2)}{=} \frac{1}{2} \left(\frac{d}{dt} \frac{\partial \mathbf{v}^2}{\partial \dot{q}_1} - \frac{\partial \mathbf{v}^2}{\partial q_1} \right) \tag{38}$$

and

$$\mathbf{v}^2 \underset{(34)}{=} \sum_{r=1}^{3} (f_r \dot{q}_r)^2 \tag{39}$$

so that

$$\frac{\partial \mathbf{v}^2}{\partial \dot{q}_1} \underset{(39)}{=} 2 f_1{}^2 \dot{q}_1 \qquad \frac{\partial \mathbf{v}^2}{\partial q_1} \underset{(39)}{=} 2 \sum_{s=1}^{3} f_s \frac{\partial f_s}{\partial q_1} \dot{q}_s{}^2 \tag{40}$$

and

$$\mathbf{v}_1 \cdot \mathbf{a} \underset{(38,\,40)}{=} \frac{d}{dt}(f_1{}^2 \dot{q}_1) - \sum_{s=1}^{3} f_s \frac{\partial f_s}{\partial q_1} \dot{q}_s{}^2 \tag{41}$$

Substituting from Eq. (41) into Eq. (37) and solving for a_1, one thus finds that a_1 is given by

$$a_1 = \frac{1}{f_1}\left[\frac{d}{dt}(f_1{}^2 \dot{q}_1) - \sum_{s=1}^{3} f_s \frac{\partial f_s}{\partial q_1} \dot{q}_s{}^2\right] \tag{42}$$

and, after using similar processes in connection with a_2 and a_3, one can conclude that

$$a_r = \frac{1}{f_r}\left[\frac{d}{dt}(f_r{}^2 \dot{q}_r) - \sum_{s=1}^{3} f_s \frac{\partial f_s}{\partial q_r} \dot{q}_s{}^2\right] \qquad (r = 1,2,3) \tag{43}$$

To illustrate the use of this quite general formula, we return to the orthogonal curvilinear coordinates q_1, q_2, q_3 introduced in Eq. (25). Expressions for f_r $(r = 1,2,3)$ available in Eqs. (29)–(31) permit us to write

$$\frac{d}{dt}(f_1{}^2 \dot{q}_1) \underset{(29)}{=} \ddot{q}_1 \qquad \frac{\partial f_1}{\partial q_1} \underset{(29)}{=} 0 \qquad \frac{\partial f_2}{\partial q_1} \underset{(30)}{=} 1 \qquad \frac{\partial f_3}{\partial q_1} \underset{(31)}{=} 0 \tag{44}$$

and, therefore,

$$a_1 \underset{(43,\,44)}{=} \ddot{q}_1 - q_1 \dot{q}_2{}^2 \tag{45}$$

Similarly,

$$\frac{d}{dt}(f_2{}^2 \dot{q}_2) \underset{(30)}{=} \frac{d}{dt}(q_1{}^2 \dot{q}_2) \qquad \frac{\partial f_s}{\partial q_2} \underset{(29\text{–}31)}{=} 0 \qquad (s = 1,2,3) \tag{46}$$

so that

$$a_2 \underset{(43,\,46)}{=} \frac{1}{q_1}\frac{d}{dt}(q_1{}^2 \dot{q}_2) \tag{47}$$

and

$$\frac{d}{dt}(f_3{}^2 \dot{q}_3) \underset{(31)}{=} \ddot{q}_3 \qquad \frac{\partial f_s}{\partial q_3} \underset{(29\text{–}31)}{=} 0 \qquad (s = 1,2,3) \tag{48}$$

which means that

$$a_3 \underset{(43,\,48)}{=} \ddot{q}_3 \tag{49}$$

Thus, we now can express the acceleration of P in A as

$$\mathbf{a} \underset{(33)}{=} \underset{(45)}{(\ddot{q}_1 - q_1 \dot{q}_2{}^2)}\, \hat{\mathbf{n}}_1 + \underset{(47)}{\frac{1}{q_1}\frac{d}{dt}(q_1{}^2 \dot{q}_2)}\hat{\mathbf{n}}_2 + \underset{(49)}{\ddot{q}_3}\, \hat{\mathbf{n}}_3 \tag{50}$$

4 MASS DISTRIBUTION

The motion that results when forces act on a material system depends not only on the forces but also on the constitution of the system. In particular, the manner in which mass is distributed throughout a system generally affects the behavior of the system. For example, suppose that a rod is supported at one end by a fixed horizontal pin and that a relatively heavy particle is attached at a point of the rod, so that together the rod and the particle form a pendulum. The frequency of the oscillations that ensue when the pendulum is released from rest after having been displaced from the vertical depends on the location of the particle along the rod, that is, on the manner in which mass is distributed throughout the pendulum.

For the purpose of certain analyses, it is unnecessary to know in detail how mass is distributed throughout each of the bodies forming a system; all one needs to know for each body is the location of the mass center, as well as the values of six quantities called inertia scalars. The subject of mass center location is considered in Secs. 4.1 and 4.2. Products of inertia and moments of inertia, which are inertia scalars, are defined in Sec. 4.3 in terms of quantities called inertia vectors. Sections 4.4–4.7 deal with the evaluation of inertia scalars, in connection with which inertia matrices and inertia dyadics are discussed. A special kind of moment of inertia, called a principal moment of inertia, is introduced in Sec. 4.8. The chapter concludes with an examination of the relationship between principal moments of inertia, on the one hand, and maximum and minimum moments of inertia, on the other hand.

4.1 MASS CENTER

If S is a set of particles $P_1, \ldots, P_\nu$ of masses $m_1, \ldots, m_\nu$, respectively, there exists a unique point $S^\star$ such that

$$\sum_{i=1}^{\nu} m_i \mathbf{r}_i = \mathbf{0} \tag{1}$$

where $\mathbf{r}_i$ is the position vector from $S^\star$ to P_i $(i = 1, \ldots, \nu)$. $S^\star$, called the *mass center* of S, can be located as follows. Let O be any point whatsoever, and let $\mathbf{p}_i$ be the position vector from O to P_i $(i = 1, \ldots, \nu)$. Then $\mathbf{p}^\star$, the position vector from O to $S^\star$, is given by

$$\mathbf{p}^\star = \frac{\sum_{i=1}^{\nu} m_i \mathbf{p}_i}{\sum_{i=1}^{\nu} m_i} \tag{2}$$

Derivation With $\mathbf{p}_i$ as defined, introduce a point $\widetilde{S}$, let $\widetilde{\mathbf{p}}$ be the position vector from O to $\widetilde{S}$, and let $\widetilde{\mathbf{r}}_i$ be the position vector from $\widetilde{S}$ to P_i $(i = 1, \ldots, \nu)$. Then

$$\widetilde{\mathbf{r}}_i = \mathbf{p}_i - \widetilde{\mathbf{p}} \qquad (i = 1, \ldots, \nu) \tag{3}$$

and

$$\begin{aligned} \sum_{i=1}^{\nu} m_i \widetilde{\mathbf{r}}_i &\underset{(3)}{=} \sum_{i=1}^{\nu} m_i \mathbf{p}_i - \sum_{i=1}^{\nu} m_i \widetilde{\mathbf{p}} \\ &= \sum_{i=1}^{\nu} m_i \mathbf{p}_i - \left(\sum_{i=1}^{\nu} m_i \right) \widetilde{\mathbf{p}} \end{aligned} \tag{4}$$

Set the right-hand member of this equation equal to zero, solve the resulting equation for $\widetilde{\mathbf{p}}$, and call the value of $\widetilde{\mathbf{p}}$ thus obtained $\mathbf{p}^\star$. This produces Eq. (2). Next, replace $\widetilde{\mathbf{p}}$ with $\mathbf{p}^\star$ in Eq. (3) and let $\mathbf{r}_i$ denote the resulting value of $\widetilde{\mathbf{r}}_i$. Then $\widetilde{\mathbf{r}}_i$ may be replaced with $\mathbf{r}_i$ in Eq. (4) whenever $\widetilde{\mathbf{p}}$ is replaced with $\mathbf{p}^\star$, and under these circumstances Eq. (4) reduces to Eq. (1).

Example A process called "static balancing" consists of adding matter to, or removing matter from, an object (for example, an automobile wheel) in such a way as to minimize the distance from the mass center of the new object thus created to a specified line (for example, the axle of the wheel). Consider, for instance, a set of three particles P_1, P_2, P_3 situated at corners of a cube as shown in Fig. 4.1.1 and having

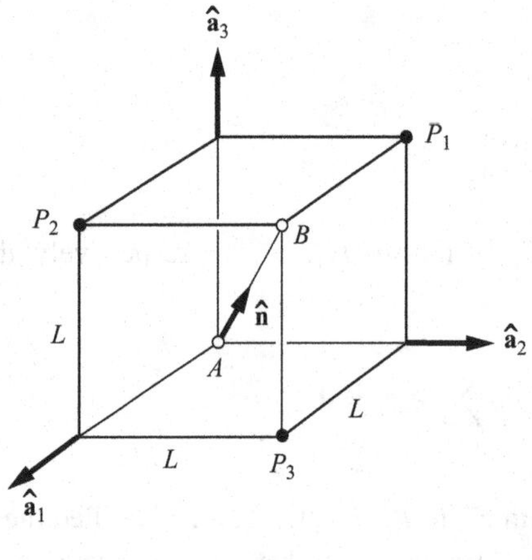

Figure 4.1.1

masses m, $2m$, $3m$, respectively. The mass center of this set of particles does not lie on line AB, but, by replacing P_3 with a particle Q of mass μ and choosing μ suitably, one can minimize the distance from line AB to the mass center $S^\star$ of the set S of particles P_1, P_2, and Q. To determine μ, introduce $\mathbf{p}^\star$ as the position vector from A to the mass center $S^\star$ of S, let $\hat{\mathbf{n}}$ be a unit vector directed as shown in Fig. 4.1.1, and note that D^2, the square of the distance D from $S^\star$ to line AB, is given by

$$D^2 = (\mathbf{p}^\star \times \hat{\mathbf{n}})^2 \tag{5}$$

Now, if $\hat{\mathbf{a}}_1$, $\hat{\mathbf{a}}_2$, $\hat{\mathbf{a}}_3$ are unit vectors directed as in Fig. 4.1.1, then

$$\mathbf{p}^\star \underset{(2)}{=} \frac{m(\hat{\mathbf{a}}_2 + \hat{\mathbf{a}}_3) + 2m(\hat{\mathbf{a}}_3 + \hat{\mathbf{a}}_1) + \mu(\hat{\mathbf{a}}_1 + \hat{\mathbf{a}}_2)}{3m + \mu} L \tag{6}$$

while

$$\hat{\mathbf{n}} = \frac{\hat{\mathbf{a}}_1 + \hat{\mathbf{a}}_2 + \hat{\mathbf{a}}_3}{\sqrt{3}} \tag{7}$$

Consequently,

$$D^2 \underset{(5-7)}{=} \frac{2L^2(\mu^2 - 3\mu m + 3m^2)}{3(3m + \mu)^2} \tag{8}$$

and, since a value of μ that minimizes D must satisfy the requirement

$$\frac{dD^2}{d\mu} = 0 \tag{9}$$

it may be verified that $\mu = 5m/3$. In accordance with Eq. (8), the associated (minimum) distance from line AB to $S^\star$ is equal to $L/\sqrt{42}$.

4.2 CURVES, SURFACES, AND SOLIDS

When a body B is modeled as matter distributed along a curve, over a surface, or throughout a solid, there exists a unique point $B^\star$ such that

$$\int_F \rho \mathbf{r}\, d\tau = \mathbf{0} \tag{1}$$

where ρ is the mass density (that is, the mass per unit of length, area, or volume) of B at a generic point P of B, $\mathbf{r}$ is the position vector from $B^\star$ to P; $d\tau$ is the length, area, or volume of a differential element of the figure F (curve, surface, or solid) occupied by B; and the integration is extended throughout F. $B^\star$, called the *mass center* of B, can be located as follows. Let O be any point whatsoever, and let $\mathbf{p}$ be the position vector from O to P. Then $\mathbf{p}^\star$, the position vector from O to $B^\star$, is given by

$$\mathbf{p}^\star = \frac{\displaystyle\int_F \rho \mathbf{p}\, d\tau}{\displaystyle\int_F \rho\, d\tau} \tag{2}$$

The lines of reasoning leading to Eqs. (1) and (2) are analogous to those followed in connection with Eqs. (4.1.1) and (4.1.2).

When ρ varies from point to point of B, the integrals appearing in Eq. (2) generally must be worked out by the analyst who wishes to locate $B^\star$; but when B is a *uniform* body, that is, when ρ is independent of the position of P in B, then the desired information frequently is readily available, for $B^\star$ then coincides with the *centroid* of the figure F occupied by B; the centroids of many figures have been found (by integration), and the results have been recorded, as in Appendix III. This information makes it possible to locate, without performing any integrations, the mass center of any body B that can be regarded as composed solely of uniform bodies $B_1, \ldots, B_\nu$ whose masses and mass center locations are known. Under these circumstances, the mass center of B coincides with the mass center of a (fictitious) set of particles (see Sec. 4.1) whose masses are those of $B_1, \ldots, B_\nu$, and which are situated at the mass centers of $B_1, \ldots, B_\nu$, respectively.

Example Figure 4.2.1 shows a body B consisting of a wire, EFG, attached to a piece of sheet metal, EGH. (The lines X_1, X_2, X_3 are mutually perpendicular.) The mass of the sheet metal is 10 times that of the wire.

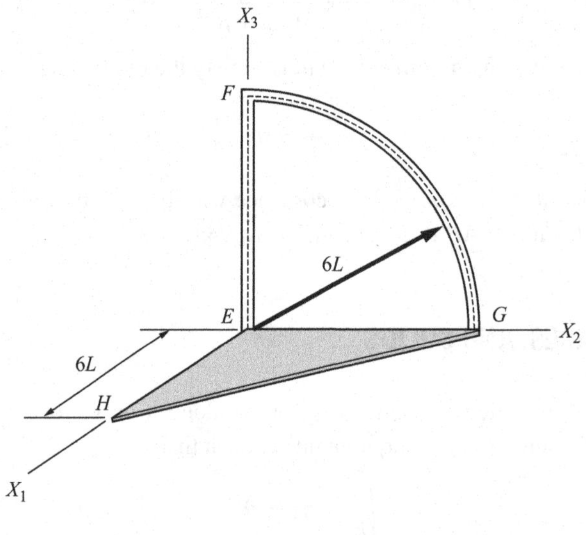

Figure 4.2.1

To locate the mass center $B^\star$ of B, we let B_1 and B_2 be bodies formed by matter distributed uniformly along the straight line EF and the circular curve FG, respectively, and model B_3, the sheet metal portion of B, as matter distributed uniformly over the plane triangular surface EGH. The mass centers of B_1, B_2, and B_3, found by reference to Appendix III, are the points $B_1{}^\star$, $B_2{}^\star$, and $B_3{}^\star$ in Fig. 4.2.2, and the masses m_1, m_2, and m_3 of B_1, B_2, and B_3 are taken to be

$$m_1 = m \qquad m_2 = \left(\frac{\pi}{2}\right) m \qquad m_3 = 10\left(1 + \frac{\pi}{2}\right) m \tag{3}$$

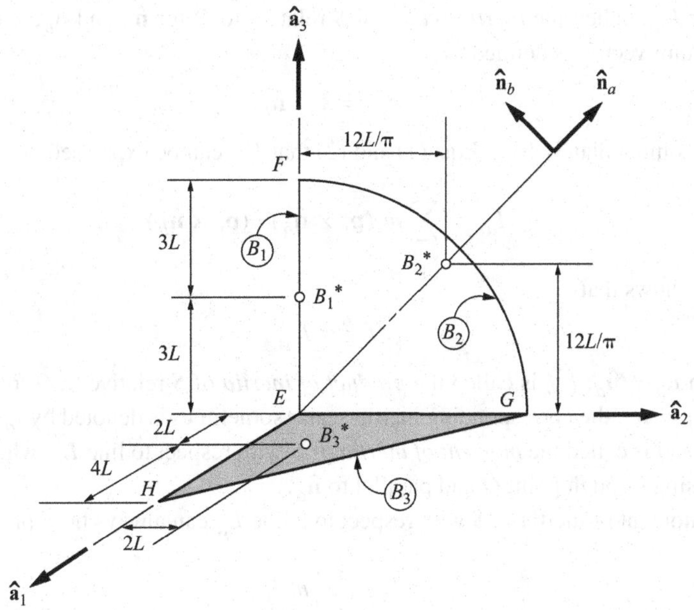

Figure 4.2.2

where m is arbitrary. In accordance with Eq. (4.1.2), the position vector $\mathbf{p}^\star$ from point E to the mass center of three particles situated at $B_1{}^\star$, $B_2{}^\star$, and $B_3{}^\star$, and having masses m_1, m_2, and m_3, respectively, is given by

$$
\begin{aligned}
\mathbf{p}^\star &= \frac{m_1(3L\hat{\mathbf{a}}_3) + m_2[(12L/\pi)\hat{\mathbf{a}}_2 + (12L/\pi)\hat{\mathbf{a}}_3] + m_3(2L\hat{\mathbf{a}}_1 + 2L\hat{\mathbf{a}}_2)}{m_1 + m_2 + m_3} \\
&\underset{(3)}{=} L\frac{20(1+\pi/2)\hat{\mathbf{a}}_1 + [6 + 20(1+\pi/2)]\hat{\mathbf{a}}_2 + 9\hat{\mathbf{a}}_3}{11(1+\pi/2)} \\
&= (1.82\hat{\mathbf{a}}_1 + 2.03\hat{\mathbf{a}}_2 + 0.318\hat{\mathbf{a}}_3)L
\end{aligned} \tag{4}
$$

The vector $\mathbf{p}^\star$ is the position vector from E to $B^\star$, the mass center of B, and Eq. (4) shows that $B^\star$ lies neither on the wire nor on the sheet metal portion of B.

4.3 INERTIA VECTOR, INERTIA SCALARS

If S is a set of particles $P_1, \ldots, P_\nu$ of masses $m_1, \ldots, m_\nu$, respectively, $\mathbf{p}_i$ is the position vector from a point O to P_i $(i = 1, \ldots, \nu)$, and $\hat{\mathbf{n}}_a$ is a unit vector, then a vector $\mathbf{I}_a$, called the *inertia vector* of S relative to O for $\hat{\mathbf{n}}_a$, is defined as

$$
\mathbf{I}_a \triangleq \sum_{i=1}^{\nu} m_i \mathbf{p}_i \times (\hat{\mathbf{n}}_a \times \mathbf{p}_i) \tag{1}
$$

A scalar I_{ab}, called the *inertia scalar* of S relative to O for $\hat{\mathbf{n}}_a$ and $\hat{\mathbf{n}}_b$, where $\hat{\mathbf{n}}_b$, like $\hat{\mathbf{n}}_a$, is a unit vector, is defined as

$$I_{ab} \triangleq \mathbf{I}_a \cdot \hat{\mathbf{n}}_b \tag{2}$$

It follows immediately from Eqs. (1) and (2) that I_{ab} can be expressed as

$$I_{ab} = \sum_{i=1}^{\nu} m_i(\mathbf{p}_i \times \hat{\mathbf{n}}_a) \cdot (\mathbf{p}_i \times \hat{\mathbf{n}}_b) \tag{3}$$

and this shows that

$$I_{ab} = I_{ba} \tag{4}$$

When $\hat{\mathbf{n}}_b \neq \hat{\mathbf{n}}_a$, I_{ab} is called the *product of inertia* of S relative to O for $\hat{\mathbf{n}}_a$ and $\hat{\mathbf{n}}_b$. When $\hat{\mathbf{n}}_b = \hat{\mathbf{n}}_a$, the corresponding inertia scalar sometimes is denoted by I_a (rather than by I_{aa}) and is called the *moment of inertia* of S with respect to line L_a, where L_a is the line passing through point O and parallel to $\hat{\mathbf{n}}_a$.

The moment of inertia of S with respect to a line L_a can always be expressed both as

$$I_a = \sum_{i=1}^{\nu} m_i {l_i}^2 \tag{5}$$

where l_i is the distance from P_i to line L_a, and as

$$I_a = m{k_a}^2 \tag{6}$$

where m is the total mass of S, and k_a is a real, nonnegative quantity called the *radius of gyration* of S with respect to line L_a. Equation (5) follows from the fact that

$$I_a \underset{(3)}{=} \sum_{i=1}^{\nu} m_i(\mathbf{p}_i \times \hat{\mathbf{n}}_a)^2 \tag{7}$$

and that $\mathbf{p}_i \times \hat{\mathbf{n}}_a$ has the magnitude l_i, the vector $\mathbf{p}_i$ being the position vector from a point on L_a to P_i.

Inertia vectors, products of inertia, moments of inertia, and radii of gyration of a body B modeled as matter distributed along a curve, over a surface, or throughout a solid are defined analogously. Specifically, if ρ is the mass density (that is, the mass per unit of length, area, or volume) of B at a generic point P of B, $\mathbf{p}$ is the position vector from a point O to P, $d\tau$ is the length, area, or volume of a differential element of the figure F (curve, surface, or solid) occupied by B, and $\hat{\mathbf{n}}_a$ and $\hat{\mathbf{n}}_b$ are unit vectors, then Eqs. (1), (2), and (5) give way, respectively, to

$$\mathbf{I}_a \triangleq \int_F \rho\mathbf{p} \times (\hat{\mathbf{n}}_a \times \mathbf{p})d\tau \tag{8}$$

$$I_{ab} \triangleq \mathbf{I}_a \cdot \hat{\mathbf{n}}_b \underset{(8)}{=} \int_F \rho(\mathbf{p} \times \hat{\mathbf{n}}_a) \cdot (\mathbf{p} \times \hat{\mathbf{n}}_b)d\tau \tag{9}$$

$$I_a = \int_F \rho l^2 d\tau \tag{10}$$

where l is the distance from P to line L_a; and Eq. (6) applies to B as well as to S.

Example In Fig. 4.3.1, B designates a thin, uniform, rectangular plate of mass m. When B is modeled as matter distributed over a rectangular surface R, then ρ, the mass per unit of area, is given by

$$\rho = \frac{m}{L_1 L_2} \tag{11}$$

while $\mathbf{p}$, the position vector from point O to a generic point P of B, can be expressed as

$$\mathbf{p} = x_1 \hat{\mathbf{n}}_1 + x_2 \hat{\mathbf{n}}_2 \tag{12}$$

where x_1 and x_2 are distances, and $\hat{\mathbf{n}}_1$ and $\hat{\mathbf{n}}_2$ are unit vectors, as shown in Fig. 4.3.2.

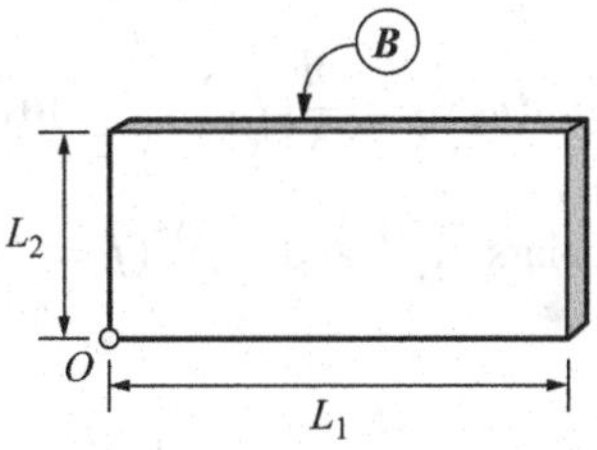

Figure 4.3.1

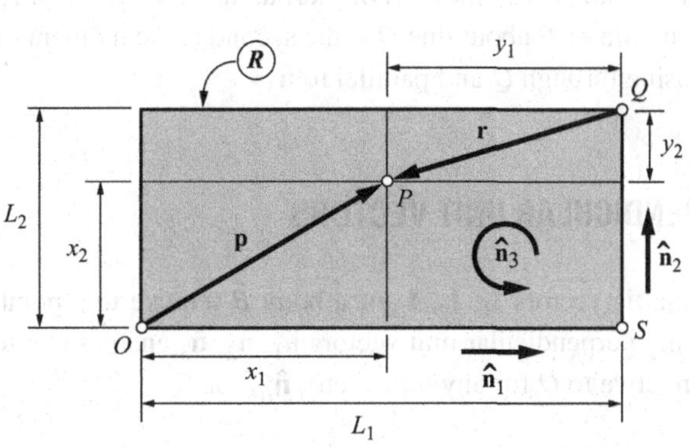

Figure 4.3.2

The inertia vector of B with respect to O for $\hat{\mathbf{n}}_1$ is thus

$$
\begin{aligned}
\mathbf{I}_1{}^O &\underset{(8)}{=} \int_R \rho \mathbf{p} \times (\hat{\mathbf{n}}_1 \times \mathbf{p}) d\tau \\
&\underset{(11,\,12)}{=} \int_0^{L_2} \int_0^{L_1} \frac{m}{L_1 L_2} (x_1 \hat{\mathbf{n}}_1 + x_2 \hat{\mathbf{n}}_2) \times [\hat{\mathbf{n}}_1 \times (x_1 \hat{\mathbf{n}}_1 + x_2 \hat{\mathbf{n}}_2)] \, dx_1 \, dx_2 \\
&= \frac{m}{L_1 L_2} \int_0^{L_2} \int_0^{L_1} (x_2{}^2 \hat{\mathbf{n}}_1 - x_1 x_2 \hat{\mathbf{n}}_2) \, dx_1 \, dx_2 \\
&= \frac{m}{L_1 L_2} \left(\hat{\mathbf{n}}_1 \int_0^{L_2} \int_0^{L_1} x_2{}^2 \, dx_1 \, dx_2 - \hat{\mathbf{n}}_2 \int_0^{L_2} \int_0^{L_1} x_1 x_2 \, dx_1 \, dx_2 \right) \\
&= \frac{m}{L_1 L_2} \left(\hat{\mathbf{n}}_1 \frac{L_1 L_2{}^3}{3} - \hat{\mathbf{n}}_2 \frac{L_1{}^2 L_2{}^2}{4} \right) = m L_2 \left(\frac{L_2}{3} \hat{\mathbf{n}}_1 - \frac{L_1}{4} \hat{\mathbf{n}}_2 \right) \qquad (13)
\end{aligned}
$$

Similarly, $\mathbf{I}_3{}^Q$, the inertia vector of B with respect to point Q for $\hat{\mathbf{n}}_3$, can be written (see Fig. 4.3.2 for $\mathbf{r}$)

$$
\begin{aligned}
\mathbf{I}_3{}^Q &\underset{(8)}{=} \int_R \rho \mathbf{r} \times (\hat{\mathbf{n}}_3 \times \mathbf{r}) d\tau \\
&= \int_0^{L_2} \int_0^{L_1} \frac{m}{L_1 L_2} (-y_1 \hat{\mathbf{n}}_1 - y_2 \hat{\mathbf{n}}_2) \times [\hat{\mathbf{n}}_3 \times (-y_1 \hat{\mathbf{n}}_1 - y_2 \hat{\mathbf{n}}_2)] \, dy_1 \, dy_2 \\
&= \frac{m}{L_1 L_2} \int_0^{L_2} \int_0^{L_1} (y_1{}^2 + y_2{}^2) \, dy_1 \, dy_2 \, \hat{\mathbf{n}}_3 = \frac{m}{3} (L_1{}^2 + L_2{}^2) \hat{\mathbf{n}}_3 \qquad (14)
\end{aligned}
$$

With the aid of these results, the inertia scalars $I_{1j}{}^O$ and $I_{3j}{}^Q$ ($j = 1,2,3$) can be formed as

$$
I_{11}{}^O \underset{(2)}{=} \mathbf{I}_1{}^O \cdot \hat{\mathbf{n}}_1 \underset{(13)}{=} \frac{m L_2{}^2}{3} \qquad I_{12}{}^O \underset{(2,\,13)}{=} -\frac{m L_1 L_2}{4} \qquad I_{13}{}^O \underset{(2,\,13)}{=} 0 \qquad (15)
$$

$$
I_{31}{}^Q \underset{(2)}{=} \mathbf{I}_3{}^Q \cdot \hat{\mathbf{n}}_1 \underset{(14)}{=} 0 \qquad I_{32}{}^Q \underset{(2,\,14)}{=} 0 \qquad I_{33}{}^Q \underset{(2,\,14)}{=} \frac{m}{3} (L_1{}^2 + L_2{}^2) \qquad (16)
$$

Two of these inertia scalars are moments of inertia, namely, $I_{11}{}^O$ and $I_{33}{}^Q$. The first is the moment of inertia of B about line OS; the second is the moment of inertia of B about the line passing through Q and parallel to $\hat{\mathbf{n}}_3$.

4.4 MUTUALLY PERPENDICULAR UNIT VECTORS

Knowledge of the inertia vectors $\mathbf{I}_1$, $\mathbf{I}_2$, $\mathbf{I}_3$ of a body B relative to a point O (see Sec. 4.3) for three mutually perpendicular unit vectors $\hat{\mathbf{n}}_1$, $\hat{\mathbf{n}}_2$, $\hat{\mathbf{n}}_3$ enables one to find $\mathbf{I}_a$, the inertia vector of B relative to O for any unit vector $\hat{\mathbf{n}}_a$, for

$$
\mathbf{I}_a = \sum_{j=1}^{3} a_j \mathbf{I}_j \qquad (1)
$$

where a_1, a_2, a_3 are defined as

$$a_j \triangleq \hat{\mathbf{n}}_a \cdot \hat{\mathbf{n}}_j \qquad (j = 1,2,3) \tag{2}$$

Similarly, I_{ab}, the inertia scalar of B relative to O for $\hat{\mathbf{n}}_a$ and $\hat{\mathbf{n}}_b$ (see Sec. 4.3), can be found easily when the inertia scalars I_{jk} $(j,k = 1,2,3)$ are known, for

$$I_{ab} = \sum_{j=1}^{3}\sum_{k=1}^{3} a_j I_{jk} b_k \tag{3}$$

where

$$b_k \triangleq \hat{\mathbf{n}}_b \cdot \hat{\mathbf{n}}_k \qquad (k = 1,2,3) \tag{4}$$

Derivations It follows from Eq. (2) that (see Sec. 1.6)

$$\hat{\mathbf{n}}_a = \sum_{j=1}^{3} a_j \hat{\mathbf{n}}_j \tag{5}$$

Consequently,

$$\begin{aligned}
\mathbf{I}_a \underset{(4.3.1)}{=} & \sum_{i=1}^{\nu} m_i \mathbf{p}_i \times \left(\sum_{j=1}^{3} a_j \underset{(5)}{\hat{\mathbf{n}}_j} \times \mathbf{p}_i \right) \\
= & \sum_{j=1}^{3} a_j \sum_{i=1}^{\nu} m_i \mathbf{p}_i \times (\hat{\mathbf{n}}_j \times \mathbf{p}_i) \\
= & \sum_{j=1}^{3} a_j \underset{(4.3.1)}{\mathbf{I}_j}
\end{aligned} \tag{6}$$

which establishes the validity of Eq. (1). As for Eq. (3), note that Eq. (4) implies that (see Sec. 1.6)

$$\hat{\mathbf{n}}_b = \sum_{k=1}^{3} b_k \hat{\mathbf{n}}_k \tag{7}$$

Hence,

$$\begin{aligned}
I_{ab} \underset{(4.3.2)}{=} & \underset{(1)}{\left(\sum_{j=1}^{3} a_j \mathbf{I}_j \right)} \cdot \underset{(7)}{\sum_{k=1}^{3} b_k \hat{\mathbf{n}}_k} \\
= & \sum_{j=1}^{3}\sum_{k=1}^{3} a_j \mathbf{I}_j \cdot \hat{\mathbf{n}}_k b_k = \sum_{j=1}^{3}\sum_{k=1}^{3} a_j \underset{(4.3.2)}{I_{jk}} b_k
\end{aligned} \tag{8}$$

Example Table 4.4.1 shows the inertia scalars I_{jk} of B relative to O for $\hat{\mathbf{n}}_j$ and $\hat{\mathbf{n}}_k$ $(j,k = 1,2,3)$, where B is the rectangular plate considered in the example in Sec. 4.3.

To find the moment of inertia of B with respect to line OQ in Fig. 4.3.2, let $\hat{\mathbf{n}}_a$ be a unit vector parallel to this line, so that

$$\hat{\mathbf{n}}_a = \frac{L_1\hat{\mathbf{n}}_1 + L_2\hat{\mathbf{n}}_2}{(L_1{}^2 + L_2{}^2)^{1/2}} \tag{9}$$

which means that, in accordance with Eq. (2),

$$a_1 = \frac{L_1}{(L_1{}^2 + L_2{}^2)^{1/2}} \qquad a_2 = \frac{L_2}{(L_1{}^2 + L_2{}^2)^{1/2}} \qquad a_3 = 0 \tag{10}$$

With $b_k = a_k$ $(k = 1,2,3)$, Eq. (3) then yields [see also Eq. (4.3.4)]

$$\begin{aligned} I_a &= a_1{}^2 I_{11} + a_2{}^2 I_{22} + a_3{}^2 I_{33} + 2(a_1a_2I_{12} + a_2a_3I_{23} + a_3a_1I_{31}) \\ &\underset{(10)}{=} \frac{L_1{}^2 I_{11} + L_2{}^2 I_{22} + 2L_1L_2I_{12}}{L_1{}^2 + L_2{}^2} \end{aligned} \tag{11}$$

and use of Table 4.4.1 leads to

$$I_a \underset{(11)}{=} \frac{mL_1{}^2L_2{}^2/3 + mL_1{}^2L_2{}^2/3 - mL_1{}^2L_2{}^2/2}{L_1{}^2 + L_2{}^2} = \frac{m(L_1L_2)^2}{6(L_1{}^2 + L_2{}^2)} \tag{12}$$

Table 4.4.1

I_{jk}	1	2	3
1	$mL_2{}^2/3$	$-mL_1L_2/4$	0
2	$-mL_1L_2/4$	$mL_1{}^2/3$	0
3	0	0	$m(L_1{}^2 + L_2{}^2)/3$

4.5 INERTIA MATRIX, INERTIA DYADIC

The inertia scalars I_{jk} of a set S of particles relative to a point O for unit vectors $\hat{\mathbf{n}}_j$ and $\hat{\mathbf{n}}_k$ $(j, k = 1,2,3)$ can be used to define a square matrix I, called the *inertia matrix* of S relative to O for $\hat{\mathbf{n}}_1$, $\hat{\mathbf{n}}_2$, $\hat{\mathbf{n}}_3$, as follows:

$$I \triangleq \begin{bmatrix} I_{11} & I_{12} & I_{13} \\ I_{21} & I_{22} & I_{23} \\ I_{31} & I_{32} & I_{33} \end{bmatrix} \tag{1}$$

Suppose that $\hat{\mathbf{n}}_1$, $\hat{\mathbf{n}}_2$, $\hat{\mathbf{n}}_3$ are mutually perpendicular and that row matrices a and b are defined as

$$a \triangleq \lfloor a_1 \quad a_2 \quad a_3 \rfloor \qquad b \triangleq \lfloor b_1 \quad b_2 \quad b_3 \rfloor \tag{2}$$

where a_1, a_2, a_3 and b_1, b_2, b_3 are given by Eqs. (4.4.2) and (4.4.4), respectively. Then it follows immediately from Eq. (4.4.3) and the rules for multiplication of matrices that I_{ab}, the inertia scalar of S relative to O for $\hat{\mathbf{n}}_a$ and $\hat{\mathbf{n}}_b$, is given by

$$I_{ab} = aIb^T \tag{3}$$

where b^T is the transpose of b, that is, the column matrix having b_i as the element in the ith row ($i = 1, 2, 3$). Equation (3) is useful when inertia scalars are evaluated by means of machine computations and matrix multiplication routines are readily available.

The set S does not possess a unique inertia matrix relative to O, for, if $\hat{\mathbf{n}}_1{}'$, $\hat{\mathbf{n}}_2{}'$, $\hat{\mathbf{n}}_3{}'$ are mutually perpendicular unit vectors other than $\hat{\mathbf{n}}_1$, $\hat{\mathbf{n}}_2$, $\hat{\mathbf{n}}_3$, and I' is defined as

$$I' \triangleq \begin{bmatrix} I_{11}' & I_{12}' & I_{13}' \\ I_{21}' & I_{22}' & I_{23}' \\ I_{31}' & I_{32}' & I_{33}' \end{bmatrix} \tag{4}$$

where I_{jk}' ($j, k = 1, 2, 3$) are the inertia scalars of S relative to O for $\hat{\mathbf{n}}_j{}'$ and $\hat{\mathbf{n}}_k{}'$, then I', like I, is an inertia matrix of S relative to O, but I and I' are by no means equal to each other. Hence, when working with inertia matrices, one must keep in mind that each such matrix is associated with a specific vector basis. By way of contrast, the use of *dyadics* enables one to deal with certain topics involving inertia vectors and/or inertia scalars in a *basis-independent* way. To acquaint the reader with dyadics, we begin by focusing attention on two vectors, $\mathbf{u}$ and $\mathbf{v}$, given by

$$\mathbf{u} = \mathbf{w} \cdot \mathbf{ab} + \mathbf{w} \cdot \mathbf{cd} + \cdots \tag{5}$$

and

$$\mathbf{v} = \mathbf{ab} \cdot \mathbf{w} + \mathbf{cd} \cdot \mathbf{w} + \cdots \tag{6}$$

respectively, where $\mathbf{a}, \mathbf{b}, \mathbf{c}, \mathbf{d}, \ldots$, and $\mathbf{w}$ are any vectors whatsoever. Equations (5) and (6) can be rewritten as

$$\mathbf{u} = \mathbf{w} \cdot (\mathbf{ab} + \mathbf{cd} + \cdots) \tag{7}$$

and

$$\mathbf{v} = (\mathbf{ab} + \mathbf{cd} + \cdots) \cdot \mathbf{w} \tag{8}$$

if it is understood that the right-hand members of Eqs. (7) and (8) have the same meanings as those of Eqs. (5) and (6), respectively. Furthermore, if the quantity within parentheses in Eqs. (7) and (8) is denoted by $\underline{\mathbf{Q}}$, that is, if $\underline{\mathbf{Q}}$ is defined as

$$\underline{\mathbf{Q}} \triangleq \mathbf{ab} + \mathbf{cd} + \cdots \tag{9}$$

then Eqs. (7) and (8) give way to

$$\mathbf{u} = \mathbf{w} \cdot \underline{\mathbf{Q}} \tag{10}$$

and

$$\mathbf{v} = \underline{\mathbf{Q}} \cdot \mathbf{w} \tag{11}$$

respectively; $\underline{\mathbf{Q}}$ is called a dyadic; and Eqs. (5), (9), and (10) constitute a definition of *scalar premultiplication* of a dyadic with a vector, while Eqs. (6), (9), and (11) define the operation of *scalar postmultiplication* of a dyadic with a vector. In summary, then, a dyadic is a juxtaposition of vectors as in the right-hand member of Eq. (9), and scalar multiplication (pre- or post-) of a dyadic with a vector produces a vector.

A rather special dyadic $\underline{\mathbf{U}}$, called the *unit dyadic*, comes to light in connection with Eq. (1.6.3), which suggests the definition

$$\underline{\mathbf{U}} \triangleq \hat{\mathbf{a}}_1\hat{\mathbf{a}}_1 + \hat{\mathbf{a}}_2\hat{\mathbf{a}}_2 + \hat{\mathbf{a}}_3\hat{\mathbf{a}}_3 \tag{12}$$

where $\hat{\mathbf{a}}_1$, $\hat{\mathbf{a}}_2$, $\hat{\mathbf{a}}_3$ are mutually perpendicular unit vectors. In accordance with Eqs. (5), (9), and (10),

$$\mathbf{v}\cdot\underline{\mathbf{U}} \underset{(12)}{=} \mathbf{v}\cdot\hat{\mathbf{a}}_1\hat{\mathbf{a}}_1 + \mathbf{v}\cdot\hat{\mathbf{a}}_2\hat{\mathbf{a}}_2 + \mathbf{v}\cdot\hat{\mathbf{a}}_3\hat{\mathbf{a}}_3 \underset{(1.6.3)}{=} \mathbf{v} \tag{13}$$

Moreover, it follows from Eqs. (6), (9), and (11) that

$$\underline{\mathbf{U}}\cdot\mathbf{v} \underset{(12)}{=} \hat{\mathbf{a}}_1\hat{\mathbf{a}}_1\cdot\mathbf{v} + \hat{\mathbf{a}}_2\hat{\mathbf{a}}_2\cdot\mathbf{v} + \hat{\mathbf{a}}_3\hat{\mathbf{a}}_3\cdot\mathbf{v} \underset{(1.6.3)}{=} \mathbf{v} \tag{14}$$

In other words, $\underline{\mathbf{U}}$ is a dyadic whose scalar product (pre- or post-) with any vector is equal to the vector itself.

Returning to the subject of inertia vectors and inertia scalars, let us consider the inertia vector $\mathbf{I}_a$, defined as in Eq. (4.3.1). This can be expressed as

$$\begin{aligned}\mathbf{I}_a &\underset{(4.3.1)}{=} \sum_{i=1}^{\nu} m_i(\hat{\mathbf{n}}_a\mathbf{p}_i{}^2 - \hat{\mathbf{n}}_a\cdot\mathbf{p}_i\mathbf{p}_i)\\ &= \sum_{i=1}^{\nu} m_i(\underbrace{\hat{\mathbf{n}}_a\cdot\underline{\mathbf{U}}}_{(13)}\mathbf{p}_i{}^2 - \hat{\mathbf{n}}_a\cdot\mathbf{p}_i\mathbf{p}_i) = \sum_{i=1}^{\nu} m_i(\underbrace{\underline{\mathbf{U}}\cdot\hat{\mathbf{n}}_a}_{(14)}\mathbf{p}_i{}^2 - \mathbf{p}_i\mathbf{p}_i\cdot\hat{\mathbf{n}}_a)\end{aligned} \tag{15}$$

Hence, if a dyadic $\underline{\mathbf{I}}$ is defined as

$$\underline{\mathbf{I}} \triangleq \sum_{i=1}^{\nu} m_i(\underline{\mathbf{U}}\mathbf{p}_i{}^2 - \mathbf{p}_i\mathbf{p}_i) \tag{16}$$

then $\mathbf{I}_a$ can be expressed as

$$\mathbf{I}_a = \hat{\mathbf{n}}_a\cdot\underline{\mathbf{I}} = \underline{\mathbf{I}}\cdot\hat{\mathbf{n}}_a \tag{17}$$

The dyadic $\underline{\mathbf{I}}$ is called the *inertia dyadic* of S relative to O. When a body B is modeled as matter occupying a figure F (a curve, surface, or solid), then Eq. (16) is replaced with

$$\underline{\mathbf{I}} \triangleq \int_F \rho(\underline{\mathbf{U}}\mathbf{p}^2 - \mathbf{p}\mathbf{p})d\tau \tag{18}$$

where ρ, $\mathbf{p}$, and $d\tau$ have the same meanings as in connection with Eq. (4.3.8).

Dot multiplication of Eq. (17) with a unit vector $\hat{\mathbf{n}}_b$ leads to

$$\mathbf{I}_a\cdot\hat{\mathbf{n}}_b = (\hat{\mathbf{n}}_a\cdot\mathbf{I})\cdot\hat{\mathbf{n}}_b \tag{19}$$

In view of Eq. (4.3.2) and the fact that $(\hat{\mathbf{n}}_a\cdot\underline{\mathbf{I}})\cdot\hat{\mathbf{n}}_b = \hat{\mathbf{n}}_a\cdot(\underline{\mathbf{I}}\cdot\hat{\mathbf{n}}_b)$, so that the parentheses in Eq. (19) are unnecessary, one thus finds that the inertia scalar I_{ab} can be expressed as

$$I_{ab} = \hat{\mathbf{n}}_a\cdot\underline{\mathbf{I}}\cdot\hat{\mathbf{n}}_b \tag{20}$$

The inertia dyadic $\underline{\mathbf{I}}$ is said to be basis independent because its definition, Eq. (16), does not involve any basis vectors. However, it can be expressed in various basis-dependent forms. For example, if $\hat{\mathbf{n}}_1$, $\hat{\mathbf{n}}_2$, $\hat{\mathbf{n}}_3$ are mutually perpendicular unit vectors, $\mathbf{I}_j$ is the inertia vector of S relative to O for $\hat{\mathbf{n}}_j$ $(j = 1,2,3)$, and I_{jk} is the inertia scalar of S relative to O for $\hat{\mathbf{n}}_j$ and $\hat{\mathbf{n}}_k$ $(j,k = 1,2,3)$, then $\underline{\mathbf{I}}$ is given both by

$$\underline{\mathbf{I}} = \sum_{j=1}^{3} \mathbf{I}_j \hat{\mathbf{n}}_j \tag{21}$$

and by

$$\underline{\mathbf{I}} = \sum_{j=1}^{3} \sum_{k=1}^{3} I_{jk} \hat{\mathbf{n}}_j \hat{\mathbf{n}}_k \tag{22}$$

Derivation It may be verified that a relationship analogous to Eq. (1.6.3) applies to any dyadic $\underline{\mathbf{Q}}$; that is,

$$\underline{\mathbf{Q}} = \underline{\mathbf{Q}} \cdot \hat{\mathbf{n}}_1 \hat{\mathbf{n}}_1 + \underline{\mathbf{Q}} \cdot \hat{\mathbf{n}}_2 \hat{\mathbf{n}}_2 + \underline{\mathbf{Q}} \cdot \hat{\mathbf{n}}_3 \hat{\mathbf{n}}_3 \tag{23}$$

Consequently,

$$\begin{aligned} \underline{\mathbf{I}} &= \underline{\mathbf{I}} \cdot \hat{\mathbf{n}}_1 \hat{\mathbf{n}}_1 + \underline{\mathbf{I}} \cdot \hat{\mathbf{n}}_2 \hat{\mathbf{n}}_2 + \underline{\mathbf{I}} \cdot \hat{\mathbf{n}}_3 \hat{\mathbf{n}}_3 \\ &\underset{(17)}{=} \mathbf{I}_1 \hat{\mathbf{n}}_1 + \mathbf{I}_2 \hat{\mathbf{n}}_2 + \mathbf{I}_3 \hat{\mathbf{n}}_3 \end{aligned} \tag{24}$$

in agreement with Eq. (21). As for Eq. (22), refer to Eq. (1.6.3) to write

$$\begin{aligned} \mathbf{I}_j &= \mathbf{I}_j \cdot \hat{\mathbf{n}}_1 \hat{\mathbf{n}}_1 + \mathbf{I}_j \cdot \hat{\mathbf{n}}_2 \hat{\mathbf{n}}_2 + \mathbf{I}_j \cdot \hat{\mathbf{n}}_3 \hat{\mathbf{n}}_3 \\ &\underset{(4.3.2)}{=} I_{j1} \hat{\mathbf{n}}_1 + I_{j2} \hat{\mathbf{n}}_2 + I_{j3} \hat{\mathbf{n}}_3 \qquad (j = 1,2,3) \end{aligned} \tag{25}$$

and substitute these expressions for $\mathbf{I}_1$, $\mathbf{I}_2$, and $\mathbf{I}_3$ into Eq. (24) to obtain

$$\begin{aligned} \underline{\mathbf{I}} = {} & (I_{11} \hat{\mathbf{n}}_1 + I_{12} \hat{\mathbf{n}}_2 + I_{13} \hat{\mathbf{n}}_3) \hat{\mathbf{n}}_1 \\ & + (I_{21} \hat{\mathbf{n}}_1 + I_{22} \hat{\mathbf{n}}_2 + I_{23} \hat{\mathbf{n}}_3) \hat{\mathbf{n}}_2 \\ & + (I_{31} \hat{\mathbf{n}}_1 + I_{32} \hat{\mathbf{n}}_2 + I_{33} \hat{\mathbf{n}}_3) \hat{\mathbf{n}}_3 \end{aligned} \tag{26}$$

which, in view of Eq. (4.3.4), is seen to agree with Eq. (22).

Example If S is a set of particles $P_1, \ldots, P_\nu$ of masses $m_1, \ldots, m_\nu$, respectively, moving in a reference frame A with velocities ${}^A\mathbf{v}^{P_1}, \ldots, {}^A\mathbf{v}^{P_\nu}$ (see Sec. 2.6), then a vector ${}^A\mathbf{H}^{S/O}$, called the *angular momentum* of S relative to O in A, is defined as

$${}^A\mathbf{H}^{S/O} \triangleq \sum_{i=1}^{\nu} m_i \mathbf{p}_i \times {}^A\mathbf{v}^{P_i} \tag{27}$$

where $\mathbf{p}_i$ is the position vector from a point O to P_i $(i = 1, \ldots, \nu)$. The point O need not be fixed in A; for example, it can be the mass center $S^\star$ of S, in which case ${}^A\mathbf{H}^{S/O}$ becomes ${}^A\mathbf{H}^{S/S^\star}$ and is called the *central angular momentum* of S in A.

If the particles of S form a rigid body B, then $\mathbf{H}$, the central angular momentum of B in A, can be expressed as

$$\mathbf{H} = \underline{\mathbf{I}} \cdot \boldsymbol{\omega} \tag{28}$$

where $\underline{\mathbf{I}}$, called the *central inertia dyadic* of B, is the inertia dyadic of B relative to the mass center $B^\star$ of B, and $\boldsymbol{\omega}$ is the angular velocity of B in A. To verify that $\mathbf{H}$ can be written as in Eq. (28), let $\mathbf{r}_i$ be the position vector from $B^\star$ to P_i and refer to Eqs. (27) and (2.7.1) to write

$$\begin{aligned} \mathbf{H} &= \sum_{i=1}^{\nu} m_i \mathbf{r}_i \times ({}^A\mathbf{v}^{B^\star} + \boldsymbol{\omega} \times \mathbf{r}_i) \\ &= \left(\sum_{i=1}^{\nu} m_i \mathbf{r}_i \right) \times {}^A\mathbf{v}^{B^\star} + \sum_{i=1}^{\nu} m_i \mathbf{r}_i \times (\boldsymbol{\omega} \times \mathbf{r}_i) \end{aligned} \tag{29}$$

Then note that

$$\sum_{i=1}^{\nu} m_i \mathbf{r}_i \underset{(4.1.1)}{=} \mathbf{0} \tag{30}$$

while

$$\begin{aligned} \sum_{i=1}^{\nu} m_i \mathbf{r}_i \times (\boldsymbol{\omega} \times \mathbf{r}_i) &= \sum_{i=1}^{\nu} m_i (\mathbf{r}_i{}^2 \boldsymbol{\omega} - \mathbf{r}_i \mathbf{r}_i \cdot \boldsymbol{\omega}) \\ &= \sum_{i=1}^{\nu} m_i (\mathbf{r}_i{}^2 \underline{\mathbf{U}} - \mathbf{r}_i \mathbf{r}_i) \cdot \boldsymbol{\omega} \underset{(16)}{=} \underline{\mathbf{I}} \cdot \boldsymbol{\omega} \end{aligned} \tag{31}$$

If $\omega_1, \omega_2, \omega_3$ are defined as

$$\omega_i \triangleq \boldsymbol{\omega} \cdot \hat{\mathbf{n}}_i \qquad (i = 1,2,3) \tag{32}$$

then it follows directly from Eqs. (28), (22), and (32) that

$$\begin{aligned} \mathbf{H} &= \sum_{j=1}^{3} \sum_{k=1}^{3} I_{jk} \hat{\mathbf{n}}_j \hat{\mathbf{n}}_k \cdot \sum_{i=1}^{3} \omega_i \hat{\mathbf{n}}_i \\ &= \sum_{i=1}^{3} \sum_{j=1}^{3} \sum_{k=1}^{3} I_{jk}\, \omega_i\, \hat{\mathbf{n}}_j \hat{\mathbf{n}}_k \cdot \hat{\mathbf{n}}_i \end{aligned} \tag{33}$$

so that, since $\hat{\mathbf{n}}_k \cdot \hat{\mathbf{n}}_i$ vanishes except when $i = k$, in which case it is equal to unity, $\mathbf{H}$ can be written

$$\mathbf{H} \underset{(33)}{=} \sum_{j=1}^{3} \sum_{k=1}^{3} I_{jk}\, \omega_k\, \hat{\mathbf{n}}_j \tag{34}$$

This is a useful relationship. It reduces to the convenient form given in Eq. (2.3.5) if $\hat{\mathbf{n}}_1$, $\hat{\mathbf{n}}_2$, $\hat{\mathbf{n}}_3$ are chosen in such a way that I_{12}, I_{23}, and I_{31} vanish. The fact that it is always possible to choose $\hat{\mathbf{n}}_1$, $\hat{\mathbf{n}}_2$, $\hat{\mathbf{n}}_3$ in this way is established in Sec. 4.8.

4.6 PARALLEL AXES THEOREMS

The inertia dyadic $\underline{\mathbf{I}}^{S/O}$ of a set S of ν particles $P_1, \ldots, P_\nu$ relative to a point O (see Sec. 4.5) is related in a simple way to the *central inertia dyadic* $\underline{\mathbf{I}}^{S/S^\star}$ of S, that is, the inertia dyadic of S relative to the mass center $S^\star$ of S. Specifically,

$$\underline{\mathbf{I}}^{S/O} = \underline{\mathbf{I}}^{S/S^\star} + \underline{\mathbf{I}}^{S^\star/O} \tag{1}$$

where $\underline{\mathbf{I}}^{S^\star/O}$ denotes the inertia dyadic relative to O of a (fictitious) particle situated at $S^\star$ and having a mass equal to the total mass of S. Similarly, if $I^{S/O}$ and $I^{S/S^\star}$ are inertia matrices of S relative to O and $S^\star$, respectively, for unit vectors $\hat{\mathbf{n}}_1$, $\hat{\mathbf{n}}_2$, $\hat{\mathbf{n}}_3$, and $I^{S^\star/O}$ is the inertia matrix relative to O of a particle having the same mass as S and situated at $S^\star$, also for $\hat{\mathbf{n}}_1$, $\hat{\mathbf{n}}_2$, $\hat{\mathbf{n}}_3$, then

$$I^{S/O} = I^{S/S^\star} + I^{S^\star/O} \tag{2}$$

Moreover, analogous relationships apply to inertia vectors, products of inertia, and moments of inertia (see Sec. 4.3); that is,

$$\mathbf{I}_a{}^{S/O} = \mathbf{I}_a{}^{S/S^\star} + \mathbf{I}_a{}^{S^\star/O} \tag{3}$$

$$I_{ab}{}^{S/O} = I_{ab}{}^{S/S^\star} + I_{ab}{}^{S^\star/O} \tag{4}$$

and

$$I_a{}^{S/O} = I_a{}^{S/S^\star} + I_a{}^{S^\star/O} \tag{5}$$

The quantities $I_{ab}{}^{S/S^\star}$ and $I_a{}^{S/S^\star}$ are called *central inertia scalars*. Equations (3)–(5) are referred to as parallel axes theorems because one can associate certain parallel lines with each of these equations, such as, in the case of Eq. (5), the two lines that are parallel to $\hat{\mathbf{n}}_a$ and pass through O and $S^\star$.

Derivations Let $\mathbf{p}_i$ and $\mathbf{r}_i$ be the position vectors from O to P_i and from $S^\star$ to P_i ($i = 1, \ldots, \nu$), respectively; note that Eq. (4.1.1) is satisfied if m_i denotes the mass of P_i, and that

$$\mathbf{p}_i = \mathbf{p}^\star + \mathbf{r}_i \tag{6}$$

where $\mathbf{p}^\star$ is the position vector from O to $S^\star$. Then $\underline{\mathbf{I}}^{S/O}$, $\underline{\mathbf{I}}^{S/S^\star}$, and $\underline{\mathbf{I}}^{S^\star/O}$ are given by

$$\underline{\mathbf{I}}^{S/O} \underset{(4.5.16)}{=} \sum_{i=1}^{\nu} m_i (\underline{\mathbf{U}}\mathbf{p}_i{}^2 - \mathbf{p}_i\mathbf{p}_i) \tag{7}$$

$$\underline{\mathbf{I}}^{S/S^\star} \underset{(4.5.16)}{=} \sum_{i=1}^{\nu} m_i (\underline{\mathbf{U}}\mathbf{r}_i{}^2 - \mathbf{r}_i\mathbf{r}_i) \tag{8}$$

and

$$\underline{\mathbf{I}}^{S^\star/O} \underset{(4.5.16)}{=} \left(\sum_{i=1}^{\nu} m_i\right) (\underline{\mathbf{U}}\mathbf{p}^{\star 2} - \mathbf{p}^\star\mathbf{p}^\star) \tag{9}$$

Hence,

$$\begin{aligned}\underline{\mathbf{I}}^{S/O} &\underset{(7)}{=} \sum_{i=1}^{\nu} m_i[\underline{\mathbf{U}}(\underset{(6)}{\mathbf{p}^\star + \mathbf{r}_i})^2 - (\underset{(6)}{\mathbf{p}^\star + \mathbf{r}_i})(\underset{(6)}{\mathbf{p}^\star + \mathbf{r}_i})] \\ &= \sum_{i=1}^{\nu} m_i[\underline{\mathbf{U}}(\mathbf{p}^{\star 2} + 2\mathbf{p}^\star \cdot \mathbf{r}_i + \mathbf{r}_i{}^2) - \mathbf{p}^\star\mathbf{p}^\star \\ &\quad - \mathbf{r}_i\mathbf{p}^\star - \mathbf{p}^\star\mathbf{r}_i - \mathbf{r}_i\mathbf{r}_i] \\ &= \sum_{i=1}^{\nu} m_i(\underline{\mathbf{U}}\mathbf{r}_i{}^2 - \mathbf{r}_i\mathbf{r}_i) + \left(\sum_{i=1}^{\nu} m_i\right)(\underline{\mathbf{U}}\mathbf{p}^{\star 2} - \mathbf{p}^\star\mathbf{p}^\star) \\ &\quad + 2\underline{\mathbf{U}}\mathbf{p}^\star \cdot \left(\sum_{i=1}^{\nu} m_i\mathbf{r}_i\right) - \left(\sum_{i=1}^{\nu} m_i\mathbf{r}_i\right)\mathbf{p}^\star - \mathbf{p}^\star\left(\sum_{i=1}^{\nu} m_i\mathbf{r}_i\right)\end{aligned} \tag{10}$$

and use of Eqs. (8), (9), and (4.1.1) leads directly to Eq. (1).

Premultiplication of Eq. (1) with a unit vector $\hat{\mathbf{n}}_a$ yields Eq. (3) when Eq. (4.5.17) is taken into account; similarly, use of Eq. (4.3.2) after dot multiplication of Eq. (3) with $\hat{\mathbf{n}}_b$ and $\hat{\mathbf{n}}_a$ leads to Eqs. (4) and (5), respectively. Finally, Eq. (2) is an immediate consequence of the fact that the elements of $I^{S/O}$, $I^{S/S^\star}$, and $I^{S^\star/O}$ satisfy Eqs. (4) and (5).

Example In Fig. 4.6.1, C represents a wing of a delta-wing aircraft, modeled as a uniform, thin, right-triangular plate of mass m. The central inertia matrix of C for the unit vectors $\hat{\mathbf{n}}_1$, $\hat{\mathbf{n}}_2$, $\hat{\mathbf{n}}_3$ is given by

$$I^{C/C^\star} = mc^2 \begin{bmatrix} \frac{9}{2} & \frac{3}{4} & 0 \\ \frac{3}{4} & \frac{1}{2} & 0 \\ 0 & 0 & 5 \end{bmatrix} \tag{11}$$

The product of inertia of C relative to O for $\hat{\mathbf{n}}_a$ and $\hat{\mathbf{n}}_b$ is to be determined (see Fig. 4.6.1 for point O and the unit vectors $\hat{\mathbf{n}}_a$ and $\hat{\mathbf{n}}_b$).

The unit vectors $\hat{\mathbf{n}}_a$ and $\hat{\mathbf{n}}_b$ can be expressed, respectively, as

$$\hat{\mathbf{n}}_a = \frac{-\hat{\mathbf{n}}_1 + 3\hat{\mathbf{n}}_2}{\sqrt{10}} \qquad \hat{\mathbf{n}}_b = -\hat{\mathbf{n}}_1 \tag{12}$$

Hence, if matrices a and b are defined as

$$a \triangleq [-1/\sqrt{10} \quad 3/\sqrt{10} \quad 0] \qquad b \triangleq [-1 \quad 0 \quad 0] \tag{13}$$

then the central product of inertia of C for $\hat{\mathbf{n}}_a$ and $\hat{\mathbf{n}}_b$ is given by

$$I_{ab}{}^{C/C^\star} \underset{(4.5.3)}{=} aI^{C/C^\star}b^T \underset{(11,\ 13)}{=} \frac{9mc^2}{4\sqrt{10}} \tag{14}$$

The position vector $\mathbf{p}^\star$ from O to $C^\star$ is

$$\mathbf{p}^\star = c\hat{\mathbf{n}}_1 + 3c\hat{\mathbf{n}}_2 \tag{15}$$

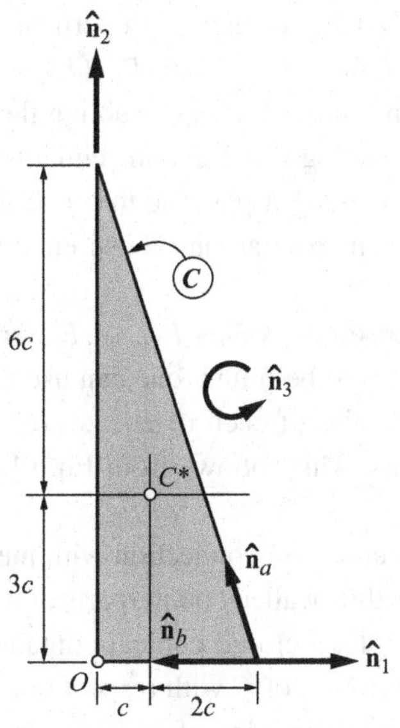

Figure 4.6.1

Consequently, the product of inertia of a particle of mass m situated at $C^\star$, relative to O, for $\hat{\mathbf{n}}_a$ and $\hat{\mathbf{n}}_b$, is

$$I_{ab}{}^{C^\star/O} \underset{(4.3.3)}{=} m(\mathbf{p}^\star \times \hat{\mathbf{n}}_a) \cdot (\mathbf{p}^\star \times \hat{\mathbf{n}}_b) \underset{(12,\ 15)}{=} \frac{18mc^2}{\sqrt{10}} \tag{16}$$

and

$$I_{ab}{}^{C/O} \underset{(4)}{=} I_{ab}{}^{C/C^\star} + I_{ab}{}^{C^\star/O} \underset{(14,\ 16)}{=} \frac{81mc^2}{4\sqrt{10}} \tag{17}$$

4.7 EVALUATION OF INERTIA SCALARS

Generally speaking, the most direct way to find inertia scalars (see Sec. 4.3) of a set of particles (including sets consisting of a single particle) is to use Eq. (4.3.3). Exceptions to this rule arise when the value of an appropriate central inertia scalar (see Sec. 4.6) is known, in which event one can appeal to Eq. (4.6.4), or when a suitable inertia vector (see Sec. 4.3), inertia matrix, or inertia dyadic (see Sec. 4.5) is available, making it possible to use Eqs. (4.3.2), (4.5.3), or (4.5.20). In a sense, the opposite applies to the evaluation of inertia scalars of a body modeled as matter distributed along a curve, over a surface, or throughout a solid. Here, one should use Eq. (4.3.9), which is analogous to Eq. (4.3.3), only when one cannot find any tabulated information regarding the inertia properties of the body under consideration, which is likely to be the case only when the body has a variable mass density or occupies an irregular figure. When, as happens more often in engineering practice, the mass density is uniform and the figure in question is one of those considered in an available table of inertia properties, the best way to proceed is to use, first, Eq. (4.4.3), (4.5.3), or (4.5.20) with I_{jk} $(j,k = 1,2,3)$ representing central inertia scalars, and then turn to Eq. (4.6.4), evaluating the last term in this equation with the aid of Eq. (4.3.3) after setting $\nu = 1$ and m_1 equal to the mass of the body. Appendix III contains information making it possible to proceed in this way in connection with a number of figures chosen from among those encountered most frequently in engineering practice.

When an object can be regarded as composed of bodies $B_1, \ldots, B_\nu$, and an inertia scalar of this object with respect to a point P is to be found, one can use the foregoing procedure to find the corresponding inertia scalar of each of $B_1, \ldots, B_\nu$ with respect to P and then simply add these inertia scalars. This follows from Eq. (4.3.3) and the associativity of scalar addition.

Differentials sometimes can be used to advantage in connection with inertia calculations involving objects that can be regarded as thin-walled counterparts of bodies having known inertia properties. Consider, for example, a closed cubical container C having sides of length L, and let k be the radius of gyration of C with respect to a line passing through $C^\star$, the mass center of C, and parallel to an edge of C. Suppose, further, that the walls of C are thin. To determine k, let ρ be the mass density of a uniform solid cube S having sides of length L. Then m, the mass of S, and I, the moment of inertia of S about a line parallel to an edge of S and passing through $S^\star$, the mass center of S, are given by (see Appendix III)

$$m = \rho L^3 \tag{1}$$

and

$$I = \frac{\rho L^5}{6} \tag{2}$$

respectively, and the differentials of m and I are

$$dm \underset{(1)}{=} 3\rho L^2\, dL \tag{3}$$

and

$$dI \underset{(2)}{=} \frac{5\rho L^4}{6}\, dL \tag{4}$$

Consequently, attributing to C the wall thickness $dL/2$, and hence the mass dm and the moment of inertia dI, one can write

$$\underset{(4)}{\frac{5\rho L^4}{6}\, dL} \underset{(4.3.6)}{=} \underset{(3)}{3\rho L^2\, dL}\, k^2 \tag{5}$$

from which it follows immediately that

$$k = \frac{1}{3}\sqrt{\frac{5}{2}}L \tag{6}$$

The same result is obtained with somewhat more effort, but quite revealingly, by reasoning as follows.

When C is regarded as formed by removing from a solid cube S_1 of mass density ρ and having sides of length L_1 a solid cube S_2 of the same mass density and having sides of length L_2, the mass m and moment of inertia I of C about a line passing through $C^\star$ and parallel to an edge of C can be expressed as

$$m = \rho L_1{}^3 - \rho L_2{}^3 = \rho(L_1{}^3 - L_2{}^3) \tag{7}$$

and

$$I = \rho\frac{L_1{}^5}{6} - \rho\frac{L_2{}^5}{6} = \frac{\rho}{6}(L_1{}^5 - L_2{}^5) \tag{8}$$

Consequently, $\overline{k}$, the radius of gyration of interest, is given by

$$\overline{k} \underset{(4.3.6)}{=} \left(\frac{I}{m}\right)^{1/2} \underset{(7,\,8)}{=} \left[\frac{L_1{}^5 - L_2{}^5}{6(L_1{}^3 - L_2{}^3)}\right]^{1/2} \tag{9}$$

Furthermore, in order for C to have sides of length L and walls of thickness $t/2$, L_1 and L_2 must be given by

$$L_1 = L \qquad L_2 = L - t \tag{10}$$

which leads to

$$\begin{aligned}\overline{k} &\underset{(9,\,10)}{=} \frac{1}{\sqrt{6}}\left[\frac{L^5 - (L-t)^5}{L^3 - (L-t)^3}\right]^{1/2} \\ &= \frac{1}{\sqrt{6}}\left(\frac{5L^4 - 10L^3t + 10L^2t^2 - 5Lt^3 + t^4}{3L^2 - 3Lt + t^2}\right)^{1/2}\end{aligned} \tag{11}$$

and

$$\lim_{t\to 0} \overline{k} \underset{(11)}{=} \frac{1}{3}\sqrt{\frac{5}{2}}L \tag{12}$$

Example The radius of gyration k of the body B described in the example in Sec. 4.2 and shown in Fig. 4.2.1 is to be found with respect to a line that is parallel to X_2 and passes through $B^\star$, the mass center of B.

Regarding B as composed of the three bodies B_1, B_2, B_3 introduced previously, and noting that the masses of B_1, B_2, B_3 are [see Eqs. (4.2.3)]

$$m_1 = m \qquad m_2 = \left(\frac{\pi}{2}\right)m \qquad m_3 = 10\left(1 + \frac{\pi}{2}\right)m \tag{13}$$

while $\mathbf{p}^\star$, the position vector from E to $B^\star$, is given by

$$\mathbf{p}^\star \underset{(4.2.4)}{=} (1.82\hat{\mathbf{a}}_1 + 2.03\hat{\mathbf{a}}_2 + 0.318\hat{\mathbf{a}}_3)L \tag{14}$$

and the mass centers of B_1, B_2, B_3 are the points $B_1{}^\star$, $B_2{}^\star$, $B_3{}^\star$ shown in Fig. 4.2.2, begin by forming expressions for the moment of inertia of B_i with respect to a line passing through $B_i{}^\star$ ($i = 1, 2, 3$) and parallel to $\hat{\mathbf{a}}_2$.

For B_1, from Eqs. (13), Fig. 4.2.2, and Appendix III,

$$I_2{}^{B_1/B_1{}^\star} \underset{(13)}{=} \frac{m(6L)^2}{12} = 3mL^2 \tag{15}$$

To deal with B_2, introduce unit vectors $\hat{\mathbf{n}}_a$ and $\hat{\mathbf{n}}_b$ as shown in Fig. 4.2.2, and refer to Appendix III to write

$$I_a{}^{B_2/B_2\star} = \frac{(\pi/2)m(6L)^2}{2}\left(1 - \frac{1}{\pi/2}\right) = 10.3mL^2 \tag{16}$$

$$I_b{}^{B_2/B_2\star} = \frac{(\pi/2)m(6L)^2}{2}\left[1 + \frac{1}{\pi/2} - \frac{1}{(\pi/4)^2}\right] = 0.438mL^2 \tag{17}$$

and then appeal to Eq. (4.4.3) to obtain

$$I_2{}^{B_2/B_2\star} = \left(\frac{1}{\sqrt{2}}\right)^2 I_a{}^{B_2/B_2\star} + \left(\frac{1}{\sqrt{2}}\right)^2 I_b{}^{B_2/B_2\star} \underset{(16,\ 17)}{=} 5.36mL^2 \tag{18}$$

Finally, refer to Appendix III to verify that

$$I_2{}^{B_3/B_3\star} = 10\left(1 + \frac{\pi}{2}\right)m\frac{(6L)^2}{18} = 51.4mL^2 \tag{19}$$

Next, determine $I_2{}^{B_i\star/B\star}$ $(i = 1,2,3)$ by using Eq. (4.3.3) with $\nu = 1$, that is, by expressing $I_2{}^{B_i\star/B\star}$ as

$$I_2{}^{B_i\star/B\star} = m_i[(\mathbf{p}^\star - \mathbf{p}_i) \times \hat{\mathbf{a}}_2]^2 \qquad (i = 1,2,3) \tag{20}$$

where $\mathbf{p}_i$ is the position vector from E (see Fig. 4.2.2) to $B_i{}^\star$ $(i = 1,2,3)$. Specifically,

$$\begin{aligned} I_2{}^{B_1\star/B\star} &\underset{(20)}{=} m[(\mathbf{p}^\star - 3L\hat{\mathbf{a}}_3) \times \hat{\mathbf{a}}_2]^2 \\ &\underset{(14)}{=} mL^2(1.82\hat{\mathbf{a}}_3 + 2.68\hat{\mathbf{a}}_1)^2 = 10.5mL^2 \end{aligned} \tag{21}$$

$$I_2{}^{B_2\star/B\star} \underset{(20)}{=} \left(\frac{\pi}{2}\right)m\left\{\left[\mathbf{p}^\star - \left(\frac{12L}{\pi}\right)(\hat{\mathbf{a}}_2 + \hat{\mathbf{a}}_3)\right] \times \hat{\mathbf{a}}_2\right\}^2 \underset{(14)}{=} 24.5mL^2 \tag{22}$$

$$I_2{}^{B_3\star/B\star} \underset{(20)}{=} 10\left(1 + \frac{\pi}{2}\right)m\{[\mathbf{p}^\star - 2L(\hat{\mathbf{a}}_1 + \hat{\mathbf{a}}_2)] \times \hat{\mathbf{a}}_2\}^2 \underset{(14)}{=} 3.43mL^2 \tag{23}$$

Now refer to Eq. (4.6.5) to evaluate $I_2{}^{B_i/B\star}$ $(i = 1,2,3)$ as

$$I_2{}^{B_i/B\star} = I_2{}^{B_i/B_i\star} + I_2{}^{B_i\star/B\star} \qquad (i = 1,2,3) \tag{24}$$

which gives

$$I_2{}^{B_1/B\star} \underset{(24)}{=} \underset{(15)}{3mL^2} + \underset{(21)}{10.5mL^2} = 13.5mL^2 \tag{25}$$

$$I_2{}^{B_2/B\star} \underset{(24)}{=} \underset{(18)}{5.36mL^2} + \underset{(22)}{24.5mL^2} = 29.9mL^2 \tag{26}$$

$$I_2{}^{B_3/B\star} \underset{(24)}{=} \underset{(19)}{51.4mL^2} + \underset{(23)}{3.43mL^2} = 54.8mL^2 \tag{27}$$

Consequently,

$$I_2{}^{B/B\star} = I_2{}^{B_1/B\star} + I_2{}^{B_2/B\star} + I_2{}^{B_3/B\star} \underset{(25\text{–}27)}{=} 98.2mL^2 \tag{28}$$

and

$$k \underset{(4.3.6)}{=} \left(\frac{I_2^{B/B^\star}}{m_1 + m_2 + m_3} \right)^{1/2} \underset{(13,\ 28)}{=} 1.86L \tag{29}$$

4.8 PRINCIPAL MOMENTS OF INERTIA

In general, the inertia vector $\mathbf{I}_a$ (see Sec. 4.3) is not parallel to $\hat{\mathbf{n}}_a$. When $\hat{\mathbf{n}}_z$ is a unit vector such that $\mathbf{I}_z$ is parallel to $\hat{\mathbf{n}}_z$, the line L_z passing through O and parallel to $\hat{\mathbf{n}}_z$ is called a *principal axis* of S for O, the plane P_z passing through O and normal to $\hat{\mathbf{n}}_z$ is called a *principal plane* of S for O, the moment of inertia I_z of S with respect to L_z is called a *principal moment of inertia* of S for O, and the radius of gyration of S with respect to L_z is called a *principal radius of gyration* of S for O. When the point O under consideration is the mass center of S, one speaks of *central* principal axes, central principal planes, central principal moments of inertia, and central principal radii of gyration.

When $\hat{\mathbf{n}}_z$ is parallel to a principal axis of S for O, the inertia vector $\mathbf{I}_z$ of S relative to O for $\hat{\mathbf{n}}_z$ can be expressed as

$$\mathbf{I}_z = I_z \hat{\mathbf{n}}_z \tag{1}$$

and, if $\hat{\mathbf{n}}_y$ is any unit vector perpendicular to $\hat{\mathbf{n}}_z$, then the product of inertia of S relative to O for $\hat{\mathbf{n}}_y$ and $\hat{\mathbf{n}}_z$ vanishes; that is,

$$I_{yz} = 0 \tag{2}$$

Suppose that $\hat{\mathbf{n}}_1$, $\hat{\mathbf{n}}_2$, $\hat{\mathbf{n}}_3$ are mutually perpendicular unit vectors, each parallel to a principal axis of S for O, and $\hat{\mathbf{n}}_a$ and $\hat{\mathbf{n}}_b$ are any two unit vectors. Then, if a_i and b_i are defined as

$$a_i \triangleq \hat{\mathbf{n}}_a \cdot \hat{\mathbf{n}}_i \qquad b_i \triangleq \hat{\mathbf{n}}_b \cdot \hat{\mathbf{n}}_i \qquad (i = 1,2,3) \tag{3}$$

the inertia scalar I_{ab} of S relative to O for $\hat{\mathbf{n}}_a$ and $\hat{\mathbf{n}}_b$ is given by

$$I_{ab} = a_1 I_1 b_1 + a_2 I_2 b_2 + a_3 I_3 b_3 \tag{4}$$

where I_1, I_2, I_3 are the principal moments of inertia of S for O associated with $\hat{\mathbf{n}}_1$, $\hat{\mathbf{n}}_2$, $\hat{\mathbf{n}}_3$, respectively. This relationship is considerably simpler than its more general counterpart, Eq. (4.4.3), which applies even when $\hat{\mathbf{n}}_1$, $\hat{\mathbf{n}}_2$, $\hat{\mathbf{n}}_3$ are not parallel to principal axes of S for O. Similarly, Eqs. (4.5.1) and (4.5.22) reduce to

$$I = \begin{bmatrix} I_1 & 0 & 0 \\ 0 & I_2 & 0 \\ 0 & 0 & I_3 \end{bmatrix} \tag{5}$$

and

$$\underline{\mathbf{I}} = I_1 \hat{\mathbf{n}}_1 \hat{\mathbf{n}}_1 + I_2 \hat{\mathbf{n}}_2 \hat{\mathbf{n}}_2 + I_3 \hat{\mathbf{n}}_3 \hat{\mathbf{n}}_3 \tag{6}$$

respectively, when $\hat{\mathbf{n}}_1$, $\hat{\mathbf{n}}_2$, $\hat{\mathbf{n}}_3$ are parallel to principal axes of S for O.

For every set of particles there exists at least one set of three mutually perpendicular principal axes for every point in space. To locate principal axes of a set S of particles for a point O, and to determine the associated principal moments of inertia, one exploits the following facts.

If $\hat{\mathbf{n}}_1$, $\hat{\mathbf{n}}_2$, $\hat{\mathbf{n}}_3$ are any mutually perpendicular unit vectors, I_{jk} $(j, k = 1,2,3)$ are the associated inertia scalars of S for O, and I_z is a principal moment of inertia of S for O, then I_z satisfies the cubic *characteristic equation*

$$\begin{vmatrix} (I_1 - I_z) & I_{12} & I_{13} \\ I_{21} & (I_2 - I_z) & I_{23} \\ I_{31} & I_{32} & (I_3 - I_z) \end{vmatrix} = 0 \tag{7}$$

A unit vector $\hat{\mathbf{n}}_z$ is parallel to the principal axis associated with the principal moment of inertia I_z if

$$\hat{\mathbf{n}}_z = z_1 \hat{\mathbf{n}}_1 + z_2 \hat{\mathbf{n}}_2 + z_3 \hat{\mathbf{n}}_3 \tag{8}$$

where z_1, z_2, z_3 satisfy the four equations

$$\sum_{j=1}^{3} z_j I_{jk} = I_z z_k \qquad (k = 1,2,3) \tag{9}$$

$$z_1{}^2 + z_2{}^2 + z_3{}^2 = 1 \tag{10}$$

When Eq. (7) has precisely two equal roots, every line passing through O and lying in the principal plane of S for O corresponding to the remaining root of Eq. (7) is a principal axis of S for O; when Eq. (7) has three equal roots, every line passing through O is a principal axis of S for O.

Once a principal plane P_z of S for O has been identified, principal axes of S for O lying in P_z, as well as the associated principal moments of inertia of S for O, can be found without solving a cubic equation. If $\hat{\mathbf{n}}_a$ and $\hat{\mathbf{n}}_b$ are any two unit vectors parallel to P_z and perpendicular to each other, while I_a, I_b, and I_{ab} are the associated moments of inertia and product of inertia of S for O, then the angle θ between $\hat{\mathbf{n}}_a$ and each of the principal axes in question satisfies the equation

$$\tan 2\theta = \frac{2I_{ab}}{I_a - I_b} \qquad (I_a \neq I_b) \tag{11}$$

and the associated principal moments of inertia of S for O, say, I_x and I_y, are given by

$$I_x, I_y = \frac{I_a + I_b}{2} \pm \left[\left(\frac{I_a - I_b}{2} \right)^2 + I_{ab}{}^2 \right]^{1/2} \tag{12}$$

If $I_a = I_b$, then every line passing through O and lying in P_z is a principal axis of S for O, and the associated principal moments of inertia of S for O all have the value I_a.

Frequently, principal planes can be located by means of symmetry considerations. For example, when all particles of S lie in a plane, then this plane is a principal plane of S for every point of the plane, for the inertia vector of S relative to any point of the plane is then normal to the plane.

The eigenvalues of the matrix I defined in Eq. (4.5.1) are principal moments of inertia of S for O because these eigenvalues satisfy Eq. (7); and, if a matrix $\lfloor e_1 \quad e_2 \quad e_3 \rfloor$ is an eigenvector of I, and $e_1{}^2 + e_2{}^2 + e_3{}^2 = 1$, then $e_1\hat{\mathbf{n}}_1 + e_2\hat{\mathbf{n}}_2 + e_3\hat{\mathbf{n}}_3$ is a unit vector parallel to a principal axis of S for O because e_1, e_2, e_3 then satisfy Eqs. (9) whenever I_z is an eigenvalue of I. Hence, when computer routines for finding eigenvalues and eigenvectors of a symmetric matrix are available, these can be used directly to determine principal moments of inertia and to locate the associated principal axes.

Derivations When $\hat{\mathbf{n}}_z$ is parallel to a principal axis of S for O, and hence $\mathbf{I}_z$ is parallel to $\hat{\mathbf{n}}_z$, then there exists a quantity λ such that

$$\lambda\hat{\mathbf{n}}_z = \mathbf{I}_z \tag{13}$$

Dot multiplication of this equation with $\hat{\mathbf{n}}_z$ gives

$$\lambda = \mathbf{I}_z \cdot \hat{\mathbf{n}}_z \underset{(4.3.2)}{=} I_z \tag{14}$$

and substitution from this equation into Eq. (13) yields Eq. (1). Furthermore, if $\hat{\mathbf{n}}_y$ is any unit vector perpendicular to $\hat{\mathbf{n}}_z$, then dot multiplication of Eq. (1) with $\hat{\mathbf{n}}_y$ and use of Eqs. (4.3.2) and (4.3.4) lead to Eq. (2).

If $\hat{\mathbf{n}}_1$, $\hat{\mathbf{n}}_2$, $\hat{\mathbf{n}}_3$ are mutually perpendicular unit vectors, each parallel to a principal axis of S for O, then, in accordance with Eq. (2),

$$I_{12} = I_{23} = I_{31} = 0 \tag{15}$$

and substitution from these equations into Eqs. (4.4.3), (4.5.1), and (4.5.22) establishes the validity of Eqs. (4), (5), and (6), respectively.

If $\hat{\mathbf{n}}_1$, $\hat{\mathbf{n}}_2$, $\hat{\mathbf{n}}_3$ are any mutually perpendicular unit vectors, I_{jk} $(j,k = 1,2,3)$ are the associated inertia scalars of S for O, $\hat{\mathbf{n}}_z$ is any unit vector, and z_1, z_2, z_3 are defined as

$$z_i \triangleq \hat{\mathbf{n}}_z \cdot \hat{\mathbf{n}}_i \qquad (i = 1,2,3) \tag{16}$$

then

$$\hat{\mathbf{n}}_z \underset{(1.6.1)}{=} \sum_{k=1}^{3} z_k\hat{\mathbf{n}}_k \tag{17}$$

and

$$\mathbf{I}_z \underset{(4.4.1)}{=} \sum_{j=1}^{3} z_j\mathbf{I}_j \tag{18}$$

Now

$$\mathbf{I}_j \underset{(1.6.3)}{=} \sum_{k=1}^{3} \mathbf{I}_j \cdot \hat{\mathbf{n}}_k\hat{\mathbf{n}}_k \underset{(4.3.2)}{=} \sum_{k=1}^{3} I_{jk}\hat{\mathbf{n}}_k \qquad (j = 1,2,3) \tag{19}$$

Hence,

$$\mathbf{I}_z \underset{(18,\ 19)}{=} \sum_{j=1}^{3}\sum_{k=1}^{3} z_j I_{jk}\hat{\mathbf{n}}_k \tag{20}$$

and $\hat{\mathbf{n}}_z$ is parallel to a principal axis of S for O if

$$\sum_{j=1}^{3}\sum_{k=1}^{3} z_j I_{jk}\hat{\mathbf{n}}_k \underset{(20)}{} \underset{(1)}{=} I_z \sum_{k=1}^{3} z_k \hat{\mathbf{n}}_k \underset{(17)}{} \tag{21}$$

This vector equation is equivalent to the three scalar equations

$$\sum_{j=1}^{3} z_j I_{jk} = I_z z_k \qquad (k = 1,2,3) \tag{22}$$

Also, z_1, z_2, z_3 satisfy the equation

$$z_1^2 + z_2^2 + z_3^2 = 1 \tag{23}$$

because $\hat{\mathbf{n}}_z$ is a unit vector. [Equations (22) and (23) are Eqs. (9) and (10), respectively.] As Eqs. (22) are linear and homogeneous in z_1, z_2, z_3, and as these three quantities cannot all vanish, for this would violate Eq. (23), Eqs. (22) can be satisfied only if the determinant of the coefficients of z_1, z_2, z_3 vanishes, that is, if Eq. (7) is satisfied. Now, Eq. (7) is cubic in I_z. Hence, there exist three values of I_z (not necessarily distinct) that satisfy Eq. (7). It will now be shown that all such values of I_z are real.

Let A, B, α_j and β_j $(j = 1,2,3)$ be real quantities such that

$$I_z = A + iB \tag{24}$$

and

$$z_j = \alpha_j + i\beta_j \qquad (j = 1,2,3) \tag{25}$$

where $i \triangleq \sqrt{-1}$. Then

$$\sum_{j=1}^{3} \underset{(25)}{(\alpha_j + i\beta_j)} I_{jk} \underset{(22)}{=} \underset{(24)}{(A + iB)} \underset{(25)}{(\alpha_k + i\beta_k)} \qquad (k = 1,2,3) \tag{26}$$

and, after separating the real and imaginary parts of this equation, one has

$$\sum_{j=1}^{3} \alpha_j I_{jk} \underset{(26)}{=} A\alpha_k - B\beta_k \qquad (k = 1,2,3) \tag{27}$$

$$\sum_{j=1}^{3} \beta_j I_{jk} \underset{(26)}{=} B\alpha_k + A\beta_k \qquad (k = 1,2,3) \tag{28}$$

Multiply Eq. (28) by α_k and Eq. (27) by β_k, and subtract, obtaining

$$\sum_{j=1}^{3} \alpha_k \beta_j I_{jk} - \sum_{j=1}^{3} \alpha_j \beta_k I_{jk} = B(\alpha_k^2 + \beta_k^2) \qquad (k = 1,2,3) \tag{29}$$

and add these three equations, which yields

$$\sum_{k=1}^{3}\sum_{j=1}^{3} \alpha_k \beta_j I_{jk} - \sum_{k=1}^{3}\sum_{j=1}^{3} \alpha_j \beta_k I_{jk} = B \sum_{k=1}^{3} (\alpha_k^2 + \beta_k^2) \tag{30}$$

Now,

$$\sum_{k=1}^{3}\sum_{j=1}^{3}\alpha_k\beta_j I_{jk} \equiv \sum_{k=1}^{3}\sum_{j=1}^{3}\alpha_j\beta_k I_{kj} \underset{(4.3.4)}{=} \sum_{k=1}^{3}\sum_{j=1}^{3}\alpha_j\beta_k I_{jk} \tag{31}$$

Consequently, the left-hand member of Eq. (30) is equal to zero, and Eq. (30) reduces to

$$B\sum_{k=1}^{3}(\alpha_k{}^2 + \beta_k{}^2) = 0 \tag{32}$$

The quantities α_k and β_k $(k = 1,2,3)$ cannot all vanish, for this would mean [see Eqs. (25)] that z_1, z_2, z_3 all vanish, which is ruled out by Eq. (23). Hence, the only way Eq. (32) can be satisfied is for B to be equal to zero, and Eq. (24) thus shows that I_z is real.

Suppose that two roots of Eq. (7), say, I_x and I_y, are distinct from each other. Then, with self-explanatory notation, we can write

$$\mathbf{I}_x \underset{(1)}{=} I_x\hat{\mathbf{n}}_x \tag{33}$$

$$\mathbf{I}_y \underset{(1)}{=} I_y\hat{\mathbf{n}}_y \tag{34}$$

whereupon dot mutliplication of Eq. (33) with $\hat{\mathbf{n}}_y$, Eq. (34) with $\hat{\mathbf{n}}_x$, and subtraction of the resulting equations produces

$$\mathbf{I}_x \cdot \hat{\mathbf{n}}_y - \mathbf{I}_y \cdot \hat{\mathbf{n}}_x = (I_x - I_y)\hat{\mathbf{n}}_x \cdot \hat{\mathbf{n}}_y \tag{35}$$

or, in view of Eq. (4.3.2),

$$I_{xy} - I_{yx} \underset{(4.3.4)}{=} 0 \underset{(35)}{=} (I_x - I_y)\hat{\mathbf{n}}_x \cdot \hat{\mathbf{n}}_y \tag{36}$$

Since I_x differs from I_y by hypothesis, it follows that

$$\hat{\mathbf{n}}_x \cdot \hat{\mathbf{n}}_y = 0 \tag{37}$$

which proves that, whenever two principal moments of inertia of S for O are unequal, the corresponding principal axes of S for O are perpendicular to each other. Consequently, when Eq. (7) has three distinct roots, a unique principal axis of S for O corresponds to each root, and these three principal axes of S for O are mutually perpendicular. What remains to be done, is to deal with the matter of repeated roots of Eq. (7).

Let I_c denote one of the roots of Eq. (7), let $\hat{\mathbf{n}}_c$ be a unit vector parallel to the associated principal axis of S for O, and suppose that the remaining two roots of Eq. (7) are equal to each other (and possibly to I_c). Let $\hat{\mathbf{n}}_a$ and $\hat{\mathbf{n}}_b$ be any unit vectors perpendicular to $\hat{\mathbf{n}}_c$ and to each other, and note that, in accordance with Eq. (2),

$$I_{ac} \underset{(4.3.4)}{=} I_{ca} = 0 \qquad I_{bc} \underset{(4.3.4)}{=} I_{cb} = 0 \tag{38}$$

In Eq. (7), the subscripts 1, 2, and 3 may be replaced with a, b, and c, respectively. When this is done and Eqs. (38) are used, Eq. (7) reduces to

$$(I_c - I_z)[I_z{}^2 - (I_a + I_b)I_z + I_aI_b - I_{ab}{}^2] = 0 \tag{39}$$

which has the three roots

$$I_z = \frac{I_a + I_b}{2} \pm \left[\left(\frac{I_a - I_b}{2} \right)^2 + I_{ab}{}^2 \right]^{1/2}, \quad I_c \tag{40}$$

The first two of these can be equal to each other only if

$$\left(\frac{I_a - I_b}{2} \right)^2 + I_{ab}{}^2 = 0 \tag{41}$$

which is possible only if

$$I_a = I_b \tag{42}$$

and

$$I_{ab} = 0 \tag{43}$$

But Eqs. (40) show that, under these circumstances, I_a is one of the values of I_z; that is, I_a is a principal moment of inertia of S for O, and $\hat{\mathbf{n}}_a$ is thus parallel to a principal axis of S for O. Since $\hat{\mathbf{n}}_a$ was restricted only to the extent of being required to be perpendicular to $\hat{\mathbf{n}}_c$, this means that every line passing through O and perpendicular to $\hat{\mathbf{n}}_c$ is a principal axis of S for O; it follows from this that, when Eq. (7) has three equal roots, every line passing through O is a principal axis of S for O.

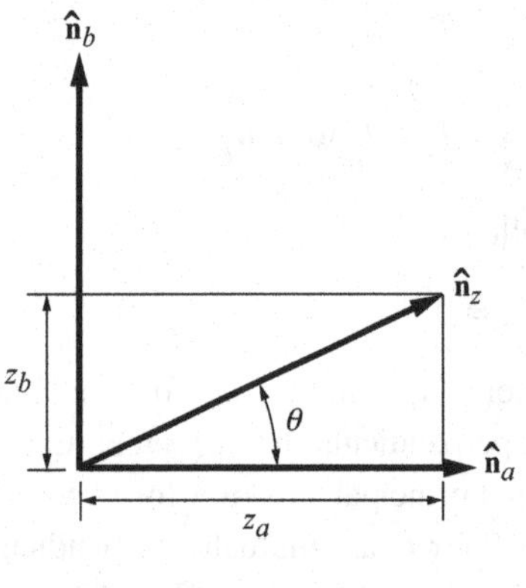

Figure 4.8.1

The validity of Eqs. (12) is established by Eqs. (40). Finally, Eq. (11) may be derived from Eqs. (9) by writing the first of these with the subscripts 1, 2, and 3 replaced with a, b, and c, respectively, and with I_{ca} set equal to zero in accordance with Eqs. (38), which gives

$$z_a I_a + z_b I_{ab} = I_z z_a \tag{44}$$

where z_a and z_b are then the $\hat{\mathbf{n}}_a$ and $\hat{\mathbf{n}}_b$ measure numbers of a unit vector $\hat{\mathbf{n}}_z$ that is parallel to one of the principal axes of S for O associated with the first two values of I_z

in Eqs. (40). The unit vectors $\hat{\mathbf{n}}_a$, $\hat{\mathbf{n}}_b$, and $\hat{\mathbf{n}}_z$ are shown in Fig. 4.8.1, as is the angle θ, which is seen to satisfy the equation

$$\tan\theta = \frac{z_b}{z_a} \tag{45}$$

Now,

$$\tan 2\theta = \frac{2\tan\theta}{1-\tan^2\theta} \underset{(45)}{=} \frac{2(z_b/z_a)}{1-(z_b/z_a)^2} \tag{46}$$

and

$$\frac{z_b}{z_a} \underset{(44)}{=} \frac{I_z - I_a}{I_{ab}} \underset{(40)}{=} -\frac{1}{I_{ab}}\left\{\frac{I_a - I_b}{2} \pm \left[\left(\frac{I_a - I_b}{2}\right)^2 + {I_{ab}}^2\right]^{1/2}\right\} \tag{47}$$

so that

$$1 - \left(\frac{z_b}{z_a}\right)^2 \underset{(47)}{=} -\frac{I_a - I_b}{{I_{ab}}^2}\left\{\frac{I_a - I_b}{2} \pm \left[\left(\frac{I_a - I_b}{2}\right)^2 + {I_{ab}}^2\right]^{1/2}\right\} \tag{48}$$

Thus, Eq. (11) is obtained by substituting from Eqs. (47) and (48) into Eq. (46).

Example A uniform rectangular plate B of mass m has the dimensions shown in Fig. 4.8.2. Two sets of principal moments of inertia and associated principal axes of B are to be found, namely, the principal moments of inertia I_x, I_y, I_z of B for O and the associated principal axes X, Y, Z, and the principal moments of inertia I_x', I_y', I_z' of B for O' and the associated principal axes X', Y', Z', where O' is a point situated at a distance $3L$ from O on a line passing through O and parallel to $\hat{\mathbf{n}}_3$, as shown in Fig. 4.8.2, and $\hat{\mathbf{n}}_3 = \hat{\mathbf{n}}_1 \times \hat{\mathbf{n}}_2$.

When, as in the example in Sec. 4.3, B is modeled as matter distributed over a rectangular surface, this surface is a principal plane of B for O, the line Z passing through O and parallel to $\hat{\mathbf{n}}_3$ is a principal axis of S for O, and the corresponding principal moment of inertia of B for O, I_z, is equal to the moment of inertia of B about Z. Referring to Table 4.8.1, where the inertia scalars I_{jk} of B relative to O for $\hat{\mathbf{n}}_j$ and $\hat{\mathbf{n}}_k$ $(j, k = 1,2,3)$ are recorded in accordance with Table 4.4.1, one thus finds that

$$I_z = I_{33} = \frac{5mL^2}{3} = 1.67mL^2 \tag{49}$$

and the remaining principal moments of inertia of B for O, I_x, and I_y, found with

Table 4.8.1

I_{jk}	1	2	3
1	$4mL^2/3$	$-mL^2/2$	0
2	$-mL^2/2$	$mL^2/3$	0
3	0	0	$5mL^2/3$

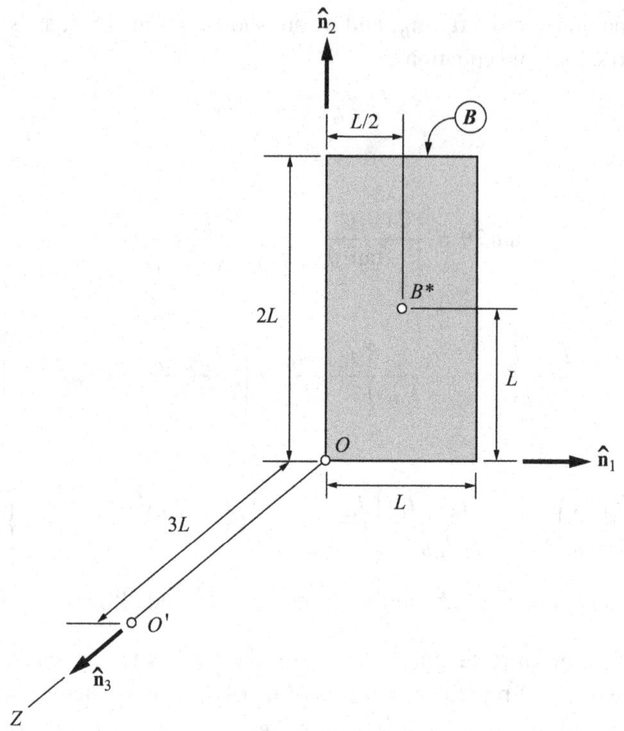

Figure 4.8.2

the aid of Eq. (12), where the subscripts a and b may be replaced with 1 and 2, respectively, are given by

$$I_x, I_y = \frac{5mL^2}{6} \pm \left[\left(\frac{mL^2}{2}\right)^2 + \left(\frac{mL^2}{2}\right)^2\right]^{1/2}$$

$$= mL^2\left(\frac{5}{6} \pm \frac{1}{\sqrt{2}}\right) = 0.126mL^2,\ 1.54mL^2 \tag{50}$$

while the angle θ between the associated principal axes and $\hat{\mathbf{n}}_1$ satisfies the equation

$$\tan 2\theta \underset{(11)}{=} \frac{-2(mL^2/2)}{(4mL^2/3) - (mL^2/3)} = -1 \tag{51}$$

so that θ can have the values

$$\theta = 67.5°,\ -22.5° \tag{52}$$

Consequently, the principal axes X, Y, and Z for O are oriented as shown in Fig. 4.8.3.

Determining I_x', I_y', I_z' and locating X', Y', Z' is somewhat more difficult, for here one cannot appeal to Eqs. (11) and (12) but must use Eqs. (7), (9), and (10), instead. Furthermore, the inertia scalars appearing in Eqs. (7) and (9) must be the

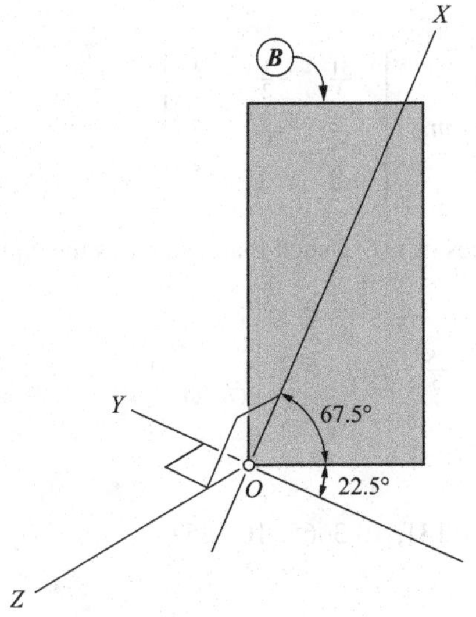

Figure 4.8.3

inertia scalars $I_{jk}{}'$ of B relative to point O' for $\hat{\mathbf{n}}_j$ and $\hat{\mathbf{n}}_k$ $(j,k = 1,2,3)$, which are not, as yet, available. To find them, let $I^{B/B^\star}$ denote the inertia matrix of B relative to $B^\star$, the mass center of B, for $\hat{\mathbf{n}}_1$, $\hat{\mathbf{n}}_2$, $\hat{\mathbf{n}}_3$, and refer to Eq. (4.6.2) to write

$$I^{B/B^\star} = I^{B/O} - I^{B^\star/O} \tag{53}$$

$$I' = I^{B/B^\star} + I^{B^\star/O'}$$

$$\underset{(53)}{=} I^{B/O} - I^{B^\star/O} + I^{B^\star/O'} \tag{54}$$

where I' is the inertia matrix of B relative to O' for $\hat{\mathbf{n}}_1$, $\hat{\mathbf{n}}_2$, $\hat{\mathbf{n}}_3$, and $I^{B/O}$, the inertia matrix of B relative to O for $\hat{\mathbf{n}}_1$, $\hat{\mathbf{n}}_2$, $\hat{\mathbf{n}}_3$, then can be written (see Table 4.8.1)

$$I^{B/O} = mL^2 \begin{bmatrix} \frac{4}{3} & -\frac{1}{2} & 0 \\ -\frac{1}{2} & \frac{1}{3} & 0 \\ 0 & 0 & \frac{5}{3} \end{bmatrix} \tag{55}$$

while $I^{B^\star/O}$ and $I^{B^\star/O'}$, found with the aid of Eqs. (4.3.3), are given by

$$I^{B^\star/O} = mL^2 \begin{bmatrix} 1 & -\frac{1}{2} & 0 \\ -\frac{1}{2} & \frac{1}{4} & 0 \\ 0 & 0 & \frac{5}{4} \end{bmatrix} \tag{56}$$

and

$$I^{B^\star/O'} = mL^2 \begin{bmatrix} 10 & -\frac{1}{2} & \frac{3}{2} \\ -\frac{1}{2} & \frac{37}{4} & 3 \\ \frac{3}{2} & 3 & \frac{5}{4} \end{bmatrix} \tag{57}$$

Consequently,

$$I' \underset{(54\text{–}57)}{=} mL^2 \begin{bmatrix} \frac{31}{3} & -\frac{1}{2} & \frac{3}{2} \\ -\frac{1}{2} & \frac{28}{3} & 3 \\ \frac{3}{2} & 3 & \frac{5}{3} \end{bmatrix} \tag{58}$$

The eigenvalues of I' are values of $mL^2\lambda$ such that λ satisfies the equation

$$\begin{vmatrix} \frac{31}{3} - \lambda & -\frac{1}{2} & \frac{3}{2} \\ -\frac{1}{2} & \frac{28}{3} - \lambda & 3 \\ \frac{3}{2} & 3 & \frac{5}{3} - \lambda \end{vmatrix} \underset{(7,\ 58)}{=} 0 \tag{59}$$

which is the case for

$$\lambda = 0.381131,\ 10.3665,\ 10.5857 \tag{60}$$

Hence,

$$I_x{}' = 0.381131mL^2 \qquad I_y{}' = 10.3665mL^2 \qquad I_z{}' = 10.5857mL^2 \tag{61}$$

To locate X', the principal axis of B for O' associated with $I_x{}'$, refer to Eqs. (9) to write

$$z_1 I_{11}{}' + z_2 I_{21}{}' + z_3 I_{31}{}' = I_x{}' z_1 \tag{62}$$

$$z_1 I_{12}{}' + z_2 I_{22}{}' + z_3 I_{32}{}' = I_x{}' z_2 \tag{63}$$

or, by reference to Eqs. (58) and (61),

$$z_1(\tfrac{31}{3} - 0.381131) + z_2(-\tfrac{1}{2}) \underset{(62)}{=} z_3(-\tfrac{3}{2}) \tag{64}$$

$$z_1(-\tfrac{1}{2}) + z_2(\tfrac{28}{3} - 0.381131) \underset{(63)}{=} z_3(-3) \tag{65}$$

Solution of these two equations for z_1 and z_2 yields

$$z_1 = -0.1680z_3 \qquad z_2 = -0.3445z_3 \tag{66}$$

and substitution into Eq. (10) shows that z_3 must satisfy the equation

$$[(-0.1680)^2 + (-0.3445)^2 + 1]z_3{}^2 = 1 \tag{67}$$

which is satisfied by

$$z_3 = \pm 0.9338 \tag{68}$$

Hence,

$$z_1 \underset{(66,\ 68)}{=} \mp 0.1569 \qquad z_2 = \mp 0.3217 \tag{69}$$

and X' is parallel to the unit vector $\hat{\mathbf{x}}'$ defined as

$$\hat{\mathbf{x}}' \triangleq \underset{(69)}{\mp 0.157}\hat{\mathbf{n}}_1 \underset{(69)}{\mp 0.322}\hat{\mathbf{n}}_2 \underset{(68)}{\pm 0.934}\hat{\mathbf{n}}_3 \tag{70}$$

Similarly, to locate Y' and Z', one needs only to replace 0.381131 with 10.3665 and 10.5857, respectively [see Eqs. (61)], in Eqs. (64) and (65), solve the resulting equations for z_1 and z_2 in terms of z_3, and then use Eq. (10) to find that Y' and Z' are parallel to the unit vectors $\hat{\mathbf{y}}'$ and $\hat{\mathbf{z}}'$ defined as

$$\hat{\mathbf{y}}' \triangleq \mp 0.0724\hat{\mathbf{n}}_1 \pm 0.947\hat{\mathbf{n}}_2 \pm 0.314\hat{\mathbf{n}}_3 \tag{71}$$

and

$$\hat{\mathbf{z}}' \triangleq \mp 0.985\hat{\mathbf{n}}_1 \mp 0.0185\hat{\mathbf{n}}_2 \mp 0.172\hat{\mathbf{n}}_3 \tag{72}$$

4.9 MAXIMUM AND MINIMUM MOMENTS OF INERTIA

The locus E of points whose distance R from a point O is inversely proportional to the square root of the moment of inertia of a set S of particles about line OP (see Fig. 4.9.1) is an ellipsoid, called an *inertia ellipsoid* of S for O, whose axes are parallel to the principal axes of S for O (see Sec. 4.8). It follows that, of all lines passing through O, those with respect to which S has a larger, or smaller, moment of inertia than it has with respect to all other lines passing through O are principal axes of S for O; and this, in turn, means that no moment of inertia of S is smaller than the smallest central principal moment of inertia of S.

Derivations In Fig. 4.9.1, X, Y, Z designate mutually perpendicular coordinate axes passing through O, chosen in such a way that each is a principal axis of S for O; x, y, z are the coordinates of P; $\hat{\mathbf{n}}_x$, $\hat{\mathbf{n}}_y$, $\hat{\mathbf{n}}_z$ are unit vectors parallel to X, Y, Z, respectively; and $\hat{\mathbf{n}}_a$ is a unit vector parallel to OP. It is to be shown that x, y, z satisfy the equation of an ellipsoid whenever

$$R = \lambda I_a^{-1/2} \tag{1}$$

where I_a is the moment of inertia of S with respect to line OP, and λ is any constant.

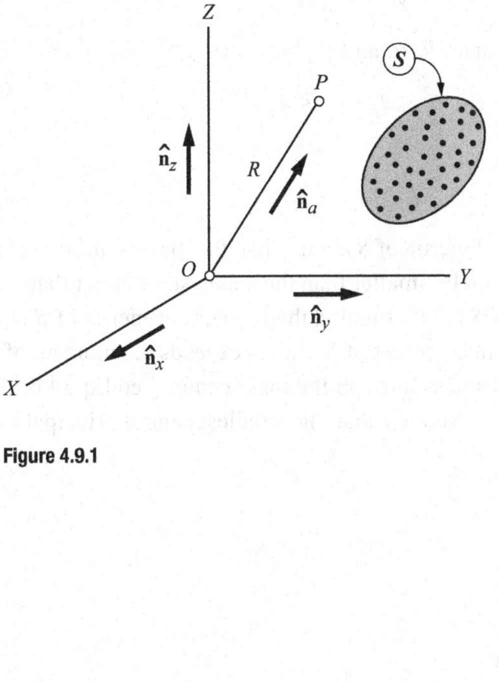

Figure 4.9.1

Let

$$a_x \triangleq \hat{\mathbf{n}}_a \cdot \hat{\mathbf{n}}_x \qquad a_y \triangleq \hat{\mathbf{n}}_a \cdot \hat{\mathbf{n}}_y \qquad a_z \triangleq \hat{\mathbf{n}}_a \cdot \hat{\mathbf{n}}_z \tag{2}$$

so that

$$\hat{\mathbf{n}}_a = a_x\hat{\mathbf{n}}_x + a_y\hat{\mathbf{n}}_y + a_z\hat{\mathbf{n}}_z \tag{3}$$

and note that the position vector from O to P can be expressed both as $R\hat{\mathbf{n}}_a$ and as $x\hat{\mathbf{n}}_x + y\hat{\mathbf{n}}_y + z\hat{\mathbf{n}}_z$, so that, in view of Eq. (3),

$$a_x = \frac{x}{R} \qquad a_y = \frac{y}{R} \qquad a_z = \frac{z}{R} \tag{4}$$

Also,

$$I_a \underset{(4.8.4)}{=} a_x{}^2 I_x + a_y{}^2 I_y + a_z{}^2 I_z \tag{5}$$

where I_x, I_y, I_z are the principal moments of inertia of S for O. Hence,

$$I_a \underset{(5,\,4)}{=} \frac{1}{R^2}(x^2 I_x + y^2 I_y + z^2 I_z) \tag{6}$$

and elimination of R by reference to Eq. (1) leads to

$$\frac{x^2}{\alpha^2} + \frac{y^2}{\beta^2} + \frac{z^2}{\gamma^2} = 1 \tag{7}$$

where α, β, γ are defined as

$$\alpha \triangleq \lambda I_x{}^{-1/2} \qquad \beta \triangleq \lambda I_y{}^{-1/2} \qquad \gamma \triangleq \lambda I_z{}^{-1/2} \tag{8}$$

Equation (7) is the equation of an ellipsoid whose axes are X, Y, Z, that is, the principal axes of S for O, and whose semidiameters have lengths α, β, γ given by Eqs. (8). Now, the naming of the axes always can be accomplished in such a way that $I_z \geq I_y \geq I_x$ or, in view of Eqs. (8), that $\gamma \leq \beta \leq \alpha$, and the distance R from the center O to any point P of an ellipsoid is never smaller than the smallest semidiameter, and never larger than the largest semidiameter, which then means that

$$\gamma \leq R \leq \alpha \tag{9}$$

Using Eqs. (1) and (8) to eliminate R, α, and γ gives

$$I_z{}^{-1/2} \leq I_a{}^{-1/2} \leq I_x{}^{-1/2} \tag{10}$$

or, equivalently,

$$I_z \geq I_a \geq I_x \tag{11}$$

which shows that the moment of inertia of S about a line that passes through O and is not a principal axis of S for O cannot be smaller than the smallest, or larger than the largest, principal moment of inertia of S for O. Finally, the moment of inertia of S about a line that does not pass through the mass center of S always exceeds the moment of inertia of S about a parallel line that does pass through the mass center [see Eq. (4.6.5)]. Hence, no moment of inertia of S can be smaller than the smallest central principal moment of inertia of S.

Example When a uniform rectangular plate B of mass m has the dimensions shown in Fig. 4.8.2, its principal moments of inertia for point O have the values $0.126mL^2$, $1.54mL^2$, and $1.67mL^2$ [see Eqs. (4.8.49) and (4.8.50)], and X, Y, and Z, the respective principal axes of B for O, are oriented as shown in Fig. 4.8.3. Hence, if the constant λ in Eqs. (8) is assigned the value

$$\lambda = \frac{L^2\sqrt{m}}{2} \tag{12}$$

then the principal semidiameters of the associated inertia ellipsoid of B for O have lengths α, β, and γ given, in accordance with Eqs. (8), by

$$\alpha = \frac{L}{2\sqrt{0.126}} = 1.4L \qquad \beta = 0.40L \qquad \gamma = 0.39L \tag{13}$$

and the ellipsoid appears as shown in Fig. 4.9.2. Furthermore, the moment of inertia of B about line X is smaller than the moment of inertia of B about any other line passing through point O. This fact is of practical interest in the following situation. Suppose that a shaft S is fixed in a reference frame A, and it is desired to mount B on S in such a way that the axis of S passes through O while the magnitude of a force applied to B normal to the plane of B in order to impart to B any angular acceleration in A is as small as possible. Since the magnitude of the force is proportional to the moment of inertia of B about the axis of S, this axis must be made to coincide with X.

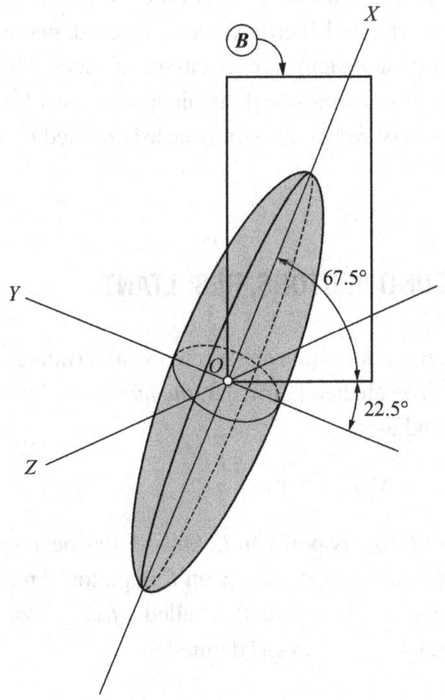

Figure 4.9.2

5 GENERALIZED FORCES

The necessity to cross-multiply a vector $\mathbf{v}$ with the position vector $\mathbf{p}^{AB}$ from a point A to a point B arises frequently [for example, see Eqs. (2.7.1) and (4.5.27)]. Now, $\mathbf{p}^{AB} \times \mathbf{v} = \mathbf{p}^{AC} \times \mathbf{v}$, where $\mathbf{p}^{AC}$ is the position vector from A to *any* point C of the line L that is parallel to $\mathbf{v}$ and passes through B; and, when C is chosen properly, it may be easier to evaluate $\mathbf{p}^{AC} \times \mathbf{v}$ than $\mathbf{p}^{AB} \times \mathbf{v}$. This fact provides the motivation for introducing the concepts of "bound" vectors and "moments" of such vectors as in Sec. 5.1. The terms "couple" and "torque," which have to do with special sets of bound vectors, are defined in Sec. 5.2, and the concepts of "equivalence" and "replacement," each of which involves two sets of bound vectors, are discussed in Sec. 5.3. This material then is used throughout the rest of the chapter to facilitate the forming of expressions for quantities that play a preeminent role in connection with dynamical equations of motion, namely, two kinds of generalized forces. Sections 5.4–5.8 deal with generalized active forces, which come into play whenever the particles of a system are subject to the actions of contact and/or distance forces. Generalized inertia forces, which depend on both the motion and the mass distribution of a system, are discussed in Sec. 5.9. Mastery of the material brings one into position to formulate dynamical equations for any system possessing a finite number of degrees of freedom, as may be ascertained by reading Sec. 8.1.

5.1 MOMENT ABOUT A POINT, BOUND VECTORS, RESULTANT

Of the infinitely many lines that are parallel to every vector $\mathbf{v}$, a particular one, say, L, called the *line of action of* $\mathbf{v}$, must be selected before $\mathbf{M}$, *the moment of* $\mathbf{v}$ *about a point* P, can be evaluated, for $\mathbf{M}$ is defined as

$$\mathbf{M} \triangleq \mathbf{p} \times \mathbf{v} \tag{1}$$

where $\mathbf{p}$ is the position vector from P to *any* point on L. Once L has been specified, $\mathbf{v}$ is said to be a *bound vector*, and it is customary to show $\mathbf{v}$ on L in pictorial representations of $\mathbf{v}$. A vector for which no line of action is specified is called a *free vector*.

The *resultant* $\mathbf{R}$ of a set S of vectors $\mathbf{v}_1, \ldots, \mathbf{v}_\nu$ is defined as

$$\mathbf{R} \triangleq \sum_{i=1}^{\nu} \mathbf{v}_i \tag{2}$$

and, if $\mathbf{v}_1, \ldots, \mathbf{v}_\nu$ are bound vectors, the sum of their moments about a point P is called *the moment of S about P*.

At times, it is convenient to regard the resultant $\mathbf{R}$ of a set S of bound vectors $\mathbf{v}_1, \ldots, \mathbf{v}_\nu$ as a bound vector. Suppose, for example, that $\mathbf{M}^{S/P}$ and $\mathbf{M}^{S/Q}$ denote the moments of S about points P and Q, respectively, and $\mathbf{R}$ is regarded as a bound vector whose line of action passes through Q. Then one can find $\mathbf{M}^{S/P}$ simply by adding to $\mathbf{M}^{S/Q}$ the moment of $\mathbf{R}$ about P, for

$$\mathbf{M}^{S/P} = \mathbf{M}^{S/Q} + \mathbf{r}^{PQ} \times \mathbf{R} \tag{3}$$

where $\mathbf{r}^{PQ}$ is the position vector from P to Q.

Derivation Let $\mathbf{p}_i$ and $\mathbf{q}_i$ be the position vectors from P and Q, respectively, to a point on the line of action L_i of $\mathbf{v}_i$ $(i = 1, \ldots, \nu)$, and let $\mathbf{r}^{PQ}$ be the position vector from P to Q, as shown in Fig. 5.1.1. Then, by definition,

$$\mathbf{M}^{S/P} = \sum_{i=1}^{\nu} \mathbf{p}_i \times \mathbf{v}_i \tag{4}$$

and

$$\mathbf{M}^{S/Q} = \sum_{i=1}^{\nu} \mathbf{q}_i \times \mathbf{v}_i \tag{5}$$

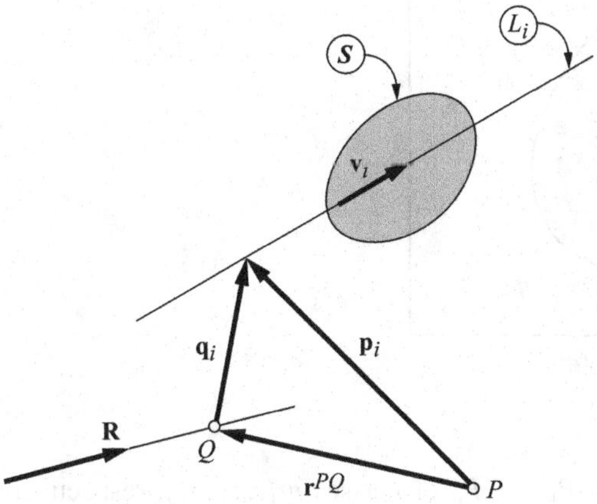

Figure 5.1.1

Now (see Fig. 5.1.1),

$$\mathbf{p}_i = \mathbf{r}^{PQ} + \mathbf{q}_i \qquad (i = 1, \ldots, \nu) \tag{6}$$

Hence,

$$\mathbf{M}^{S/P} \underset{(4)}{=} \sum_{i=1}^{\nu} (\mathbf{r}^{PQ} \underset{(6)}{+} \mathbf{q}_i) \times \mathbf{v}_i = \mathbf{r}^{PQ} \times \sum_{i=1}^{\nu} \mathbf{v}_i + \sum_{i=1}^{\nu} \mathbf{q}_i \times \mathbf{v}_i$$
$$= \mathbf{r}^{PQ} \times \underset{(2)}{\mathbf{R}} + \underset{(5)}{\mathbf{M}^{S/Q}} \qquad (7)$$

This establishes the validity of Eq. (3). If **R** is regarded as a bound vector whose line of action passes through Q, then, in accordance with Eq. (1), $\mathbf{r}^{PQ} \times \mathbf{R}$ is the moment of **R** about P.

Examples In Fig. 5.1.2, P and Q are two points of a rigid body B that is moving in a reference frame A. If the angular velocity ${}^A\boldsymbol{\omega}^B$ of B in A is regarded as a bound vector whose line of action passes through Q, then **M**, the moment of ${}^A\boldsymbol{\omega}^B$ about point P, is given by

$$\mathbf{M} \underset{(1)}{=} -\mathbf{r} \times {}^A\boldsymbol{\omega}^B \qquad (8)$$

where **r** is the position vector from Q to P, as shown in Fig. 5.1.2. This moment vector is equal to the difference between the velocities of P and Q in A, for

$$\mathbf{M} \underset{(8)}{=} {}^A\boldsymbol{\omega}^B \times \mathbf{r} \underset{(2.7.1)}{=} {}^A\mathbf{v}^P - {}^A\mathbf{v}^Q \qquad (9)$$

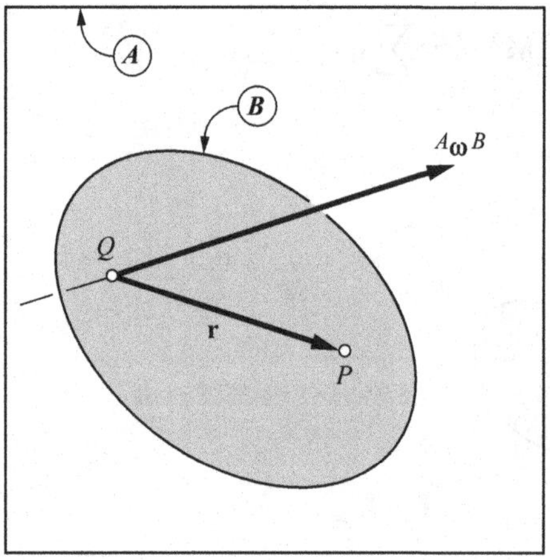

Figure 5.1.2

If S is a set of particles $P_1, \ldots, P_\nu$ of masses $m_1, \ldots, m_\nu$, respectively, moving in a reference frame A with velocities ${}^A\mathbf{v}^{P_1}, \ldots, {}^A\mathbf{v}^{P_\nu}$, then the vector $m_i {}^A\mathbf{v}^{P_i}$ is called the *linear momentum* of P_i in A ($i = 1, \ldots, \nu$), and, if this vector is assigned a line of action passing through P_i, the sum of the moments of the linear momenta of $P_1, \ldots, P_\nu$ in A about any point O is equal to the angular momentum of S relative to O in A, a conclusion that follows from Eq. (4.5.27) together with the definition of the moment of a set of bound vectors about a point.

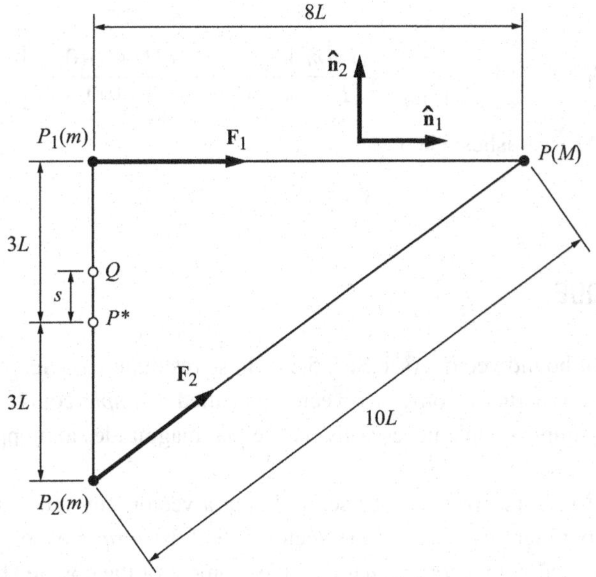

Figure 5.1.3

The observation just made sheds light on the usage of the phrase "moment of momentum" in place of angular momentum, and it enables one to appeal to Eq. (3) to establish effortlessly the following useful proposition. If the linear momentum of S in A, defined as the resultant of the linear momenta of $P_1, \ldots, P_\nu$ in A, is regarded as a bound vector whose line of action passes through an arbitrary point Q, then the angular momentum of S relative to any other point O is equal to the sum of the angular momentum of S relative to Q in A and the moment, about point O, of the linear momentum of S in A.

As a final example, consider the set of three particles P_1, P_2, and P shown in Fig. 5.1.3, and let P_1 and P_2 each have a mass m while P has a mass M. In accordance with Newton's law of gravitation, the gravitational forces, $\mathbf{F}_1$ and $\mathbf{F}_2$, exerted, respectively, on P_1 and P_2 by P have the lines of action shown in Fig. 5.1.3 and are given by

$$\mathbf{F}_1 = \frac{GMm}{64L^2}\hat{\mathbf{n}}_1 \qquad \mathbf{F}_2 = \frac{GMm}{100L^2}(0.8\hat{\mathbf{n}}_1 + 0.6\hat{\mathbf{n}}_2) \tag{10}$$

where G is the universal gravitational constant,† and $\hat{\mathbf{n}}_1$ and $\hat{\mathbf{n}}_2$ are unit vectors directed as shown. Under these circumstances, there exists a point Q on line $P^\star P_1$, where $P^\star$ is the mass center of P_1 and P_2, such that $\mathbf{M}^Q$, the resultant of the moments of $\mathbf{F}_1$ and $\mathbf{F}_2$ about Q, is equal to zero. To locate Q, let s be the distance from $P^\star$ to Q, as shown in Fig. 5.1.3, and let $\mathbf{p}_1$ and $\mathbf{p}_2$ be the position vectors from Q to P_1 and P_2, respectively. Then

$$\mathbf{p}_1 = (3L - s)\hat{\mathbf{n}}_2 \qquad \mathbf{p}_2 = -(3L + s)\hat{\mathbf{n}}_2 \tag{11}$$

† $G \approx 6.6732 \times 10^{-11}$ N m^2 kg^{-2} (see E. A. Mechtly, "The International System of Units: Physical Constants and Conversion Factors," NASA SP-7012, revised, 1969).

and

$$\mathbf{M}^Q = \mathbf{p}_1 \times \mathbf{F}_1 + \mathbf{p}_2 \times \mathbf{F}_2 \underset{(11,\,10)}{=} -\frac{GMm}{L^2}\left[\frac{3L-s}{64} - \frac{(3L+s)(0.8)}{100}\right]\hat{\mathbf{n}}_1 \times \hat{\mathbf{n}}_2 \quad (12)$$

Consequently, $\mathbf{M}^Q$ vanishes when

$$s = 0.968L \quad (13)$$

5.2 COUPLES, TORQUE

A *couple* is a set of bound vectors (see Sec. 5.1) whose resultant (see Sec. 5.1) is equal to zero. A couple consisting of only two vectors is called a *simple* couple. Hence, the vectors forming a simple couple necessarily have equal magnitudes and opposite directions.

Couples are not vectors, for a set of vectors is not a vector, any more than a set of points is a point; but there exists a unique vector, called the *torque* of the couple, that is intimately associated with a couple, namely, the moment of the couple about a point. It is unique because, as can be seen by reference to Eq. (5.1.3), a couple has the same moment about *all* points.

Example

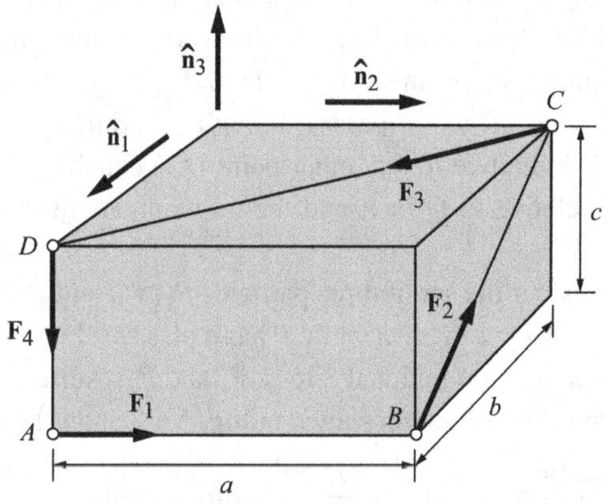

Figure 5.2.1

Four forces, $\mathbf{F}_1, \ldots, \mathbf{F}_4$, have the lines of action shown in Fig. 5.2.1, and the magnitudes of $\mathbf{F}_1, \ldots, \mathbf{F}_4$ are proportional to the lengths of the lines AB, BC, CD, and DA, respectively; that is,

$$\mathbf{F}_1 = ka\hat{\mathbf{n}}_2 \qquad \mathbf{F}_2 = k(-b\hat{\mathbf{n}}_1 + c\hat{\mathbf{n}}_3) \quad (1)$$

$$\mathbf{F}_3 = k(b\hat{\mathbf{n}}_1 - a\hat{\mathbf{n}}_2) \qquad \mathbf{F}_4 = -kc\hat{\mathbf{n}}_3 \quad (2)$$

where k is a constant and $\hat{\mathbf{n}}_1$, $\hat{\mathbf{n}}_2$, $\hat{\mathbf{n}}_3$ are mutually perpendicular unit vectors. The forces $\mathbf{F}_1,\ldots,\mathbf{F}_4$ form a couple since their resultant, $\mathbf{F}_1 + \mathbf{F}_2 + \mathbf{F}_3 + \mathbf{F}_4$, is equal to zero. The torque $\mathbf{T}$ of the couple, found, for example, by adding the moments of $\mathbf{F}_1,\ldots,\mathbf{F}_4$ about point C, is given by (note that the moments of $\mathbf{F}_2$ and $\mathbf{F}_3$ about C are equal to zero)

$$\begin{aligned}\mathbf{T} &= (-c\hat{\mathbf{n}}_3 + b\hat{\mathbf{n}}_1) \times \mathbf{F}_1 + (b\hat{\mathbf{n}}_1 - a\hat{\mathbf{n}}_2) \times \mathbf{F}_4 \\ &\underset{(1,\,2)}{=} k(2ca\hat{\mathbf{n}}_1 + bc\hat{\mathbf{n}}_2 + ab\hat{\mathbf{n}}_3) \qquad (3)\end{aligned}$$

5.3 EQUIVALENCE, REPLACEMENT

Two sets of bound vectors are said to be *equivalent* when they have equal resultants and equal moments about one point; either set then is called a *replacement* of the other.

Two couples having equal torques are equivalent, for they necessarily have equal resultants (zero; see Sec. 5.2) and their respective moments about every point are equal to each other because each such moment is equal to the torque of the corresponding couple.

When two sets of bound vectors, say, S and S', are equivalent, they have equal moments about *every* point. To see this, let S and S' have equal resultants, and let Q be *one* point about which S and S' have equal moments; then note that, in accordance with Eq. (5.1.3), the moments of S and S' about any point P other than Q depend solely on the moments of S and S' about Q, equal by hypothesis, and on the resultants of S and S', also equal by hypothesis.

If S is any set of bound vectors while S' is a set of bound vectors consisting of a couple C of torque $\mathbf{T}$ together with a single bound vector $\mathbf{v}$ whose line of action passes through a point P selected arbitrarily, then the following two requirements must be satisfied in order for S' to be a replacement of S: $\mathbf{T}$ is equal to the moment of S about P, and $\mathbf{v}$ is equal to the resultant of S. For, when one of these requirements is violated, S and S' either have unequal resultants or there exists no point about which S and S' have equal moments. Conversely, satisfying both requirements guarantees the equivalence of S and S'. These facts enable one to deal in simple analytical terms with sets of bound vectors, such as contact forces exerted by one body on another, in situations in which little is known about the individual vectors of such a set.

Example Figure 5.3.1 is a schematic representation of a device known as Hooke's joint, described as follows. Two shafts, S and S', are mounted in bearings, B and B', which are fixed in a reference frame R. The axes of S and S' are parallel to unit vectors $\hat{\mathbf{n}}$ and $\hat{\mathbf{n}}'$, respectively, and intersect at a point A. Each shaft terminates in a "yoke," and these yokes, Y and Y', are connected to each other by a rigid cross C, one of whose arms is supported by bearings at D and E in Y, the other by bearings at D' and E' in Y'. The two arms of C have equal lengths and form a right angle with each other. Furthermore, the arm supported by Y is perpendicular to $\hat{\mathbf{n}}$, while the one supported by Y' is perpendicular to $\hat{\mathbf{n}}'$. Finally, circular disks G and G' having radii r and r' are attached to S and S', respectively, and each of these is subjected to the

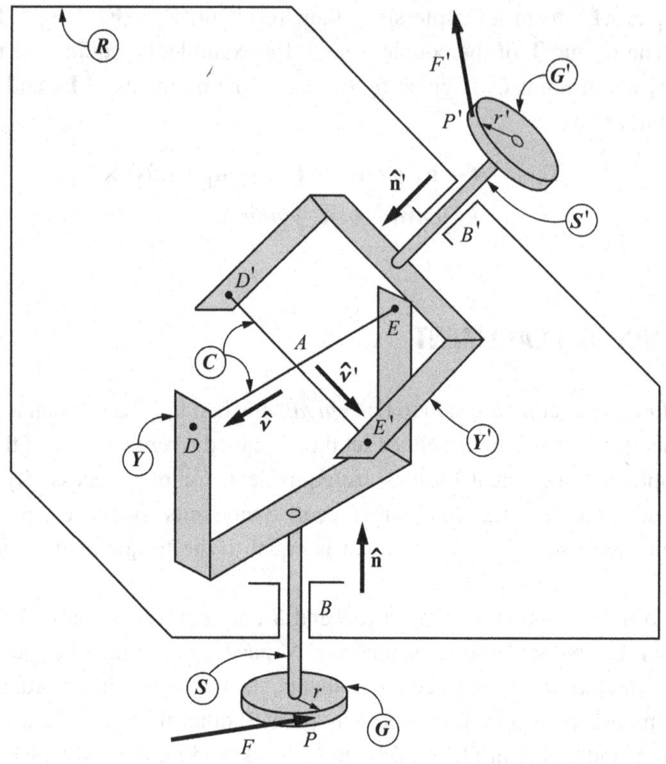

Figure 5.3.1

action of a tangential force, one having a magnitude F and point of application P, the other a magnitude of F' and point of application P'.

In order for the system formed by G, G', S, S', Y, Y', and C to be in equilibrium, the ratio F/F' must be related suitably to r, r', $\hat{\mathbf{n}}$, $\hat{\mathbf{n}}'$, $\hat{\nu}$, and $\hat{\nu}'$, where $\hat{\nu}$ and $\hat{\nu}'$ are unit vectors parallel to the arms of C, as shown in Fig. 5.3.1. To determine this relationship, one may consider the equilibrium of each of three rigid bodies, namely, the rigid body Z formed by G, S, and Y, the rigid body Z' consisting of G', S', and Y', and the body C. The first of these is depicted in Fig. 5.3.2, where the vectors $\boldsymbol{\alpha}$, $\mathbf{A}$, $\boldsymbol{\beta}$, and $\mathbf{B}$ are associated with replacements of sets of contact forces exerted on Z by C and by the bearing at B. Specifically, the set of contact forces exerted on Z by C across the bearing surfaces at D and E is replaced with a couple of torque $\boldsymbol{\alpha}$ together with a force $\mathbf{A}$ whose line of action passes through point A, and the set of contact forces exerted on Z by the bearing at B is replaced with a couple of torque $\boldsymbol{\beta}$ together with a force $\mathbf{B}$ whose line of action passes through the center of G. The vectors $\boldsymbol{\alpha}$, $\mathbf{A}$, $\boldsymbol{\beta}$, and $\mathbf{B}$ are unknown, but, assuming that all bearing surfaces are so smooth that the associated contact forces have lines of action normal to the surfaces,

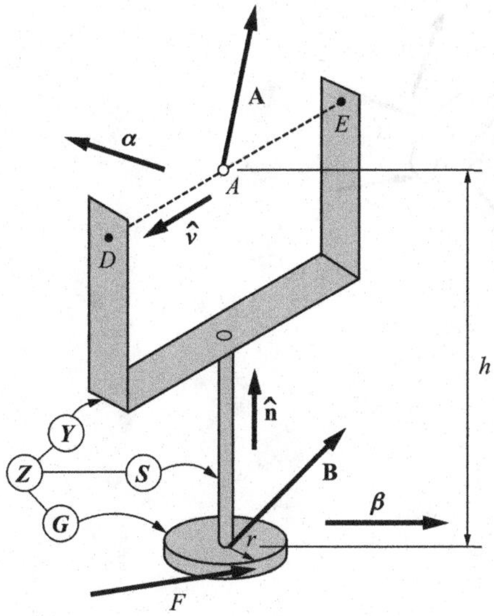

Figure 5.3.2

one can conclude that $\boldsymbol{\alpha}$ must be perpendicular to $\hat{\nu}$, and $\boldsymbol{\beta}$ to $\hat{\mathbf{n}}$, so that

$$\hat{\nu} \cdot \boldsymbol{\alpha} = 0 \tag{1}$$

and

$$\hat{\mathbf{n}} \cdot \boldsymbol{\beta} = 0 \tag{2}$$

Now, when Z is in equilibrium, the sum of the moments of all forces acting on Z about the center of G is equal to zero. Hence, treating gravitational forces as negligible, one can write (see Fig. 5.3.2 for h)

$$rF\hat{\mathbf{n}} + \boldsymbol{\alpha} + \boldsymbol{\beta} + h\hat{\mathbf{n}} \times \mathbf{A} = \mathbf{0} \tag{3}$$

and dot multiplication of this equation with $\hat{\mathbf{n}}$ yields [when Eq. (2) is taken into account]

$$rF + \hat{\mathbf{n}} \cdot \boldsymbol{\alpha} = 0 \tag{4}$$

The body Z' and vectors $\boldsymbol{\alpha}'$, $\mathbf{A}'$, $\boldsymbol{\beta}'$, and $\mathbf{B}'$ respectively analogous to $\boldsymbol{\alpha}$, $\mathbf{A}$, $\boldsymbol{\beta}$, and $\mathbf{B}$ are shown in Fig. 5.3.3. Reasoning as before, one finds that

$$\hat{\nu}' \cdot \boldsymbol{\alpha}' = 0 \tag{5}$$

$$\hat{\mathbf{n}}' \cdot \boldsymbol{\beta}' = 0 \tag{6}$$

and

$$r'F' + \hat{\mathbf{n}}' \cdot \boldsymbol{\alpha}' = 0 \tag{7}$$

Finally, in Fig. 5.3.4, C is shown together with vectors representing the sets of

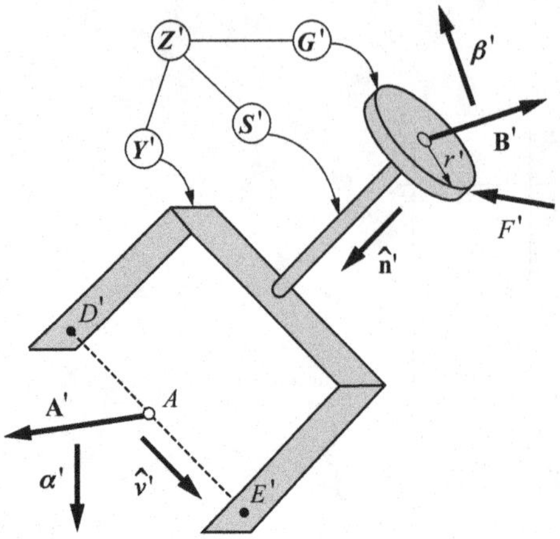

Figure 5.3.3

contact forces exerted on C by Y and Y'. These vectors are labeled $-\boldsymbol{\alpha}$, $-\boldsymbol{\alpha}'$, $-\mathbf{A}$, and $-\mathbf{A}'$ in accordance with the law of action and reaction; that is, since to every force exerted on Y by C there corresponds a force exerted on C by Y, these two forces

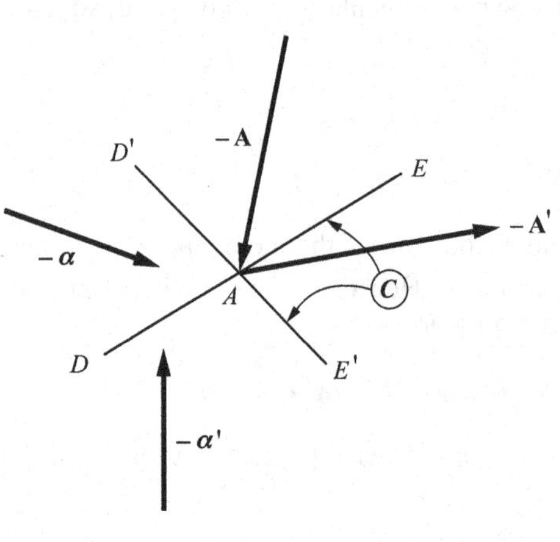

Figure 5.3.4

having equal magnitudes, the same line of action, but opposite senses, it follows from the definition of a replacement of a set of bound vectors that, if the set of forces exerted on Y by C is replaced with a couple of torque $\boldsymbol{\alpha}$ together with a force $\mathbf{A}$ whose line of action passes through point A, while the set of forces exerted on C by Y is replaced with a couple of torque $\tilde{\boldsymbol{\alpha}}$ together with a force $\tilde{\mathbf{A}}$ whose line of action passes through point A, then $\tilde{\boldsymbol{\alpha}} = -\boldsymbol{\alpha}$ and $\tilde{\mathbf{A}} = -\mathbf{A}$. Similar considerations apply to the interaction of C with Y'.

Because C is presumed to be in equilibrium, the sum of the moments about point A of all forces acting on C may be set equal to zero, so that, in view of Fig. 5.3.4, one can write

$$-\boldsymbol{\alpha} - \boldsymbol{\alpha}' = \mathbf{0} \tag{8}$$

Using this equation to eliminate $\boldsymbol{\alpha}'$ from Eqs. (5) and (7), one arrives at

$$\hat{\nu}' \cdot \boldsymbol{\alpha} \underset{(8)\ (5)}{=} 0 \tag{9}$$

and

$$r'F' - \hat{\mathbf{n}}' \cdot \boldsymbol{\alpha} \underset{(8)\ (7)}{=} 0 \tag{10}$$

Furthermore, Eqs. (1) and (9) show that $\boldsymbol{\alpha}$ is perpendicular both to $\hat{\nu}$ and to $\hat{\nu}'$, which means that there exists a quantity λ such that

$$\boldsymbol{\alpha} = \lambda \hat{\nu} \times \hat{\nu}' \tag{11}$$

Consequently,

$$rF \underset{(4,\ 11)}{=} -\lambda \hat{\mathbf{n}} \cdot \hat{\nu} \times \hat{\nu}' \tag{12}$$

while

$$r'F' \underset{(10,\ 11)}{=} \lambda \hat{\mathbf{n}}' \cdot \hat{\nu} \times \hat{\nu}' \tag{13}$$

Hence, the desired expression for F/F' is

$$\frac{F}{F'} \underset{(12,\ 13)}{=} -\frac{r'\hat{\mathbf{n}} \cdot \hat{\nu} \times \hat{\nu}'}{r\hat{\mathbf{n}}' \cdot \hat{\nu} \times \hat{\nu}'} \tag{14}$$

What is most important here is to realize that it was possible to determine the relationship between F/F' and r, r', $\hat{\mathbf{n}}$, $\hat{\mathbf{n}}'$, $\hat{\nu}$, and $\hat{\nu}'$ despite the fact that relatively little is known about the contact forces exerted on S by the bearing B, on Y by C, and so forth, and that the introduction of torques together with forces having well-defined lines of action greatly facilitated the solution process.

5.4 GENERALIZED ACTIVE FORCES

Let S denote a system composed of particles $P_1, \ldots, P_\nu$. When S is a holonomic system possessing n degrees of freedom in a reference frame A (see Sec. 3.5), and $u_1, \ldots, u_n$ are generalized velocities for S in A, then n quantities $F_1, \ldots, F_n$ called *holonomic generalized active forces* for S in A are defined as

$$F_r \triangleq \sum_{i=1}^{\nu} \mathbf{v}_r^{P_i} \cdot \mathbf{R}_i \qquad (r = 1, \ldots, n) \tag{1}$$

where $\mathbf{v}_r^{P_i}$ is a holonomic partial velocity of P_i in A (see Sec. 3.6), and $\mathbf{R}_i$ is the resultant (see Sec. 5.1) of all contact forces (for example, friction forces) and distance forces (for example, gravitational forces, magnetic forces, and so forth) acting on P_i.

When S is a simple nonholonomic system possessing p degrees of freedom in A

(see Sec. 3.5), and $u_1, \ldots, u_p$ are generalized velocities for S in A, then p quantities $\tilde{F}_1, \ldots, \tilde{F}_p$ called *nonholonomic generalized active forces* for S in A are defined as

$$\tilde{F}_r \triangleq \sum_{i=1}^{\nu} \tilde{\mathbf{v}}_r^{P_i} \cdot \mathbf{R}_i \qquad (r = 1, \ldots, p) \tag{2}$$

where $\tilde{\mathbf{v}}_r^{P_i}$ is a nonholonomic partial velocity of P_i in A (see Sec. 3.6).

When S is a complex nonholonomic system possessing c degrees of freedom in A (see Sec. 3.7), and $\dot{u}_1, \ldots, \dot{u}_c$ are time derivatives of motion variables for S in A, then c quantities $\tilde{\tilde{F}}_1, \ldots, \tilde{\tilde{F}}_c$ are defined as

$$\tilde{\tilde{F}}_r \triangleq \sum_{i=1}^{\nu} \tilde{\mathbf{a}}_r^{P_i} \cdot \mathbf{R}_i \qquad (r = 1, \ldots, c) \tag{3}$$

where $\tilde{\mathbf{a}}_r^{P_i}$ is a nonholonomic partial acceleration of P_i in A (see Sec. 3.8). The term "nonholonomic generalized active force" may also be used in connection with $\tilde{\tilde{F}}_1, \ldots, \tilde{\tilde{F}}_c$.

As in the case of holonomic and nonholonomic partial angular velocities and partial velocities, one can generally omit the adjectives "holonomic" and "nonholonomic" when speaking of generalized forces, but notation involving tildes should be used in writing to distinguish the three kinds of generalized active forces from one another.

The generalized active forces for S in A defined in Eqs. (1) and (2) are related to each other and to the quantities A_{rs} $(s = 1, \ldots, p;\ r = p + 1, \ldots, n)$ introduced in Eqs. (3.5.2), as follows:

$$\tilde{F}_r = F_r + \sum_{s=p+1}^{n} F_s A_{sr} \qquad (r = 1, \ldots, p) \tag{4}$$

Likewise, there exist relationships involving the generalized active forces for S in A defined in Eqs. (2) and (3), and the quantities $\tilde{A}_{rs}$ $(s = 1, \ldots, c;\ r = c + 1, \ldots, p)$ appearing in Eqs. (3.7.1):

$$\tilde{\tilde{F}}_r = \tilde{F}_r + \sum_{s=c+1}^{p} \tilde{F}_s \tilde{A}_{sr} \qquad (r = 1, \ldots, c) \tag{5}$$

Derivation Referring to Eq. (3.6.17) to express $\tilde{\mathbf{v}}_r^{P_i}$ $(r = 1, \ldots, p)$ in terms of $\mathbf{v}_s^{P_i}$ $(s = p + 1, \ldots, n)$, one has

$$\begin{aligned} \tilde{F}_r &\underset{(2)}{=} \sum_{i=1}^{\nu} \left(\mathbf{v}_r^{P_i} + \sum_{s=p+1}^{n} \mathbf{v}_s^{P_i} A_{sr} \right) \cdot \mathbf{R}_i \\ &= \sum_{i=1}^{\nu} \mathbf{v}_r^{P_i} \cdot \mathbf{R}_i + \sum_{s=p+1}^{n} \left(\sum_{i=1}^{\nu} \mathbf{v}_s^{P_i} \cdot \mathbf{R}_i \right) A_{sr} \qquad (r = 1, \ldots, p) \end{aligned} \tag{6}$$

and use of Eqs. (1) then leads immediately to Eqs. (4).

By appealing to Eqs. (3.8.15) and (3.8.9), one can write

$$\tilde{\mathbf{a}}_r^{P_i} \underset{(3.8.15)}{=} \mathbf{a}_r^{P_i} + \sum_{s=c+1}^{p} \mathbf{a}_s^{P_i} \tilde{A}_{sr} \underset{(3.8.9)}{=} \tilde{\mathbf{v}}_r^{P_i} + \sum_{s=c+1}^{p} \tilde{\mathbf{v}}_s^{P_i} \tilde{A}_{sr} \qquad (r = 1, \ldots, c) \tag{7}$$

so that

$$\underset{(3)}{\tilde{\tilde{F}}_r =} \sum_{i=1}^{\nu} \left(\tilde{\mathbf{v}}_r^{P_i} + \underset{(7)}{\sum_{s=c+1}^{p}} \tilde{\mathbf{v}}_s^{P_i} \tilde{A}_{sr} \right) \cdot \mathbf{R}_i$$

$$= \sum_{i=1}^{\nu} \tilde{\mathbf{v}}_r^{P_i} \cdot \mathbf{R}_i + \sum_{s=c+1}^{p} \left(\sum_{i=1}^{\nu} \tilde{\mathbf{v}}_s^{P_i} \cdot \mathbf{R}_i \right) \tilde{A}_{sr} \qquad (r = 1, \ldots, c) \tag{8}$$

Substitution from Eqs. (2) allows one to proceed directly to Eqs. (5).

Example In Fig. 5.4.1, P_1 and P_2 designate particles of masses m_1 and m_2 that can slide freely in a smooth tube T and are attached to light linear springs σ_1 and σ_2 having spring constants k_1 and k_2 and "natural" lengths L_1 and L_2. T is made to rotate about a fixed horizontal axis passing through one end of T, in such a way that the angle between the vertical and the axis of T is a prescribed function $\theta(t)$ of the time t. Generalized active forces F_1 and F_2 associated with generalized velocities u_1 and u_2 defined as

$$u_r \triangleq \dot{q}_r \qquad (r = 1, 2) \tag{9}$$

are to be determined for the system S formed by the two particles P_1 and P_2, with q_1 and q_2 (see Fig. 5.4.1) designating the displacements of P_1 and P_2 from the positions occupied by P_1 and P_2 in T when σ_1 and σ_2 are undeformed.

The velocities $\mathbf{v}^{P_1}$ and $\mathbf{v}^{P_2}$ of P_1 and P_2 can be expressed as (see Fig. 5.4.1 for the unit vectors $\hat{\mathbf{t}}_1$ and $\hat{\mathbf{t}}_2$)

$$\mathbf{v}^{P_1} \underset{(2.8.1)}{=} u_1 \hat{\mathbf{t}}_1 + (L_1 + q_1)\dot{\theta}\,\hat{\mathbf{t}}_2 \tag{10}$$

$$\mathbf{v}^{P_2} \underset{(2.8.1)}{=} u_2 \hat{\mathbf{t}}_1 + (L_1 + L_2 + q_2)\dot{\theta}\,\hat{\mathbf{t}}_2 \tag{11}$$

S is a holonomic system, and the partial velocities of P_1 and P_2 are

$$\mathbf{v}_1^{P_1} \underset{(10)}{=} \hat{\mathbf{t}}_1 \qquad \mathbf{v}_2^{P_1} \underset{(10)}{=} \mathbf{0} \tag{12}$$

$$\mathbf{v}_1^{P_2} \underset{(11)}{=} \mathbf{0} \qquad \mathbf{v}_2^{P_2} \underset{(11)}{=} \hat{\mathbf{t}}_1 \tag{13}$$

Contact forces are applied to P_1 by σ_1, σ_2, and T, and a gravitational force is exerted on P_1 by the Earth. The forces applied to P_1 by σ_1 and σ_2 can be written $-k_1 q_1 \hat{\mathbf{t}}_1$ and $k_2(q_2 - q_1)\hat{\mathbf{t}}_1$, respectively, and the force exerted on P_1 by T can be expressed as $T_{12}\hat{\mathbf{t}}_2 + T_{13}\hat{\mathbf{t}}_3$, where T_{12} and T_{13} are unknown scalars. No component parallel to $\hat{\mathbf{t}}_1$ is included because T is presumed to be smooth. Finally, the gravitational force exerted on P_1 by the Earth is $m_1 g \hat{\mathbf{k}}$, where g is the local gravitational force per unit mass and $\hat{\mathbf{k}}$ is a unit vector directed vertically downward. Hence, if the gravitational force exerted on P_1 by P_2 is ignored, $\mathbf{R}_1$, the resultant of all contact and distance forces acting on P_1, is given by

$$\mathbf{R}_1 = -k_1 q_1 \hat{\mathbf{t}}_1 + k_2(q_2 - q_1)\hat{\mathbf{t}}_1 + T_{12}\hat{\mathbf{t}}_2 + T_{13}\hat{\mathbf{t}}_3 + m_1 g \hat{\mathbf{k}} \tag{14}$$

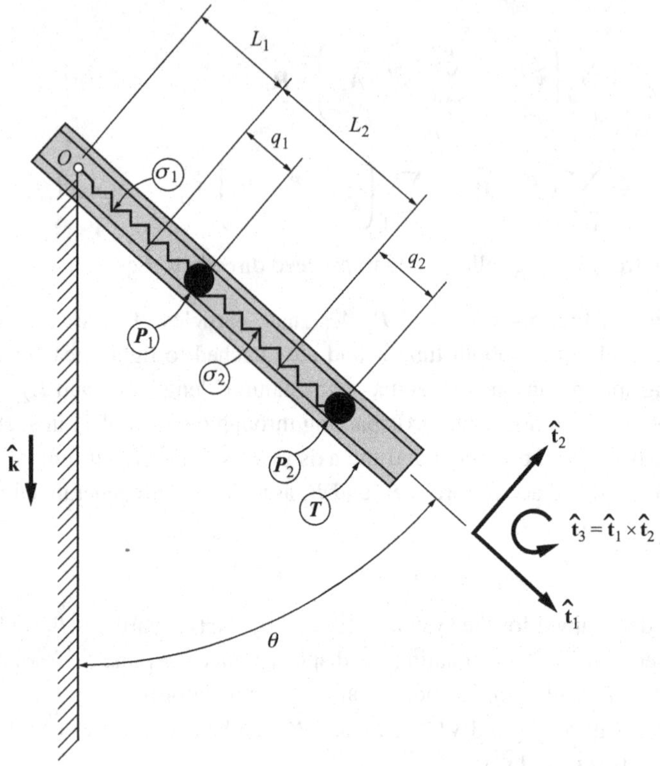

Figure 5.4.1

and $(F_1)_{P_1}$, the contribution to the generalized active force F_1 of all contact and distance forces acting on P_1, is

$$(F_1)_{P_1} \underset{(1)}{=} \mathbf{v}_1^{P_1} \cdot \mathbf{R}_1 \underset{(12,\ 14)}{=} -k_1 q_1 + k_2(q_2 - q_1) + m_1 g \cos\theta \tag{15}$$

Similarly, the resultant $\mathbf{R}_2$ of all contact and distance forces acting on P_2 is

$$\mathbf{R}_2 = -k_2(q_2 - q_1)\hat{\mathbf{t}}_1 + T_{22}\hat{\mathbf{t}}_2 + T_{23}\hat{\mathbf{t}}_3 + m_2 g\hat{\mathbf{k}} \tag{16}$$

so that $(F_1)_{P_2}$, the contribution to the generalized active force F_1 of all contact and distance forces acting on P_2, is

$$(F_1)_{P_2} \underset{(1)}{=} \mathbf{v}_1^{P_2} \cdot \mathbf{R}_2 \underset{(13,\ 16)}{=} 0 \tag{17}$$

and, proceeding in the same way, one finds that $(F_2)_{P_1}$ and $(F_2)_{P_2}$, the contributions to the generalized active force F_2 of all contact and distance forces acting on P_1 and P_2, respectively, are

$$(F_2)_{P_1} \underset{(1)}{=} \mathbf{v}_2^{P_1} \cdot \mathbf{R}_1 \underset{(12,\ 14)}{=} 0 \tag{18}$$

and

$$(F_2)_{P_2} \underset{(1)}{=} \mathbf{v}_2^{P_2} \cdot \mathbf{R}_2 \underset{(13,\,16)}{=} -k_2(q_2 - q_1) + m_2 g \cos\theta \tag{19}$$

The desired generalized active forces are thus

$$F_1 = (F_1)_{P_1} + (F_1)_{P_2} \underset{(15,\,17)}{=} -k_1 q_1 + k_2(q_2 - q_1) + m_1 g \cos\theta \tag{20}$$

and

$$F_2 = (F_2)_{P_1} + (F_2)_{P_2} \underset{(18,\,19)}{=} -k_2(q_2 - q_1) + m_2 g \cos\theta \tag{21}$$

It is worth noting that the (unknown) contact forces exerted on P_1 and P_2 by T contribute nothing to the generalized active forces F_1 and F_2.

5.5 FORCES ACTING ON A RIGID BODY

Suppose B is a rigid body belonging to a system S, and a set of contact and/or distance forces acting on B is equivalent (see Sec. 5.3) to a couple of torque $\mathbf{T}$ (see Sec. 5.2) together with a force $\mathbf{R}$ whose line of action passes through a point Q of B. When S is a holonomic, simple nonholonomic, or complex nonholonomic system possessing, respectively, n, p, or c degrees of freedom in a reference frame A, the set of forces acting on B contributes to the generalized active forces defined in Eqs. (5.4.1), (5.4.2), and (5.4.3), respectively, as follows:

$$(F_r)_B = {}^A\boldsymbol{\omega}_r^B \cdot \mathbf{T} + {}^A\mathbf{v}_r^Q \cdot \mathbf{R} \qquad (r = 1,\ldots,n) \tag{1}$$

where ${}^A\boldsymbol{\omega}_r^B$ is a holonomic partial angular velocity of B in A (see Sec. 3.6), and ${}^A\mathbf{v}_r^Q$ is a holonomic partial velocity of Q in A,

$$(\tilde{F}_r)_B = {}^A\tilde{\boldsymbol{\omega}}_r^B \cdot \mathbf{T} + {}^A\tilde{\mathbf{v}}_r^Q \cdot \mathbf{R} \qquad (r = 1,\ldots,p) \tag{2}$$

where ${}^A\tilde{\boldsymbol{\omega}}_r^B$ is a nonholonomic partial angular velocity of B in A (see Sec. 3.6), and ${}^A\tilde{\mathbf{v}}_r^Q$ is a nonholonomic partial velocity of Q in A, and

$$(\tilde{\tilde{F}}_r)_B = {}^A\tilde{\boldsymbol{\alpha}}_r^B \cdot \mathbf{T} + {}^A\tilde{\mathbf{a}}_r^Q \cdot \mathbf{R} \qquad (r = 1,\ldots,c) \tag{3}$$

where ${}^A\tilde{\boldsymbol{\alpha}}_r^B$ is a nonholonomic partial angular acceleration of B in A (see Sec. 3.8), and ${}^A\tilde{\mathbf{a}}_r^Q$ is a nonholonomic partial acceleration of Q in A.

Derivation Let $\mathbf{K}_1,\ldots,\mathbf{K}_\beta$ be the contact and/or distance forces acting on particles $P_1,\ldots,P_\beta$ of B, and let $\mathbf{p}_1,\ldots,\mathbf{p}_\beta$ be the position vectors from Q to $P_1,\ldots,P_\beta$, respectively. Then, by definition of equivalence, the resultant of $\mathbf{K}_1,\ldots,\mathbf{K}_\beta$ is equal to $\mathbf{R}$, that is

$$\sum_{i=1}^{\beta} \mathbf{K}_i \underset{(5.1.2)}{=} \mathbf{R} \tag{4}$$

and the sum of the moments of $\mathbf{K}_1, \ldots, \mathbf{K}_\beta$ about Q is equal to $\mathbf{T}$, so that

$$\sum_{i=1}^{\beta} \mathbf{p}_i \times \mathbf{K}_i \underset{(5.1.1)}{=} \mathbf{T} \tag{5}$$

Also by definition, the contribution of $\mathbf{K}_1, \ldots, \mathbf{K}_\beta$ to F_r is

$$(F_r)_B \underset{(5.4.1)}{=} \sum_{i=1}^{\beta} {}^A\mathbf{v}_r^{P_i} \cdot \mathbf{K}_i \qquad (r = 1, \ldots, n) \tag{6}$$

where, with the aid of Eqs. (2.7.1), (3.6.2), and (3.6.1), one can express ${}^A\mathbf{v}_r^{P_i}$ as

$${}^A\mathbf{v}_r^{P_i} = {}^A\mathbf{v}_r^{Q} + {}^A\boldsymbol{\omega}_r^{B} \times \mathbf{p}_i \qquad (r = 1, \ldots, n;\ i = 1, \ldots, \beta) \tag{7}$$

Hence,

$$\begin{aligned}(F_r)_B &\underset{(6)}{=} \sum_{i=1}^{\beta} ({}^A\mathbf{v}_r^{Q} + {}^A\boldsymbol{\omega}_r^{B} \times \mathbf{p}_i) \cdot \mathbf{K}_i \underset{(7)}{} = {}^A\mathbf{v}_r^{Q} \cdot \sum_{i=1}^{\beta} \mathbf{K}_i + {}^A\boldsymbol{\omega}_r^{B} \cdot \sum_{i=1}^{\beta} \mathbf{p}_i \times \mathbf{K}_i \\ &= {}^A\mathbf{v}_r^{Q} \underset{(4)}{\cdot} \mathbf{R} + {}^A\boldsymbol{\omega}_r^{B} \underset{(5)}{\cdot} \mathbf{T} \qquad (r = 1, \ldots, n)\end{aligned} \tag{8}$$

in agreement with Eq. (1).

Equations (2) can be obtained similarly, by first appealing to Eqs. (5.4.2) and then allowing ${}^A\tilde{\mathbf{v}}_r^{P_i}$, ${}^A\tilde{\mathbf{v}}_r^{Q}$, and ${}^A\tilde{\boldsymbol{\omega}}_r^{B}$ $(r = 1, \ldots, p)$ to play the roles of ${}^A\mathbf{v}_r^{P_i}$, ${}^A\mathbf{v}_r^{Q}$, and ${}^A\boldsymbol{\omega}_r^{B}$ $(r = 1, \ldots, n)$, respectively, in Eqs. (7).

Finally, to derive Eqs. (3), one may begin by noting that the contribution of $\mathbf{K}_1, \ldots, \mathbf{K}_\beta$ to $\tilde{\tilde{F}}_r$ is, by definition,

$$(\tilde{\tilde{F}}_r)_B \underset{(5.4.3)}{=} \sum_{i=1}^{\beta} {}^A\tilde{\mathbf{a}}_r^{P_i} \cdot \mathbf{K}_i \qquad (r = 1, \ldots, c) \tag{9}$$

A relationship for ${}^A\tilde{\mathbf{a}}_r^{P_i}$ similar in form to Eqs. (7) can be obtained from Eq. (2.7.2) after expressing ${}^A\mathbf{a}^{P_i}$ and ${}^A\mathbf{a}^{Q}$ as in Eq. (3.8.4), ${}^A\boldsymbol{\alpha}^{B}$ as in Eq. (3.8.3), observing that motion-variable time derivatives are absent from ${}^A\boldsymbol{\omega}^{B}$ [see Eqs. (3.6.1) and (3.6.3)], and subsequently equating coefficients of $\dot{u}_r$ on both sides of the resulting expression. Hence,

$${}^A\tilde{\mathbf{a}}_r^{P_i} = {}^A\tilde{\mathbf{a}}_r^{Q} + {}^A\tilde{\boldsymbol{\alpha}}_r^{B} \times \mathbf{p}_i \qquad (r = 1, \ldots, c;\ i = 1, \ldots, \beta) \tag{10}$$

Consequently,

$$\begin{aligned}(\tilde{\tilde{F}}_r)_B &\underset{(9)}{=} \sum_{i=1}^{\beta} ({}^A\tilde{\mathbf{a}}_r^{Q} + {}^A\tilde{\boldsymbol{\alpha}}_r^{B} \times \mathbf{p}_i) \cdot \mathbf{K}_i \underset{(10)}{} = {}^A\tilde{\mathbf{a}}_r^{Q} \cdot \sum_{i=1}^{\beta} \mathbf{K}_i + {}^A\tilde{\boldsymbol{\alpha}}_r^{B} \cdot \sum_{i=1}^{\beta} \mathbf{p}_i \times \mathbf{K}_i \\ &= {}^A\tilde{\mathbf{a}}_r^{Q} \underset{(4)}{\cdot} \mathbf{R} + {}^A\tilde{\boldsymbol{\alpha}}_r^{B} \underset{(5)}{\cdot} \mathbf{T} \qquad (r = 1, \ldots, c)\end{aligned} \tag{11}$$

Example Figure 5.5.1 shows a uniform rod B of mass M and length L. B is free to move in a plane A, and P is a particle of mass m, fixed in A.

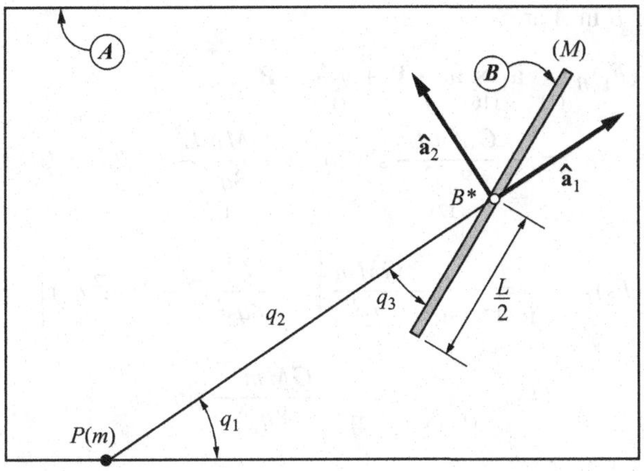

Figure 5.5.1

It can be shown† that the set of gravitational forces exerted by P on B is approximately equivalent to a couple of torque $\mathbf{T}$ together with a force $\mathbf{R}$ whose line of action passes through the mass center $B^\star$ of B, with $\mathbf{T}$ and $\mathbf{R}$ given by

$$\mathbf{T} = -\frac{GMmL^2}{8{q_2}^3} \sin 2q_3 \, \hat{\mathbf{a}}_1 \times \hat{\mathbf{a}}_2 \tag{12}$$

and

$$\mathbf{R} = -\frac{GMm}{{q_2}^2} \left\{ \hat{\mathbf{a}}_1 \left[1 + \frac{L^2}{8{q_2}^2}(2 - 3\sin^2 q_3) \right] - \hat{\mathbf{a}}_2 \frac{L^2}{8{q_2}^2} \sin 2q_3 \right\} \tag{13}$$

where G is the universal gravitational constant, $\hat{\mathbf{a}}_1$ and $\hat{\mathbf{a}}_2$ are unit vectors directed as shown in Fig. 5.5.1, and q_1, q_2, q_3 are generalized coordinates characterizing the configuration of B in A (see Fig. 5.5.1). If motion variables u_1, u_2, u_3 are introduced as

$$u_r \triangleq \dot{q}_r \qquad (r = 1,2,3) \tag{14}$$

then $\boldsymbol{\omega}$, the angular velocity of B in A, and $\mathbf{v}$, the velocity of $B^\star$ in A, are given by

$$\boldsymbol{\omega} = (u_1 + u_3)\hat{\mathbf{a}}_1 \times \hat{\mathbf{a}}_2 \qquad \mathbf{v} = u_2\hat{\mathbf{a}}_1 + u_1 q_2 \hat{\mathbf{a}}_2 \tag{15}$$

so that the partial angular velocities of B in A and the partial velocities of $B^\star$ in A are

$$\boldsymbol{\omega}_1 = \hat{\mathbf{a}}_1 \times \hat{\mathbf{a}}_2 \qquad \boldsymbol{\omega}_2 = \mathbf{0} \qquad \boldsymbol{\omega}_3 = \hat{\mathbf{a}}_1 \times \hat{\mathbf{a}}_2 \tag{16}$$

$$\mathbf{v}_1 = q_2\hat{\mathbf{a}}_2 \qquad \mathbf{v}_2 = \hat{\mathbf{a}}_1 \qquad \mathbf{v}_3 = \mathbf{0} \tag{17}$$

The contributions of the gravitational forces exerted by P on B to the generalized

† T. R. Kane, P. W. Likins, and D. A. Levinson, *Spacecraft Dynamics* (New York: McGraw-Hill, 1983), Secs. 2.3, 2.6.

active forces for B in A are

$$(F_1)_B \underset{(1)}{=} \underset{(16)}{\hat{\mathbf{a}}_1 \times \hat{\mathbf{a}}_2 \cdot \mathbf{T}} + \underset{(17)}{q_2\hat{\mathbf{a}}_2 \cdot \mathbf{R}}$$

$$= -\underset{(12)}{\frac{GMmL^2}{8{q_2}^3}\sin 2q_3} + \underset{(13)}{\frac{GMmL^2}{8{q_2}^3}\sin 2q_3} = 0 \tag{18}$$

$$(F_2)_B \underset{(1,\ 16,\ 17,\ 13)}{=} -\frac{GMm}{{q_2}^2}\left[1 + \frac{L^2}{8{q_2}^2}(2 - 3\sin^2 q_3)\right] \tag{19}$$

$$(F_3)_B \underset{(1,\ 16,\ 17,\ 12)}{=} -\frac{GMmL^2}{8{q_2}^3}\sin 2q_3 \tag{20}$$

5.6 CONTRIBUTING INTERACTION FORCES

In Sec. 6.6 it will be shown that certain interaction forces, that is, forces exerted by one part of a system on another, make no contributions to generalized active forces. In some situations, forces of interaction *do* contribute to generalized active forces. For example, whenever two particles of a system are not rigidly connected to each other, the gravitational forces exerted by the particles on each other can make such contributions. Bodies connected to each other by certain energy storage or energy dissipation devices furnish additional examples.

Example Figure 5.6.1 shows a double pendulum consisting of two rigid rods, A and B. Rod A is pinned to a fixed support, and A and B are pin-connected. Relative motion of A and B is resisted by a light torsion spring of modulus σ and by a viscous fluid damper with a damping constant δ. In other words, the set of forces exerted on A by B through the spring and damper is equivalent (see Sec. 5.3) to a couple whose torque $\mathbf{T}_A$ is given by

$$\mathbf{T}_A = (\sigma q_2 + \delta\dot{q}_2)\hat{\mathbf{n}} \tag{1}$$

where q_2 is the angle between A and B, as shown in Fig. 5.6.1, and the set of forces exerted by A on B through the spring and damper is equivalent to a couple whose torque $\mathbf{T}_B$ can be written

$$\mathbf{T}_B = -\mathbf{T}_A \tag{2}$$

Suppose now that motion variables u_1 and u_2 are defined as (see Fig. 5.6.1 for q_1)

$$u_r = \dot{q}_r \qquad (r = 1,2) \tag{3}$$

and let f_r $(r = 1,2)$ be the contribution of the spring and damper forces to the generalized active force F_r $(r = 1,2)$. Then

$$f_r \underset{(5.5.1)}{=} \boldsymbol{\omega}_r^A \cdot \mathbf{T}_A + \boldsymbol{\omega}_r^B \cdot \mathbf{T}_B \qquad (r = 1,2) \tag{4}$$

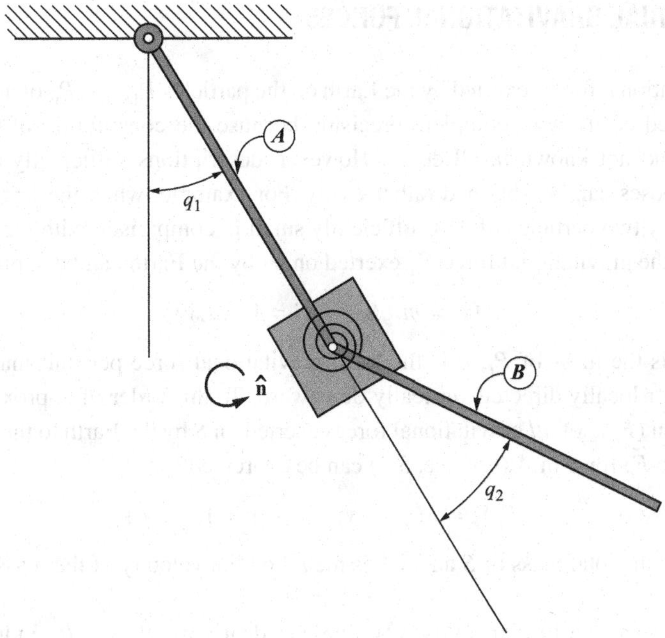

Figure 5.6.1

Now,

$$\boldsymbol{\omega}^A = \dot{q}_1\hat{\mathbf{n}} \underset{(3)}{=} u_1\hat{\mathbf{n}} \tag{5}$$

$$\boldsymbol{\omega}^B = (\dot{q}_1 + \dot{q}_2)\hat{\mathbf{n}} \underset{(3)}{=} (u_1 + u_2)\hat{\mathbf{n}} \tag{6}$$

so that

$$\boldsymbol{\omega}_1^A \underset{(5)}{=} \hat{\mathbf{n}} \qquad \boldsymbol{\omega}_2^A \underset{(5)}{=} \mathbf{0} \qquad \boldsymbol{\omega}_1^B \underset{(6)}{=} \hat{\mathbf{n}} \qquad \boldsymbol{\omega}_2^B \underset{(6)}{=} \hat{\mathbf{n}} \tag{7}$$

Hence,

$$f_1 \underset{(4)}{=} \underset{(7)}{\hat{\mathbf{n}}} \cdot \mathbf{T}_A + \underset{(7)}{\hat{\mathbf{n}}} \cdot \underset{(2)}{(-\mathbf{T}_A)} = 0 \tag{8}$$

and

$$f_2 \underset{(4)}{=} \underset{(7)}{\mathbf{0}} \cdot \mathbf{T}_A + \hat{\mathbf{n}} \cdot \underset{(2)}{(-\mathbf{T}_A)} \underset{(1)}{=} -(\sigma q_2 + \delta \dot{q}_2) \tag{9}$$

Thus, the interaction forces associated with the spring and damper here contribute to F_2 but not to F_1. The reader should verify that, if u_1 and u_2 are defined as $u_1 = \dot{q}_1$, $u_2 = \dot{q}_1 + \dot{q}_2$, rather than as in Eqs. (3), then the spring forces and damper forces contribute to both F_1 and F_2.

5.7 TERRESTRIAL GRAVITATIONAL FORCES

The gravitational forces exerted by the Earth on the particles $P_1, \ldots, P_\nu$ of a set S cannot be evaluated easily with complete precision because the constitution of the Earth is complex and not known in all detail. However, descriptions sufficiently accurate for many purposes can be obtained rather easily. For example, when the largest distance between any two particles of S is sufficiently small in comparison with the diameter of the Earth, the gravitational force $\mathbf{G}_i$ exerted on P_i by the Earth can be approximated as

$$\mathbf{G}_i = m_i g \hat{\mathbf{k}} \qquad (i = 1, \ldots, \nu) \tag{1}$$

where m_i is the mass of P_i, g is the local gravitational force per unit mass, and $\hat{\mathbf{k}}$ is a unit vector locally directed vertically downward. To this order of approximation, the contribution $(\tilde{F}_r)_\gamma$ of *all* gravitational forces exerted on S by the Earth to the generalized active force $\tilde{F}_r$ for S in A (see Sec. 5.4) can be expressed as

$$(\tilde{F}_r)_\gamma = Mg\hat{\mathbf{k}} \cdot \tilde{\mathbf{v}}_r^\star \qquad (r = 1, \ldots, p) \tag{2}$$

where M is the total mass of S and $\tilde{\mathbf{v}}_r^\star$ is the rth partial velocity of the mass center of S in A.

As an alternative to using Eqs. (2), one can deal with $P_1, \ldots, P_\nu$ individually by expressing $(\tilde{F}_r)_\gamma$ as

$$(\tilde{F}_r)_\gamma = \sum_{i=1}^{\nu} \tilde{\mathbf{v}}_r^{P_i} \cdot \mathbf{G}_i \qquad (r = 1, \ldots, p) \tag{3}$$

Whether it is more convenient to use Eqs. (2) or Eqs. (3) depends on the relative ease of finding $\tilde{\mathbf{v}}_r^\star$ $(r = 1, \ldots, p)$, on the one hand, and $\tilde{\mathbf{v}}_r^{P_i}$ $(r = 1, \ldots, p;\ i = 1, \ldots, \nu)$, on the other hand.

Derivation The position vector $\mathbf{p}^\star$ from a point O fixed in A to the mass center of S is related to the position vectors $\mathbf{p}_1, \ldots, \mathbf{p}_\nu$ from O to the particles of S by

$$M\mathbf{p}^\star \underset{(4.1.2)}{=} \sum_{i=1}^{\nu} m_i \mathbf{p}_i \tag{4}$$

Differentiation with respect to t in A yields

$$\underset{(2.6.1)}{M\mathbf{v}^\star} \underset{(4)}{=} \sum_{i=1}^{\nu} \underset{(2.6.1)}{m_i \mathbf{v}^{P_i}} \tag{5}$$

Consequently, the partial velocities $\tilde{\mathbf{v}}_r^\star$ and $\tilde{\mathbf{v}}_r^{P_i}$ $(r = 1, \ldots, p)$ are related to each other by [see Eq. (3.6.4)]

$$M\, \tilde{\mathbf{v}}_r^\star \underset{(5)}{=} \sum_{i=1}^{\nu} m_i\, \tilde{\mathbf{v}}_r^{P_i} \qquad (r = 1, \ldots, p) \tag{6}$$

Now,

$$(\tilde{F}_r)_\gamma \underset{(3)}{=} \sum_{i=1}^{\nu} \tilde{\mathbf{v}}_r^{P_i} \cdot \mathbf{G}_i \underset{(1)}{=} g\hat{\mathbf{k}} \cdot \sum_{i=1}^{\nu} m_i\, \tilde{\mathbf{v}}_r^{P_i} \qquad (r = 1, \ldots, p) \tag{7}$$

Hence,

$$(\tilde{F}_r)_\gamma \underset{(7)}{=} g\hat{\mathbf{k}} \cdot (M \underset{(6)}{\tilde{\mathbf{v}}_r^\star}) \qquad (r = 1, \ldots, p) \tag{8}$$

in agreement with Eqs. (2).

Example Figure 5.7.1 shows a system S formed by a rigid frame A that carries two sharp-edged circular disks, B and C, each of radius R. Point $S^\star$ is the mass center of S, and Q is a point of A that comes into contact with a plane P that supports S. Generalized active forces are to be determined on the basis of the assumptions that B and C roll on P without slipping and are completely free to rotate relative to A, while P is inclined to the horizontal at an angle θ.

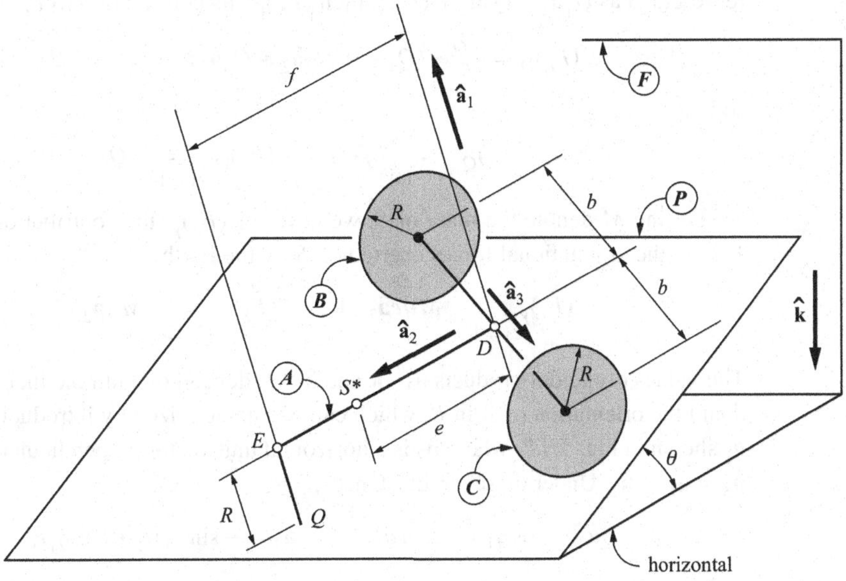

Figure 5.7.1

Five generalized coordinates are required to specify the configuration of S in a reference frame F in which P is fixed. Of the five associated motion variables, three are dependent on the remaining two when B and C roll on P without slipping. In other words, if u_1 and u_2 are defined as

$$u_1 \triangleq \boldsymbol{\omega}^A \cdot \hat{\mathbf{a}}_1 \qquad u_2 \triangleq \mathbf{v}^D \cdot \hat{\mathbf{a}}_2 \tag{9}$$

where $\hat{\mathbf{a}}_1$ and $\hat{\mathbf{a}}_2$ are unit vectors directed as shown in Fig. 5.7.1, $\boldsymbol{\omega}^A$ is the angular velocity of A in F, and $\mathbf{v}^D$ is the velocity in F of the midpoint D of the axle that carries B and C, then the velocity in F of every point of S can be expressed as a linear function of u_1 and u_2. For example, $\mathbf{v}^\star$, the velocity of $S^\star$ in F, becomes (see Fig. 5.7.1 for $\hat{\mathbf{a}}_3$)

$$\mathbf{v}^\star \underset{(2.7.1)}{=} u_2\hat{\mathbf{a}}_2 + eu_1\hat{\mathbf{a}}_3 \tag{10}$$

and $\mathbf{v}^Q$, the velocity in F of the point Q of A that comes into contact with P, is

$$\mathbf{v}^Q \underset{(2.7.1)}{=} u_2\hat{\mathbf{a}}_2 + f u_1 \hat{\mathbf{a}}_3 \tag{11}$$

The associated partial velocities are

$$\tilde{\mathbf{v}}_1^\star \underset{(10)}{=} e\hat{\mathbf{a}}_3 \qquad \tilde{\mathbf{v}}_2^\star \underset{(10)}{=} \hat{\mathbf{a}}_2 \tag{12}$$

$$\tilde{\mathbf{v}}_1^Q \underset{(11)}{=} f\hat{\mathbf{a}}_3 \qquad \tilde{\mathbf{v}}_2^Q \underset{(11)}{=} \hat{\mathbf{a}}_2 \tag{13}$$

The only contact force that contributes to the generalized active forces $\tilde{F}_1$ and $\tilde{F}_2$ is (as will be seen in Sec. 6.6) the force exerted on A by P at Q. When this force is represented as $Q_1\hat{\mathbf{a}}_1 + Q_2\hat{\mathbf{a}}_2 + Q_3\hat{\mathbf{a}}_3$, then $(\tilde{F}_r)_Q$, its contribution to $\tilde{F}_r$, is given by

$$(\tilde{F}_r)_Q = \tilde{\mathbf{v}}_r^Q \cdot (Q_1\hat{\mathbf{a}}_1 + Q_2\hat{\mathbf{a}}_2 + Q_3\hat{\mathbf{a}}_3) \qquad (r = 1,2) \tag{14}$$

so that

$$(\tilde{F}_1)_Q \underset{(13,\ 14)}{=} fQ_3 \qquad (\tilde{F}_2)_Q \underset{(13,\ 14)}{=} Q_2 \tag{15}$$

Letting M denote the mass of S, we have for $(\tilde{F}_r)_\gamma$, the contribution to $\tilde{F}_r$ $(r = 1,2)$ of the gravitational forces exerted on S by the Earth,

$$(\tilde{F}_1)_\gamma \underset{(2,\ 12)}{=} Mge\hat{\mathbf{a}}_3 \cdot \hat{\mathbf{k}} \qquad (\tilde{F}_2)_\gamma \underset{(2,\ 12)}{=} Mg\hat{\mathbf{a}}_2 \cdot \hat{\mathbf{k}} \tag{16}$$

The values of the dot products $\hat{\mathbf{a}}_3 \cdot \hat{\mathbf{k}}$ and $\hat{\mathbf{a}}_2 \cdot \hat{\mathbf{k}}$ depend on both the inclination angle θ and the orientation of A in P, which can be characterized by introducing the angle q_1 shown in Fig. 5.7.2, where $\hat{\mathbf{n}}_2$ is a horizontal unit vector perpendicular to $\hat{\mathbf{a}}_1$, while $\hat{\mathbf{n}}_3 = \hat{\mathbf{a}}_1 \times \hat{\mathbf{n}}_2$. Under these circumstances,

$$\hat{\mathbf{a}}_2 = \cos q_1 \hat{\mathbf{n}}_2 + \sin q_1 \hat{\mathbf{n}}_3 \qquad \hat{\mathbf{a}}_3 = -\sin q_1 \hat{\mathbf{n}}_2 + \cos q_1 \hat{\mathbf{n}}_3 \tag{17}$$

and

$$\hat{\mathbf{k}} = -\cos\theta \hat{\mathbf{a}}_1 + \sin\theta \hat{\mathbf{n}}_3 \tag{18}$$

Hence,

$$(\tilde{F}_1)_\gamma \underset{(16–18)}{=} Mge \sin\theta \cos q_1 \qquad (\tilde{F}_2)_\gamma \underset{(16–18)}{=} Mg \sin\theta \sin q_1 \tag{19}$$

and the desired generalized active forces are

$$\tilde{F}_1 = (\tilde{F}_1)_Q + (\tilde{F}_1)_\gamma \underset{(15,\ 19)}{=} fQ_3 + Mge \sin\theta \cos q_1 \tag{20}$$

and

$$\tilde{F}_2 = (\tilde{F}_2)_Q + (\tilde{F}_2)_\gamma \underset{(15,\ 19)}{=} Q_2 + Mg \sin\theta \sin q_1 \tag{21}$$

The reason for using Eqs. (2) rather than Eqs. (3) to find $(\tilde{F}_1)_\gamma$ and $(\tilde{F}_2)_\gamma$ [see Eqs. (16)] is that it would be very laborious to deal individually with each of the particles forming A, B, and C. By way of contrast, consider once again the contributions

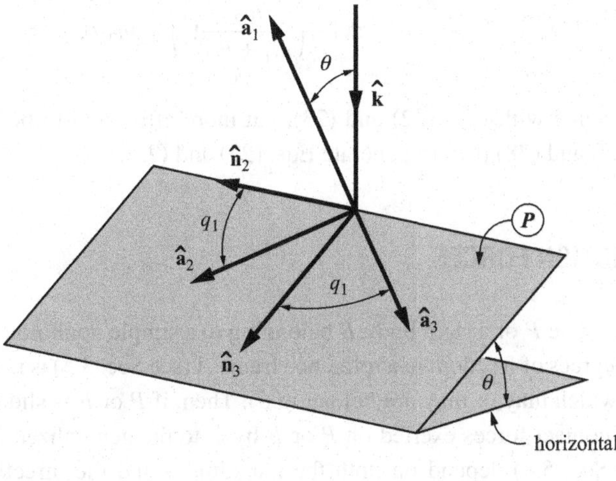

Figure 5.7.2

of gravitational forces to the generalized active forces F_1 and F_2 in the example in Sec. 5.4. Since the partial velocities of P_1 and P_2 are available in Eqs. (5.4.12) and (5.4.13), $(F_1)_\gamma$ and $(F_2)_\gamma$ are formed easily as

$$(F_1)_\gamma \underset{(3)}{=} \mathbf{v}_1^{P_1} \cdot (m_1 g\hat{\mathbf{k}}) + \mathbf{v}_1^{P_2} \cdot (m_2 g\hat{\mathbf{k}}) = \underset{(5.4.12)}{m_1 g \cos\theta} + \underset{(5.4.13)}{0} \tag{22}$$

$$(F_2)_\gamma \underset{(3)}{=} \mathbf{v}_2^{P_1} \cdot (m_1 g\hat{\mathbf{k}}) + \mathbf{v}_2^{P_2} \cdot (m_2 g\hat{\mathbf{k}}) = \underset{(5.4.12)}{0} + \underset{(5.4.13)}{m_2 g \cos\theta} \tag{23}$$

whereas, to use Eqs. (2), one must first locate the mass center $S^\star$ of P_1 and P_2, determine its velocity, and use this to form partial velocities. Specifically, letting $\mathbf{p}^\star$ be the position vector from point O in Fig. 5.4.1 to $S^\star$, one has

$$\mathbf{p}^\star = \frac{m_1(L_1 + q_1) + m_2(L_1 + L_2 + q_2)}{m_1 + m_2}\hat{\mathbf{t}}_1 \tag{24}$$

so that $\mathbf{v}^\star$, the velocity of $S^\star$, is given by

$$\mathbf{v}^\star \underset{(24)}{=} \frac{m_1 u_1 + m_2 u_2}{m_1 + m_2}\hat{\mathbf{t}}_1 \underset{(5.4.9)}{+} \frac{m_1(L_1 + q_1) + m_2(L_1 + L_2 + q_2)}{m_1 + m_2}\dot{\theta}\hat{\mathbf{t}}_2 \tag{25}$$

and the partial velocities of $S^\star$ are

$$\mathbf{v}_1^\star = \frac{m_1}{m_1 + m_2}\hat{\mathbf{t}}_1 \qquad \mathbf{v}_2^\star = \frac{m_2}{m_1 + m_2}\hat{\mathbf{t}}_1 \tag{26}$$

Now one can refer to Eqs. (2) to write

$$(F_1)_\gamma \underset{(2)}{=} [(m_1 + m_2)g\hat{\mathbf{k}}] \cdot \underset{(26)}{\left(\frac{m_1}{m_1 + m_2}\hat{\mathbf{t}}_1\right)} = m_1 g \cos\theta \tag{27}$$

$$(F_2)_\gamma \underset{(2)}{=} [(m_1 + m_2)g\hat{\mathbf{k}}] \cdot \underset{(26)}{\left(\frac{m_2}{m_1 + m_2}\hat{\mathbf{t}}_1\right)} = m_2 g \cos\theta \tag{28}$$

These results agree with Eqs. (22) and (23), but more effort had to be expended to derive Eqs. (27) and (28) than to generate Eqs. (22) and (23).

5.8 COULOMB FRICTION FORCES

Suppose that a particle P or a rigid body B belonging to a simple nonholonomic system S possessing p degrees of freedom in a reference frame A (see Sec. 3.5) is in contact with a rigid body C (which may or may not belong to S). Then, if P or B is sliding on C, the contributions of contact forces exerted on P or B by C to the generalized active forces $\tilde{F}_1, \ldots, \tilde{F}_p$ (see Sec. 5.4) depend on both the magnitudes and the directions of such contact forces. When contact takes place across dry, clean surfaces, certain information regarding the magnitudes and the directions of contact forces can be obtained from the *laws of Coulomb friction*, which will now be stated.

When a particle P that is in contact with a rigid body C is at rest relative to C, then C exerts on P a contact force $\mathbf{C}$ that can be expressed as

$$\mathbf{C} = N\hat{\nu} + T\hat{\tau} \tag{1}$$

where $\hat{\nu}$ is a unit vector normal to the surface Σ of C at P and directed from C toward P, $\hat{\tau}$ is a unit vector perpendicular to $\hat{\nu}$, N is nonnegative, and T satisfies the inequality

$$|T| \leq \mu N \tag{2}$$

in which μ, called the *coefficient of static friction* for P and C, is a quantity whose value depends solely on the materials of which P and C are made. Typical values are 0.2 for metal on metal, 0.6 for metal on wood.

When P is in a state of impending tangential motion relative to C, that is, when P is on the verge of moving tangentially relative to C, then

$$|T| = \mu N \tag{3}$$

and the vector $T\hat{\tau}$ appearing in Eq. (1) points in the direction opposite to that in which P is about to move relative to C.

When P is sliding relative to C, Eq. (1) remains in force, but the inequality (2) gives way to the equality

$$|T| = \mu' N \tag{4}$$

where μ', called the *coefficient of kinetic friction* for P and C, has a value generally smaller than that of μ; and the vector $T\hat{\tau}$ in Eq. (1) now is directed oppositely to ${}^C\mathbf{v}^P$, the velocity of P in C.

When a rigid body B, rather than a particle P, is in contact with C, the same laws apply if the surface Σ over which B and C are in contact has an area so small that Σ can be regarded as a point. Otherwise, that is, when Σ has an area that cannot be regarded as negligibly small, the laws already stated apply in connection with every differential element of Σ. More specifically, if P is a point of B within a portion $\overline{\Sigma}$ of Σ that has an area $\overline{A}$, then the set of contact forces exerted on B by C across $\overline{\Sigma}$ can be replaced with a couple of torque $\mathbf{M}$ together with a force $\mathbf{C}$ whose line of action passes through P; $\mathbf{M}$ and $\mathbf{C}$ depend on $\overline{A}$, and both approach zero as $\overline{A}$ approaches zero, but

$$\lim_{\overline{A}\to 0} \frac{\mathbf{M}}{\overline{A}} = \mathbf{0} \tag{5}$$

whereas $\mathbf{C}/\overline{A}$ has a nonzero limit that can be expressed as

$$\lim_{\overline{A}\to 0} \frac{\mathbf{C}}{\overline{A}} = n\hat{\nu} + t\hat{\tau} \tag{6}$$

where n, called the *pressure* at point P, and t, called the *shear* at P, depend on the position of P within Σ, n is nonnegative, and $\hat{\nu}$ and $\hat{\tau}$ have the same meanings as before. Equations (5) and (6) imply that, if P is a point of B lying within a differential element $d\Sigma$ of Σ having an area dA, the set of contact forces exerted on B by C across $d\Sigma$ is equivalent to a force $d\mathbf{C}$ whose line of action passes through P and that is given by

$$d\mathbf{C} = (n\hat{\nu} + t\hat{\tau})dA \tag{7}$$

This equation takes the place of Eq. (1) when B and C are in contact over an extended surface, and the relationships (2)–(4) then are replaced with, respectively,

$$|t| \le \mu n \tag{8}$$

when P is at rest relative to C,

$$|t| = \mu n \tag{9}$$

when P is in a state of impending tangential motion relative to C, and

$$|t| = \mu' n \tag{10}$$

when P is sliding relative to C.

When a particle P is in contact with a rigid body C modeled as matter distributed along a curve Γ (for example, a thin rod modeled as matter distributed along a straight line), then Eqs. (1)–(4) and the statements made in connection with these equations remain in force, provided that $\hat{\tau}$ is regarded as a unit vector tangent to Γ at P while $\hat{\nu}$ is simply a unit vector perpendicular to $\hat{\tau}$. Similarly, when contact between two rigid bodies B and C is regarded as taking place along a curve Γ (for example, when a generator of a right-circular cylinder is in contact with a plane), then Eq. (7) applies after dA has been replaced with dL, the length of a differential element $d\Gamma$ of Γ, and $\hat{\nu}$ and $\hat{\tau}$ then have their original meanings.

Example In Fig. 5.8.1, P is a particle of mass m that can slide freely on a smooth, uniform rod R of mass M and length $2L$. R is rigidly attached at an angle β to a light sleeve S that is supported by a smooth, fixed vertical shaft V and a smooth bearing surface B, and S is subjected to the action of a couple whose torque $\mathbf{T}$ is given by

$$\mathbf{T} = T\hat{\mathbf{s}}_1 \tag{11}$$

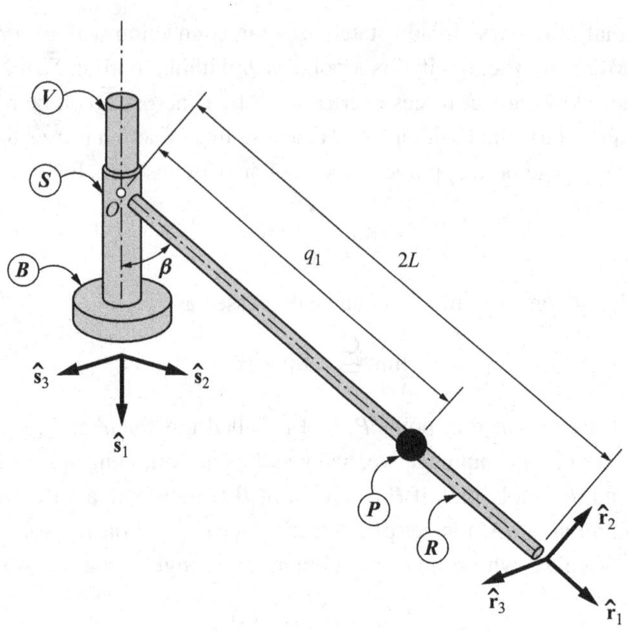

Figure 5.8.1

where T is time dependent and $\hat{\mathbf{s}}_1$ is a unit vector directed vertically downward. The unit vectors $\hat{\mathbf{s}}_2$, $\hat{\mathbf{s}}_3$, $\hat{\mathbf{r}}_1$, $\hat{\mathbf{r}}_2$, and $\hat{\mathbf{r}}_3$ in Fig. 5.8.1 are defined as follows: $\hat{\mathbf{s}}_2$ is perpendicular to $\hat{\mathbf{s}}_1$ and parallel to the plane determined by the axes of S and R; $\hat{\mathbf{s}}_3 = \hat{\mathbf{s}}_1 \times \hat{\mathbf{s}}_2$; and $\hat{\mathbf{r}}_1$, $\hat{\mathbf{r}}_2$, $\hat{\mathbf{r}}_3$ form a dextral set of mutually perpendicular unit vectors, with $\hat{\mathbf{r}}_1$ parallel to the axis of R and $\hat{\mathbf{r}}_2$ parallel to the plane determined by the axes of S and R.

The system formed by P, R, and S possesses two degrees of freedom and, if generalized velocities u_1 and u_2 are defined as

$$u_1 \triangleq \dot{q}_1 \qquad u_2 \triangleq {}^B\boldsymbol{\omega}^R \cdot \hat{\mathbf{s}}_1 \tag{12}$$

where q_1 is the distance from O to P and ${}^B\boldsymbol{\omega}^R$ is the angular velocity of R in B, then the associated partial angular velocities of R in B and partial velocities of P in B are

$${}^B\boldsymbol{\omega}_1^R = \mathbf{0} \qquad {}^B\mathbf{v}_1^P = \hat{\mathbf{r}}_1 \tag{13}$$

$${}^B\boldsymbol{\omega}_2^R = \hat{\mathbf{s}}_1 \qquad {}^B\mathbf{v}_2^P = q_1 \sin\beta\, \hat{\mathbf{r}}_3 \tag{14}$$

Hence, the generalized active forces F_1 and F_2, found by substituting from Eqs. (11), (13), and (14) into [see Eqs. (5.5.1) and (5.7.1)]

$$F_r = {}^B\boldsymbol{\omega}_r^R \cdot \mathbf{T} + {}^B\mathbf{v}_r^P \cdot (mg\hat{\mathbf{s}}_1) \qquad (r = 1, 2) \tag{15}$$

are

$$F_1 = mg\cos\beta \tag{16}$$

$$F_2 = T \tag{17}$$

Now, suppose that the contact between S and the bearing surface B, as well as that between P and R, takes place across a rough rather than a smooth surface, but that the vertical shaft V can be regarded as smooth, as heretofore. Then, when S is moving relative to B, and P relative to R, contact forces that contribute to the generalized active forces F_1 and F_2 come into play. $(F_r)_C$, the contribution to F_r $(r = 1,2)$ of the contact forces, will now be determined.

The contact force $\boldsymbol{\rho}$ exerted by R on P can be expressed as

$$\boldsymbol{\rho} = \rho_1 \hat{\mathbf{r}}_1 + \rho_2 \hat{\mathbf{r}}_2 + \rho_3 \hat{\mathbf{r}}_3 \tag{18}$$

and the contact force $\overline{\boldsymbol{\rho}}$ exerted by P on R is given by

$$\overline{\boldsymbol{\rho}} = -\boldsymbol{\rho} \tag{19}$$

Hence, by letting ${}^B\mathbf{v}_r^{\overline{R}}$ $(r = 1,2)$ denote the partial velocities in B of the point of R at which $\overline{\boldsymbol{\rho}}$ is applied to R, so that

$${}^B\mathbf{v}_1^{\overline{R}} = \mathbf{0} \qquad {}^B\mathbf{v}_2^{\overline{R}} = q_1 \sin\beta \hat{\mathbf{r}}_3 \tag{20}$$

one can express the contributions of $\boldsymbol{\rho}$ and $\overline{\boldsymbol{\rho}}$ to the generalized active forces F_1 and F_2 as

$${}^B\mathbf{v}_1^P \cdot \boldsymbol{\rho} + {}^B\mathbf{v}_1^{\overline{R}} \cdot \overline{\boldsymbol{\rho}} \underset{(13)}{=} \hat{\mathbf{r}}_1 \cdot \boldsymbol{\rho} + 0 \underset{(20)\ (18)}{=} \rho_1 \tag{21}$$

and

$${}^B\mathbf{v}_2^P \cdot \boldsymbol{\rho} + {}^B\mathbf{v}_2^{\overline{R}} \cdot \overline{\boldsymbol{\rho}} \underset{(19)}{=} ({}^B\mathbf{v}_2^P - {}^B\mathbf{v}_2^{\overline{R}}) \cdot \boldsymbol{\rho} \underset{(14,\ 20)}{=} 0 \tag{22}$$

The laws of friction make it possible to express ρ_1 in terms of ρ_2 and ρ_3, for $\rho_1 \hat{\mathbf{r}}_1$ in Eq. (18) corresponds to the second term of Eq. (1), while $\rho_2 \hat{\mathbf{r}}_2 + \rho_3 \hat{\mathbf{r}}_3$ plays the part of the first term. Accordingly,

$$|\rho_1| \underset{(4)}{=} \mu_1' (\rho_2{}^2 + \rho_3{}^2)^{1/2} \underset{(18)}{} \tag{23}$$

where μ_1' is the coefficient of kinetic friction for P and R. Furthermore, $\rho_1 \hat{\mathbf{r}}_1$ must have a direction opposite to that of ${}^R\mathbf{v}^P$, the velocity of P in R, which means that $\rho_1 \hat{\mathbf{r}}_1$ can be written

$$\begin{aligned} \rho_1 \hat{\mathbf{r}}_1 = -|\rho_1| \frac{{}^R\mathbf{v}^P}{|{}^R\mathbf{v}^P|} &\underset{(23)}{=} -\mu_1' (\rho_2{}^2 + \rho_3{}^2)^{1/2} \underset{(12)}{\frac{u_1 \hat{\mathbf{r}}_1}{|u_1|}} \\ &= -\mu_1' (\rho_2{}^2 + \rho_3{}^2)^{1/2} \operatorname{sgn} u_1 \hat{\mathbf{r}}_1 \end{aligned} \tag{24}$$

from which it follows that

$$\rho_1 = -\mu_1' (\rho_2{}^2 + \rho_3{}^2)^{1/2} \operatorname{sgn} u_1 \tag{25}$$

so that

$${}^B\mathbf{v}_1^P \cdot \boldsymbol{\rho} + {}^B\mathbf{v}_1^{\overline{R}} \cdot \overline{\boldsymbol{\rho}} \underset{(21,\ 25)}{=} -\mu_1' (\rho_2{}^2 + \rho_3{}^2)^{1/2} \operatorname{sgn} u_1 \tag{26}$$

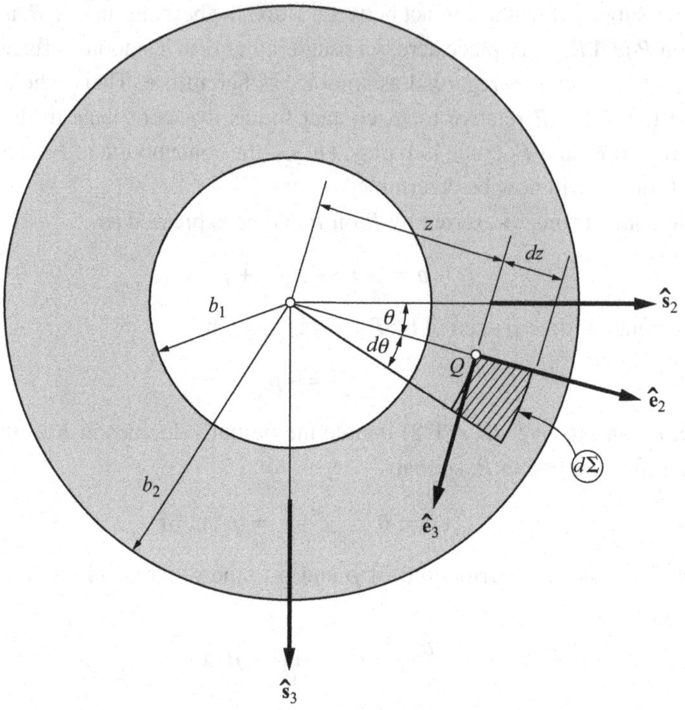

Figure 5.8.2

To deal with the contributions to F_1 and F_2 of the contact forces exerted on S by B, we let b_1 and b_2 be the inner and outer radii of S, respectively, and consider a generic differential element $d\Sigma$ of the surface Σ of S that is in contact with B, $d\Sigma$ having the dimensions indicated in Fig. 5.8.2, where θ and z are variables used to locate one corner, Q, of $d\Sigma$. The force $d\boldsymbol{\sigma}$ exerted on S by B across $d\Sigma$ then can be written

$$d\boldsymbol{\sigma} \underset{(7)}{=} (-n\hat{\mathbf{s}}_1 + t_2\hat{\mathbf{s}}_2 + t_3\hat{\mathbf{s}}_3)z\, dz\, d\theta \tag{27}$$

where $-\hat{\mathbf{s}}_1$, which points from B toward S, corresponds to $\hat{\boldsymbol{\nu}}$ in Eq. (7) since Q is a point of $d\Sigma$, and ${}^B\mathbf{v}^Q$, the velocity of Q in B, is given by (see Fig. 5.8.2 for the unit vectors $\hat{\mathbf{e}}_2$ and $\hat{\mathbf{e}}_3$)

$${}^B\mathbf{v}^Q = {}^B\boldsymbol{\omega}^R \times (z\hat{\mathbf{e}}_2) \underset{(12)}{=} (u_2\hat{\mathbf{s}}_1) \times (z\hat{\mathbf{e}}_2) = u_2 z\hat{\mathbf{e}}_3 \tag{28}$$

so that the partial velocities of Q are

$${}^B\mathbf{v}_1^Q = \mathbf{0} \qquad {}^B\mathbf{v}_2^Q = z\hat{\mathbf{e}}_3 \tag{29}$$

We now can express the contributions of $d\boldsymbol{\sigma}$ to F_1 and F_2 as

$${}^B\mathbf{v}_1^Q \cdot d\boldsymbol{\sigma} \underset{(29)}{=} 0 \tag{30}$$

and

$$ {}^B\mathbf{v}_2^Q \cdot d\boldsymbol{\sigma} \underset{(29,\,27)}{=} z^2(-t_2 \sin\theta + t_3 \cos\theta)\, dz\, d\theta \tag{31} $$

In addition, we have the relationship

$$ (t_2^{\,2} + t_3^{\,2})^{1/2} \underset{(10)}{=} \mu_2{}' n \tag{32} $$

where $\mu_2{}'$ is the coefficient of kinetic friction for B and S, and the requirement that $t_2\hat{\mathbf{s}}_2 + t_3\hat{\mathbf{s}}_3$ be directed oppositely to ${}^B\mathbf{v}^Q$ implies that

$$ t_2\hat{\mathbf{s}}_2 + t_3\hat{\mathbf{s}}_3 = -(t_2^{\,2} + t_3^{\,2})^{1/2} \frac{{}^B\mathbf{v}^Q}{|{}^B\mathbf{v}^Q|} \tag{33} $$

from which it follows that

$$ t_2\hat{\mathbf{s}}_2 + t_3\hat{\mathbf{s}}_3 \underset{(33)}{=} -\underset{(32)}{\mu_2{}' n}\ \underset{(28)}{\operatorname{sgn} u_2 \hat{\mathbf{e}}_3} \tag{34} $$

or, upon dot multiplication with $\hat{\mathbf{e}}_3$, that

$$ -t_2 \sin\theta + t_3 \cos\theta = -\mu_2{}' n \operatorname{sgn} u_2 \tag{35} $$

Hence,

$$ {}^B\mathbf{v}_2^Q \cdot d\boldsymbol{\sigma} \underset{(31,\,35)}{=} -\mu_2{}' \operatorname{sgn} u_2\, n z^2\, dz\, d\theta \tag{36} $$

$(F_1)_C$ and $(F_2)_C$, the contributions to F_1 and F_2, respectively, of the contact forces exerted on P by R, on R by P, and on S by B are

$$ \begin{aligned} (F_1)_C &= {}^B\mathbf{v}_1^P \cdot \boldsymbol{\rho} + {}^B\mathbf{v}_1^{\overline{R}} \cdot \overline{\boldsymbol{\rho}} + \int {}^B\mathbf{v}_1^Q \cdot d\boldsymbol{\sigma} \\ &\underset{(26,\,30)}{=} -\mu_1{}'(\rho_2^{\,2} + \rho_3^{\,2})^{1/2} \operatorname{sgn} u_1 \end{aligned} \tag{37} $$

and

$$ \begin{aligned} (F_2)_C &= {}^B\mathbf{v}_2^P \cdot \boldsymbol{\rho} + {}^B\mathbf{v}_2^{\overline{R}} \cdot \overline{\boldsymbol{\rho}} + \int {}^B\mathbf{v}_2^Q \cdot d\boldsymbol{\sigma} \\ &\underset{(22,\,36)}{=} -\mu_2{}' \operatorname{sgn} u_2 \int_0^{2\pi}\int_{b_1}^{b_2} n z^2\, dz\, d\theta \end{aligned} \tag{38} $$

The definite integral in Eq. (38) can be evaluated only when the pressure n is known as a function of z and θ. To discover this functional relationship, one must use the methods of the theory of elasticity. However, satisfactory results often can be obtained in this sort of situation by making a relatively simple assumption regarding the pressure distribution in question, such as, for example, that n is independent of z and θ and thus has a value n^* that depends solely on the time t. Under these circumstances,

$$ \begin{aligned} (F_2)_C &\underset{(38)}{=} -\mu_2{}' \operatorname{sgn} u_2\, n^* \int_0^{2\pi}\int_{b_1}^{b_2} z^2\, dz\, d\theta \\ &= -\frac{2\pi n^*}{3}(b_2^{\,3} - b_1^{\,3})\mu_2{}' \operatorname{sgn} u_2 \end{aligned} \tag{39} $$

When the contributions recorded in Eqs. (37) and (39) are added to those of all gravitational forces acting on S, R, and P and of the torque $\mathbf{T}$ described by Eq. (11), the complete generalized active forces are given by

$$F_1 = \underset{(16)}{mg\cos\beta} - \underset{(37)}{\mu_1'(\rho_2{}^2 + \rho_3{}^2)^{1/2}\,\text{sgn}\,u_1} \tag{40}$$

$$F_2 = \underset{(17)}{T} - \underset{(39)}{\frac{2\pi n^*}{3}(b_2{}^3 - b_1{}^3)\mu_2'\,\text{sgn}\,u_2} \tag{41}$$

As will be seen later, ρ_2, ρ_3, and n^* ultimately can be determined by resorting to the procedure employed in Sec. 6.7, that is, by introducing three suitable motion variables in addition to the generalized velocities u_1 and u_2.

5.9 GENERALIZED INERTIA FORCES

Let S denote a system composed of particles $P_1,\ldots,P_\nu$. When S is a holonomic system possessing n degrees of freedom in a reference frame A (see Sec. 3.5), and $u_1,\ldots,u_n$ are generalized velocities for S in A, then n quantities $F_1^\star,\ldots,F_n^\star$ called *holonomic generalized inertia forces* for S in A are defined as

$$F_r^\star \triangleq \sum_{i=1}^{\nu} {}^A\mathbf{v}_r^{P_i} \cdot \mathbf{R}_i^\star \qquad (r = 1,\ldots,n) \tag{1}$$

where ${}^A\mathbf{v}_r^{P_i}$ is a holonomic partial velocity of P_i in A (see Sec. 3.6).

When S is a simple nonholonomic system possessing p degrees of freedom in A (see Sec. 3.5), and $u_1,\ldots,u_p$ are generalized velocities for S in A, then p quantities $\tilde{F}_1^\star,\ldots,\tilde{F}_p^\star$ called *nonholonomic generalized inertia forces* for S in A are defined as

$$\tilde{F}_r^\star \triangleq \sum_{i=1}^{\nu} {}^A\tilde{\mathbf{v}}_r^{P_i} \cdot \mathbf{R}_i^\star \qquad (r = 1,\ldots,p) \tag{2}$$

where ${}^A\tilde{\mathbf{v}}_r^{P_i}$ is a nonholonomic partial velocity of P_i in A (see Sec. 3.6).

When S is a complex nonholonomic system possessing c degrees of freedom in A (see Sec. 3.7), and $\dot{u}_1,\ldots,\dot{u}_c$ are time derivatives of motion variables for S in A, then c quantities $\tilde{\tilde{F}}_1^\star,\ldots,\tilde{\tilde{F}}_c^\star$ are defined as

$$\tilde{\tilde{F}}_r^\star \triangleq \sum_{i=1}^{\nu} {}^A\tilde{\mathbf{a}}_r^{P_i} \cdot \mathbf{R}_i^\star \qquad (r = 1,\ldots,c) \tag{3}$$

where ${}^A\tilde{\mathbf{a}}_r^{P_i}$ is a nonholonomic partial acceleration of P_i in A (see Sec. 3.8). The term "nonholonomic generalized inertia force" may also be used in connection with $\tilde{\tilde{F}}_1^\star,\ldots,\tilde{\tilde{F}}_c^\star$.

The vector $\mathbf{R}_i^\star$ that appears in Eqs. (1)–(3) is the *inertia force* for P_i in A— that is,

$$\mathbf{R}_i^\star \triangleq -m_i{}^A\mathbf{a}^{P_i} \qquad (i = 1,\ldots,\nu) \tag{4}$$

where m_i is the mass of P_i, and ${}^A\mathbf{a}^{P_i}$ is the acceleration of P_i in A.

The generalized inertia forces for S in A defined in Eqs. (1) and (2) are related to each other and to the quantities A_{rs} ($s = 1,\ldots,p$; $r = p+1,\ldots,n$) introduced in Eqs. (3.5.2), as follows:

$$\tilde{F}_r^\star = F_r^\star + \sum_{s=p+1}^{n} F_s^\star A_{sr} \qquad (r = 1,\ldots,p) \tag{5}$$

Likewise, there exist relationships involving the generalized inertia forces for S in A defined in Eqs. (2) and (3), and the quantities $\tilde{A}_{rs}$ ($s = 1,\ldots,c$; $r = c+1,\ldots,p$) appearing in Eqs. (3.7.1):

$$\tilde{\tilde{F}}_r^\star = \tilde{F}_r^\star + \sum_{s=c+1}^{p} \tilde{F}_s^\star \tilde{A}_{sr} \qquad (r = 1,\ldots,c) \tag{6}$$

When B is a rigid body belonging to S, all inertia forces for the particles of B make contributions denoted by $(F_r^\star)_B$, $(\tilde{F}_r^\star)_B$, and $(\tilde{\tilde{F}}_r^\star)_B$, respectively, to the generalized inertia forces defined in Eqs. (1)–(3), as follows:

$$(F_r^\star)_B = {}^A\boldsymbol{\omega}_r^B \cdot \mathbf{T}^\star + {}^A\mathbf{v}_r^{B^\star} \cdot \mathbf{R}^\star \qquad (r = 1,\ldots,n) \tag{7}$$

where ${}^A\boldsymbol{\omega}_r^B$ is a holonomic partial angular velocity of B in A (see Sec. 3.6), and ${}^A\mathbf{v}_r^{B^\star}$ is a holonomic partial velocity of $B^\star$, the mass center of B, in A,

$$(\tilde{F}_r^\star)_B = {}^A\tilde{\boldsymbol{\omega}}_r^B \cdot \mathbf{T}^\star + {}^A\tilde{\mathbf{v}}_r^{B^\star} \cdot \mathbf{R}^\star \qquad (r = 1,\ldots,p) \tag{8}$$

where ${}^A\tilde{\boldsymbol{\omega}}_r^B$ is a nonholonomic partial angular velocity of B in A (see Sec. 3.6), and ${}^A\tilde{\mathbf{v}}_r^{B^\star}$ is a nonholonomic partial velocity of $B^\star$ in A, and

$$(\tilde{\tilde{F}}_r^\star)_B = {}^A\tilde{\boldsymbol{\alpha}}_r^B \cdot \mathbf{T}^\star + {}^A\tilde{\mathbf{a}}_r^{B^\star} \cdot \mathbf{R}^\star \qquad (r = 1,\ldots,c) \tag{9}$$

where ${}^A\tilde{\boldsymbol{\alpha}}_r^B$ is a nonholonomic partial angular acceleration of B in A (see Sec. 3.8), and ${}^A\tilde{\mathbf{a}}_r^{B^\star}$ is a nonholonomic partial acceleration of $B^\star$ in A.

The vector $\mathbf{T}^\star$ appearing in Eqs. (7)–(9) is called the *inertia torque* for B in A; it is defined as

$$\mathbf{T}^\star \triangleq -\sum_{i=1}^{\beta} m_i \mathbf{r}_i \times {}^A\mathbf{a}^{P_i} \tag{10}$$

where β is the number of particles forming B, m_i is the mass of a generic particle P_i of B, $\mathbf{r}_i$ is the position vector from $B^\star$ to P_i, and ${}^A\mathbf{a}^{P_i}$ is the acceleration of P_i in A. $\mathbf{R}^\star$ is called the *inertia force* for B in A, and it is defined as

$$\mathbf{R}^\star \triangleq -M\,{}^A\mathbf{a}^{B^\star} \tag{11}$$

where M is the total mass of B, and ${}^A\mathbf{a}^{B^\star}$ is the acceleration of $B^\star$ in A.

To use Eqs. (7)–(9) effectively, one must take advantage of the fact that $\mathbf{T}^\star$ can be expressed in a number of ways making it unnecessary to perform explicitly the summation indicated in Eq. (10). For example,

$$\mathbf{T}^\star = -{}^A\boldsymbol{\alpha}^B \cdot \underline{\mathbf{I}} - {}^A\boldsymbol{\omega}^B \times \underline{\mathbf{I}} \cdot {}^A\boldsymbol{\omega}^B \tag{12}$$

where ${}^A\boldsymbol{\alpha}^B$ and ${}^A\boldsymbol{\omega}^B$ are, respectively, the angular acceleration of B in A and the angular velocity of B in A, and $\underline{\mathbf{I}}$ is the central inertia dyadic of B (see Sec. 4.5). If $\hat{\mathbf{c}}_1$, $\hat{\mathbf{c}}_2$, $\hat{\mathbf{c}}_3$ form a dextral set of mutually perpendicular unit vectors, each parallel to a central principal axis of B (see Sec. 4.8), but not necessarily fixed in B, and α_j, ω_j, and I_j are defined as

$$\alpha_j \triangleq {}^A\boldsymbol{\alpha}^B \cdot \hat{\mathbf{c}}_j \qquad \omega_j \triangleq {}^A\boldsymbol{\omega}^B \cdot \hat{\mathbf{c}}_j \qquad I_j \triangleq \hat{\mathbf{c}}_j \cdot \underline{\mathbf{I}} \cdot \hat{\mathbf{c}}_j \qquad (j = 1,2,3) \tag{13}$$

then Eq. (12) can be replaced with

$$\begin{aligned}\mathbf{T}^\star = &-[\alpha_1 I_1 - \omega_2\omega_3(I_2 - I_3)]\hat{\mathbf{c}}_1 \\ &-[\alpha_2 I_2 - \omega_3\omega_1(I_3 - I_1)]\hat{\mathbf{c}}_2 \\ &-[\alpha_3 I_3 - \omega_1\omega_2(I_1 - I_2)]\hat{\mathbf{c}}_3\end{aligned} \tag{14}$$

Derivations To establish the validity of Eqs. (5) and (6), one can proceed as in the derivations of Eqs. (5.4.4) and (5.4.5).

With β, m_i, and $\mathbf{r}_i$ as defined,

$$\sum_{i=1}^{\beta} m_i \mathbf{r}_i \underset{(4.1.1)}{=} \mathbf{0} \tag{15}$$

and ${}^A\mathbf{a}^{P_i}$ can be expressed as

$${}^A\mathbf{a}^{P_i} \underset{(2.7.2)}{=} {}^A\mathbf{a}^{B^\star} + {}^A\boldsymbol{\alpha}^B \times \mathbf{r}_i + {}^A\boldsymbol{\omega}^B \times ({}^A\boldsymbol{\omega}^B \times \mathbf{r}_i) \qquad (i = 1,\ldots,\beta) \tag{16}$$

Consequently,

$$\begin{aligned}\sum_{i=1}^{\beta} m_i {}^A\mathbf{a}^{P_i} &\underset{(16)}{=} \sum_{i=1}^{\beta} m_i {}^A\mathbf{a}^{B^\star} + {}^A\boldsymbol{\alpha}^B \times \sum_{i=1}^{\beta} m_i \mathbf{r}_i + {}^A\boldsymbol{\omega}^B \times \left({}^A\boldsymbol{\omega}^B \times \sum_{i=1}^{\beta} m_i \mathbf{r}_i\right) \\ &\underset{(15)}{=} M {}^A\mathbf{a}^{B^\star} \underset{(11)}{=} -\mathbf{R}^\star\end{aligned} \tag{17}$$

Now,

$$(F_r^\star)_B \underset{(1,\,4)}{=} -\sum_{i=1}^{\beta} m_i {}^A\mathbf{v}_r^{P_i} \cdot {}^A\mathbf{a}^{P_i} \qquad (r = 1,\ldots,n) \tag{18}$$

and, since the velocity ${}^A\mathbf{v}^{P_i}$ of P_i in A and the velocity ${}^A\mathbf{v}^{B^\star}$ of $B^\star$ in A are related by

$${}^A\mathbf{v}^{P_i} \underset{(2.7.1)}{=} {}^A\mathbf{v}^{B^\star} + {}^A\boldsymbol{\omega}^B \times \mathbf{r}_i \qquad (i = 1,\ldots,\beta) \tag{19}$$

it follows from Eqs. (3.6.2) and (3.6.1) that

$${}^A\mathbf{v}_r^{P_i} = {}^A\mathbf{v}_r^{B^\star} + {}^A\boldsymbol{\omega}_r^B \times \mathbf{r}_i \qquad (i = 1,\ldots,\beta;\ r = 1,\ldots,n) \tag{20}$$

Hence,

$$
\begin{aligned}
(F_r^{\star})_B &\underset{(18)}{=} -\sum_{i=1}^{\beta} m_i({}^A\mathbf{v}_r^{B^\star} \underset{(20)}{+} {}^A\boldsymbol{\omega}_r^B \times \mathbf{r}_i)\cdot {}^A\mathbf{a}^{P_i} \\
&= -{}^A\mathbf{v}_r^{B^\star}\cdot \sum_{i=1}^{\beta} m_i {}^A\mathbf{a}^{P_i} - {}^A\boldsymbol{\omega}_r^B \cdot \sum_{i=1}^{\beta} m_i \mathbf{r}_i \times {}^A\mathbf{a}^{P_i} \\
&= {}^A\mathbf{v}_r^{B^\star}\cdot \underset{(17)}{\mathbf{R}^\star} + {}^A\boldsymbol{\omega}_r^B \cdot \underset{(10)}{\mathbf{T}^\star} \qquad (r = 1,\ldots,n)
\end{aligned} \tag{21}
$$

which establishes the validity of Eqs. (7). Equations (8) can be obtained similarly, by first appealing to Eqs. (2) and then allowing ${}^A\tilde{\mathbf{v}}_r^{P_i}$, ${}^A\tilde{\mathbf{v}}_r^{B^\star}$, and ${}^A\tilde{\boldsymbol{\omega}}_r^B$ $(r = 1,\ldots,p)$ to play the roles of ${}^A\mathbf{v}_r^{P_i}$, ${}^A\mathbf{v}_r^{B^\star}$, and ${}^A\boldsymbol{\omega}_r^B$ $(r = 1,\ldots,n)$, respectively, in Eqs. (20). Finally, one can proceed in much the same way from Eqs. (3) to (9) by permitting $B^\star$ to play the part of Q in Eqs. (5.5.10), and using these relationships in place of (20).

With the aid of Eqs. (16), one can express $\mathbf{T}^\star$ as

$$
\begin{aligned}
\mathbf{T}^\star &\underset{(10)}{=} -\sum_{i=1}^{\beta} m_i\mathbf{r}_i \times [{}^A\mathbf{a}^{B^\star} + {}^A\boldsymbol{\alpha}^B \times \mathbf{r}_i + {}^A\boldsymbol{\omega}^B \times ({}^A\boldsymbol{\omega}^B \times \mathbf{r}_i)] \\
&\underset{(15)}{=} -\sum_{i=1}^{\beta} m_i\mathbf{r}_i \times ({}^A\boldsymbol{\alpha}^B \times \mathbf{r}_i) - \sum_{i=1}^{\beta} m_i\mathbf{r}_i \times [{}^A\boldsymbol{\omega}^B \times ({}^A\boldsymbol{\omega}^B \times \mathbf{r}_i)]
\end{aligned} \tag{22}
$$

Now,

$$
\begin{aligned}
\sum_{i=1}^{\beta} m_i\mathbf{r}_i \times ({}^A\boldsymbol{\alpha}^B \times \mathbf{r}_i) &= \sum_{i=1}^{\beta} m_i(\mathbf{r}_i^{2\,A}\boldsymbol{\alpha}^B - {}^A\boldsymbol{\alpha}^B \cdot \mathbf{r}_i\mathbf{r}_i) \\
&= {}^A\boldsymbol{\alpha}^B \cdot \sum_{i=1}^{\beta} m_i(\underline{\mathbf{U}}\mathbf{r}_i^2 - \mathbf{r}_i\mathbf{r}_i) \underset{(4.5.16)}{=} {}^A\boldsymbol{\alpha}^B \cdot \underline{\mathbf{I}}
\end{aligned} \tag{23}
$$

and

$$
\begin{aligned}
\sum_{i=1}^{\beta} m_i\mathbf{r}_i \times [{}^A\boldsymbol{\omega}^B \times ({}^A\boldsymbol{\omega}^B \times \mathbf{r}_i)] &= -\sum_{i=1}^{\beta} m_i\mathbf{r}_i \cdot {}^A\boldsymbol{\omega}^{B\,A}\boldsymbol{\omega}^B \times \mathbf{r}_i \\
&= -{}^A\boldsymbol{\omega}^B \times \left[\left(\sum_{i=1}^{\beta} m_i\mathbf{r}_i\mathbf{r}_i\right)\cdot {}^A\boldsymbol{\omega}^B\right] \\
&= -{}^A\boldsymbol{\omega}^B \times \left\{\left[\sum_{i=1}^{\beta} m_i(\mathbf{r}_i\mathbf{r}_i - \mathbf{r}_i^2\underline{\mathbf{U}})\right]\cdot {}^A\boldsymbol{\omega}^B\right\} \\
&\underset{(4.5.16)}{=} {}^A\boldsymbol{\omega}^B \times \underline{\mathbf{I}} \cdot {}^A\boldsymbol{\omega}^B
\end{aligned} \tag{24}
$$

Substitution from Eqs. (23) and (24) into Eq. (22) produces Eq. (12).

With α_j, ω_j, and I_j as defined in Eqs. (13),

$$
{}^A\boldsymbol{\alpha}^B = \alpha_1\hat{\mathbf{c}}_1 + \alpha_2\hat{\mathbf{c}}_2 + \alpha_3\hat{\mathbf{c}}_3 \tag{25}
$$

$$
{}^A\boldsymbol{\omega}^B = \omega_1\hat{\mathbf{c}}_1 + \omega_2\hat{\mathbf{c}}_2 + \omega_3\hat{\mathbf{c}}_3 \tag{26}
$$

and

$$\underline{\mathbf{I}} = I_1\hat{\mathbf{c}}_1\hat{\mathbf{c}}_1 + I_2\hat{\mathbf{c}}_2\hat{\mathbf{c}}_2 + I_3\hat{\mathbf{c}}_3\hat{\mathbf{c}}_3 \tag{27}$$

so that

$${}^A\boldsymbol{\alpha}^B \cdot \underline{\mathbf{I}} \underset{(25,\ 27)}{=} \alpha_1 I_1\hat{\mathbf{c}}_1 + \alpha_2 I_2\hat{\mathbf{c}}_2 + \alpha_3 I_3\hat{\mathbf{c}}_3 \tag{28}$$

and

$${}^A\boldsymbol{\omega}^B \times \underline{\mathbf{I}} \cdot {}^A\boldsymbol{\omega}^B = -\omega_2\omega_3(I_2 - I_3)\hat{\mathbf{c}}_1 - \omega_3\omega_1(I_3 - I_1)\hat{\mathbf{c}}_2 - \omega_1\omega_2(I_1 - I_2)\hat{\mathbf{c}}_3 \tag{29}$$

Equation (14) thus follows directly from Eqs. (12), (28), and (29).

Example For the system considered in the example in Sec. 5.8, the generalized inertia forces corresponding to u_1 and u_2 are given by

$$F_r^\star \underset{(1,\ 4,\ 7,\ 11)}{=} {}^B\mathbf{v}_r^P \cdot (-m\,{}^B\mathbf{a}^P) + {}^B\boldsymbol{\omega}_r^R \cdot \mathbf{T}^\star + {}^B\mathbf{v}_r^{R^\star} \cdot (-M\,{}^B\mathbf{a}^{R^\star}) \quad (r = 1,2) \tag{30}$$

where ${}^B\mathbf{v}_r^P$ and ${}^B\boldsymbol{\omega}_r^R$ $(r = 1,2)$ are given in Eqs. (5.8.13) and (5.8.14), ${}^B\mathbf{v}_r^{R^\star}$ $(r = 1,2)$ are partial velocities in B of the mass center of R, ${}^B\mathbf{a}^P$ is the acceleration in B of P, ${}^B\mathbf{a}^{R^\star}$ is the acceleration in B of the mass center of R, and $\mathbf{T}^\star$ is the inertia torque for R. To construct the necessary expressions for ${}^B\mathbf{v}_r^{R^\star}$ $(r = 1,2)$, ${}^B\mathbf{a}^P$, ${}^B\mathbf{a}^{R^\star}$, and $\mathbf{T}^\star$, begin by noting that ${}^B\mathbf{v}^{R^\star}$, the velocity in B of the mass center of R, is given by

$${}^B\mathbf{v}^{R^\star} = u_2\hat{\mathbf{s}}_1 \times (L\hat{\mathbf{r}}_1) = u_2 L \sin\beta\,\hat{\mathbf{r}}_3 \tag{31}$$

so that

$${}^B\mathbf{v}_1^{R^\star} \underset{(31)}{=} \mathbf{0} \qquad {}^B\mathbf{v}_2^{R^\star} \underset{(31)}{=} L\sin\beta\,\hat{\mathbf{r}}_3 \tag{32}$$

while

$${}^B\mathbf{a}^{R^\star} = \frac{{}^B d\,{}^B\mathbf{v}^{R^\star}}{dt} \underset{(31)}{=} L\sin\beta(\dot{u}_2\hat{\mathbf{r}}_3 - {u_2}^2\hat{\mathbf{s}}_2) \tag{33}$$

Next, write

$$\begin{aligned}
{}^B\mathbf{a}^P &= \frac{{}^B d\,{}^B\mathbf{v}^P}{dt} = \frac{{}^B d}{dt}(u_1\hat{\mathbf{r}}_1 + u_2 q_1 \sin\beta\,\hat{\mathbf{r}}_3) \\
&\underset{(5.8.12)}{=} \dot{u}_1\hat{\mathbf{r}}_1 + (\dot{u}_2 q_1 + u_2\dot{q}_1)\sin\beta\,\hat{\mathbf{r}}_3 + (u_2\hat{\mathbf{s}}_1) \times (u_1\hat{\mathbf{r}}_1 + u_2 q_1 \sin\beta\,\hat{\mathbf{r}}_3) \\
&= \dot{u}_1\hat{\mathbf{r}}_1 + [(\dot{u}_2 q_1 + 2u_1u_2)\hat{\mathbf{r}}_3 - {u_2}^2 q_1\hat{\mathbf{s}}_2]\sin\beta
\end{aligned} \tag{34}$$

and express ω_j, α_j, and I_j $(j = 1,2,3)$, needed for substitution into Eq. (14), as

$$\omega_1 = {}^B\boldsymbol{\omega}^R \cdot \hat{\mathbf{r}}_1 = u_2\hat{\mathbf{s}}_1 \cdot \hat{\mathbf{r}}_1 = u_2\cos\beta \qquad \omega_2 = -u_2\sin\beta \qquad \omega_3 = 0 \tag{35}$$

$$\alpha_1 = \dot{\omega}_1 = \dot{u}_2\cos\beta \qquad \alpha_2 = -\dot{u}_2\sin\beta \qquad \alpha_3 = 0 \tag{36}$$

$$I_1 = 0 \qquad I_2 = \frac{ML^2}{3} \qquad I_3 = \frac{ML^2}{3} \tag{37}$$

so that, from Eq. (14),

$$\mathbf{T}^\star = \frac{ML^2}{3}\dot{u}_2 \sin\beta\hat{\mathbf{r}}_2 + \frac{ML^2}{3}{u_2}^2 \sin\beta\cos\beta\hat{\mathbf{r}}_3 \tag{38}$$

Finally, substitute into Eqs. (30) to obtain

$$F_1^\star = -m\hat{\mathbf{r}}_1 \cdot \left\{\dot{u}_1\hat{\mathbf{r}}_1 + [(\dot{u}_2 q_1 + 2u_1u_2)\hat{\mathbf{r}}_3 - {u_2}^2 q_1\hat{\mathbf{s}}_2]\sin\beta\right\}$$

(5.8.13, 34, 32)

$$= -m(\dot{u}_1 - {u_2}^2 q_1 \sin^2\beta) \tag{39}$$

$$F_2^\star = -\left[\left(m{q_1}^2 + \frac{4ML^2}{3}\right)\dot{u}_2 + 2mq_1u_1u_2\right]\sin^2\beta \tag{40}$$

(5.8.14, 34, 38, 32, 33)

Equations (39) and (40) furnish expressions for the generalized inertia forces that correspond to the generalized active forces F_1 and F_2 given in Eqs. (5.8.16) and (5.8.17), respectively.

6 CONSTRAINT FORCES, CONSTRAINT TORQUES

Satisfaction of the constraints discussed in Chapter 3 requires that the particles of a system be subject to the actions of certain forces known as *constraint forces*. When a set of such forces constitutes a couple, the associated torque is referred to as a *constraint torque*. One may express constraint equations in terms of accelerations of points and/or angular accelerations of rigid bodies. Such expressions then can be employed to determine the directions of constraint forces and the points to which they must be applied; likewise, one can identify the directions of constraint torques and the rigid bodies upon which they must act. Procedures for utilizing constraint equations in this manner are presented early in the chapter. The forces and torques needed to bring about constraints contribute to certain types of generalized active forces but not to other types, depending on the nature of the constraint. This fact permits the analyst to bring constraint forces and constraint torques into evidence in equations of motion or to leave them out, depending on whether such forces and torques are, or are not, of interest.

6.1 CONSTRAINT EQUATIONS, ACCELERATION, FORCE

When a configuration constraint is imposed on a set S of ν particles $P_1, \ldots, P_\nu$ moving in a reference frame A, the constraint can be described in the manner of Eq. (3.1.1) with a holonomic relationship involving position vectors. As will be demonstrated in Sec. 6.2, the time derivative of such an equation can be expressed as

$$\sum_{i=1}^{\nu} {}^A\mathbf{v}^{P_i} \cdot \mathbf{w}_i + Y = 0 \tag{1}$$

where Y is a scalar, ${}^A\mathbf{v}^{P_i}$ is the velocity of P_i in A, and the vector $\mathbf{w}_i$ is not the velocity of any particle in any reference frame, which means that Eq. (1) is linear in velocity. Differentiation once more with respect to time yields

$$\sum_{i=1}^{\nu} {}^A\mathbf{a}^{P_i} \cdot \mathbf{w}_i + Z = 0 \tag{2}$$

where Z is a scalar and ${}^A\mathbf{a}^{P_i}$ is the acceleration of P_i in A. We will establish in Sec. 6.3 that a motion constraint can often be expressed with a relationship having the form of Eq. (1), in which case differentiation again leads to Eq. (2). In fact, even when a motion constraint is described with a nonholonomic equation having the general form of Eq.

(3.5.1), differentiation leads once more to Eq. (2) as will be seen in Sec. 6.4. In each of these cases $\mathbf{w}_i$ is not the acceleration of any particle in any reference frame, which means that Eq. (2) is linear in acceleration.

There exist certain reference frames N that can be regarded as Newtonian or inertial, as discussed in detail in Secs. 8.1 and 8.2. When A can be regarded as such a reference frame, ${}^N\mathbf{v}^{P_i}$ and ${}^N\mathbf{a}^{P_i}$ play the respective roles of ${}^A\mathbf{v}^{P_i}$ and ${}^A\mathbf{a}^{P_i}$ in constraint equations (1) and (2).

The motion of P_i is governed by Newton's second law, which asserts

$$\mathbf{R}_i = m_i {}^N\mathbf{a}^{P_i} \qquad (i = 1,\ldots,\nu) \tag{3}$$

where $\mathbf{R}_i$ is the resultant of all contact forces and distance forces applied to P_i, m_i is the mass of P_i, and ${}^N\mathbf{a}^{P_i}$ is the acceleration of P_i in N. $\mathbf{R}_i$ can be regarded as the sum

$$\mathbf{R}_i = \mathbf{f}_i + \mathbf{C}_i \qquad (i = 1,\ldots,\nu) \tag{4}$$

where the force $\mathbf{f}_i$ acts on P_i regardless of whether Eq. (2) is satisfied, and where $\mathbf{C}_i$ is a constraint force applied to P_i so as to ensure satisfaction of Eq. (2). Substitution from Eqs. (4) and (3) into (2) yields

$$\sum_{i=1}^{\nu} \frac{(\mathbf{f}_i + \mathbf{C}_i)}{m_i} \cdot \mathbf{w}_i + Z = 0 \tag{5}$$

Because any component of $\mathbf{C}_i$ perpendicular to $\mathbf{w}_i$ will not play a part in Eq. (5), it is sufficient for $\mathbf{C}_i$ to be parallel to $\mathbf{w}_i$, a condition expressed as $\mathbf{C}_i = \lambda_i \mathbf{w}_i$, where the scalar λ_i scales the vector $\mathbf{w}_i$ $(i = 1,\ldots,\nu)$. The scalars are chosen such that $\lambda_1 = \lambda_2 = \ldots = \lambda_\nu = \lambda$ for two reasons. First, this leads to a pair of constraint forces related by the law of action and reaction when two particles exert constraint forces on each other. Second, this results in one scalar being associated with each constraint equation having the form of (2), which in turn facilitates the process of solving equations for the scalars. Therefore, constraint forces that ensure satisfaction of Eq. (2) are written as

$$\mathbf{C}_i = \lambda \mathbf{w}_i \qquad (i = 1,\ldots,\nu) \tag{6}$$

When a constraint equation having the form of (2) is in hand and A can be regarded as a Newtonian reference frame, one may inspect the equation for the presence of dot products ${}^N\mathbf{a}^{P_i} \cdot \mathbf{w}_i$ and proceed immediately to write Eqs. (6). In doing so, one determines that the constraint can be satisfied by applying to P_i a constraint force parallel to $\mathbf{w}_i$ $(i = 1,\ldots,\nu)$. In the event that a constraint equation in the form of (1) is available, one may instead inspect such a relationship. The appearance of dot products ${}^N\mathbf{v}^{P_i} \cdot \mathbf{w}_i$ allows one to write Eqs. (6) and reach the same conclusions regarding the point of application and direction of the constraint force $\mathbf{C}_i$.

In general, one cannot expect Eq. (5) to be satisfied in the absence of the constraint forces $\mathbf{C}_1,\ldots,\mathbf{C}_\nu$. The value of λ that does permit Eq. (5) [and, hence, Eq. (2)] to be satisfied evidently depends on the mass of each particle as well as the nature of the force $\mathbf{f}_i$ applied to it. An understanding of the material presented in Secs. 6.7, 8.3, and 9.7 will bring one into position to determine λ.

6.2 HOLONOMIC CONSTRAINT EQUATIONS

When a set S of ν particles $P_1,\ldots,P_\nu$ is subject to configuration constraints (see Sec. 3.1), the associated holonomic constraint equations can be expressed in the form of Eq. (3.1.1). In particular, when M such independent equations are under consideration, one can write

$$g_s(\mathbf{p}_1,\ldots,\mathbf{p}_\nu,t) = 0 \qquad (s = 1,\ldots,M) \tag{1}$$

where g_s is a scalar function of the time t and of $\mathbf{p}_1,\ldots,\mathbf{p}_\nu$, the position vectors to $P_1,\ldots,P_\nu$, respectively, from a point O fixed in a reference frame A.

The ordinary derivative of the scalar function g_s with respect to t can be written as

$$\frac{dg_s}{dt} = \sum_{i=1}^{\nu} {}^A\mathbf{v}^{P_i} \cdot \mathbf{w}_{is} + Y_s = 0 \qquad (s = 1,\ldots,M) \tag{2}$$

where ${}^A\mathbf{v}^{P_i}$ is the velocity of P_i in A, $\mathbf{w}_{is}$ is a vector function in A of t and generalized coordinates $q_1,\ldots,q_n$ for S in A, and Y_s is a scalar function of the same variables. It is important to note that $\mathbf{w}_{is}$ is not a vector function in A of any generalized coordinate time derivatives; therefore, $\mathbf{w}_{is}$ is not a velocity of any particle in any reference frame. Consequently, Eqs. (2) are said to be linear in velocity.

Differentiation once more with respect to time yields

$$\frac{d^2g_s}{dt^2} = \sum_{i=1}^{\nu} {}^A\mathbf{a}^{P_i} \cdot \mathbf{w}_{is} + Z_s = 0 \qquad (s = 1,\ldots,M) \tag{3}$$

where ${}^A\mathbf{a}^{P_i}$ is the acceleration of P_i in A and Z_s is a scalar function of $q_1,\ldots,q_n$, $\dot{q}_1,\ldots,\dot{q}_n$, and t. When A can be regarded as a Newtonian reference frame, in view of Sec. 6.1, one may inspect either Eqs. (2) or (3) and conclude that constraint forces given by

$$\mathbf{C}_{is} = \lambda_s \mathbf{w}_{is} \qquad (i = 1,\ldots,\nu;\ s = 1,\ldots,M) \tag{4}$$

must be applied to P_i in order for the configuration constraints to be satisfied.

Derivation The ordinary derivative of g_s with respect to time can be obtained with the aid of Eq. (1.13.1) by regarding g_s as a scalar function of the scalar variable t and ν vector variables $\mathbf{p}_1,\ldots,\mathbf{p}_\nu$, each of which in turn is regarded as a function of t. Accordingly,

$$\frac{dg_s}{dt} \underset{(1.13.1)}{=} \sum_{i=1}^{\nu} \frac{\partial g_s}{\partial \mathbf{p}_i} \cdot \frac{{}^A d\mathbf{p}_i}{dt} + \frac{\partial g_s}{\partial t} = 0 \qquad (s = 1,\ldots,M) \tag{5}$$

Equations (2) then are obtained from these relationships after defining $\mathbf{w}_{is}$, ${}^A\mathbf{v}^{P_i}$, and Y_s,

$$\mathbf{w}_{is} \triangleq \frac{\partial g_s}{\partial \mathbf{p}_i} \qquad {}^A\mathbf{v}^{P_i} \underset{(2.6.1)}{\triangleq} \frac{{}^A d\mathbf{p}_i}{dt} \qquad Y_s \triangleq \frac{\partial g_s}{\partial t} \qquad (i = 1,\ldots,\nu;\ s = 1,\ldots,M) \tag{6}$$

Finally, time differentiation of Eqs. (2) leads to

$$\sum_{i=1}^{\nu}\left(\frac{{}^{A}d\,{}^{A}\mathbf{v}^{P_i}}{dt}\cdot\mathbf{w}_{is}+{}^{A}\mathbf{v}^{P_i}\cdot\frac{{}^{A}d\mathbf{w}_{is}}{dt}\right)+\frac{dY_s}{dt}=0 \qquad (s=1,\ldots,M) \tag{7}$$

in agreement with Eqs. (3) after ${}^{A}\mathbf{a}^{P_i}$ and Z_s are defined as

$$ {}^{A}\mathbf{a}^{P_i}\underset{(2.6.2)}{\triangleq}\frac{{}^{A}d\,{}^{A}\mathbf{v}^{P_i}}{dt} \qquad Z_s\triangleq\frac{dY_s}{dt}+\sum_{i=1}^{\nu}{}^{A}\mathbf{v}^{P_i}\cdot\frac{{}^{A}d\mathbf{w}_{is}}{dt} \qquad (s=1,\ldots,M) \tag{8}$$

Example The set S of two particles considered in the example in Sec. 3.1 is subject to configuration constraints described by three holonomic constraint equations that have the form of Eqs. (1). Requirements for P_1 and P_2 to remain between two panes of glass are described, respectively, by the two relationships

$$g_s=\mathbf{p}_s\cdot\hat{\mathbf{b}}_z\underset{(3.1.4)}{=}0 \qquad (s=1,2) \tag{9}$$

where $\hat{\mathbf{b}}_z$ is a unit vector normal to the plane determined by the panes of glass, directed as in Fig. 3.1.2. The thin rigid rod of constant length L that connects P_1 and P_2 imposes a third requirement expressed as

$$g_3=(\mathbf{p}_1-\mathbf{p}_2)\cdot(\mathbf{p}_1-\mathbf{p}_2)-L^2\underset{(3.1.5)}{=}0 \tag{10}$$

If reference frame A (see Figs. 3.1.1 and 3.1.2) can be regarded as Newtonian, constraint forces needed to satisfy Eqs. (9)–(10) are identified as follows. First, Eqs. (9) are differentiated with respect to time t,

$$\begin{aligned}\frac{dg_s}{dt}&\underset{(9)}{=}\frac{{}^{A}d\mathbf{p}_s}{dt}\cdot\hat{\mathbf{b}}_z+\mathbf{p}_s\cdot\frac{{}^{A}d\hat{\mathbf{b}}_z}{dt}\\&\underset{(2.6.1)}{=}{}^{A}\mathbf{v}^{P_s}\cdot\hat{\mathbf{b}}_z+\mathbf{p}_s\cdot\underset{(2.3.1)}{(\omega\hat{\mathbf{b}}_y\times\hat{\mathbf{b}}_z)}\\&={}^{A}\mathbf{v}^{P_s}\cdot\hat{\mathbf{b}}_z+\omega\mathbf{p}_s\cdot\hat{\mathbf{b}}_x=0 \qquad (s=1,2)\end{aligned} \tag{11}$$

The vectors $\mathbf{w}_{is}$ and scalars Y_s ($i=1,2;\ s=1,2$) are obtained by comparing each of these relationships with Eqs. (2). First, with $s=1$,

$$\mathbf{w}_{11}=\hat{\mathbf{b}}_z \qquad \mathbf{w}_{21}=\mathbf{0} \qquad Y_1=\omega\mathbf{p}_1\cdot\hat{\mathbf{b}}_x \tag{12}$$

Consequently, constraint forces to be applied to P_1 and P_2 are identified as

$$\mathbf{C}_{11}\underset{(4,\,12)}{=}\lambda_1\hat{\mathbf{b}}_z \qquad \mathbf{C}_{21}\underset{(4,\,12)}{=}\mathbf{0} \tag{13}$$

Similarly, with $s=2$ inspection reveals that

$$\mathbf{w}_{12}=\mathbf{0} \qquad \mathbf{w}_{22}=\hat{\mathbf{b}}_z \qquad Y_2=\omega\mathbf{p}_2\cdot\hat{\mathbf{b}}_x \tag{14}$$

and the associated constraint forces are given by

$$\mathbf{C}_{12}\underset{(4,\,14)}{=}\mathbf{0} \qquad \mathbf{C}_{22}\underset{(4,\,14)}{=}\lambda_2\hat{\mathbf{b}}_z \tag{15}$$

Next, time differentiation of Eq. (10) yields

$$\frac{dg_3}{dt} = \left(\frac{{}^A d\mathbf{p}_1}{dt} - \frac{{}^A d\mathbf{p}_2}{dt}\right) \cdot (\mathbf{p}_1 - \mathbf{p}_2) + (\mathbf{p}_1 - \mathbf{p}_2) \cdot \left(\frac{{}^A d\mathbf{p}_1}{dt} - \frac{{}^A d\mathbf{p}_2}{dt}\right) + 0$$
$$= 2\underset{(2.6.1)}{\left({}^A\mathbf{v}^{P_1} - {}^A\mathbf{v}^{P_2}\right)} \cdot (\mathbf{p}_1 - \mathbf{p}_2) = 0 \tag{16}$$

A comparison of this relationship with Eqs. (2) leads, with $s = 3$, to

$$\mathbf{w}_{13} = 2(\mathbf{p}_1 - \mathbf{p}_2) \qquad \mathbf{w}_{23} = -2(\mathbf{p}_1 - \mathbf{p}_2) \qquad Y_3 = 0 \tag{17}$$

and subsequently to the conclusion that constraint forces are given by

$$\mathbf{C}_{13} \underset{(4,\,17)}{=} 2\lambda_3(\mathbf{p}_1 - \mathbf{p}_2) \qquad \mathbf{C}_{23} \underset{(4,\,17)}{=} -2\lambda_3(\mathbf{p}_1 - \mathbf{p}_2) \tag{18}$$

This result, $\mathbf{C}_{23} = -\mathbf{C}_{13}$, is seen to be in accordance with the law of action and reaction.

The resultant of all constraint forces applied to P_1 is given by

$$\mathbf{C}_1 = \mathbf{C}_{11} + \mathbf{C}_{12} + \mathbf{C}_{13} = \underset{(13)}{\lambda_1 \hat{\mathbf{b}}_z} + \underset{(18)}{2\lambda_3(\mathbf{p}_1 - \mathbf{p}_2)} \tag{19}$$

whereas the resultant constraint force applied to P_2 is given by

$$\mathbf{C}_2 = \mathbf{C}_{21} + \mathbf{C}_{22} + \mathbf{C}_{23} = \underset{(15)}{\lambda_2 \hat{\mathbf{b}}_z} - \underset{(18)}{2\lambda_3(\mathbf{p}_1 - \mathbf{p}_2)} \tag{20}$$

6.3 LINEAR NONHOLONOMIC CONSTRAINT EQUATIONS

Let S be a set of ν particles $P_1, \ldots, P_\nu$ whose configuration in a reference frame A is described by generalized coordinates $q_1, \ldots, q_n$ (see Sec. 3.2). When S is subject to motion constraints (see Sec. 3.5), the associated nonholonomic constraint equations can be expressed in the form of Eq. (3.5.1). In many cases the equations are linear in velocity; the condition of rolling (which is the absence of slipping) and the restriction on velocity imposed by a sharp-edged blade furnish two examples of motion constraints described by such equations. Nonholonomic constraint equations that are linear in velocity can be expressed in vector form as

$$\sum_{i=1}^{\nu} {}^A\mathbf{v}^{P_i} \cdot \mathbf{w}_{is} + Y_s = 0 \qquad (s = 1, \ldots, m) \tag{1}$$

where ${}^A\mathbf{v}^{P_i}$ is the velocity of P_i in A, $\mathbf{w}_{is}$ are vector functions of $q_1, \ldots, q_n$ and t in A, and Y_s are scalar functions of the same variables. Although Eqs. (6.2.2) bear a resemblance to Eqs. (1), the latter cannot be obtained by differentiating relationships that involve position vectors, such as Eqs. (6.2.1); hence, Eqs. (1) are referred to as nonholonomic, and $\mathbf{w}_{is}$ and Y_s are *not* defined in terms of partial derivatives as in Eqs. (6.2.6). When S is subject to motion constraints characterized by m independent relationships (1), it is referred to as a simple nonholonomic system (see Sec. 3.5).

Differentiation of Eqs. (1) with respect to time yields

$$\sum_{i=1}^{\nu} {}^{A}\mathbf{a}^{P_i} \cdot \mathbf{w}_{is} + Z_s = 0 \qquad (s = 1,\dots,m) \tag{2}$$

where ${}^{A}\mathbf{a}^{P_i}$ is the acceleration of P_i in A and Z_s is a scalar function of $q_1,\dots,q_n$, $\dot{q}_1,\dots,\dot{q}_n$, and t. When A can be regarded as a Newtonian reference frame, according to the material in Sec. 6.1, one may inspect either Eqs. (1) or (2) and conclude that constraint forces given by

$$\mathbf{C}_{is} = \lambda_s \mathbf{w}_{is} \qquad (i = 1,\dots,\nu;\ s = 1,\dots,m) \tag{3}$$

must be applied to P_i in order for the motion constraints to be satisfied.

Derivation S is referred to as a simple nonholonomic system when motion variables $u_1,\dots,u_n$ for S in A are related to each other by Eqs. (3.5.2), which can be obtained from Eqs. (1) as follows. First, the velocity of P_i in A is expressed in terms of motion variables,

$$ {}^{A}\mathbf{v}^{P_i} \underset{(3.6.2)}{=} \sum_{r=1}^{n} {}^{A}\mathbf{v}_r^{P_i} u_r + {}^{A}\mathbf{v}_t^{P_i} \qquad (i = 1,\dots,\nu) \tag{4}$$

where ${}^{A}\mathbf{v}_r^{P_i}$ $(r = 1,\dots,n)$ and ${}^{A}\mathbf{v}_t^{P_i}$ are functions of $q_1,\dots,q_n$ and the time t, and hence have the same functional character as the vectors $\mathbf{w}_{is}$ in Eqs. (1). Substitution from Eqs. (4) into (1) yields

$$\sum_{i=1}^{\nu}\left(\sum_{r=1}^{n} {}^{A}\mathbf{v}_r^{P_i} u_r + {}^{A}\mathbf{v}_t^{P_i}\right) \cdot \mathbf{w}_{is} + Y_s =$$

$$\sum_{r=1}^{n}\left(\sum_{i=1}^{\nu} {}^{A}\mathbf{v}_r^{P_i} \cdot \mathbf{w}_{is}\right) u_r + \sum_{i=1}^{\nu} {}^{A}\mathbf{v}_t^{P_i} \cdot \mathbf{w}_{is} + Y_s = 0 \qquad (s = 1,\dots,m) \tag{5}$$

The coefficients of u_r and the remaining terms can be abbreviated respectively by means of two definitions,

$$\alpha_{sr} \triangleq \sum_{i=1}^{\nu} {}^{A}\mathbf{v}_r^{P_i} \cdot \mathbf{w}_{is} \qquad (s = 1,\dots,m;\ r = 1,\dots,n) \tag{6}$$

and

$$\beta_s \triangleq Y_s + \sum_{i=1}^{\nu} {}^{A}\mathbf{v}_t^{P_i} \cdot \mathbf{w}_{is} \qquad (s = 1,\dots,m) \tag{7}$$

Use of these definitions in Eqs. (5) leads to relationships that are linear in the motion variables,

$$\sum_{r=1}^{n} \underset{(6)}{\alpha_{sr}}\, u_r + \underset{(7)}{\beta_s} \underset{(5)}{=} 0 \qquad (s = 1,\dots,m) \tag{8}$$

As long as these relationships are independent of one another, they will give way to expressions having the form of Eqs. (3.5.2) in which m of the motion variables, say

$u_{p+1},\ldots,u_n$, are each written in terms of the remaining motion variables (the generalized velocities) $u_1,\ldots,u_p$, where $p = n - m$. Equations (1) are thus shown to describe motion constraints imposed on a simple nonholonomic system.

Finally, time differentiation of Eqs. (1) leads to Eqs. (2) after ${}^A\mathbf{a}^{P_i}$ and Z_s are defined as

$$ {}^A\mathbf{a}^{P_i} \underset{(2.6.2)}{\triangleq} \frac{{}^A d\,{}^A\mathbf{v}^{P_i}}{dt} \qquad Z_s \triangleq \frac{dY_s}{dt} + \sum_{i=1}^{\nu} {}^A\mathbf{v}^{P_i} \cdot \frac{{}^A d\mathbf{w}_{is}}{dt} \qquad (s = 1,\ldots,m) \tag{9} $$

Example The example in Sec. 3.5 involves a system S modeled as a particle P_1 rigidly connected to a second particle $D^\star$ located at the center of a sharp-edged circular disk. The sharp edge constitutes a motion constraint that prevents $D^\star$ from moving in reference frame B in the direction of unit vector $\hat{\mathbf{e}}_y$ (see Fig. 3.5.1). The constraint can be expressed as

$$ \begin{aligned} ({}^A\mathbf{v}^{D^\star} - {}^A\mathbf{v}^{\overline{B}}) \cdot \hat{\mathbf{e}}_y &\underset{(3.5.6)}{=} {}^A\mathbf{v}^{D^\star} \cdot \hat{\mathbf{e}}_y + \omega(q_1 + Lc_3)\hat{\mathbf{e}}_z \underset{(3.5.9)}{\cdot} \hat{\mathbf{e}}_y \\ &= {}^A\mathbf{v}^{D^\star} \cdot \hat{\mathbf{e}}_y = 0 \end{aligned} \tag{10} $$

where $\overline{B}$, the point of B with which $D^\star$ coincides, does not belong to S. This relationship can be compared to Eqs. (1) to identify

$$ \mathbf{w} = \hat{\mathbf{e}}_y \qquad Y = 0 \tag{11} $$

where the subscripts $i = 2$ and $s = 1$ have been omitted for convenience. If A can be regarded as a Newtonian reference frame, then it can be concluded that a constraint force given by

$$ \mathbf{C} \underset{(3,\ 11)}{=} \lambda\hat{\mathbf{e}}_y \tag{12} $$

must be applied to $D^\star$ to prevent it from moving in the direction of $\hat{\mathbf{e}}_y$.

6.4 NONLINEAR NONHOLONOMIC CONSTRAINT EQUATIONS

When a set S of ν particles $P_1,\ldots,P_\nu$ is subject to one or more motion constraints (see Sec. 3.5) described by relationships that are inherently nonlinear in the velocities ${}^A\mathbf{v}^{P_1},\ldots,{}^A\mathbf{v}^{P_\nu}$ of $P_1,\ldots,P_\nu$ in a reference frame A, S is referred to as a complex nonholonomic system (see Sec. 3.7). The nonlinear nonholonomic constraint equations can be expressed in terms of ℓ scalar functions

$$ f_s({}^A\mathbf{v}^{P_1},\ldots,{}^A\mathbf{v}^{P_\nu},t) = 0 \qquad (s = 1,\ldots,\ell) \tag{1} $$

Differentiation of these relationships with respect to time t yields equations that are linear in acceleration,

$$ \sum_{i=1}^{\nu} {}^A\mathbf{a}^{P_i} \cdot \mathbf{W}_{is} + Z_s = 0 \qquad (s = 1,\ldots,\ell) \tag{2} $$

where $\mathbf{W}_{is}$ are vector functions of $q_1,\ldots,q_n$, $u_1,\ldots,u_p$, and t in A, and Z_s are scalar functions of the same variables. The acceleration of P_i in A is denoted by ${}^A\mathbf{a}^{P_i}$. Provided that Eqs. (2) are independent of one another, they can be brought into the form of Eqs. (3.7.1).

When A can be regarded as a Newtonian reference frame, in view of Sec. 6.1, one may inspect Eqs. (2) and conclude that constraint forces given by

$$\mathbf{C}_{is} = \mu_s \mathbf{W}_{is} \qquad (i = 1,\ldots,\nu;\ s = 1,\ldots,\ell) \tag{3}$$

must be applied to P_i in order for the motion constraints to be satisfied.

Derivations The ordinary derivative of f_s with respect to time can be obtained from Eqs. (1) with the aid of Eq. (1.13.1) by regarding f_s as a scalar function of the scalar variable t and ν vector variables ${}^A\mathbf{v}^{P_1},\ldots,{}^A\mathbf{v}^{P_\nu}$, each of which in turn is regarded as a function of t. Accordingly,

$$\frac{df_s}{dt} \underset{(1.13.1)}{=} \sum_{i=1}^{\nu} \frac{\partial f_s}{\partial\, {}^A\mathbf{v}^{P_i}} \cdot \frac{{}^A d\, {}^A\mathbf{v}^{P_i}}{dt} + \frac{\partial f_s}{\partial t} = 0 \qquad (s = 1,\ldots,\ell) \tag{4}$$

Equations (2) then are obtained from these relationships after defining $\mathbf{W}_{is}$, ${}^A\mathbf{a}^{P_i}$, and Z_s, as

$$\mathbf{W}_{is} \triangleq \frac{\partial f_s}{\partial\, {}^A\mathbf{v}^{P_i}} \qquad {}^A\mathbf{a}^{P_i} \underset{(2.6.2)}{\triangleq} \frac{{}^A d\, {}^A\mathbf{v}^{P_i}}{dt} \qquad Z_s \triangleq \frac{\partial f_s}{\partial t} \qquad (i = 1,\ldots,\nu;\ s = 1,\ldots,\ell) \tag{5}$$

To see that Eqs. (2) lead to (3.7.1), one can proceed as follows. First, express the acceleration of P_i in A in terms of motion-variable time derivatives,

$${}^A\mathbf{a}^{P_i} \underset{(3.8.2)}{=} \sum_{r=1}^{p} {}^A\mathbf{a}_r^{P_i}\, \dot{u}_r + {}^A\mathbf{a}_t^{P_i} \qquad (i = 1,\ldots,\nu) \tag{6}$$

where the partial accelerations ${}^A\mathbf{a}_r^{P_i}$ are functions of $q_1,\ldots,q_n$ and t, and where ${}^A\mathbf{a}_t^{P_i}$ are functions of $q_1,\ldots,q_n$, $u_1,\ldots,u_p$, and t. Next, substitute from Eqs. (6) into (2) to obtain

$$\sum_{i=1}^{\nu}\left(\sum_{r=1}^{p} {}^A\mathbf{a}_r^{P_i}\, \dot{u}_r + {}^A\mathbf{a}_t^{P_i}\right) \cdot \mathbf{W}_{is} + Z_s =$$

$$\sum_{r=1}^{p}\left(\sum_{i=1}^{\nu} {}^A\mathbf{a}_r^{P_i} \cdot \mathbf{W}_{is}\right)\dot{u}_r + \sum_{i=1}^{\nu} {}^A\mathbf{a}_t^{P_i} \cdot \mathbf{W}_{is} + Z_s = 0 \qquad (s = 1,\ldots,\ell) \tag{7}$$

The coefficients of $\dot{u}_r$ and the remaining terms can be abbreviated respectively by means of two definitions,

$$\widetilde{\alpha}_{sr} \triangleq \sum_{i=1}^{\nu} {}^A\mathbf{a}_r^{P_i} \cdot \mathbf{W}_{is} \qquad (s = 1,\ldots,\ell;\ r = 1,\ldots,p) \tag{8}$$

and

$$\widetilde{\beta}_s \triangleq Z_s + \sum_{i=1}^{\nu} {}^A\mathbf{a}_t^{P_i} \cdot \mathbf{W}_{is} \qquad (s = 1,\ldots,\ell) \tag{9}$$

Use of these definitions in Eqs. (7) leads to relationships that are linear in motion-variable time derivatives,

$$\sum_{r=1}^{p} \underset{(8)}{\tilde{\alpha}_{sr}} \dot{u}_r + \underset{(9)}{\tilde{\beta}_s} \underset{(7)}{=} 0 \qquad (s = 1, \ldots, \ell) \tag{10}$$

As long as these relationships are independent of one another, they will give way to expressions having the form of Eqs. (3.7.1) in which ℓ of the motion-variable time derivatives, say $\dot{u}_{c+1}, \ldots, \dot{u}_p$, are each written in terms of the remaining motion-variable time derivatives $\dot{u}_1, \ldots, \dot{u}_c$, where $c = p - \ell$.

Example The complex nonholonomic system considered in the example in Sec. 3.7 consists of two particles P_1 and P_2 whose velocities in a reference frame A are required to be perpendicular to each other. This restriction can be expressed by the nonholonomic constraint equation

$$^A\mathbf{v}^{P_2} \cdot {}^A\mathbf{v}^{P_1} \underset{(3.7.4)}{=} 0 \tag{11}$$

that is evidently nonlinear in velocity. Differentiation with respect to time yields

$$^A\mathbf{a}^{P_2} \cdot {}^A\mathbf{v}^{P_1} + {}^A\mathbf{v}^{P_2} \cdot {}^A\mathbf{a}^{P_1} \underset{(3.7.7)}{=} 0 \tag{12}$$

which is linear in acceleration. Substitution into the left-hand member of this relationship from Eqs. (3.7.3), (3.8.17), and (3.8.18) leads directly to the middle member of Eq. (3.7.8). Equation (3.7.9) then follows immediately and that expression has the form of Eqs. (3.7.1) with $p = 4$, $\ell = 1$, and $c = 3$.

Inspection of Eq. (12) with Eqs. (2) in mind enables one to identify the vectors $\mathbf{W}_{is}$ and scalar Z_s ($i = 1,2$; $s = 1$) and record them as

$$\mathbf{W}_{11} = {}^A\mathbf{v}^{P_2} \qquad \mathbf{W}_{21} = {}^A\mathbf{v}^{P_1} \qquad Z_1 = 0 \tag{13}$$

If A can be regarded as a Newtonian reference frame, then constraint forces given by

$$\mathbf{C}_{11} \underset{(3,\ 13)}{=} \mu_1 {}^A\mathbf{v}^{P_2} \qquad \mathbf{C}_{21} \underset{(3,\ 13)}{=} \mu_1 {}^A\mathbf{v}^{P_1} \tag{14}$$

can be applied to P_1 and P_2, respectively, to ensure orthogonality of their velocities.

6.5 CONSTRAINT FORCES ACTING ON A RIGID BODY

If a rigid body B belongs to a system S that is subject to configuration constraints and/or motion constraints described by nonholonomic equations that are linear in velocity, then Eqs. (6.2.2) and (6.3.1), taken together, can be put in the form

$$^A\mathbf{v}^Q \cdot \mathbf{w}_s + \cdots + {}^A\boldsymbol{\omega}^B \cdot \boldsymbol{\tau}_s + \cdots + Y_s = 0 \qquad (s = 1, \ldots, M + m) \tag{1}$$

where $^A\mathbf{v}^Q$ is the velocity in A of a point Q fixed in B and $^A\boldsymbol{\omega}^B$ is the angular velocity of B in A. When A can be regarded as a Newtonian reference frame, in view of Sec. 6.1, one may inspect Eqs. (1) and conclude that constraint forces given by

$$\mathbf{C}_s^Q = \lambda_s \mathbf{w}_s \qquad (s = 1, \ldots, M + m) \tag{2}$$

must be applied to B at Q in order for the constraints to be satisfied. Additionally, one can conclude that B must be acted upon by couples whose constraint torques are

$$\mathbf{T}_s^B = \lambda_s \boldsymbol{\tau}_s \qquad (s = 1, \ldots, M + m) \tag{3}$$

The ellipsis that follows the term ${}^A\mathbf{v}^Q \cdot \mathbf{w}_s$ in Eqs. (1) indicates there may be other such terms involving velocities in A of other points fixed in other rigid bodies, or possibly particles in S that do not belong to a rigid body. Likewise, there may be terms similar to ${}^A\boldsymbol{\omega}^B \cdot \boldsymbol{\tau}_s$ involving angular velocities in A of other rigid bodies.

The technique of inspecting constraint equations having the form of (1) and subsequently expressing constraint forces and constraint torques according to Eqs. (2) and (3) yields several pieces of useful information. In connection with each constraint force, one is able to identify its direction, the rigid body or particle on which it acts, and the point at which it is applied; similarly, one determines the direction of a constraint torque, together with the rigid body to which it is applied.

If a system S contains a rigid body B and is subject to motion constraints described by nonholonomic equations that are inherently nonlinear in velocity, then Eqs. (6.4.2) can be expressed as

$${}^A\mathbf{a}^Q \cdot \mathbf{W}_s + \cdots + {}^A\boldsymbol{\alpha}^B \cdot \boldsymbol{\tau}_s + \cdots + Z_s' = 0 \qquad (s = 1, \ldots, \ell) \tag{4}$$

where ${}^A\mathbf{a}^Q$ is the acceleration of Q in A and ${}^A\boldsymbol{\alpha}^B$ is the angular acceleration of B in A. When A can be regarded as a Newtonian reference frame, in view of Sec. 6.1, one may inspect Eqs. (4) and conclude that constraint forces

$$\mathbf{C}_s^Q = \mu_s \mathbf{W}_s \qquad (s = 1, \ldots, \ell) \tag{5}$$

must be applied to B at Q in order for the motion constraints to be satisfied. Further, one can deduce that B must be acted upon by couples whose constraint torques are

$$\mathbf{T}_s^B = \mu_s \boldsymbol{\tau}_s \qquad (s = 1, \ldots, \ell) \tag{6}$$

Derivations Because Eqs. (6.3.1) have the same form as Eqs. (6.2.2), it is convenient to consider a single set of equations that are independent of one another and write

$$\sum_{i=1}^{\nu} {}^A\mathbf{v}^{P_i} \cdot \mathbf{w}_{is} + Y_s = 0 \qquad (s = 1, \ldots, M + m) \tag{7}$$

When particles $P_1, \ldots, P_\beta$, a subset of the particles $P_1, \ldots, P_\nu$ belonging to S, form a rigid body B, the portions of constraint equations (7) dealing with the particles of B can be written in terms of the velocity ${}^A\mathbf{v}^Q$ in A of a point Q fixed in B, and the angular velocity ${}^A\boldsymbol{\omega}^B$ of B in A, so that

$$\sum_{i=1}^{\beta} {}^A\mathbf{v}^{P_i} \cdot \mathbf{w}_{is} \underset{(2.7.1)}{=} \sum_{i=1}^{\beta} ({}^A\mathbf{v}^Q + {}^A\boldsymbol{\omega}^B \times \mathbf{r}_i) \cdot \mathbf{w}_{is}$$

$$= {}^A\mathbf{v}^Q \cdot \sum_{i=1}^{\beta} \mathbf{w}_{is} + {}^A\boldsymbol{\omega}^B \cdot \sum_{i=1}^{\beta} \mathbf{r}_i \times \mathbf{w}_{is}$$

$$\triangleq {}^A\mathbf{v}^Q \cdot \mathbf{w}_s + {}^A\boldsymbol{\omega}^B \cdot \boldsymbol{\tau}_s \qquad (s = 1, \ldots, M + m) \tag{8}$$

where $\mathbf{r}_i$ is the position vector from Q to P_i.

As stated in connection with Eqs. (6.2.2) and (6.3.1), the appearance of the vector $\mathbf{w}_{is}$ in Eqs. (8) requires the application of a constraint force $\mathbf{C}_{is} = \lambda_s \mathbf{w}_{is}$ to P_i. After selecting the line of action of $\mathbf{w}_{is}$ such that it passes through P_i, and defining the resultants

$$\mathbf{w}_s \triangleq \sum_{i=1}^{\beta} \mathbf{w}_{is} \qquad \mathbf{C}_s^Q \triangleq \sum_{i=1}^{\beta} \mathbf{C}_{is} \qquad (s = 1, \ldots, M + m) \tag{9}$$

the set of forces $\mathbf{C}_{1s}, \ldots, \mathbf{C}_{\beta s}$ applied to B is regarded as equivalent to a single force $\mathbf{C}_s^Q$ whose line of action passes through Q, together with a couple whose torque is $\mathbf{T}_s^B$. The resultant $\mathbf{C}_s^Q$ is given by

$$\mathbf{C}_s^Q = \sum_{i=1}^{\beta} \mathbf{C}_{is} = \sum_{i=1}^{\beta} \lambda_s \mathbf{w}_{is} = \lambda_s \mathbf{w}_s \qquad (s = 1, \ldots, M + m) \tag{10}$$

in agreement with Eqs. (2), and the torque $\mathbf{T}_s^B$ is equal to the moment of $\mathbf{C}_{1s}, \ldots, \mathbf{C}_{\beta s}$ about Q,

$$\mathbf{T}_s^B = \sum_{i=1}^{\beta} \mathbf{r}_i \times \mathbf{C}_{is} = \sum_{i=1}^{\beta} \mathbf{r}_i \times \lambda_s \mathbf{w}_{is} = \lambda_s \boldsymbol{\tau}_s \qquad (s = 1, \ldots, M + m) \tag{11}$$

in agreement with Eqs. (3). The vector $\boldsymbol{\tau}_s$ is the moment of $\mathbf{w}_{1s}, \ldots, \mathbf{w}_{\beta s}$ about Q,

$$\boldsymbol{\tau}_s \triangleq \sum_{i=1}^{\beta} \mathbf{r}_i \times \mathbf{w}_{is} \qquad (s = 1, \ldots, M + m) \tag{12}$$

A similar exercise leads from Eqs. (6.4.2) to (4)–(6). The terms in Eqs. (6.4.2) associated with the particles of B can be rewritten as

$$
\begin{aligned}
\sum_{i=1}^{\beta} {}^{A}\mathbf{a}^{P_i} \cdot \mathbf{W}_{is}
&\underset{(2.7.2)}{=} \sum_{i=1}^{\beta} [{}^{A}\mathbf{a}^{Q} + {}^{A}\boldsymbol{\alpha}^{B} \times \mathbf{r}_i + {}^{A}\boldsymbol{\omega}^{B} \times ({}^{A}\boldsymbol{\omega}^{B} \times \mathbf{r}_i)] \cdot \mathbf{W}_{is} \\
&= {}^{A}\mathbf{a}^{Q} \cdot \sum_{i=1}^{\beta} \mathbf{W}_{is} + {}^{A}\boldsymbol{\alpha}^{B} \cdot \sum_{i=1}^{\beta} \mathbf{r}_i \times \mathbf{W}_{is} + \sum_{i=1}^{\beta} [{}^{A}\boldsymbol{\omega}^{B} \times ({}^{A}\boldsymbol{\omega}^{B} \times \mathbf{r}_i)] \cdot \mathbf{W}_{is} \\
&\triangleq {}^{A}\mathbf{a}^{Q} \cdot \mathbf{W}_s + {}^{A}\boldsymbol{\alpha}^{B} \cdot \boldsymbol{\tau}_s + \sum_{i=1}^{\beta} [{}^{A}\boldsymbol{\omega}^{B} \times ({}^{A}\boldsymbol{\omega}^{B} \times \mathbf{r}_i)] \cdot \mathbf{W}_{is} \qquad (s = 1, \ldots, \ell)
\end{aligned}
\tag{13}
$$

One then can proceed to develop relationships analogous to Eqs. (9)–(12) by letting $\mathbf{W}$, ℓ, and μ play the roles of $\mathbf{w}$, $M + m$, and λ, respectively. Hence, the dot products in Eqs. (13) involving the acceleration ${}^{A}\mathbf{a}^{Q}$ of Q in A, and the angular acceleration ${}^{A}\boldsymbol{\alpha}^{B}$ of B in A, give rise to the two dot products in Eqs. (4). The term Z_s' in Eqs. (4) is the sum of the scalar Z_s appearing in Eqs. (6.4.2) and the term $\sum_{i=1}^{\beta} [{}^{A}\boldsymbol{\omega}^{B} \times ({}^{A}\boldsymbol{\omega}^{B} \times \mathbf{r}_i)] \cdot \mathbf{W}_{is}$ contributed by each rigid body belonging to S.

Example Consider a mechanical system S composed of two rigid spheres, A and B, moving in a reference frame E as shown in Fig. 6.5.1. The mass of each sphere is uniformly distributed, and the radii of A and B are denoted by R_A and R_B, respectively. Suppose that the spheres move on a rigid, planar, and horizontal surface fixed in E, and that terrestrial gravitational forces (see Sec. 5.7) act on A and B. The surface imposes a configuration constraint on each sphere; the mass center $A^\star$ of A must remain a distance R_A above the surface, whereas the mass center $B^\star$ of B is required to be at a height of R_B. Additionally, suppose that each sphere rolls on the surface. One point, fixed in a sphere, is in contact with another point, fixed in E. The condition of rolling, a motion constraint, requires identical velocities in any reference frame of the two points in question; in this case, the velocities in E are zero. A final motion constraint is to be imposed on S; the velocities in E of $A^\star$ and $B^\star$ are mandated to be perpendicular to each other. Thus, S is a complex nonholonomic system (see Sec. 3.7).

The configuration constraint imposed on A restricts the position that $A^\star$ may occupy. A holonomic constraint equation having the form of Eqs. (6.2.1) can be written as

$$
\mathbf{p}^{OA^\star} \cdot \hat{\mathbf{e}}_3 - R_A = 0 \tag{14}
$$

where $\mathbf{p}^{OA^\star}$ is the position vector to $A^\star$ from O, a point fixed in E, on the planar surface supporting A. Unit vector $\hat{\mathbf{e}}_3$ is fixed in E and perpendicular to the horizontal surface as indicated in Fig. 6.5.1. Time differentiation of Eq. (14), with all vectors differentiated with respect to t in E, yields

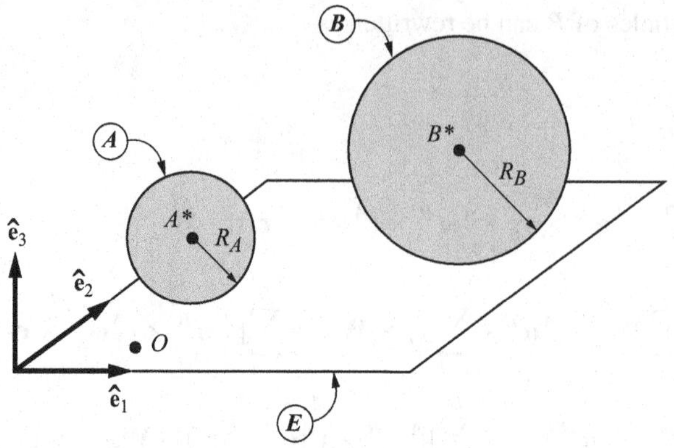

Figure 6.5.1

$$\frac{{}^E d\mathbf{p}^{OA^\star}}{dt} \cdot \hat{\mathbf{e}}_3 + \mathbf{p}^{OA^\star} \cdot \frac{{}^E d\hat{\mathbf{e}}_3}{dt} + \frac{dR_A}{dt} \underset{(2.6.1)}{=} {}^E\mathbf{v}^{A^\star} \cdot \hat{\mathbf{e}}_3 + \mathbf{p}^{OA^\star} \cdot \mathbf{0} + 0$$
$$= {}^E\mathbf{v}^{A^\star} \cdot \hat{\mathbf{e}}_3 = 0 \tag{15}$$

where ${}^E\mathbf{v}^{A^\star}$ is the velocity of $A^\star$ in E. This relationship has the vector form of Eqs. (1). In this instance $\hat{\mathbf{e}}_3$ plays the part of $\mathbf{w}_s$, and $Y_s = 0$ $(s = 1)$. No angular velocity of a rigid body in E appears in the equation; therefore, no constraint torque is needed to enforce the restriction. If E can be regarded as a Newtonian reference frame, one may inspect Eq. (15) and determine that a single constraint force given by

$$\mathbf{C}_1^{A^\star} = \lambda_1 \hat{\mathbf{e}}_3 \tag{16}$$

must be applied to $A^\star$, as indicated in Eqs. (2).

Equations (14)–(16) can be suitably modified in order to represent the situation in the case of B. The relevant holonomic constraint equation is stated as

$$\mathbf{p}^{OB^\star} \cdot \hat{\mathbf{e}}_3 - R_B = 0 \tag{17}$$

and time differentiation results in

$${}^E\mathbf{v}^{B^\star} \cdot \hat{\mathbf{e}}_3 = 0 \tag{18}$$

Inspection of this relationship leads to the conclusion that a constraint force

$$\mathbf{C}_2^{B^\star} = \lambda_2 \hat{\mathbf{e}}_3 \tag{19}$$

must be applied to $B^\star$.

The motion constraint involving rolling is treated as follows. Let P be the point of A that is in rolling contact with E. Two nonholonomic constraint equations describing rolling at P can be written as

$${}^E\mathbf{v}^P \cdot \hat{\mathbf{e}}_{s-2} = 0 \qquad (s = 3, 4) \tag{20}$$

where ${}^E\mathbf{v}^P$ is the velocity of P in E and unit vectors $\hat{\mathbf{e}}_1$ and $\hat{\mathbf{e}}_2$ are parallel to the horizontal surface as shown in Fig. 6.5.1. Together, $\hat{\mathbf{e}}_1$, $\hat{\mathbf{e}}_2$, and $\hat{\mathbf{e}}_3$ constitute a dextral set of mutually perpendicular unit vectors. Inspection of Eqs. (20) indicates that constraint forces

$$\mathbf{C}_3^P = \lambda_3\hat{\mathbf{e}}_1 \qquad \mathbf{C}_4^P = \lambda_4\hat{\mathbf{e}}_2 \tag{21}$$

must be applied to P. Alternatively, ${}^E\mathbf{v}^P$ can be expressed in terms of ${}^E\mathbf{v}^{A^\star}$ and ${}^E\boldsymbol{\omega}^A$, the angular velocity of A in E. Hence, Eqs. (20) can be rewritten as

$$[{}^E\mathbf{v}^{A^\star} + {}^E\boldsymbol{\omega}^A \times (-R_A\hat{\mathbf{e}}_3)] \cdot \hat{\mathbf{e}}_{s-2} \underset{(2.7.1)}{=}$$

$${}^E\mathbf{v}^{A^\star} \cdot \hat{\mathbf{e}}_{s-2} + {}^E\boldsymbol{\omega}^A \cdot (-R_A\hat{\mathbf{e}}_3 \times \hat{\mathbf{e}}_{s-2}) = 0 \qquad (s = 3,4) \tag{22}$$

In the case of these equations, the part of $\mathbf{w}_s$ in Eqs. (1) is played by $\hat{\mathbf{e}}_{s-2}$, whereas the role of $\boldsymbol{\tau}_s$ is played by $-R_A\hat{\mathbf{e}}_3 \times \hat{\mathbf{e}}_{s-2}$. Hence, with Eqs. (2) and (3) in mind, inspection reveals that constraint forces $\mathbf{C}_3^{A^\star} = \mathbf{C}_3^P$ and $\mathbf{C}_4^{A^\star} = \mathbf{C}_4^P$ must be applied to A at $A^\star$, together with constraint torques

$$\mathbf{T}_3^A = \lambda_3(-R_A\hat{\mathbf{e}}_3 \times \hat{\mathbf{e}}_1) = -\lambda_3 R_A\hat{\mathbf{e}}_2 \qquad \mathbf{T}_4^A = \lambda_4(-R_A\hat{\mathbf{e}}_3 \times \hat{\mathbf{e}}_2) = \lambda_4 R_A\hat{\mathbf{e}}_1 \tag{23}$$

that are applied to A. The equivalence of Eqs. (20) and (22) is thus mirrored in equivalent sets of constraint forces obtained by the process of inspection. Clearly, a constraint force $\mathbf{C}_3^P$ applied at P is equivalent to $\mathbf{C}_3^P$ applied at $A^\star$, together with a couple whose torque is $\mathbf{T}_3^A$. Likewise, a constraint force $\mathbf{C}_4^P$ applied at P is equivalent to $\mathbf{C}_4^P$ applied at $A^\star$, together with a couple whose torque is $\mathbf{T}_4^A$. A similar exercise leads to the conclusion that rolling requires constraint forces

$$\mathbf{C}_5^{B^\star} = \lambda_5\hat{\mathbf{e}}_1 \qquad \mathbf{C}_6^{B^\star} = \lambda_6\hat{\mathbf{e}}_2 \tag{24}$$

applied to B at $B^\star$, together with couples whose constraint torques act on B and are given by

$$\mathbf{T}_5^B = -\lambda_5 R_B\hat{\mathbf{e}}_2 \qquad \mathbf{T}_6^B = \lambda_6 R_B\hat{\mathbf{e}}_1 \tag{25}$$

This brings us to the final motion constraint, which requires perpendicular velocities in E of $A^\star$ and $B^\star$. The constraint can be expressed by the relationship

$${}^E\mathbf{v}^{B^\star} \cdot {}^E\mathbf{v}^{A^\star} = 0 \tag{26}$$

This constraint equation is nonlinear in the velocity vectors and thus does not have the form of Eqs. (1). Consequently, differentiation with respect to t is performed to obtain a relationship having the form of Eqs. (4), one that is linear in the acceleration vectors.

$${}^E\mathbf{a}^{B^\star} \cdot {}^E\mathbf{v}^{A^\star} + {}^E\mathbf{a}^{A^\star} \cdot {}^E\mathbf{v}^{B^\star} = 0 \tag{27}$$

With Eqs. (4)–(6) in mind, it can be concluded that the constraint necessitates application of the forces

$$\mathbf{C}_7^{B^\star} = \mu_7\,{}^E\mathbf{v}^{A^\star} \qquad \mathbf{C}_7^{A^\star} = \mu_7\,{}^E\mathbf{v}^{B^\star} \tag{28}$$

to $B^\star$ and $A^\star$, respectively. Two observations now can be made regarding the constraint forces $\mathbf{C}_7^{B\star}$ and $\mathbf{C}_7^{A\star}$. First, they need not be of equal magnitudes because the constraint does not require ${}^E\mathbf{v}^{B\star}$ and ${}^E\mathbf{v}^{A\star}$ to be equal in magnitude. Second, $\mathbf{C}_7^{B\star}$ is perpendicular to $\mathbf{C}_7^{A\star}$ when the constraint is satisfied.

The set of constraint forces acting on A is therefore equivalent to a force

$$\mathbf{C}^{A\star} = \mathbf{C}_1^{A\star} + \mathbf{C}_3^{A\star} + \mathbf{C}_4^{A\star} + \mathbf{C}_7^{A\star} = \lambda_1\hat{\mathbf{e}}_3 + \lambda_3\hat{\mathbf{e}}_1 + \lambda_4\hat{\mathbf{e}}_2 + \mu_7\,{}^E\mathbf{v}^{B\star} \tag{29}$$

applied at $A^\star$, together with a couple whose torque is

$$\mathbf{T}^A = \mathbf{T}_3^A + \mathbf{T}_4^A = R_A(\lambda_4\hat{\mathbf{e}}_1 - \lambda_3\hat{\mathbf{e}}_2) \tag{30}$$

Similarly, the constraint forces acting on B can be replaced by a force

$$\mathbf{C}^{B\star} = \mathbf{C}_2^{B\star} + \mathbf{C}_5^{B\star} + \mathbf{C}_6^{B\star} + \mathbf{C}_7^{B\star} = \lambda_2\hat{\mathbf{e}}_3 + \lambda_5\hat{\mathbf{e}}_1 + \lambda_6\hat{\mathbf{e}}_2 + \mu_7\,{}^E\mathbf{v}^{A\star} \tag{31}$$

applied at $B^\star$, together with a couple whose torque is

$$\mathbf{T}^B = \mathbf{T}_5^B + \mathbf{T}_6^B = R_B(\lambda_6\hat{\mathbf{e}}_1 - \lambda_5\hat{\mathbf{e}}_2) \tag{32}$$

6.6 NONCONTRIBUTING FORCES

Some forces that contribute to $\mathbf{R}_i$ as defined in Sec. 5.4 make no contributions to certain generalized active forces defined in Eqs. (5.4.1)–(5.4.3). Indeed, this is the principal motivation for introducing generalized active forces. Consider the following examples. If B is a rigid body belonging to S, the total contribution to F_r of all contact forces and distance forces exerted by all particles of B on each other is equal to zero. The total contribution to F_r of all contact forces exerted on particles of S across smooth surfaces of rigid bodies vanishes. When B rolls without slipping on a rigid body C, the total contribution to $\tilde{F}_r$ of all contact forces exerted on B by C is equal to zero if C is not a part of S, and the total contribution to $\tilde{F}_r$ of all contact forces exerted by B and C on each other is equal to zero if C is a part of S.

Derivations In Fig. 6.6.1, P_i and P_j designate any two particles of a rigid body B belonging to S. The rigid nature of B is a consequence of restrictions placed on the positions of P_i and P_j; the distance between them is required to remain constant (see Sec. 3.3). A holonomic constraint equation can be expressed in terms of $\mathbf{p}_{ij}$, the position vector from P_i to P_j, and L, the distance between the two particles,

$$\mathbf{p}_{ij} \cdot \mathbf{p}_{ij} - L^2 = 0 \tag{1}$$

Now, $\mathbf{p}_{ij}$ can be written as the difference

$$\mathbf{p}_{ij} = \mathbf{p}_j - \mathbf{p}_i \tag{2}$$

where $\mathbf{p}_i$ and $\mathbf{p}_j$ are, respectively, the position vectors to P_i and P_j from a point $\tilde{A}$ fixed in a reference frame A. After substituting from Eq. (2) into (1), performing time

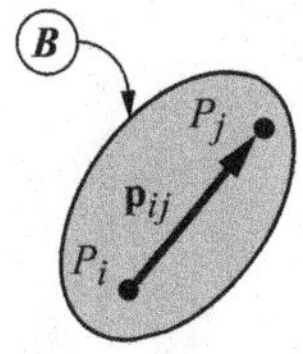

Figure 6.6.1

differentiation on the result, and differentiating all vectors with respect to t in A, one obtains

$$\left(\frac{{}^A d\mathbf{p}_j}{dt} - \frac{{}^A d\mathbf{p}_i}{dt}\right) \cdot (\mathbf{p}_j - \mathbf{p}_i) + (\mathbf{p}_j - \mathbf{p}_i) \cdot \left(\frac{{}^A d\mathbf{p}_j}{dt} - \frac{{}^A d\mathbf{p}_i}{dt}\right) - \frac{dL^2}{dt} = 2\,({}^A\mathbf{v}^{P_j} - {}^A\mathbf{v}^{P_i}) \cdot (\mathbf{p}_j - \mathbf{p}_i) - 0 \underset{(2.6.1)}{=} 2({}^A\mathbf{v}^{P_j} - {}^A\mathbf{v}^{P_i}) \cdot \mathbf{p}_{ij} \underset{(2)}{=} 0 \tag{3}$$

It is convenient to eliminate the factor of 2 before comparing the result with Eqs. (6.5.1), in which case Eqs. (6.5.2) indicate that application of constraint forces given by

$$\mathbf{C}^{P_j} = \lambda \mathbf{p}_{ij} \qquad \mathbf{C}^{P_i} = -\lambda \mathbf{p}_{ij} \tag{4}$$

to P_j and to P_i, respectively, will ensure satisfaction of the configuration constraint whenever A can be regarded as a Newtonian reference frame. The results are in accord with the law of action and reaction, which asserts that $\mathbf{C}^{P_i}$ and $\mathbf{C}^{P_j}$ have equal magnitudes, opposite directions, and coincident lines of action.

We shall show that the total contribution of $\mathbf{C}^{P_i}$ and $\mathbf{C}^{P_j}$ to F_r $(r = 1,\ldots,n)$ is equal to zero, from which it follows that the total contribution to F_r of all contact forces and distance forces exerted by all particles of B on each other is equal to zero.

The partial velocities ${}^A\mathbf{v}_r^{P_i}$ and ${}^A\mathbf{v}_r^{P_j}$ of P_i and P_j in A are related to the partial angular velocity ${}^A\boldsymbol{\omega}_r^B$ of B in A by [see Eqs. (2.7.1), (3.6.1), and (3.6.2)]

$${}^A\mathbf{v}_r^{P_j} = {}^A\mathbf{v}_r^{P_i} + {}^A\boldsymbol{\omega}_r^B \times \mathbf{p}_{ij} \qquad (r = 1,\ldots,n) \tag{5}$$

Consequently, the total contribution of $\mathbf{C}^{P_i}$ and $\mathbf{C}^{P_j}$ to F_r $(r = 1,\ldots,n)$ is [see Eqs. (5.4.1)]

$$\begin{aligned} {}^A\mathbf{v}_r^{P_i} \cdot \mathbf{C}^{P_i} + {}^A\mathbf{v}_r^{P_j} \cdot \mathbf{C}^{P_j} &\underset{(4)}{=} ({}^A\mathbf{v}_r^{P_j} - {}^A\mathbf{v}_r^{P_i}) \cdot \lambda \mathbf{p}_{ij} \\ &\underset{(5)}{=} {}^A\boldsymbol{\omega}_r^B \times \mathbf{p}_{ij} \cdot \lambda \mathbf{p}_{ij} = 0 \qquad (r = 1,\ldots,n) \end{aligned} \tag{6}$$

Suppose P, a particle of S, is in contact with a rigid body B whose surface is smooth. Let $\tilde{A}$ be a point fixed in A, $\tilde{B}$ be a particle belonging to B, and $\overline{B}$ be a point fixed in B that coincides with P at the instant under consideration. Further, let $\mathbf{p}$ denote the position vector from $\tilde{A}$ to P, $\mathbf{r}$ the position vector from $\tilde{A}$ to $\tilde{B}$, and $\mathbf{q}$ the position vector

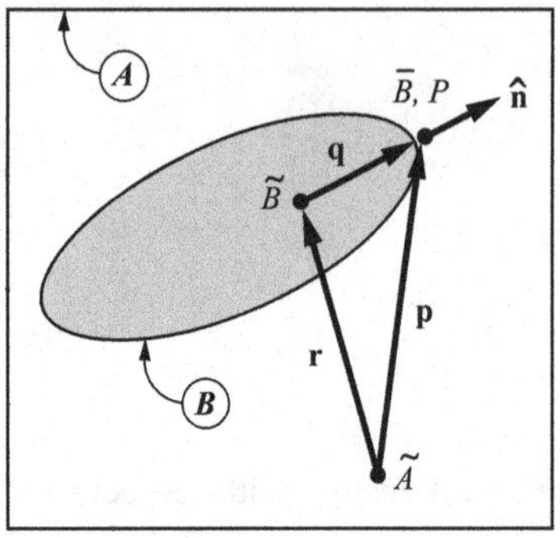

Figure 6.6.2

from $\tilde{B}$ to P, as shown in Fig. 6.6.2. Hence,

$$\mathbf{p} = \mathbf{r} + \mathbf{q} \tag{7}$$

Then, if $\hat{\mathbf{n}}$ is a unit vector normal to the surface of B at P, $\tilde{B}$ may always be taken such that

$$\mathbf{q} = q\hat{\mathbf{n}} \tag{8}$$

Moreover, if P is neither to lose contact with B nor to penetrate B, then the $\hat{\mathbf{n}}$ measure number of $\mathbf{q}$ must remain constant; this configuration constraint can be expressed as

$$(\mathbf{p} - \mathbf{r}) \cdot \hat{\mathbf{n}} - q\hat{\mathbf{n}} \cdot \hat{\mathbf{n}} = (\mathbf{p} - \mathbf{r}) \cdot \hat{\mathbf{n}} - q \underset{(7,\ 8)}{=} 0 \tag{9}$$

Differentiation with respect to time yields

$$\left(\frac{{}^A d\mathbf{p}}{dt} - \frac{{}^A d\mathbf{r}}{dt}\right) \cdot \hat{\mathbf{n}} + (\mathbf{p} - \mathbf{r}) \cdot \frac{{}^A d\hat{\mathbf{n}}}{dt} - \frac{dq}{dt} =$$

$$(\underset{(2.8.3)}{{}^A\mathbf{v}^P} - \underset{(2.8.5)}{{}^A\mathbf{v}^{\tilde{B}}}) \cdot \hat{\mathbf{n}} + (\mathbf{p} - \mathbf{r}) \cdot \underset{(2.1.2)}{({}^A\boldsymbol{\omega}^B \times \hat{\mathbf{n}})} - 0 =$$

$$({}^A\mathbf{v}^P - {}^A\mathbf{v}^{\tilde{B}}) \cdot \hat{\mathbf{n}} + \underset{(7,\ 8)}{q\hat{\mathbf{n}}} \cdot \left({}^A\boldsymbol{\omega}^B \times \hat{\mathbf{n}}\right) =$$

$$({}^A\mathbf{v}^P - {}^A\mathbf{v}^{\tilde{B}}) \cdot \hat{\mathbf{n}} = 0 \tag{10}$$

Now, taking $\tilde{B}$ as the point $\overline{B}$ of B that coincides with P at the instant under consideration, so that Eqs. (2.8.8) are in effect, means that Eq. (10) gives way to

$$({}^A\mathbf{v}^P - {}^A\mathbf{v}^{\overline{B}}) \cdot \hat{\mathbf{n}} = 0 \tag{11}$$

With the velocities in A of P and $\overline{B}$ in evidence, this relationship takes on the form of Eqs. (6.5.1). When A can be regarded as a Newtonian reference frame, Eqs. (6.5.2) facilitate identification of the constraint forces

$$\mathbf{C}^P = \lambda\hat{\mathbf{n}} \qquad \mathbf{C}^{\overline{B}} = -\lambda\hat{\mathbf{n}} \tag{12}$$

where λ is some scalar. $\mathbf{C}^P$ and $\mathbf{C}^{\overline{B}}$ have equal magnitudes, opposite directions, and coincident lines of action, and therefore obey the law of action and reaction. Because the surface of B is presumed to be smooth, the contact forces exerted by B and P on each other have no component perpendicular to $\hat{\mathbf{n}}$.

The velocities ${}^A\mathbf{v}^P$ and ${}^A\mathbf{v}^{\overline{B}}$ are related to each other,

$$ {}^A\mathbf{v}^P \underset{(2.8.1)}{=} {}^A\mathbf{v}^{\overline{B}} + {}^B\mathbf{v}^P \tag{13} $$

Now ${}^B\mathbf{v}^P$, the velocity of P in B, must be perpendicular to $\hat{\mathbf{n}}$ in view of Eqs. (11) and (13), and this means that

$$ {}^B\mathbf{v}_r^P \cdot \hat{\mathbf{n}} = 0 \qquad (r = 1, \ldots, n) \tag{14} $$

because otherwise there can exist values of $u_1, \ldots, u_n$ such that ${}^B\mathbf{v}^P$ is not perpendicular to $\hat{\mathbf{n}}$. Suppose that B is part of S. Then [see Eq. (3.6.2)]

$$ {}^A\mathbf{v}_r^P \underset{(13)}{=} {}^A\mathbf{v}_r^{\overline{B}} + {}^B\mathbf{v}_r^P \qquad (r = 1, \ldots, n) \tag{15} $$

Consequently,

$$ ({}^A\mathbf{v}_r^P - {}^A\mathbf{v}_r^{\overline{B}}) \cdot \hat{\mathbf{n}} \underset{(15)}{=} {}^B\mathbf{v}_r^P \cdot \hat{\mathbf{n}} \underset{(14)}{=} 0 \qquad (r = 1, \ldots, n) \tag{16} $$

and the contribution to F_r of the forces exerted on each other by P and B is [see Eqs. (5.4.1)]

$$ {}^A\mathbf{v}_r^P \cdot \mathbf{C}^P + {}^A\mathbf{v}_r^{\overline{B}} \cdot \mathbf{C}^{\overline{B}} \underset{(12)}{=} ({}^A\mathbf{v}_r^P - {}^A\mathbf{v}_r^{\overline{B}}) \cdot \lambda\hat{\mathbf{n}} \underset{(16)}{=} 0 \qquad (r = 1, \ldots, n) \tag{17} $$

Alternatively, if B is not part of S, then $u_1, \ldots, u_n$ always can be chosen in such a way that ${}^A\mathbf{v}^{\overline{B}}$ is independent of $u_1, \ldots, u_n$, in which event

$$ {}^A\mathbf{v}_r^P \underset{(13)}{=} {}^B\mathbf{v}_r^P \qquad (r = 1, \ldots, n) \tag{18} $$

and the contribution to F_r of the contact force exerted by B on P is [see Eqs. (5.4.1)]

$$ {}^A\mathbf{v}_r^P \cdot \mathbf{C}^P \underset{(18)}{=} {}^B\mathbf{v}_r^P \underset{(12)}{\cdot} \lambda\hat{\mathbf{n}} \underset{(14)}{=} 0 \qquad (r = 1, \ldots, n) \tag{19} $$

In both cases it thus follows that the total contribution to F_r of all contact forces exerted on particles of S across smooth surfaces of rigid bodies vanishes.

Finally, to deal with contact forces that come into play when B rolls on a rigid body C, we let $\overline{B}$ be a point of B, and $\overline{C}$ a point of C, choosing these such that $\overline{B}$ is in rolling contact with $\overline{C}$, which means that

$$ {}^A\mathbf{v}^{\overline{C}} = {}^A\mathbf{v}^{\overline{B}} \tag{20} $$

where A is any reference frame.

Because B and C are rigid bodies, a configuration constraint at the point of contact prohibits them from penetrating each other. If they are also to be prevented from losing contact, $\overline{C}$ can play the role of P in the analysis just concluded in which case Eq. (11) gives way to

$$ ({}^A\mathbf{v}^{\overline{C}} - {}^A\mathbf{v}^{\overline{B}}) \cdot \hat{\mathbf{n}} = 0 \tag{21} $$

where $\hat{\mathbf{n}}$ denotes a unit vector that is normal to the surface of B at $\overline{B}$, and normal to the surface of C at $\overline{C}$. The constraint forces required to prevent penetration or loss of contact are given by

$$\mathbf{N}^{\overline{C}} = \lambda\hat{\mathbf{n}} \qquad \mathbf{N}^{\overline{B}} = -\lambda\hat{\mathbf{n}} \tag{22}$$

$\mathbf{N}^{\overline{C}}$ and $\mathbf{N}^{\overline{B}}$ do not contribute to F_r if C is part of S, and $\mathbf{N}^{\overline{B}}$ does not contribute to F_r if C is not a part of S.

The condition of rolling constitutes a motion constraint and requires that, for any unit vector $\hat{\boldsymbol{\tau}}$ perpendicular to $\hat{\mathbf{n}}$,

$$^A\mathbf{v}^{\overline{C}} \cdot \hat{\boldsymbol{\tau}} = {}^A\mathbf{v}^{\overline{B}} \cdot \hat{\boldsymbol{\tau}} \tag{23}$$

because otherwise slipping occurs. Rearrangement of this relationship yields the nonholonomic constraint equation

$$({}^A\mathbf{v}^{\overline{C}} - {}^A\mathbf{v}^{\overline{B}}) \cdot \hat{\boldsymbol{\tau}} = 0 \tag{24}$$

In view of Eqs. (6.5.1) and (6.5.2), Eq. (24) can be inspected to identify constraint forces that prevent slipping,

$$\mathbf{T}^{\overline{C}} = \sigma\hat{\boldsymbol{\tau}} \qquad \mathbf{T}^{\overline{B}} = -\sigma\hat{\boldsymbol{\tau}} \tag{25}$$

where σ is some scalar.

The combination of a configuration constraint described by Eq. (21) and a motion constraint expressed as in Eq. (24) results in Eq. (20). When rolling takes place, S is a simple nonholonomic system (see Sec. 3.5); we now examine the contributions of contact forces to nonholonomic generalized active forces $\tilde{F}_r$.

First, consider the case in which C belongs to S. Nonholonomic partial velocities of $\overline{C}$ and $\overline{B}$ then are identical,

$$^A\tilde{\mathbf{v}}_r^{\overline{C}} \underset{(20)}{=} {}^A\tilde{\mathbf{v}}_r^{\overline{B}} \qquad (r = 1,\ldots,p) \tag{26}$$

The contribution of $\mathbf{T}^{\overline{C}}$ and $\mathbf{T}^{\overline{B}}$ to $\tilde{F}_r$ is given by [see Eqs. (5.4.2)]

$$^A\tilde{\mathbf{v}}_r^{\overline{C}} \cdot \mathbf{T}^{\overline{C}} + {}^A\tilde{\mathbf{v}}_r^{\overline{B}} \cdot \mathbf{T}^{\overline{B}} \underset{(26)}{=} {}^A\tilde{\mathbf{v}}_r^{\overline{C}} \cdot \underset{(25)}{(\sigma\hat{\boldsymbol{\tau}} - \sigma\hat{\boldsymbol{\tau}})} = 0 \tag{27}$$

The contribution of $\mathbf{N}^{\overline{C}}$ and $\mathbf{N}^{\overline{B}}$ to $\tilde{F}_r$ likewise vanishes,

$$^A\tilde{\mathbf{v}}_r^{\overline{C}} \cdot \mathbf{N}^{\overline{C}} + {}^A\tilde{\mathbf{v}}_r^{\overline{B}} \cdot \mathbf{N}^{\overline{B}} \underset{(26)}{=} {}^A\tilde{\mathbf{v}}_r^{\overline{C}} \cdot \underset{(22)}{(\lambda\hat{\mathbf{n}} - \lambda\hat{\mathbf{n}})} = 0 \tag{28}$$

Thus, when B and C roll on each other and C belongs to S, the total contribution to $\tilde{F}_r$ of all contact forces exerted by B and C on each other is equal to zero.

Alternatively, when C does not belong to S, motion variables always can be chosen such that $^A\mathbf{v}^{\overline{C}}$ is independent of all of them, so that

$$^A\tilde{\mathbf{v}}_r^{\overline{C}} = \mathbf{0} \qquad (r = 1,\ldots,p) \tag{29}$$

and, consequently,

$$^A\tilde{\mathbf{v}}_r^{\overline{B}} \underset{(20,\ 29)}{=} \mathbf{0} \qquad (r = 1,\ldots,p) \tag{30}$$

from which it follows that ${}^A\tilde{\mathbf{v}}_r^{\overline{B}} \cdot \mathbf{T}^{\overline{B}}$ and ${}^A\tilde{\mathbf{v}}_r^{\overline{B}} \cdot \mathbf{N}^{\overline{B}}$, the contributions to $\widetilde{F}_r$ of the contact forces exerted on B by C at $\overline{B}$, are equal to zero.

Example In the example in Sec. 6.5 involving rigid spheres A and B, a force $\mathbf{C}_1^P = \lambda_1 \hat{\mathbf{e}}_3$ applied to P, the contact point of A, prevents A from penetrating the horizontal surface fixed in E. That force is equivalent (see Sec. 5.3) to $\mathbf{C}_1^{A^\star} = \lambda_1 \hat{\mathbf{e}}_3$ applied at $A^\star$, the mass center of A. Likewise, penetration of E by B is prevented by a force equivalent to $\mathbf{C}_2^{B^\star} = \lambda_2 \hat{\mathbf{e}}_3$ applied at $B^\star$.

Suppose that A and B are each permitted to slip on E, and that the velocities in E of $A^\star$ and $B^\star$ are not required always to be perpendicular to each other. In that event, the system S made up of A and B is holonomic and possesses ten degrees of freedom in E. The velocities in E of the mass centers of A and B can be written

$$ {}^E\mathbf{v}^{A^\star} = u_1 \hat{\mathbf{e}}_1 + u_2 \hat{\mathbf{e}}_2 \qquad {}^E\mathbf{v}^{B^\star} = u_4 \hat{\mathbf{e}}_1 + u_6 \hat{\mathbf{e}}_2 \tag{31} $$

and the angular velocities in E of A and B can be expressed as

$$ {}^E\boldsymbol{\omega}^A = u_9 \hat{\mathbf{e}}_1 + u_{10} \hat{\mathbf{e}}_2 + u_3 \hat{\mathbf{e}}_3 \qquad {}^E\boldsymbol{\omega}^B = u_7 \hat{\mathbf{e}}_1 + u_8 \hat{\mathbf{e}}_2 + u_5 \hat{\mathbf{e}}_3 \tag{32} $$

Holonomic partial velocities ${}^E\mathbf{v}_r^{A^\star}$ and ${}^E\mathbf{v}_r^{B^\star}$ then are identified as

$$ {}^E\mathbf{v}_1^{A^\star} = \hat{\mathbf{e}}_1 \qquad {}^E\mathbf{v}_2^{A^\star} = \hat{\mathbf{e}}_2 \qquad {}^E\mathbf{v}_r^{A^\star} = \mathbf{0} \quad (r = 3, \dots, 10) \tag{33} $$

$$ {}^E\mathbf{v}_4^{B^\star} = \hat{\mathbf{e}}_1 \qquad {}^E\mathbf{v}_6^{B^\star} = \hat{\mathbf{e}}_2 \qquad {}^E\mathbf{v}_r^{B^\star} = \mathbf{0} \quad (r = 1,2,3,5,7, \dots, 10) \tag{34} $$

The forces $\mathbf{C}_1^{A^\star}$ and $\mathbf{C}_2^{B^\star}$ individually contribute nothing to the holonomic generalized active forces F_r $(r = 1, \dots, 10)$. The contribution of $\mathbf{C}_1^{A^\star}$ [see Eqs. (5.5.1)] is given by

$$ \begin{aligned} {}^E\mathbf{v}_1^{A^\star} \cdot \mathbf{C}_1^{A^\star} &\underset{(33)}{=} \hat{\mathbf{e}}_1 \cdot \lambda_1 \hat{\mathbf{e}}_3 = 0 \\ {}^E\mathbf{v}_2^{A^\star} \cdot \mathbf{C}_1^{A^\star} &\underset{(33)}{=} \hat{\mathbf{e}}_2 \cdot \lambda_1 \hat{\mathbf{e}}_3 = 0 \\ {}^E\mathbf{v}_r^{A^\star} \cdot \mathbf{C}_1^{A^\star} &\underset{(33)}{=} \mathbf{0} \cdot \lambda_1 \hat{\mathbf{e}}_3 = 0 \qquad (r = 3, \dots, 10) \end{aligned} \tag{35} $$

Similarly, the contribution of $\mathbf{C}_2^{B^\star}$ vanishes:

$$ \begin{aligned} {}^E\mathbf{v}_4^{B^\star} \cdot \mathbf{C}_2^{B^\star} &\underset{(34)}{=} \hat{\mathbf{e}}_1 \cdot \lambda_2 \hat{\mathbf{e}}_3 = 0 \\ {}^E\mathbf{v}_6^{B^\star} \cdot \mathbf{C}_2^{B^\star} &\underset{(34)}{=} \hat{\mathbf{e}}_2 \cdot \lambda_2 \hat{\mathbf{e}}_3 = 0 \\ {}^E\mathbf{v}_r^{B^\star} \cdot \mathbf{C}_2^{B^\star} &\underset{(34)}{=} \mathbf{0} \cdot \lambda_2 \hat{\mathbf{e}}_3 = 0 \qquad (r = 1,2,3,5,7, \dots, 10) \end{aligned} \tag{36} $$

The condition of rolling now can be brought into the picture. In the case of A, this means that Eqs. (6.5.22) are in effect; the first of these leads to

$$ \underset{(31)}{(u_1 \hat{\mathbf{e}}_1 + u_2 \hat{\mathbf{e}}_2)} \cdot \hat{\mathbf{e}}_1 + \underset{(32)}{(u_9 \hat{\mathbf{e}}_1 + u_{10} \hat{\mathbf{e}}_2 + u_3 \hat{\mathbf{e}}_3)} \cdot (-R_A \hat{\mathbf{e}}_3 \times \hat{\mathbf{e}}_1) \underset{(6.5.22)}{=} 0 \tag{37} $$

and thereupon to

$$ u_1 - R_A u_{10} = 0 \tag{38} $$

whereas the consequences of the second relationship are

$$\underset{(31)}{(u_1\hat{\mathbf{e}}_1 + u_2\hat{\mathbf{e}}_2)} \cdot \hat{\mathbf{e}}_2 + \underset{(32)}{(u_9\hat{\mathbf{e}}_1 + u_{10}\hat{\mathbf{e}}_2 + u_3\hat{\mathbf{e}}_3)} \cdot \underset{(6.5.22)}{(-R_A\hat{\mathbf{e}}_3 \times \hat{\mathbf{e}}_2)} = 0 \tag{39}$$

and

$$u_2 + R_A u_9 = 0 \tag{40}$$

After employing analagous reasoning the conditions for rolling of B upon E are stated as

$$u_4 - R_B u_8 = 0 \tag{41}$$

and

$$u_6 + R_B u_7 = 0 \tag{42}$$

In view of the four relationships involving u_7, u_8, u_9, and u_{10}, S can be regarded as a simple nonholonomic system possessing six degrees of freedom in E. Although equations (31) are unaffected, u_7, u_8, u_9, and u_{10} can be eliminated from Eqs. (32) so that

$$^E\boldsymbol{\omega}^A = -\frac{u_2\hat{\mathbf{e}}_1}{R_A} + \frac{u_1\hat{\mathbf{e}}_2}{R_A} + u_3\hat{\mathbf{e}}_3 \qquad ^E\boldsymbol{\omega}^B = -\frac{u_6\hat{\mathbf{e}}_1}{R_B} + \frac{u_4\hat{\mathbf{e}}_2}{R_B} + u_5\hat{\mathbf{e}}_3 \tag{43}$$

Nonholonomic partial velocities $^E\tilde{\mathbf{v}}_r^{A\star}$ and $^E\tilde{\mathbf{v}}_r^{B\star}$ can be identified by inspecting Eqs. (31),

$$^E\tilde{\mathbf{v}}_1^{A\star} = \hat{\mathbf{e}}_1 \qquad ^E\tilde{\mathbf{v}}_2^{A\star} = \hat{\mathbf{e}}_2 \qquad ^E\tilde{\mathbf{v}}_r^{A\star} = \mathbf{0} \quad (r = 3,4,5,6) \tag{44}$$

$$^E\tilde{\mathbf{v}}_r^{B\star} = \mathbf{0} \quad (r = 1,2,3,5) \qquad ^E\tilde{\mathbf{v}}_4^{B\star} = \hat{\mathbf{e}}_1 \qquad ^E\tilde{\mathbf{v}}_6^{B\star} = \hat{\mathbf{e}}_2 \tag{45}$$

and nonholonomic partial angular velocities $^E\tilde{\boldsymbol{\omega}}_r^A$ and $^E\tilde{\boldsymbol{\omega}}_r^B$ are obtained from Eqs. (43),

$$^E\tilde{\boldsymbol{\omega}}_1^A = \frac{\hat{\mathbf{e}}_2}{R_A} \qquad ^E\tilde{\boldsymbol{\omega}}_2^A = -\frac{\hat{\mathbf{e}}_1}{R_A} \qquad ^E\tilde{\boldsymbol{\omega}}_3^A = \hat{\mathbf{e}}_3 \qquad ^E\tilde{\boldsymbol{\omega}}_r^A = \mathbf{0} \quad (r = 4,5,6) \tag{46}$$

$$^E\tilde{\boldsymbol{\omega}}_r^B = \mathbf{0} \quad (r = 1,2,3) \qquad ^E\tilde{\boldsymbol{\omega}}_4^B = \frac{\hat{\mathbf{e}}_2}{R_B} \qquad ^E\tilde{\boldsymbol{\omega}}_5^B = \hat{\mathbf{e}}_3 \qquad ^E\tilde{\boldsymbol{\omega}}_6^B = -\frac{\hat{\mathbf{e}}_1}{R_B} \tag{47}$$

The set of forces applied to A in order to prevent slipping is equivalent to a force with resultant $\mathbf{C}_3^{A\star} + \mathbf{C}_4^{A\star} = \lambda_3\hat{\mathbf{e}}_1 + \lambda_4\hat{\mathbf{e}}_2$ applied at $A^\star$, together with a couple of torque $\mathbf{T}_3^A + \mathbf{T}_4^A = R_A(\lambda_4\hat{\mathbf{e}}_1 - \lambda_3\hat{\mathbf{e}}_2)$ [see Eqs. (6.5.23)]. The contribution of this set of forces to the nonholonomic generalized active force $\tilde{F}_r$ can be determined with the aid of Eqs. (5.5.2), first with $r = 1$,

$$\begin{aligned} &^E\tilde{\mathbf{v}}_1^{A\star} \cdot (\mathbf{C}_3^{A\star} + \mathbf{C}_4^{A\star}) + {}^E\tilde{\boldsymbol{\omega}}_1^A \cdot (\mathbf{T}_3^A + \mathbf{T}_4^A) \\ &= \underset{(44)}{\hat{\mathbf{e}}_1} \cdot (\lambda_3\hat{\mathbf{e}}_1 + \lambda_4\hat{\mathbf{e}}_2) + \underset{(46)}{(\hat{\mathbf{e}}_2/R_A)} \cdot [R_A(\lambda_4\hat{\mathbf{e}}_1 - \lambda_3\hat{\mathbf{e}}_2)] \\ &= \lambda_3 - \lambda_3 = 0 \end{aligned} \tag{48}$$

then with $r = 2$,

$$
\begin{aligned}
{}^E\widetilde{\mathbf{v}}_2^{A\star} \cdot (\mathbf{C}_3^{A\star} + \mathbf{C}_4^{A\star}) + {}^E\widetilde{\boldsymbol{\omega}}_2^{A} \cdot (\mathbf{T}_3^A + \mathbf{T}_4^A) \\
&= \underset{(44)}{\hat{\mathbf{e}}_2} \cdot (\lambda_3\hat{\mathbf{e}}_1 + \lambda_4\hat{\mathbf{e}}_2) + \underset{(46)}{(-\hat{\mathbf{e}}_1/R_A)} \cdot [R_A(\lambda_4\hat{\mathbf{e}}_1 - \lambda_3\hat{\mathbf{e}}_2)] \\
&= \lambda_4 - \lambda_4 = 0
\end{aligned}
\tag{49}
$$

next with $r = 3$,

$$
\begin{aligned}
{}^E\widetilde{\mathbf{v}}_3^{A\star} \cdot (\mathbf{C}_3^{A\star} + \mathbf{C}_4^{A\star}) + {}^E\widetilde{\boldsymbol{\omega}}_3^{A} \cdot (\mathbf{T}_3^A + \mathbf{T}_4^A) \\
&= \underset{(44)}{\mathbf{0}} \cdot (\lambda_3\hat{\mathbf{e}}_1 + \lambda_4\hat{\mathbf{e}}_2) + \underset{(46)}{\hat{\mathbf{e}}_3} \cdot [R_A(\lambda_4\hat{\mathbf{e}}_1 - \lambda_3\hat{\mathbf{e}}_2)] \\
&= 0
\end{aligned}
\tag{50}
$$

and, finally,

$$
\begin{aligned}
{}^E\widetilde{\mathbf{v}}_r^{A\star} \cdot (\mathbf{C}_3^{A\star} + \mathbf{C}_4^{A\star}) + {}^E\widetilde{\boldsymbol{\omega}}_r^{A} \cdot (\mathbf{T}_3^A + \mathbf{T}_4^A) \\
&= \underset{(44)}{\mathbf{0}} \cdot (\lambda_3\hat{\mathbf{e}}_1 + \lambda_4\hat{\mathbf{e}}_2) + \underset{(46)}{\mathbf{0}} \cdot [R_A(\lambda_4\hat{\mathbf{e}}_1 - \lambda_3\hat{\mathbf{e}}_2)] \\
&= 0 \qquad (r = 4,5,6)
\end{aligned}
\tag{51}
$$

Likewise, B is prevented from slipping on E by a set of forces, described in Eqs. (6.5.24) and (6.5.25), equivalent to a single force $\mathbf{C}_5^{B\star} + \mathbf{C}_6^{B\star} = \lambda_5\hat{\mathbf{e}}_1 + \lambda_6\hat{\mathbf{e}}_2$ applied at $B^\star$, and a couple of torque $\mathbf{T}_5^B + \mathbf{T}_6^B = R_B(\lambda_6\hat{\mathbf{e}}_1 - \lambda_5\hat{\mathbf{e}}_2)$ acting on B. One can demonstrate that this set of forces contributes nothing to $\widetilde{F}_r$ $(r = 1,\ldots,6)$ by proceeding with steps completely analogous to those just taken in connection with A.

Equations (35) show that F_r $(r = 1,\ldots,10)$ receive no contribution from the force $\mathbf{C}_1^{A\star} = \lambda_1\hat{\mathbf{e}}_3$ that prevents A from penetrating E. Moreover, $\mathbf{C}_1^{A\star}$ contributes nothing to $\widetilde{F}_r$ $(r = 1,\ldots,6)$; it is easily seen that

$$
{}^E\widetilde{\mathbf{v}}_r^{A\star} \cdot \mathbf{C}_1^{A\star} = {}^E\widetilde{\mathbf{v}}_r^{A\star} \cdot \lambda_1\hat{\mathbf{e}}_3 \underset{(44)}{=} 0 \qquad (r = 1,\ldots,6) \tag{52}
$$

The same can be said of $\mathbf{C}_2^{B\star} = \lambda_2\hat{\mathbf{e}}_3$,

$$
{}^E\widetilde{\mathbf{v}}_r^{B\star} \cdot \mathbf{C}_2^{B\star} = {}^E\widetilde{\mathbf{v}}_r^{B\star} \cdot \lambda_2\hat{\mathbf{e}}_3 \underset{(45)}{=} 0 \qquad (r = 1,\ldots,6) \tag{53}
$$

The final constraint considered in the example of Sec. 6.5 requires perpendicular velocities in E of $A^\star$ and $B^\star$. Before we leave this example, it is worth noting that the set of forces consisting of $\mathbf{C}_7^{A\star}$ and $\mathbf{C}_7^{B\star}$ contributes nothing to $\approx{F}_r$ $(r = 1,\ldots,5)$. In fact, $\approx{F}_r$ receive no contributions from *any* of the constraint forces under consideration in this example.

After developing expressions for the accelerations in E of $A^\star$ and $B^\star$,

$$
{}^E\mathbf{a}^{A\star} \underset{(2.6.2)}{=} \frac{{}^E d\,{}^E\mathbf{v}^{A\star}}{dt} \underset{(31)}{=} \dot{u}_1\hat{\mathbf{e}}_1 + \dot{u}_2\hat{\mathbf{e}}_2 \tag{54}
$$

$$
{}^E\mathbf{a}^{B\star} \underset{(2.6.2)}{=} \frac{{}^E d\,{}^E\mathbf{v}^{B\star}}{dt} \underset{(31)}{=} \dot{u}_4\hat{\mathbf{e}}_1 + \dot{u}_6\hat{\mathbf{e}}_2 \tag{55}
$$

one can evaluate the constraint equation (6.5.27) to obtain

$$\underset{(55)}{(\dot{u}_4\hat{\mathbf{e}}_1 + \dot{u}_6\hat{\mathbf{e}}_2)} \cdot \underset{(31)}{(u_1\hat{\mathbf{e}}_1 + u_2\hat{\mathbf{e}}_2)} + \underset{(54)}{(\dot{u}_1\hat{\mathbf{e}}_1 + \dot{u}_2\hat{\mathbf{e}}_2)} \cdot \underset{(31)}{(u_4\hat{\mathbf{e}}_1 + u_6\hat{\mathbf{e}}_2)}$$
$$= u_4\dot{u}_1 + u_6\dot{u}_2 + u_1\dot{u}_4 + u_2\dot{u}_6 \underset{(6.5.27)}{=} 0 \tag{56}$$

This relationship can be solved for $\dot{u}_6$,

$$\dot{u}_6 = -(u_4\dot{u}_1 + u_6\dot{u}_2 + u_1\dot{u}_4)/u_2 \tag{57}$$

When this constraint equation is satisfied, S is a complex nonholonomic system (see Sec. 3.7) possessing five degrees of freedom in E. The expression for ${}^E\mathbf{a}^{A^\star}$ in Eq. (54) remains unaltered because it does not involve $\dot{u}_6$; however, ${}^E\mathbf{a}^{B^\star}$ now becomes

$${}^E\mathbf{a}^{B^\star} \underset{(55)}{=} \dot{u}_4\hat{\mathbf{e}}_1 \underset{(57)}{-} [u_4\dot{u}_1 + u_6\dot{u}_2 + u_1\dot{u}_4)/u_2]\,\hat{\mathbf{e}}_2 \tag{58}$$

Nonholonomic partial accelerations (see Sec. 3.8) ${}^E\tilde{\mathbf{a}}_r^{A^\star}$ $(r = 1, \ldots, 5)$ are identified by inspecting Eq. (54),

$${}^E\tilde{\mathbf{a}}_1^{A^\star} = \hat{\mathbf{e}}_1 \qquad {}^E\tilde{\mathbf{a}}_2^{A^\star} = \hat{\mathbf{e}}_2 \qquad {}^E\tilde{\mathbf{a}}_r^{A^\star} = \mathbf{0} \quad (r = 3, 4, 5) \tag{59}$$

and an examination of Eq. (58) produces ${}^E\tilde{\mathbf{a}}_r^{B^\star}$, where

$${}^E\tilde{\mathbf{a}}_1^{B^\star} = -\frac{u_4}{u_2}\hat{\mathbf{e}}_2 \quad {}^E\tilde{\mathbf{a}}_2^{B^\star} = -\frac{u_6}{u_2}\hat{\mathbf{e}}_2 \quad {}^E\tilde{\mathbf{a}}_4^{B^\star} = \hat{\mathbf{e}}_1 - \frac{u_1}{u_2}\hat{\mathbf{e}}_2 \quad {}^E\tilde{\mathbf{a}}_r^{B^\star} = \mathbf{0} \ (r = 3, 5) \tag{60}$$

It now can be shown that the set of forces $\mathbf{C}_7^{A^\star}$ and $\mathbf{C}_7^{B^\star}$ [see Eqs. (6.5.28)], which is required to keep ${}^E\mathbf{v}^{A^\star}$ perpendicular to ${}^E\mathbf{v}^{B^\star}$, contributes nothing to the nonholonomic generalized active forces $\tilde{\tilde{F}}_r$ $(r = 1, \ldots, 5)$. Upon referring to Eqs. (5.5.3) and setting $r = 1$, the contribution to $\tilde{\tilde{F}}_1$ is found to be

$${}^E\tilde{\mathbf{a}}_1^{A^\star} \cdot \mathbf{C}_7^{A^\star} + {}^E\tilde{\mathbf{a}}_1^{B^\star} \cdot \mathbf{C}_7^{B^\star} = \underset{(59)}{\hat{\mathbf{e}}_1} \cdot \underset{(6.5.28)}{\mu_7\,{}^E\mathbf{v}^{B^\star}} - \underset{(60)}{\frac{u_4}{u_2}\hat{\mathbf{e}}_2} \cdot \underset{(6.5.28)}{\mu_7\,{}^E\mathbf{v}^{A^\star}}$$
$$= \hat{\mathbf{e}}_1 \cdot \mu_7\,\underset{(31)}{(u_4\hat{\mathbf{e}}_1 + u_6\hat{\mathbf{e}}_2)} - \frac{u_4}{u_2}\hat{\mathbf{e}}_2 \cdot \mu_7\,\underset{(31)}{(u_1\hat{\mathbf{e}}_1 + u_2\hat{\mathbf{e}}_2)}$$
$$= \mu_7(u_4 - u_4) = 0 \tag{61}$$

Similarly, the contribution for $r = 2, 3, 4, 5$ is observed to vanish:

$${}^E\tilde{\mathbf{a}}_2^{A^\star} \cdot \mathbf{C}_7^{A^\star} + {}^E\tilde{\mathbf{a}}_2^{B^\star} \cdot \mathbf{C}_7^{B^\star} = \hat{\mathbf{e}}_2 \cdot \mu_7(u_4\hat{\mathbf{e}}_1 + u_6\hat{\mathbf{e}}_2) - \frac{u_6}{u_2}\hat{\mathbf{e}}_2 \cdot \mu_7(u_1\hat{\mathbf{e}}_1 + u_2\hat{\mathbf{e}}_2)$$
$$= \mu_7(u_6 - u_6) = 0 \tag{62}$$

$${}^E\tilde{\mathbf{a}}_r^{A^\star} \cdot \mathbf{C}_7^{A^\star} + {}^E\tilde{\mathbf{a}}_r^{B^\star} \cdot \mathbf{C}_7^{B^\star} = \mathbf{0} \cdot \mu_7(u_4\hat{\mathbf{e}}_1 + u_6\hat{\mathbf{e}}_2) + \mathbf{0} \cdot \mu_7(u_1\hat{\mathbf{e}}_1 + u_2\hat{\mathbf{e}}_2)$$
$$= 0 \qquad (r = 3, 5) \tag{63}$$

$${}^E\tilde{\mathbf{a}}_4^{A^\star} \cdot \mathbf{C}_7^{A^\star} + {}^E\tilde{\mathbf{a}}_4^{B^\star} \cdot \mathbf{C}_7^{B^\star} = \mathbf{0} \cdot \mathbf{C}_7^{A^\star} + \left(\hat{\mathbf{e}}_1 - \frac{u_1}{u_2}\hat{\mathbf{e}}_2\right) \cdot \mu_7(u_1\hat{\mathbf{e}}_1 + u_2\hat{\mathbf{e}}_2)$$
$$= \mu_7(u_1 - u_1) = 0 \tag{64}$$

Expressions for angular accelerations in E of A and B can be obtained as

$$^E\boldsymbol{\alpha}^A \underset{(2.5.1)}{=} \frac{^E d\,^E\boldsymbol{\omega}^A}{dt} \underset{(43)}{=} -\frac{\dot{u}_2}{R_A}\hat{\mathbf{e}}_1 + \frac{\dot{u}_1}{R_A}\hat{\mathbf{e}}_2 + \dot{u}_3\hat{\mathbf{e}}_3 \tag{65}$$

and

$$\begin{aligned} ^E\boldsymbol{\alpha}^B &\underset{(2.5.1)}{=} \frac{^E d\,^E\boldsymbol{\omega}^B}{dt} \underset{(43)}{=} -\frac{\dot{u}_6}{R_B}\hat{\mathbf{e}}_1 + \frac{\dot{u}_4}{R_B}\hat{\mathbf{e}}_2 + \dot{u}_5\hat{\mathbf{e}}_3 \\ &\underset{(57)}{=} \frac{(u_4\dot{u}_1 + u_6\dot{u}_2 + u_1\dot{u}_4)}{R_B u_2}\hat{\mathbf{e}}_1 + \frac{\dot{u}_4}{R_B}\hat{\mathbf{e}}_2 + \dot{u}_5\hat{\mathbf{e}}_3 \end{aligned} \tag{66}$$

Subsequent inspection of these relationships enables the identification of nonholonomic partial angular accelerations (see Sec. 3.8),

$$^E\tilde{\boldsymbol{\alpha}}_1^A = \frac{\hat{\mathbf{e}}_2}{R_A} \quad ^E\tilde{\boldsymbol{\alpha}}_2^A = -\frac{\hat{\mathbf{e}}_1}{R_A} \quad ^E\tilde{\boldsymbol{\alpha}}_3^A = \hat{\mathbf{e}}_3 \quad ^E\tilde{\boldsymbol{\alpha}}_r^A = \mathbf{0} \quad (r = 4,5) \tag{67}$$

$$^E\tilde{\boldsymbol{\alpha}}_1^B = \frac{u_4\hat{\mathbf{e}}_1}{u_2 R_B} \qquad ^E\tilde{\boldsymbol{\alpha}}_2^B = \frac{u_6\hat{\mathbf{e}}_1}{u_2 R_B} \qquad ^E\tilde{\boldsymbol{\alpha}}_3^B = \mathbf{0}$$

$$^E\tilde{\boldsymbol{\alpha}}_4^B = \frac{1}{R_B}\left(\frac{u_1}{u_2}\hat{\mathbf{e}}_1 + \hat{\mathbf{e}}_2\right) \qquad ^E\tilde{\boldsymbol{\alpha}}_5^B = \hat{\mathbf{e}}_3 \tag{68}$$

It has been pointed out previously that $\tilde{F}_r$ $(r = 1,\ldots,6)$ receive no contribution from the set of forces applied to A in order to prevent slipping [see Eqs. (48)–(51)], or from the corresponding set of forces applied to B. It can be shown that, individually, each of these sets makes no contribution to $\tilde{\tilde{F}}_r$ $(r = 1,\ldots,5)$. For example, the contribution to $\tilde{\tilde{F}}_1$ from the set applied to B is found to be

$$\begin{aligned} &^E\tilde{\mathbf{a}}_1^{B\star} \cdot (\mathbf{C}_5^{B\star} + \mathbf{C}_6^{B\star}) + {^E\tilde{\boldsymbol{\alpha}}_1^B} \cdot (\mathbf{T}_5^B + \mathbf{T}_6^B) \\ &= \underset{(60)}{-\frac{u_4}{u_2}\hat{\mathbf{e}}_2} \cdot (\lambda_5\hat{\mathbf{e}}_1 + \lambda_6\hat{\mathbf{e}}_2) + \underset{(68)}{\frac{u_4}{u_2 R_B}\hat{\mathbf{e}}_1} \cdot [R_B(\lambda_6\hat{\mathbf{e}}_1 - \lambda_5\hat{\mathbf{e}}_2)] \\ &= \frac{u_4}{u_2}(-\lambda_6 + \lambda_6) = 0 \end{aligned} \tag{69}$$

A contribution from this set of forces to $\tilde{\tilde{F}}_r$ is shown to be absent by employing the same approach, using $^E\tilde{\mathbf{a}}_r^{B\star}$ and $^E\tilde{\boldsymbol{\alpha}}_r^B$ $(r = 2,3,4,5)$ in place of $^E\tilde{\mathbf{a}}_1^{B\star}$ and $^E\tilde{\boldsymbol{\alpha}}_1^B$, respectively. Furthermore, $\tilde{\tilde{F}}_1,\ldots,\tilde{\tilde{F}}_5$ contain no evidence of the force $\mathbf{C}_3^{A\star} + \mathbf{C}_4^{A\star}$ applied to $A^\star$, together with a couple whose torque is $\mathbf{T}_3^A + \mathbf{T}_4^A$.

Heretofore it has been shown that the force preventing A from penetrating E, $\mathbf{C}_1^{A\star} = \lambda_1\hat{\mathbf{e}}_3$, makes no contribution to F_r $(r = 1,\ldots,10)$ [see Eqs. (35)] or to $\tilde{F}_r$ $(r = 1,\ldots,6)$ [see Eqs. (52)]. In fact, it is readily apparent that the force contributes nothing to $\tilde{\tilde{F}}_r$ $(r = 1,\ldots,5)$

$$^E\tilde{\mathbf{a}}_r^{A\star} \cdot \mathbf{C}_1^{A\star} = {^E\tilde{\mathbf{a}}_r^{A\star}} \cdot \lambda_1\hat{\mathbf{e}}_3 \underset{(59)}{=} 0 \qquad (r = 1,\ldots,5) \tag{70}$$

Likewise, the force $\mathbf{C}_2^{B\star}$ that prevents B from penetrating E makes no contribution to F_r $(r = 1,\ldots,10)$ [see Eqs. (36)], $\tilde{F}_r$ $(r = 1,\ldots,6)$ [see Eqs. (53)], or $\tilde{\tilde{F}}_r$ $(r = 1,\ldots,5)$,

$$^{E}\tilde{\mathbf{a}}_r^{B\star} \cdot \mathbf{C}_2^{B\star} = {}^{E}\tilde{\mathbf{a}}_r^{B\star} \cdot \lambda_2\hat{\mathbf{e}}_3 \underset{(60)}{=} 0 \qquad (r = 1,\ldots,5) \tag{71}$$

6.7 BRINGING NONCONTRIBUTING FORCES INTO EVIDENCE

As was mentioned in Sec. 6.6, the fact that certain forces acting on the particles of a system make no contributions to generalized active forces usually is helpful. But it can occur that precisely such a noncontributing force, or a torque of a couple formed by noncontributing forces, is of interest in its own right. In that event, one can bring this force or torque into evidence through the introduction of a motion variable properly related to the force or torque in question, that is, a motion variable that gives rise to a partial velocity of the point of application of the force, or a partial angular velocity of the rigid body on which the couple acts, such that the dot product of the partial velocity and the force, or the dot product of the partial angular velocity and the torque, does not vanish. The force or torque of interest then comes into evidence in the generalized active forces associated with a new set composed of original and additional motion variables.

The introduction of a suitable additional motion variable is accomplished by permitting points to have certain velocities, or rigid bodies to have certain angular velocities, which they cannot, in fact possess, doing so *without* introducing additional generalized coordinates. When forming expressions for velocities and/or angular velocities, one then takes the additional motion variable into account but uses the same generalized coordinates as before. One identifies partial velocities and partial angular velocities corresponding to all of the motion variables by inspection, as always. The partial velocities and partial angular velocities corresponding to the original motion variables, as well as the associated original generalized active forces and generalized inertia forces, may change and should therefore be reevaluated subsequent to the introduction of additional motion variables.

In general, during the motion of interest the additional motion variables are in fact related to the original motion variables by expressions having the form of Eqs. (3.5.2). When *all* of the additional motion variables are identically *zero* during the motion of interest, the partial velocities, partial angular velocities, generalized active forces, and generalized inertia forces corresponding to the original motion variables will remain unaltered subsequent to the introduction of the additional motion variables.

Derivations Let S be a simple nonholonomic system possessing generalized coordinates $q_1,\ldots,q_n$, and p degrees of freedom in a reference frame A (see Sec. 3.5), so that its motion can be characterized by a set of generalized velocities $u_1,\ldots,u_p$. Further, let P_i be a typical particle of S and denote its partial velocities in A by ${}^{A}\tilde{\mathbf{v}}_1^{P_i},\ldots,{}^{A}\tilde{\mathbf{v}}_p^{P_i}$ $(i = 1,\ldots,\nu)$. The generalized forces for S in A corresponding to the original motion variables are $\tilde{F}_1,\ldots,\tilde{F}_p$. Now suppose that additional motion variables $u_{p+1},\ldots,u_n$ are introduced for the purpose of bringing into evidence $n-p$ measure numbers of noncontributing forces and/or noncontributing torques. Let partial velocities and generalized active forces associated with the new set of motion variables, $u_1,\ldots,u_n$, be denoted, respectively, by ${}^A\mathbf{v}_1^{P_i},\ldots,{}^A\mathbf{v}_n^{P_i}$ $(i = 1,\ldots,\nu)$ and $F_1,\ldots,F_n$. Of these, ${}^A\mathbf{v}_{p+1}^{P_i},\ldots,{}^A\mathbf{v}_n^{P_i}$ are the partial velocities associated with the additional motion variables $u_{p+1},\ldots,u_n$, and $F_{p+1},\ldots,F_n$ are the corresponding generalized active forces. That leaves ${}^A\mathbf{v}_1^{P_i},\ldots,$ ${}^A\mathbf{v}_p^{P_i}$, together with $F_1,\ldots,F_p$, as the partial velocities and generalized active forces associated with $u_1,\ldots,u_p$ subsequent to the introduction of the additional motion variables.

During the motion of interest, $u_{p+1},\ldots,u_n$ are related to $u_1,\ldots,u_p$ by expressions having the form of Eqs. (3.5.2); therefore, Eqs. (3.6.17) are in effect, in which case

$${}^A\mathbf{v}_r^{P_i} = {}^A\tilde{\mathbf{v}}_r^{P_i} - \sum_{s=p+1}^{n} {}^A\mathbf{v}_s^{P_i} A_{sr} \qquad (r = 1,\ldots,p) \tag{1}$$

Thus, the partial velocity ${}^A\tilde{\mathbf{v}}_r^{P_i}$ obtained before introducing additional motion variables becomes ${}^A\mathbf{v}_r^{P_i}$ subsequent to the introduction. However, when all of the additional motion variables are in fact zero during the motion of interest, all of the coefficients A_{sr} vanish, in which case the partial velocities ${}^A\tilde{\mathbf{v}}_r^{P_i}$ $(r = 1,\ldots,p)$ remain unaltered by the introduction of those additional motion variables. The same conclusions are reached with regard to partial angular velocities [see Eqs. (3.6.15)] and generalized inertia forces [see Eqs. (5.9.5)].

As for generalized active forces, Eqs. (5.4.4) are applicable during the motion of interest so that

$$F_r = \tilde{F}_r - \sum_{s=p+1}^{n} F_s A_{sr} \qquad (r = 1,\ldots,p) \tag{2}$$

The goal of introducing additional motion variables is to bring noncontributing forces and/or noncontributing torques into evidence in $F_{p+1},\ldots,F_n$. Generalized active forces $\tilde{F}_1,\ldots,\tilde{F}_p$ obtained prior to introducing additional motion variables are afterward replaced with $F_1,\ldots,F_p$, which may then contain evidence of constraint forces and constraint torques.

Relationships similar to Eqs. (1) and (2) apply in the case of a holonomic system. A complex nonholonomic system (see Sec. 3.7) can be dealt with by introducing additional time derivatives of motion variables and by referring to Eqs. (3.7.1), (3.8.15), and (5.4.5).

The following example involves additional motion variables that all vanish identically during the motion of interest. In contrast, Problems 9.10 and 15.1 furnish illustrative examples in which additional motion variables are in fact functions of the original motion variables during actual motions of the system.

Example Consider once again the system formed by the sleeve S, rod R, and particle P depicted in Fig. 5.8.1 and considered previously in the example in Sec. 5.8. When all contact surfaces are treated as smooth, the two generalized active forces F_1 and F_2 [see Eqs. (5.8.16) and (5.8.17)] contain no information about either the forces exerted on the sleeve S by the supporting shaft V and bearing surface B or any forces exerted by P and R on each other. To bring such forces into evidence, we begin by replacing the set of forces exerted on S by B and V with a couple of torque $\boldsymbol{\tau}$, expressed as

$$\boldsymbol{\tau} = \tau_2 \hat{\mathbf{s}}_2 + \tau_3 \hat{\mathbf{s}}_3 \tag{3}$$

together with a force $\boldsymbol{\sigma}$ applied to S at point O, with $\boldsymbol{\sigma}$ given by

$$\boldsymbol{\sigma} = \sigma_1 \hat{\mathbf{s}}_1 + \sigma_2 \hat{\mathbf{s}}_2 + \sigma_3 \hat{\mathbf{s}}_3 \tag{4}$$

and we let $\boldsymbol{\rho}$ be the force exerted on P by R, expressing $\boldsymbol{\rho}$ as

$$\boldsymbol{\rho} = \rho_2 \hat{\mathbf{r}}_2 + \rho_3 \hat{\mathbf{r}}_3 \tag{5}$$

[The reason for omitting an $\hat{\mathbf{s}}_1$-component from Eq. (3) and an $\hat{\mathbf{r}}_1$-component from Eq. (5) is that all contact surfaces are presumed to be smooth.] Suppose now that one is interested in, say, σ_1, τ_2, and ρ_3. To bring these into evidence in expressions for generalized forces, one can proceed as follows.

The constraint force $\sigma_1 \hat{\mathbf{s}}_1$ applied to S at O prevents S from moving in a direction parallel to $\hat{\mathbf{s}}_1$; therefore, in view of Eqs. (6.5.1) and (6.5.2), it can be identified by inspecting the constraint equation

$$^B\mathbf{v}^O \cdot \hat{\mathbf{s}}_1 = 0 \tag{6}$$

provided B can be regarded as an inertial reference frame. Introduction of an additional motion variable

$$u_3 \triangleq {}^B\mathbf{v}^O \cdot \hat{\mathbf{s}}_1 \tag{7}$$

is thus tantamount to permitting O, regarded as a point of R, to have velocity in B that it cannot in fact possess, in a direction parallel to the constraint force of interest. The constraint torque $\tau_2 \hat{\mathbf{s}}_2$ prevents S (and therefore R) from having an angular velocity in B in a direction parallel to $\hat{\mathbf{s}}_2$. The constraint equation

$$^B\boldsymbol{\omega}^R \cdot \hat{\mathbf{s}}_2 = 0 \tag{8}$$

gives rise, by inspection, to a constraint torque $\tau_2 \hat{\mathbf{s}}_2$. Therefore, introducing an additional motion variable as

$$u_4 \triangleq {}^B\boldsymbol{\omega}^R \cdot \hat{\mathbf{s}}_2 \tag{9}$$

permits R to have an angular velocity in B that it cannot possess, in a direction parallel to the constraint torque of interest. The constraint force $\rho_3 \hat{\mathbf{r}}_3$ prevents P from losing contact with R or penetrating R by moving in a direction parallel to $\hat{\mathbf{r}}_3$. The law of action and reaction asserts that a constraint force $-\rho_3 \hat{\mathbf{r}}_3$ is exerted by P on R at $\overline{R}$, the point of R in contact with P. This pair of constraint forces can be identified by inspecting the constraint equation

$$({}^B\mathbf{v}^P - {}^B\mathbf{v}^{\overline{R}}) \cdot \hat{\mathbf{r}}_3 \underset{(2.8.1)}{=} {}^R\mathbf{v}^P \cdot \hat{\mathbf{r}}_3 = 0 \tag{10}$$

where ${}^B\mathbf{v}^P$ is the velocity of P in B and where ${}^B\mathbf{v}^{\overline{R}}$ is the velocity of $\overline{R}$ in B. Hence, a suitable additional motion variable is introduced as

$$u_5 \triangleq {}^R\mathbf{v}^P \cdot \hat{\mathbf{r}}_3 \tag{11}$$

The original motion variables u_1 and u_2 [see Eqs. (5.8.12)], as well as the additional motion variables u_3, u_4, and u_5, are used to form expressions for the angular velocity of R in B, and the velocities in B of O, P, $\overline{R}$, and $R^\star$, the mass center of R.

$$^B\boldsymbol{\omega}^R = u_2\hat{\mathbf{s}}_1 + u_4\hat{\mathbf{s}}_2 \tag{12}$$

$$^B\mathbf{v}^O = u_3\hat{\mathbf{s}}_1 \tag{13}$$

$$^B\mathbf{v}^{\overline{R}} \underset{(2.7.1)}{=} {}^B\mathbf{v}^O + {}^B\boldsymbol{\omega}^R \times (q_1\hat{\mathbf{r}}_1) \underset{(12,\ 13)}{=} u_3\hat{\mathbf{s}}_1 + q_1(u_2 \sin\beta - u_4 \cos\beta)\hat{\mathbf{r}}_3 \tag{14}$$

$$^R\mathbf{v}^P = \dot{q}_1\hat{\mathbf{r}}_1 + u_5\hat{\mathbf{r}}_3 \underset{(5.8.12)}{=} u_1\ \hat{\mathbf{r}}_1 + u_5\hat{\mathbf{r}}_3 \tag{15}$$

$$^B\mathbf{v}^P \underset{(2.8.1)}{=} {}^B\mathbf{v}^{\overline{R}} + {}^R\mathbf{v}^P \underset{(14,\ 15)}{=} u_1\hat{\mathbf{r}}_1 + u_2q_1 \sin\beta\hat{\mathbf{r}}_3 + u_3\hat{\mathbf{s}}_1 - u_4q_1 \cos\beta\hat{\mathbf{r}}_3 + u_5\hat{\mathbf{r}}_3 \tag{16}$$

After noting that $R^\star$ now has a velocity $^B\mathbf{v}^{R^\star}$ given by

$$^B\mathbf{v}^{R^\star} = {}^B\mathbf{v}^O + {}^B\boldsymbol{\omega}^R \times (L\hat{\mathbf{r}}_1) \underset{(12,\ 13)}{=} u_3\hat{\mathbf{s}}_1 + L(u_2 \sin\beta - u_4 \cos\beta)\hat{\mathbf{r}}_3 \tag{17}$$

one then can record the partial angular velocities of R and the partial velocities of O, P, $\overline{R}$, and $R^\star$ as in Table 6.7.1, and this puts one into position to form F_r ($r = 1,\ldots,5$) by substituting from Eqs. (5.8.11) and (3)–(5) into

$$F_r = {}^B\boldsymbol{\omega}_r^R \cdot (\mathbf{T} + \boldsymbol{\tau}) + {}^B\mathbf{v}_r^O \cdot \boldsymbol{\sigma} + {}^B\mathbf{v}_r^P \cdot (mg\hat{\mathbf{s}}_1 + \boldsymbol{\rho}) + {}^B\mathbf{v}_r^{\overline{R}} \cdot (-\boldsymbol{\rho}) + {}^B\mathbf{v}_r^{R^\star}(Mg\hat{\mathbf{s}}_1) \qquad (r = 1,\ldots,5) \tag{18}$$

which leads to

$$F_1 = mg\cos\beta \tag{19}$$
$$F_2 = T \tag{20}$$
$$F_3 - \sigma_1 + (m + M)g \tag{21}$$
$$F_4 = \tau_2 \tag{22}$$
$$F_5 = \rho_3 \tag{23}$$

Table 6.7.1

	$r = 1$	$r = 2$	$r = 3$	$r = 4$	$r = 5$	Reference
$^B\boldsymbol{\omega}_r^R$	$\mathbf{0}$	$\hat{\mathbf{s}}_1$	$\mathbf{0}$	$\hat{\mathbf{s}}_2$	$\mathbf{0}$	Eq. (12)
$^B\mathbf{v}_r^O$	$\mathbf{0}$	$\mathbf{0}$	$\hat{\mathbf{s}}_1$	$\mathbf{0}$	$\mathbf{0}$	Eq. (13)
$^B\mathbf{v}_r^P$	$\hat{\mathbf{r}}_1$	$q_1 \sin\beta\hat{\mathbf{r}}_3$	$\hat{\mathbf{s}}_1$	$-q_1 \cos\beta\hat{\mathbf{r}}_3$	$\hat{\mathbf{r}}_3$	Eq. (16)
$^B\mathbf{v}_r^{\overline{R}}$	$\mathbf{0}$	$q_1 \sin\beta\hat{\mathbf{r}}_3$	$\hat{\mathbf{s}}_1$	$-q_1 \cos\beta\hat{\mathbf{r}}_3$	$\mathbf{0}$	Eq. (14)
$^B\mathbf{v}_r^{R^\star}$	$\mathbf{0}$	$L \sin\beta\hat{\mathbf{r}}_3$	$\hat{\mathbf{s}}_1$	$-L \cos\beta\hat{\mathbf{r}}_3$	$\mathbf{0}$	Eq. (17)

It is worth noting that, in this case, additional motion variables u_3, u_4, u_5 are introduced such that their actual values are all zero; consequently, the original partial velocities, partial angular velocities, and generalized active forces, remain unaltered. The expressions for ${}^B\mathbf{v}_r^P$ and ${}^B\boldsymbol{\omega}_r^R$ ($r = 1,2$) recorded in Table 6.7.1 are the same as those in Eqs. (5.8.13) and (5.8.14). The generalized active forces F_1 and F_2 reported in Eqs. (19) and (20) are identical to those in Eqs. (5.8.16) and (5.8.17). On the other hand, when an additional motion variable is introduced such that the actual value is a function of the original motion variables [as in Eqs. (3.5.2)], there will be changes to some or all of the original partial velocities, partial angular velocities, and generalized active forces.

The quantities σ_2, σ_3, τ_3, and ρ_2, which are absent from $F_1, \ldots, F_5$, can be brought into evidence similarly; that is, if $u_6, \ldots, u_9$ are introduced such that

$$^B\mathbf{v}^O = u_3\hat{\mathbf{s}}_1 + u_6\hat{\mathbf{s}}_2 + u_7\hat{\mathbf{s}}_3 \tag{24}$$

$$^B\boldsymbol{\omega}^R = u_2\hat{\mathbf{s}}_1 + u_4\hat{\mathbf{s}}_2 + u_8\hat{\mathbf{s}}_3 \tag{25}$$

and

$$^R\mathbf{v}^P = u_1\hat{\mathbf{r}}_1 + u_9\hat{\mathbf{r}}_2 + u_5\hat{\mathbf{r}}_3 \tag{26}$$

then the generalized active forces corresponding to $u_6, \ldots, u_9$ are

$$F_6 = \sigma_2 \tag{27}$$

$$F_7 = \sigma_3 \tag{28}$$

$$F_8 = \tau_3 - (mq_1 + ML)g \sin\beta \tag{29}$$

$$F_9 = \rho_2 - mg \sin\beta \tag{30}$$

7 ENERGY FUNCTIONS

The use of potential energy functions and kinetic energy functions sometimes enables one to construct integrals of equations of motion (see Secs. 9.1 and 9.2). In addition, potential energy functions can be helpful when one seeks to form expressions for generalized active forces, and expressions for generalized inertia forces can be formed with the aid of kinetic energy functions. Hence, familiarity with these functions is certainly desirable. However, since one can readily formulate equations of motion and extract information from such equations without invoking energy concepts, one need not master the material in the present chapter before moving on to Chapter 8.

7.1 POTENTIAL ENERGY

If S is a holonomic system (see Sec. 3.5) possessing generalized coordinates $q_1,\ldots,q_n$ (see Sec. 3.2) and generalized velocities $u_1,\ldots,u_n$ (see Sec. 3.4) in a reference frame A, and the generalized velocities are defined as

$$u_r \triangleq \dot{q}_r \qquad (r = 1,\ldots,n) \tag{1}$$

then there may exist functions V of $q_1,\ldots,q_n$ and the time t that satisfy *all* of the equations

$$F_r = -\frac{\partial V}{\partial q_r} \qquad (r = 1,\ldots,n) \tag{2}$$

where $F_1,\ldots,F_n$ are generalized active forces for S in A (see Sec. 5.4) associated with $u_1,\ldots,u_n$, respectively. Any such function V is called a *potential energy* of S in A. [One speaks of *a* potential energy, rather than *the* potential energy because, if V satisfies Eqs. (2), then $V + C$, where C is any function of t, also satisfies Eqs. (2) and is, therefore, a potential energy of S in A.]

When a potential energy V of S satisfies the equation

$$\frac{\partial V}{\partial t} = 0 \tag{3}$$

then $\dot{V}$, the total time-derivative of V, is given by

$$\dot{V} = -\sum_{r=1}^{n} F_r \dot{q}_r \tag{4}$$

It is by virtue of this fact that potential energy plays an important part in the construction of integrals of equations of motion and checking functions, as will be shown in Secs. 9.2 and 9.3, respectively.

Given generalized active forces F_r $(r = 1,\dots,n)$, all of which can be regarded as functions of $q_1,\dots,q_n$, and t (but not of $u_1,\dots,u_n$), one can either prove that V does not exist or find $V(q_1,\dots,q_n;t)$ explicitly, as follows: Determine whether or not *all* of the equations

$$\frac{\partial F_r}{\partial q_s} = \frac{\partial F_s}{\partial q_r} \qquad (r,s = 1,\dots,n) \tag{5}$$

are satisfied. If one or more of Eqs. (5) are violated, then V does not exist; if all of Eqs. (5) are satisfied, then V exists and is given by

$$V = \int_{\alpha_1}^{q_1} \frac{\partial}{\partial q_1} V(\zeta,\alpha_2,\dots,\alpha_n;t)d\zeta + \int_{\alpha_2}^{q_2} \frac{\partial}{\partial q_2} V(q_1,\zeta,\alpha_3,\dots,\alpha_n;t)d\zeta$$
$$+ \cdots + \int_{\alpha_n}^{q_n} \frac{\partial}{\partial q_n} V(q_1,\dots,q_{n-1},\zeta;t)d\zeta + C \tag{6}$$

where $\alpha_1,\dots,\alpha_n$ and C are *any* functions of t. [It is advantageous to set as many of $\alpha_1,\dots,\alpha_n$ equal to zero as is possible without rendering any of the integrals in Eq. (6) improper.]

When S is holonomic and $u_1,\dots,u_n$ are defined as

$$u_r \underset{(3.4.1)}{\triangleq} \sum_{s=1}^{n} Y_{rs}\,\dot{q}_s + Z_r \qquad (r = 1,\dots,n) \tag{7}$$

rather than as in Eqs. (1), so that

$$\dot{q}_s \underset{(3.6.5)}{=} \sum_{r=1}^{n} W_{sr}u_r + X_s \qquad (s = 1,\dots,n) \tag{8}$$

where W_{sr} and X_s are functions of $q_1,\dots,q_n$, and t, then Eqs. (2)–(4) give way to

$$F_r = -\sum_{s=1}^{n} \frac{\partial V}{\partial q_s} W_{sr} \qquad (r = 1,\dots,n) \tag{9}$$

$$\frac{\partial V}{\partial t} + \sum_{s=1}^{n} \frac{\partial V}{\partial q_s} X_s = 0 \tag{10}$$

and

$$\dot{V} = -\sum_{r=1}^{n} F_r u_r \tag{11}$$

respectively. Under these circumstances, one can either prove that V does not exist, or find $V(q_1,\dots,q_n;t)$ explicitly, as follows: Solve Eqs. (9) for $\partial V/\partial q_s$ $(s = 1,\dots,n)$†

† The solution of Eqs. (9) can be written

$$\frac{\partial V}{\partial q_s} \underset{(7\text{–}9)}{=} -\sum_{k=1}^{n} Y_{ks}F_k \qquad (s = 1,\dots,n).$$

and determine whether or not *all* of the equations

$$\frac{\partial}{\partial q_s}\left(\frac{\partial V}{\partial q_r}\right) = \frac{\partial}{\partial q_r}\left(\frac{\partial V}{\partial q_s}\right) \qquad (r, s = 1, \dots, n) \tag{12}$$

are satisfied. If one or more of Eqs. (12) are violated, then V does not exist; if all of Eqs. (12) are satisfied, then V exists and can be found by using Eq. (6).

When S is a simple nonholonomic system possessing p degrees of freedom in A (see Sec. 3.5), $u_1, \dots, u_n$ are defined as in Eqs. (1), and the motion constraint equations relating $\dot{q}_{p+1}, \dots, \dot{q}_n$ to $\dot{q}_1, \dots, \dot{q}_p$ are [this is a special case of Eqs. (3.5.2)]

$$\dot{q}_k = \sum_{r=1}^{p} C_{kr}\dot{q}_r + D_k \qquad (k = p+1, \dots, n) \tag{13}$$

where C_{kr} and D_k are functions of $q_1, \dots, q_n$, and t, then Eqs. (2), (3), and (4) are replaced with

$$\widetilde{F}_r = -\left(\frac{\partial V}{\partial q_r} + \sum_{s=p+1}^{n} \frac{\partial V}{\partial q_s} C_{sr}\right) \qquad (r = 1, \dots, p) \tag{14}$$

$$\frac{\partial V}{\partial t} + \sum_{s=p+1}^{n} \frac{\partial V}{\partial q_s} D_s = 0 \tag{15}$$

and

$$\dot{V} = -\sum_{r=1}^{p} \widetilde{F}_r \dot{q}_r \tag{16}$$

respectively, whereas, when $u_1, \dots, u_n$ are defined as in Eqs. (7), so that Eqs. (8) apply, while the motion constraint equations relating $u_{p+1}, \dots, u_n$ to $u_1, \dots, u_p$ are

$$u_k \underset{(3.5.2)}{=} \sum_{r-1}^{p} A_{kr}u_r + B_k \qquad (k = p+1, \dots, n) \tag{17}$$

where A_{kr} and B_k are functions of $q_1, \dots, q_n$, and t, then Eqs. (2), (3), and (4) are replaced with

$$\widetilde{F}_r = -\sum_{s=1}^{n} \frac{\partial V}{\partial q_s}\left(W_{sr} + \sum_{k=p+1}^{n} W_{sk}A_{kr}\right) \qquad (r = 1, \dots, p) \tag{18}$$

$$\frac{\partial V}{\partial t} + \sum_{s=1}^{n} \frac{\partial V}{\partial q_s}\left(X_s + \sum_{r=p+1}^{n} W_{sr}B_r\right) = 0 \tag{19}$$

and

$$\dot{V} = -\sum_{r=1}^{p} \widetilde{F}_r u_r \tag{20}$$

respectively. In both cases, the procedure for either proving that V does not exist or finding V explicitly is more complicated than in the two cases considered previously, the underlying reason for this being that the n partial derivatives $\partial V/\partial q_1, \ldots, \partial V/\partial q_n$ needed in Eqs. (6) appear in only p equations, namely, Eqs. (14) or (18). What follows is a seven-step procedure for surmounting this hurdle.

Step 1 Introduce $m \triangleq n - p$ quantities $f_1, \ldots, f_m$ as

$$f_{s-p} \triangleq \frac{\partial V}{\partial q_s} \qquad (s = p + 1, \ldots, n) \tag{21}$$

and regard each of these as a function of $q_1, \ldots, q_n$, and t, except when *both* of the following conditions are fulfilled for some value of r, say, $r = i$: (1) the generalized active force $\widetilde{F}_i$ is a function of q_i only; (2) the right-hand members of Eqs. (14) or (18) reduce to $-\partial V/\partial q_i$. In that event, regard each of $f_1, \ldots, f_m$ as a function of t and all of $q_1, \ldots, q_n$ except q_i. [Unless this is done, Eqs. (21) and the now applicable relationship $\widetilde{F}_i = -\partial V/\partial q_i$ lead to conflicting expressions for $\partial^2 V/\partial q_s \partial q_i$ $(s = p + 1, \ldots, n; s \neq i)$, namely, $\partial f_{s-p}/\partial q_i \neq 0$ and $\partial \widetilde{F}_i/\partial q_s = 0$, respectively.]

Step 2 In accordance with Eqs. (21), replace $\partial V/\partial q_s$ with f_{s-p} $(s = p + 1, \ldots, n)$ in Eqs. (14) or (18), and solve the resulting p equations for $\partial V/\partial q_r$ $(r = 1, \ldots, p)$.

Step 3 Using the expressions obtained in Step 2 for $\partial V/\partial q_r$ $(r = 1, \ldots, p)$, form $p(n - 1)$ expressions for $\partial(\partial V/\partial q_r)/\partial q_j$ $(r = 1, \ldots, p; j = 1, \ldots, n; j \neq r)$. Referring to Eqs. (21), form the $m(n - 1)$ equations $\partial(\partial V/\partial q_s)/\partial q_j = \partial f_{s-p}/\partial q_j$ $(s = p + 1, \ldots, n; j = 1, \ldots, n; j \neq s)$. Substitute into Eqs. (12) to obtain $n(n - 1)/2$ linear algebraic equations in the mn quantities $\partial f_i/\partial q_j$ $(i = 1, \ldots, m; j = 1, \ldots, n)$.

Step 4 Identify an $n(n - 1)/2 \times mn$ matrix $[Z]$ and an $n(n - 1)/2 \times 1$ matrix $\{Y\}$ such that the set of equations written in Step 3 is equivalent to the matrix equation $[Z]\{X\} = \{Y\}$, where $\{X\}$ is an $mn \times 1$ matrix having $\partial f_1/\partial q_1, \ldots, \partial f_1/\partial q_n, \ldots, \partial f_m/\partial q_1, \ldots, \partial f_m/\partial q_n$ as successive elements.

Step 5 Determine the rank ρ of $[Z]$. If $\rho = n(n - 1)/2$, then V may exist, but cannot be found by the application of a straightforward procedure. If $\rho \neq n(n - 1)/2$, use any ρ rows of $[Z]$, hereafter called independent rows, to express each of the remaining rows of $[Z]$, hereafter called the dependent rows, as a weighted, linear combination of the ρ independent rows; and solve the resulting set of equations simultaneously to determine the weighting factors.

Step 6 Express each element of $\{Y\}$ corresponding to a dependent row of $[Z]$ as a weighted, linear combination of the ρ elements of $\{Y\}$ corresponding to the independent rows of $[Z]$, using the weighting factors found in Step 5, and solve the resulting set of equations for $f_1, \ldots, f_m$. If this cannot be done uniquely, or if one or more of $f_1, \ldots, f_m$ turn out to be functions of a generalized coordinate of which they should be independent in accordance with Step 1, then a potential energy V of S in A does not exist.

Step 7 Substitute the functions $f_1,\ldots,f_m$ found in Step 6 into Eqs. (21) and into the expressions for $\partial V/\partial q_1,\ldots,\partial V/\partial q_p$ formed in Step 2, thus obtaining expressions for $\partial V/\partial q_1,\ldots,\partial V/\partial q_n$ as explicit functions of $q_1,\ldots,q_n$, and t. Finally, form V in accordance with Eq. (6).

Derivations Multiplication of both sides of Eqs. (18) with u_r and subsequent summation yields

$$\sum_{r=1}^{p}\tilde{F}_r u_r = -\sum_{r=1}^{p}\left[\sum_{s=1}^{n}\frac{\partial V}{\partial q_s}\left(W_{sr} + \sum_{k=p+1}^{n} W_{sk}A_{kr}\right)\right]u_r \tag{22}$$

or, equivalently,

$$-\sum_{r=1}^{p}\tilde{F}_r u_r = \sum_{s=1}^{n}\left[\frac{\partial V}{\partial q_s}\sum_{r=1}^{p}\left(W_{sr} + \sum_{k=p+1}^{n} W_{sk}A_{kr}\right)u_r\right] \tag{23}$$

Now,

$$\begin{aligned}\sum_{r=1}^{p}\left(W_{sr} + \sum_{k=p+1}^{n} W_{sk}A_{kr}\right)u_r &= \sum_{r=1}^{p} W_{sr}u_r + \sum_{k=p+1}^{n} W_{sk}\sum_{r=1}^{p} A_{kr}u_r \\ &\underset{(17)}{=} \sum_{r=1}^{p} W_{sr}u_r + \sum_{k=p+1}^{n} W_{sk}\left(u_k - B_k\right) \\ &= \sum_{r=1}^{n} W_{sr}u_r - \sum_{k=p+1}^{n} W_{sk}B_k \\ &\underset{(8)}{=} \dot{q}_s - \left(X_s + \sum_{k=p+1}^{n} W_{sk}B_k\right) \qquad (s = 1,\ldots,n)\end{aligned} \tag{24}$$

Consequently,

$$\begin{aligned}-\sum_{r=1}^{p}\tilde{F}_r u_r &\underset{(23)}{=} \sum_{s=1}^{n}\frac{\partial V}{\partial q_s}\left[\dot{q}_s - \left(X_s + \sum_{\underset{(24)}{k=p+1}}^{n} W_{sk}B_k\right)\right] \\ &= \dot{V} - \left[\frac{\partial V}{\partial t} + \sum_{s=1}^{n}\frac{\partial V}{\partial q_s}\left(X_s + \sum_{k=p+1}^{n} W_{sk}B_k\right)\right] \\ &\underset{(19)}{=} \dot{V}\end{aligned} \tag{25}$$

which is Eq. (20).

One can obtain Eqs. (1) by taking $W_{sr} = \delta_{sr}$, the Kronecker delta, and $X_s = 0$ $(r,s = 1,\ldots,n)$ in Eqs. (8); and, setting $A_{kr} = C_{kr}$ and $B_k = D_k$ $(r = 1,\ldots,p; k = p+1,\ldots,n)$, one then finds that Eqs. (17)–(20) lead to Eqs. (13)–(16), respectively. When $p = n$, in which event Eqs. (17) drop out of the picture, then Eqs. (18)–(20) become Eqs. (9)–(11), respectively. Finally, when Eqs. (8) reduce to Eqs. (1) and $p = n$, then Eqs. (18)–(20) reduce to Eqs. (2)–(4), respectively.

The rationale underlying the seven-step procedure for the construction of potential energy functions is the following. Step 1 is taken in recognition of the fact that Eqs. (18) or (14) form a set of p equations in the n partial derivatives $\partial V/\partial q_1, \ldots, \partial V/\partial q_n$, so that, since $n = p+m$, m additional relationships are required for the determination of all of these n partial derivatives. In Step 2, the task begun in Step 1, that is, the constructing of a set of expressions for the partial derivatives $\partial V/\partial q_1, \ldots, \partial V/\partial q_n$, is brought to completion. Step 3 consists of imposing requirements that must be satisfied in order that V possess continuous first partial derivatives. Steps 4–6 allow one to determine f_i $(i = 1, \ldots, m)$ by exploiting the fact† that the matrix equation $[Z]\{X\} = \{Y\}$ can be solved uniquely for $\{X\}$ if and only if the rank of $[Z]$ is equal to that of the matrix $[[Z] \mid \{Y\}]$. Finally, in Step 7 the desired expression for V is formed. [Partial differentiations of Eq. (6) with respect to $q_1, \ldots, q_n$ show that this equation is a generally valid relationship between a function V of $q_1, \ldots, q_n$, and t and the partial derivatives $\partial V/\partial q_1, \ldots, \partial V/\partial q_n$ when Eqs. (12) are satisfied.]

Example Suppose that the system S formed by the particle P_1 and the sharp-edged circular disk D considered in the example in Sec. 3.5 and shown in Fig. 3.5.1 is subjected to the action of a contact force $\mathbf{K}$ applied to P_1, with $\mathbf{K}$ given by

$$\mathbf{K} = k\hat{\mathbf{e}}_x - \frac{k}{L}\mathbf{p} \cdot \hat{\mathbf{e}}_y \hat{\mathbf{e}}_y \tag{26}$$

where k is a constant, $\hat{\mathbf{e}}_x$ and $\hat{\mathbf{e}}_y$ are unit vectors directed as shown in Fig. 3.5.1, and $\mathbf{p}$ is the position vector from point O to P_1. Furthermore, let m_1 and m_2 be the masses of P_1 and D, respectively; regard the rod R connecting P_1 and D as having a negligible mass; assume that Y, the axis of rotation of B, is vertical; and, designating as E a rigid body in which $\hat{\mathbf{e}}_x$, $\hat{\mathbf{e}}_y$, and $\hat{\mathbf{e}}_z = \hat{\mathbf{e}}_x \times \hat{\mathbf{e}}_y$ are fixed, define motion variables u_1, u_2, and u_3 as

$$u_1 \triangleq {}^{A}\mathbf{v}^{P_1} \cdot \hat{\mathbf{e}}_x \qquad u_2 \triangleq {}^{A}\mathbf{v}^{P_1} \cdot \hat{\mathbf{e}}_y \qquad u_3 \triangleq {}^{B}\boldsymbol{\omega}^{E} \cdot \hat{\mathbf{e}}_z \tag{27}$$

so that, in accordance with Eqs. (3.4.8),

$$\dot{q}_1 = u_1 c_3 - u_2 s_3 \qquad \dot{q}_2 = u_1 s_3 + u_2 c_3 \qquad \dot{q}_3 \triangleq u_3 \tag{28}$$

Then S is subject to the motion constraint

$$u_3 \underset{(3.5.15)}{=} -\frac{u_2}{L} \tag{29}$$

arising from the requirement that ${}^{B}\mathbf{v}^{D^\star} \cdot \hat{\mathbf{e}}_y$ be equal to zero, where ${}^{B}\mathbf{v}^{D^\star}$ is the velocity in B of the center $D^\star$ of D, and Eqs. (28) and (29) play the roles of Eqs. (8) and (17), respectively; that is,

$$n = 3 \qquad m = 1 \qquad p = n - m = 2 \tag{30}$$

and

$$W_{11} = c_3 \qquad W_{12} = -s_3 \qquad W_{13} = 0 \tag{31}$$

† M. H. Protter and C. B. Morrey, Jr., *Modern Mathematical Analysis* (Reading, Mass.: Addison-Wesley, 1966), p. 300.

$$W_{21} = s_3 \qquad W_{22} = c_3 \qquad W_{23} = 0 \tag{32}$$

$$W_{31} = 0 \qquad W_{32} = 0 \qquad W_{33} = 1 \tag{33}$$

$$X_1 = X_2 = X_3 = 0 \tag{34}$$

while

$$A_{31} = 0 \qquad A_{32} = -\frac{1}{L} \qquad B_3 = 0 \tag{35}$$

Moreover, the generalized active forces $\widetilde{F}_1$ and $\widetilde{F}_2$ for S in A are given by

$$\widetilde{F}_r \underset{(5.4.2)}{=} {}^A\widetilde{\mathbf{v}}_r^{P_1} \cdot (\mathbf{K} + m_1 g\hat{\mathbf{z}}) + {}^A\widetilde{\mathbf{v}}_r^{D^\star} \cdot (m_2 g\hat{\mathbf{z}}) \qquad (r = 1,2) \tag{36}$$

where $\hat{\mathbf{z}}$ is a unit vector directed vertically downward; the partial velocities ${}^A\widetilde{\mathbf{v}}_r^{P_1}$ and ${}^A\widetilde{\mathbf{v}}_r^{D^\star}$ $(r = 1,2)$ are available in Eqs. (3.6.25) and (3.6.31), respectively. Referring to Eq. (26) and noting that the position vector from O to P_1 can be written

$$\mathbf{p} = (q_1 c_3 + q_2 s_3)\hat{\mathbf{e}}_x - (q_1 s_3 - q_2 c_3)\hat{\mathbf{e}}_y \tag{37}$$

while

$$\hat{\mathbf{z}} = -(s_3\hat{\mathbf{e}}_x + c_3\hat{\mathbf{e}}_y) \tag{38}$$

one thus has

$$\widetilde{F}_1 \underset{(36)}{=} k - (m_1 + m_2)g s_3 \qquad \widetilde{F}_2 \underset{(36)}{=} \frac{k}{L}(q_1 s_3 - q_2 c_3) - m_1 g c_3 \tag{39}$$

A potential energy V of S now will be found by following the seven-step procedure after noting that substitution from Eqs. (31)–(33), (35), and (39) into Eqs. (18) yields

$$k - (m_1 + m_2)g s_3 = -\left(\frac{\partial V}{\partial q_1}c_3 + \frac{\partial V}{\partial q_2}s_3\right) \tag{40}$$

$$\frac{k}{L}(q_1 s_3 - q_2 c_3) - m_1 g c_3 = \frac{\partial V}{\partial q_1}s_3 - \frac{\partial V}{\partial q_2}c_3 + \frac{1}{L}\frac{\partial V}{\partial q_3} \tag{41}$$

In accordance with Eqs. (21), f_1 is introduced as

$$f_1 \triangleq \frac{\partial V}{\partial q_3} \tag{42}$$

and is regarded as a function of q_1, q_2, and q_3 because, although $\widetilde{F}_1$ [see Eqs. (39)] is a function of q_3 only, the right-hand member of Eq. (40) is not $-\partial V/\partial q_3$.

Step 2 Elimination of $\partial V/\partial q_3$ from Eq. (41) with the aid of Eq. (42) and simultaneous solution of the resulting equation and Eq. (40) for $\partial V/\partial q_1$ and $\partial V/\partial q_2$, yield

$$\frac{\partial V}{\partial q_1} = -kc_3 + \frac{k}{L}(q_1 s_3 - q_2 c_3)s_3 + g m_2 s_3 c_3 - \frac{f_1}{L}s_3 \tag{43}$$

$$\frac{\partial V}{\partial q_2} = -ks_3 - \frac{k}{L}(q_1 s_3 - q_2 c_3)c_3 + g(m_1 + m_2 {s_3}^2) + \frac{f_1}{L}c_3 \tag{44}$$

Step 3 Differentiations of Eqs. (42)–(44) yield the following mixed partial derivatives of V:

$$\frac{\partial^2 V}{\partial q_1 \partial q_2} \underset{(44)}{=} -\frac{k}{L}s_3c_3 + \frac{1}{L}\frac{\partial f_1}{\partial q_1}c_3 \tag{45}$$

$$\frac{\partial^2 V}{\partial q_2 \partial q_1} \underset{(43)}{=} -\frac{k}{L}c_3s_3 - \frac{1}{L}\frac{\partial f_1}{\partial q_2}s_3 \tag{46}$$

$$\frac{\partial^2 V}{\partial q_2 \partial q_3} \underset{(42)}{=} \frac{\partial f_1}{\partial q_2} \tag{47}$$

$$\begin{aligned}\frac{\partial^2 V}{\partial q_3 \partial q_2} &\underset{(44)}{=} -kc_3 - \frac{k}{L}[q_1(c_3{}^2 - s_3{}^2) + 2q_2s_3c_3] \\ &\quad + 2gm_2s_3c_3 + \frac{1}{L}\left(\frac{\partial f_1}{\partial q_3}c_3 - f_1s_3\right)\end{aligned} \tag{48}$$

$$\begin{aligned}\frac{\partial^2 V}{\partial q_3 \partial q_1} &\underset{(43)}{=} ks_3 + \frac{k}{L}[2q_1s_3c_3 - q_2(c_3{}^2 - s_3{}^2)] + gm_2(c_3{}^2 - s_3{}^2) \\ &\quad - \frac{1}{L}\left(\frac{\partial f_1}{\partial q_3}s_3 + f_1c_3\right)\end{aligned} \tag{49}$$

$$\frac{\partial^2 V}{\partial q_1 \partial q_3} \underset{(42)}{=} \frac{\partial f_1}{\partial q_1} \tag{50}$$

Equations (12) thus lead to the following three equations in $\partial f_1/\partial q_1, \partial f_1/\partial q_2, \partial f_1/\partial q_3$:

$$c_3\frac{\partial f_1}{\partial q_1} + s_3\frac{\partial f_1}{\partial q_2} \underset{(45,\,46)}{=} 0 \tag{51}$$

$$\begin{aligned}\frac{\partial f_1}{\partial q_2} - \frac{c_3}{L}\frac{\partial f_1}{\partial q_3} &\underset{(47,\,48)}{=} -\frac{s_3}{L}f_1 + 2gm_2s_3c_3 - kc_3 \\ &\quad - \frac{k}{L}[q_1(c_3{}^2 - s_3{}^2) + 2q_2s_3c_3]\end{aligned} \tag{52}$$

$$\begin{aligned}\frac{\partial f_1}{\partial q_1} + \frac{s_3}{L}\frac{\partial f_1}{\partial q_3} &\underset{(49,\,50)}{=} -\frac{c_3}{L}f_1 + gm_2(c_3{}^2 - s_3{}^2) + ks_3 \\ &\quad + \frac{k}{L}[2q_1s_3c_3 - q_2(c_3{}^2 - s_3{}^2)]\end{aligned} \tag{53}$$

Step 4 Inspection of Eqs. (51)–(53) reveals that this set of equations is equivalent to $[Z]\{X\} = \{Y\}$ if $\{X\}$, $\{Y\}$, and $[Z]$ are defined as

$$\{X\} \triangleq \begin{Bmatrix} \dfrac{\partial f_1}{\partial q_1} \\[2ex] \dfrac{\partial f_1}{\partial q_2} \\[2ex] \dfrac{\partial f_1}{\partial q_3} \end{Bmatrix} \tag{54}$$

$$\{Y\} \triangleq \left\{ \begin{array}{c} 0 \\ -\dfrac{s_3}{L}f_1 + 2gm_2s_3c_3 - kc_3 - \dfrac{k}{L}[q_1(c_3{}^2 - s_3{}^2) + 2q_2s_3c_3] \\ -\dfrac{c_3}{L}f_1 + gm_2(c_3{}^2 - s_3{}^2) + ks_3 + \dfrac{k}{L}[2q_1s_3c_3 - q_2(c_3{}^2 - s_3{}^2)] \end{array} \right\} \tag{55}$$

and

$$[Z] \triangleq \begin{bmatrix} c_3 & s_3 & 0 \\ 0 & 1 & -\dfrac{c_3}{L} \\ 1 & 0 & \dfrac{s_3}{L} \end{bmatrix} \tag{56}$$

Step 5 The matrix $[Z]$ is singular, but it possesses a nonvanishing determinant of order 2. Hence, $\rho = 2$. Selecting the first two rows of $[Z]$ as the ones to be treated as independent, one can express the third row as

$$\lfloor 1 \quad 0 \quad s_3/L \rfloor = w_1 \lfloor c_3 \quad s_3 \quad 0 \rfloor + w_2 \lfloor 0 \quad 1 \quad -c_3/L \rfloor \tag{57}$$

where w_1 and w_2 are weighting factors. Equating the first elements on the right-hand and left-hand sides of Eq. (57), one finds that $w_1 = 1/c_3$, and equating the second elements then leads to $w_2 = -s_3/c_3$.

Step 6 Expressing Y_3, the element in the third row of $\{Y\}$ in Eq. (55), as $w_1Y_1 + w_2Y_2$, where Y_1 and Y_2 are the elements in the first two rows of $\{Y\}$, one has

$$\begin{aligned} -\frac{c_3}{L}f_1 + gm_2(c_3{}^2 - s_3{}^2) + ks_3 + \frac{k}{L}[2q_1s_3c_3 - q_2(c_3{}^2 - s_3{}^2)] \\ = -\frac{s_3}{c_3}\left\{-\frac{s_3}{L}f_1 + 2gm_2s_3c_3 - kc_3 - \frac{k}{L}[q_1(c_3{}^2 - s_3{}^2) + 2q_2s_3c_3]\right\} \end{aligned} \tag{58}$$

and, solving this equation for f_1, one finds that

$$f_1 = m_2gLc_3 + k(q_1s_3 - q_2c_3) \tag{59}$$

Step 7 Substituting f_1 as given in Eq. (59) into Eqs. (42)–(44), one arrives at

$$\left.\frac{\partial V}{\partial q_1}\right|_{(43)} = -kc_3 \tag{60}$$

$$\left.\frac{\partial V}{\partial q_2}\right|_{(44)} = -ks_3 + g(m_1 + m_2) \tag{61}$$

$$\left.\frac{\partial V}{\partial q_3}\right|_{(42)} = m_2gLc_3 + k(q_1s_3 - q_2c_3) \tag{62}$$

and, proceeding in accordance with Eq. (6) after setting $\alpha_1 = \alpha_2 = \alpha_3 = C = 0$, one can thus write

$$V = \int_0^{q_1} \underset{(60)}{[-k\cos(0)]}\, d\zeta + \int_0^{q_2} \underset{(61)}{[-k\sin(0) + g(m_1 + m_2)]}\, d\zeta$$
$$+ \int_0^{q_3} \underset{(62)}{[m_2 gL\cos\zeta + k(q_1\sin\zeta - q_2\cos\zeta)]}\, d\zeta \tag{63}$$

so that, after performing the indicated integrations, one has

$$V = -k(q_1 c_3 + q_2 s_3) + g[(m_1 + m_2)q_2 + m_2 L s_3] \tag{64}$$

For the problem at hand, Eq. (19) reduces to

$$\frac{\partial V}{\partial t} = 0 \tag{65}$$

by virtue of Eqs. (34) and the last of Eqs. (35). Since V as given by Eq. (64) satisfies Eq. (65), Eq. (20) should be satisfied. To see that this is, in fact, the case, note that

$$\dot{V} \underset{(64)}{=} -k(\dot{q}_1 c_3 - q_1\dot{q}_3 s_3 + \dot{q}_2 s_3 + q_2\dot{q}_3 c_3) + g[(m_1 + m_2)\dot{q}_2 + m_2 L\dot{q}_3 c_3] \tag{66}$$

and that, when Eqs. (28) are used to eliminate $\dot{q}_1$, $\dot{q}_2$, and $\dot{q}_3$ from Eq. (66), one obtains, with the aid of Eq. (29),

$$\dot{V} \underset{(66)}{=} -[k - (m_1 + m_2)g s_3]u_1 - \left[\frac{k}{L}(q_1 s_3 - q_2 c_3) - m_1 g c_3\right] u_2 \tag{67}$$

or, in view of Eqs. (39),

$$\dot{V} = -(\widetilde{F}_1 u_1 + \widetilde{F}_2 u_2) \tag{68}$$

in agreement with Eq. (20).

To illustrate the possibility of nonexistence of a potential energy, suppose that the disk D is replaced with a particle P_2 of mass m_2, and let P_2 be free to slide in plane B (see Fig. 3.5.1), so that the motion constraint expressed by Eq. (29) no longer applies and the system S formed by P_1 and P_2 is a holonomic system possessing three degrees of freedom in A. The associated generalized active forces then are given by

$$F_r \underset{(5.4.1)}{=} {}^A\mathbf{v}_r^{P_1} \cdot (\mathbf{K} + m_1 g\hat{\mathbf{z}}) + {}^A\mathbf{v}_r^{P_2} \cdot (m_2 g\hat{\mathbf{z}}) \qquad (r = 1,2,3) \tag{69}$$

where the partial velocities ${}^A\mathbf{v}_r^{P_1}$ $(r = 1,2,3)$ are available in Eqs. (3.6.20) and the partial velocities ${}^A\mathbf{v}_r^{P_2}$ $(r = 1,2,3)$ are [see Eqs. (3.6.30)]

$${}^A\mathbf{v}_1^{P_2} = \hat{\mathbf{e}}_x \qquad {}^A\mathbf{v}_2^{P_2} = \hat{\mathbf{e}}_y \qquad {}^A\mathbf{v}_3^{P_2} = L\hat{\mathbf{e}}_y \tag{70}$$

Consequently [see Eq. (26) for $\mathbf{K}$, Eq. (37) for $\mathbf{p}$, and Eq. (38) for $\hat{\mathbf{z}}$],

$$F_1 = k - (m_1 + m_2)g s_3 \tag{71}$$

$$F_2 = \frac{k}{L}(q_1 s_3 - q_2 c_3) - (m_1 + m_2)g c_3 \tag{72}$$

$$F_3 = -m_2 g L c_3 \tag{73}$$

and Eqs. (9) can be written, with the aid of Eqs. (31)–(33), as

$$k - (m_1 + m_2)g s_3 = -\left(\frac{\partial V}{\partial q_1} c_3 + \frac{\partial V}{\partial q_2} s_3\right) \tag{74}$$

$$\frac{k}{L}(q_1 s_3 - q_2 c_3) - (m_1 + m_2)g c_3 = -\left(-\frac{\partial V}{\partial q_1} s_3 + \frac{\partial V}{\partial q_2} c_3\right) \tag{75}$$

$$-m_2 g L c_3 = -\frac{\partial V}{\partial q_3} \tag{76}$$

Solving these equations for $\partial V/\partial q_r$ $(r = 1,2,3)$, one has

$$\frac{\partial V}{\partial q_1} = -k c_3 + \frac{k}{L}(q_1 s_3 - q_2 c_3) s_3 \tag{77}$$

$$\frac{\partial V}{\partial q_2} = -k s_3 - \frac{k}{L}(q_1 s_3 - q_2 c_3) c_3 + (m_1 + m_2) g \tag{78}$$

$$\frac{\partial V}{\partial q_3} = m_2 g L c_3 \tag{79}$$

and differentiation of these relationships yields

$$\frac{\partial^2 V}{\partial q_1 \partial q_2} \underset{(78)}{=} -\frac{k}{L} s_3 c_3 \qquad \frac{\partial^2 V}{\partial q_2 \partial q_1} \underset{(77)}{=} -\frac{k}{L} c_3 s_3 \tag{80}$$

$$\frac{\partial^2 V}{\partial q_2 \partial q_3} \underset{(79)}{=} 0 \qquad \frac{\partial^2 V}{\partial q_3 \partial q_2} \underset{(78)}{=} -k c_3 - \frac{k}{L}[q_1({c_3}^2 - {s_3}^2) + 2 q_2 s_3 c_3] \tag{81}$$

$$\frac{\partial^2 V}{\partial q_3 \partial q_1} \underset{(77)}{=} k s_3 + \frac{k}{L}[2 q_1 s_3 c_3 - q_2({c_3}^2 - {s_3}^2)] \qquad \frac{\partial^2 V}{\partial q_1 \partial q_3} \underset{(79)}{=} 0 \tag{82}$$

Hence, unless $k = 0$,

$$\frac{\partial^2 V}{\partial q_2 \partial q_3} \underset{(81)}{\neq} \frac{\partial^2 V}{\partial q_3 \partial q_2} \tag{83}$$

and

$$\frac{\partial^2 V}{\partial q_3 \partial q_1} \underset{(82)}{\neq} \frac{\partial^2 V}{\partial q_1 \partial q_3} \tag{84}$$

so that not all of Eqs. (12) are satisfied. Therefore, there exists no potential energy.

7.2 POTENTIAL ENERGY CONTRIBUTIONS

Referring to Sec. 7.1, divide the set of all contact and/or distance forces contributing to $\tilde{F}_r$ $(r = 1,\ldots,p)$ into subsets $\alpha, \beta, \ldots$ associated with particular sets of contact and/or distance forces. Furthermore, let $(\tilde{F}_r)_\alpha, (\tilde{F}_r)_\beta, \ldots$ denote the contributions of $\alpha, \beta, \ldots$ respectively, to $\tilde{F}_r$ $(r = 1,\ldots,p)$, and let $V_\alpha, V_\beta, \ldots$ be functions of $q_1,\ldots,q_n$, and t such that Eqs. (7.1.18) are satisfied when $\tilde{F}_r$ and V are replaced with $(\tilde{F}_r)_\alpha$ and V_α, respectively, and similarly for $\beta, \gamma, \ldots$. The functions $V_\alpha, V_\beta, \ldots$ are called *potential energy contributions* of $\alpha, \beta, \ldots$ for S, and the function V of $q_1, \ldots, q_n$, and t defined as

$$V \triangleq V_\alpha + V_\beta + \cdots \tag{1}$$

is a potential energy of S (see Sec. 7.1).

Considering the set γ of all gravitational forces exerted on particles of S by the Earth E, and assuming that these forces can be treated as in Sec. 5.7, let

$$V_\gamma \triangleq -Mg\hat{\mathbf{k}} \cdot \mathbf{p}^\star \tag{2}$$

where M is the total mass of S, g is the local gravitational force per unit mass, $\hat{\mathbf{k}}$ is a unit vector locally directed vertically downward, and $\mathbf{p}^\star$ is the position vector from any point fixed in E to $S^\star$. Then V_γ is a potential energy contribution of γ for S.

When one end of a spring is fixed in a reference frame A while the other end is attached to a particle of S, or when a spring connects two particles of S (the spring, itself, being in neither case a part of S), let σ be the set of forces exerted by the spring on particles of S, and define V_σ as

$$V_\sigma \triangleq \int_0^x f(\zeta)d\zeta \tag{3}$$

where x is a function of $q_1, \ldots, q_n$, and t that measures the *extension* of the spring, that is, the difference between the spring's current length and the spring's natural length; $f(x)$ defines the spring's elastic characteristics. For instance, $f(x)$ may be given by

$$f(x) = kx \tag{4}$$

where k is a constant called the *spring constant* or *spring modulus*. Under these circumstances, the spring is said to be a *linear* spring, and Eq. (3) leads to

$$V_\sigma = \tfrac{1}{2}kx^2 \tag{5}$$

In any event, V_σ as defined in Eqs. (3) [and, hence, V_σ as given in Eq. (5)] is a potential energy contribution of σ for S.

Equations (3)–(5) apply also when one end of a torsion spring is fixed in a reference frame A while the other end is attached to a rigid body B belonging to S and free to rotate relative to A about an axis fixed in both A and B, or when a torsion spring connects two rigid bodies belonging to S and free to rotate relative to each other about an axis fixed in both bodies. Under these circumstances, x measures the rotational deformation of the spring.

Derivations It follows from the definitions of $(\tilde{F}_r)_\alpha, (\tilde{F}_r)_\beta, \ldots$ and $V_\alpha, V_\beta, \ldots$ that

$$\tilde{F}_r = (\tilde{F}_r)_\alpha + (\tilde{F}_r)_\beta + \cdots \qquad (r = 1, \ldots, p) \tag{6}$$

and that

$$(\tilde{F}_r)_\alpha \underset{(7.1.18)}{=} -\sum_{s=1}^{n} \frac{\partial V_\alpha}{\partial q_s}\left(W_{sr} + \sum_{k=p+1}^{n} W_{sk}A_{kr}\right) \qquad (r = 1, \ldots, p) \tag{7}$$

$$(\widetilde{F}_r)_\beta \underset{(7.1.18)}{=} -\sum_{s=1}^{n} \frac{\partial V_\beta}{\partial q_s}\left(W_{sr} + \sum_{k=p+1}^{n} W_{sk}A_{kr}\right) \qquad (r = 1, \ldots, p) \tag{8}$$

and so forth. Consequently,

$$(\widetilde{F}_r)_\alpha + (\widetilde{F}_r)_\beta + \cdots \underset{(7,\,8)}{=} -\sum_{s=1}^{n}\left[\frac{\partial}{\partial q_s}(V_\alpha + V_\beta + \cdots)\right]\left(W_{sr} + \sum_{k=p+1}^{n} W_{sk}A_{kr}\right) \qquad (r = 1, \ldots, p) \tag{9}$$

which reduces to Eq. (7.1.18) when Eqs. (6) and (1) are brought into play. Hence, V as defined in Eq. (1) is a potential energy of S.

In order to show that V_γ and V_σ are potential energy contributions of γ and σ, respectively, it is necessary to make use of the following kinematical proposition. If $\mathbf{p}$, regarded as a function of $q_1, \ldots, q_n$, and t in A, is the position vector from a point fixed in A to a point P of S, and $\mathbf{v}$, the velocity of P in A, is expressed as in Eq. (3.6.4), then

$$\widetilde{\mathbf{v}}_r = \sum_{s=1}^{n} \frac{\partial \mathbf{p}}{\partial q_s}\left(W_{sr} + \sum_{k=p+1}^{n} W_{sk}A_{kr}\right) \qquad (r = 1, \ldots, p) \tag{10}$$

To see this, note that

$$\widetilde{\mathbf{v}}_r \underset{(3.6.17)}{=} \sum_{s=1}^{n} \frac{\partial \mathbf{p}}{\partial q_s} \underset{(3.6.11)}{W_{sr}} + \sum_{k=p+1}^{n} \mathbf{v}_k A_{kr}$$
$$= \sum_{s=1}^{n} \frac{\partial \mathbf{p}}{\partial q_s} W_{sr} + \sum_{k=p+1}^{n} \sum_{s=1}^{n} \underset{(3.6.11)}{\frac{\partial \mathbf{p}}{\partial q_s} W_{sk}} A_{kr} \qquad (r = 1, \ldots, p) \tag{11}$$

which is equivalent to Eqs. (10).

Now consider V_γ as defined in Eq. (2). Partial differentiations with respect to q_s $(s = 1, \ldots, n)$ give

$$\frac{\partial V_\gamma}{\partial q_s} \underset{(2)}{=} -Mg\hat{\mathbf{k}} \cdot \frac{\partial \mathbf{p}^\star}{\partial q_s} \qquad (s = 1, \ldots, n) \tag{12}$$

Hence,

$$-\sum_{s=1}^{n} \frac{\partial V_\gamma}{\partial q_s}\left(W_{sr} + \sum_{k=p+1}^{n} W_{sk}A_{kr}\right) \underset{(12)}{=} Mg\hat{\mathbf{k}} \cdot \sum_{s=1}^{n} \frac{\partial \mathbf{p}^\star}{\partial q_s}\left(W_{sr} + \sum_{k=p+1}^{n} W_{sk}A_{kr}\right)$$
$$\underset{(10)}{=} Mg\hat{\mathbf{k}} \cdot \widetilde{\mathbf{v}}_r^{\,\star} \underset{(5.7.2)}{=} (\widetilde{F}_r)_\gamma \qquad (r = 1, \ldots, p) \tag{13}$$

which shows that Eqs. (7.1.18) are satisfied when $\widetilde{F}_r$ and V are replaced with $(\widetilde{F}_r)_\gamma$ and V_γ, respectively, and this means that V_γ is a potential energy contribution of γ for S.

As for V_σ, defined in Eq. (3), we begin once more by forming partial derivatives with respect to q_s $(s = 1,\dots,n)$, obtaining

$$\frac{\partial V_\sigma}{\partial q_s} \underset{(3)}{=} f(x)\frac{\partial x}{\partial q_s} \qquad (s = 1,\dots,n) \tag{14}$$

whereupon we can write

$$-\sum_{s=1}^{n} \frac{\partial V_\sigma}{\partial q_s}\left(W_{sr} + \sum_{k=p+1}^{n} W_{sk}A_{kr}\right) \underset{(14)}{=} -f(x)\sum_{s=1}^{n}\frac{\partial x}{\partial q_s}\left(W_{sr} + \sum_{k=p+1}^{n} W_{sk}A_{kr}\right)$$
$$(r = 1,\dots,p) \tag{15}$$

Next, we construct an expression for $(\tilde{F}_r)_\sigma$, the contribution to $\tilde{F}_r$ of the force exerted by a spring on a particle P of S when one end of the spring is attached to a point O fixed in A. To facilitate this task, we introduce a unit vector $\hat{\mathbf{e}}$ directed from O to P, and express $\mathbf{p}$, the position vector from O to P, as

$$\mathbf{p} = (L + x)\hat{\mathbf{e}} \tag{16}$$

where L is the natural length of the spring. The holonomic partial velocities of P in A then can be written

$$\mathbf{v}_r \underset{(3.6.11)}{=} \sum_{s=1}^{n}\frac{\partial \mathbf{p}}{\partial q_s}W_{sr} \underset{(16)}{=} \sum_{s=1}^{n}\left[\frac{\partial x}{\partial q_s}\hat{\mathbf{e}} + (L + x)\frac{\partial \hat{\mathbf{e}}}{\partial q_s}\right]W_{sr} \qquad (r = 1,\dots,n) \tag{17}$$

and, with $\mathbf{T}$, the (tensile) force exerted on P by the spring, expressed as

$$\mathbf{T} = -f(x)\hat{\mathbf{e}} \tag{18}$$

we find that $(F_r)_\sigma$, the contribution of $\mathbf{T}$ to the *holonomic* generalized active force F_r, is given by

$$(F_r)_\sigma = \mathbf{v}_r \cdot \mathbf{T} \underset{(17,\,18)}{=} -f(x)\sum_{s=1}^{n}\frac{\partial x}{\partial q_s}W_{sr} \qquad (r = 1,\dots,n) \tag{19}$$

because $2\hat{\mathbf{e}} \cdot \partial\hat{\mathbf{e}}/\partial q_s = \partial(\hat{\mathbf{e}} \cdot \hat{\mathbf{e}})/\partial q_s = \partial(1)/\partial q_s = 0$ $(s = 1,\dots,n)$; and $(\tilde{F}_r)_\sigma$, the contribution of $\mathbf{T}$ to the *nonholonomic* generalized active force $\tilde{F}_r$, now can be formed as

$$\begin{aligned}(\tilde{F}_r)_\sigma &\underset{(5.4.4)}{=} (F_r)_\sigma + \sum_{k=p+1}^{n}(F_k)_\sigma A_{kr} \\ &\underset{(19)}{=} -f(x)\left(\sum_{s=1}^{n}\frac{\partial x}{\partial q_s}W_{sr} + \sum_{k=p+1}^{n}\sum_{s=1}^{n}\frac{\partial x}{\partial q_s}W_{sk}A_{kr}\right) \\ &= -f(x)\sum_{s=1}^{n}\frac{\partial x}{\partial q_s}\left(W_{sr} + \sum_{k=p+1}^{n}W_{sk}A_{kr}\right) \qquad (r = 1,\dots,p)\end{aligned} \tag{20}$$

The right-hand members of Eqs. (20) and (15) are identical. Consequently,

$$(\tilde{F}_r)_\sigma = -\sum_{s=1}^{n}\frac{\partial V_\sigma}{\partial q_s}\left(W_{sr} + \sum_{k=p+1}^{n}W_{sk}A_{kr}\right) \qquad (r = 1,\dots,p) \tag{21}$$

This is precisely Eq. (7.1.18) when $\widetilde{F}_r$ and V are replaced with $(\widetilde{F}_r)_\sigma$ and V_σ, respectively, which means that V_σ is a potential energy contribution of σ for S. A parallel proof shows that this conclusion is also valid when the spring connects two particles of S.

Examples Referring to the example in Sec. 5.4, one can express $\mathbf{p}^\star$, the position vector from O to the mass center of the system S formed by P_1 and P_2, as

$$\mathbf{p}^\star = \frac{m_1(L_1 + q_1) + m_2(L_1 + L_2 + q_2)}{m_1 + m_2}\hat{\mathbf{t}}_1 \tag{22}$$

Forming, with the aid of Eq. (2), a potential energy contribution V_γ of the set γ of gravitational forces acting on S, one has

$$V_\gamma \underset{(2,\ 22)}{=} -[m_1(L_1 + q_1) + m_2(L_1 + L_2 + q_2)]g \cos\theta \tag{23}$$

The extensions of the springs σ_1 and σ_2 are q_1 and $q_2 - q_1$, respectively. Hence, if potential energy contributions of the forces exerted by σ_1 and σ_2 are denoted by V_{σ_1} and V_{σ_2}, respectively, then, in accordance with Eq. (5),

$$V_{\sigma_1} = \tfrac{1}{2}k_1 {q_1}^2 \qquad V_{\sigma_2} = \tfrac{1}{2}k_2(q_2 - q_1)^2 \tag{24}$$

Now consider the function V defined as

$$\begin{aligned} V &\triangleq V_\gamma + V_{\sigma_1} + V_{\sigma_2} \\ &\underset{(23,\ 24)}{=} -[m_1(L_1 + q_1) + m_2(L_1 + L_2 + q_2)]g \cos\theta \\ &\quad + \tfrac{1}{2}[k_1 {q_1}^2 + k_2(q_2 - q_1)^2] \end{aligned} \tag{25}$$

Since gravitational forces and forces exerted on P_1 and P_2 by σ_1 and σ_2 are the only forces contributing to generalized active forces for S in A, V as given by Eq. (25) is a potential energy of S [see Eq. (1)]. It follows that the generalized active forces F_1 and F_2 for S in A can be found by substituting from Eq. (25) into Eqs. (7.1.2), which leads to

$$F_1 = m_1 g \cos\theta - k_1 q_1 + k_2(q_2 - q_1) \tag{26}$$

and

$$F_2 = m_2 g \cos\theta - k_2(q_2 - q_1) \tag{27}$$

in agreement with Eqs. (5.4.20) and (5.4.21), respectively.

When $\theta(t)$ is a constant, the potential energy V as given by Eq. (25) satisfies Eq. (7.1.3) and, therefore, Eq. (7.1.4). Conversely, when $\theta(t)$ is not a constant, then V does not satisfy Eq. (7.1.3), and it may be verified with the aid of Eqs. (26) and (27) that Eq. (7.1.4) is violated. As will be seen in Sec. 9.2, these facts play a decisive role in connection with the formulation of integrals of the equations of motion of S.

Suppose that generalized velocities u_1 and u_2 are defined as

$$u_1 \triangleq \mathbf{v}^{P_1} \cdot \hat{\mathbf{k}} = \dot{q}_1 \cos\theta - (L_1 + q_1)\dot{\theta} \sin\theta \tag{28}$$

$$u_2 \triangleq \mathbf{v}^{P_2} \cdot \hat{\mathbf{k}} = \dot{q}_2 \cos\theta - (L_1 + L_2 + q_2)\dot{\theta} \sin\theta \tag{29}$$

rather than as in Eqs. (5.4.9). The relevant equations of Sec. 7.1 then are Eqs. (7.1.8) and (7.1.9), with [solve Eqs. (28) and (29) for $\dot{q}_1$ and $\dot{q}_2$ and compare the resulting equations with Eqs. (7.1.8)]

$$W_{11} = \sec\theta \qquad W_{12} = 0 \qquad X_1 = (L_1 + q_1)\dot{\theta}\tan\theta \tag{30}$$

$$W_{21} = 0 \qquad W_{22} = \sec\theta \qquad X_2 = (L_1 + L_2 + q_2)\dot{\theta}\tan\theta \tag{31}$$

Consequently, F_1 and F_2 are now given by

$$F_1 \underset{(7.1.9)}{=} -\left(\frac{\partial V}{\partial q_1}W_{11} + \frac{\partial V}{\partial q_2}W_{21}\right) \underset{(25,\,30,\,31)}{=} m_1 g - [k_1 q_1 - k_2(q_2 - q_1)]\sec\theta \tag{32}$$

$$F_2 \underset{(7.1.9)}{=} -\left(\frac{\partial V}{\partial q_1}W_{12} + \frac{\partial V}{\partial q_2}W_{22}\right) \underset{(25,\,30,\,31)}{=} m_2 g - k_2(q_2 - q_1)\sec\theta \tag{33}$$

rather than by Eqs. (26) and (27). Moreover, solving Eqs. (32) and (33) [with the aid of Eqs. (30) and (31)] for $\partial V/\partial q_1$ and $\partial V/\partial q_2$ and then using Eq. (7.1.6), one recovers V as given by Eq. (25), thus verifying that the choice of motion variables affects the generalized active forces but not the potential energy of S.

7.3 DISSIPATION FUNCTIONS

If S is a simple nonholonomic system (see Sec. 3.5) possessing generalized coordinates $q_1, \ldots, q_n$ (see Sec. 3.2) and motion variables $u_1, \ldots, u_n$ (see Sec. 3.4) in a reference frame A, with $u_{p+1}, \ldots, u_n$ dependent upon $u_1, \ldots, u_p$ in accordance with Eqs. (3.5.2), and $(\tilde{F}_1)_C, \ldots, (\tilde{F}_p)_C$ are the contributions to the generalized active forces $\tilde{F}_1, \ldots, \tilde{F}_p$ (see Sec. 5.4), respectively, of a set C of contact forces acting on particles of S, there may exist a function $\mathscr{F}$ of $q_1, \ldots, q_n$, $u_1, \ldots, u_p$, and t such that

$$(\tilde{F}_r)_C = -\frac{\partial \mathscr{F}}{\partial u_r} \qquad (r = 1, \ldots, p) \tag{1}$$

Under these circumstances, $\mathscr{F}$ is called a *dissipation function* for C.

Example Referring to the example in Sec. 5.6, note that $n = 2$ and $p = n$ (because there are no motion constraints), and let C be the set of contact forces exerted on the rods A and B by the viscous fluid. Then $(F_1)_C$ and $(F_2)_C$, the contributions of C to the generalized active forces F_1 and F_2, respectively, are given by

$$(F_1)_C \underset{(5.6.8)}{=} 0 \qquad (F_2)_C \underset{(5.6.9,\,5.6.3)}{=} -\delta u_2 \tag{2}$$

and $\mathscr{F}$, a function of q_1, q_2, u_1, and u_2 defined as

$$\mathscr{F} \triangleq \tfrac{1}{2}\delta {u_2}^2 \tag{3}$$

is a dissipation function for C because

$$\frac{\partial \mathscr{F}}{\partial u_1} \underset{(3)}{=} 0 \underset{(2)}{=} -(F_1)_C \tag{4}$$

and

$$\frac{\partial \mathscr{F}}{\partial u_2} \underset{(3)}{=} \delta u_2 \underset{(2)}{=} -(F_2)_C \tag{5}$$

so that Eqs. (1) are satisfied for $r = 1, \ldots, p$.

7.4 KINETIC ENERGY

The *kinetic energy* K of a set S of ν particles $P_1, \ldots, P_\nu$ in a reference frame A is defined as

$$K \triangleq \frac{1}{2} \sum_{i=1}^{\nu} m_i (\mathbf{v}^{P_i})^2 \tag{1}$$

where m_i is the mass of P_i and $\mathbf{v}^{P_i}$ is the velocity of P_i in A.

When a subset of S forms a rigid body B, then K_B, the contribution of B to K, can be expressed as

$$K_B = K_\omega + K_v \tag{2}$$

where K_ω, called the *rotational* kinetic energy of B in A, and K_v, called the *translational* kinetic energy of B in A, depend, respectively, on the angular velocity $\boldsymbol{\omega}$ of B in A and the central inertia dyadic $\underline{\mathbf{I}}$ of B, and on the velocity $\mathbf{v}$ in A of the mass center $B^\star$ of B and the mass m of B. Specifically,

$$K_\omega \triangleq \tfrac{1}{2}\boldsymbol{\omega} \cdot \underline{\mathbf{I}} \cdot \boldsymbol{\omega} \tag{3}$$

and

$$K_v \triangleq \tfrac{1}{2} m \mathbf{v}^2 \tag{4}$$

Furthermore, the kinetic energy of rotation of B in A is given also by

$$K_\omega = \tfrac{1}{2} I \boldsymbol{\omega}^2 \tag{5}$$

where I is the moment of inertia of B about the line that passes through $B^\star$ and is parallel to $\boldsymbol{\omega}$ (in general, I is time dependent), and by

$$K_\omega = \frac{1}{2} \sum_{j=1}^{3} \sum_{k=1}^{3} \omega_j I_{jk} \omega_k \tag{6}$$

where I_{jk} $(j,k = 1,2,3)$ are inertia scalars of B relative to $B^\star$ (see Sec. 4.3) for any three mutually perpendicular unit vectors (in general, I_{jk} is time-dependent), and ω_1, ω_2, ω_3 are the associated measure numbers of $\boldsymbol{\omega}$. Finally, if I_1, I_2, I_3 are central principal moments of inertia of B (see Sec. 4.8), and ω_1, ω_2, ω_3 are the associated measure numbers of $\boldsymbol{\omega}$, then

$$K_\omega = \tfrac{1}{2}(I_1 {\omega_1}^2 + I_2 {\omega_2}^2 + I_3 {\omega_3}^2) \tag{7}$$

Derivations Letting $\mathbf{r}_i$ be the position vector from $B^\star$ to P_i, a generic particle of B, one can write

$$\mathbf{v}^{P_i} \underset{(2.7.1)}{=} \mathbf{v} + \boldsymbol{\omega} \times \mathbf{r}_i \tag{8}$$

where $\mathbf{v}$ is the velocity of $B^\star$ in A. Hence,

$$\begin{aligned} K &\underset{(1,\,8)}{=} \frac{1}{2}\sum_{i=1}^{\nu} m_i[\mathbf{v}^2 + 2\mathbf{v}\cdot\boldsymbol{\omega}\times\mathbf{r}_i + (\boldsymbol{\omega}\times\mathbf{r}_i)\cdot(\boldsymbol{\omega}\times\mathbf{r}_i)] \\ &= \frac{1}{2}\left(\sum_{i=1}^{\nu} m_i\right)\mathbf{v}^2 + \mathbf{v}\cdot\boldsymbol{\omega}\times\sum_{i=1}^{\nu} m_i\mathbf{r}_i + \tfrac{1}{2}\boldsymbol{\omega}\cdot\sum_{i=1}^{\nu} m_i\mathbf{r}_i\times(\boldsymbol{\omega}\times\mathbf{r}_i) \\ &= \tfrac{1}{2}m\mathbf{v}^2 + \underset{(4.1.1)}{0} + \underset{(4.5.31)}{\tfrac{1}{2}\boldsymbol{\omega}\cdot\underline{\mathbf{I}}\cdot\boldsymbol{\omega}} = \underset{(4)}{K_v} + \underset{(3)}{K_\omega} \end{aligned} \tag{9}$$

in agreement with Eq. (2).

Let $\hat{\mathbf{n}}_\omega$ and ω be a unit vector and a scalar, respectively, such that

$$\boldsymbol{\omega} = \hat{\mathbf{n}}_\omega\,\omega \tag{10}$$

Then

$$K_\omega \underset{(3,\,10)}{=} \tfrac{1}{2}\omega^2\hat{\mathbf{n}}_\omega\cdot\underline{\mathbf{I}}\cdot\hat{\mathbf{n}}_\omega \tag{11}$$

Now,

$$\omega^2 \underset{(10)}{=} \boldsymbol{\omega}^2 \tag{12}$$

and

$$\hat{\mathbf{n}}_\omega\cdot\underline{\mathbf{I}}\cdot\hat{\mathbf{n}}_\omega \underset{(4.5.20)}{=} I \tag{13}$$

Substitution from Eqs. (12) and (13) into Eq. (11) thus leads to Eq. (5).

When $\boldsymbol{\omega}$ is expressed as

$$\boldsymbol{\omega} = \sum_{i=1}^{3} \omega_i\hat{\mathbf{n}}_i \tag{14}$$

where $\hat{\mathbf{n}}_1, \hat{\mathbf{n}}_2, \hat{\mathbf{n}}_3$ are any mutually perpendicular unit vectors, then

$$\begin{aligned} K_\omega &\underset{(3,\,14)}{=} \frac{1}{2}\sum_{i=1}^{3}\omega_i\hat{\mathbf{n}}_i\cdot\underset{(4.5.22)}{\sum_{j=1}^{3}\sum_{k=1}^{3} I_{jk}\hat{\mathbf{n}}_j\hat{\mathbf{n}}_k}\cdot\sum_{i=1}^{3}\omega_i\hat{\mathbf{n}}_i \\ &= \frac{1}{2}\left(\sum_{i=1}^{3}\sum_{j=1}^{3}\sum_{k=1}^{3}\omega_i\hat{\mathbf{n}}_i\cdot\hat{\mathbf{n}}_j I_{jk}\hat{\mathbf{n}}_k\right)\cdot\sum_{i=1}^{3}\omega_i\hat{\mathbf{n}}_i \end{aligned} \tag{15}$$

or, since $\hat{\mathbf{n}}_i \cdot \hat{\mathbf{n}}_j$ vanishes except when $i = j$, and is equal to unity when $i = j$,

$$K_\omega \underset{(15)}{=} \frac{1}{2}\left(\sum_{j=1}^{3}\sum_{k=1}^{3}\omega_j I_{jk}\hat{\mathbf{n}}_k\right) \cdot \sum_{i=1}^{3}\omega_i\hat{\mathbf{n}}_i$$

$$= \frac{1}{2}\sum_{j=1}^{3}\sum_{k=1}^{3}\sum_{i=1}^{3}\omega_j I_{jk}\omega_i\hat{\mathbf{n}}_k \cdot \hat{\mathbf{n}}_i \qquad (16)$$

But $\hat{\mathbf{n}}_k \cdot \hat{\mathbf{n}}_i = 0$ except when $i = k$, and is equal to unity when $i = k$. Hence, Eq. (16) reduces to Eq. (6). Finally, when $\hat{\mathbf{n}}_1$, $\hat{\mathbf{n}}_2$, $\hat{\mathbf{n}}_3$ are parallel to central principal axes of B, then I_{jk} vanishes except for $j = k$ [see Eq. (4.8.2)], and Eq. (6) yields

$$K_\omega = \tfrac{1}{2}(I_{11}{\omega_1}^2 + I_{22}{\omega_2}^2 + I_{33}{\omega_3}^2) \qquad (17)$$

which establishes the validity of Eq. (7).

Example Suppose that the system S considered in the example in Sec. 5.7 has the following inertia properties: The frame A has a mass m_A, and $A^\star$, the mass center of A, is situated on line DE, at a distance a from D; the moment of inertia of A about a line passing through $A^\star$ and parallel to $\hat{\mathbf{a}}_1$ has the value I_A; wheels B and C are identical, uniform, thin disks of mass m_B and radius R. The kinetic energy K of S in reference frame F is to be expressed in terms of the parameters a, b, R, m_A, m_B, I_A, the generalized coordinates $q_1, \ldots, q_5$ defined in Problem 10.8, and the time derivatives of these coordinates.

The contribution K_A of A to K is

$$K_A \underset{(2)}{=} \tfrac{1}{2}I_A\underset{(5)}{(\boldsymbol{\omega}^A)^2} + \tfrac{1}{2}m_A\underset{(4)}{(\mathbf{v}^{A^\star})^2} \qquad (18)$$

with

$$\boldsymbol{\omega}^A = \dot{q}_1\hat{\mathbf{a}}_1 \qquad (19)$$

and (see Fig. 5.7.2 for $\hat{\mathbf{a}}_2$, $\hat{\mathbf{n}}_2$, and $\hat{\mathbf{n}}_3$)

$$\mathbf{v}^{A^\star} = \mathbf{v}^D + \boldsymbol{\omega}^A \times (a\hat{\mathbf{a}}_2)$$

$$= (-as_1\dot{q}_1 + \dot{q}_2)\hat{\mathbf{n}}_2 + (ac_1\dot{q}_1 + \dot{q}_3)\hat{\mathbf{n}}_3 \qquad (20)$$

Thus,

$$K_A \underset{(18\text{–}20)}{=} \tfrac{1}{2}I_A{\dot{q}_1}^2 + \tfrac{1}{2}m_A[a^2{\dot{q}_1}^2 - 2a\dot{q}_1(s_1\dot{q}_2 - c_1\dot{q}_3) + {\dot{q}_2}^2 + {\dot{q}_3}^2] \qquad (21)$$

K_B, the contribution of B to K, can be expressed as

$$K_B \underset{(2)}{=} \tfrac{1}{2}\underset{(7)}{[{I_1}^B({\omega_1}^B)^2 + {I_2}^B({\omega_2}^B)^2 + {I_3}^B({\omega_3}^B)^2]} + \tfrac{1}{2}m_B\underset{(4)}{(\mathbf{v}^{B^\star})^2} \qquad (22)$$

where

$${I_1}^B = {I_2}^B = \frac{m_B R^2}{4} \qquad {I_3}^B = \frac{m_B R^2}{2} \qquad (23)$$

$${\omega_1}^B = \dot{q}_1 \qquad {\omega_2}^B = 0 \qquad {\omega_3}^B = \dot{q}_4 \qquad (24)$$

and

$$\begin{aligned}\mathbf{v}^{B^\star} &= \mathbf{v}^D + \boldsymbol{\omega}^A \times (-b\hat{\mathbf{a}}_3)\\ &= (\dot{q}_2 + b\dot{q}_1 c_1)\hat{\mathbf{n}}_2 + (\dot{q}_3 + b\dot{q}_1 s_1)\hat{\mathbf{n}}_3\end{aligned} \tag{25}$$

Hence,

$$K_B \underset{(22\text{–}25)}{=} \frac{m_B}{2}\left[\left(\frac{R^2}{4} + b^2\right)\dot{q}_1^{\,2} + \dot{q}_2^{\,2} + \dot{q}_3^{\,2} + \frac{R^2}{2}\dot{q}_4^{\,2} + 2b\dot{q}_1(c_1\dot{q}_2 + s_1\dot{q}_3)\right] \tag{26}$$

Similarly,

$$K_C = \frac{m_B}{2}\left[\left(\frac{R^2}{4} + b^2\right)\dot{q}_1^{\,2} + \dot{q}_2^{\,2} + \dot{q}_3^{\,2} + \frac{R^2}{2}\dot{q}_5^{\,2} - 2b\dot{q}_1(c_1\dot{q}_2 + s_1\dot{q}_3)\right] \tag{27}$$

Consequently, the desired expression for K is

$$\begin{aligned}K &= K_A + K_B + K_C\\ &\underset{(21,\,26,\,27)}{=} \frac{1}{2}\Bigg\{\left[I_A + m_A a^2 + 2m_B\left(\frac{R^2}{4} + b^2\right)\right]\dot{q}_1^{\,2}\\ &\quad + (m_A + 2m_B)(\dot{q}_2^{\,2} + \dot{q}_3^{\,2}) + \frac{m_B R^2}{2}(\dot{q}_4^{\,2} + \dot{q}_5^{\,2})\\ &\quad - 2m_A a\dot{q}_1(s_1\dot{q}_2 - c_1\dot{q}_3)\Bigg\}\end{aligned} \tag{28}$$

7.5 HOMOGENEOUS KINETIC ENERGY FUNCTIONS

If S is a simple nonholonomic system (see Sec. 3.5) possessing n generalized coordinates $q_1,\ldots,q_n$ (see Sec. 3.2), n motion variables $u_1,\ldots,u_n$ (see Sec. 3.4), and p degrees of freedom (see Sec. 3.5) in a reference frame A, and Eqs. (3.5.2) and (3.6.5) apply, then the kinetic energy K of S in A (see Sec. 7.4) can be expressed as

$$K = K_0 + K_1 + K_2 \tag{1}$$

where K_i is a function of $q_1,\ldots,q_n$, $u_1,\ldots,u_p$, and the time t, and is *homogeneous and of degree i* $(i = 0, 1, 2)$ in $u_1,\ldots,u_p$.

The function K_2 is given by

$$K_2 = \frac{1}{2}\sum_{r=1}^{p}\sum_{s=1}^{p} m_{rs} u_r u_s \tag{2}$$

where m_{rs}, called an *inertia coefficient* of S in A, is defined in terms of the masses $m_1,\ldots,m_\nu$ and partial velocities $\tilde{\mathbf{v}}_r^{P_i}$ $(i = 1,\ldots,\nu;\ r = 1,\ldots,p)$ in A (see Sec. 3.6) of the ν particles $P_1,\ldots,P_\nu$ forming S as

$$m_{rs} \triangleq \sum_{i=1}^{\nu} m_i\, \tilde{\mathbf{v}}_r^{P_i} \cdot \tilde{\mathbf{v}}_s^{P_i} \qquad (r, s = 1,\ldots,p) \tag{3}$$

so that

$$m_{rs} = m_{sr} \qquad (r, s = 1, \ldots, p) \tag{4}$$

Inertia coefficients† figure prominently in the theory of small vibrations of mechanical systems. Generally, the most convenient way to determine inertia coefficients is simply to inspect a kinetic energy expression.

Derivations When $\mathbf{v}^{P_i}$, the velocity in A of P_i, a generic particle of S, is expressed as

$$\mathbf{v}^{P_i} \underset{(3.6.4)}{=} \sum_{r=1}^{p} \tilde{\mathbf{v}}_r^{P_i} u_r + \tilde{\mathbf{v}}_t^{P_i} \qquad (i = 1, \ldots, \nu) \tag{5}$$

then Eq. (7.4.1) leads to

$$K = \frac{1}{2} \sum_{i=1}^{\nu} \sum_{r=1}^{p} \sum_{s=1}^{p} m_i \, \tilde{\mathbf{v}}_r^{P_i} \cdot \tilde{\mathbf{v}}_s^{P_i} u_r u_s + \sum_{i=1}^{\nu} \sum_{r=1}^{p} m_i \, \tilde{\mathbf{v}}_r^{P_i} \cdot \tilde{\mathbf{v}}_t^{P_i} u_r + \frac{1}{2} \sum_{i=1}^{\nu} m_i (\tilde{\mathbf{v}}_t^{P_i})^2 \tag{6}$$

and, if K_0, K_1, and K_2 are defined as

$$K_0 \triangleq \frac{1}{2} \sum_{i=1}^{\nu} m_i (\tilde{\mathbf{v}}_t^{P_i})^2 \tag{7}$$

$$K_1 \triangleq \sum_{i=1}^{\nu} \sum_{r=1}^{p} m_i \, \tilde{\mathbf{v}}_r^{P_i} \cdot \tilde{\mathbf{v}}_t^{P_i} u_r \tag{8}$$

and

$$K_2 \triangleq \frac{1}{2} \sum_{i=1}^{\nu} \sum_{r=1}^{p} \sum_{s=1}^{p} m_i \, \tilde{\mathbf{v}}_r^{P_i} \cdot \tilde{\mathbf{v}}_s^{P_i} u_r u_s \tag{9}$$

respectively, then Eq. (1) follows directly from Eqs. (7)–(9). Moreover, K_0, K_1, and K_2 can be seen to be homogeneous functions of, respectively, degree 0, 1, and 2 in $u_1, \ldots, u_p$. Furthermore, since the order in which the summations in Eq. (9) are performed is immaterial, Eqs. (9) and (3) yield Eq. (2).

Example When Eqs. (3.4.7) are used to define u_1, u_2, and u_3, the kinetic energy K in A of the system S formed by the particle P_1 of mass m_1 and the sharp-edged circular disk D of mass m_2 considered in the example in Sec. 3.5 becomes

$$K = \tfrac{1}{2}[(m_1 + m_2){u_1}^2 + m_1 {u_2}^2] + \tfrac{1}{2}\omega^2[m_1 {q_1}^2 + m_2(q_1 + Lc_3)^2] \tag{10}$$

K_0, the portion of K that is of degree zero in u_1 and u_2, is thus seen to be given by

$$K_0 = \tfrac{1}{2}\omega^2[m_1 {q_1}^2 + m_2(q_1 + Lc_3)^2] \tag{11}$$

† Strictly speaking, the quantities defined in Eqs. (3) should be called *nonholonomic* inertia coefficients to distinguish them from the quantities obtained when $\tilde{\mathbf{v}}_r^{P_i}$ and $\tilde{\mathbf{v}}_s^{P_i}$ are replaced with $\mathbf{v}_r^{P_i}$ and $\mathbf{v}_s^{P_i}$, respectively, and p is replaced with n. The latter quantities would be called *holonomic* inertia coefficients of the nonholonomic system S in A. As we shall have no occasion to use them, we leave them undefined. When S is a holonomic system, the distinction between the two kinds of inertia coefficients disappears in any event.

Since K contains no terms of degree 1 in u_1 and u_2, the function K_1 is equal to zero. Finally, K_2, the *quadratic* part of K, as it is frequently called, is

$$K_2 = \frac{1}{2}[(m_1 + m_2)u_1{}^2 + m_1 u_2{}^2] \tag{12}$$

For $p = 2$, as is the case here, Eq. (2) yields

$$K_2 = \frac{1}{2}[m_{11}u_1{}^2 + (m_{12} + m_{21})u_1 u_2 + m_{22}u_2{}^2] \tag{13}$$

Comparing Eqs. (12) and (13), one finds that the inertia coefficients m_{11}, m_{12}, m_{21}, and m_{22} are

$$m_{11} = m_1 + m_2 \qquad m_{12} \underset{(4)}{=} m_{21} = 0 \qquad m_{22} = m_1 \tag{14}$$

7.6 KINETIC ENERGY AND GENERALIZED INERTIA FORCES

If S is a set of ν particles $P_1, \ldots, P_\nu$ of masses $m_1, \ldots, m_\nu$, respectively, possessing p degrees of freedom in a reference frame A, a quantity σ of interest in connection with generalized inertia forces (see Sec. 5.9) is defined as

$$\sigma \triangleq \sum_{i=1}^{\nu} m_i \mathbf{v}^{P_i} \cdot \frac{d\,\widetilde{\mathbf{v}}_t^{P_i}}{dt} \tag{1}$$

where $\mathbf{v}^{P_i}$ is the velocity of P_i in A, and $d\,\widetilde{\mathbf{v}}_t^{P_i}/dt$ is the time derivative in A of $\widetilde{\mathbf{v}}_t^{P_i}$ [see Eq. (3.6.4)]. σ_B, the contribution to σ of the particles of a rigid body B that belongs to S, is given by

$$\sigma_B = m\mathbf{v} \cdot \frac{d\,\widetilde{\mathbf{v}}_t}{dt} + \boldsymbol{\omega} \cdot \underline{\mathbf{I}} \cdot \frac{d\,\widetilde{\boldsymbol{\omega}}_t}{dt} \tag{2}$$

where m is the mass of B, $\mathbf{v}$ is the velocity in A of the mass center of B, $d\,\widetilde{\mathbf{v}}_t/dt$ is the time derivative in A of $\widetilde{\mathbf{v}}_t$ [see Eq. (3.6.4)], $\boldsymbol{\omega}$ is the angular velocity of B in A, $\underline{\mathbf{I}}$ is the central inertia dyadic of B, and $d\,\widetilde{\boldsymbol{\omega}}_t/dt$ is the time derivative in A of $\widetilde{\boldsymbol{\omega}}_t$ [see Eq. (3.6.3)].

If and only if

$$\sigma = \sum_{i=1}^{\nu} m_i \mathbf{v}^{P_i} \cdot \frac{d\,\widetilde{\mathbf{v}}_t^{P_i}}{dt} = 0 \tag{3}$$

then

$$\dot{K}_2 - \dot{K}_0 = -\sum_{r=1}^{p} \widetilde{F}_r^\star u_r \tag{4}$$

where $\widetilde{F}_r^\star$ is the rth nonholonomic generalized inertia force for S in A (see Sec. 5.9), and all other symbols have the same meanings as in Sec. 7.5. It is by virtue of these facts that kinetic energy plays an important part in the construction of integrals of equations of motion, as will be shown in Sec. 9.2.

When K is regarded as a function of $q_1,\ldots,q_n$, $\dot{q}_1,\ldots,\dot{q}_n$, and t, then $\widetilde{F}_r^\star$ can be expressed as

$$\widetilde{F}_r^\star = -\sum_{s=1}^{n}\left(\frac{d}{dt}\frac{\partial K}{\partial \dot{q}_s} - \frac{\partial K}{\partial q_s}\right)\left(W_{sr} + \sum_{k=p+1}^{n} W_{sk}A_{kr}\right) \qquad (r = 1,\ldots,p) \qquad (5)$$

where W_{sr} and A_{kr} have the same meaning as in Eqs. (3.6.5) and (3.5.2), respectively. These relationships can be simplified when $u_r = \dot{q}_r$ $(r = 1,\ldots,n)$ and/or when S is a holonomic system. When $u_r = \dot{q}_r$ $(r = 1,\ldots,n)$,

$$\widetilde{F}_r^\star = -\left[\frac{d}{dt}\frac{\partial K}{\partial \dot{q}_r} - \frac{\partial K}{\partial q_r} + \sum_{s=p+1}^{n}\left(\frac{d}{dt}\frac{\partial K}{\partial \dot{q}_s} - \frac{\partial K}{\partial q_s}\right)C_{sr}\right] \qquad (r = 1,\ldots,p) \qquad (6)$$

where C_{sr} has the same meaning as in Eqs. (7.1.13). When S is a holonomic system, but Eqs. (3.6.5) apply, $\widetilde{F}_r^\star$ is replaced with $F_r^\star$ and

$$F_r^\star = -\sum_{s=1}^{n}\left(\frac{d}{dt}\frac{\partial K}{\partial \dot{q}_s} - \frac{\partial K}{\partial q_s}\right)W_{sr} \qquad (r = 1,\ldots,n) \qquad (7)$$

Finally, when $u_r = \dot{q}_r$ $(r = 1,\ldots,n)$ and S is a holonomic system, then

$$F_r^\star = -\left(\frac{d}{dt}\frac{\partial K}{\partial \dot{q}_r} - \frac{\partial K}{\partial q_r}\right) \qquad (r = 1,\ldots,n) \qquad (8)$$

One can use Eqs. (5)–(8) to find expressions for generalized inertia forces, but frequently it is inefficient to use this approach.

Derivations When a simple nonholonomic system S contains a subset of particles $P_1,\ldots,P_\beta$ that constitute a rigid body B, the contribution of B to the right-hand member of Eq. (1) can be written as

$$\sigma_B = \sum_{i=1}^{\beta} m_i \mathbf{v}^{P_i} \cdot \frac{d\,\widetilde{\mathbf{v}}_t^{P_i}}{dt} \qquad (9)$$

With the aid of Eq. (2.7.1), one can write

$$\mathbf{v}^{P_i} = \mathbf{v} + \boldsymbol{\omega} \times \mathbf{r}_i \qquad (i = 1,\ldots,\beta) \qquad (10)$$

where $\mathbf{v}$ is the velocity in A of $B^\star$, the mass center of B; $\boldsymbol{\omega}$ is the angular velocity of B in A; and $\mathbf{r}_i$ is the position vector from $B^\star$ to P_i $(i = 1,\ldots,\beta)$. Moreover, after referring to Eqs. (3.6.3) and (3.6.4), one can express $\widetilde{\mathbf{v}}_t^{P_i}$ as

$$\widetilde{\mathbf{v}}_t^{P_i} = \widetilde{\mathbf{v}}_t + \widetilde{\boldsymbol{\omega}}_t \times \mathbf{r}_i \qquad (i = 1,\ldots,\beta) \qquad (11)$$

One may then substitute from these relationships into Eq. (9) to obtain

$$\begin{aligned}\sigma_B &= \sum_{i=1}^{\beta} m_i\,(\mathbf{v} + \boldsymbol{\omega} \times \mathbf{r}_i) \cdot \frac{d}{dt}(\widetilde{\mathbf{v}}_t + \widetilde{\boldsymbol{\omega}}_t \times \mathbf{r}_i) \\ &= \sum_{i=1}^{\beta} m_i\,(\mathbf{v} + \boldsymbol{\omega} \times \mathbf{r}_i) \cdot \left[\frac{d\,\widetilde{\mathbf{v}}_t}{dt} + \frac{d\,\widetilde{\boldsymbol{\omega}}_t}{dt} \times \mathbf{r}_i + \widetilde{\boldsymbol{\omega}}_t \times (\boldsymbol{\omega} \times \mathbf{r}_i)\right] \qquad (12)\end{aligned}$$

where d/dt denotes time differentiation in A. Now,

$$\sum_{i=1}^{\beta} m_i \mathbf{v} \cdot \left[\frac{d\,\tilde{\mathbf{v}}_t}{dt} + \frac{d\,\tilde{\boldsymbol{\omega}}_t}{dt} \times \mathbf{r}_i + \tilde{\boldsymbol{\omega}}_t \times (\boldsymbol{\omega} \times \mathbf{r}_i)\right]$$
$$= \mathbf{v} \cdot \left[\frac{d\,\tilde{\mathbf{v}}_t}{dt} \sum_{i=1}^{\beta} m_i + \frac{d\,\tilde{\boldsymbol{\omega}}_t}{dt} \times \sum_{i=1}^{\beta} m_i \mathbf{r}_i + \tilde{\boldsymbol{\omega}}_t \times \left(\boldsymbol{\omega} \times \sum_{i=1}^{\beta} m_i \mathbf{r}_i\right)\right]$$
$$= \mathbf{v} \cdot \left[\frac{d\,\tilde{\mathbf{v}}_t}{dt} m + \frac{d\,\tilde{\boldsymbol{\omega}}_t}{dt} \times \underset{(4.1.1)}{\mathbf{0}} + \tilde{\boldsymbol{\omega}}_t \times \left(\boldsymbol{\omega} \times \underset{(4.1.1)}{\mathbf{0}}\right)\right]$$
$$= m\mathbf{v} \cdot \frac{d\,\tilde{\mathbf{v}}_t}{dt} \tag{13}$$

where m is the mass of B. Also,

$$\sum_{i=1}^{\beta} m_i (\boldsymbol{\omega} \times \mathbf{r}_i) \cdot \frac{d\,\tilde{\mathbf{v}}_t}{dt} = \left(\boldsymbol{\omega} \times \sum_{i=1}^{\beta} m_i \mathbf{r}_i\right) \cdot \frac{d\,\tilde{\mathbf{v}}_t}{dt} = \left(\boldsymbol{\omega} \times \underset{(4.1.1)}{\mathbf{0}}\right) \cdot \frac{d\,\tilde{\mathbf{v}}_t}{dt}$$
$$= 0 \tag{14}$$

Next, by letting $d\,\tilde{\boldsymbol{\omega}}_t/dt$ play the role of $\boldsymbol{\omega}$ in Eq. (4.5.31), one can write

$$\sum_{i=1}^{\beta} m_i (\boldsymbol{\omega} \times \mathbf{r}_i) \cdot \left(\frac{d\,\tilde{\boldsymbol{\omega}}_t}{dt} \times \mathbf{r}_i\right) = \boldsymbol{\omega} \cdot \sum_{i=1}^{\beta} m_i \mathbf{r}_i \times \left(\frac{d\,\tilde{\boldsymbol{\omega}}_t}{dt} \times \mathbf{r}_i\right) = \boldsymbol{\omega} \cdot \underline{\mathbf{I}} \cdot \frac{d\,\tilde{\boldsymbol{\omega}}_t}{dt} \tag{15}$$

where $\underline{\mathbf{I}}$ is the central inertia dyadic of B. Finally,

$$\sum_{i=1}^{\beta} m_i (\boldsymbol{\omega} \times \mathbf{r}_i) \cdot [\tilde{\boldsymbol{\omega}}_t \times (\boldsymbol{\omega} \times \mathbf{r}_i)] = 0 \tag{16}$$

because each term in the sum involves a scalar triple product in which a single vector, $\boldsymbol{\omega} \times \mathbf{r}_i$, appears twice. The validity of Eq. (2) is established by substituting from Eqs. (13)–(16) into Eq. (12).

The acceleration $\mathbf{a}^{P_i}$ of a generic particle P_i of S in A, found by differentiating Eq. (7.5.5) with respect to t in A, is

$$\mathbf{a}^{P_i} = \sum_{s=1}^{p} \left(\frac{d\,\tilde{\mathbf{v}}_s^{P_i}}{dt} u_s + \tilde{\mathbf{v}}_s^{P_i} \dot{u}_s\right) + \frac{d\,\tilde{\mathbf{v}}_t^{P_i}}{dt} \qquad (i = 1, \ldots, \nu) \tag{17}$$

Hence, with the aid of Eqs. (5.9.2) and (5.9.4), one can express the right-hand member of Eq. (4) as

$$-\sum_{r=1}^{p} \tilde{F}_r^{\star} u_r = \sum_{r=1}^{p} \sum_{i=1}^{\nu} m_i\, \tilde{\mathbf{v}}_r^{P_i} \cdot \left[\sum_{s=1}^{p} \left(\frac{d\,\tilde{\mathbf{v}}_s^{P_i}}{dt} u_s + \tilde{\mathbf{v}}_s^{P_i} \dot{u}_s\right) + \frac{d\,\tilde{\mathbf{v}}_t^{P_i}}{dt}\right] u_r \tag{18}$$

As for the left-hand member, we note that

$$\dot{K}_0 \underset{(7.5.7)}{=} \sum_{i=1}^{\nu} m_i\, \tilde{\mathbf{v}}_t^{P_i} \cdot \frac{d\,\tilde{\mathbf{v}}_t^{P_i}}{dt} \tag{19}$$

and that

$$\dot{K}_2 \underset{(7.5.9)}{=} \sum_{i=1}^{\nu}\sum_{r=1}^{p}\sum_{s=1}^{p} m_i \left(\tilde{\mathbf{v}}_r^{P_i} \cdot \frac{d\,\tilde{\mathbf{v}}_s^{P_i}}{dt} u_r u_s + \tilde{\mathbf{v}}_r^{P_i} \cdot \tilde{\mathbf{v}}_s^{P_i} u_r \dot{u}_s \right)$$

$$\underset{(18)}{=} -\sum_{r=1}^{p} \tilde{F}_r^\star u_r - \sum_{i=1}^{\nu}\sum_{r=1}^{p} m_i\, \tilde{\mathbf{v}}_r^{P_i} \cdot \frac{d\,\tilde{\mathbf{v}}_t^{P_i}}{dt} u_r \tag{20}$$

so that

$$\dot{K}_2 - \dot{K}_0 \underset{(19,\,20)}{=} -\sum_{r=1}^{p} \tilde{F}_r^\star u_r - \sum_{i=1}^{\nu} m_i \left(\sum_{r=1}^{p} \tilde{\mathbf{v}}_r^{P_i} u_r + \tilde{\mathbf{v}}_t^{P_i} \right) \cdot \frac{d\,\tilde{\mathbf{v}}_t^{P_i}}{dt} \tag{21}$$

or, in view of Eqs. (7.5.5),

$$\dot{K}_2 - \dot{K}_0 \underset{(21)}{=} -\sum_{r=1}^{p} \tilde{F}_r^\star u_r - \sum_{i=1}^{\nu} m_i \mathbf{v}^{P_i} \cdot \frac{d\,\tilde{\mathbf{v}}_t^{P_i}}{dt} = -\sum_{r=1}^{p} \tilde{F}_r^\star u_r \underset{(1)}{-\sigma} \tag{22}$$

Hence, when Eq. (3) is satisfied, then so is Eq. (4), and vice versa.

The validity of Eqs. (5) is established by writing

$$\tilde{F}_r^\star \underset{(5.9.2,\,5.9.4)}{=} -\sum_{i=1}^{\nu} m_i\, \tilde{\mathbf{v}}_r^{P_i} \cdot \mathbf{a}^{P_i} \tag{23}$$

$$\underset{(3.9.8)}{=} -\sum_{i=1}^{\nu} m_i \frac{1}{2} \sum_{s=1}^{n} \left\{ \left[\frac{d}{dt}\frac{\partial(\mathbf{v}^{P_i})^2}{\partial \dot{q}_s} - \frac{\partial(\mathbf{v}^{P_i})^2}{\partial q_s} \right] \left(W_{sr} + \sum_{k=p+1}^{n} W_{sk} A_{kr} \right) \right\}$$

$$= -\sum_{s=1}^{n} \left\langle \frac{d}{dt} \left\{ \frac{\partial}{\partial \dot{q}_s} \left[\frac{1}{2} \sum_{i=1}^{\nu} m_i (\mathbf{v}^{P_i})^2 \right] \right\} - \frac{\partial}{\partial q_s} \left[\frac{1}{2} \sum_{i=1}^{\nu} m_i (\mathbf{v}^{P_i})^2 \right] \right\rangle \left(W_{sr} + \sum_{k=p+1}^{n} W_{sk} A_{kr} \right)$$

$$\underset{(7.4.1)}{=} -\sum_{s=1}^{n} \left(\frac{d}{dt}\frac{\partial K}{\partial \dot{q}_s} - \frac{\partial K}{\partial q_s} \right) \left(W_{sr} + \sum_{k=p+1}^{n} W_{sk} A_{kr} \right) \qquad (r = 1, \ldots, p)$$

Finally, Eqs. (6)–(8) are special cases of Eqs. (5).

Example For the example in Sec. 7.5, n and p have the values 3 and 2, respectively. To see that Eq. (3) is satisfied, proceed as follows. With motion variables u_1, u_2, and u_3 chosen according to Eqs. (3.4.7), the velocity ${}^A\mathbf{v}^{P_1}$ of P_1 in A is given by

$${}^A\mathbf{v}^{P_1} \underset{(3.4.14)}{=} u_1 \hat{\mathbf{e}}_x + u_2 \hat{\mathbf{e}}_y - \omega q_1 \hat{\mathbf{e}}_z \tag{24}$$

where unit vectors $\hat{\mathbf{e}}_x$, $\hat{\mathbf{e}}_y$, and $\hat{\mathbf{e}}_z$ are directed as shown in Fig. 3.5.1, and related as indicated in Table 2.6.1 to unit vectors $\hat{\mathbf{b}}_x$, $\hat{\mathbf{b}}_y$, and $\hat{\mathbf{b}}_z$ (see Fig. 2.6.1). Inspection of Eq. (24) yields

$${}^A\tilde{\mathbf{v}}_t^{P_1} \underset{(24)}{=} -\omega q_1 \hat{\mathbf{e}}_z \underset{\text{(Table 2.6.1)}}{=} -\omega q_1 \hat{\mathbf{b}}_z \tag{25}$$

and the time derivative in A of this vector is given by

$$\frac{{}^A d\, {}^A\tilde{\mathbf{v}}_t^{P_1}}{dt} = -\omega\dot{q}_1\hat{\mathbf{b}}_z - \omega q_1 \underset{(2.1.2,\ 2.8.18)}{(\omega\hat{\mathbf{b}}_y \times \hat{\mathbf{b}}_z)} = -\omega\dot{q}_1\hat{\mathbf{b}}_z - \omega^2 q_1\hat{\mathbf{b}}_x \tag{26}$$

Two dot products are needed to form σ according to Eq. (1); the first of these is given by

$$\begin{aligned} {}^A\mathbf{v}^{P_1} \cdot \frac{{}^A d\, {}^A\tilde{\mathbf{v}}_t^{P_1}}{dt} &= \underset{(24)}{(u_1\hat{\mathbf{e}}_x + u_2\hat{\mathbf{e}}_y - \omega q_1\hat{\mathbf{e}}_z)} \cdot \underset{(26)}{(-\omega\dot{q}_1\hat{\mathbf{b}}_z - \omega^2 q_1\hat{\mathbf{b}}_x)} \\ &\underset{(\text{Table 2.6.1})}{=} \omega^2 q_1\dot{q}_1 - \omega^2 q_1[u_1 c_3 + u_2(-s_3) + 0] \\ &= \omega^2 q_1\dot{q}_1 - \omega^2 q_1 \underset{(3.4.8)}{\dot{q}_1} = 0 \end{aligned} \tag{27}$$

The velocity in A of $D^\star$, the center of the sharp-edged circular disk D, is given by

$$ {}^A\mathbf{v}^{D^\star} \underset{(3.6.29)}{=} u_1\hat{\mathbf{e}}_x - \omega(q_1 + Lc_3)\hat{\mathbf{e}}_z \tag{28}$$

in which case

$$ {}^A\tilde{\mathbf{v}}_t^{D^\star} \underset{(28)}{=} -\omega(q_1 + Lc_3)\hat{\mathbf{e}}_z \underset{(\text{Table 2.6.1})}{=} -\omega(q_1 + Lc_3)\hat{\mathbf{b}}_z \tag{29}$$

and

$$\begin{aligned} \frac{{}^A d\, {}^A\tilde{\mathbf{v}}_t^{D^\star}}{dt} &= -\omega(\dot{q}_1 - Ls_3\dot{q}_3)\hat{\mathbf{b}}_z - \omega(q_1 + Lc_3)(\omega\hat{\mathbf{b}}_y \times \hat{\mathbf{b}}_z) \\ &= -\omega(\dot{q}_1 - Ls_3\dot{q}_3)\hat{\mathbf{b}}_z - \omega^2(q_1 + Lc_3)\hat{\mathbf{b}}_x \\ &= -\omega \underset{(3.4.8,\ 3.5.15)}{(\dot{q}_1 + u_2 s_3)} \hat{\mathbf{b}}_z - \omega^2(q_1 + Lc_3)\hat{\mathbf{b}}_x \end{aligned} \tag{30}$$

Thus, the final dot product needed to evaluate σ is found to be

$$\begin{aligned} {}^A\mathbf{v}^{D^\star} \cdot \frac{{}^A d\, {}^A\tilde{\mathbf{v}}_t^{D^\star}}{dt} &= \underset{(28)}{[u_1\hat{\mathbf{e}}_x - \omega(q_1 + Lc_3)\hat{\mathbf{e}}_z]} \cdot \\ &\quad \underset{(30)}{[-\omega(\dot{q}_1 + u_2 s_3)\hat{\mathbf{b}}_z - \omega^2(q_1 + Lc_3)\hat{\mathbf{b}}_x]} \\ &\underset{(\text{Table 2.6.1})}{=} \omega^2(q_1 + Lc_3)(\dot{q}_1 + u_2 s_3) - \omega^2(q_1 + Lc_3)(u_1 c_3 + 0) \\ &= \omega^2(q_1 + Lc_3)[\underset{(3.4.8)}{(u_1 c_3 - u_2 s_3)} + u_2 s_3 - u_1 c_3] = 0 \end{aligned} \tag{31}$$

Satisfaction of Eq. (3) is demonstrated by substitution from Eqs. (27) and (31).

To verify that Eq. (4) is satisfied, we first evaluate $\dot{K}_2 - \dot{K}_0$ by reference to Eqs. (7.5.11) and (7.5.12), obtaining

$$\begin{aligned} \dot{K}_2 - \dot{K}_0 &= (m_1 + m_2)u_1\dot{u}_1 + m_1 u_2\dot{u}_2 \\ &\quad - \omega^2[m_1 q_1\dot{q}_1 + m_2(q_1 + Lc_3)(\dot{q}_1 - Ls_3\dot{q}_3)] \\ &\underset{(3.4.8,\ 3.5.15)}{=} \{(m_1 + m_2)\dot{u}_1 - \omega^2[m_1 q_1 + m_2(q_1 + Lc_3)]c_3\}u_1 \\ &\quad + m_1(\dot{u}_2 + \omega^2 q_1 s_3)u_2 \end{aligned} \tag{32}$$

and then, to illustrate the use of Eqs. (5), form $\widetilde{F}_1^\star$ as

$$\begin{aligned}\widetilde{F}_1^\star &= -\left(\frac{d}{dt}\frac{\partial K}{\partial \dot{q}_1} - \frac{\partial K}{\partial q_1}\right)(W_{11} + W_{13}A_{31}) \\ &\quad -\left(\frac{d}{dt}\frac{\partial K}{\partial \dot{q}_2} - \frac{\partial K}{\partial q_2}\right)(W_{21} + W_{23}A_{31}) \\ &\quad -\left(\frac{d}{dt}\frac{\partial K}{\partial \dot{q}_3} - \frac{\partial K}{\partial q_3}\right)(W_{31} + W_{33}A_{31}) \\ &\underset{(7.1.31\text{–}7.1.33,\ 7.1.35)}{=} -\left(\frac{d}{dt}\frac{\partial K}{\partial \dot{q}_1} - \frac{\partial K}{\partial q_1}\right)c_3 - \left(\frac{d}{dt}\frac{\partial K}{\partial \dot{q}_2} - \frac{\partial K}{\partial q_2}\right)s_3\end{aligned} \tag{33}$$

Before performing the differentiations indicated in Eq. (33), one must express K as a function of q_1, q_2, q_3, $\dot{q}_1$, $\dot{q}_2$, $\dot{q}_3$, and t, leaving the motion constraint equation $u_3 = -u_2/L$ out of account. The required expression for K, available in Problem 11.1, is

$$\begin{aligned}K &= \tfrac{1}{2}\omega^2[m_1 q_1^{\,2} + m_2(q_1 + Lc_3)^2] + \tfrac{1}{2}(m_1 + m_2)(\dot{q}_1^{\,2} + \dot{q}_2^{\,2}) \\ &\quad - m_2 L\left(\dot{q}_1 s_3 - \dot{q}_2 c_3 - \frac{L}{2}\dot{q}_3\right)\dot{q}_3\end{aligned} \tag{34}$$

Hence,

$$\frac{\partial K}{\partial \dot{q}_1} \underset{(34)}{=} (m_1 + m_2)\dot{q}_1 - m_2 L s_3 \dot{q}_3 \tag{35}$$

$$\frac{\partial K}{\partial \dot{q}_2} \underset{(34)}{=} (m_1 + m_2)\dot{q}_2 + m_2 L c_3 \dot{q}_3 \tag{36}$$

$$\frac{\partial K}{\partial q_1} \underset{(34)}{=} \omega^2[m_1 q_1 + m_2(q_1 + Lc_3)] \tag{37}$$

$$\frac{\partial K}{\partial q_2} \underset{(34)}{=} 0 \tag{38}$$

and substitution into Eq. (33) produces

$$\begin{aligned}\widetilde{F}_1^\star &= -\{\underset{(35)}{(m_1 + m_2)\ddot{q}_1 - m_2 L(c_3\dot{q}_3^{\,2} + s_3\ddot{q}_3)} \\ &\quad \underset{(37)}{- \omega^2[m_1 q_1 + m_2(q_1 + Lc_3)]}\}c_3 \\ &\quad - [\underset{(36,\ 38)}{(m_1 + m_2)\ddot{q}_2 - m_2 L(s_3\dot{q}_3^{\,2} - c_3\ddot{q}_3)}]s_3 \\ &= -(m_1 + m_2)(\ddot{q}_1 c_3 + \ddot{q}_2 s_3) + m_2 L\dot{q}_3^{\,2} \\ &\quad + \omega^2[m_1 q_1 + m_2(q_1 + Lc_3)]c_3\end{aligned} \tag{39}$$

from which we must eliminate the time derivatives of q_1, q_2, and q_3 to accomplish our ultimate objective. This can be done by using Eqs. (3.4.8), which permits one to write

$$\begin{aligned}\ddot{q}_1 c_3 + \ddot{q}_2 s_3 &= (\dot{u}_1 c_3 - u_1 \dot{q}_3 s_3 - \dot{u}_2 s_3 - u_2 \dot{q}_3 c_3) c_3 \\ &\quad + (\dot{u}_1 s_3 + u_1 \dot{q}_3 c_3 + \dot{u}_2 c_3 - u_2 \dot{q}_3 s_3) s_3 \\ &= \dot{u}_1 - u_2 u_3 \end{aligned} \tag{40}$$

and $\widetilde{F}_1^\star$ then is seen to be given by

$$\widetilde{F}_1^\star \underset{(39,\,40)}{=} -(m_1 + m_2)(\dot{u}_1 - u_2 u_3) + m_2 L u_3{}^2 + \omega^2 [m_1 q_1 + m_2 (q_1 + L c_3)] c_3 \tag{41}$$

Finally, after eliminating u_3 with the aid of Eq. (3.5.15), we arrive at

$$\widetilde{F}_1^\star = -(m_1 + m_2)\dot{u}_1 - m_1 \frac{u_2{}^2}{L} + \omega^2 [m_1 q_1 + m_2 (q_1 + L c_3)] c_3 \tag{42}$$

(in agreement with the results in Problem 8.14).

Using the expression for $\widetilde{F}_2^\star$ available in the results in Problem 8.14, we can write the right-hand member of Eq. (4) as

$$\begin{aligned}-(\widetilde{F}_1^\star u_1 + \widetilde{F}_2^\star u_2) = \Bigg\{ & (m_1 + m_2)\dot{u}_1 + m_1 \frac{u_2{}^2}{L} \\ & - \omega^2 [m_1 q_1 + m_2 (q_1 + L c_3)] c_3 \Bigg\} u_1 \\ & + m_1 \left(\dot{u}_2 - \frac{u_1 u_2}{L} + \omega^2 q_1 s_3 \right) u_2 \end{aligned} \tag{43}$$

which reduces to the right-hand member of Eq. (32) in conformity with Eq. (4).

8 FORMULATION OF EQUATIONS OF MOTION

In Sec. 8.1, the notion of a Newtonian reference frame is presented, and it is shown that dynamical differential equations can be formulated easily, once expressions for generalized active forces and generalized inertia forces are in hand. Questions regarding reference frame choices are examined in Sec. 8.2, and Sec. 8.3 deals with the formulation of equations intended for the determination of forces and/or torques that do not come into evidence explicitly in equations of motion unless special measures are taken. Section 8.4 contains a detailed exposition of a method for generating linearized forms of dynamical equations. Next, consideration is given to three kinds of motion that deserve attention because of their practical importance, namely rest, steady motion, and motions resembling states of rest, which are treated in Secs. 8.5, 8.6, and 8.7, respectively. The concepts of generalized impulse and generalized momentum are introduced in Sec. 8.8 in preparation for the presentation, in Sec. 8.9, of a method for analyzing phenomena involving collisions.

8.1 DYNAMICAL EQUATIONS

There exist reference frames N such that, if S is a holonomic system possessing n degrees of freedom in N, and F_r and $F_r^\star$ $(r = 1, \ldots, n)$ are, respectively, the holonomic generalized active forces (see Sec. 5.4) and the holonomic generalized inertia forces (see Sec. 5.9) for S in N, then the equations

$$F_r + F_r^\star = 0 \qquad (r = 1, \ldots, n) \tag{1}$$

govern all motions of S in *any* reference frame. The reference frames N are called *Newtonian* or *inertial* reference frames. If S is a simple nonholonomic system possessing p degrees of freedom in N (see Sec. 3.5), Eqs. (1) are replaced with

$$\tilde{F}_r + \tilde{F}_r^\star = 0 \qquad (r = 1, \ldots, p) \tag{2}$$

where $\tilde{F}_r$ and $\tilde{F}_r^\star$ are defined in Secs. 5.4 and 5.9, respectively. Equations (1) and (2) are known as *Kane's dynamical equations*. If S is a complex nonholonomic system possessing c degrees of freedom in N (see Sec. 3.7), Eqs. (2) give way to

$$\approx{F}_r + \approx{F}_r^\star = 0 \qquad (r = 1, \ldots, c) \tag{3}$$

where the generalized forces $\approx{F}_r$ and $\approx{F}_r^\star$ are presented in Secs. 5.4 and 5.9, respectively.

Ultimately, the justification for regarding a particular reference frame as Newtonian can come only from experiments. One such experiment, first performed by Foucault in 1851, is discussed in the example that follows the derivation of Eqs. (1).

Derivations If $\mathbf{R}_i$ is the resultant of all contact forces and distance forces acting on a typical particle P_i of S, and ${}^N\mathbf{a}^{P_i}$ is the acceleration of P_i in a Newtonian reference frame N, then, in accordance with Newton's second law,

$$\mathbf{R}_i - m_i{}^N\mathbf{a}^{P_i} = \mathbf{0} \qquad (i = 1,\ldots,\nu) \tag{4}$$

where m_i is the mass of P_i and ν is the number of particles of S. Dot multiplication of Eqs. (4) with the partial velocities ${}^N\mathbf{v}_r^{P_i}$ of P_i in N (see Sec. 3.6) and subsequent summation yields

$$\sum_{i=1}^{\nu} {}^N\mathbf{v}_r^{P_i} \cdot \mathbf{R}_i + \sum_{i=1}^{\nu} {}^N\mathbf{v}_r^{P_i} \cdot (-m_i{}^N\mathbf{a}^{P_i}) \underset{(4)}{=} 0 \qquad (r = 1,\ldots,n) \tag{5}$$

In accordance with Eq. (5.4.1), the first sum in this equation is F_r; the second sum is $F_r^\star$, as may be seen by reference to Eqs. (5.9.1) and (5.9.4). Thus, Eqs. (5) lead directly to Eqs. (1). Equations (2) are obtained in a similar manner by using the nonholonomic partial velocities ${}^N\tilde{\mathbf{v}}_r^{P_i}$ $(r = 1,\ldots,p)$ in place of ${}^N\mathbf{v}_r^{P_i}$ in Eqs. (5); Eqs. (3) are a consequence of employing nonholonomic partial accelerations ${}^N\tilde{\mathbf{a}}_r^{P_i}$ $(r = 1,\ldots,c)$ (see Sec. 3.8).

Example In Fig. 8.1.1, E represents the Earth, modeled as a sphere centered at a point $E^\star$, and P is a particle of mass m, suspended by means of a light, inextensible string of length L from a point Q that is fixed relative to E. Point O is the intersection of line $E^\star Q$ with the surface of E; h is the distance from O to Q; $\hat{\mathbf{k}}$ is a unit vector parallel to line K, the Earth's polar axis; $\hat{\mathbf{e}}_1$, $\hat{\mathbf{e}}_2$, and $\hat{\mathbf{e}}_3$ are unit vectors pointing southward, eastward, and upward at O, respectively. Finally, F represents a reference frame in which $E^\star$ is fixed and in which E has a simple angular velocity (see Sec. 2.2) ${}^F\boldsymbol{\omega}^E$ given by

$$ {}^F\boldsymbol{\omega}^E = \omega\hat{\mathbf{k}} \tag{6}$$

where $\omega = 7.29 \times 10^{-5}$ rad/s, so that E performs in F one rotation per sidereal day (24 h of sidereal time or 23 h 56 min 4.09054 s of mean solar time). (F differs from the so-called astronomical reference frame primarily in that one point of F coincides permanently with a point of E, the point $E^\star$. No point of E is fixed in the astronomical reference frame.)

To bring P into a general position, one can proceed as follows: Place P on line OQ, subject line QP to a rotation characterized by the vector $q_1\hat{\mathbf{e}}_1$ rad, and let $\hat{\mathbf{a}}_1$, $\hat{\mathbf{a}}_2$, $\hat{\mathbf{a}}_3$ be unit vectors directed as shown in Fig. 8.1.2; subject line QP to a rotation characterized by the vector $q_2\hat{\mathbf{a}}_2$ rad, thus bringing line QP into the position shown in Fig. 8.1.3. The quantities q_1 and q_2 can be used as generalized coordinates of P in E or in F, since the motion of E in F is specified as taking place in accordance with Eq. (6).

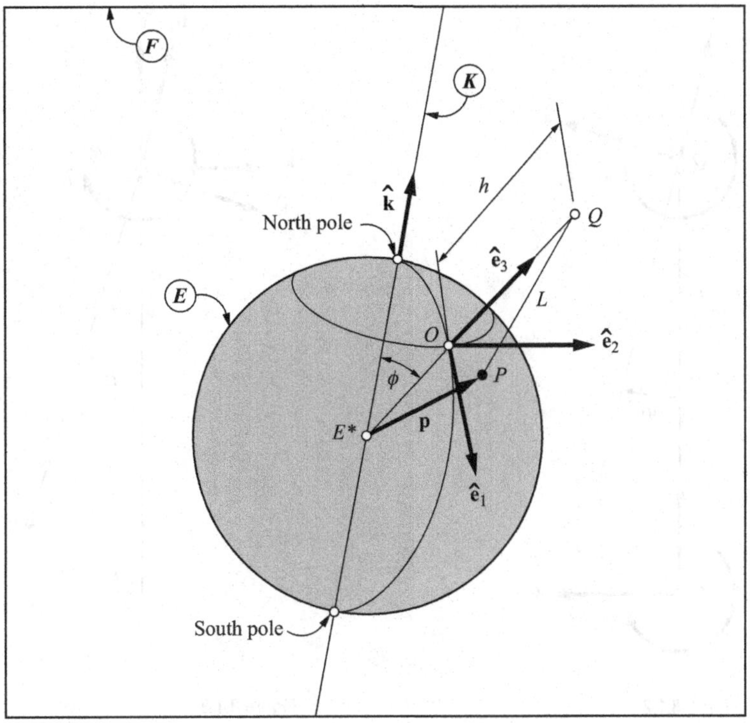

Figure 8.1.1

With a view to writing the dynamical equations governing all motions of P, we introduce motion variables u_1 and u_2 as

$$u_r \triangleq {}^E\mathbf{v}^P \cdot \hat{\mathbf{b}}_r \qquad (r = 1,2) \tag{7}$$

where ${}^E\mathbf{v}^P$ is the velocity of P in E, and $\hat{\mathbf{b}}_1$ and $\hat{\mathbf{b}}_2$ are unit vectors directed as shown in Fig. 8.1.3. Next, we begin the task of formulating expressions for partial velocities of P in F by writing

$$ {}^F\mathbf{v}^P \underset{(2.8.1)}{=} {}^F\mathbf{v}^{\overline{E}} + {}^E\mathbf{v}^P \tag{8}$$

where $\overline{E}$ denotes that point of E with which P coincides, so that, if $\mathbf{p}$ is the position vector from $E^\star$ to P, as shown in Fig. 8.1.1,

$$ {}^F\mathbf{v}^{\overline{E}} \underset{(2.7.1)}{=} {}^F\mathbf{v}^{E^\star} + {}^F\boldsymbol{\omega}^E \times \mathbf{p} \tag{9}$$

or, since

$$ {}^F\mathbf{v}^{E^\star} = \mathbf{0} \tag{10}$$

then, by hypothesis, we have

$$ {}^F\mathbf{v}^{\overline{E}} \underset{(9,\,10)}{=} {}^F\boldsymbol{\omega}^E \times \mathbf{p} \underset{(6)}{=} \omega\hat{\mathbf{k}} \times \mathbf{p} \tag{11}$$

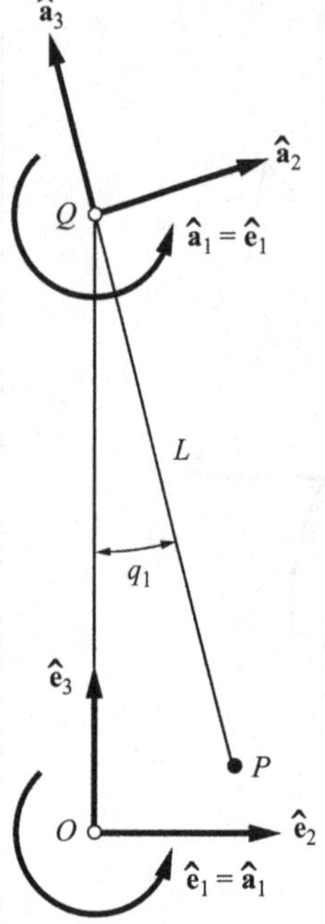

Figure 8.1.2

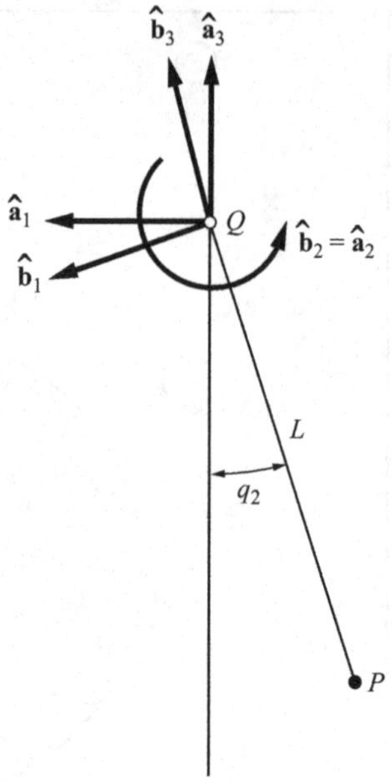

Figure 8.1.3

As for ${}^E\mathbf{v}^P$, it follows directly from Eqs. (7) together with the fact that ${}^E\mathbf{v}^P \cdot \hat{\mathbf{b}}_3 = 0$ (because ${}^E\mathbf{v}^P$ must be perpendicular to $\hat{\mathbf{b}}_3$) that (see Sec. 1.6)

$$ {}^E\mathbf{v}^P = u_1\hat{\mathbf{b}}_1 + u_2\hat{\mathbf{b}}_2 \tag{12} $$

Thus,

$$ {}^F\mathbf{v}^P \underset{(8)}{=} \underset{(11)}{\omega\hat{\mathbf{k}} \times \mathbf{p}} + \underset{(12)}{u_1\hat{\mathbf{b}}_1 + u_2\hat{\mathbf{b}}_2} \tag{13} $$

and the partial velocities of P in F are simply

$$ {}^F\mathbf{v}_1^P \underset{(13)}{=} \hat{\mathbf{b}}_1 \qquad {}^F\mathbf{v}_2^P \underset{(13)}{=} \hat{\mathbf{b}}_2 \tag{14} $$

The acceleration of P in F, required in connection with the generalized inertia forces for P in F, is given by

$$ {}^F\mathbf{a}^P \underset{(2.8.2)}{=} {}^F\mathbf{a}^{\overline{E}} + {}^E\mathbf{a}^P + 2\,{}^F\boldsymbol{\omega}^E \times {}^E\mathbf{v}^P \tag{15} $$

with

$$ \begin{aligned} {}^F\mathbf{a}^{\overline{E}} &\underset{(2.7.2)}{=} {}^F\mathbf{a}^{E^\star} + {}^F\boldsymbol{\omega}^E \times ({}^F\boldsymbol{\omega}^E \times \mathbf{p}) + {}^F\boldsymbol{\alpha}^E \times \mathbf{p} \\ &= \underset{(6)}{\mathbf{0}} + \omega^2\hat{\mathbf{k}} \times (\hat{\mathbf{k}} \times \mathbf{p}) + \underset{(6)}{\mathbf{0}} \end{aligned} \tag{16} $$

$$ {}^E\mathbf{a}^P \underset{(2.6.2)}{=} \frac{{}^E d}{dt}({}^E\mathbf{v}^P) \tag{17} $$

$$ \underset{(12)}{=} \left(\dot{u}_1 - \frac{u_2{}^2 s_2}{Lc_2}\right)\hat{\mathbf{b}}_1 + \left(\dot{u}_2 + \frac{u_1 u_2 s_2}{Lc_2}\right)\hat{\mathbf{b}}_2 + \frac{1}{L}(u_1{}^2 + u_2{}^2)\hat{\mathbf{b}}_3 $$

and

$$ 2\,{}^F\boldsymbol{\omega}^E \times {}^E\mathbf{v}^P = 2\underset{(6)}{\omega\hat{\mathbf{k}}} \times \underset{(12)}{(u_1\hat{\mathbf{b}}_1 + u_2\hat{\mathbf{b}}_2)} \tag{18} $$

so that

$$ {}^F\mathbf{a}^P \underset{(15)}{=} \underset{(16)}{\omega^2\hat{\mathbf{k}} \times (\hat{\mathbf{k}} \times \mathbf{p})} + \underset{(18)}{2\omega\hat{\mathbf{k}} \times (u_1\hat{\mathbf{b}}_1 + u_2\hat{\mathbf{b}}_2)} \tag{19} $$

$$ + \underset{(17)}{\left(\dot{u}_1 - \frac{u_2{}^2 s_2}{Lc_2}\right)\hat{\mathbf{b}}_1 + \left(\dot{u}_2 + \frac{u_1 u_2 s_2}{Lc_2}\right)\hat{\mathbf{b}}_2 + \frac{1}{L}(u_1{}^2 + u_2{}^2)\hat{\mathbf{b}}_3} $$

Hence, the generalized inertia forces for P in F, given by

$$ F_r^\star \underset{(5.9.1)}{=} {}^F\mathbf{v}_r^P \cdot \underset{(5.9.4)}{(-m\,{}^F\mathbf{a}^P)} \qquad (r = 1,2) \tag{20} $$

where m is the mass of P, are

$$ F_1^\star \underset{(20,\ 14,\ 19)}{=} -m\left[\omega^2\hat{\mathbf{k}} \times (\hat{\mathbf{k}} \times \mathbf{p}) \cdot \hat{\mathbf{b}}_1 + \dot{u}_1 - \frac{u_2{}^2 s_2}{Lc_2} - 2\omega u_2\hat{\mathbf{k}} \cdot \hat{\mathbf{b}}_3\right] \tag{21} $$

and

$$ F_2^\star \underset{(20,\ 14,\ 19)}{=} -m\left[\omega^2\hat{\mathbf{k}} \times (\hat{\mathbf{k}} \times \mathbf{p}) \cdot \hat{\mathbf{b}}_2 + \dot{u}_2 + \frac{u_1 u_2 s_2}{Lc_2} + 2\omega u_1\hat{\mathbf{k}} \cdot \hat{\mathbf{b}}_3\right] \tag{22} $$

The only contact force acting on P is the force exerted on P by the string connecting P to Q. Since the string is presumed to be "light," the line of action of this force is regarded as parallel to $\hat{\mathbf{b}}_3$, and the force contributes nothing to the generalized active forces F_1 and F_2 for S in F because the partial velocities ${}^F\mathbf{v}_1^P$ and ${}^F\mathbf{v}_2^P$ [see Eqs. (14)] are perpendicular to $\hat{\mathbf{b}}_3$. By way of contrast, the gravitational force $\mathbf{G}$ exerted on P by E, given by

$$ \mathbf{G} \underset{(5.7.1)}{=} -mg\hat{\mathbf{e}}_3 \tag{23} $$

where g is the gravitational force per unit mass at O, does contribute to F_1 and F_2. Hence,

$$ F_r = \mathbf{G} \cdot {}^F\mathbf{v}_r^P \underset{(23)}{=} -mg\hat{\mathbf{e}}_3 \cdot {}^F\mathbf{v}_r^P \qquad (r = 1,2) \tag{24} $$

so that

$$ F_1 = -mg\hat{\mathbf{e}}_3 \cdot \hat{\mathbf{b}}_1 \qquad F_2 = -mg\hat{\mathbf{e}}_3 \cdot \hat{\mathbf{b}}_2 \tag{25} $$

and substitution from Eqs. (25), (21), and (22) into Eqs. (1) yields

$$ \dot{u}_1 = -\omega^2\hat{\mathbf{k}} \times (\hat{\mathbf{k}} \times \mathbf{p}) \cdot \hat{\mathbf{b}}_1 + \frac{u_2{}^2 s_2}{Lc_2} + 2\omega u_2\hat{\mathbf{k}} \cdot \hat{\mathbf{b}}_3 - g\hat{\mathbf{e}}_3 \cdot \hat{\mathbf{b}}_1 \tag{26} $$

$$\dot{u}_2 = -\omega^2\hat{\mathbf{k}} \times (\hat{\mathbf{k}} \times \mathbf{p}) \cdot \hat{\mathbf{b}}_2 - \frac{u_1 u_2 \, s_2}{L c_2} - 2\omega u_1 \hat{\mathbf{k}} \cdot \hat{\mathbf{b}}_3 - g\hat{\mathbf{e}}_3 \cdot \hat{\mathbf{b}}_2 \tag{27}$$

The leading term in each of these equations may be dropped because it is approximately four orders of magnitude smaller than the last term in each equation. As for the dot products $\hat{\mathbf{e}}_3 \cdot \hat{\mathbf{b}}_1$ and $\hat{\mathbf{e}}_3 \cdot \hat{\mathbf{b}}_2$, reference to Figs. 8.1.2 and 8.1.3 permits one to express these as

$$\hat{\mathbf{e}}_3 \cdot \hat{\mathbf{b}}_1 = -c_1 s_2 \qquad \hat{\mathbf{e}}_3 \cdot \hat{\mathbf{b}}_2 = s_1 \tag{28}$$

and, if ϕ is the angle between line K and line $E^\star O$ (see Fig. 8.1.1), then $\hat{\mathbf{k}} \cdot \hat{\mathbf{b}}_3$ is given by

$$\hat{\mathbf{k}} \cdot \hat{\mathbf{b}}_3 = c_1 c_2 c_\phi - s_2 s_\phi \tag{29}$$

Hence, if F is a Newtonian reference frame, the dynamical equations governing all motions of P are

$$\dot{u}_1 \underset{(26)}{=} \frac{u_2{}^2 \, s_2}{L c_2} + 2\omega u_2 \underset{(29)}{(c_1 c_2 c_\phi - s_2 s_\phi)} + \underset{(28)}{g c_1 s_2} \tag{30}$$

$$\dot{u}_2 \underset{(27)}{=} -\frac{u_1 u_2 \, s_2}{L c_2} - 2\omega u_1 \underset{(29)}{(c_1 c_2 c_\phi - s_2 s_\phi)} - \underset{(28)}{g s_1} \tag{31}$$

Kinematical equations, needed in addition to Eqs. (30) and (31), are obtained by taking advantage of the fact that ${}^E\mathbf{v}^P$ is given not only by Eq. (12), but also by (see Fig. 8.1.3)

$${}^E\mathbf{v}^P \underset{(2.6.1)}{=} \frac{{}^E d}{dt}(-L\hat{\mathbf{b}}_3) \underset{(2.1.2)}{=} -L\,{}^E\boldsymbol{\omega}^B \times \hat{\mathbf{b}}_3 \tag{32}$$

where ${}^E\boldsymbol{\omega}^B$ is the angular velocity in E of a reference frame B in which $\hat{\mathbf{b}}_1$, $\hat{\mathbf{b}}_2$, and $\hat{\mathbf{b}}_3$ are fixed; that is (see Figs. 8.1.2 and 8.1.3),

$${}^E\boldsymbol{\omega}^B = \dot{q}_1\hat{\mathbf{a}}_1 + \dot{q}_2\hat{\mathbf{b}}_2 = \dot{q}_1(c_2\hat{\mathbf{b}}_1 + s_2\hat{\mathbf{b}}_3) + \dot{q}_2\hat{\mathbf{b}}_2 \tag{33}$$

Consequently,

$${}^E\mathbf{v}^P \underset{(32,\ 33)}{=} -L(\dot{q}_2\hat{\mathbf{b}}_1 - \dot{q}_1 c_2 \hat{\mathbf{b}}_2) \tag{34}$$

and, comparing this equation with Eq. (12), one concludes that the needed equations are

$$\dot{q}_1 = \frac{u_2}{L c_2} \qquad \dot{q}_2 = -\frac{u_1}{L} \tag{35}$$

Since Eqs. (30), (31), and (35) form a set of coupled, *nonlinear* differential equations, they cannot be solved in closed form. However, particular solutions, corresponding to specific initial conditions, can be obtained by solving the equations numerically (a subject discussed in its own right in Sec. 9.6). Suppose, for instance, that $L = 10$ m, $\phi = 45°$, and q_1, q_2, u_1, u_2 have the following initial values:

$$q_1(0) = 10° \qquad q_2(0) = u_1(0) = u_2(0) = 0 \tag{36}$$

In other words, P is initially displaced toward the east in such a way that the string

Table 8.1.1

	F regarded as Newtonian		*E* regarded as Newtonian	
1	2	3	4	5
t (s)	q_1 (deg)	q_2 (deg)	q_1 (deg)	q_2 (deg)
0.0	10.00000	0.00000	10.00000	0.00000
2.0	−3.95464	0.00088	−3.95464	0.00000
4.0	−6.87783	0.00103	−6.87783	0.00000
6.0	9.38835	−0.00303	9.38835	0.00000
8.0	−0.54578	0.00074	−0.54578	0.00000
10.0	−8.95759	0.00432	−8.95760	0.00000
12.0	7.62758	−0.00498	7.62758	0.00000
14.0	2.93015	−0.00160	2.93015	0.00000
16.0	−9.94057	0.00801	−9.94057	0.00000
18.0	4.93192	−0.00497	4.93193	0.00000
20.0	6.04611	−0.00574	6.04611	0.00000
22.0	−9.70745	0.01096	−9.70745	0.00000
24.0	1.63083	−0.00250	1.63083	0.00000
26.0	8.42027	−0.01084	8.42028	0.00000
28.0	−8.28649	0.01207	−8.28650	0.00000
30.0	−1.87067	0.00235	−1.87067	0.00000
32.0	9.76298	−0.01573	9.76300	0.00000
34.0	−5.85037	0.01054	−5.85038	0.00000
36.0	−5.14229	0.00898	−5.14230	0.00000
38.0	9.91113	−0.01917	9.91115	0.00000
40.0	−2.69639	0.00599	−2.69640	0.00000
42.0	−7.78272	0.01628	−7.78273	0.00000
44.0	8.84674	−0.02000	8.84677	0.00000
46.0	0.78885	−0.00134	0.78885	0.00000
48.0	−9.46933	0.02289	−9.46936	0.00000
50.0	6.69904	−0.01741	6.69906	0.00000
52.0	4.17714	−0.01059	4.17715	0.00000
54.0	−9.99700	0.02739	−9.99704	0.00000
56.0	3.72975	−0.01112	3.72977	0.00000
58.0	7.05248	−0.02043	7.05251	0.00000
60.0	−9.30173	0.02851	−9.30178	0.00000

makes an angle of 10° with the local vertical, and *P* is then released from a state of rest in *E* [see Eqs. (12) and (36)]. The numerical solution of Eqs. (30), (31), and (35) corresponding to these initial conditions (and $\omega = 7.29 \times 10^{-5}$ rad/s, $g = 9.81$ m/s^2) leads to values of q_1 and q_2 such as those recorded in columns 2 and 3 of Table 8.1.1. Columns 4 and 5 show the values one obtains when one assumes that *E* rather than *F*

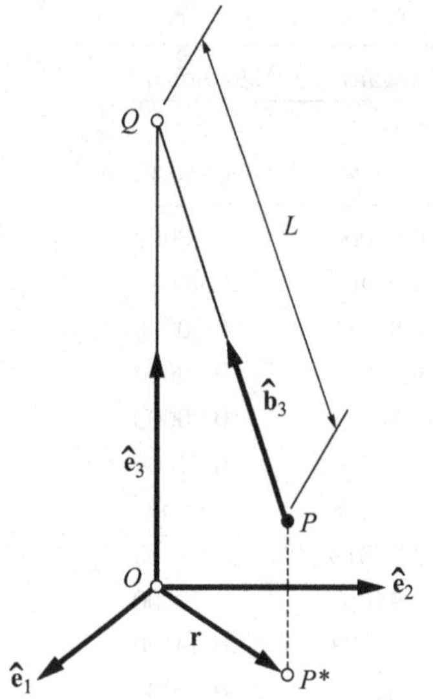

Figure 8.1.4

is a Newtonian reference frame, results that can be generated by using Eqs. (30) and (31) with $\omega = 0$. The two sets of results can be seen to agree with each other rather well as regards q_1, but differ from each other markedly as regards q_2. To determine whether column 3 or column 5 corresponds more nearly to reality, it is helpful to proceed as follows.

Let $\mathbf{r}$ be the position vector from O to $P^\star$, the orthogonal projection of P on the horizontal plane passing through O, as shown in Fig. 8.1.4. Then

$$\mathbf{r} = L\hat{\mathbf{e}}_3 \times (\hat{\mathbf{e}}_3 \times \hat{\mathbf{b}}_3) \tag{37}$$

or, since $\hat{\mathbf{b}}_3 = s_2\hat{\mathbf{e}}_1 - s_1c_2\hat{\mathbf{e}}_2 + c_1c_2\hat{\mathbf{e}}_3$,

$$\mathbf{r} = L(-s_2\hat{\mathbf{e}}_1 + s_1c_2\hat{\mathbf{e}}_2) \tag{38}$$

Hence, if E_1 and E_2 are Cartesian coordinate axes passing through O and parallel to $\hat{\mathbf{e}}_1$ and $\hat{\mathbf{e}}_2$, respectively, then the E_1 and E_2 coordinates of $P^\star$ are $-Ls_2$ and Ls_1c_2, respectively, and one can plot the path traced out by $P^\star$ in the E_1–E_2 plane, as has been done in Fig. 8.1.5 with data corresponding to columns 2 and 3 of Table 8.1.1; a portion of the E_2 axis represents columns 4 and 5 since, with $q_2 = 0$, Eq. (38) reduces to

$$\mathbf{r} = Ls_1\hat{\mathbf{e}}_2 \tag{39}$$

Figure 8.1.5 shows that the assumption that F is a Newtonian reference frame leads to the prediction that $P^\star$ must trace out a complicated curve in the E_1–E_2 plane. If it is assumed that E is a Newtonian reference frame, then the path predicted for $P^\star$ is simply a straight line. Foucault's experiments, performed at the Panthéon in Paris and subsequently duplicated in numerous other locations, revealed that $P^\star$ moves on

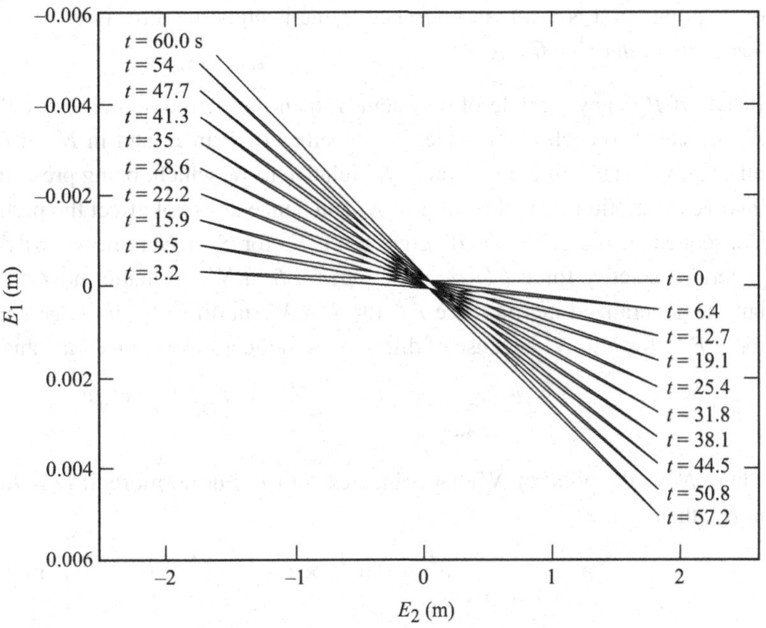

Figure 8.1.5

a curve such as the one in Fig. 8.1.5. Moreover, good quantitative agreement was obtained between experimental data and values predicted mathematically. Hence, regarding F as a Newtonian reference frame is more realistic than assuming that E is a Newtonian reference frame. But this does not mean that F is, in fact, a Newtonian reference frame or that one may not regard E as such a reference frame in certain contexts.

8.2 SECONDARY NEWTONIAN REFERENCE FRAMES

Astronomical observations furnish a large body of data showing that the reference frame F of the example in Sec. 8.1 is not a Newtonian reference frame, because motion predictions based on Eqs. (8.1.1) or (8.1.2) together with the assumption that F is such a reference frame conflict with the data. This fact suggests the following question: Why does the hypothesis that F is a Newtonian reference frame lead to a satisfactory description of the motion of a pendulum relative to the Earth, but to an incorrect description of, for example, the motion of the Moon relative to the Earth? This question and a number of related ones can be answered in the light of the following theorem:

If N' is a reference frame performing a prescribed motion relative to a Newtonian reference frame N, then N' is a Newtonian reference frame throughout some time interval if and only if throughout this time interval the acceleration in N of every point of N' is equal to zero. When a reference frame moves in such a way that the acceleration of each of its points in a Newtonian reference frame is equal to zero, it is called a *secondary Newtonian reference frame*.

Proof If P is any particle of a system S, then the partial velocities of P in N and in N' are equal to each other since the velocities of P in N and in N' differ from each other only because of the motion of N' relative to N, which, being prescribed, does not involve the motion variables for S in N, and hence does not affect the partial velocities. Consequently, the generalized active force $\widetilde{F}_r$ for S in N is necessarily equal to the generalized active force $\widetilde{F}_r'$ $(r = 1,\ldots,p)$ for S in N', no matter how N' moves in N; but the generalized inertia force $\widetilde{F}_r^\star$ for S in N can differ from the generalized inertia force $\widetilde{F}_r^{\star\prime}$ for S in N' because of differences in the accelerations ${}^N\mathbf{a}^P$ and ${}^{N'}\mathbf{a}^P$. Now,

$$ {}^N\mathbf{a}^P \underset{(2.8.2)}{=} {}^{N'}\mathbf{a}^P + {}^N\mathbf{a}^{\overline{N}'} + 2\,{}^N\boldsymbol{\omega}^{N'} \times {}^{N'}\mathbf{v}^P \tag{1} $$

where $\overline{N}'$ is the point of N' that coincides with P. Furthermore, if O is any point fixed in N', then

$$ {}^N\mathbf{a}^{\overline{N}'} \underset{(2.7.2)}{=} {}^N\mathbf{a}^O + {}^N\boldsymbol{\alpha}^{N'} \times \mathbf{r} + {}^N\boldsymbol{\omega}^{N'} \times ({}^N\boldsymbol{\omega}^{N'} \times \mathbf{r}) \tag{2} $$

where $\mathbf{r}$ is the position vector from O to $\overline{N}'$. Hence, if the acceleration in N of every point of N' is equal to zero, so that

$$ {}^N\mathbf{a}^{\overline{N}'} = {}^N\mathbf{a}^O = \mathbf{0} \tag{3} $$

then

$$ {}^N\boldsymbol{\alpha}^{N'} \times \mathbf{r} + {}^N\boldsymbol{\omega}^{N'} \times ({}^N\boldsymbol{\omega}^{N'} \times \mathbf{r}) \underset{(2,\,3)}{=} \mathbf{0} \tag{4} $$

and this can be satisfied for all $\mathbf{r}$ only if

$$ {}^N\boldsymbol{\alpha}^{N'} = \mathbf{0} \tag{5} $$

and

$$ {}^N\boldsymbol{\omega}^{N'} = \mathbf{0} \tag{6} $$

in which event

$$ {}^N\mathbf{a}^P \underset{(1,\,3,\,6)}{=} {}^{N'}\mathbf{a}^P \tag{7} $$

and, therefore [see Eqs. (5.9.2) and (5.9.4)],

$$ \widetilde{F}_r^\star = \widetilde{F}_r^{\star\prime} \qquad (r = 1,\ldots,p) \tag{8} $$

which means that whenever $\widetilde{F}_r + \widetilde{F}_r^\star = 0$ $(r = 1,\ldots,p)$, then $\widetilde{F}_r' + \widetilde{F}_r^{\star\prime} = 0$ $(r = 1,\ldots,p)$ and, consequently, that N' is a Newtonian reference frame.

To show that N' is not a Newtonian reference frame unless the acceleration in N of every point of N' is equal to zero, let O be a point of N' such that

$$ {}^N\mathbf{a}^O \neq \mathbf{0} \tag{9} $$

and let S consist of a single particle P situated permanently at O. Then

$$ {}^{N'}\mathbf{a}^P = \mathbf{0} \tag{10} $$

while, since now $\mathbf{r} = \mathbf{0}$ and ${}^{N'}\mathbf{v}^P = \mathbf{0}$,

$$ {}^{N}\mathbf{a}^P \underset{(1,\,2,\,10)}{=} {}^{N}\mathbf{a}^O \underset{(9)}{\neq} \mathbf{0} \tag{11} $$

so that

$$ {}^{N}\mathbf{a}^P \underset{(10,\,11)}{\neq} {}^{N'}\mathbf{a}^P \tag{12} $$

and, therefore,

$$ \widetilde{F}_r^{\star} \neq \widetilde{F}_r^{\star\prime} \qquad (r = 1, \ldots, p) \tag{13} $$

which means that whenever $\widetilde{F}_r + \widetilde{F}_r^{\star} = 0$ $(r = 1, \ldots, p)$, then $\widetilde{F}_r' + \widetilde{F}_r^{\star\prime} \neq 0$ $(r = 1, \ldots, p)$ and, consequently, that N' is not a Newtonian reference frame.

Returning to the question raised at the beginning of the section, we let A be a reference frame in which the Sun S remains fixed and relative to which the orientation of the reference frame F introduced in the example in Sec. 8.1 does not vary, so that ${}^{A}\boldsymbol{\omega}^F = \mathbf{0}$, but in which the center $E^{\star}$ of the Earth E moves on a plane curve. A, S, F, and E are depicted in Fig. 8.2.1, which also shows the Moon M.

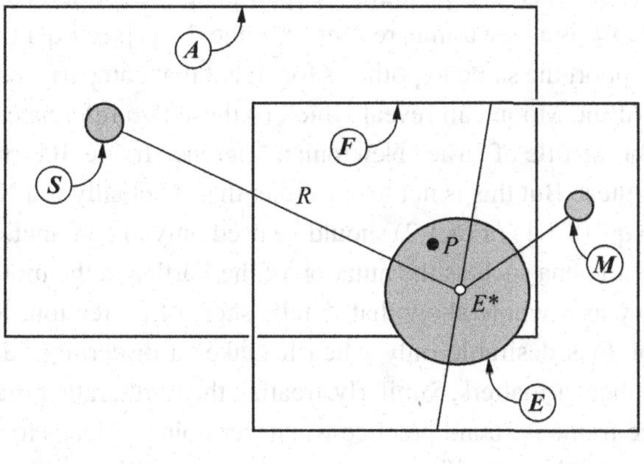

Figure 8.2.1

Suppose that A is a Newtonian reference frame. Then F cannot possibly be a Newtonian reference frame, because there exists no point of F whose acceleration in A is equal to zero. In fact, the acceleration in A of every point that is fixed in F is equal to the acceleration ${}^{A}\mathbf{a}^{E^{\star}}$ of $E^{\star}$ in A, and this acceleration has a magnitude that can be estimated with sufficient accuracy for present purposes by assuming that $E^{\star}$ moves in A on a circle of radius $R \approx 1.5 \times 10^{11}$ m, traced out once per year. How important is this acceleration? That depends on the acceleration in F of the particles of the system under consideration. As the angular velocity of F in A is equal to zero, and, hence, the angular acceleration of F in A is also equal to zero, the acceleration ${}^{A}\mathbf{a}^P$ of a particle P in A differs from the acceleration ${}^{F}\mathbf{a}^P$ of P in F by precisely ${}^{A}\mathbf{a}^{E^{\star}}$. Errors resulting from regarding F, rather than A, as a Newtonian reference frame therefore grow in importance as the ratio of $|{}^A\mathbf{a}^{E^\star}|$ to $|{}^F\mathbf{a}^P|$ increases. Hence, in connection with studies of motions of the Moon, an object whose particles move in F, essentially, on circles having radii of approximately 4.0×10^8 m, each particle completing about 12 such orbits per year, it must make a substantial difference whether A or F is assigned the role of Newtonian reference frame, because the ratio in question here has the value (let ${}^F\mathbf{a}^M$ be the acceleration in F of a typical particle of the Moon)

$$\frac{|{}^A\mathbf{a}^{E^\star}|}{|{}^F\mathbf{a}^M|} = \frac{1.5 \times 10^{11}}{4.0 \times 10^8}\left(\frac{1}{12}\right)^2 \approx 2.6 \tag{14}$$

which means that $|{}^A\mathbf{a}^{E^\star}|$ cannot be regarded as negligible in comparison with $|{}^F\mathbf{a}^M|$. By way of contrast, for the pendulum considered in Sec. 8.1, $g\alpha$ is a reasonable upper bound for $|{}^F\mathbf{a}^P|$, where α is the maximum value of the angle between lines OQ and PQ in Fig. 8.1.1, so that, with $g = 9.81$ m/s^2 and α even as small as 0.01 rad,

$$\frac{|{}^A\mathbf{a}^{E^\star}|}{|{}^F\mathbf{a}^P|} \approx 0.06 \tag{15}$$

Thus, so far as numerical results are concerned, it here matters far less whether A or F is regarded as a Newtonian reference frame.

As was pointed out in Sec. 8.1, the pendulum experiments performed by Foucault support the hypothesis that F is a Newtonian reference frame. Now [see Eq. (15)] we see that these experiments support the same hypothesis for A, but that comparisons of actual with predicted motions of the Moon can reveal which of these two reference frames is the stronger contender for the title of "true" Newtonian reference frame. It turns out that A is the winner of this contest. But this is not to say either that A actually is a Newtonian reference frame or that Eqs. (8.1.1) or (8.1.2) should be used only in conjunction with A (rather than with F). Phenomena such as the nutation of the Earth and the motion of the Sun relative to the galaxy as a whole show that A falls short of perfection. Moreover, the use of A in place of F is desirable only when it makes a discernible difference; otherwise it merely complicates matters. Similarly, treating the Earth, rather than A or F, as a Newtonian reference frame is sound practice whenever doing so leads to analytical simplifications unaccompanied by significant losses in accuracy, as is the case in a large number of situations encountered in engineering. Hence, unless explicitly exempted, every use of Eqs. (8.1.1) or (8.1.2) in the remainder of this book will be predicated on the assumption that all reference frames rigidly attached to the Earth, as well as all reference frames all of whose points have zero acceleration relative to the Earth, may be regarded as Newtonian reference frames.

8.3 ADDITIONAL DYNAMICAL EQUATIONS

When motion variables are introduced in addition to $u_1, \ldots, u_p$ for the purpose of bringing into evidence forces and/or torques that contribute nothing to the generalized active forces $\widetilde{F}_1, \ldots, \widetilde{F}_p$ or, if S is a holonomic system, to $F_1, \ldots, F_n$, then the dynamical equations corresponding to the new set of motion variables furnish the needed information about the forces and/or torques in question. To generate these dynamical equations, one first follows the procedure set forth in Sec. 6.7 to form expressions for the generalized active forces corresponding to the new set of motion variables, which are used in Eqs. (8.1.1) or (8.1.2) together with generalized inertia forces constructed according to instructions that will now be given.

Generalized inertia forces corresponding to the new set of motion variables are found by using in Eqs. (5.9.1), (5.9.2), (5.9.7), and (5.9.8) partial velocities and partial angular velocities formed as explained in Sec. 6.7. The expressions to be used here for the vectors ${}^A\mathbf{a}^{P_i}$, ${}^A\mathbf{a}^{B\star}$, ${}^A\boldsymbol{\alpha}^B$, and ${}^A\boldsymbol{\omega}^B$ appearing variously in Eqs. (5.9.4), (5.9.10), (5.9.11), and (5.9.12) are precisely those employed in forming the generalized inertia forces $\widetilde{F}_1^\star, \ldots, \widetilde{F}_p^\star$, or, if S is a holonomic system, $F_1^\star, \ldots, F_n^\star$; in other words, the expressions do not contain the additional motion variables or their time derivatives.

As discussed in Sec. 6.7, generalized active forces and generalized inertia forces associated with the original motion variables will, in general, require revision subsequent to the introduction of additional motion variables.

Example When the bearing surface B and the rod R of the system introduced in the example in Sec. 5.8 and depicted in Fig. 5.8.1 both are perfectly smooth, the dynamical equations governing the motion of the system, written by substituting from Eqs. (5.8.16), (5.8.17), (5.9.39), and (5.9.40) into Eqs. (8.1.1), are

$$\dot{u}_1 = g \cos\beta + q_1 (u_2 \sin\beta)^2 \tag{1}$$

$$\dot{u}_2 = \frac{T \csc^2\beta - 2mq_1u_1u_2}{mq_1{}^2 + 4ML^2/3} \tag{2}$$

and these, together with the kinematical equation

$$\dot{q}_1 \underset{(5.8.12)}{=} u_1 \tag{3}$$

permit one to determine q_1, u_1, and u_2 for $t > 0$ when the values of these variables are known for $t = 0$ (and T has been specified as a function of q_1, u_1, u_2, and t). But if, as subsequently supposed in the example in Sec. 5.8, the contact between the sleeve S and the bearing surface B, as well as the contact between P and R, is presumed to take place across a rough surface, then Eqs. (5.8.40) and (5.8.41) replace Eqs. (5.8.16) and (5.8.17), respectively, with the result that the dynamical equations become

$$\dot{u}_1 = g \cos\beta - \left(\frac{\mu_1{}'}{m}\right)(\rho_2{}^2 + \rho_3{}^2)^{1/2} \operatorname{sgn} u_1 + q_1 (u_2 \sin\beta)^2 \tag{4}$$

$$\dot{u}_2 = \frac{[T - (2\pi n^*/3)(b_2{}^3 - b_1{}^3)\mu_2{}' \operatorname{sgn} u_2] \csc^2\beta - 2mq_1u_1u_2}{mq_1{}^2 + 4ML^2/3} \tag{5}$$

and, since these contain ρ_2, ρ_3, and n^*, they do not suffice [together with Eq. (3)] for the determination of q_1, u_1, and u_2. However, one can supplement Eqs. (3)–(5) with the dynamical equations corresponding to u_3, u_5, and u_9 as introduced in Sec. 6.7 [see Eqs. (6.7.13), (6.7.15), and (6.7.26), respectively].

After introducing u_3 as in Eq. (6.7.13), one finds that $(F_3)_C$, the contribution to the generalized active force F_3 of all contact forces exerted by B and V on S, and by R and P on each other, can be written

$$(F_3)_C = {}^B\mathbf{v}_3^P \cdot \boldsymbol{\rho} + {}^B\mathbf{v}_3^{\bar{R}} \cdot \bar{\boldsymbol{\rho}} + \int {}^B\mathbf{v}_3^Q \cdot d\boldsymbol{\sigma} \tag{6}$$

where (see Table 6.7.1)

$$ {}^B\mathbf{v}_3^P = {}^B\mathbf{v}_3^{\bar{R}} \tag{7}$$

If h is the distance from O to Σ, then

$$ {}^B\mathbf{v}^Q \underset{(2.7.1)}{=} {}^B\mathbf{v}^O + {}^B\boldsymbol{\omega}^R \times (h\hat{\mathbf{s}}_1 + z\hat{\mathbf{e}}_2) \underset{(5.8.12)}{=} u_3\hat{\mathbf{s}}_1 + u_2 z\hat{\mathbf{e}}_3 \tag{8}$$

and the partial velocity ${}^B\mathbf{v}_3^Q$ is given by

$$ {}^B\mathbf{v}_3^Q = \hat{\mathbf{s}}_1 \tag{9}$$

Consequently [see Eqs. (5.8.19) and (5.8.27)],

$$\begin{aligned}(F_3)_C &\underset{(6,\,7,\,9)}{=} -n^* \int_0^{2\pi} \int_{b_1}^{b_2} z\, dz\, d\theta \\ &= -\pi n^* ({b_2}^2 - {b_1}^2)\end{aligned} \tag{10}$$

The quantities ρ_2 and ρ_3 appearing in Eq. (4) come into evidence also when one forms the contributions $(F_5)_C$ and $(F_9)_C$ of contact forces to the generalized active forces F_5 and F_9 corresponding to u_5 and u_9, respectively, where u_5 and u_9 have the same meaning as in Eq. (6.7.26). Specifically, in agreement with Eqs. (6.7.23) and (6.7.30),

$$(F_5)_C = \rho_3 \qquad (F_9)_C = \rho_2 \tag{11}$$

Expressions for the contributions to F_r $(r = 3,5,9)$ of all gravitational forces acting on S, R, and P and of the torque $\mathbf{T}$ described by Eq. (5.8.11) can be formulated as they were in connection with Eqs. (6.7.21), (6.7.23), and (6.7.30). Hence, the complete generalized active forces are

$$F_3 = \underset{(6.7.21)}{(m+M)g} - \underset{(10)}{\pi n^* ({b_2}^2 - {b_1}^2)} \tag{12}$$

$$F_5 = \underset{(6.7.23,\,11)}{\rho_3} \tag{13}$$

$$F_9 = \underset{(6.7.30)}{-mg \sin\beta} + \underset{(11)}{\rho_2} \tag{14}$$

To form the generalized inertia forces $F_3^\star$ and $F_5^\star$ corresponding to the generalized active forces F_3 and F_5 in Eqs. (12) and (13), all one needs to do is use Eqs. (5.9.30) with $r = 3,5$ and with ${}^B\mathbf{v}_r^P$, ${}^B\boldsymbol{\omega}_r^R$, and ${}^B\mathbf{v}_r^{R\star}$, $(r = 3,5)$ as given in Table

6.7.1, while ${}^{B}\mathbf{a}^{R^\star}$, ${}^{B}\mathbf{a}^{P}$, and $\mathbf{T}^\star$ are given by Eqs. (5.9.33), (5.9.34), and (5.9.38), respectively, as heretofore. Thus one finds that

$$F_3^\star = -m\dot{u}_1 \cos\beta \tag{15}$$

$$F_5^\star = -m(\dot{u}_2 q_1 + 2u_1 u_2)\sin\beta \tag{16}$$

By proceeding similarly, one can form an expression for $F_9^\star$ corresponding to the generalized active force in Eq. (14),

$$F_9^\star = m{u_2}^2 q_1 \sin\beta\cos\beta \tag{17}$$

Finally, the required additional dynamical equations are obtained,

$$\underset{(12)}{(m+M)g - \pi n^*({b_2}^2 - {b_1}^2)} \underset{(15)}{- m\dot{u}_1\cos\beta} \underset{(8.1.1)}{=} 0 \tag{18}$$

$$\underset{(13)}{\rho_3} \underset{(16)}{- m(\dot{u}_2 q_1 + 2u_1 u_2)\sin\beta} \underset{(8.1.1)}{=} 0 \tag{19}$$

$$\underset{(14)}{-mg\sin\beta + \rho_2} \underset{(17)}{+ m{u_2}^2 q_1 \sin\beta\cos\beta} \underset{(8.1.1)}{=} 0 \tag{20}$$

and with their aid one can eliminate n^*, ρ_2, and ρ_3 from Eqs. (4) and (5) and thus come into position to determine q_1, u_1, and u_2 as functions of t.

8.4 LINEARIZATION OF DYNAMICAL EQUATIONS

Frequently, one can obtain much useful information about the behavior of a system S from *linearized* forms of kinematical and/or dynamical equations, that is, equations derived from nonlinear equations by omitting all terms of second or higher degree in *perturbations* of some (or all) of the motion variables $u_1, \ldots, u_n$, and generalized coordinates $q_1, \ldots, q_n$. This is true primarily because linear differential equations generally can be solved more easily than can nonlinear differential equations. Of course, the solutions of such linearized equations lead one only to approximations of the solutions of the corresponding full, nonlinear equations; and they may be rather poor approximations. In any event, however, the approximations become ever better as the perturbations involved in the linearization take on ever smaller values.

When nonlinear kinematical and/or dynamical equations are in hand, one forms their linearized counterparts by expanding all functions of the perturbations involved in the linearization into power series in these perturbations and dropping all nonlinear terms. To formulate linearized dynamical equations directly, that is, without first writing exact dynamical equations, proceed as follows:

Develop fully nonlinear expressions for angular velocities of rigid bodies belonging to S, for velocities of mass centers of such bodies, and for velocities of particles of S to which contact and/or distance forces contributing to generalized active forces are applied. Use these nonlinear expressions to determine partial angular velocities and partial velocities by inspection. Linearize all angular velocities of rigid bodies and velocities of particles, and use the linearized forms to construct linearized angular accelerations and accelerations. Linearize all partial angular velocities and partial velocities. Form linearized generalized active forces and linearized generalized inertia forces, and substitute into Eqs. (8.1.1) or (8.1.2).

Examples As was shown in the example in Sec. 8.1, all motions of the Foucault pendulum are governed by the equations

$$\dot{u}_1 \underset{(8.1.30)}{=} \frac{u_2{}^2\, s_2}{L c_2} + 2\omega u_2 (c_1 c_2 c_\phi - s_2 s_\phi) + g c_1 s_2 \tag{1}$$

$$\dot{u}_2 \underset{(8.1.31)}{=} -\frac{u_1 u_2\, s_2}{L c_2} - 2\omega u_1 (c_1 c_2 c_\phi - s_2 s_\phi) - g s_1 \tag{2}$$

$$\dot{q}_1 \underset{(8.1.35)}{=} \frac{u_2}{L c_2} \qquad \dot{q}_2 \underset{(8.1.35)}{=} -\frac{u_1}{L} \tag{3}$$

As may be verified by inspection, these equations possess the solution

$$u_1 = u_2 = q_1 = q_2 = 0 \tag{4}$$

Hence, one may hope to obtain useful information regarding the motion of the pendulum by employing equations resulting from linearization of Eqs. (1)–(3) "about this solution," that is, by replacing u_1, u_2, q_1, and q_2 with perturbation functions $u_1{}^*$, $u_2{}^*$, $q_1{}^*$, and $q_2{}^*$, respectively, and then linearizing in these perturbations. For example, when $u_1{}^*$ and $u_2{}^*$ are written in place of u_1 and u_2, respectively, in Eq. (1), and c_1, c_2, and s_2 are replaced with 1, 1, and $q_2{}^*$, respectively, these being the terms of degree less than two in the expansions of $\cos q_1{}^*$, $\cos q_2{}^*$, and $\sin q_2{}^*$, respectively, then $\dot{u}_1{}^*$ is given by

$$\dot{u}_1{}^* \underset{(1)}{=} 2\omega u_2{}^* (c_\phi - q_2{}^* s_\phi) + g q_2{}^* \tag{5}$$

This, however, is an incompletely linearized equation because $u_2{}^* q_2{}^*$ is a second-degree term. Dropping this term, one arrives at the linear equation corresponding to Eq. (1), namely,

$$\dot{u}_1{}^* = 2\omega u_2{}^* c_\phi + g q_2{}^* \tag{6}$$

Similarly, the linear equations corresponding to Eqs. (2) and (3) are found to be

$$\dot{u}_2{}^* \underset{(2)}{=} -2\omega u_1{}^* c_\phi - g q_1{}^* \tag{7}$$

and

$$\dot{q}_1{}^* \underset{(3)}{=} \frac{u_2{}^*}{L} \qquad \dot{q}_2{}^* \underset{(3)}{=} -\frac{u_1{}^*}{L} \tag{8}$$

To solve Eqs. (6)–(8), and thus to obtain a detailed, approximate, analytical description of the motion of the pendulum, it is helpful to introduce $\boldsymbol{\rho}$ as the linearized form of the vector $\mathbf{r}$ shown in Fig. 8.1.4, that is, to let

$$\boldsymbol{\rho} \underset{(8.1.38)}{=} L(-q_2{}^*\, \hat{\mathbf{e}}_1 + q_1{}^*\, \hat{\mathbf{e}}_2) \tag{9}$$

for it may be verified by carrying out the indicated differentiations that Eqs. (6)–(8) are together equivalent to the single linear vector differential equation

$$\frac{^E d^2 \boldsymbol{\rho}}{dt^2} + 2\omega c_\phi \hat{\mathbf{e}}_3 \times \frac{^E d\boldsymbol{\rho}}{dt} + \frac{g}{L}\boldsymbol{\rho} = \mathbf{0} \tag{10}$$

and this equation can be replaced with an even simpler one through the introduction of a reference frame R whose angular velocity in E is taken to be

$$^E\boldsymbol{\omega}^R \triangleq -\omega c_\phi \hat{\mathbf{e}}_3 \tag{11}$$

which permits one to write

$$\frac{^E d\boldsymbol{\rho}}{dt} \underset{(2.3.1)}{=} \frac{^R d\boldsymbol{\rho}}{dt} + {^E\boldsymbol{\omega}^R} \times \boldsymbol{\rho} \underset{(11)}{=} \frac{^R d\boldsymbol{\rho}}{dt} - \omega c_\phi \hat{\mathbf{e}}_3 \times \boldsymbol{\rho} \tag{12}$$

$$\frac{^E d^2\boldsymbol{\rho}}{dt^2} \underset{(12)}{=} \frac{^R d^2\boldsymbol{\rho}}{dt^2} - \omega c_\phi \hat{\mathbf{e}}_3 \times \frac{^R d\boldsymbol{\rho}}{dt} \underset{(11)}{-} \omega c_\phi \hat{\mathbf{e}}_3 \times \frac{^R d\boldsymbol{\rho}}{dt} \underset{(11)}{+} (\omega c_\phi)^2 \hat{\mathbf{e}}_3 \times (\hat{\mathbf{e}}_3 \times \boldsymbol{\rho}) \tag{13}$$

Substitution from Eqs. (12) and (13) into Eq. (10) yields

$$\frac{^R d^2\boldsymbol{\rho}}{dt^2} + \frac{g}{L}\boldsymbol{\rho} = \mathbf{0} \tag{14}$$

if, as in Sec. 8.1, terms involving ω^2 are dropped. Now, Eq. (14) possesses the general solution

$$\boldsymbol{\rho} = \mathbf{A}\cos pt + \mathbf{B}\sin pt \tag{15}$$

where $\mathbf{A}$ and $\mathbf{B}$ are vectors fixed in reference frame R, and p, known as the *circular natural frequency* of the pendulum, is defined as

$$p = \left(\frac{g}{L}\right)^{1/2} \tag{16}$$

The vectors $\mathbf{A}$ and $\mathbf{B}$ depend on initial conditions. Specifically, denoting the initial values of $\boldsymbol{\rho}$ and $^R d\boldsymbol{\rho}/dt$ by $\boldsymbol{\rho}(0)$ and $^R d\boldsymbol{\rho}(0)/dt$, respectively, one can write

$$\mathbf{A} \underset{(15)}{=} \boldsymbol{\rho}(0) \qquad \mathbf{B} \underset{(15)}{=} \frac{1}{p}\frac{^R d\boldsymbol{\rho}(0)}{dt} \tag{17}$$

Suppose, for example, that P is initially displaced toward the east in such a way that the string makes an angle α with the vertical, and that P is then released from a state of rest in E. Correspondingly,

$$q_1(0) = \alpha \qquad q_2(0) = 0 \qquad u_1(0) = u_2(0) = 0 \tag{18}$$

Hence,

$$\mathbf{r}(0) \underset{(8.1.38,\ 18)}{=} L s_\alpha \hat{\mathbf{e}}_2(0) \tag{19}$$

so that linearization in α yields

$$\boldsymbol{\rho}(0) = L\alpha\hat{\mathbf{e}}_2(0) \tag{20}$$

where $\hat{\mathbf{e}}_2(0)$ denotes the value of $\hat{\mathbf{e}}_2$ in R at $t = 0$. Similarly, one can write

$$\begin{aligned}\frac{{}^R d\mathbf{r}(0)}{dt} &= \frac{{}^E d\mathbf{r}(0)}{dt} + {}^R\boldsymbol{\omega}^E(0) \times \mathbf{r}(0) \\ &= {}^E\mathbf{v}^P(0) + \underset{(11)}{\omega c_\phi \hat{\mathbf{e}}_3(0)} \times \underset{(19)}{[L s_\alpha \hat{\mathbf{e}}_2(0)]} \\ &= \underset{(8.1.34,\ 8.1.35,\ 18)}{\mathbf{0}} - L\omega c_\phi s_\alpha \hat{\mathbf{e}}_1(0)\end{aligned} \tag{21}$$

and, linearizing in α, one has

$$\frac{{}^R d\boldsymbol{\rho}(0)}{dt} \underset{(21)}{=} -L\omega c_\phi \alpha \hat{\mathbf{e}}_1(0) \tag{22}$$

Thus, $\boldsymbol{\rho}$ can be written

$$\begin{aligned}\boldsymbol{\rho} &\underset{(15)}{=} \underset{(17,\ 20)}{L\alpha\hat{\mathbf{e}}_2(0)\cos pt} - \underset{(17,\ 22)}{L\left(\frac{\omega}{p}\right) c_\phi \alpha \hat{\mathbf{e}}_1(0)\sin pt} \\ &= L\alpha\left[-\left(\frac{\omega}{p}\right) c_\phi \sin pt\, \hat{\mathbf{e}}_1(0) + \cos pt\, \hat{\mathbf{e}}_2(0)\right]\end{aligned} \tag{23}$$

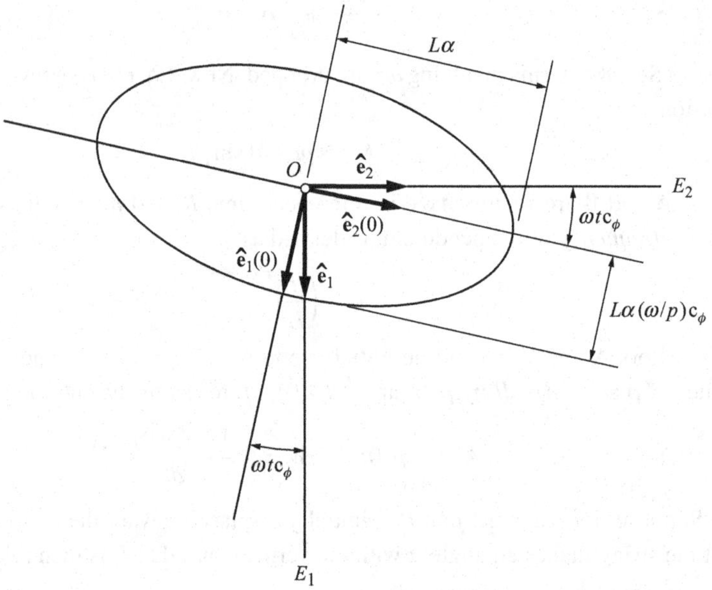

Figure 8.4.1

According to Eq. (23), $P^\star$ (see Fig. 8.1.4) moves in R on an elliptical path having the proportions shown in Fig. 8.4.1, and $P^\star$ traverses this path once every $2\pi/p$ units of time. The ellipse, being fixed in R, rotates in E at a rate of ωc_ϕ radians per unit of time [see Eq. (11)], the rotation being clockwise as seen by an observer looking from point Q toward point O (see Fig. 8.1.4), provided that O is situated in the northern hemisphere. These predictions agree qualitatively with those obtained in Sec. 8.1 by using the full, nonlinear equations of motion, Eqs. (8.1.30), (8.1.31), and (8.1.35). To assess the merits of the linear theory in quantitative terms, one can compare $|\mathbf{r}|$ — calculated by using Eq. (8.1.38) after integrating Eqs. (8.1.30), (8.1.31), and (8.1.35) numerically with $\phi = 45°$, $L = 10$ m, $\omega = 7.29 \times 10^{-5}$ rad/s, $q_1(0) = \alpha$, $q_2(0) = u_1(0) = u_2(0) = 0$— with $|\rho|$ as obtained by using Eq. (23). Plots resulting from such calculations are shown in Fig. 8.4.2 for $\alpha = 10°$, $30°$, and $60°$. The solid curves represent values of $|\mathbf{r}|$, whereas the broken curves correspond to values of $|\rho|$. As was to be expected, the agreement between the two curves associated with a given value of α is best for $\alpha = 10°$ and worst for $\alpha = 60°$. Of course, whether or not any given result obtained from linearized equations can be regarded as "good" depends on the user's needs.

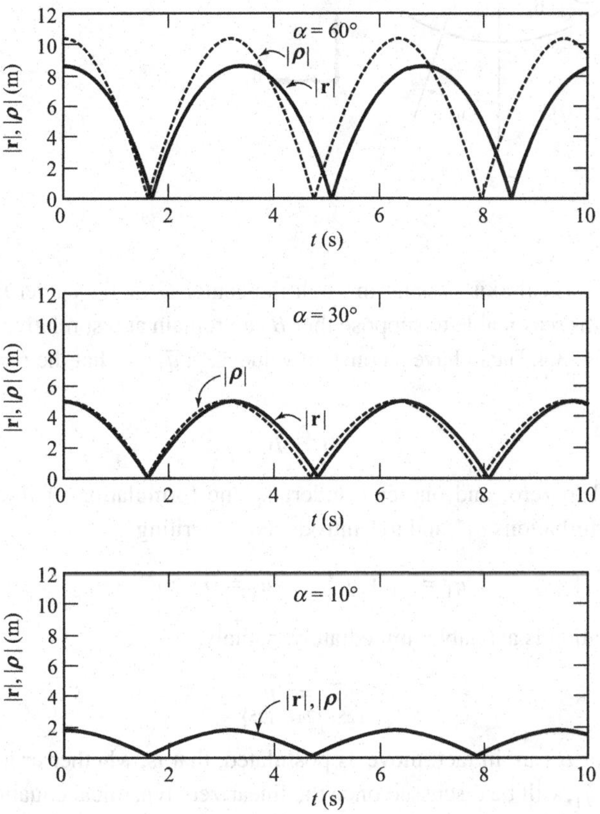

Figure 8.4.2

To illustrate the process of formulating linearized equations of motion without first constructing fully nonlinear equations, we consider a uniform bar B of mass m supported by frictionless sliders on a circular wire W of radius R, as shown in Fig. 8.4.3, and suppose that W is being made to rotate with a constant angular speed Ω

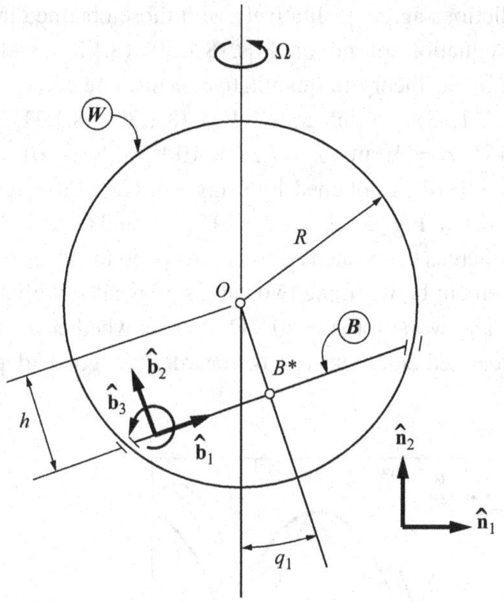

Figure 8.4.3

about a fixed vertical axis passing through the center O of W. Under these circumstances, it seems reasonable to suppose that B can remain at rest relative to W, that is, that q_1 (see Fig. 8.4.3) can have a constant value, say, $\overline{q}_1$, so that the motion variable u_1 defined as

$$u_1 \triangleq \dot{q}_1 \tag{24}$$

remains equal to zero; and one can undertake the formulation of linear equations governing perturbations $q_1{}^*$ and $u_1{}^*$ introduced by writing

$$q_1 = \overline{q}_1 + q_1{}^* \qquad u_1 = 0 + u_1{}^* \tag{25}$$

One such equation is available immediately, namely,

$$\underset{(25)}{\dot{q}_1{}^*} \underset{(24)}{=} \underset{(25)}{u_1{}^*} \tag{26}$$

Whether or not B can, in fact, move as postulated, that is, whether or not there exist real values of $\overline{q}_1$, will be discussed once the linearized dynamical equation governing $u_1{}^*$ and $q_1{}^*$ has been formulated.

The angular velocity $\boldsymbol{\omega}$ of B and the velocity $\mathbf{v}^\star$ of the center $B^\star$ of B are given by (see Fig. 8.4.3 for the unit vectors $\hat{\mathbf{b}}_1$, $\hat{\mathbf{b}}_2$, and $\hat{\mathbf{b}}_3$)

$$\boldsymbol{\omega} = \Omega(s_1\hat{\mathbf{b}}_1 + c_1\hat{\mathbf{b}}_2) + u_1\hat{\mathbf{b}}_3 \tag{27}$$

and

$$\mathbf{v}^\star = h(u_1\hat{\mathbf{b}}_1 - \Omega s_1\hat{\mathbf{b}}_3) \tag{28}$$

respectively. Hence, the partial angular velocity $\boldsymbol{\omega}_1$ and partial velocity $\mathbf{v}_1^\star$ are

$$\boldsymbol{\omega}_1 \underset{(27)}{=} \hat{\mathbf{b}}_3 \qquad \mathbf{v}_1^\star \underset{(28)}{=} h\hat{\mathbf{b}}_1 \tag{29}$$

To linearize Eqs. (27) and (28) in $q_1{}^*$ and $u_1{}^*$, we observe that

$$\mathrm{s}_1 \underset{(25)}{=} \sin(\overline{q}_1 + q_1{}^*) = \sin\overline{q}_1 \cos q_1{}^* + \cos\overline{q}_1 \sin q_1{}^* \tag{30}$$

so that, if $\overline{\mathrm{s}}_1$ and $\overline{\mathrm{c}}_1$ are written in place of $\sin\overline{q}_1$ and $\cos\overline{q}_1$, respectively, and $\cos q_1{}^*$ and $\sin q_1{}^*$ are replaced with unity and $q_1{}^*$, respectively, then

$$\mathrm{s}_1 \underset{(30)}{\approx} \overline{\mathrm{s}}_1 + \overline{\mathrm{c}}_1 q_1{}^* \tag{31}$$

Similarly,

$$\mathrm{c}_1 \approx \overline{\mathrm{c}}_1 - \overline{\mathrm{s}}_1 q_1{}^* \tag{32}$$

Hence, the linearized forms of $\boldsymbol{\omega}$ and $\mathbf{v}^\star$ are

$$\boldsymbol{\omega} \underset{(27)}{\approx} \Omega[\underset{(31)}{(\overline{\mathrm{s}}_1 + \overline{\mathrm{c}}_1 q_1{}^*)}\hat{\mathbf{b}}_1 + \underset{(32)}{(\overline{\mathrm{c}}_1 - \overline{\mathrm{s}}_1 q_1{}^*)}\hat{\mathbf{b}}_2] + \underset{(25)}{u_1{}^*}\hat{\mathbf{b}}_3 \tag{33}$$

and

$$\mathbf{v}^\star \underset{(28)}{\approx} h[\underset{(25)}{u_1{}^*}\hat{\mathbf{b}}_1 - \Omega\underset{(31)}{(\overline{\mathrm{s}}_1 + \overline{\mathrm{c}}_1 q_1{}^*)}\hat{\mathbf{b}}_3)] \tag{34}$$

As for $\boldsymbol{\omega}_1$ and $\mathbf{v}_1^\star$, these, as written in Eqs. (29), contain no terms that are nonlinear in $q_1{}^*$. Note, however, that $\mathbf{v}_1^\star$ can be written

$$\mathbf{v}_1^\star \underset{(29)}{=} h(\mathrm{c}_1\hat{\mathbf{n}}_1 + \mathrm{s}_1\hat{\mathbf{n}}_2) \tag{35}$$

where $\hat{\mathbf{n}}_1$ and $\hat{\mathbf{n}}_2$ are unit vectors directed as shown in Fig. 8.4.3, and that the linearized form of $\mathbf{v}_1^\star$ then is

$$\mathbf{v}_1^\star \underset{(35)}{\approx} h[\underset{(32)}{(\overline{\mathrm{c}}_1 - \overline{\mathrm{s}}_1 q_1{}^*)}\hat{\mathbf{n}}_1 + \underset{(31)}{(\overline{\mathrm{s}}_1 + \overline{\mathrm{c}}_1 q_1{}^*)}\hat{\mathbf{n}}_2] \tag{36}$$

To show that, in general, one must form partial velocities (and partial angular velocities) *before* linearizing, we rewrite Eq. (34) as

$$\mathbf{v}^\star \approx h[u_1{}^*(\overline{\mathrm{c}}_1 - \overline{\mathrm{s}}_1 q_1{}^*)\hat{\mathbf{n}}_1 + u_1{}^*(\overline{\mathrm{s}}_1 + \overline{\mathrm{c}}_1 q_1{}^*)\hat{\mathbf{n}}_2 - \Omega(\overline{\mathrm{s}}_1 + \overline{\mathrm{c}}_1 q_1{}^*)\hat{\mathbf{b}}_3] \tag{37}$$

or, after completing the linearization process,

$$\mathbf{v}^\star \underset{(37)}{\approx} h[u_1{}^*\overline{\mathrm{c}}_1\hat{\mathbf{n}}_1 + u_1{}^*\overline{\mathrm{s}}_1\hat{\mathbf{n}}_2 - \Omega(\overline{\mathrm{s}}_1 + \overline{\mathrm{c}}_1 q_1{}^*)\hat{\mathbf{b}}_3] \tag{38}$$

Starting with this equation, one would conclude that $\mathbf{v}_1^\star \approx h(\overline{\mathrm{c}}_1\hat{\mathbf{n}}_1 + \overline{\mathrm{s}}_1\hat{\mathbf{n}}_2)$, which conflicts with Eq. (36) and is incorrect.

Linearized expressions for the angular acceleration $\boldsymbol{\alpha}$ of B and the acceleration $\mathbf{a}^\star$ of $B^\star$ are formed by working with the already linearized Eqs. (33) and (34), respectively. Specifically,

$$\boldsymbol{\alpha} \underset{(33)}{\approx} \Omega \underset{(26)}{u_1{}^*} (\overline{\mathrm{c}}_1\hat{\mathbf{b}}_1 - \overline{\mathrm{s}}_1\hat{\mathbf{b}}_2) + \dot{u}_1{}^*\hat{\mathbf{b}}_3 \tag{39}$$

and

$$\mathbf{a}^\star \underset{(34)}{\approx} h(\dot{u}_1{}^*\hat{\mathbf{b}}_1 - \Omega \bar{c}_1 u_1{}^* \hat{\mathbf{b}}_3) + \underset{(26)}{\boldsymbol{\omega} \times \mathbf{v}^\star} \tag{40}$$

Now,

$$\boldsymbol{\omega} \times \mathbf{v}^\star \underset{(33,\ 34)}{\approx} -h\Omega^2(\bar{c}_1 - \bar{s}_1 q_1{}^*)(\bar{s}_1 + \bar{c}_1 q_1{}^*)\hat{\mathbf{b}}_1 + \cdots$$

$$\approx -h\Omega^2[\bar{c}_1\bar{s}_1 + (\bar{c}_1{}^2 - \bar{s}_1{}^2)q_1{}^*]\hat{\mathbf{b}}_1 + \cdots \tag{41}$$

where only the $\hat{\mathbf{b}}_1$ component has been worked out in detail because $\mathbf{v}_1^\star$ [see Eqs. (29)] is parallel to $\hat{\mathbf{b}}_1$, and $\mathbf{v}_1^\star$ presently will be dot-multiplied with $\mathbf{a}^\star$ [see Eqs. (5.9.7) and (5.9.11)]. Hence, we have

$$\mathbf{a}^\star \underset{(40,\ 41)}{\approx} h\{\dot{u}_1{}^* - \Omega^2[\bar{c}_1\bar{s}_1 + (\bar{c}_1{}^2 - \bar{s}_1{}^2)q_1{}^*]\}\hat{\mathbf{b}}_1 + \cdots \tag{42}$$

The inertia torque $\mathbf{T}^\star$ for B is given by

$$\mathbf{T}^\star \underset{(5.9.14)}{\approx} -\frac{m(R^2 - h^2)}{3}[\underset{(39)}{\dot{u}_1{}^*} + \Omega^2 (\bar{s}_1 + \bar{c}_1 q_1{}^*)\underset{(33)}{(\bar{c}_1 - \bar{s}_1 q_1{}^*)}]\hat{\mathbf{b}}_3 + \cdots$$

$$\approx -\frac{m(R^2 - h^2)}{3}\{\dot{u}_1{}^* + \Omega^2[\bar{c}_1\bar{s}_1 + (\bar{c}_1{}^2 - \bar{s}_1{}^2)q_1{}^*]\}\hat{\mathbf{b}}_3 + \cdots \tag{43}$$

where, once again, only the term that will survive [after dot multiplication, as per Eq. (5.9.7), with $\boldsymbol{\omega}_1$ as given in Eqs. (29)] has been written in detail. Now we are in position to form the generalized inertia force $F_1^\star$ as

$$F_1^\star \underset{(5.9.7)}{=} \boldsymbol{\omega}_1 \cdot \mathbf{T}^\star + \mathbf{v}_1^\star \cdot \underset{(5.9.11)}{(-m\mathbf{a}^\star)}$$

$$\approx -\frac{m(R^2 - h^2)}{3}\underset{(29,\ 43)}{\{\dot{u}_1{}^* + \Omega^2[\bar{c}_1\bar{s}_1 + (\bar{c}_1{}^2 - \bar{s}_1{}^2)q_1{}^*]\}}$$

$$- mh^2\underset{(29,\ 42)}{\{\dot{u}_1{}^* - \Omega^2[\bar{c}_1\bar{s}_1 + (\bar{c}_1{}^2 - \bar{s}_1{}^2)q_1{}^*]\}}$$

$$= -m\left\{\left(\frac{R^2 - h^2}{3} + h^2\right)\dot{u}_1{}^* + \Omega^2\left(\frac{R^2 - h^2}{3} - h^2\right)[\bar{c}_1\bar{s}_1 + (\bar{c}_1{}^2 - \bar{s}_1{}^2)q_1{}^*]\right\} \tag{44}$$

and the generalized active force F_1 is given by

$$F_1 \underset{(5.7.2)}{=} -mg(s_1\hat{\mathbf{b}}_1 + c_1\hat{\mathbf{b}}_2) \cdot \mathbf{v}_1^\star \underset{(29)}{=} -mghs_1$$

$$\approx -mgh\underset{(31)}{(\bar{s}_1 + \bar{c}_1 q_1{}^*)} \tag{45}$$

Finally, substitution from Eqs. (44) and (45) into Eqs. (8.1.1) yields

$$(R^2 + 2h^2)\dot{u}_1{}^* + \Omega^2(R^2 - 4h^2)[\bar{c}_1\bar{s}_1 + (\bar{c}_1{}^2 - \bar{s}_1{}^2)q_1{}^*] + 3gh(\bar{s}_1 + \bar{c}_1 q_1{}^*) \approx 0 \tag{46}$$

Now, this equation must be satisfied by $q_1{}^* = u_1{}^* = 0$. Otherwise, $q_1 = \bar{q}_1$ and $u_1 = 0$ cannot be solutions of the full, nonlinear equations governing q_1 and u_1. Hence, by setting $q_1{}^* = u_1{}^* = 0$ in Eq. (46), we arrive at the conclusion that $\bar{q}_1$ must satisfy the equation

$$\Omega^2(R^2 - 4h^2)\bar{c}_1\bar{s}_1 + 3gh\bar{s}_1 = 0 \tag{47}$$

Moreover, when this equation is satisfied, Eq. (46) reduces to

$$\dot{u}_1{}^* + \frac{\Omega^2(R^2 - 4h^2)(\bar{c}_1{}^2 - \bar{s}_1{}^2) + 3gh\bar{c}_1}{R^2 + 2h^2} q_1{}^* \approx 0 \tag{48}$$

This is the desired linearized dynamical equation.

Equation (47) possesses the physically distinct solutions $\bar{q}_1 = 0$ and $\bar{q}_1 = \pi$ regardless of the values of Ω, R, and h; when $\bar{q}_1 \neq 0$ and $\bar{q}_1 \neq \pi$, Eq. (47) requires that the cosine of $\bar{q}_1$ be given by

$$\bar{c}_1 = \frac{3gh}{\Omega^2(4h^2 - R^2)} \tag{49}$$

Now, real values of $\bar{q}_1$ satisfying this equation exist if and only if

$$-1 \leq \frac{3gh}{\Omega^2(4h^2 - R^2)} \leq 1 \tag{50}$$

Thus, B can move as postulated, either with $\bar{q}_1 = 0$ or $\bar{q}_1 = \pi$ and Ω, R, and h unrestricted, or with $\bar{q}_1$ governed by Eq. (49) and Ω, R, and h restricted by Eq. (50).

8.5 SYSTEMS AT REST IN A NEWTONIAN REFERENCE FRAME

Since rest is a special form of motion, Eqs. (8.1.1) and (8.1.2) apply to any system S at rest in a Newtonian reference frame N, reducing under these circumstances to

$$F_r = 0 \qquad (r = 1, \ldots, n) \tag{1}$$

and

$$\tilde{F}_r = 0 \qquad (r = 1, \ldots, p) \tag{2}$$

respectively. Moreover, if S possesses a potential energy V in N (see Sec. 7.1), then Eqs. (2) may be replaced with [see Eqs. (7.1.18)]

$$\sum_{s=1}^{n} \frac{\partial V}{\partial q_s}\left(W_{sr} + \sum_{k=p+1}^{n} W_{sk}A_{kr}\right) = 0 \qquad (r = 1, \ldots, p) \tag{3}$$

or with [see Eqs. (7.1.14)]

$$\frac{\partial V}{\partial q_r} + \sum_{s=p+1}^{n} \frac{\partial V}{\partial q_s} C_{sr} = 0 \qquad (r = 1, \ldots, p) \tag{4}$$

whereas Eqs. (1) are equivalent to [see Eqs. (7.1.9)]

$$\sum_{s=1}^{n} \frac{\partial V}{\partial q_s} W_{sr} = 0 \qquad (r = 1, \ldots, n) \tag{5}$$

or to [see Eqs. (7.1.2)]

$$\frac{\partial V}{\partial q_r} = 0 \qquad (r = 1, \ldots, n) \tag{6}$$

The last set of equations expresses the *principle of stationary potential energy*; Eqs. (3)–(5) represent generalizations of this principle.

Example Figure 8.5.1 shows a frame F supported by wheels W_1, W_2, and W_3, this assembly resting on a plane P that is inclined at an angle θ to the horizontal. W_1 can rotate freely relative to F about the axis of W_1; rotation of W_2 relative to F is resisted by a braking couple whose torque has a magnitude T; and W_3 is mounted in a fork that can rotate freely relative to F about the axis of the fork, which is normal to P, while W_3 is free to rotate relative to the fork about the axis of W_3.

The configuration of the system S formed by the frame, the fork, and the wheels is characterized completely by wheel rotation angles q_1, q_2, and q_3, a steering angle

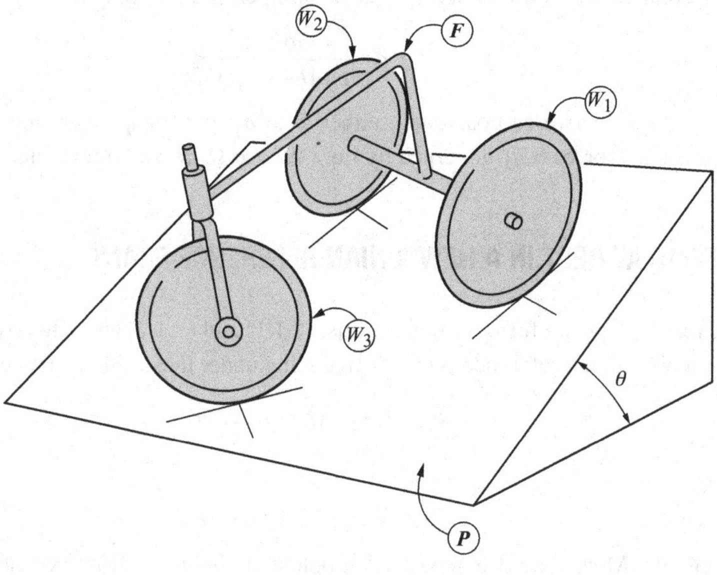

Figure 8.5.1

q_4, a frame orientation angle q_5, and two position coordinates, q_6 and q_7, defined as

$$q_6 \triangleq \mathbf{p}^\star \cdot \hat{\mathbf{n}}_1 \qquad q_7 \triangleq \mathbf{p}^\star \cdot \hat{\mathbf{n}}_2 \tag{7}$$

where $\mathbf{p}^\star$ is the position vector from a point O fixed in P to the mass center $S^\star$ of S, and $\hat{\mathbf{n}}_1$ and $\hat{\mathbf{n}}_2$ are, respectively, a horizontal unit vector and a unit vector pointing in the direction of steepest descent on P, as indicated in Fig. 8.5.2. Motion variables $u_1, \ldots, u_7$ may be defined as

$$u_1 \triangleq \dot{q}_3 \qquad u_2 \triangleq \dot{q}_4 \qquad u_3 \triangleq \dot{q}_5 \tag{8}$$

$$u_4 \triangleq \dot{q}_1 \qquad u_5 \triangleq \dot{q}_2 \qquad u_6 \triangleq \mathbf{v}^\star \cdot \hat{\mathbf{f}}_1 \qquad u_7 \triangleq \mathbf{v}^\star \cdot \hat{\mathbf{f}}_2 \tag{9}$$

where $\mathbf{v}^\star$ is the velocity of $S^\star$, and $\hat{\mathbf{f}}_1$ and $\hat{\mathbf{f}}_2$ are unit vectors directed as indicated in Fig. 8.5.2.

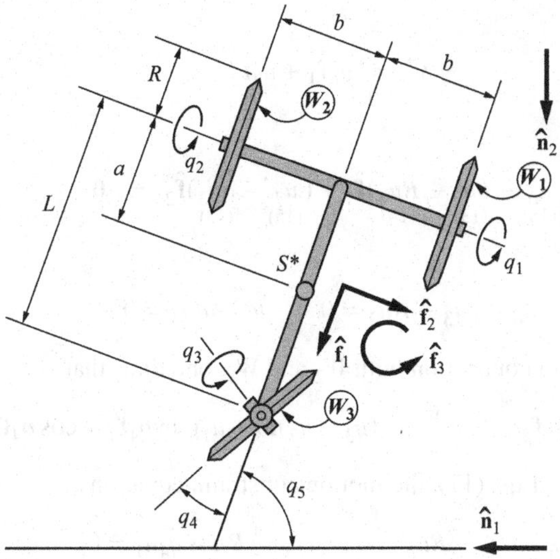

Figure 8.5.2

On the basis of the assumption that W_1, W_2, and W_3 roll, rather than slip, on P, equations are to be formulated for the purpose of determining how T is related to the inclination angle θ, the total mass M of S, the dimensions R, a, b, and L (see Fig. 8.5.2), and the generalized coordinates $q_1, \ldots, q_7$ when S is at rest.

The assumption that W_1, W_2, and W_3 do not slip on P leads to five motion constraint equations (see Sec. 3.5). Hence, S possesses two degrees of freedom, and Eqs. (2) can be written with $p = 2$. To form $\widetilde{F}_r$ $(r = 1, 2)$, we write

$$\widetilde{F}_r \underset{(5.5.2)}{=} Mg\hat{\mathbf{k}} \cdot \widetilde{\mathbf{v}}_r^\star \underset{(5.7.2)}{+} T\hat{\mathbf{f}}_2 \cdot \widetilde{\boldsymbol{\omega}}_r^F - T\hat{\mathbf{f}}_2 \cdot \widetilde{\boldsymbol{\omega}}_r^{W_2} \qquad (r = 1, 2) \tag{10}$$

where $\hat{\mathbf{k}}$ is a unit vector directed vertically downward; that is, with $\hat{\mathbf{f}}_3 \triangleq \hat{\mathbf{f}}_1 \times \hat{\mathbf{f}}_2$,

$$\hat{\mathbf{k}} = \sin q_5 \sin\theta \hat{\mathbf{f}}_1 + \cos q_5 \sin\theta \hat{\mathbf{f}}_2 - \cos\theta \hat{\mathbf{f}}_3 \tag{11}$$

The partial velocities and partial angular velocities appearing in Eqs. (10) are found as follows.

The velocities of the points of W_1, W_2, and W_3 that are in contact with P must be equal to zero. In the case of W_1 this requirement can be expressed as [use Eq. (2.7.1) twice]

$$\mathbf{v}^\star + \boldsymbol{\omega}^F \times (-a\hat{\mathbf{f}}_1 + b\hat{\mathbf{f}}_2) + \boldsymbol{\omega}^{W_1} \times (-R\hat{\mathbf{f}}_3) = \mathbf{0} \tag{12}$$

where

$$\boldsymbol{\omega}^F \underset{(8)}{=} u_3\hat{\mathbf{f}}_3 \tag{13}$$

and

$$\boldsymbol{\omega}^{W_1} \underset{(2.4.1)}{=} \boldsymbol{\omega}^F + \dot{q}_1\hat{\mathbf{f}}_2 = \underset{(13)}{u_3}\hat{\mathbf{f}}_3 + \underset{(9)}{u_4}\hat{\mathbf{f}}_2 \tag{14}$$

Moreover,

$$\mathbf{v}^\star \underset{(9)}{=} u_6\hat{\mathbf{f}}_1 + u_7\hat{\mathbf{f}}_2 \tag{15}$$

Hence,

$$(\underset{(15)}{u_6} - \underset{(13)}{bu_3} - \underset{(14)}{Ru_4})\hat{\mathbf{f}}_1 + (\underset{(15)}{u_7} - \underset{(13)}{au_3})\hat{\mathbf{f}}_2 \underset{(12)}{=} \mathbf{0} \tag{16}$$

which means that

$$u_6 - bu_3 - Ru_4 = 0 \qquad u_7 - au_3 = 0 \tag{17}$$

Proceeding similarly in connection with W_2 and W_3, one finds that

$$\boldsymbol{\omega}^{W_2} = u_3\hat{\mathbf{f}}_3 + u_5\hat{\mathbf{f}}_2 \qquad \boldsymbol{\omega}^{W_3} = (u_3 - u_2)\hat{\mathbf{f}}_3 + u_1(\sin q_4\hat{\mathbf{f}}_1 + \cos q_4\hat{\mathbf{f}}_2) \tag{18}$$

and that, in addition to Eqs. (17), the motion constraint equations

$$u_6 + bu_3 - Ru_5 = 0 \qquad u_6 - R\cos q_4 u_1 = 0 \tag{19}$$

and

$$u_7 + (L - a)u_3 + R\sin q_4 u_1 = 0 \tag{20}$$

must be satisfied. Solved simultaneously for u_3 and u_7, Eq. (20) and the second of Eqs. (17) yield

$$u_3 = -\frac{R}{L}\sin q_4 u_1 \qquad u_7 = -\frac{aR}{L}\sin q_4 u_1 \tag{21}$$

The second of Eqs. (19) shows that

$$u_6 = R\cos q_4 u_1 \tag{22}$$

and, solving the first of Eqs. (19) for u_5 with the aid of Eq. (22) and the first of Eqs. (21), one obtains

$$u_5 = \left(\cos q_4 - \frac{b}{L}\sin q_4\right)u_1 \tag{23}$$

Hence,

$$\mathbf{v}^\star \underset{(15)}{=} \underset{(22)}{R\cos q_4 u_1}\hat{\mathbf{f}}_1 - \underset{(21)}{\frac{aR}{L}\sin q_4 u_1}\hat{\mathbf{f}}_2 \tag{24}$$

$$\boldsymbol{\omega}^F \underset{(13)}{=} -\underset{(21)}{\frac{R}{L}\sin q_4 u_1}\hat{\mathbf{f}}_3 \tag{25}$$

and

$$\boldsymbol{\omega}^{W_2} = \underset{(21)}{-\frac{R}{L}\sin q_4 u_1}\hat{\mathbf{f}}_3 + \underset{(23)}{\left(\cos q_4 - \frac{b}{L}\sin q_4\right)u_1}\hat{\mathbf{f}}_2 \tag{26}$$

The partial velocities of $S^\star$ and the partial angular velocities of F and W_2 needed for substitution into Eqs. (10) thus are given by [see Eqs. (24)–(26)]

$$\tilde{\mathbf{v}}_1^\star = R\cos q_4\hat{\mathbf{f}}_1 - \frac{aR}{L}\sin q_4\hat{\mathbf{f}}_2 \tag{27}$$

$$\tilde{\boldsymbol{\omega}}_1{}^F = -\frac{R}{L}\sin q_4\hat{\mathbf{f}}_3 \tag{28}$$

$$\tilde{\boldsymbol{\omega}}_1{}^{W_2} = \left(\cos q_4 - \frac{b}{L}\sin q_4\right)\hat{\mathbf{f}}_2 - \frac{R}{L}\sin q_4\hat{\mathbf{f}}_3 \tag{29}$$

and

$$\tilde{\mathbf{v}}_2^\star = \tilde{\boldsymbol{\omega}}_2{}^F = \tilde{\boldsymbol{\omega}}_2{}^{W_2} = \mathbf{0} \tag{30}$$

Substituting these into Eqs. (10) with $r = 1$, and setting the result equal to zero in accordance with Eqs. (2), one finds with the aid of Eq. (11) that

$$MgR\sin\theta\left(\cos q_4\sin q_5 - \frac{a}{L}\sin q_4\cos q_5\right) - T\left(\cos q_4 - \frac{b}{L}\sin q_4\right) = 0 \tag{31}$$

With $r = 2$, Eqs. (10) yield $\tilde{F}_2 = 0$, so that Eqs. (2) are satisfied identically in this case. Equation (31) is the desired relationship between T, θ, M, R, a, b, L, and the generalized coordinates $q_1,\ldots,q_7$.

8.6 STEADY MOTION

A simple nonholonomic system S possessing p degrees of freedom in a Newtonian reference frame N is said to be in a state of *steady motion* in N when the generalized velocities $u_1,\ldots,u_p$ have constant values, say, $\bar{u}_1,\ldots,\bar{u}_p$, respectively. To determine the conditions under which steady motions can occur, use Eqs. (8.1.1) or (8.1.2), proceeding as follows: Form expressions for angular velocities of rigid bodies belonging to S, velocities of mass centers of these bodies, and so forth, *without* regard to the fact that $u_1,\ldots,u_p$ are to remain constant, and use these expressions to construct partial angular velocities and partial velocities. Set $u_r = \bar{u}_r$ $(r = 1,\ldots,p)$ in angular velocity and velocity expressions, then differentiate with respect to time to generate needed angular accelerations of rigid bodies and accelerations of various points. Formulate expressions for $\tilde{F}_r$ and $\tilde{F}_r^\star$ $(r = 1,\ldots,p)$ in the case of Eqs. (8.1.2), or F_r and $F_r^\star$ $(r = 1,\ldots,n)$ in the case of Eqs. (8.1.1), and substitute into Eqs. (8.1.2) or (8.1.1).

Example Figure 8.6.1 shows a right-circular, uniform, solid cone C in contact with a fixed, horizontal plane P. The motion that C performs —when C rolls on P in such a way that the mass center $C^\star$ of C (see Fig. 8.6.1) remains fixed while the plane determined by the axis of C and a vertical line passing through $C^\star$ has an angular velocity $-\Omega\hat{\mathbf{k}}$ (Ω constant)— is a steady motion, as will be shown presently. But this motion can take place only if Ω, the radius R of the base of C, the height $4h$ of C, and the inclination angle θ (see Fig. 8.6.1) are related to each other suitably. To determine the conditions under which the motion is possible, we begin by noting that C has three degrees of freedom, and introduce generalized velocities u_1, u_2, and u_3 as

$$u_r \triangleq \boldsymbol{\omega} \cdot \hat{\mathbf{b}}_r \qquad (r = 1,2,3) \tag{1}$$

where $\boldsymbol{\omega}$ is the angular velocity of C, and $\hat{\mathbf{b}}_1$, $\hat{\mathbf{b}}_2$, and $\hat{\mathbf{b}}_3$ are mutually perpendicular unit vectors directed as indicated in Fig. 8.6.1. (Note that $\hat{\mathbf{b}}_2$ and $\hat{\mathbf{b}}_3$ are *not* fixed in C.)

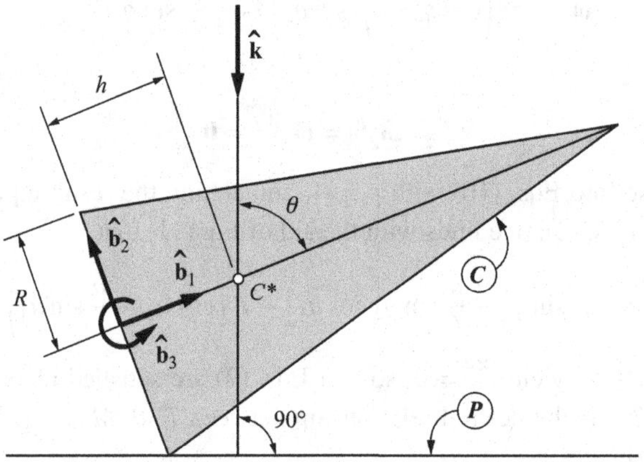

Figure 8.6.1

As is pointed out in Problem 3.12, $\boldsymbol{\omega}$ is given by

$$\boldsymbol{\omega} = \Omega s_\theta \left(\frac{h}{R}\hat{\mathbf{b}}_1 + \hat{\mathbf{b}}_2 \right) \tag{2}$$

when C moves as required. Hence, u_1, u_2, and u_3 then have the constant values

$$\bar{u}_1 \underset{(1)}{=} \Omega s_\theta \underset{(2)}{\frac{h}{R}} \qquad \bar{u}_2 \underset{(1)}{=} \underset{(2)}{\Omega s_\theta} \qquad \bar{u}_3 \underset{(1)}{=} \underset{(2)}{0} \tag{3}$$

respectively.

The generalized active forces $\tilde{F}_r$ $(r = 1,2,3)$ are expressed most conveniently as

$$\tilde{F}_r \underset{(5.7.2)}{=} Mg\hat{\mathbf{k}} \cdot \tilde{\mathbf{v}}_r^\star \qquad (r = 1,2,3) \tag{4}$$

where M is the mass of C and

$$\hat{\mathbf{k}} = -(c_\theta \hat{\mathbf{b}}_1 + s_\theta \hat{\mathbf{b}}_2) \tag{5}$$

As for the generalized inertia forces $\tilde{F}_r^\star$ $(r = 1,2,3)$, we write

$$\tilde{F}_r^\star \underset{(5.9.8)}{=} \tilde{\boldsymbol{\omega}}_r \cdot \mathbf{T}^\star + \tilde{\mathbf{v}}_r^\star \cdot \mathbf{R}^\star \qquad (r = 1,2,3) \tag{6}$$

and defer detailed consideration of $\mathbf{T}^\star$ and $\mathbf{R}^\star$ until after expressions for $\tilde{\boldsymbol{\omega}}_r$ and $\tilde{\mathbf{v}}_r^\star$ have been constructed.

From Eqs. (1),

$$\boldsymbol{\omega} = u_1\hat{\mathbf{b}}_1 + u_2\hat{\mathbf{b}}_2 + u_3\hat{\mathbf{b}}_3 \tag{7}$$

and, when C rolls on P, the velocity $\mathbf{v}^\star$ of $C^\star$ is given by

$$\begin{aligned}\mathbf{v}^\star &= \boldsymbol{\omega} \times (h\hat{\mathbf{b}}_1 + R\hat{\mathbf{b}}_2) \\ &\underset{(7)}{=} -Ru_3\hat{\mathbf{b}}_1 + hu_3\hat{\mathbf{b}}_2 + (Ru_1 - hu_2)\hat{\mathbf{b}}_3\end{aligned} \tag{8}$$

The partial angular velocities and partial velocities obtained by inspection of Eqs. (7) and (8), respectively, are recorded in Table 8.6.1.

Table 8.6.1

r	$\tilde{\boldsymbol{\omega}}_r$	$\tilde{\mathbf{v}}_r^\star$
1	$\hat{\mathbf{b}}_1$	$R\hat{\mathbf{b}}_3$
2	$\hat{\mathbf{b}}_2$	$-h\hat{\mathbf{b}}_3$
3	$\hat{\mathbf{b}}_3$	$-R\hat{\mathbf{b}}_1 + h\hat{\mathbf{b}}_2$

The steady motion form of $\boldsymbol{\omega}$ is available in Eq. (2). Differentiating both sides of this equation with respect to time in order to find $\boldsymbol{\alpha}$, the angular acceleration of C, we obtain

$$\boldsymbol{\alpha} = \boldsymbol{\omega}^B \times \left[\Omega s_\theta \left(\frac{h}{R}\hat{\mathbf{b}}_1 + \hat{\mathbf{b}}_2\right)\right] \tag{9}$$

where $\boldsymbol{\omega}^B$, the angular velocity of a reference frame in which $\hat{\mathbf{b}}_1$, $\hat{\mathbf{b}}_2$, and $\hat{\mathbf{b}}_3$ are fixed, is given by

$$\boldsymbol{\omega}^B = \Omega(c_\theta\hat{\mathbf{b}}_1 + s_\theta\hat{\mathbf{b}}_2) \tag{10}$$

Consequently,

$$\boldsymbol{\alpha} \underset{(9,\,10)}{=} \Omega^2 s_\theta \left(c_\theta - \frac{h}{R}s_\theta\right)\hat{\mathbf{b}}_3 \tag{11}$$

The velocity $\mathbf{v}^\star$ is equal to zero, by hypothesis, during the steady motion of interest, a result one can verify by setting $u_r = \bar{u}_r$ $(r = 1, 2, 3)$ in Eq. (8) and then using Eqs. (3). It follows that $\mathbf{a}^\star$, the acceleration of $C^\star$, vanishes, and this means that the inertia force $\mathbf{R}^\star$ appearing in Eqs. (6) also vanishes. As for the inertia torque $\mathbf{T}^\star$, one can express this as

$$\mathbf{T}^\star = \Omega^2 s_\theta \left(I_1 \frac{h}{R}s_\theta - I_2 c_\theta\right)\hat{\mathbf{b}}_3 \tag{12}$$

by substituting from Eqs. (11) and (2) into Eq. (5.9.12), with

$$\underline{\mathbf{I}} = I_1\hat{\mathbf{b}}_1\hat{\mathbf{b}}_1 + I_2(\hat{\mathbf{b}}_2\hat{\mathbf{b}}_2 + \hat{\mathbf{b}}_3\hat{\mathbf{b}}_3) \tag{13}$$

Expressed in terms of M, R, and h, the moments of inertia I_1 and I_2 are given by (see

Appendix III)

$$I_1 = \frac{3MR^2}{10} \qquad I_2 = \frac{3M(R^2 + 4h^2)}{20} \tag{14}$$

Referring to Eq. (5) and Table 8.6.1, one finds by substitution into Eqs. (4) that

$$\widetilde{F}_1 = \widetilde{F}_2 = 0 \qquad \widetilde{F}_3 = Mg(Rc_\theta - hs_\theta) \tag{15}$$

Similarly, Eqs. (6) lead to

$$\widetilde{F}_1^\star = \widetilde{F}_2^\star = 0 \qquad \widetilde{F}_3^\star = \Omega^2 s_\theta \left(I_1 \frac{h}{R} s_\theta - I_2 c_\theta \right) \tag{16}$$

Consequently, Eqs. (8.1.2) are satisfied identically when $r = 1$ and $r = 2$; for $r = 3$, substitution from Eqs. (15), (16), and (14) into Eqs. (8.1.2) leads to the conclusion that the steady motion under consideration can take place only when

$$\frac{3}{20}\left(\frac{R\Omega^2}{g}\right)\left\{\left[1 + 4\left(\frac{h}{R}\right)^2\right] c_\theta - 2\left(\frac{h}{R}\right) s_\theta \right\} s_\theta + \frac{h}{R} s_\theta - c_\theta = 0 \tag{17}$$

8.7 MOTIONS RESEMBLING STATES OF REST

A simple nonholonomic system S possessing p degrees of freedom in a Newtonian reference frame N is said to be performing a *motion resembling a state of rest* when the generalized coordinates $q_1, \ldots, q_n$ have constant values, say, $\overline{q}_1, \ldots, \overline{q}_n$, respectively. To determine the conditions under which such motions are possible, one can use Eqs. (8.1.1) or (8.1.2), employing a procedure analogous to that set forth in Sec. 8.6. Alternatively, *if the kinetic energy* K, when regarded as a function of $q_1, \ldots, q_n, \dot{q}_1, \ldots, \dot{q}_n$, and t, *does not involve* t *explicitly*, then one can use the equations

$$\sum_{s=1}^{n} \frac{\partial K}{\partial q_s}\left(W_{sr} + \sum_{k=p+1}^{n} W_{sk} A_{kr} \right) + \widetilde{F}_r = 0 \qquad (r = 1, \ldots, p) \tag{1}$$

where W_{sr} $(s = 1, \ldots, n;\ r = 1, \ldots, p)$ and A_{kr} $(k = p+1, \ldots, n;\ r = 1, \ldots, p)$ have the same meanings as in Eqs. (3.6.5) and (3.5.2), respectively, and $\widetilde{F}_r$ $(r = 1, \ldots, p)$ are defined in Eqs. (5.4.2). If $u_1, \ldots, u_n$ are defined as $u_r = \dot{q}_r$ $(r = 1, \ldots, n)$, then Eqs. (1) give way to

$$\frac{\partial K}{\partial q_r} + \sum_{s=p+1}^{n} \frac{\partial K}{\partial q_s} C_{sr} + \widetilde{F}_r = 0 \qquad (r = 1, \ldots, p) \tag{2}$$

where C_{sr} $(s = p+1, \ldots, n;\ r = 1, \ldots, p)$ has the same meaning as in Eq. (7.1.13).

If S is a holonomic system with u_r defined as in Eqs. (3.4.1), so that Eqs. (3.6.5) apply, then

$$\sum_{s=1}^{n} \frac{\partial K}{\partial q_s} W_{sr} + F_r = 0 \qquad (r = 1, \ldots, n) \tag{3}$$

during motions resembling states of rest; and if $u_r = \dot{q}_r$ $(r = 1,\ldots,n)$, then Eqs. (3) are replaced with

$$\frac{\partial K}{\partial q_r} + F_r = 0 \qquad (r = 1,\ldots,n) \tag{4}$$

When using Eqs. (1)–(4), one can work with expressions for K, $\widetilde{F}_r$ $(r = 1,\ldots,p)$, or F_r $(r = 1,\ldots,n)$ that apply only when $\dot{q}_1 = \ldots = \dot{q}_n = 0$, rather than during the most general motion of S; generally, this simplifies matters considerably.

If S possesses a potential energy V in N, and $\mathscr{L}$ is defined as

$$\mathscr{L} \triangleq K - V \tag{5}$$

then Eqs. (1)–(4) can be replaced with

$$\sum_{s=1}^{n} \frac{\partial \mathscr{L}}{\partial q_s}\left(W_{sr} + \sum_{k=p+1}^{n} W_{sk}A_{kr}\right) = 0 \qquad (r = 1,\ldots,p) \tag{6}$$

$$\frac{\partial \mathscr{L}}{\partial q_r} + \sum_{s=p+1}^{n} \frac{\partial \mathscr{L}}{\partial q_s} C_{sr} = 0 \qquad (r = 1,\ldots,p) \tag{7}$$

$$\sum_{s=1}^{n} \frac{\partial \mathscr{L}}{\partial q_s} W_{sr} = 0 \qquad (r = 1,\ldots,n) \tag{8}$$

$$\frac{\partial \mathscr{L}}{\partial q_r} = 0 \qquad (r = 1,\ldots,n) \tag{9}$$

respectively. Use of these equations obviates forming expressions for accelerations.

Derivations If K, when regarded as a function of $q_1,\ldots,q_n$, $\dot{q}_1,\ldots,\dot{q}_n$, and t, does not involve t explicitly, then the partial derivatives $\partial K/\partial \dot{q}_r$ $(r = 1,\ldots,n)$ are functions f_r $(r = 1,\ldots,n)$ of precisely the $2n$ variables q_r and $\dot{q}_r$ $(r = 1,\ldots,n)$, so that one can write

$$\frac{\partial K}{\partial \dot{q}_r} = f_r(q_1,\ldots,q_n, \dot{q}_1,\ldots,\dot{q}_n) \qquad (r = 1,\ldots,n) \tag{10}$$

and

$$\frac{d}{dt}\left(\frac{\partial K}{\partial \dot{q}_r}\right) = \sum_{s=1}^{n}\left(\frac{\partial f_r}{\partial q_s}\dot{q}_s + \frac{\partial f_r}{\partial \dot{q}_s}\ddot{q}_s\right) \qquad (r = 1,\ldots,n) \tag{11}$$

But, since $q_s = \overline{q}_s$ $(s = 1,\ldots,n)$ by hypothesis,

$$\dot{q}_s = \ddot{q}_s = 0 \qquad (s = 1,\ldots,n) \tag{12}$$

so that

$$\frac{d}{dt}\left(\frac{\partial K}{\partial \dot{q}_r}\right) \underset{(11,\ 12)}{=} 0 \qquad (r = 1,\ldots,n) \tag{13}$$

and use of these relationships in conjunction with the four equations in Problem 12.14 (taken in reverse order) leads directly to Eqs. (1)–(4).

When S possesses a potential energy V in N, $\widetilde{F}_r$ in Eqs. (1) can be replaced with the right-hand member of Eq. (7.1.18), which shows that

$$\sum_{s=1}^{n}\left(\frac{\partial K}{\partial q_s}-\frac{\partial V}{\partial q_s}\right)\left(W_{sr}+\sum_{k=p+1}^{n}W_{sk}A_{kr}\right)=0 \qquad (r=1,\ldots,p) \tag{14}$$

under these circumstances. Now,

$$\frac{\partial K}{\partial q_s}-\frac{\partial V}{\partial q_s}=\frac{\partial}{\partial q_s}(K-V)\underset{(5)}{=}\frac{\partial \mathscr{L}}{\partial q_s} \qquad (s=1,\ldots,n) \tag{15}$$

Substitution from Eqs. (15) into Eqs. (14) produces Eqs. (6). Similarly, Eqs. (2), (15), and (7.1.14) lead to Eqs. (7); Eqs. (3), (15), and (7.1.9) underlie Eqs. (8); and Eqs. (4), (15), and (7.1.2) can be used to establish the validity of Eqs. (9).

Example Two uniform bars, A and B, each of mass m and length L, are connected by a pin, and A is pinned to a vertical shaft that is made to rotate with constant angular speed Ω, as indicated in Fig. 8.7.1. (The axes of the pins are parallel to each other and horizontal.) This system can move in such a way that q_1 and q_2 (see Fig. 8.7.1) remain constant. Equations (9) will be used to formulate equations governing q_1 and q_2 during such motions.

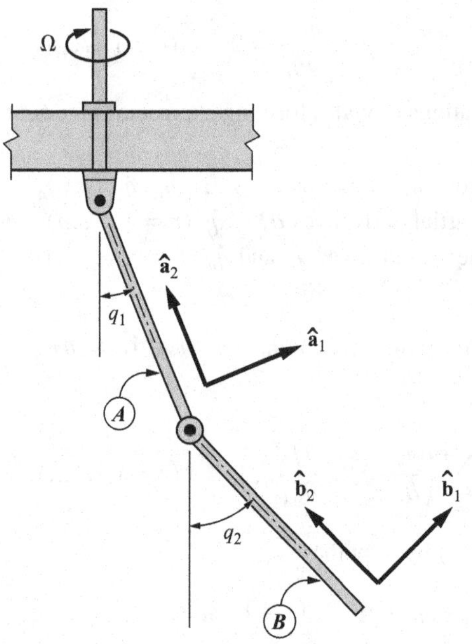

Figure 8.7.1

When $\dot{q}_1=\dot{q}_2=0$, the angular velocities of A and B can be written

$$\boldsymbol{\omega}^A=-\Omega(s_1\hat{\mathbf{a}}_1+c_1\hat{\mathbf{a}}_2) \qquad \boldsymbol{\omega}^B=-\Omega(s_2\hat{\mathbf{b}}_1+c_2\hat{\mathbf{b}}_2) \tag{16}$$

where $\hat{\mathbf{a}}_i$ and $\hat{\mathbf{b}}_i$ $(i = 1,2)$ are unit vectors directed as shown in Fig. 8.7.1; the velocity of $B^\star$, the mass center of B, is

$$\mathbf{v}^{B^\star} = L\Omega\left(s_1 + \frac{s_2}{2}\right)\hat{\mathbf{b}}_3 \tag{17}$$

and the kinetic energy K is thus given by

$$K = \underset{\text{(Problem 11.4, 16) (7.4.2)}}{\frac{1}{2}\frac{mL^2}{3}\Omega^2 {s_1}^2} + \underset{(7.4.7,\ 16)}{\frac{1}{2}\frac{mL^2}{12}\Omega^2 {s_2}^2} + \underset{(7.4.4,\ 17)}{\frac{1}{2}mL^2\Omega^2\left(s_1 + \frac{s_2}{2}\right)^2}$$

$$= \frac{mL^2\Omega^2}{2}\left(\frac{4}{3}{s_1}^2 + s_1 s_2 + \frac{1}{3}{s_2}^2\right) \tag{18}$$

The potential energy V can be expressed as

$$V = -mgL\left(\frac{1}{2}c_1 + c_1 + \frac{1}{2}c_2\right) = -\frac{mgL}{2}(3c_1 + c_2) \tag{19}$$

Consequently,

$$\mathscr{L} \underset{(5)}{=} \frac{mL^2\Omega^2}{2}\underset{(18)}{\left(\frac{4}{3}{s_1}^2 + s_1 s_2 + \frac{1}{3}{s_2}^2\right)} + \underset{(19)}{\frac{mgL}{2}(3c_1 + c_2)} \tag{20}$$

and, setting $\partial\mathscr{L}/\partial q_r = 0$ $(r = 1,2)$, one obtains the desired equations, namely,

$$\left(\frac{L\Omega^2}{g}\right)c_1(8s_1 + 3s_2) - 9s_1 = 0 \tag{21}$$

and

$$\left(\frac{L\Omega^2}{g}\right)c_2(3s_1 + 2s_2) - 3s_2 = 0 \tag{22}$$

8.8 GENERALIZED IMPULSE, GENERALIZED MOMENTUM

When a system S is subjected to the action of forces that become very large during a very short time interval, the velocities of certain particles of S may change substantially during this time interval while the configuration of S in a Newtonian reference frame N remains essentially unaltered. This happens, for example, when two relatively inflexible bodies collide with each other. Although such phenomena may appear to be more complex than motions that proceed more smoothly, they frequently can be treated analytically with comparatively simple methods because the presumption that the configuration of S does not change enables one to integrate Eqs. (8.1.1) or (8.1.2) in general terms and thus to construct a theory involving algebraic rather than differential equations. To this end, two sets of quantities, called generalized impulses and generalized momenta, are defined as follows.

Suppose that S is a nonholonomic system possessing n generalized coordinates $q_1, \ldots, q_n$ (see Sec. 3.2) and $n - m$ independent motion variables $u_1, \ldots, u_{n-m}$ (see Sec. 3.4) in N. Let $\tilde{\mathbf{v}}_r^{P_i}$ be the rth nonholonomic partial velocity in N of a generic particle P_i of S (see Sec. 3.6); let $\mathbf{R}_i$ be the resultant of all contact forces and distance forces acting on P_i; and let t_1 and t_2 be the initial and final instants of a time interval such that $q_1, \ldots, q_n$ can be regarded as constant throughout this interval. The *generalized impulse* I_r is defined as

$$I_r \triangleq \sum_{i=1}^{\nu} \tilde{\mathbf{v}}_r^{P_i}(t_1) \cdot \int_{t_1}^{t_2} \mathbf{R}_i \, dt \qquad (r = 1, \ldots, n - m) \tag{1}$$

where ν is the number of particles of S, and the *generalized momentum* p_r is defined as

$$p_r(t) \triangleq \sum_{i=1}^{\nu} m_i \, \tilde{\mathbf{v}}_r^{P_i}(t) \cdot \mathbf{v}^{P_i}(t) \qquad (r = 1, \ldots, n - m) \tag{2}$$

where m_i is the mass of P_i $(i = 1, \ldots, \nu)$.

When forming expressions for I_r $(r = 1, \ldots, n - m)$ in accordance with Eqs. (1), one regards as negligible the contributions to the time integral of $\mathbf{R}_i$ of all forces that remain constant during the time interval beginning at t_1 and ending at t_2. Moreover, I_r can be expressed as

$$I_r \approx \int_{t_1}^{t_2} \tilde{F}_r \, dt \qquad (r = 1, \ldots, n - m) \tag{3}$$

where $\tilde{F}_1, \ldots, \tilde{F}_{n-m}$ are the nonholonomic generalized active forces for S in N (see Sec. 5.4). Consequently, forces that make no contributions to generalized active forces (see Sec. 6.6) also make no contributions to generalized impulses.

The constructing of expressions for generalized momenta often is simplified by the following fact. If K, the kinetic energy of S in N (see Sec. 7.4), is regarded as a function of $q_1, \ldots, q_n, u_1, \ldots, u_{n-m}$, and t (see Sec. 7.5), then

$$p_r = \frac{\partial K}{\partial u_r} \qquad (r = 1, \ldots, n - m) \tag{4}$$

The aforementioned integration of Eqs. (8.1.2) from t_1 to t_2 results in the relationships

$$I_r \approx p_r(t_2) - p_r(t_1) \qquad (r = 1, \ldots, n - m) \tag{5}$$

When S is a holonomic system, m is set equal to zero, tildes are omitted from Eqs. (1)–(3), and integration of Eqs. (8.1.1), rather than (8.1.2), leads to Eqs. (5).

Derivations In accordance with Eqs. (5.4.2),

$$\int_{t_1}^{t_2} \tilde{F}_r \, dt = \int_{t_1}^{t_2} \sum_{i=1}^{\nu} \tilde{\mathbf{v}}_r^{P_i} \cdot \mathbf{R}_i \, dt = \sum_{i=1}^{\nu} \int_{t_1}^{t_2} \tilde{\mathbf{v}}_r^{P_i} \cdot \mathbf{R}_i \, dt \qquad (r = 1, \ldots, n - m) \tag{6}$$

In general, $\tilde{\mathbf{v}}_r^{P_i}$ is a function of t in N. But if $t_2 \approx t_1$ and $q_r(t_2) \approx q_r(t_1)$, then $\tilde{\mathbf{v}}_r^{P_i}$ is nearly fixed in N (and nearly equal to its value at time t_1) throughout the time interval beginning at t_1 and ending at t_2, and

$$\int_{t_1}^{t_2} \tilde{F}_r \, dt \underset{(6)}{\approx} \sum_{i=1}^{\nu} \tilde{\mathbf{v}}_r^{P_i}(t_1) \cdot \int_{t_1}^{t_2} \mathbf{R}_i \, dt \underset{(1)}{=} I_r \qquad (r = 1, \ldots, n - m) \tag{7}$$

in agreement with Eqs. (3).

To establish the validity of Eqs. (4), one can write

$$\frac{\partial K}{\partial u_r} \underset{(7.4.1)}{=} \frac{\partial}{\partial u_r}\left[\frac{1}{2}\sum_{i=1}^{\nu} m_i (\mathbf{v}^{P_i})^2\right] = \sum_{i=1}^{\nu} m_i \frac{\partial \mathbf{v}^{P_i}}{\partial u_r} \cdot \mathbf{v}^{P_i}$$

$$\underset{(3.6.4)}{=} \sum_{i=1}^{\nu} m_i \, \tilde{\mathbf{v}}_r^{P_i} \cdot \mathbf{v}^{P_i} \underset{(2)}{=} p_r \qquad (r = 1, \ldots, n-m) \tag{8}$$

Finally, integration of Eqs. (5.9.2) produces

$$\int_{t_1}^{t_2} \tilde{F}_r^\star \, dt = \int_{t_1}^{t_2} \sum_{i=1}^{\nu} \tilde{\mathbf{v}}_r^{P_i} \underset{(5.9.4)}{\cdot} (-m_i \mathbf{a}_i) \, dt$$

$$= -\sum_{i=1}^{\nu} m_i \int_{t_1}^{t_2} \tilde{\mathbf{v}}_r^{P_i} \cdot \frac{{}^N d\mathbf{v}^{P_i}}{dt} \, dt$$

$$\approx -\sum_{i=1}^{\nu} m_i \, \tilde{\mathbf{v}}_r^{P_i}(t_1) \cdot \int_{t_1}^{t_2} \frac{{}^N d\mathbf{v}^{P_i}}{dt} \, dt$$

$$= -\sum_{i=1}^{\nu} m_i \, \tilde{\mathbf{v}}_r^{P_i}(t_1) \cdot [\mathbf{v}^{P_i}(t_2) - \mathbf{v}^{P_i}(t_1)]$$

$$\underset{(2)}{=} -[p_r(t_2) - p_r(t_1)] \qquad (r = 1, \ldots, n-m) \tag{9}$$

Consequently, integration of Eqs. (8.1.2) yields

$$\int_{t_1}^{t_2} \tilde{F}_r \, dt + \int_{t_1}^{t_2} \tilde{F}_r^\star \, dt \approx \underset{(7)}{I_r} - \underset{(9)}{[p_r(t_2) - p_r(t_1)]} \underset{(8.1.2)}{=} 0 \qquad (r = 1, \ldots, n-m) \tag{10}$$

or, equivalently, Eqs. (5).

Example Figure 8.8.1 shows a gear train consisting of three identical gears G_1, G_2, and G_3, each having a radius b and a moment of inertia J about its axis. The angular speed Ω of G_1 is to be determined on the basis of the assumption that the gears are set into motion suddenly when G_1 is meshed with a gear G' of radius b' and axial moment of inertia J', G' having an angular speed Ω' at the instant of first contact between G_1 and G'.

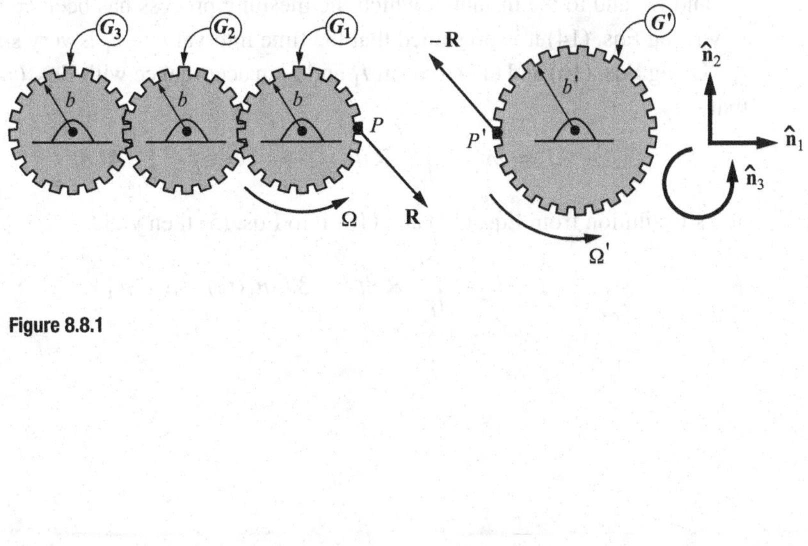

Figure 8.8.1

Before G_1 and G' are brought into contact, the system S formed by G_1, G_2, G_3, and G' possesses two degrees of freedom, and, if motion variables u_1 and u_2 are defined in terms of $\boldsymbol{\omega}^{G_1}$ and $\boldsymbol{\omega}^{G'}$, the angular velocities of G_1 and G', respectively, as

$$u_1 \triangleq \boldsymbol{\omega}^{G_1} \cdot \hat{\mathbf{n}}_3 \qquad u_2 \triangleq \boldsymbol{\omega}^{G'} \cdot \hat{\mathbf{n}}_3 \tag{11}$$

where $\hat{\mathbf{n}}_3$ is one of three mutually perpendicular unit vectors directed as shown in Fig. 8.8.1, then the kinetic energy K of S is given by

$$K \underset{(7.4.5)}{=} \tfrac{3}{2} J u_1{}^2 + \tfrac{1}{2} J' u_2{}^2 \tag{12}$$

Consequently, the generalized momenta p_1 and p_2 are

$$p_1 \underset{(4)}{=} \frac{\partial K}{\partial u_1} \underset{(12)}{=} 3Ju_1 \qquad p_2 \underset{(4)}{=} \frac{\partial K}{\partial u_2} \underset{(12)}{=} J'u_2 \tag{13}$$

When G_1 and G' are meshed suddenly, large forces come into play at the points P and P' (see Fig. 8.8.1) where G_1 and G' come into contact with each other, as well as at the points of contact between G_1 and G_2, G_2 and G_3, and each of G_1, G_2, G_3, and G' with its supports. Of all of these forces, the only ones that make contributions to generalized active forces (see Sec. 5.4), and hence to generalized impulses, are the interaction forces $\mathbf{R}$ and $-\mathbf{R}$ exerted on G_1 at P and on G' at P', respectively. The generalized impulses I_1 and I_2 are, therefore, given by

$$I_r \underset{(1)}{=} \mathbf{v}_r^P \cdot \int_{t_1}^{t_2} \mathbf{R}\, dt + \mathbf{v}_r^{P'} \cdot \int_{t_1}^{t_2} (-\mathbf{R})\, dt \qquad (r = 1, 2) \tag{14}$$

where the partial velocities $\mathbf{v}_r^P$ and $\mathbf{v}_r^{P'}$ $(r = 1, 2)$, found by inspection of the velocity expressions

$$\mathbf{v}^P = bu_1\hat{\mathbf{n}}_2 \qquad \mathbf{v}^{P'} = -b'u_2\hat{\mathbf{n}}_2 \tag{15}$$

are

$$\mathbf{v}_1^P = b\hat{\mathbf{n}}_2 \qquad \mathbf{v}_1^{P'} = \mathbf{0} \tag{16}$$

and

$$\mathbf{v}_2^P = \mathbf{0} \qquad \mathbf{v}_2^{P'} = -b'\hat{\mathbf{n}}_2 \tag{17}$$

The symbols t_1 and t_2 in Eqs. (14) refer, respectively, to the instant of first contact of G_1 and G' and to the instant at which the meshing process has been completed, and in writing Eqs. (14) it is presumed that the time interval $t_2 - t_1$ is very small.

Using Eqs. (16) and (17) to form I_1 and I_2 in accordance with Eqs. (14), one finds that

$$I_1 = b\hat{\mathbf{n}}_2 \cdot \int_{t_1}^{t_2} \mathbf{R}\, dt \qquad I_2 = b'\hat{\mathbf{n}}_2 \cdot \int_{t_1}^{t_2} \mathbf{R}\, dt \tag{18}$$

and substitution from Eqs. (13) and (18) into Eqs. (5) then yields

$$\underset{(18)}{b\hat{\mathbf{n}}_2 \cdot \int_{t_1}^{t_2} \mathbf{R}\, dt} \underset{(5)}{\approx} \underset{(13)}{3J\,[u_1(t_2) - u_1(t_1)]} \tag{19}$$

and

$$b'\hat{\mathbf{n}}_2 \cdot \underset{(18)}{\int_{t_1}^{t_2} \mathbf{R}\, dt} \underset{(5)}{\approx} \underset{(13)}{J'\,[u_2(t_2) - u_2(t_1)]} \tag{20}$$

or, after elimination of $\hat{\mathbf{n}}_2 \cdot \int_{t_1}^{t_2} \mathbf{R}\, dt$,

$$\frac{3J}{b}[u_1(t_2) - u_1(t_1)] \underset{(19,\ 20)}{\approx} \frac{J'}{b'}[u_2(t_2) - u_2(t_1)] \tag{21}$$

The quantities $u_1(t_1)$ and $u_2(t_1)$ have known values, namely [see Eqs. (11)],

$$u_1(t_1) = 0 \qquad u_2(t_1) = \Omega' \tag{22}$$

At time t_2, the meshing processing having been completed, points P and P' have equal velocities, which means that

$$bu_1(t_2) \underset{(15)}{=} -b'u_2(t_2) \tag{23}$$

or, if Ω is defined as

$$\Omega \triangleq u_1(t_2) \tag{24}$$

that

$$u_2(t_2) \underset{(23)}{=} -\frac{b}{b'}u_1(t_2) \underset{(24)}{=} -\frac{b}{b'}\Omega \tag{25}$$

Consequently,

$$\frac{3J}{b}(\underset{(24)}{\Omega} - \underset{(22)}{0}) \underset{(21)}{\approx} \frac{J'}{b'}\left(\underset{(25)}{-\frac{b}{b'}\Omega} - \underset{(22)}{\Omega'}\right) \tag{26}$$

and Ω, the angular speed that was to be determined, is seen to be given by

$$\Omega \underset{(26)}{=} \frac{-(J'/b')\Omega'}{3(J/b) + (b/b')J'/b'} \tag{27}$$

Before leaving this example, we examine the changes that take place, during the time interval beginning at t_1 and ending at t_2, in the kinetic energy of S and in the angular momentum of S with respect to any fixed point.

The kinetic energies of S at t_1 and at t_2 are given by

$$K(t_1) \underset{(12,\ 22)}{=} \tfrac{1}{2}J'(\Omega')^2 \tag{28}$$

and

$$K(t_2) \underset{(12,\ 24,\ 25)}{=} \frac{\Omega^2}{2}\left[J'\left(\frac{b}{b'}\right)^2 + 3J\right]$$

$$\underset{(27)}{=} \frac{(\Omega' J'/b')^2[J'(b/b')^2 + 3J]}{2[3(J/b) + (b/b')J'/b']^2} \tag{29}$$

respectively. The ratio of these two kinetic energies thus can be expressed as

$$\frac{K(t_2)}{K(t_1)} \underset{(28,\ 29)}{=} \frac{1}{1 + 3(J/J')(b'/b)^2} \tag{30}$$

Since the right-hand member of this equation is smaller than unity, it is evident that the kinetic energy of S decreases during the meshing process under consideration. As for the angular momenta $\mathbf{H}(t_1)$ and $\mathbf{H}(t_2)$ of S relative to any fixed point, these can be written

$$\mathbf{H}(t_1) \underset{(22)}{=} J'\Omega'\hat{\mathbf{n}}_3 \tag{31}$$

and

$$\mathbf{H}(t_2) = \Big(\underset{(25)}{-J'\frac{b}{b'}\Omega} + J\Omega - J\Omega + J\Omega\Big)\hat{\mathbf{n}}_3 \tag{32}$$

where the last three terms in the parentheses can be seen to reflect the contributions of G_1, G_2, and G_3 when kinematical relationships are taken into account. From Eqs. (32) and (27) it thus follows that

$$\mathbf{H}(t_2) = -\frac{(\Omega'J'/b')(J - J'b/b')}{3(J/b) + (b/b')J'/b'}\hat{\mathbf{n}}_3 \tag{33}$$

and the ratio of the magnitudes of $\mathbf{H}(t_2)$ and $\mathbf{H}(t_1)$ is

$$\frac{|\mathbf{H}(t_2)|}{|\mathbf{H}(t_1)|} \underset{(33,\ 31)}{=} \frac{|1 - (J/J')(b'/b)|}{1 + 3(J/J')(b'/b)^2} \tag{34}$$

Here we have a quantity that can be smaller or larger than unity, so that angular momentum can decrease or increase.

It is tempting to "explain" the decrease in kinetic energy by referring to the sound and heat generation known to accompany events of the kind under consideration. However, this is seen to be unsound when one realizes that the method at hand can also lead to kinetic energy increases (see, for example, Problem 13.19). What must be remembered is that Eqs. (5) are *approximate* relationships, which means that results obtained by using these equations can be somewhat, or even totally, unrealistic. Ultimately, only experiments can reveal the degree of utility of a given approximate solution of a problem. Hence, energy decreases should not be regarded as reassuring, or energy increases as alarming, in the present context. Most importantly, *one should not attempt to base a solution of a problem involving sudden velocity changes on an energy conservation principle.*

The change in angular momentum magnitude manifested in Eq. (34) *can* be explained readily. During the time interval beginning at t_1 and ending at t_2, forces are exerted on S at points of contact between G' and its support and, similarly, on points of G_1, G_2, and G_3 by their supports. There is no reason to think that the sum of the moments of all of these forces about any fixed point vanishes. Hence, in accordance with the angular momentum principle [see Eq. (9.4.9)], the angular momentum of S with respect to any fixed point must be expected to change during the time interval beginning at t_1 and ending at t_2. The principle of conservation of angular momentum applies to any system undergoing abrupt velocity changes, just as it applies to any system having other motions, only when the resultant moment about the system's mass center (or about a point fixed in a Newtonian reference frame) of all forces acting on the system is equal to zero.

8.9 COLLISIONS

When a system S is involved in a collision beginning at time t_1 and ending at time t_2, the motion of S at time t_2 frequently cannot be determined solely by use of Eqs. (8.8.5) together with a complete description of the motion of S at time t_1. Generally, some information about the velocity of one or more particles of S at time t_2 must be used in addition to Eqs. (8.8.5), and this information must be expressed in mathematical form [see, for example, Eq. (8.8.23)]. What follows is an attempt to come to grips with this problem by formulating two assumptions that, as experiments have shown, are valid in many situations of practical interest.

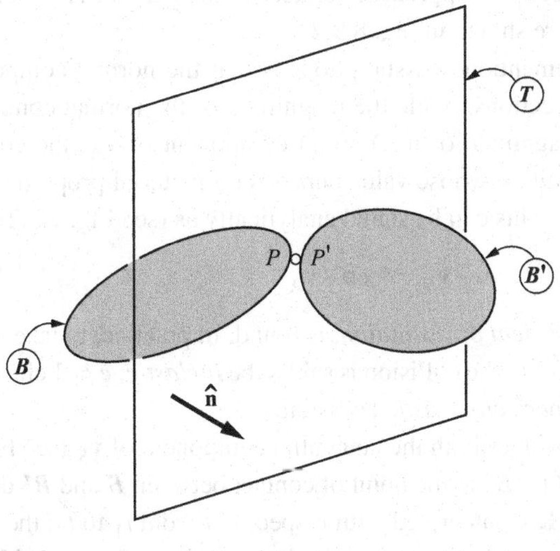

Figure 8.9.1

In Fig. 8.9.1, P and P' designate points that come into contact with each other during a collision of two bodies B and B'. (P and P' are points of B and B' respectively.) T is the plane that is tangent to the surfaces of B and B' at their point of contact, and $\hat{\mathbf{n}}$ is a unit vector perpendicular to T.

If $\mathbf{v}^P(t)$ and $\mathbf{v}^{P'}(t)$ denote the velocities of P and P', respectively, at time t, then $\mathbf{v}_A$, defined as

$$\mathbf{v}_A \triangleq \mathbf{v}^P(t_1) - \mathbf{v}^{P'}(t_1) \tag{1}$$

is called the *velocity of approach* of B and B', and $\mathbf{v}_S$, defined as

$$\mathbf{v}_S \triangleq \mathbf{v}^P(t_2) - \mathbf{v}^{P'}(t_2) \tag{2}$$

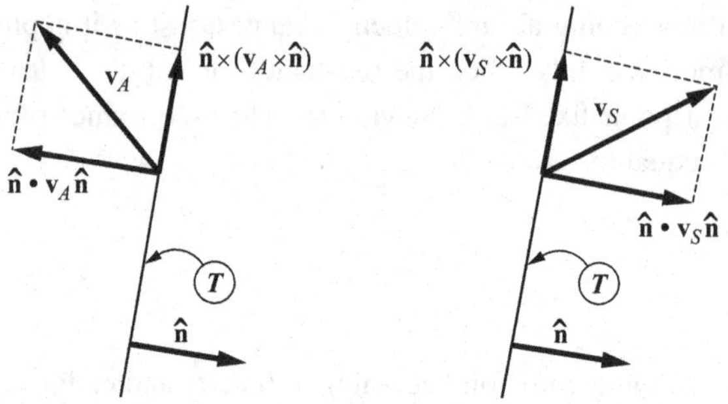

Figure 8.9.2

is called the *velocity of separation* of B and B'. Each of the vectors $\mathbf{v}_A$ and $\mathbf{v}_S$ can be resolved into two components, one parallel to $\hat{\mathbf{n}}$, called the *normal component*, the other perpendicular to $\hat{\mathbf{n}}$, called the *tangential component*, and the normal components can be expressed as $\hat{\mathbf{n}} \cdot \mathbf{v}_A\hat{\mathbf{n}}$ and $\hat{\mathbf{n}} \cdot \mathbf{v}_S\hat{\mathbf{n}}$, while the tangential components are $\hat{\mathbf{n}} \times (\mathbf{v}_A \times \hat{\mathbf{n}})$ and $\hat{\mathbf{n}} \times (\mathbf{v}_S \times \hat{\mathbf{n}})$. The velocity of approach, velocity of separation, and their normal and tangential components are shown in Fig. 8.9.2.

The first of the two aforementioned assumptions is that the normal components of $\mathbf{v}_A$ and $\mathbf{v}_S$ have opposite directions, while the magnitude of the normal component of $\mathbf{v}_S$ is proportional to the magnitude of the normal component of $\mathbf{v}_A$, the constant of proportionality being a quantity e whose value depends on material properties, but not on the motions, of B and B'. This can be stated analytically as (see Fig. 8.9.2)

$$\hat{\mathbf{n}} \cdot \mathbf{v}_S = -e\hat{\mathbf{n}} \cdot \mathbf{v}_A \tag{3}$$

The constant e, called a *coefficient of restitution*, is found, in practice, to take on values such that $0 \le e \le 1$. When $e = 0$, the collision is said to be *inelastic*; $e = 1$ characterizes an idealized event termed a *perfectly elastic* collision.

The second assumption involves both the tangential component of $\mathbf{v}_S$ (see Fig. 8.9.2) and the force $\mathbf{R}$ exerted on B by B' at the point of contact between B and B' during the collision. More specifically, $\mathbf{R}$ is integrated with respect to t from t_1 to t_2; the resulting vector is resolved into two components, one normal to the plane T, denoted by ν, and called the *normal impulse*, the other parallel to T, denoted by $\boldsymbol{\tau}$, and called the *tangential impulse*, as indicated in Fig. 8.9.3. The assumption is this: If and only if the inequality

$$|\boldsymbol{\tau}| < \mu|\nu| \tag{4}$$

is satisfied, where μ is the coefficient of static friction for B and B' (see Sec. 5.8), then there is no slipping at t_2, which means that (see Fig. 8.9.2)

$$\hat{\mathbf{n}} \times (\mathbf{v}_S \times \hat{\mathbf{n}}) = \mathbf{0} \tag{5}$$

If the inequality (4) is violated, then

$$\boldsymbol{\tau} = -\mu'|\nu|\frac{\hat{\mathbf{n}} \times (\mathbf{v}_S \times \hat{\mathbf{n}})}{|\hat{\mathbf{n}} \times (\mathbf{v}_S \times \hat{\mathbf{n}})|} \tag{6}$$

where μ' is the coefficient of kinetic friction for B and B' (see Sec. 5.8), and there is slipping at t_2, so that Eq. (5) does not apply.

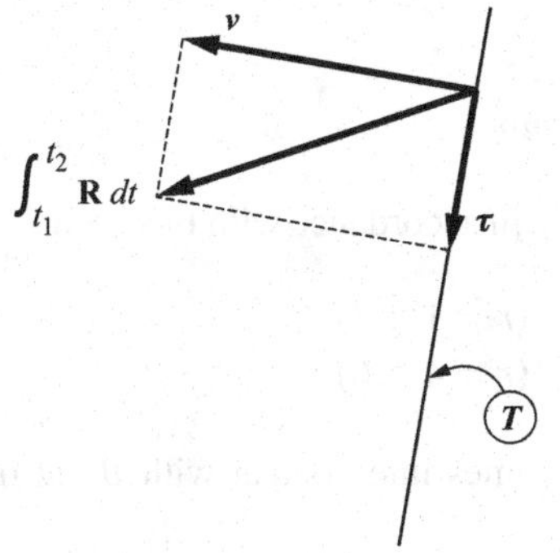

Figure 8.9.3

Example Consider the collision of a uniform sphere B of mass m and radius b with a fixed body B' that is bounded by a horizontal plane, as indicated in Fig. 8.9.4. The angular velocity $\boldsymbol{\omega}$ of B and the velocity $\mathbf{v}^\star$ of the center $B^\star$ of B can be expressed in terms of generalized velocities $u_1, \ldots, u_6$ as

$$\boldsymbol{\omega} = u_1\hat{\mathbf{n}}_1 + u_2\hat{\mathbf{n}}_2 + u_3\hat{\mathbf{n}}_3 \tag{7}$$

and

$$\mathbf{v}^\star = u_4\hat{\mathbf{n}}_1 + u_5\hat{\mathbf{n}}_2 + u_6\hat{\mathbf{n}}_3 \tag{8}$$

The values of $u_1, \ldots, u_6$ at the instant t_1 at which B comes into contact with B' are presumed to be known. It is desired to determine the values of $u_1, \ldots, u_6$ at time t_2, the instant at which B loses contact with B'.

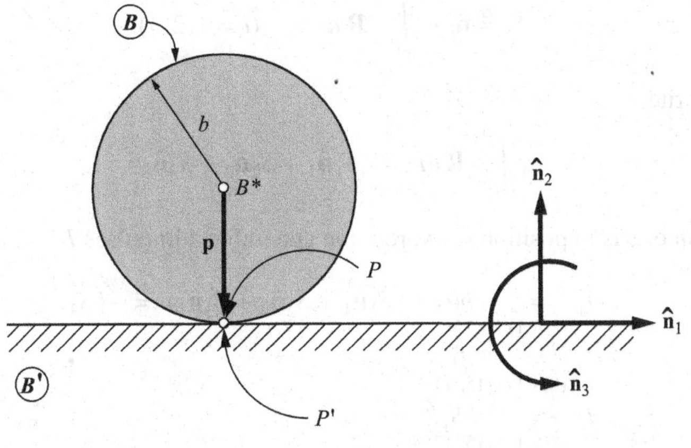

Figure 8.9.4

The kinetic energy K of B is given by [see Eqs. (7.4.2), (7.4.4), and (7.4.7)]

$$K = \tfrac{1}{2}J(u_1^2 + u_2^2 + u_3^2) + \tfrac{1}{2}m(u_4^2 + u_5^2 + u_6^2) \tag{9}$$

where

$$J \triangleq \tfrac{2}{5}mb^2 \tag{10}$$

The generalized momenta $p_1, \ldots, p_6$, formed in accordance with Eqs. (8.8.4), are thus

$$p_r = \begin{cases} Ju_r & (r = 1,2,3) \\ mu_r & (r = 4,5,6) \end{cases} \tag{11}$$

The velocity of the point P of B that comes into contact with B' at time t_1 is given at all times by

$$\mathbf{v}^P = \mathbf{v}^\star + \boldsymbol{\omega} \times \mathbf{p} \tag{12}$$

where $\mathbf{p}$ is the position vector from $B^\star$ to P. Hence,

$$\mathbf{v}^P \underset{(12,\ 7,\ 8)}{=} u_4\hat{\mathbf{n}}_1 + u_5\hat{\mathbf{n}}_2 + u_6\hat{\mathbf{n}}_3 + (u_1\hat{\mathbf{n}}_1 + u_2\hat{\mathbf{n}}_2 + u_3\hat{\mathbf{n}}_3) \times \mathbf{p} \tag{13}$$

and the partial velocities of P at any time t are

$$\mathbf{v}_r^P(t) \underset{(13)}{=} \begin{cases} \hat{\mathbf{n}}_r \times \mathbf{p} & (r = 1,2,3) \\ \hat{\mathbf{n}}_{r-3} & (r = 4,5,6) \end{cases} \tag{14}$$

At time t_1, $\mathbf{p}$ is given by $\mathbf{p} = -b\hat{\mathbf{n}}_2$, and the first three of Eqs. (14) thus lead to

$$\mathbf{v}_1^P(t_1) = -b\hat{\mathbf{n}}_3 \qquad \mathbf{v}_2^P(t_1) = \mathbf{0} \qquad \mathbf{v}_3^P(t_1) = b\hat{\mathbf{n}}_1 \tag{15}$$

Letting $\mathbf{R}$ be the force exerted on B by B' at their point of contact during the time interval beginning at t_1 and ending at t_2, and defining S_i as

$$S_i \triangleq \hat{\mathbf{n}}_i \cdot \int_{t_1}^{t_2} \mathbf{R}\, dt \qquad (i = 1,2,3) \tag{16}$$

one can write

$$\int_{t_1}^{t_2} \mathbf{R}\, dt \underset{(16)}{=} S_1\hat{\mathbf{n}}_1 + S_2\hat{\mathbf{n}}_2 + S_3\hat{\mathbf{n}}_3 \tag{17}$$

whereupon one is in position to express the generalized impulses $I_1, \ldots, I_6$ as

$$I_1 \underset{(8.8.1)}{=} \underset{(15)}{-b\hat{\mathbf{n}}_3} \cdot \underset{(17)}{(S_1\hat{\mathbf{n}}_1 + S_2\hat{\mathbf{n}}_2 + S_3\hat{\mathbf{n}}_3)} = -bS_3 \tag{18}$$

$$I_2 \underset{(8.8.1)}{=} \underset{(15,\ 17)}{0} \tag{19}$$

$$I_3 \underset{(8.8.1)}{=} \underset{(15,\ 17)}{bS_1} \tag{20}$$

$$I_r \underset{(8.8.1)}{=} \underset{(14,\ 17)}{S_{r-3}} \qquad (r = 4,5,6) \tag{21}$$

so that substitution from Eqs. (11) and (18)–(21) into Eqs. (8.8.5) produces

$$-bS_3 \approx J[u_1(t_2) - u_1(t_1)] \tag{22}$$

$$0 \approx J[u_2(t_2) - u_2(t_1)] \tag{23}$$

$$bS_1 \approx J[u_3(t_2) - u_3(t_1)] \tag{24}$$

$$S_1 \approx m[u_4(t_2) - u_4(t_1)] \tag{25}$$

$$S_2 \approx m[u_5(t_2) - u_5(t_1)] \tag{26}$$

$$S_3 \approx m[u_6(t_2) - u_6(t_1)] \tag{27}$$

One of the quantities to be determined, namely, $u_2(t_2)$, now can be found immediately, for Eq. (23) yields

$$u_2(t_2) \approx u_2(t_1) \tag{28}$$

As for the rest, information in addition to that furnished by Eqs. (22) and (24)–(27) is required because these five equations involve the eight unknowns $u_1(t_2), u_3(t_2), \ldots, u_6(t_2), S_1, S_2$, and S_3.

The velocity of the point P' of B' (see Fig. 8.9.4) with which B comes into contact during the collision is equal to zero at all times if B' is fixed, as is being presumed. Hence,

$$\mathbf{v}^{P'}(t_1) = \mathbf{v}^{P'}(t_2) = \mathbf{0} \tag{29}$$

and $\mathbf{v}_A$, the velocity of approach, can be expressed as

$$\mathbf{v}_A \underset{(1)}{=} \mathbf{v}^P(t_1) - \underset{(29)}{\mathbf{0}} \tag{30}$$

or, since $\mathbf{p} = -b\hat{\mathbf{n}}_2$ at t_1, as

$$\mathbf{v}_A \underset{(30,\ 13)}{=} [u_4(t_1) + bu_3(t_1)]\hat{\mathbf{n}}_1 + u_5(t_1)\hat{\mathbf{n}}_2 + [u_6(t_1) - bu_1(t_1)]\hat{\mathbf{n}}_3 \tag{31}$$

Similarly, $\mathbf{v}_S$, the velocity of separation, formed in accordance with Eq. (2), is given by

$$\mathbf{v}_S \underset{(2,\ 29,\ 13)}{=} [u_4(t_2) + bu_3(t_2)]\hat{\mathbf{n}}_1 + u_5(t_2)\hat{\mathbf{n}}_2 + [u_6(t_2) - bu_1(t_2)]\hat{\mathbf{n}}_3 \tag{32}$$

and, with $\hat{\mathbf{n}}_2$ playing the role of $\hat{\mathbf{n}}$ (see Figs. 8.9.1 and 8.9.4), one can thus write one further equation by appealing to Eqs. (3), namely,

$$u_5(t_2) \underset{(3,\ 32,\ 31)}{=} -eu_5(t_1) \tag{33}$$

Furthermore, Eq. (26) may be replaced with

$$S_2 \approx -m(1+e)u_5(t_1) \tag{34}$$

Four equations, namely, Eqs. (22), (24), (25), and (27), now are available for the determination of the remaining six unknowns, $u_1(t_2)$, $u_3(t_2)$, $u_4(t_2)$, $u_6(t_2)$, S_1, and S_3. To supplement these, we note that $\boldsymbol{\nu}$, the normal impulse, and $\boldsymbol{\tau}$, the tangential impulse, are given by [see Fig. 8.9.3, Eq. (17), and Fig. 8.9.4]

$$\boldsymbol{\nu} = S_2\hat{\mathbf{n}}_2 \tag{35}$$

and

$$\boldsymbol{\tau} = S_1 \hat{\mathbf{n}}_1 + S_3 \hat{\mathbf{n}}_3 \tag{36}$$

respectively, and we form the tangential component of the velocity of separation as (see Figs. 8.9.2 and 8.9.4)

$$\hat{\mathbf{n}}_2 \times (\mathbf{v}_S \times \hat{\mathbf{n}}_2) \underset{(32)}{=} [u_4(t_2) + bu_3(t_2)]\hat{\mathbf{n}}_1 + [u_6(t_2) - bu_1(t_2)]\hat{\mathbf{n}}_3 \tag{37}$$

Now there are two possibilities: There is no slipping at t_2, in which case

$$u_4(t_2) + bu_3(t_2) \underset{(5,\ 37)}{=} 0 \tag{38}$$

$$u_6(t_2) - bu_1(t_2) \underset{(5,\ 37)}{=} 0 \tag{39}$$

and

$$\underset{(36)}{(S_1{}^2 + S_3{}^2)^{1/2}} < \underset{(4)}{\mu} \underset{(35)}{|S_2|} \tag{40}$$

Alternatively, there is slipping at t_2, and Eq. (6) requires that

$$\underset{(36)}{S_1} \underset{(6)}{=} -\mu' \underset{(35)}{|S_2|} \underset{(37)}{\frac{u_4(t_2) + bu_3(t_2)}{\{[u_4(t_2) + bu_3(t_2)]^2 + [u_6(t_2) - bu_1(t_2)]^2\}^{1/2}}} \tag{41}$$

and

$$\underset{(36)}{S_3} \underset{(6)}{=} -\mu' \underset{(35)}{|S_2|} \underset{(37)}{\frac{u_6(t_2) - bu_1(t_2)}{\{[u_4(t_2) + bu_3(t_2)]^2 + [u_6(t_2) - bu_1(t_2)]^2\}^{1/2}}} \tag{42}$$

We shall examine these two possibilities separately. Before doing so, however, we establish two relationships that apply in both cases. Specifically, we eliminate S_1 from Eqs. (24) and (25), obtaining

$$bm[u_4(t_2) - u_4(t_1)] \approx J[u_3(t_2) - u_3(t_1)] \tag{43}$$

Similarly, eliminating S_3 from Eqs. (22) and (27), we find that

$$-\,bm[u_6(t_2) - u_6(t_1)] \approx J[u_1(t_2) - u_1(t_1)] \tag{44}$$

If there is no slipping at t_2, then Eq. (38) applies, and elimination of $u_4(t_2)$ from Eqs. (38) and (43) reveals that

$$u_3(t_2) \approx \frac{Ju_3(t_1) - mbu_4(t_1)}{mb^2 + J} \tag{45}$$

Once $u_3(t_2)$ has been evaluated, one can find $u_4(t_2)$ by using the relationship

$$u_4(t_2) \underset{(38)}{=} -bu_3(t_2) \tag{46}$$

As for $u_1(t_2)$, elimination of $u_6(t_2)$ from Eqs. (39) and (44) results in

$$u_1(t_2) \approx \frac{Ju_1(t_1) + mbu_6(t_1)}{mb^2 + J} \tag{47}$$

and Eq. (39) then permits one to evaluate $u_6(t_2)$ as

$$u_6(t_2) = bu_1(t_2) \tag{48}$$

Successive use of Eqs. (28), (33), and (45)–(48) thus yields a set of values of $u_1, \ldots, u_6$ at time t_2, and these values apply if and only if the inequality (40) is satisfied when S_1, S_2, and S_3 have the values given by Eqs. (25), (26), and (27), respectively.

If there is slipping at t_2, then, as before, $u_2(t_2)$ and $u_5(t_2)$ are given by Eqs. (28) and (33), respectively, and S_2 can be found with the aid of Eq. (26). To determine $u_1(t_2)$, $u_3(t_2)$, $u_4(t_2)$, and $u_6(t_2)$, one can begin by using Eqs. (24) and (22) to express S_1 and S_3 as

$$S_1 \approx \frac{J}{b}[u_3(t_2) - u_3(t_1)] \tag{49}$$

and

$$S_3 \approx -\frac{J}{b}[u_1(t_2) - u_1(t_1)] \tag{50}$$

Next,

$$u_3(t_2) \underset{(49)}{\approx} u_3(t_1) + \frac{bS_1}{J} \tag{51}$$

$$u_1(t_2) \underset{(50)}{\approx} u_1(t_1) - \frac{bS_3}{J} \tag{52}$$

$$u_4(t_2) \underset{(43)}{\approx} u_4(t_1) + \underset{(51)}{\frac{S_1}{m}} \tag{53}$$

$$u_6(t_2) \underset{(44)}{\approx} u_6(t_1) + \underset{(52)}{\frac{S_3}{m}} \tag{54}$$

Hence,

$$u_4(t_2) + bu_3(t_2) \underset{(51,\,53)}{\approx} u_4(t_1) + bu_3(t_1) + \left(\frac{1}{m} + \frac{b^2}{J}\right) S_1 \tag{55}$$

$$u_6(t_2) - bu_1(t_2) \underset{(52,\,54)}{\approx} u_6(t_1) - bu_1(t_1) + \left(\frac{1}{m} + \frac{b^2}{J}\right) S_3 \tag{56}$$

and, if α, γ, and k are defined as

$$\alpha \triangleq u_4(t_1) + bu_3(t_1) \qquad \gamma \triangleq u_6(t_1) - bu_1(t_1) \tag{57}$$

and

$$k \triangleq \frac{1}{m} + \frac{b^2}{J} \tag{58}$$

respectively, then Eqs. (41) and (42) lead to

$$S_1 \underset{(41)}{\approx} -\mu'|S_2| \underset{(55-58)}{\frac{\alpha + kS_1}{[(\alpha + kS_1)^2 + (\gamma + kS_3)^2]^{1/2}}} \tag{59}$$

and

$$S_3 \underset{(42)}{\approx} -\mu'|S_2|\frac{\gamma + kS_3}{[(\alpha + kS_1)^2 + (\gamma + kS_3)^2]^{1/2}} \tag{60}$$

$\qquad\qquad\qquad\qquad\qquad\qquad\qquad\qquad (55-58)$

from which it follows that

$$\frac{S_1}{S_3} \underset{(59,\,60)}{\approx} \frac{\alpha + kS_1}{\gamma + kS_3} \tag{61}$$

or, equivalently, that

$$S_3 \approx \frac{\gamma}{\alpha}S_1 \tag{62}$$

which makes it possible to replace Eq. (59) with

$$\begin{aligned} S_1 &\underset{(59)}{\approx} -\mu'|S_2|\frac{\alpha + kS_1}{\{(\alpha + kS_1)^2 + [\gamma + k(\gamma/\alpha)S_1]^2\}^{1/2}} \\ &= -\mu'|S_2|\frac{\alpha + kS_1}{|\alpha + kS_1|[1 + (\gamma/\alpha)^2]^{1/2}} \end{aligned} \tag{63}$$

Now, $(\alpha + kS_1)/|\alpha + kS_1|$ has the value 1 when $\alpha + kS_1 > 0$, and the value -1 when $\alpha + kS_1 < 0$. In the first case, therefore,

$$S_1 \underset{(63)}{\approx} -\frac{\mu'|S_2|}{[1 + (\gamma/\alpha)^2]^{1/2}} \tag{64}$$

so that

$$\alpha + kS_1 \underset{(64)}{\approx} \alpha - \frac{k\mu'|S_2|}{[1 + (\gamma/\alpha)]^{1/2}} > 0 \tag{65}$$

which implies that

$$\alpha > 0 \tag{66}$$

In the second case,

$$S_1 \underset{(63)}{\approx} \frac{\mu'|S_2|}{[1 + (\gamma/\alpha)^2]^{1/2}} \tag{67}$$

so that

$$\alpha + kS_1 \underset{(67)}{\approx} \alpha + \frac{k\mu'|S_2|}{[1 + (\gamma/\alpha)^2]^{1/2}} < 0 \tag{68}$$

which implies that

$$\alpha < 0 \tag{69}$$

Since the inequalities (66) and (69) are mutually exclusive, the sign of α is sufficient to settle the question of whether Eq. (64) or Eq. (67) should be used, and one can accommodate both equations by writing

$$S_1 \approx -\frac{\mu'\alpha|S_2|}{|\alpha|[1 + (\gamma/\alpha)^2]^{1/2}} \tag{70}$$

which brings one into position to evaluate S_3 by reference to Eq. (62), whereupon one can find $u_1(t_2)$, $u_3(t_2)$, $u_4(t_2)$, and $u_6(t_2)$ with the aid of Eqs. (52), (51), (53), and (54), respectively.

In Fig. 8.9.5, a complete algorithm for the evaluation of $u_1, \ldots, u_6$ at time t_2 is set forth in the form of a flow chart, in which numbers in parentheses refer to corresponding equations. The values of the physical parameters b, m, J, e, μ, and μ' and the generalized velocities $u_1, \ldots, u_6$ at time t_1 are presumed to be known, and $u_5(t_1)$ must be negative, since no collision will occur otherwise. The physical significance of the preceding analysis is brought to light with the aid of a numerical example.

Suppose that $e = 0.8$, $\mu = 0.25$, $\mu' = 0.20$, and B has a "topspin" when it strikes B', which is the case, for instance, if [see Eqs. (7) and (8) and Fig. 8.9.4]

$$u_1(t_1) = u_2(t_1) = 0 \qquad u_3(t_1) = -\frac{3V}{b} \tag{71}$$

$$u_4(t_1) = -u_5(t_1) = V \qquad u_6(t_1) = 0 \tag{72}$$

where V is any (positive) speed. Then, following the steps indicated in Fig. 8.9.5, one finds that $S_1 \approx 4mV/7$, $S_2 \approx 1.8mV$, $S_3 \approx 0$. Hence, the inequality (40) is violated, which means that there is slipping at t_2. Proceeding in accordance with Fig. 8.9.5, one obtains

$$u_1(t_2) = u_2(t_2) = 0 \qquad u_3(t_2) = -\frac{2.1V}{b} \tag{73}$$

$$u_4(t_2) = 1.36V \qquad u_5(t_2) = 0.8V \qquad u_6(t_2) = 0 \tag{74}$$

The fact that $u_6(t_2) = 0$ shows that $B^\star$ moves in the same plane subsequent to the collision of B with B' as it does prior to the collision. The angles θ_1 and θ_2 that the vector $\mathbf{v}^\star$ [see Eq. (8)] makes with the vertical before and after impact, respectively, can be compared with each other when it is noted that

$$\theta_i = \tan^{-1}\left|\frac{u_4(t_i)}{u_5(t_i)}\right| \qquad (i = 1,2) \tag{75}$$

so that [see Eqs. (72)] $\theta_1 = 45°$ while [see Eqs. (74)] $\theta_2 = 59.5°$. This means that the topspin initially imparted to B has the effect of producing a "drop," a fact that will not surprise tennis players.

Finally, one can estimate how long B will continue to bounce. The maximum height reached by $B^\star$ during any bounce is, of course, smaller than that attained during the preceding one, so that this height approaches zero (and bouncing ceases) when the number n of bounces approaches infinity. The time T_n required for n bounces to occur is found by noting that the time that elapses between two successive impacts depends only on the value of u_5 at the end of the first of these; that is, if τ_n denotes the time required for the nth bounce, then τ_1 is given by

$$\tau_1 = \frac{2u_5(t_2)}{g} \underset{(33)}{=} \frac{-2eu_5(t_1)}{g} \tag{76}$$

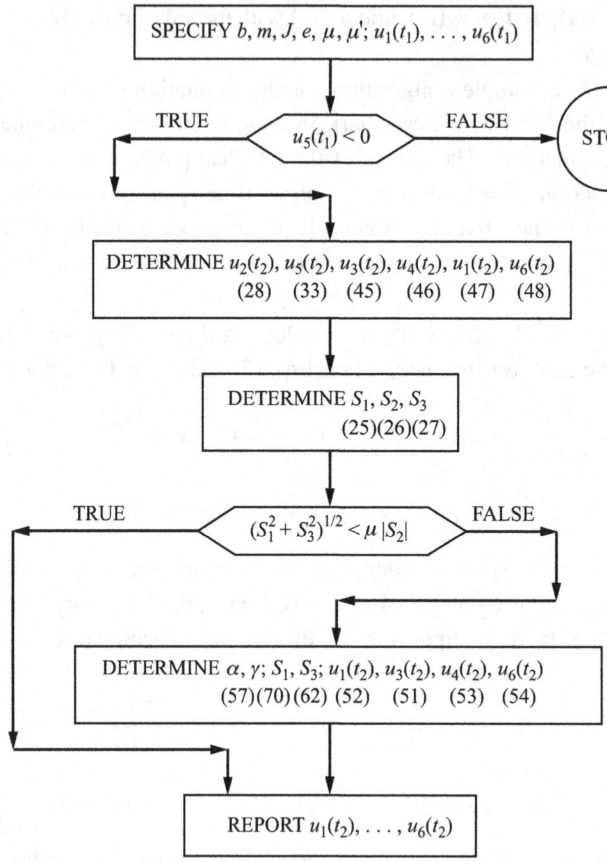

Figure 8.9.5

and, similarly,

$$\tau_2 = \frac{2u_5(t_4)}{g} = \frac{-2eu_5(t_3)}{g} \tag{77}$$

But

$$u_5(t_3) = -u_5(t_2) \underset{(33)}{=} eu_5(t_1) \tag{78}$$

Hence,

$$\tau_2 \underset{(77,\ 78)}{=} \frac{-2e^2u_5(t_1)}{g} \tag{79}$$

Similarly,

$$\tau_3 = \frac{-2e^3u_5(t_1)}{g} \tag{80}$$

and

$$\tau_n = \frac{-2e^n u_5(t_1)}{g} \tag{81}$$

Consequently,

$$T_n = \tau_1 + \cdots + \tau_n = \frac{-2(e + e^2 + \cdots + e^n)u_5(t_1)}{g} \tag{82}$$

Now,

$$e + e^2 + \cdots + e^n = \frac{e(1 - e^n)}{1 - e} \tag{83}$$

Thus,

$$T_n \underset{(82,\ 83)}{=} \frac{-2e(1 - e^n)u_5(t_1)}{(1 - e)g} \tag{84}$$

and the total time T required for infinitely many bounces is given by (for $0 < e < 1$)

$$T = \lim_{n\to\infty} T_n \underset{(84)}{=} \frac{-2eu_5(t_1)}{(1 - e)g} \tag{85}$$

Since this analysis does not account for the time consumed by the collisions (infinitely many), Eq. (85) should be regarded as a lower bound on the time required for bouncing to cease subsequent to time t_1. Applied to the numerical example considered previously, Eq. (85) yields

$$T = \frac{-2(0.8)(-V)}{(1 - 0.8)g} = \frac{8V}{g} \tag{86}$$

Hence, if $V = 10$ m/s, B may be expected to bounce for a little longer than 8 s.

9 EXTRACTION OF INFORMATION FROM EQUATIONS OF MOTION

This chapter is intended to bring to fruition the effort that has been expended by the reader in learning the material covered in the first eight chapters. The chapter begins in Sec. 9.1 with an introduction of terminology employed in connection with solutions of equations of motion. The construction of an energy integral, and the circumstances under which such an integral of equations of motion exists, are taken up in Sec. 9.2. When an energy integral cannot be formed, it is possible to construct a checking function instead, as explained in Sec. 9.3. The existence of momentum integrals, and instructions for forming them, are presented in Sec. 9.4. Exact closed-form solutions of equations of motion are considered in Sec. 9.5, and means for obtaining numerical results when closed-form solutions are unavailable are discussed in Sec. 9.6. These techniques, used in conjunction with material in Sec. 8.3, permit the evaluation of constraint forces and constraint torques, as is shown in Sec. 9.7. Section 9.8 deals with a purely mathematical problem that arises frequently in dynamics, namely, finding real solutions of a set of nonlinear, algebraic equations. Finally, Sec. 9.9 contains an exposition of the principal concepts underlying the theory of small vibrations of mechanical systems.

9.1 INTEGRALS OF EQUATIONS OF MOTION

The behavior of a simple nonholonomic system S possessing p degrees of freedom in a Newtonian reference frame N (see Sec. 3.5) is governed by $2n$ equations called, collectively, the *equations of motion* of S in N. This set of equations consists of three subsets: the n *kinematical differential equations*, Eqs. (3.4.1) or, equivalently, Eqs. (3.6.5); the $n - p$ *nonholonomic constraint equations*, Eqs. (3.5.2); and the p *dynamical differential equations*, Eqs. (8.1.2). If S is a holonomic system, then $n = p$, the nonholonomic constraint equations are absent, and Eqs. (8.1.1) furnish n dynamical equations.

A solution of the equations of motion is said to be in hand when the generalized coordinates $q_1, \ldots, q_n$ (see Sec. 3.2) and motion variables $u_1, \ldots, u_n$ (see Sec. 3.4) are known as functions of the time t. The *general solution* of the equations of motion consists of $2n$ independent equations of the form

$$f_i(q_1, \ldots, q_n, u_1, \ldots, u_n, t) = C_i \qquad (i = 1, \ldots, 2n) \tag{1}$$

where f_i $(i = 1, \ldots, 2n)$ is a function whose total time-derivative vanishes, whenever $q_1, \ldots, q_n$ and $u_1, \ldots, u_n$ satisfy all equations of motion, and $C_1, \ldots, C_{2n}$ are *arbitrary* constants. An equation having the form of Eqs. (1) is called an *integral of the equations of motion*. Thus, the general solution of the equations of motion consists of $2n$ integrals.

Examples For the nonholonomic system S formed by the particle P_1 and the sharp-edged circular disk D connected by a rigid rod R as in the example in Sec. 3.5, $p = 2$ and $n = 3$. When the kinematical differential equations are taken to be [see Eqs. (3.4.7)]

$$u_1 = \dot{q}_1 c_3 + \dot{q}_2 s_3 \qquad u_2 = -\dot{q}_1 s_3 + \dot{q}_2 c_3 \qquad u_3 = \dot{q}_3 \tag{2}$$

then the nonholonomic constraint equation expressing the restriction that the center of D may not move perpendicularly to R is [see Eq. (3.5.15)]

$$u_3 = -\frac{u_2}{L} \tag{3}$$

Furthermore, if line Y in Fig. 3.5.1 is vertical, P_1 and D have masses m_1 and m_2, respectively, and P_1 is subjected to the action of a contact force $\mathbf{K}$ given by Eq. (7.1.26), then the generalized active forces for S are given by [see Eqs. (7.1.39)]

$$\tilde{F}_1 = k - (m_1 + m_2) g s_3 \tag{4}$$

$$\tilde{F}_2 = \frac{k}{L}(q_1 s_3 - q_2 c_3) - m_1 g c_3 \tag{5}$$

while the generalized inertia forces can be written (see Problem 8.14)

$$\tilde{F}_1^\star = (m_1 + m_2)(\omega^2 q_1 c_3 - \dot{u}_1) - \frac{m_1 {u_2}^2}{L} + m_2 L \omega^2 {c_3}^2 \tag{6}$$

$$\tilde{F}_2^\star = -m_1 \left(\dot{u}_2 + \omega^2 q_1 s_3 - \frac{u_1 u_2}{L} \right) \tag{7}$$

Substitution from Eqs. (4)–(7) into Eqs. (8.1.2) thus yields the dynamical differential equations

$$\underbrace{k - (m_1 + m_2) g s_3}_{(4)} + \underbrace{(m_1 + m_2)(\omega^2 q_1 c_3 - \dot{u}_1) - \frac{m_1 {u_2}^2}{L} + m_2 L \omega^2 {c_3}^2}_{(6)} = 0 \tag{8}$$

and

$$\underbrace{\frac{k}{L}(q_1 s_3 - q_2 c_3) - m_1 g c_3}_{(5)} \underbrace{- m_1 \left(\dot{u}_2 + \omega^2 q_1 s_3 - \frac{u_1 u_2}{L} \right)}_{(7)} = 0 \tag{9}$$

Equations (2), (3), (8), and (9) are the six equations of motion of S.

As will be shown presently, the equation

$$\begin{aligned} -k(q_1 c_3 + q_2 s_3) + g[(m_1 + m_2) q_2 + m_2 L s_3] + \tfrac{1}{2}[(m_1 + m_2){u_1}^2 + m_1 {u_2}^2] \\ - \tfrac{1}{2}\omega^2[m_1 {q_1}^2 + m_2 (q_1 + L c_3)^2] = C \end{aligned} \tag{10}$$

where C is an arbitrary constant, is an integral of the equations of motion of S. It does not matter for present purposes how this integral was found; this subject is discussed in its own right in the example in Sec. 9.2.

Denoting the left-hand member of Eq. (10) by $f(q_1,q_2,q_3,u_1,u_2,u_3,t)$, and expressing the total time-derivative of f as

$$\dot{f} = \sum_{r=1}^{3}\left(\frac{\partial f}{\partial q_r}\dot{q}_r + \frac{\partial f}{\partial u_r}\dot{u}_r\right) + \frac{\partial f}{\partial t} \tag{11}$$

one finds that

$$\begin{aligned}\dot{f} \underset{(11,\ 10)}{=} & \{-kc_3 - \omega^2[m_1 q_1 + m_2(q_1 + Lc_3)]\}\dot{q}_1 \\ & + [-ks_3 + g(m_1 + m_2)]\dot{q}_2 \\ & + [k(q_1 s_3 - q_2 c_3) + g m_2 L c_3 + \omega^2 m_2 (q_1 + Lc_3) L s_3]\dot{q}_3 \\ & + (m_1 + m_2)u_1\dot{u}_1 + m_1 u_2 \dot{u}_2\end{aligned} \tag{12}$$

Solution of the kinematical differential equations, Eqs. (2), for $\dot{q}_1$, $\dot{q}_2$, and $\dot{q}_3$ yields

$$\dot{q}_1 = u_1 c_3 - u_2 s_3 \qquad \dot{q}_2 = u_1 s_3 + u_2 c_3 \qquad \dot{q}_3 = u_3 \tag{13}$$

while solution of the dynamical differential equations for $\dot{u}_1$ and $\dot{u}_2$ leads to

$$\dot{u}_1 \underset{(8)}{=} \omega^2 c_3\left(q_1 + \frac{m_2 L c_3}{m_1 + m_2}\right) + \frac{k}{m_1 + m_2} - g s_3 - \frac{m_1 {u_2}^2}{(m_1 + m_2)L} \tag{14}$$

$$\dot{u}_2 \underset{(9)}{=} -\omega^2 q_1 s_3 + \frac{k}{m_1 L}(q_1 s_3 - q_2 c_3) - g c_3 + \frac{u_1 u_2}{L} \tag{15}$$

If u_3 in the third of Eqs. (13) now is replaced with $-u_2/L$ in accordance with the nonholonomic constraint equation, Eq. (3), and Eqs. (13)–(15) then are used to eliminate $\dot{q}_1$, $\dot{q}_2$, $\dot{q}_3$, $\dot{u}_1$, and $\dot{u}_2$ from Eq. (12), $\dot{f}$ is found to be equal to zero, and this establishes Eq. (10) as an integral of the equations of motion of S. (To obtain the general solution of the equations of motion, one must produce five more integrals.)

As a second example, we consider a system governed by differential equations whose explicit solution is well known, namely, the *harmonic oscillator*, that is, any system whose behavior is governed by the differential equation

$$\ddot{q} + \omega^2 q = 0 \tag{16}$$

where ω^2 is a constant. This equation can be replaced with the two first-order equations

$$\dot{q} = u \qquad \dot{u} = -\omega^2 q \tag{17}$$

the first of which is simply a definition of u and plays the role of a kinematical differential equation, while the second is a dynamical differential equation. Thus, $n = p = 1$.

Using the same procedure as in the preceding example, one can verify that

$$q \sin\omega t + \frac{u}{\omega}\cos\omega t = C_1 \tag{18}$$

and

$$q \cos \omega t - \frac{u}{\omega} \sin \omega t = C_2 \tag{19}$$

are integrals of the equations of motion, Eqs. (17). Together, these two integrals constitute the general solution of Eqs. (17). To bring this solution into a form that may be more familiar, one can multiply both sides of Eq. (18) by $\sin \omega t$, multiply both sides of Eq. (19) by $\cos \omega t$, and add the resulting expressions to obtain

$$q = C_1 \sin \omega t + C_2 \cos \omega t \tag{20}$$

9.2 THE ENERGY INTEGRAL

When a system S possesses a potential energy V in a Newtonian reference frame N (see Sec. 7.1), there may exist an integral of the equations of motion of S in N (see Sec. 9.1) that can be expressed as

$$H = C \tag{1}$$

where C is a constant and H, called a *Hamiltonian* of S in N, is defined as

$$H \triangleq V + K_2 - K_0 \tag{2}$$

with K_0 and K_2 given by Eqs. (7.5.7) and (7.5.9), respectively. Equation (1) is called the *energy integral* of the equations of motion of S in N. The conditions under which it

Table 9.2.1

	$u_r = \dot{q}_r$ $(X_r = 0, W_{rs} = \delta_{rs})$ $(r,s = 1,\ldots,n)$	$u_r = \sum_{s=1}^{n} Y_{rs}\dot{q}_s + Z_r$ $(r = 1,\ldots,n)$
Holonomic	Eq. (7.1.3) and Eqs. (7.6.3)	Eq. (7.1.10) and Eqs. (7.6.3)
Simple Nonholonomic	Eq. (7.1.15) and Eqs. (7.6.3)	Eq. (7.1.19) and Eqs. (7.6.3)

exists depend on the manner in which motion variables are introduced and on whether S is a holonomic or simple nonholonomic system. In Table 9.2.1, sufficient conditions for the existence of the energy integral of S in N are indicated by listing the numbers of equations satisfaction of which ensures the existence of an energy integral.

If the kinetic energy of S is a homogeneous quadratic function of $u_1,\ldots,u_n$, that is, if $K_0 = K_1 = 0$, so that $K = K_2$ [see Eq. (7.5.1)], then Eqs. (1) and (2) yield

$$V + K = E \tag{3}$$

where E is a constant. Equation (3) expresses the *principle of conservation of mechanical energy*.

Derivation Suppose that Eqs. (7.1.19) and (7.6.3) are satisfied (see Table 9.2.1). Then

$$\dot{V} \underset{(7.1.20)}{=} -\sum_{r=1}^{p} \widetilde{F}_r u_r \tag{4}$$

and

$$\dot{K}_2 - \dot{K}_0 \underset{(7.6.4)}{=} -\sum_{r=1}^{p} \widetilde{F}_r^{\star} u_r \tag{5}$$

so that

$$\dot{H} \underset{(2)}{=} \dot{V} + \dot{K}_2 - \dot{K}_0 \underset{(4,\ 5)}{=} -\sum_{r=1}^{p} (\widetilde{F}_r + \widetilde{F}_r^{\star}) u_r \underset{(8.1.2)}{=} 0 \tag{6}$$

and Eq. (1) follows immediately. The derivations of Eq. (1) for the remaining three cases proceed similarly when the remaining entries in Table 9.2.1 are taken into account.

Examples A potential energy V and the kinetic energy functions K_0 and K_2 for the system S considered in the first example in Sec. 9.1 are given by

$$V \underset{(7.1.64)}{=} -k(q_1 c_3 + q_2 s_3) + g[(m_1 + m_2)q_2 + m_2 L s_3] \tag{7}$$

$$K_0 \underset{(7.5.11)}{=} \tfrac{1}{2}\omega^2 [m_1 {q_1}^2 + m_2 (q_1 + L c_3)^2] \tag{8}$$

and

$$K_2 \underset{(7.5.12)}{=} \tfrac{1}{2}[(m_1 + m_2){u_1}^2 + m_1 {u_2}^2] \tag{9}$$

Setting $V + K_2 - K_0$ equal to an arbitrary constant, one arrives at Eq. (9.1.10). In the example in Sec. 9.1, the equations of motion were brought into play to show that Eq. (9.1.10) is an integral of these equations. Alternatively, one can establish this fact with the aid of Eqs. (7.1.19) and (7.6.3), both of which are satisfied.

For the system depicted in Fig. 5.4.1, a potential energy V is given by

$$V \underset{(7.2.25)}{=} -[m_1(L_1 + q_1) + m_2(L_1 + L_2 + q_2)]g \cos\theta + \tfrac{1}{2}[k_1 {q_1}^2 + k_2 (q_2 - q_1)^2] \tag{10}$$

and the velocities of the two particles P_1 and P_2 are, respectively,

$$\mathbf{v}^{P_1} \underset{(5.4.10)}{=} u_1 \hat{\mathbf{t}}_1 + (L_1 + q_1)\dot{\theta}\,\hat{\mathbf{t}}_2 \tag{11}$$

$$\mathbf{v}^{P_2} \underset{(5.4.11)}{=} u_2 \hat{\mathbf{t}}_1 + (L_1 + L_2 + q_2)\dot{\theta}\,\hat{\mathbf{t}}_2 \tag{12}$$

The kinetic energy K can therefore be written

$$\begin{aligned} K &\underset{(7.4.1)}{=} \tfrac{1}{2}\{m_1[u_1\hat{\mathbf{t}}_1 + (L_1 + q_1)\dot{\theta}\,\hat{\mathbf{t}}_2]^2 + m_2[u_2\hat{\mathbf{t}}_1 + (L_1 + L_2 + q_2)\dot{\theta}\,\hat{\mathbf{t}}_2]^2\} \\ &= \tfrac{1}{2}\{m_1 {u_1}^2 + m_2 {u_2}^2 + [m_1(L_1 + q_1)^2 + m_2(L_1 + L_2 + q_2)^2]\dot{\theta}^2\} \end{aligned} \tag{13}$$

so that

$$K_0 \underset{(13)}{=} \tfrac{1}{2}[m_1(L_1+q_1)^2 + m_2(L_1+L_2+q_2)^2]\dot{\theta}^2 \tag{14}$$

and

$$K_2 \underset{(13)}{=} \tfrac{1}{2}(m_1{u_1}^2 + m_2{u_2}^2) \tag{15}$$

Since the system is holonomic and, in accordance with Eqs. (5.4.9), $u_r = \dot{q}_r$ $(r = 1,2)$, Eq. (1) furnishes an integral of the equations of motion if (see Table 9.2.1) Eqs. (7.1.3) and (7.6.3) are satisfied. The first of the two conditions can be stated

$$\frac{\partial V}{\partial t} \underset{(10)}{=} [m_1(L_1+q_1) + m_2(L_1+L_2+q_2)]g\sin\theta\dot{\theta} = 0 \tag{16}$$

whereas the second condition is expressed as

$$m_1\mathbf{v}^{P_1}\cdot\frac{d\mathbf{v}_t^{P_1}}{dt} + m_2\mathbf{v}^{P_2}\cdot\frac{d\mathbf{v}_t^{P_2}}{dt} = 0 \tag{17}$$

Equations (11) and (12) can be inspected to identify $\mathbf{v}_t^{P_1}$ and $\mathbf{v}_t^{P_2}$,

$$\mathbf{v}_t^{P_1} = (L_1+q_1)\,\dot{\theta}\,\hat{\mathbf{t}}_2 \qquad \mathbf{v}_t^{P_2} = (L_1+L_2+q_2)\,\dot{\theta}\,\hat{\mathbf{t}}_2 \tag{18}$$

and, with the angular velocity of the tube T (see Fig. 5.4.1)

$$\boldsymbol{\omega}^T = \dot{\theta}\,\hat{\mathbf{t}}_3 \tag{19}$$

in hand, the time derivatives of $\mathbf{v}_t^{P_1}$ and $\mathbf{v}_t^{P_2}$ are found to be

$$\frac{d\mathbf{v}_t^{P_1}}{dt} = (L_1+\dot{q}_1)\,\dot{\theta}\,\hat{\mathbf{t}}_2 + (L_1+q_1)(\ddot{\theta}\,\hat{\mathbf{t}}_2 - \dot{\theta}^2\,\hat{\mathbf{t}}_1) \tag{20}$$

$$\frac{d\mathbf{v}_t^{P_2}}{dt} = (L_1+L_2+\dot{q}_2)\,\dot{\theta}\,\hat{\mathbf{t}}_2 + (L_1+L_2+q_2)(\ddot{\theta}\,\hat{\mathbf{t}}_2 - \dot{\theta}^2\,\hat{\mathbf{t}}_1) \tag{21}$$

Clearly, the two requirements stated in Eqs. (16) and (17) are fulfilled if $\dot{\theta} \equiv 0$, so that

$$\theta = \overline{\theta} \tag{22}$$

where $\overline{\theta}$ is a constant. Moreover, K [see Eq. (13)] then is a homogeneous quadratic function of u_1 and u_2, and Eq. (3) applies, which makes it possible to write [see Eqs. (10) and (13)]

$$\begin{aligned} &-[m_1(L_1+q_1) + m_2(L_1+L_2+q_2)]g\cos\overline{\theta} \\ &\qquad + \tfrac{1}{2}[k_1{q_1}^2 + k_2(q_2-q_1)^2 + m_1{u_1}^2 + m_2{u_2}^2] = E \end{aligned} \tag{23}$$

with E an arbitrary constant.

9.3 THE CHECKING FUNCTION

In some cases there exists no energy integral (see Sec. 9.2) of the equations of motion of a simple nonholonomic system S in a Newtonian reference frame N. This can occur, for instance, when σ fails to vanish and Eq. (7.6.3) is left unsatisfied. Existence of an energy integral can also be precluded when S does not possess a potential energy (see Sec. 7.1), in which case the nonholonomic generalized active forces $\tilde{F}_r$ $(r = 1,\ldots,p)$ cannot be obtained as indicated in Eqs. (7.1.18). In these instances it is nevertheless possible to construct a function that remains constant throughout all motions of S in N. Such a function is referred to as a *checking function*; it can be used to test the results of numerical integrations of differential equations of motion of mechanical systems.

When a potential energy contribution V_α for S (see Sec. 7.2) is available such that Eq. (7.1.19) is satisfied with V replaced by V_α,

$$\frac{\partial V_\alpha}{\partial t} + \sum_{s=1}^{n} \frac{\partial V_\alpha}{\partial q_s}\left(X_s + \sum_{k=p+1}^{n} W_{sk} B_k\right) = 0 \tag{1}$$

where B_k has the same meaning as in Eqs. (3.5.2), and where W_{sk} and X_s have the same meanings as in Eqs. (3.6.5), the rth nonholonomic generalized active force may be regarded as the sum of two parts. The first part, denoted by $(\tilde{F}_r)_\alpha$, is obtained from V_α according to Eqs. (7.2.7). The second part, G_r, is the difference between $\tilde{F}_r$ and $(\tilde{F}_r)_\alpha$; thus

$$\begin{aligned} \tilde{F}_r &= (\tilde{F}_r)_\alpha + G_r \\ &\underset{(7.2.7)}{=} -\sum_{s=1}^{n} \frac{\partial V_\alpha}{\partial q_s}\left(W_{sr} + \sum_{k=p+1}^{n} W_{sk} A_{kr}\right) + G_r \qquad (r = 1,\ldots,p) \end{aligned} \tag{2}$$

where A_{kr} has the same meaning as in Eqs. (3.5.2). If a potential energy contribution is not available, then $(\tilde{F}_r)_\alpha = 0$ and $G_r = \tilde{F}_r$. Conversely, in the event there exists a potential energy V for S in N, then $G_r = 0$, $(\tilde{F}_r)_\alpha = \tilde{F}_r$ $(r = 1,\ldots,p)$, and V_α can be replaced by V in Eqs. (1) and (2).

If S is a holonomic system, $\tilde{F}_r$ in Eq. (2) is replaced with F_r, the holonomic generalized active force associated with u_r $(r = 1,\ldots,n)$, and the second summation is omitted in each of Eqs. (1) and (2).

A checking function $\mathscr{C}$ is defined as

$$\mathscr{C} \triangleq V_\alpha + K_2 - K_0 + Z \tag{3}$$

with K_0 and K_2 given by Eqs. (7.5.7) and (7.5.9), respectively. Z is a function of the time t that satisfies a differential equation

$$\dot{Z} = \sigma - \sum_{r=1}^{p} G_r u_r \tag{4}$$

to be integrated simultaneously with n kinematical differential equations and the p dynamical differential equations governing all motions of S in N. The definition of σ is given in Eq. (7.6.1).

When numerical integration of the aforementioned differential equations yields results in which $\mathscr{C}$ does not remain constant and the variations cannot be attributed to unavoidable roundoff errors, the results must be regarded as suspect. There may be defects in the differential equations, the checking function, or the integration procedure. On the other hand, a constant value of $\mathscr{C}$ provides confidence in the validity of the results.

Derivation In accordance with Eq. (7.6.22), the nonholonomic generalized inertia forces $\tilde{F}_r^{\star}$ corresponding to u_r $(r = 1, \ldots, p)$ satisfy the equation

$$-\sum_{r=1}^{p} \tilde{F}_r^{\star} u_r = \dot{K}_2 - \dot{K}_0 + \sigma \tag{5}$$

When $(\tilde{F}_r)_\alpha$ is given by Eqs. (7.2.7), the satisfaction of Eq. (1) and reference to Eqs. (7.1.18)–(7.1.20) permits one to write

$$\dot{V}_\alpha = -\sum_{r=1}^{p} (\tilde{F}_r)_\alpha u_r \tag{6}$$

Hence,

$$-\sum_{r=1}^{p} \tilde{F}_r u_r \underset{(2)}{=} -\sum_{r=1}^{p} [(\tilde{F}_r)_\alpha + G_r]\, u_r \underset{(6)}{=} \dot{V}_\alpha - \sum_{r=1}^{p} G_r u_r \tag{7}$$

Addition of Eqs. (5) and (7) yields

$$-\sum_{r=1}^{p} (\tilde{F}_r + \tilde{F}_r^{\star}) u_r = \dot{V}_\alpha - \sum_{r=1}^{p} G_r u_r + \dot{K}_2 - \dot{K}_0 + \sigma \underset{(4)}{=} \dot{V}_\alpha + \dot{K}_2 - \dot{K}_0 + \dot{Z} \tag{8}$$

According to Eqs. (8.1.2), each term $\tilde{F}_r + \tilde{F}_r^{\star}$ is equal to zero $(r = 1, \ldots, p)$; therefore, the left-hand member of Eq. (8) vanishes. In view of Eq. (3), the right-hand member of Eq. (8) is the time derivative of $\mathscr{C}$. Consequently, $0 = \dot{\mathscr{C}}$, which means that $\mathscr{C}$ remains constant throughout all motions of S in N.

Example Undesirable attitude motions of an orbiting spacecraft can be gradually alleviated with a device consisting of a sphere surrounded by a thin layer of viscous fluid as described in Problem 8.11. The spacecraft, represented by a rigid body C, contains a cavity in which a uniform sphere D is placed, as indicated in Fig. 9.3.1. A permanent magnet is fixed in D. The mass centers $C^{\star}$ and $D^{\star}$ of C and D, respectively, are coincident; they move in a circle of radius R around a particle P whose mass is m. Let $\hat{\mathbf{a}}_1$, $\hat{\mathbf{a}}_2$, $\hat{\mathbf{a}}_3$ be a right-handed set of mutually perpendicular unit vectors fixed in a reference frame A, where $\hat{\mathbf{a}}_1$ has the same direction as the position vector from P to $C^{\star}$, and where $\hat{\mathbf{a}}_3$, fixed both in A and in a Newtonian reference frame N, is normal to the plane of the orbit of $C^{\star}$. Thus, A has a simple angular velocity in N given by

$$^{N}\boldsymbol{\omega}^{A} = \left(\frac{Gm}{R^3}\right)^{1/2} \hat{\mathbf{a}}_3 \triangleq \Omega \hat{\mathbf{a}}_3 \tag{9}$$

where G is the universal gravitational constant.

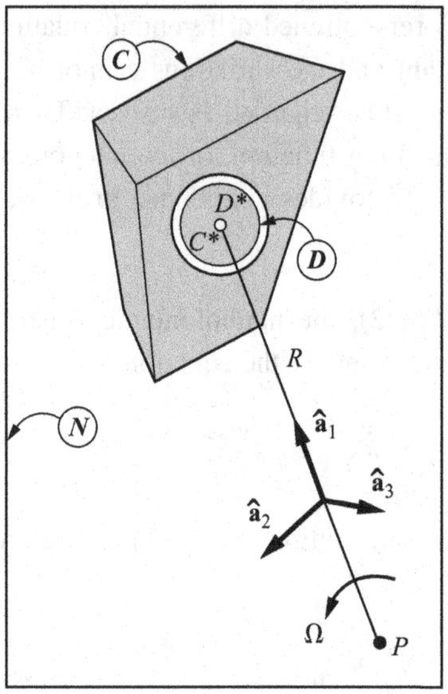

Figure 9.3.1

The set γ of all gravitational forces exerted by P on each rigid body is dealt with according to the assumptions made in Problem 10.9. For instance, the gravitational forces acting on C are taken to be equivalent to a couple of torque $(\mathbf{T}_C)_\gamma$ given by

$$(\mathbf{T}_C)_\gamma = 3\Omega^2 \hat{\mathbf{a}}_1 \times \underline{\mathbf{I}} \cdot \hat{\mathbf{a}}_1 \tag{10}$$

together with a force $(\mathbf{F}_C)_\gamma$ applied to C at $C^\star$ and expressed as

$$(\mathbf{F}_C)_\gamma = -\frac{GmM_C}{R^2}\hat{\mathbf{a}}_1 \tag{11}$$

where $\underline{\mathbf{I}}$ is the central inertia dyadic of C, and M_C is the mass of C. D is regarded as a sphere whose mass is distributed uniformly; the central inertia dyadic $\underline{\mathbf{J}}$ of D can therefore be written as

$$\underline{\mathbf{J}} = J\underline{\mathbf{U}} \tag{12}$$

where J is the central principal moment of inertia of D, and where $\underline{\mathbf{U}}$ is the unit dyadic. Consequently, Eq. (10) is in the case of D replaced by

$$(\mathbf{T}_D)_\gamma = 3\Omega^2 \hat{\mathbf{a}}_1 \times \underline{\mathbf{J}} \cdot \hat{\mathbf{a}}_1 \underset{(12)}{=} 3\Omega^2 \hat{\mathbf{a}}_1 \times J\underline{\mathbf{U}} \cdot \hat{\mathbf{a}}_1 = \mathbf{0} \tag{13}$$

while a force $(\mathbf{F}_D)_\gamma$ applied to D at $D^\star$ is given by

$$(\mathbf{F}_D)_\gamma = -\frac{GmM_D}{R^2}\hat{\mathbf{a}}_1 \tag{14}$$

where M_D is the mass of D.

The set δ of all forces exerted by the viscous fluid on C is equivalent (see Problem

8.11) to a couple of torque $(\mathbf{T}_C)_\delta$ that can be expressed in terms of ${}^C\boldsymbol{\omega}^D$, the angular velocity of D in C, as

$$(\mathbf{T}_C)_\delta = K_\delta\, {}^C\boldsymbol{\omega}^D \tag{15}$$

where K_δ is a positive constant. Moreover, the forces exerted by the fluid on D are replaced by a couple of torque $(\mathbf{T}_D)_\delta$ given by

$$(\mathbf{T}_D)_\delta = -K_\delta\, {}^C\boldsymbol{\omega}^D \tag{16}$$

A set β of distance forces is exerted on D as a result of interaction between the magnet contained within D and a magnetic field that surrounds P. The forces can be replaced by a couple whose torque $(\mathbf{T}_D)_\beta$ is given by

$$(\mathbf{T}_D)_\beta = \mathbf{m} \times \mathbf{B} \tag{17}$$

where $\mathbf{m}$ is the net dipole moment of the magnet fixed in D and where $\mathbf{B}$ is the local magnetic field vector at $D^\star$.

The system S made up of rigid bodies C and D possesses six degrees of freedom in N. Generalized velocities $u_1, \ldots, u_6$ may be introduced as

$${}^A\boldsymbol{\omega}^C \triangleq u_1\hat{\mathbf{c}}_1 + u_2\hat{\mathbf{c}}_2 + u_3\hat{\mathbf{c}}_3 \tag{18}$$

$${}^N\boldsymbol{\omega}^D \triangleq u_4\hat{\mathbf{c}}_1 + u_5\hat{\mathbf{c}}_2 + u_6\hat{\mathbf{c}}_3 \tag{19}$$

where ${}^A\boldsymbol{\omega}^C$ is the angular velocity of C in A, ${}^N\boldsymbol{\omega}^D$ is the angular velocity of D in N, and $\hat{\mathbf{c}}_1$, $\hat{\mathbf{c}}_2$, and $\hat{\mathbf{c}}_3$ form a right-handed set of mutually perpendicular unit vectors fixed in C. A checking function for S in N corresponding to $u_1, \ldots, u_6$ is to be constructed.

After noting that the angular velocity of C in N can be expressed as

$${}^N\boldsymbol{\omega}^C = {}^N\boldsymbol{\omega}^A + {}^A\boldsymbol{\omega}^C = \underset{(9)}{\Omega\hat{\mathbf{a}}_3} + \underset{(18)}{u_1\hat{\mathbf{c}}_1 + u_2\hat{\mathbf{c}}_2 + u_3\hat{\mathbf{c}}_3} \tag{20}$$

one can identify the following partial angular velocities:

$${}^N\boldsymbol{\omega}_1^C \underset{(20)}{=} \hat{\mathbf{c}}_1 \quad {}^N\boldsymbol{\omega}_2^C \underset{(20)}{=} \hat{\mathbf{c}}_2 \quad {}^N\boldsymbol{\omega}_3^C \underset{(20)}{=} \hat{\mathbf{c}}_3 \quad {}^N\boldsymbol{\omega}_r^C \underset{(20)}{=} \mathbf{0} \quad (r = 4,5,6) \tag{21}$$

$${}^N\boldsymbol{\omega}_r^D \underset{(19)}{=} \mathbf{0} \quad (r = 1,2,3) \quad {}^N\boldsymbol{\omega}_4^D \underset{(19)}{=} \hat{\mathbf{c}}_1 \quad {}^N\boldsymbol{\omega}_5^D \underset{(19)}{=} \hat{\mathbf{c}}_2 \quad {}^N\boldsymbol{\omega}_6^D \underset{(19)}{=} \hat{\mathbf{c}}_3 \tag{22}$$

Generalized active forces for S in N are formed with the aid of Eqs. (5.5.1),

$$\begin{aligned} F_r &= {}^N\boldsymbol{\omega}_r^C \cdot \left[(\mathbf{T}_C)_\gamma + (\mathbf{T}_C)_\delta\right] + {}^N\boldsymbol{\omega}_r^D \cdot \left[(\mathbf{T}_D)_\gamma + (\mathbf{T}_D)_\delta + (\mathbf{T}_D)_\beta\right] \\ &= {}^N\boldsymbol{\omega}_r^C \cdot (\underset{(10)}{3\Omega^2\hat{\mathbf{a}}_1 \times \underline{\mathbf{I}} \cdot \hat{\mathbf{a}}_1} + \underset{(15)}{K_\delta\, {}^C\boldsymbol{\omega}^D}) \\ &\quad + {}^N\boldsymbol{\omega}_r^D \cdot (\underset{(13)}{\mathbf{0}} \underset{(16)}{- K_\delta\, {}^C\boldsymbol{\omega}^D} + \underset{(17)}{\mathbf{m} \times \mathbf{B}}) \qquad (r = 1, \ldots, 6) \end{aligned} \tag{23}$$

Referring to Eqs. (18), (19), and (21)–(23), one can verify that the sum $\sum_{r=1}^{6} F_r u_r$ can be expressed as

$$\sum_{r=1}^{6} F_r u_r = {}^A\boldsymbol{\omega}^C \cdot (3\Omega^2\hat{\mathbf{a}}_1 \times \underline{\mathbf{I}} \cdot \hat{\mathbf{a}}_1 + K_\delta\, {}^C\boldsymbol{\omega}^D) + {}^N\boldsymbol{\omega}^D \cdot (-K_\delta\, {}^C\boldsymbol{\omega}^D + \mathbf{m} \times \mathbf{B}) \tag{24}$$

The inertia torque for C in N can be expressed, in view of Eq. (5.9.12), as

$$\mathbf{T}_C^{\star} = -\underline{\mathbf{I}} \cdot {}^N\boldsymbol{\alpha}^C - {}^N\boldsymbol{\omega}^C \times \underline{\mathbf{I}} \cdot {}^N\boldsymbol{\omega}^C \tag{25}$$

where ${}^N\boldsymbol{\alpha}^C$ denotes the angular acceleration of C in N. Similarly, the inertia torque for D in N is given by

$$\begin{aligned}\mathbf{T}_D^{\star} &= -\underline{\mathbf{J}} \cdot {}^N\boldsymbol{\alpha}^D - {}^N\boldsymbol{\omega}^D \times \underline{\mathbf{J}} \cdot {}^N\boldsymbol{\omega}^D \\ &= -\underline{\mathbf{J}} \cdot {}^N\boldsymbol{\alpha}^D - {}^N\boldsymbol{\omega}^D \times \underset{(12)}{J\underline{\mathbf{U}}} \cdot {}^N\boldsymbol{\omega}^D = -\underline{\mathbf{J}} \cdot {}^N\boldsymbol{\alpha}^D - \mathbf{0}\end{aligned} \tag{26}$$

where ${}^N\boldsymbol{\alpha}^D$ is the angular acceleration of D in N. Hence, generalized inertia forces for S in N assembled with the aid of Eqs. (5.9.7) are given by

$$\begin{aligned}F_r^{\star} &= {}^N\boldsymbol{\omega}_r^C \cdot \mathbf{T}_C^{\star} + {}^N\boldsymbol{\omega}_r^D \cdot \mathbf{T}_D^{\star} \\ &= -{}^N\boldsymbol{\omega}_r^C \cdot \underset{(25)}{(\underline{\mathbf{I}} \cdot {}^N\boldsymbol{\alpha}^C + {}^N\boldsymbol{\omega}^C \times \underline{\mathbf{I}} \cdot {}^N\boldsymbol{\omega}^C)} \\ &\quad - {}^N\boldsymbol{\omega}_r^D \cdot \underset{(26)}{(\underline{\mathbf{J}} \cdot {}^N\boldsymbol{\alpha}^D)} \qquad (r = 1,\ldots,6)\end{aligned} \tag{27}$$

Referring once again to Eqs. (18), (19), and (21)–(22), one can write the sum $\sum_{r=1}^{6} F_r^{\star} u_r$ as

$$\sum_{r=1}^{6} F_r^{\star} u_r \underset{(27)}{=} -{}^A\boldsymbol{\omega}^C \cdot (\underline{\mathbf{I}} \cdot {}^N\boldsymbol{\alpha}^C + {}^N\boldsymbol{\omega}^C \times \underline{\mathbf{I}} \cdot {}^N\boldsymbol{\omega}^C) - {}^N\boldsymbol{\omega}^D \cdot (\underline{\mathbf{J}} \cdot {}^N\boldsymbol{\alpha}^D) \tag{28}$$

The kinetic energy of S in N can be formed by referring to Eqs. (7.4.2)–(7.4.4) and writing

$$\begin{aligned}K &= \tfrac{1}{2}\left({}^N\boldsymbol{\omega}^C \cdot \underline{\mathbf{I}} \cdot {}^N\boldsymbol{\omega}^C + M_C\,{}^N\mathbf{v}^{C^\star} \cdot {}^N\mathbf{v}^{C^\star}\right) \\ &\quad + \tfrac{1}{2}\left({}^N\boldsymbol{\omega}^D \cdot \underline{\mathbf{J}} \cdot {}^N\boldsymbol{\omega}^D + M_D\,{}^N\mathbf{v}^{D^\star} \cdot {}^N\mathbf{v}^{D^\star}\right)\end{aligned} \tag{29}$$

Now, the mass centers $C^\star$ and $D^\star$ are coincident, fixed in A, and move on a circle of radius R; therefore, their velocities in N are identical and given by

$${}^N\mathbf{v}^{C^\star} = {}^N\mathbf{v}^{D^\star} = R\Omega\hat{\mathbf{a}}_2 \tag{30}$$

Hence, Eq. (29) can be rewritten as

$$\begin{aligned}K &= \tfrac{1}{2}\underset{(20)}{({}^N\boldsymbol{\omega}^A + {}^A\boldsymbol{\omega}^C)} \cdot \underline{\mathbf{I}} \cdot \underset{(20)}{({}^N\boldsymbol{\omega}^A + {}^A\boldsymbol{\omega}^C)} + \tfrac{1}{2}{}^N\boldsymbol{\omega}^D \cdot \underline{\mathbf{J}} \cdot {}^N\boldsymbol{\omega}^D \\ &\quad + \tfrac{1}{2}(M_C + M_D)\underset{(30)}{(R\Omega)^2} \\ &= \tfrac{1}{2}{}^N\boldsymbol{\omega}^A \cdot \underline{\mathbf{I}} \cdot {}^N\boldsymbol{\omega}^A + {}^A\boldsymbol{\omega}^C \cdot \underline{\mathbf{I}} \cdot {}^N\boldsymbol{\omega}^A + \tfrac{1}{2}{}^A\boldsymbol{\omega}^C \cdot \underline{\mathbf{I}} \cdot {}^A\boldsymbol{\omega}^C \\ &\quad + \tfrac{1}{2}{}^N\boldsymbol{\omega}^D \cdot \underline{\mathbf{J}} \cdot {}^N\boldsymbol{\omega}^D + \tfrac{1}{2}(M_C + M_D)(R\Omega)^2\end{aligned} \tag{31}$$

Each of the angular velocities ${}^A\boldsymbol{\omega}^C$ and ${}^N\boldsymbol{\omega}^D$ are homogeneous and of degree 1 in $u_1, \ldots, u_6$ [see Eqs. (18) and (19)]; therefore, the homogeneous kinetic energy functions K_0, K_1, and K_2 [see Eqs. (7.5.7)–(7.5.9)] are identified as

$$K_0 = \tfrac{1}{2}[{}^N\boldsymbol{\omega}^A \cdot \underline{\mathbf{I}} \cdot {}^N\boldsymbol{\omega}^A + (M_C + M_D)(R\Omega)^2] \tag{32}$$

$$K_1 = {}^A\boldsymbol{\omega}^C \cdot \underline{\mathbf{I}} \cdot {}^N\boldsymbol{\omega}^A \tag{33}$$

$$K_2 = \tfrac{1}{2}({}^A\boldsymbol{\omega}^C \cdot \underline{\mathbf{I}} \cdot {}^A\boldsymbol{\omega}^C + {}^N\boldsymbol{\omega}^D \cdot \underline{\mathbf{J}} \cdot {}^N\boldsymbol{\omega}^D) \tag{34}$$

When the set γ of gravitational forces exerted by a particle P on a rigid body C is given a replacement according to Eqs. (10) and (11), and the distance between P and $C^\star$ remains constant, γ makes a contribution V_γ to a potential energy of C that can be expressed as (see Problem 10.9)

$$V_\gamma = \tfrac{3}{2}\Omega^2(\hat{\mathbf{a}}_1 \cdot \underline{\mathbf{I}} \cdot \hat{\mathbf{a}}_1 - I_1) \tag{35}$$

where I_1 is a central principal moment of inertia of C. The torque $(\mathbf{T}_D)_\gamma$ vanishes as demonstrated in Eq. (13) because D is a uniform sphere; hence, the gravitational forces acting on D make no contribution to a potential energy. This can be observed formally by replacing $\underline{\mathbf{I}}$ and I_1 in the right-hand member of Eq. (35) with, respectively, $\underline{\mathbf{J}}$ and J, the value of *all* central moments of inertia of D,

$$\tfrac{3}{2}\Omega^2(\hat{\mathbf{a}}_1 \cdot \underset{(12)}{J\underline{\mathbf{U}}} \cdot \hat{\mathbf{a}}_1 - J) = \tfrac{3}{2}\Omega^2(J - J) = 0 \tag{36}$$

Consequently, the entire potential energy contribution for S is given by V_γ in Eq. (35).

The quantity σ can be formed by appealing to Eq. (7.6.2),

$$\sigma = \sigma_C + \sigma_D = M_C\,{}^N\mathbf{v}^{C\star} \cdot \frac{{}^N d\,{}^N\mathbf{v}_t^{C\star}}{dt} + {}^N\boldsymbol{\omega}^C \cdot \underline{\mathbf{I}} \cdot \frac{{}^N d\,{}^N\boldsymbol{\omega}_t^C}{dt}$$
$$+ M_D\,{}^N\mathbf{v}^{D\star} \cdot \frac{{}^N d\,{}^N\mathbf{v}_t^{D\star}}{dt} + {}^N\boldsymbol{\omega}^D \cdot \underline{\mathbf{J}} \cdot \frac{{}^N d\,{}^N\boldsymbol{\omega}_t^D}{dt} \tag{37}$$

All of the generalized velocities $u_1, \ldots, u_6$ are absent from Eqs. (30); therefore,

$${}^N\mathbf{v}_t^{C\star} = {}^N\mathbf{v}_t^{D\star} = R\Omega\hat{\mathbf{a}}_2 \tag{38}$$

Differentiation with respect to time in N yields

$$\frac{{}^N d\,{}^N\mathbf{v}_t^{C\star}}{dt} = \frac{{}^N d\,{}^N\mathbf{v}_t^{D\star}}{dt} \underset{(38)}{=} R\Omega\frac{{}^N d\hat{\mathbf{a}}_2}{dt} = R\Omega\,{}^N\boldsymbol{\omega}^A \times \hat{\mathbf{a}}_2 \underset{(9)}{=} -R\Omega^2\hat{\mathbf{a}}_1 \tag{39}$$

so that

$${}^N\mathbf{v}^{C\star} \cdot \frac{{}^N d\,{}^N\mathbf{v}_t^{C\star}}{dt} = {}^N\mathbf{v}^{D\star} \cdot \frac{{}^N d\,{}^N\mathbf{v}_t^{D\star}}{dt} = R\Omega\hat{\mathbf{a}}_2 \cdot (-R\Omega^2)\hat{\mathbf{a}}_1 = 0 \tag{40}$$

Inspection of Eq. (20) reveals that

$${}^N\boldsymbol{\omega}_t^C = \Omega\hat{\mathbf{a}}_3 \tag{41}$$

and the time derivative in N of this quantity vanishes

$$\frac{{}^N d\,{}^N\boldsymbol{\omega}_t^C}{dt} = \Omega\frac{{}^N d\hat{\mathbf{a}}_3}{dt} = \mathbf{0} \tag{42}$$

because $\hat{\mathbf{a}}_3$ is fixed in N as well as in A. In addition, after referring to Eq. (19), one can write

$$^N\boldsymbol{\omega}_t^D = \mathbf{0} \qquad \frac{^N d\,^N\boldsymbol{\omega}_t^D}{dt} = \mathbf{0} \tag{43}$$

Thus, upon substituting from Eqs. (40), (42), and (43) into (37), one finds that

$$\sigma = 0 \tag{44}$$

and is then in position to form $\dot{Z}$ according to Eq. (4) by designating the sum $\sum_{r=1}^{6} G_r u_r$ as those terms in Eq. (24) associated with the sets of forces δ and β,

$$\begin{aligned} \dot{Z} &\underset{(4)\ (44)}{=} 0 - {}^A\boldsymbol{\omega}^C \cdot (K_\delta\,{}^C\boldsymbol{\omega}^D) - {}^N\boldsymbol{\omega}^D \cdot (-K_\delta\,{}^C\boldsymbol{\omega}^D + \mathbf{m}\times\mathbf{B}) \\ &= K_\delta\,{}^C\boldsymbol{\omega}^D \cdot ({}^N\boldsymbol{\omega}^D - {}^A\boldsymbol{\omega}^C) - {}^N\boldsymbol{\omega}^D \cdot \mathbf{m}\times\mathbf{B} \end{aligned} \tag{45}$$

A checking function for S in N corresponding to $u_1, \ldots, u_6$ now can be written as

$$\begin{aligned} \mathscr{C} &\underset{(3)}{=} \tfrac{3}{2}\Omega^2(\hat{\mathbf{a}}_1 \cdot \underset{(35)}{\underline{\mathbf{I}}} \cdot \hat{\mathbf{a}}_1 - I_1) + \tfrac{1}{2}({}^A\boldsymbol{\omega}^C \cdot \underline{\mathbf{I}} \cdot {}^A\boldsymbol{\omega}^C + {}^N\boldsymbol{\omega}^D \cdot \underset{(34)}{\underline{\mathbf{J}}} \cdot {}^N\boldsymbol{\omega}^D) \\ &\quad - \tfrac{1}{2}({}^N\boldsymbol{\omega}^A \cdot \underset{(32)}{\underline{\mathbf{I}}} \cdot {}^N\boldsymbol{\omega}^A) + Z \\ &= \tfrac{3}{2}\Omega^2(\hat{\mathbf{a}}_1 \cdot \underline{\mathbf{I}} \cdot \hat{\mathbf{a}}_1 - I_1) + Z \\ &\quad + \tfrac{1}{2}({}^A\boldsymbol{\omega}^C \cdot \underline{\mathbf{I}} \cdot {}^A\boldsymbol{\omega}^C - {}^N\boldsymbol{\omega}^A \cdot \underline{\mathbf{I}} \cdot {}^N\boldsymbol{\omega}^A + {}^N\boldsymbol{\omega}^D \cdot \underline{\mathbf{J}} \cdot {}^N\boldsymbol{\omega}^D) \end{aligned} \tag{46}$$

The term $\frac{1}{2}(M_C + M_D)(R\Omega)^2$ appearing in K_0 [see Eq. (32)] is a constant, and can therefore be omitted from $\mathscr{C}$. Before leaving this example, we verify that $\mathscr{C}$ is a checking function for S by differentiating Eq. (46) with respect to time and showing that

$$\dot{\mathscr{C}} = -\sum_{r=1}^{6}(F_r + F_r^\star)u_r \underset{(8.1.1)}{=} 0 \tag{47}$$

Two identities involving a dyadic and any two vectors, $\mathbf{u}$ and $\mathbf{v}$, will prove useful in the sequel. First, for any dyadic $\underline{\mathbf{Q}} = \mathbf{ab} + \mathbf{cd}$,

$$\begin{aligned} \mathbf{u}\times\mathbf{v}\cdot\underline{\mathbf{Q}} &= \mathbf{u}\times\mathbf{v}\cdot(\mathbf{ab}+\mathbf{cd}) = \mathbf{u}\cdot\mathbf{v}\times(\mathbf{ab}+\mathbf{cd}) \\ &= \mathbf{u}\cdot\mathbf{v}\times\underline{\mathbf{Q}} \end{aligned} \tag{48}$$

Second, a *symmetric* dyadic $\underline{\mathbf{S}}$ is one that satisfies the relationship

$$\mathbf{u}\cdot\underline{\mathbf{S}}\cdot\mathbf{v} = \mathbf{v}\cdot\underline{\mathbf{S}}\cdot\mathbf{u} \tag{49}$$

Every inertia dyadic is symmetric, as can be verified by referring to Eqs. (4.3.4) and (4.5.22).

After considering the definition in Problem 5.14 of the time derivative of a dyadic in a reference frame, one finds that the time derivative in C of $\underline{\mathbf{I}}$, the central inertia dyadic of C, is the zero dyadic $\underline{\mathbf{0}}$. For this reason, time differentiation of terms involving $\underline{\mathbf{I}}$ is conveniently performed in C in what follows. The time derivative of V_γ, the first term in the right-hand member of Eq. (46), is thus obtained as

$$\begin{aligned}
\dot{V}_\gamma &\underset{(35)}{=} \tfrac{3}{2}\Omega^2\left(\frac{^C d\hat{\mathbf{a}}_1}{dt}\cdot\underline{\mathbf{I}}\cdot\hat{\mathbf{a}}_1 + \hat{\mathbf{a}}_1\cdot\frac{^C d\underline{\mathbf{I}}}{dt}\cdot\hat{\mathbf{a}}_1 + \hat{\mathbf{a}}_1\cdot\underline{\mathbf{I}}\cdot\frac{^C d\hat{\mathbf{a}}_1}{dt} - 0\right) \\
&= 3\Omega^2\left(\frac{^C d\hat{\mathbf{a}}_1}{dt}\cdot\underline{\mathbf{I}}\cdot\hat{\mathbf{a}}_1\right) + \tfrac{3}{2}\Omega^2(\hat{\mathbf{a}}_1\cdot\underset{(49)}{\underline{\mathbf{0}}}\cdot\hat{\mathbf{a}}_1) \\
&\underset{(2.3.1)}{=} 3\Omega^2\left(\frac{^A d\hat{\mathbf{a}}_1}{dt} - {}^A\boldsymbol{\omega}^C\times\hat{\mathbf{a}}_1\right)\cdot\underline{\mathbf{I}}\cdot\hat{\mathbf{a}}_1 \\
&\underset{(48)}{=} -3\Omega^{2A}\boldsymbol{\omega}^C\cdot\hat{\mathbf{a}}_1\times\underline{\mathbf{I}}\cdot\hat{\mathbf{a}}_1
\end{aligned} \tag{50}$$

This result in turn permits one to write

$$\begin{aligned}
&\frac{d}{dt}\left[\tfrac{3}{2}\Omega^2(\hat{\mathbf{a}}_1\cdot\underline{\mathbf{I}}\cdot\hat{\mathbf{a}}_1 - I_1) + Z\right] \\
&\quad= -{}^A\boldsymbol{\omega}^C\cdot\underset{(50)}{(3\Omega^2\hat{\mathbf{a}}_1\times\underline{\mathbf{I}}\cdot\hat{\mathbf{a}}_1)} \\
&\qquad - {}^A\boldsymbol{\omega}^C\cdot(K_\delta{}^C\boldsymbol{\omega}^D) - {}^N\boldsymbol{\omega}^D\cdot\underset{(45)}{(-K_\delta{}^C\boldsymbol{\omega}^D + \mathbf{m}\times\mathbf{B})} \\
&\quad\underset{(24)}{=} -\sum_{r=1}^{6} F_r u_r
\end{aligned} \tag{51}$$

Next, we differentiate the remaining terms in Eq. (46) that involve $\underline{\mathbf{I}}$, namely,

$$\begin{aligned}
&\frac{d}{dt}\left[\tfrac{1}{2}({}^A\boldsymbol{\omega}^C\cdot\underline{\mathbf{I}}\cdot{}^A\boldsymbol{\omega}^C - {}^N\boldsymbol{\omega}^A\cdot\underline{\mathbf{I}}\cdot{}^N\boldsymbol{\omega}^A)\right] \\
&= \tfrac{1}{2}\left(\frac{^C d^A\boldsymbol{\omega}^C}{dt}\cdot\underline{\mathbf{I}}\cdot{}^A\boldsymbol{\omega}^C + {}^A\boldsymbol{\omega}^C\cdot\underline{\mathbf{0}}\cdot{}^A\boldsymbol{\omega}^C + {}^A\boldsymbol{\omega}^C\cdot\underline{\mathbf{I}}\cdot\frac{^C d^A\boldsymbol{\omega}^C}{dt}\right) \\
&\quad - \tfrac{1}{2}\left(\frac{^C d^N\boldsymbol{\omega}^A}{dt}\cdot\underline{\mathbf{I}}\cdot{}^N\boldsymbol{\omega}^A + {}^N\boldsymbol{\omega}^A\cdot\underline{\mathbf{0}}\cdot{}^N\boldsymbol{\omega}^A + {}^N\boldsymbol{\omega}^A\cdot\underline{\mathbf{I}}\cdot\frac{^C d^N\boldsymbol{\omega}^A}{dt}\right) \\
&\underset{(49)}{=} {}^A\boldsymbol{\omega}^C\cdot\underline{\mathbf{I}}\cdot\frac{^C d^A\boldsymbol{\omega}^C}{dt} - {}^N\boldsymbol{\omega}^A\cdot\underline{\mathbf{I}}\cdot\frac{^C d^N\boldsymbol{\omega}^A}{dt} \\
&= {}^A\boldsymbol{\omega}^C\cdot\underline{\mathbf{I}}\cdot\frac{^C d^A\boldsymbol{\omega}^C}{dt} - ({}^N\boldsymbol{\omega}^C - {}^A\boldsymbol{\omega}^C)\cdot\underline{\mathbf{I}}\cdot\frac{^C d^N\boldsymbol{\omega}^A}{dt} \\
&= {}^A\boldsymbol{\omega}^C\cdot\underline{\mathbf{I}}\cdot\frac{^C d}{dt}({}^N\boldsymbol{\omega}^A + {}^A\boldsymbol{\omega}^C) - \frac{^C d^N\boldsymbol{\omega}^A}{dt}\underset{(49)}{\cdot}\underline{\mathbf{I}}\cdot{}^N\boldsymbol{\omega}^C \\
&= {}^A\boldsymbol{\omega}^C\cdot\underline{\mathbf{I}}\cdot\underset{(2.5.2)}{{}^N\boldsymbol{\alpha}^C} - \left(\frac{^N d^N\boldsymbol{\omega}^A}{dt}\underset{(2.3.1)}{-}{}^N\boldsymbol{\omega}^C\times{}^N\boldsymbol{\omega}^A\right)\cdot\underline{\mathbf{I}}\cdot{}^N\boldsymbol{\omega}^C \\
&= {}^A\boldsymbol{\omega}^C\cdot\underline{\mathbf{I}}\cdot{}^N\boldsymbol{\alpha}^C - (\underset{(9)}{\mathbf{0}} + {}^N\boldsymbol{\omega}^A\times{}^N\boldsymbol{\omega}^C)\cdot\underline{\mathbf{I}}\cdot{}^N\boldsymbol{\omega}^C \\
&= {}^A\boldsymbol{\omega}^C\cdot\underline{\mathbf{I}}\cdot{}^N\boldsymbol{\alpha}^C - \underset{(48)}{{}^N\boldsymbol{\omega}^A}\cdot{}^N\boldsymbol{\omega}^C\times\underline{\mathbf{I}}\cdot{}^N\boldsymbol{\omega}^C \\
&= {}^A\boldsymbol{\omega}^C\cdot\underline{\mathbf{I}}\cdot{}^N\boldsymbol{\alpha}^C - ({}^N\boldsymbol{\omega}^C - {}^A\boldsymbol{\omega}^C)\cdot{}^N\boldsymbol{\omega}^C\times\underline{\mathbf{I}}\cdot{}^N\boldsymbol{\omega}^C \\
&= {}^A\boldsymbol{\omega}^C\cdot(\underline{\mathbf{I}}\cdot{}^N\boldsymbol{\alpha}^C + {}^N\boldsymbol{\omega}^C\times\underline{\mathbf{I}}\cdot{}^N\boldsymbol{\omega}^C)
\end{aligned} \tag{52}$$

where the term that vanishes in the final step does so because $\underline{\mathbf{I}} \cdot {}^N\boldsymbol{\omega}^C$ is a vector, say, $\mathbf{H}$, and clearly ${}^N\boldsymbol{\omega}^C \cdot {}^N\boldsymbol{\omega}^C \times \mathbf{H} = 0$. The final term in Eq. (46) that requires attention is easily differentiated with respect to time,

$$\begin{aligned}
\frac{d}{dt}\left[\tfrac{1}{2}({}^N\boldsymbol{\omega}^D \cdot \underline{\mathbf{J}} \cdot {}^N\boldsymbol{\omega}^D)\right] &\underset{(12)}{=} \frac{d}{dt}\left[\tfrac{1}{2}({}^N\boldsymbol{\omega}^D \cdot J\,{}^N\boldsymbol{\omega}^D)\right] \\
&= \frac{J}{2}\left(\frac{{}^N d\,{}^N\boldsymbol{\omega}^D}{dt} \cdot {}^N\boldsymbol{\omega}^D + {}^N\boldsymbol{\omega}^D \cdot \frac{{}^N d\,{}^N\boldsymbol{\omega}^D}{dt}\right) \\
&= J\,{}^N\boldsymbol{\omega}^D \cdot {}^N\boldsymbol{\alpha}^D \underset{(2.5.1)}{=} {}^N\boldsymbol{\omega}^D \cdot J\underline{\mathbf{U}} \cdot {}^N\boldsymbol{\alpha}^D \\
&\underset{(12)}{=} {}^N\boldsymbol{\omega}^D \cdot \underline{\mathbf{J}} \cdot {}^N\boldsymbol{\alpha}^D
\end{aligned} \tag{53}$$

Therefore,

$$\begin{aligned}
&\frac{d}{dt}\left[\tfrac{1}{2}({}^A\boldsymbol{\omega}^C \cdot \underline{\mathbf{I}} \cdot {}^A\boldsymbol{\omega}^C - {}^N\boldsymbol{\omega}^A \cdot \underline{\mathbf{I}} \cdot {}^N\boldsymbol{\omega}^A + {}^N\boldsymbol{\omega}^D \cdot \underline{\mathbf{J}} \cdot {}^N\boldsymbol{\omega}^D)\right] \\
&\quad = \underbrace{{}^A\boldsymbol{\omega}^C \cdot (\underline{\mathbf{I}} \cdot {}^N\boldsymbol{\alpha}^C + {}^N\boldsymbol{\omega}^C \times \underline{\mathbf{I}} \cdot {}^N\boldsymbol{\omega}^C)}_{(52)} + \underbrace{{}^N\boldsymbol{\omega}^D \cdot \underline{\mathbf{J}} \cdot {}^N\boldsymbol{\alpha}^D}_{(53)} \\
&\quad \underset{(28)}{=} -\sum_{r=1}^{6} F_r^{\star} u_r
\end{aligned} \tag{54}$$

Finally, Eqs. (46), (51), and (54) establish the validity of Eq. (47), which shows that $\mathscr{C}$ as reported in Eq. (46) remains constant for all motions of S in N.

9.4 MOMENTUM INTEGRALS

If σ is the set of all contact forces and distance forces acting on the particles of a system S, $\mathbf{R}$ is the resultant of σ (see Sec. 5.1), $\mathbf{L}$ is the linear momentum of S in a Newtonian reference frame N, $\hat{\mathbf{n}}$ is a unit vector, and

$$\mathbf{R} \cdot \hat{\mathbf{n}} + \mathbf{L} \cdot \frac{{}^N d\hat{\mathbf{n}}}{dt} = 0 \tag{1}$$

then the equation

$$\mathbf{L} \cdot \hat{\mathbf{n}} = C \tag{2}$$

where C is any constant, is an integral of the equations of motion of S in N. Similarly, if $\mathbf{M}$ is the moment of σ about $S^{\star}$, the mass center of S, or about a point O fixed in N, $\mathbf{H}$ is the angular momentum of S about $S^{\star}$ or O in N, and

$$\mathbf{M} \cdot \hat{\mathbf{n}} + \mathbf{H} \cdot \frac{{}^N d\hat{\mathbf{n}}}{dt} = 0 \tag{3}$$

then the equation

$$\mathbf{H} \cdot \hat{\mathbf{n}} = C \tag{4}$$

where C is any constant, is an integral of the equations of motion of S in N. Finally, if S is a holonomic system possessing generalized coordinates $q_1, \ldots, q_n$ and a kinetic potential $\mathscr{L}$ in N (see Problem 12.15), and q_k is absent from $\mathscr{L}$ when $\mathscr{L}$ is expressed as a function of $q_1, \ldots, q_n, \dot{q}_1, \ldots, \dot{q}_n$, and t, then

$$\frac{\partial \mathscr{L}}{\partial \dot{q}_k} = C \tag{5}$$

where C is any constant, is an integral of the equations of motion of S in N.

Equations (2), (4), and (5) are called *momentum integrals* of the equations of motion, and q_k is referred to as a *cyclic* or *ignorable* coordinate.

A momentum integral is linear in the motion variables, and can be used to construct relationships having the form of Eqs. (3.5.2). This enables one, for example, to treat a holonomic system as a simple nonholonomic system and employ Eqs. (8.1.2) to construct dynamical equations.

Derivations In accordance with the *linear momentum principle,*

$$\frac{{}^N d\mathbf{L}}{dt} = \mathbf{R} \tag{6}$$

Hence,

$$\frac{d}{dt}(\mathbf{L} \cdot \hat{\mathbf{n}}) = \frac{{}^N d\mathbf{L}}{dt} \cdot \hat{\mathbf{n}} + \mathbf{L} \cdot \frac{{}^N d\hat{\mathbf{n}}}{dt} \underset{(6)}{=} \mathbf{R} \cdot \hat{\mathbf{n}} + \mathbf{L} \cdot \frac{{}^N d\hat{\mathbf{n}}}{dt} \tag{7}$$

and, when Eq. (1) is satisfied, then Eq. (7) yields

$$\frac{d}{dt}(\mathbf{L} \cdot \hat{\mathbf{n}}) = 0 \tag{8}$$

from which Eq. (2) follows immediately. Similarly, the *angular momentum principle* asserts that

$$\frac{{}^N d\mathbf{H}}{dt} = \mathbf{M} \tag{9}$$

so that

$$\frac{d}{dt}(\mathbf{H} \cdot \hat{\mathbf{n}}) = \frac{{}^N d\mathbf{H}}{dt} \cdot \hat{\mathbf{n}} + \mathbf{H} \cdot \frac{{}^N d\hat{\mathbf{n}}}{dt} \underset{(9)}{=} \mathbf{M} \cdot \hat{\mathbf{n}} + \mathbf{H} \cdot \frac{{}^N d\hat{\mathbf{n}}}{dt} \tag{10}$$

and this equation together with Eq. (3) leads to Eq. (4). Finally, when q_k is absent from $\mathscr{L}$, Eq. (5) is an immediate consequence of Lagrange's equations of the second kind (see Problem 12.15).

To see that Eq. (2) is linear in the motion variables, one can proceed as follows. For a system S consisting of particles $P_1, \ldots, P_\nu$ of masses $m_1, \ldots, m_\nu$, respectively, define $\mathbf{L}$ as

$$\mathbf{L} \triangleq \sum_{i=1}^{\nu} m_i {}^N\mathbf{v}^{P_i} \tag{11}$$

and substitute from Eqs. (3.6.2) to obtain

$$
\begin{aligned}
\mathbf{L}\cdot\hat{\mathbf{n}} - C &\underset{(11)}{=} \sum_{i=1}^{\nu} m_i {}^N\mathbf{v}^{P_i}\cdot\hat{\mathbf{n}} - C \\
&\underset{(3.6.2)}{=} \sum_{i=1}^{\nu} m_i \left(\sum_{r=1}^{n} {}^N\mathbf{v}_r^{P_i} u_r + {}^N\mathbf{v}_t^{P_i}\right)\cdot\hat{\mathbf{n}} - C \\
&= \sum_{r=1}^{n}\left(\sum_{i=1}^{\nu} m_i {}^N\mathbf{v}_r^{P_i}\cdot\hat{\mathbf{n}}\right) u_r + \sum_{i=1}^{\nu} m_i {}^N\mathbf{v}_t^{P_i}\cdot\hat{\mathbf{n}} - C \\
&\underset{(2)}{=} 0
\end{aligned}
\tag{12}
$$

This equation can be used to express, say, u_n, in terms of $u_1,\ldots,u_{n-1}$, thereby producing a relationship having the form of Eqs. (3.5.2). Proceeding similarly, one can demonstrate that Eq. (4) is linear in the motion variables.

$$
\begin{aligned}
\mathbf{H}\cdot\hat{\mathbf{n}} - C &\underset{(4.5.27)}{=} \sum_{i=1}^{\nu} m_i \mathbf{p}_i \times {}^N\mathbf{v}^{P_i}\cdot\hat{\mathbf{n}} - C \\
&\underset{(3.6.2)}{=} \sum_{i=1}^{\nu} m_i \mathbf{p}_i \times \left(\sum_{r=1}^{n} {}^N\mathbf{v}_r^{P_i} u_r + {}^N\mathbf{v}_t^{P_i}\right)\cdot\hat{\mathbf{n}} - C \\
&= \sum_{r=1}^{n}\left(\sum_{i=1}^{\nu} m_i \mathbf{p}_i \times {}^N\mathbf{v}_r^{P_i}\cdot\hat{\mathbf{n}}\right) u_r + \sum_{i=1}^{\nu} m_i \mathbf{p}_i \times {}^N\mathbf{v}_t^{P_i}\cdot\hat{\mathbf{n}} - C \\
&\underset{(4)}{=} 0
\end{aligned}
\tag{13}
$$

where $\mathbf{p}_i$ is the position vector from $S^\star$ or O to P_i $(i = 1,\ldots,\nu)$. Finally, referring to Problem 12.15, one can write

$$
\frac{\partial \mathscr{L}}{\partial \dot{q}_k} = \frac{\partial (K - V)}{\partial \dot{q}_k} = \frac{\partial K}{\partial \dot{q}_k}
\tag{14}
$$

because a potential energy V is not regarded as a function of $\dot{q}_1,\ldots,\dot{q}_n$ (see Sec. 7.1). Hence, Eq. (5) can be written as

$$
\begin{aligned}
\frac{\partial \mathscr{L}}{\partial \dot{q}_k} - C &\underset{(14)}{=} \frac{\partial K}{\partial \dot{q}_k} - C \\
&\underset{(8.8.8)}{=} \sum_{i=1}^{\nu} m_i {}^N\mathbf{v}_k^{P_i}\cdot {}^N\mathbf{v}^{P_i} - C \\
&\underset{(3.6.2)}{=} \sum_{i=1}^{\nu} m_i {}^N\mathbf{v}_k^{P_i}\cdot \left(\sum_{r=1}^{n} {}^N\mathbf{v}_r^{P_i} \dot{q}_r + {}^N\mathbf{v}_t^{P_i}\right) - C \\
&= \sum_{r=1}^{n}\left(\sum_{i=1}^{\nu} m_i {}^N\mathbf{v}_k^{P_i}\cdot {}^N\mathbf{v}_r^{P_i}\right) \dot{q}_r + \sum_{i=1}^{\nu} m_i {}^N\mathbf{v}_k^{P_i}\cdot {}^N\mathbf{v}_t^{P_i} - C \\
&\underset{(5)}{=} 0
\end{aligned}
\tag{15}
$$

Examples Figure 9.4.1 is a schematic representation of a *gyrostat* G consisting of a rigid body A and an axially symmetric rotor B that is constrained to rotate relative to A about an axis fixed in A. G has a mass M, and $\underline{\mathbf{I}}_G$, the central inertia dyadic of G, expressed in terms of K_1, K_2, and K_3, the central principal moments of inertia of G, is given by

$$\underline{\mathbf{I}}_G = K_1\hat{\mathbf{a}}_1\hat{\mathbf{a}}_1 + K_2\hat{\mathbf{a}}_2\hat{\mathbf{a}}_2 + K_3\hat{\mathbf{a}}_3\hat{\mathbf{a}}_3 \tag{16}$$

where $\hat{\mathbf{a}}_1$, $\hat{\mathbf{a}}_2$, $\hat{\mathbf{a}}_3$ are mutually perpendicular unit vectors fixed in A and parallel to central principal axes of G. The axis of symmetry of B is parallel to a unit vector $\hat{\mathbf{b}}_3$, and $\underline{\mathbf{I}}_B$, the central inertia dyadic of B, can be expressed as

$$\underline{\mathbf{I}}_B = I(\hat{\mathbf{b}}_1\hat{\mathbf{b}}_1 + \hat{\mathbf{b}}_2\hat{\mathbf{b}}_2) + J\hat{\mathbf{b}}_3\hat{\mathbf{b}}_3 \tag{17}$$

where $\hat{\mathbf{b}}_1$ and $\hat{\mathbf{b}}_2$ are unit vectors fixed in A, perpendicular to each other, and such that $\hat{\mathbf{b}}_3 = \hat{\mathbf{b}}_1 \times \hat{\mathbf{b}}_2$.

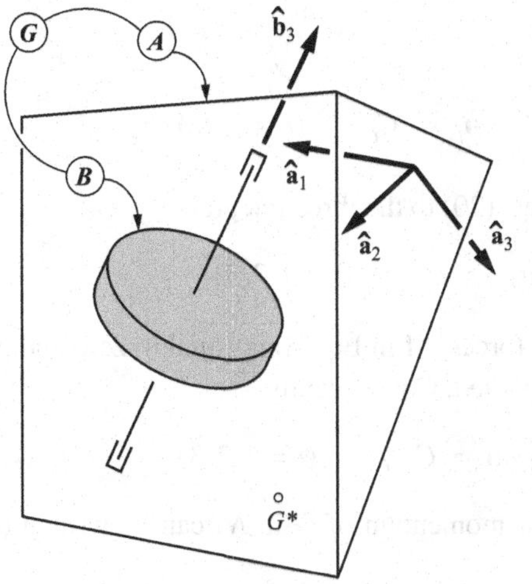

Figure 9.4.1

When G moves in a Newtonian reference frame N in the absence of external constraints, it is convenient to work with generalized velocities $u_1, \ldots, u_7$ defined as

$$u_i \triangleq {}^N\boldsymbol{\omega}^A \cdot \hat{\mathbf{a}}_i \qquad (i = 1,2,3) \tag{18}$$

$$u_4 \triangleq {}^A\boldsymbol{\omega}^B \cdot \hat{\mathbf{b}}_3 \tag{19}$$

$$u_{4+i} \triangleq {}^N\mathbf{v}^{G^\star} \cdot \hat{\mathbf{n}}_i \qquad (i = 1,2,3) \tag{20}$$

where $G^\star$ is the mass center of G and $\hat{\mathbf{n}}_1$, $\hat{\mathbf{n}}_2$, $\hat{\mathbf{n}}_3$ are mutually perpendicular unit vectors fixed in N. As generalized coordinates one can use, for example, angles q_1, q_2, and q_3 like those in Problem 1.1 to characterize the relative orientations of $\hat{\mathbf{n}}_1$, $\hat{\mathbf{n}}_2$, $\hat{\mathbf{n}}_3$ and $\hat{\mathbf{a}}_1$, $\hat{\mathbf{a}}_2$, $\hat{\mathbf{a}}_3$; an angle q_4 between two lines that are perpendicular to $\hat{\mathbf{b}}_3$, one line fixed in A, the other fixed in B; and Cartesian coordinates q_5, q_6, and q_7 of $G^\star$ relative to a set of axes fixed in N. Table 9.4.1 shows the relationship between $\hat{\mathbf{a}}_1$, $\hat{\mathbf{a}}_2$, $\hat{\mathbf{a}}_3$ and $\hat{\mathbf{n}}_1$, $\hat{\mathbf{n}}_2$, $\hat{\mathbf{n}}_3$.

Table 9.4.1

	$\hat{\mathbf{a}}_1$	$\hat{\mathbf{a}}_2$	$\hat{\mathbf{a}}_3$
$\hat{\mathbf{n}}_1$	c_2c_3	$-c_2s_3$	s_2
$\hat{\mathbf{n}}_2$	$s_1s_2c_3 + s_3c_1$	$-s_1s_2s_3 + c_3c_1$	$-s_1c_2$
$\hat{\mathbf{n}}_3$	$-c_1s_2c_3 + s_3s_1$	$c_1s_2s_3 + c_3s_1$	c_1c_2

Letting G play the role of S, suppose that G moves in N in the absence of external forces. Then $\mathbf{R}$ in Eq. (1) is equal to zero and, if one takes for $\hat{\mathbf{n}}$ in Eq. (1) any one of $\hat{\mathbf{n}}_1$, $\hat{\mathbf{n}}_2$, $\hat{\mathbf{n}}_3$, then Eq. (1) is satisfied because ${}^N d\hat{\mathbf{n}}_i/dt = \mathbf{0}$ $(i = 1,2,3)$. Since $\mathbf{L}$, the linear momentum of G in N, is given by

$$\mathbf{L} = M\,{}^N\mathbf{v}^{G^\star} \tag{21}$$

one can, therefore, write

$$ {}^N\mathbf{v}^{G^\star} \cdot \hat{\mathbf{n}}_i \underset{(2)}{=} C_i \qquad (i = 1,2,3) \tag{22}$$

which leads with the aid of Eqs. (20) to the three integrals

$$u_{4+i} = C_i \qquad (i = 1,2,3) \tag{23}$$

In the absence of external forces, $\mathbf{M}$ in Eq. (3) is equal to zero and, with $\hat{\mathbf{n}}$ equal to $\hat{\mathbf{n}}_i$ $(i = 1,2,3)$, Eq. (3) is satisfied. Consequently,

$$\mathbf{H}_G \cdot \hat{\mathbf{n}}_i = C_{3+i} \qquad (i = 1,2,3) \tag{24}$$

where $\mathbf{H}_G$, the central angular momentum of G in N, can be written (see Problem 6.5)

$$\mathbf{H}_G = \underline{\mathbf{I}}_G \cdot {}^N\boldsymbol{\omega}^A + \underline{\mathbf{I}}_B \cdot {}^A\boldsymbol{\omega}^B \tag{25}$$

or

$$\mathbf{H}_G \underset{(25)}{=} K_1u_1\hat{\mathbf{a}}_1 + \underset{(16,\ 18)}{K_2u_2\hat{\mathbf{a}}_2} + K_3u_3\hat{\mathbf{a}}_3 + \underset{(17,\ 19)}{Ju_4\hat{\mathbf{b}}_3} \tag{26}$$

Consequently,

$$K_1u_1\hat{\mathbf{a}}_1 \cdot \hat{\mathbf{n}}_i + K_2u_2\hat{\mathbf{a}}_2 \cdot \hat{\mathbf{n}}_i + K_3u_3\hat{\mathbf{a}}_3 \cdot \hat{\mathbf{n}}_i + Ju_4\hat{\mathbf{b}}_3 \cdot \hat{\mathbf{n}}_i \underset{(24,\ 26)}{=} C_{3+i} \quad (i = 1,2,3) \tag{27}$$

Furthermore, $\hat{\mathbf{b}}_3$ can be written

$$\hat{\mathbf{b}}_3 = \beta_1\hat{\mathbf{a}}_1 + \beta_2\hat{\mathbf{a}}_2 + \beta_3\hat{\mathbf{a}}_3 \tag{28}$$

where β_1, β_2, and β_3 are constants. Referring to Table 9.4.1, one thus arrives at the integrals

$$(K_1u_1 + Ju_4\beta_1)c_2c_3 - (K_2u_2 + Ju_4\beta_2)c_2s_3 + (K_3u_3 + Ju_4\beta_3)s_2 = C_4 \tag{29}$$

$$(K_1u_1 + Ju_4\beta_1)(s_1s_2c_3 + s_3c_1) + (K_2u_2 + Ju_4\beta_2)(-s_1s_2s_3 + c_3c_1) - (K_3u_3 + Ju_4\beta_3)s_1c_2 = C_5 \tag{30}$$

$$(K_1u_1 + Ju_4\beta_1)(-c_1s_2c_3 + s_3s_1) + (K_2u_2 + Ju_4\beta_2)(c_1s_2s_3 + c_3s_1) + (K_3u_3 + Ju_4\beta_3)c_1c_2 = C_6 \tag{31}$$

Now let B play the part of S in connection with Eqs. (3) and (4); that is, let $\mathbf{M}$ be the moment about $B^\star$, the mass center of B, of all contact forces and distance forces acting on B, and let $\mathbf{H}_B$ be the central angular momentum of B in N so that

$$\mathbf{H}_B \underset{(4.5.28)}{=} \underline{\mathbf{I}}_B \cdot {}^N\boldsymbol{\omega}^B = \underline{\mathbf{I}}_B \cdot ({}^N\boldsymbol{\omega}^A \underset{(2.4.1)}{+} {}^A\boldsymbol{\omega}^B) \tag{32}$$

or

$$\begin{aligned}\mathbf{H}_B &\underset{(32)}{=} [I(\hat{\mathbf{b}}_1\hat{\mathbf{b}}_1 + \hat{\mathbf{b}}_2\hat{\mathbf{b}}_2) \underset{(17)}{+} J\hat{\mathbf{b}}_3\hat{\mathbf{b}}_3] \cdot (u_1\hat{\mathbf{a}}_1 + u_2\hat{\mathbf{a}}_2 \underset{(18)}{+} u_3\hat{\mathbf{a}}_3 \underset{(19)}{+} u_4\hat{\mathbf{b}}_3) \\ &= I[(u_1\hat{\mathbf{b}}_1 \cdot \hat{\mathbf{a}}_1 + u_2\hat{\mathbf{b}}_1 \cdot \hat{\mathbf{a}}_2 + u_3\hat{\mathbf{b}}_1 \cdot \hat{\mathbf{a}}_3)\hat{\mathbf{b}}_1 \\ &\quad + (u_1\hat{\mathbf{b}}_2 \cdot \hat{\mathbf{a}}_1 + u_2\hat{\mathbf{b}}_2 \cdot \hat{\mathbf{a}}_2 + u_3\hat{\mathbf{b}}_2 \cdot \hat{\mathbf{a}}_3)\hat{\mathbf{b}}_2] \\ &\quad + J(u_1\beta_1 \underset{(28)}{+} u_2\beta_2 + u_3\beta_3 + u_4)\hat{\mathbf{b}}_3\end{aligned} \tag{33}$$

Finally, let $\hat{\mathbf{b}}_3$ play the role of $\hat{\mathbf{n}}$ in Eq. (3), and suppose that B is completely free to rotate relative to A, so that

$$\mathbf{M} \cdot \hat{\mathbf{b}}_3 = 0 \tag{34}$$

Then, since

$$\frac{{}^N d\hat{\mathbf{b}}_3}{dt} \underset{(2.1.2)}{=} {}^N\boldsymbol{\omega}^A \times \hat{\mathbf{b}}_3 \underset{(18)}{=} u_1\hat{\mathbf{a}}_1 \times \hat{\mathbf{b}}_3 + u_2\hat{\mathbf{a}}_2 \times \hat{\mathbf{b}}_3 + u_3\hat{\mathbf{a}}_3 \times \hat{\mathbf{b}}_3 \tag{35}$$

one can use the relationships

$$\hat{\mathbf{b}}_1 \cdot \hat{\mathbf{a}}_1 \times \hat{\mathbf{b}}_3 = \hat{\mathbf{b}}_3 \times \hat{\mathbf{b}}_1 \cdot \hat{\mathbf{a}}_1 = \hat{\mathbf{b}}_2 \cdot \hat{\mathbf{a}}_1 \tag{36}$$

$$\hat{\mathbf{b}}_2 \cdot \hat{\mathbf{a}}_1 \times \hat{\mathbf{b}}_3 = \hat{\mathbf{b}}_3 \times \hat{\mathbf{b}}_2 \cdot \hat{\mathbf{a}}_1 = -\hat{\mathbf{b}}_1 \cdot \hat{\mathbf{a}}_1 \tag{37}$$

and so forth, to verify that

$$\mathbf{H}_B \cdot \frac{{}^N d\hat{\mathbf{b}}_3}{dt} \underset{(33,\,35)}{=} 0 \tag{38}$$

Consequently, Eq. (3) is satisfied, and Eqs. (4) and (33) lead to the integral

$$u_1\beta_1 + u_2\beta_2 + u_3\beta_3 + u_4 = C \tag{39}$$

The same integral can be found by noting that the Lagrangian $\mathscr{L}$ is here equal to the kinetic energy of G in N, so that (see Problem 11.8)

$$\mathscr{L} = \frac{1}{2}(K_1u_1{}^2 + K_2u_2{}^2 + K_3u_3{}^2) + Ju_4\left(u_1\beta_1 + u_2\beta_2 + u_3\beta_3 + \frac{u_4}{2}\right) + \frac{1}{2}M(u_5{}^2 + u_6{}^2 + u_7{}^2) \tag{40}$$

and q_4 is absent from $\mathscr{L}$ when $u_1, \ldots, u_7$ are expressed in terms of $q_1, \ldots, q_7$ and $\dot{q}_1, \ldots, \dot{q}_7$; however, $\dot{q}_4$ appears in Eq. (40) because $u_4 = \dot{q}_4$. Consequently,

$$\frac{\partial \mathscr{L}}{\partial \dot{q}_4} = \frac{\partial \mathscr{L}}{\partial u_4} \underset{(40)}{=} J(u_1\beta_1 + u_2\beta_2 + u_3\beta_3 + u_4) \tag{41}$$

and, setting the right-hand member of this equation equal to a constant in accordance with Eq. (5), one recovers Eq. (39).

The integrals reported in Eqs. (23) and (29) also can be obtained by using Eq. (5) in conjunction with Eq. (40), but those expressed in Eqs. (30) and (31) cannot, for q_2 and q_3 appear in $\mathscr{L}$ when u_1, u_2, and u_3 are replaced in Eq. (40) with

$$u_1 = \dot{q}_1 c_2 c_3 + \dot{q}_2 s_3 \tag{42}$$

$$u_2 = -\dot{q}_1 c_2 s_3 + \dot{q}_2 c_3 \tag{43}$$

$$u_3 = \dot{q}_1 s_2 + \dot{q}_3 \tag{44}$$

respectively.

In the course of solving Problem 4.13, one may develop the following three relationships that are valid when $\hat{C}$, the contact point of circular disk C, slips on horizontal plane H,

$${}^A\boldsymbol{\omega}^C = u_1\hat{\mathbf{b}}_1 + u_2\hat{\mathbf{b}}_2 + u_3\hat{\mathbf{b}}_3 \tag{45}$$

$${}^A\mathbf{v}^{C^\star} = R(u_1\hat{\mathbf{b}}_3 - u_2 \tan q_2 \hat{\mathbf{b}}_1) + u_4\hat{\mathbf{a}}_x + u_5\hat{\mathbf{a}}_y \tag{46}$$

$${}^A\mathbf{v}^{\hat{C}} = R(u_3 - u_2 \tan q_2)\hat{\mathbf{b}}_1 + u_4\hat{\mathbf{a}}_x + u_5\hat{\mathbf{a}}_y \tag{47}$$

and inspect them to obtain the holonomic partial angular velocities, ${}^A\boldsymbol{\omega}_r^C$, and holonomic partial velocities, ${}^A\mathbf{v}_r^{C^\star}$ and ${}^A\mathbf{v}_r^{\hat{C}}$, recorded in Table P4.13 for $r = 1, \ldots, 5$. The entries in Table P4.13 then can be used to construct dynamical equations for C according to Eqs. (8.1.1). Now, when one supposes that H is perfectly smooth, as in Problem 14.5, the equations of motion of C possess a momentum integral that is linear in the motion variables; the integral can be solved for, say, u_2

$$u_2 = -2u_3 \tan q_2 + \frac{\beta}{c_2} \tag{48}$$

where β is an arbitrary constant, and this expression has the form of Eqs. (3.5.2). Substitution from Eq. (48) into Eqs. (45)–(47) yields

$${}^A\boldsymbol{\omega}^C \underset{(45)}{=} u_1\hat{\mathbf{b}}_1 + (\beta/c_2 - 2u_3 \tan q_2)\hat{\mathbf{b}}_2 + u_3\hat{\mathbf{b}}_3 \tag{49}$$

$${}^A\mathbf{v}^{C^\star} \underset{(46)}{=} R[u_1\hat{\mathbf{b}}_3 - (\beta/c_2 - 2u_3 \tan q_2) \tan q_2 \hat{\mathbf{b}}_1] + u_4\hat{\mathbf{a}}_x + u_5\hat{\mathbf{a}}_y \tag{50}$$

$${}^A\mathbf{v}^{\hat{C}} \underset{(47)}{=} R[u_3 - (\beta/c_2 - 2u_3 \tan q_2) \tan q_2]\hat{\mathbf{b}}_1 + u_4\hat{\mathbf{a}}_x + u_5\hat{\mathbf{a}}_y \tag{51}$$

Nonholonomic partial angular velocities and nonholonomic partial velocities corresponding to u_1, u_3, u_4, and u_5 then can be identified by inspection, as recorded in Table 9.4.2, and four dynamical equations can be constructed according to Eqs. (8.1.2) with $r = 1,3,4,5$.

Table 9.4.2

r	${}^A\tilde{\boldsymbol{\omega}}_r^C$	${}^A\tilde{\mathbf{v}}_r^{C^\star}$	${}^A\tilde{\mathbf{v}}_r^{\hat{C}}$
1	$\hat{\mathbf{b}}_1$	$R\hat{\mathbf{b}}_3$	$\mathbf{0}$
3	$\hat{\mathbf{b}}_3 - 2\tan q_2\hat{\mathbf{b}}_2$	$2R\tan^2 q_2\hat{\mathbf{b}}_1$	$R(1+2\tan^2 q_2)\hat{\mathbf{b}}_1$
4	$\mathbf{0}$	$\hat{\mathbf{a}}_x$	$\hat{\mathbf{a}}_x$
5	$\mathbf{0}$	$\hat{\mathbf{a}}_y$	$\hat{\mathbf{a}}_y$

9.5 EXACT CLOSED-FORM SOLUTIONS

Usually, it is difficult to find in closed form the exact, general solution of the equations of motion of a system S (see Sec. 9.1) because the differential equations governing the generalized coordinates and motion variables are both nonlinear and coupled. At times, some of the generalized coordinates and/or motion variables can be expressed as explicit, elementary functions of time, whereas the rest cannot be so expressed. Moreover, there exists no general, systematic procedure that facilitates the search for closed-form solutions. To find them, one must take advantage of special features of the problem under consideration.

Example In Fig. 9.5.1, S represents a uniform sphere of radius r that rolls on the interior surface of a fixed tube T of radius R, the axis of T being vertical. Motion variables $u_1,\ldots,u_5$ may be introduced as

$$u_i \triangleq \boldsymbol{\omega}\cdot\hat{\mathbf{e}}_i \qquad (i = 1,2,3) \tag{1}$$

$$u_4 \triangleq \mathbf{v}\cdot\hat{\mathbf{e}}_2 \qquad u_5 \triangleq \mathbf{v}\cdot\hat{\mathbf{e}}_3 \tag{2}$$

where $\boldsymbol{\omega}$ is the angular velocity of S, $\mathbf{v}$ is the velocity of the center of S, and $\hat{\mathbf{e}}_1$, $\hat{\mathbf{e}}_2$, and $\hat{\mathbf{e}}_3$ form a right-handed set of mutually perpendicular unit vectors directed as shown in Fig. 9.5.1. For generalized coordinates, one can use, in part, angles q_1, q_2, and q_3, like those in Problem 1.1, to characterize the orientation of unit vectors $\hat{\mathbf{s}}_1$, $\hat{\mathbf{s}}_2$, $\hat{\mathbf{s}}_3$ fixed in S relative to the unit vectors $\hat{\mathbf{e}}_1$, $\hat{\mathbf{e}}_2$, $\hat{\mathbf{e}}_3$, and, for the rest, the distance q_4

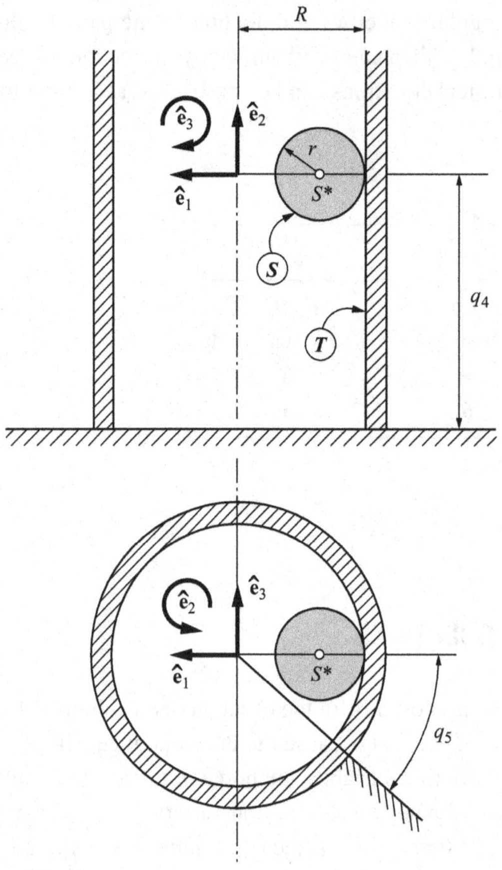

Figure 9.5.1

and the angle q_5 shown in Fig. 9.5.1. The kinematical differential equations then are

$$\dot{q}_1 = u_1 + \tan q_2 \left[\left(u_2 - \frac{u_5}{R-r}\right) s_1 - u_3 c_1\right] \tag{3}$$

$$\dot{q}_2 = \left(u_2 - \frac{u_5}{R-r}\right) c_1 + u_3 s_1 \tag{4}$$

$$\dot{q}_3 = -\sec q_2 \left[\left(u_2 - \frac{u_5}{R-r}\right) s_1 - u_3 c_1\right] \tag{5}$$

$$\dot{q}_4 = u_4 \tag{6}$$

$$\dot{q}_5 = \frac{u_5}{R-r} \tag{7}$$

The system possesses three degrees of freedom because the two nonholonomic constraint equations (see Sec. 3.5)

$$u_4 = r u_3 \tag{8}$$

and

$$u_5 = -r u_2 \tag{9}$$

must be satisfied in order that there be no slipping at the contact between S and T. Finally, with the aid of Eqs. (8.1.2), one can write the dynamical equations

$$\dot{u}_1 = \frac{r}{R-r} u_2 u_3 \tag{10}$$

$$\dot{u}_2 = 0 \tag{11}$$

$$\dot{u}_3 = -\frac{1}{7}\left(\frac{2r}{R-r} u_1 u_2 + \frac{5g}{r}\right) \tag{12}$$

In principle, Eqs. (3)–(12) suffice for the determination of $q_1, \ldots, q_5$ and $u_1, \ldots, u_5$ as functions of the time t; but, as will now be shown, it is considerably easier to find explicit expressions for $u_1, \ldots, u_5$, q_4, and q_5 than for q_1, q_2, and q_3.

Equation (11) has the general solution

$$u_2 = u_2(0) \tag{13}$$

where $u_2(0)$ denotes the value of u_2 at $t = 0$. Hence,

$$u_5 \underset{(9)}{=} -ru_2(0) \tag{14}$$

and

$$\dot{q}_5 \underset{(7,\,14)}{=} \frac{ru_2(0)}{r-R} \tag{15}$$

which has the general solution

$$q_5 = \frac{ru_2(0)t}{r-R} + q_5(0) \tag{16}$$

Next, if $u_2(0) \neq 0$ then

$$u_1 \underset{(12)}{=} \frac{r-R}{2r\,\underset{(13)}{u_2(0)}}\left(\frac{5g}{r} + 7\dot{u}_3\right) \tag{17}$$

so that

$$\dot{u}_1 \underset{(17)}{=} \frac{7(r-R)}{2ru_2(0)}\ddot{u}_3 \tag{18}$$

and

$$\underset{(18)}{\frac{7(r-R)}{2ru_2(0)}\ddot{u}_3} \underset{(10)}{=} \frac{r}{R-r}\underset{(13)}{u_2(0)}\, u_3 \tag{19}$$

which, after p has been defined as

$$p \triangleq \frac{r|u_2(0)|\sqrt{\frac{2}{7}}}{R-r} \tag{20}$$

can be written

$$\ddot{u}_3 + p^2 u_3 = 0 \tag{21}$$

The general solution of this equation is

$$u_3 = u_3(0)\cos pt + \left[\frac{\dot{u}_3(0)}{p}\right]\sin pt \tag{22}$$

where $u_3(0)$ and $\dot{u}_3(0)$ are the initial values of u_3 and $\dot{u}_3$, respectively. Moreover,

$$\dot{u}_3(0) \underset{(12)}{=} -\frac{1}{7}\left[\frac{2r}{R-r}u_1(0)u_2(0) + \frac{5g}{r}\right] \tag{23}$$

Hence,

$$u_3 \underset{(22,\ 23)}{=} u_3(0)\cos pt - \frac{1}{7p}\left[\frac{2r}{R-r}u_1(0)u_2(0) + \frac{5g}{r}\right]\sin pt \tag{24}$$

and differentiation of this equation leads to an expression for $\dot{u}_3$, with the aid of which one obtains from Eq. (17)

$$u_1 = \frac{r-R}{2ru_2(0)}\left\{\frac{5g}{r} - 7u_3(0)p\sin pt - \left[\frac{2r}{R-r}u_1(0)u_2(0) + \frac{5g}{r}\right]\cos pt\right\} \tag{25}$$

As for u_4, Eqs. (8) and (24) yield

$$u_4 = ru_3(0)\cos pt - \frac{r}{7p}\left[\frac{2r}{R-r}u_1(0)u_2(0) + \frac{5g}{r}\right]\sin pt \tag{26}$$

Finally, integrating Eq. (6) after replacing u_4 with the right-hand member of Eq. (26), one finds that

$$q_4 = \frac{r}{p}u_3(0)\sin pt + \frac{r}{7p^2}\left[\frac{2r}{R-r}u_1(0)u_2(0) + \frac{5g}{r}\right](\cos pt - 1) + q_4(0) \tag{27}$$

Equations (27), (16), (25), (13), (24), (26), and (14) provide expressions for q_4, q_5, u_1, u_2, u_3, u_4, and u_5, respectively, as explicit functions of time, and thus they furnish much physically meaningful information about the motion of S. For example, Eq. (16) reveals that $S^\star$ moves in a plane that rotates with a constant angular speed of magnitude $r|u_2(0)|/(R-r)$ about the axis of T. The motion of $S^\star$ in this rotating plane is described completely by Eq. (27), which shows that $S^\star$ performs an oscillatory vertical motion whose circular frequency, p [see Eq. (20)], depends both on the initial value of u_2 and on the geometric parameter R/r, while the amplitude of the oscillations can be adjusted to any desired value by a suitable choice of the initial values of u_1, u_2, and u_3. In particular, one can make the amplitude equal to zero —that is, one can make $S^\star$ move on a horizontal circle— by taking

$$u_3(0) = 0 \qquad u_1(0) = -\frac{5g(R-r)}{2u_2(0)r^2} \tag{28}$$

These results may conflict with one's intuition; one might expect the sphere to move downward, regardless of initial conditions. Indeed, this is what happens in reality. The reason for this discrepancy between predicted and actual motions is that certain physically unavoidable, dissipative effects, such as frictional resistance to rotation, have been left out of account in the analysis. Problem 14.12 deals with a more realistic approach.

When the expressions for u_1, u_2, and u_3 available in Eqs. (25), (13), and (24), respectively, are substituted into Eqs. (3)–(5), it can be seen that q_1, q_2, and q_3 are governed by a set of coupled, nonlinear differential equations with time-dependent coefficients. Generally, solutions of such sets of equations can be found only by numerical integration.

9.6 NUMERICAL INTEGRATION OF DIFFERENTIAL EQUATIONS OF MOTION

When the dynamical (see Sec. 8.1) and/or kinematical (see Sec. 3.4) differential equations governing the motion of a system S cannot be solved in closed form (see Sec. 9.5), one can use a numerical integration procedure to determine for time $t > t_0$, or for time $t < t_0$, the values of the dependent variables appearing in the equations, provided that the values of these variables are known for $t = t_0$, t_0 being a particular value of t. To this end, one can employ any one of many computer programs† applicable to the solution of a set of ν first-order differential equations of the form

$$\frac{dx_i}{dt} = f_i(x_1, \ldots, x_\nu; t) \qquad (i = 1, \ldots, \nu) \tag{1}$$

where $x_1, \ldots, x_\nu$ are unknown functions of t, and $f_1, \ldots, f_\nu$ are known, generally nonlinear functions of $x_1, \ldots, x_\nu$, and t. In this form, the ordinary differential equations are referred to as *explicit*.

Sometimes the differential equations of motion of a system are available precisely in the form of Eqs. (1). When this is the case, one can proceed directly to the numerical solution of the equations, taking care in programming to ensure that the evaluation of any expression occurring repeatedly in $f_1, \ldots, f_\nu$ is performed only once. More frequently, at least the dynamical differential equations (see Sec. 8.1) are not immediately available in the form of Eqs. (1), for use of Eqs. (8.1.1), (8.1.2), or (8.1.3) leads to equations of the *linearly implicit* form

$$X\dot{u} = Y \tag{2}$$

where $\dot{u}$ is a $d \times 1$ matrix having the time derivative $\dot{u}_r$ of the motion variable u_r as the rth element $(r = 1, \ldots, d)$, X is a $d \times d$ matrix, and Y is a $d \times 1$ matrix. The number of degrees of freedom, d, is equal to n, p, or c, when one employs Eqs. (8.1.1), (8.1.2), or (8.1.3), respectively. The elements of X are functions of the generalized coordinates $q_1, \ldots, q_n$, the time t, and, when Eqs. (8.1.3) apply, $u_1, \ldots, u_d$. The elements of Y are functions of $q_1, \ldots, q_n, u_1, \ldots, u_d$, and t. Under these circumstances, one must "uncouple" Eqs. (2), that is, one must solve these equations for $\dot{u}_1, \ldots, \dot{u}_d$, in order to be able to use a computer program intended for the solution of Eqs. (1). When d is sufficiently small, or when X is sufficiently sparse, one can do this in analytical terms; otherwise,

† Programs applicable to the solution of Eqs. (1) and (2) can be found in commercially available software such as IMSL® (International Mathematics and Statistics Library), Mathematica®, MATLAB®, MotionGenesis™Kane, and *Numerical Recipes: The Art of Scientific Computing* (Cambridge: Cambridge University Press, 2007).

one must resort to the use of a computer routine that solves a set of linear equations numerically, calling this routine each time $\dot{u}_1, \ldots, \dot{u}_d$ are needed.

When one undertakes the task of forming X and Y, that is, of formulating dynamical differential equations, with a view toward performing a numerical simulation of a motion of a system S, one can take certain measures that save one considerable amounts of writing and that lead to highly efficient computer codes for the evaluation of X and Y, particularly when d is large. The two guiding ideas here are that $\dot{u}_1, \ldots, \dot{u}_d$ must be kept in explicit evidence, and that repeated writing of any expression must be avoided. The introduction of auxiliary constants $k_1, k_2, \ldots$, and auxiliary functions $Z_1, Z_2, \ldots$ of $q_1, \ldots, q_n$, $u_1, \ldots, u_n$, and t, enables one to reach both of these goals, as will be illustrated in connection with an example.

Example Figure 9.6.1 is a schematic representation of a spacecraft equipped with a nutation damper. This system consists of a rigid body B and a particle P that moves in a tube T, with P attached to B by means of a light, linear spring S and a light, linear dashpot D. The axis of T is parallel to line L_1, one of the central principal axes of B; this axis is separated from L_1 by a distance b; and it lies in the L_1L_2 plane, where L_2, like L_1, is a central principal axis of B. The force $\mathbf{R}$ exerted on P by S and D is given by

$$\mathbf{R} = -(\sigma q + \delta \dot{q})\hat{\mathbf{b}}_1 \tag{3}$$

where σ and δ are constants, q is the distance from L_2 to P, and $\hat{\mathbf{b}}_1$ is a unit vector parallel to L_1.

When the spacecraft moves in a Newtonian reference frame N in the absence of external forces, then $\mathbf{H}$, the central angular momentum of the spacecraft, remains fixed in N, but θ, the angle between L_1 and $\mathbf{H}$, varies with time. The time history of θ is to be determined.

Once $\mathbf{H}$ is known as a function of t, θ can be found with the aid of the relationship

$$\theta = \cos^{-1}\left(\hat{\mathbf{b}}_1 \cdot \frac{\mathbf{H}}{|\mathbf{H}|}\right) \tag{4}$$

As for $\mathbf{H}$, Eqs. (4.5.27), (4.5.28), and the theorem stated in Problem 6.3 lead to

$$\mathbf{H} = \underline{\mathbf{I}} \cdot {}^N\boldsymbol{\omega}^B + m_B \mathbf{p}^{CB^\star} \times {}^N\mathbf{v}^{B^\star} + m_p \mathbf{p}^{CP} \times {}^N\mathbf{v}^P \tag{5}$$

where $\underline{\mathbf{I}}$ is the central inertia dyadic of B, ${}^N\boldsymbol{\omega}^B$ is the angular velocity of B in N, m_B and m_p are the masses of B and P, respectively, C denotes the mass center of the system formed by B and P, $\mathbf{p}^{CB^\star}$ and $\mathbf{p}^{CP}$ are, respectively, the position vectors from C to $B^\star$, the mass center of B, and C to P, and ${}^N\mathbf{v}^{B^\star}$ and ${}^N\mathbf{v}^P$ are, respectively, the velocities of $B^\star$ and P in N. Hence, we must determine ${}^N\boldsymbol{\omega}^B$, $\mathbf{p}^{CB^\star}$, ${}^N\mathbf{v}^{B^\star}$, $\mathbf{p}^{CP}$, and ${}^N\mathbf{v}^P$ as functions of t. To this end, we shall introduce generalized velocities $u_1, \ldots, u_7$ in such a way that each of the required vectors can be expressed as a function of $u_1, \ldots, u_7$, and q; formulate generalized inertia forces $F_1^\star, \ldots, F_7^\star$ and generalized active forces $F_1, \ldots, F_7$; use Eqs. (8.1.1) to write dynamical differential

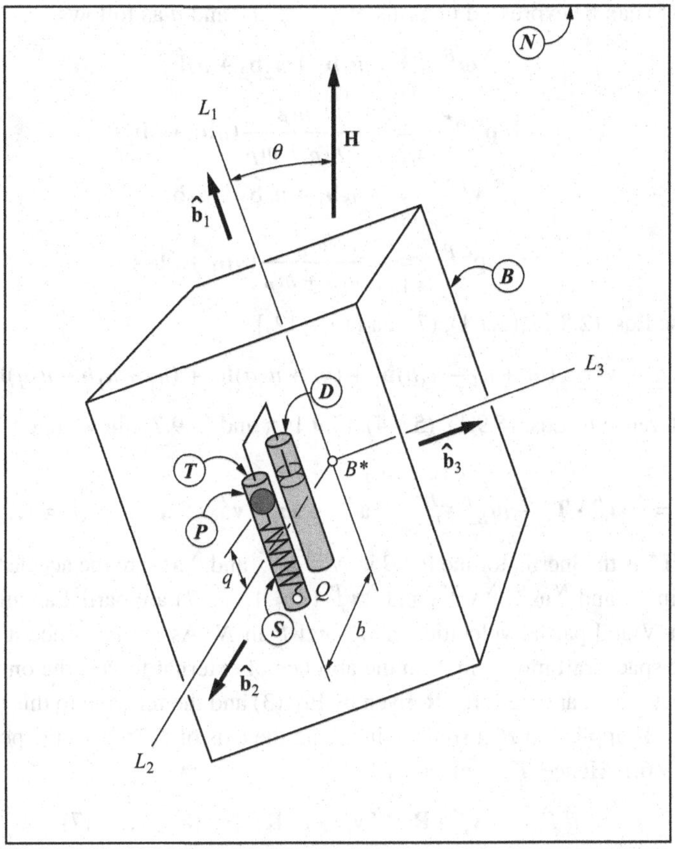

Figure 9.6.1

equations governing $u_1, \ldots, u_7$; and solve these numerically together with a kinematical differential equation that will be stated presently, thus obtaining $u_1, \ldots, u_7$, and q as functions of t.

The generalized velocities to be employed are defined as

$$u_i \triangleq {}^N\boldsymbol{\omega}^B \cdot \hat{\mathbf{b}}_i \qquad (i = 1,2,3) \tag{6}$$

$$u_4 \triangleq \dot{q} \tag{7}$$

$$u_{4+i} \triangleq {}^N\mathbf{v}^{B^\star} \cdot \hat{\mathbf{b}}_i \qquad (i = 1,2,3) \tag{8}$$

where $\hat{\mathbf{b}}_2$ and $\hat{\mathbf{b}}_3$ are unit vectors, with $\hat{\mathbf{b}}_2$ parallel to L_2, and $\hat{\mathbf{b}}_3 = \hat{\mathbf{b}}_1 \times \hat{\mathbf{b}}_2$. Equation (7) is the aforementioned kinematical differential equation, and the vectors appearing in Eq. (5) can be expressed in terms of $u_1, \ldots, u_7$, and q as follows:

$$^N\boldsymbol{\omega}^B \underset{(6)}{=} u_1\hat{\mathbf{b}}_1 + u_2\hat{\mathbf{b}}_2 + u_3\hat{\mathbf{b}}_3 \tag{9}$$

$$\mathbf{p}^{CB^\star} \underset{(4.1.2)}{=} -\frac{m_P}{m_B + m_P}(q\hat{\mathbf{b}}_1 + b\hat{\mathbf{b}}_2) \tag{10}$$

$$^N\mathbf{v}^{B^\star} \underset{(8)}{=} u_5\hat{\mathbf{b}}_1 + u_6\hat{\mathbf{b}}_2 + u_7\hat{\mathbf{b}}_3 \tag{11}$$

$$\mathbf{p}^{CP} \underset{(4.1.2)}{=} \frac{m_B}{m_B + m_P}(q\hat{\mathbf{b}}_1 + b\hat{\mathbf{b}}_2) \tag{12}$$

and [use Eqs. (2.8.1), (2.7.1), (7), and (9)–(12)]

$$^N\mathbf{v}^P = (u_5 + u_4 - u_3 b)\hat{\mathbf{b}}_1 + (u_6 + u_3 q)\hat{\mathbf{b}}_2 + (u_7 + u_1 b - u_2 q)\hat{\mathbf{b}}_3 \tag{13}$$

Reference to Eqs. (5.9.1), (5.9.4), (5.9.11), and (5.9.7) allows one to express $F_r^\star$ as

$$F_r^\star = {^N\boldsymbol{\omega}_r^B} \cdot \mathbf{T}^\star - m_B {^N\mathbf{v}_r^{B^\star}} \cdot {^N\mathbf{a}^{B^\star}} - m_p {^N\mathbf{v}_r^P} \cdot {^N\mathbf{a}^P} \qquad (r = 1, \ldots, 7) \tag{14}$$

where $\mathbf{T}^\star$ is the inertia torque for B in N, $^N\mathbf{a}^{B^\star}$ and $^N\mathbf{a}^P$ are the accelerations of $B^\star$ and P in N, and $^N\boldsymbol{\omega}_r^B$, $^N\mathbf{v}_r^{B^\star}$, and $^N\mathbf{v}_r^P$ $(r = 1, \ldots, 7)$ are partial angular velocities of B in N and partial velocities of $B^\star$ and P in N. As for F_r, since it is presumed that the spacecraft moves in N in the absence of external forces, the only forces that contribute to F_r are the force $\mathbf{R}$ given by Eq. (3) and the reaction to this force, that is, a force $-\mathbf{R}$ applied to B at some point along the axis of T, such as the point Q shown in Fig. 9.6.1. Hence, F_r is given by

$$F_r = {^N\mathbf{v}_r^P} \cdot \mathbf{R} + {^N\mathbf{v}_r^Q} \cdot (-\mathbf{R}) \qquad (r = 1, \ldots, 7) \tag{15}$$

where $^N\mathbf{v}_r^Q$ $(r = 1, \ldots, 7)$ are partial velocities of Q in N.

Equations (14) and (15) show that the partial angular velocities of B and the partial velocities of $B^\star$, P, and Q are needed. Placing Q at a distance d from L_2 (see Fig. 9.6.1), one can write

$$^N\mathbf{v}^Q = {^N\mathbf{v}^{B^\star}} + {^N\boldsymbol{\omega}^B} \times (-d\hat{\mathbf{b}}_1 + b\hat{\mathbf{b}}_2) \tag{16}$$

or, in view of Eqs. (9) and (11),

$$^N\mathbf{v}^Q = (u_5 - u_3 b)\hat{\mathbf{b}}_1 + (u_6 - u_3 d)\hat{\mathbf{b}}_2 + (u_7 + u_1 b + u_2 d)\hat{\mathbf{b}}_3 \tag{17}$$

and this brings one into position to record all necessary partial angular velocities and partial velocities as in Table 9.6.1, where numbers in parentheses in column headings refer to equations numbered correspondingly.

The next item required for substitution into Eqs. (14) is the inertia torque $\mathbf{T}^\star$, which can be written

$$\begin{aligned} \mathbf{T}^\star \underset{(5.9.14,\ 6)}{=} & -[\dot{u}_1 B_1 - u_2 u_3 (B_2 - B_3)]\hat{\mathbf{b}}_1 \\ & - [\dot{u}_2 B_2 - u_3 u_1 (B_3 - B_1)]\hat{\mathbf{b}}_2 \\ & - [\dot{u}_3 B_3 - u_1 u_2 (B_1 - B_2)]\hat{\mathbf{b}}_3 \end{aligned} \tag{18}$$

Table 9.6.1

r	${}^N\boldsymbol{\omega}_r^B$ (9)	${}^N\mathbf{v}_r^{B\star}$ (11)	${}^N\mathbf{v}_r^P$ (13)	${}^N\mathbf{v}_r^Q$ (17)
1	$\hat{\mathbf{b}}_1$	$\mathbf{0}$	$b\hat{\mathbf{b}}_3$	$b\hat{\mathbf{b}}_3$
2	$\hat{\mathbf{b}}_2$	$\mathbf{0}$	$-q\hat{\mathbf{b}}_3$	$d\hat{\mathbf{b}}_3$
3	$\hat{\mathbf{b}}_3$	$\mathbf{0}$	$-b\hat{\mathbf{b}}_1 + q\hat{\mathbf{b}}_2$	$-b\hat{\mathbf{b}}_1 - d\hat{\mathbf{b}}_2$
4	$\mathbf{0}$	$\mathbf{0}$	$\hat{\mathbf{b}}_1$	$\mathbf{0}$
5	$\mathbf{0}$	$\hat{\mathbf{b}}_1$	$\hat{\mathbf{b}}_1$	$\hat{\mathbf{b}}_1$
6	$\mathbf{0}$	$\hat{\mathbf{b}}_2$	$\hat{\mathbf{b}}_2$	$\hat{\mathbf{b}}_2$
7	$\mathbf{0}$	$\hat{\mathbf{b}}_3$	$\hat{\mathbf{b}}_3$	$\hat{\mathbf{b}}_3$

where B_1, B_2, and B_3 are the moments of inertia of B about lines L_1, L_2, and L_3 (see Fig. 9.6.1), respectively. At this point, the process of simplifying the analysis that follows can be begun through the introduction of constants k_1, k_2, k_3 as

$$k_1 \triangleq B_2 - B_3 \qquad k_2 \triangleq B_3 - B_1 \qquad k_3 \triangleq B_1 - B_2 \tag{19}$$

and three functions of u_1, u_2, and u_3 as

$$Z_1 \triangleq u_2 u_3 k_1 \qquad Z_2 \triangleq u_3 u_1 k_2 \qquad Z_3 \triangleq u_1 u_2 k_3 \tag{20}$$

which makes it possible to replace Eq. (18) with

$$\mathbf{T}^\star \underset{(18\text{–}20)}{=} (-\dot{u}_1 B_1 + Z_1)\hat{\mathbf{b}}_1 + (-\dot{u}_2 B_2 + Z_2)\hat{\mathbf{b}}_2 + (-\dot{u}_3 B_3 + Z_3)\hat{\mathbf{b}}_3 \tag{21}$$

The reason for introducing k_1, k_2, k_3 as in Eqs. (19) and then expressing Z_1, Z_2, Z_3 in terms of k_1, k_2, k_3, rather than simply defining Z_1 as $u_2 u_3(B_2 - B_3)$, and similarly for Z_2 and Z_3, is that k_1, k_2, and k_3, being constants, can be evaluated once and for all, whereas Z_1, Z_2, and Z_3, being functions of the dependent variables u_1, u_2, and u_3, must be evaluated repeatedly in the course of a numerical integration. Hence, if Z_1 were defined as $u_2 u_3(B_2 - B_3)$, the subtraction of B_3 from B_2 would have to be performed repeatedly, but if Z_1 is defined as in the first of Eqs. (20), then this subtraction takes place only once.

To form suitable expressions for ${}^N\mathbf{a}^{B\star}$ and ${}^N\mathbf{a}^P$, the two vectors in Eqs. (14) that have not yet been considered in detail, we differentiate Eqs. (11) and (13) with respect to t in N, obtaining [see Eqs. (2.6.2) and (2.3.1)] in the first case

$${}^N\mathbf{a}^{B\star} \underset{(11)}{=} \dot{u}_5\hat{\mathbf{b}}_1 + \dot{u}_6\hat{\mathbf{b}}_2 + \dot{u}_7\hat{\mathbf{b}}_3 + {}^N\boldsymbol{\omega}^B \times {}^N\mathbf{v}^{B\star} \tag{22}$$

or, after using Eq. (9),

$$\begin{aligned} {}^N\mathbf{a}^{B\star} \underset{(9,\,11)}{=} & (\dot{u}_5 + u_2 u_7 - u_3 u_6)\hat{\mathbf{b}}_1 + (\dot{u}_6 + u_3 u_5 - u_1 u_7)\hat{\mathbf{b}}_2 \\ & + (\dot{u}_7 + u_1 u_6 - u_2 u_5)\hat{\mathbf{b}}_3 \end{aligned} \tag{23}$$

which we simplify by letting

$$Z_4 \triangleq u_2 u_7 - u_3 u_6 \tag{24}$$

$$Z_5 \triangleq u_3 u_5 - u_1 u_7 \tag{25}$$

$$Z_6 \triangleq u_1 u_6 - u_2 u_5 \tag{26}$$

whereupon we have

$$^N\mathbf{a}^{B^\star} \underset{(23)}{=} (\dot{u}_5 \underset{(24)}{+} Z_4)\hat{\mathbf{b}}_1 + (\dot{u}_6 \underset{(25)}{+} Z_5)\hat{\mathbf{b}}_2 + (\dot{u}_7 \underset{(26)}{+} Z_6)\hat{\mathbf{b}}_3 \tag{27}$$

Similarly,

$$^N\mathbf{a}^P \underset{(13)}{=} (\dot{u}_5 + \dot{u}_4 - \dot{u}_3 b)\hat{\mathbf{b}}_1 + (\dot{u}_6 + \dot{u}_3 q + u_3 \dot{q})\hat{\mathbf{b}}_2 + (\dot{u}_7 + \dot{u}_1 b - \dot{u}_2 q - u_2 \dot{q})\hat{\mathbf{b}}_3 + {}^N\boldsymbol{\omega}^B \times {}^N\mathbf{v}^P \tag{28}$$

and here it is advantageous to introduce $Z_7, \ldots, Z_{13}$ that not only permit one to write the last term in a simple form, but that at the same time accommodate those portions of the $\hat{\mathbf{b}}_2$ and $\hat{\mathbf{b}}_3$ measure numbers of $^N\mathbf{a}^P$ that do not involve $\dot{u}_1, \ldots, \dot{u}_7$, that is, the two terms $u_3\dot{q}$ and $-u_2\dot{q}$, which, incidentally, are respectively equal to $u_3 u_4$ and $-u_2 u_4$, as can be seen by reference to Eq. (7). Specifically, we let [see Eqs. (13) and (9)]

$$Z_7 \triangleq u_5 + u_4 - u_3 b \tag{29}$$

$$Z_8 \triangleq u_6 + u_3 q \tag{30}$$

$$Z_9 \triangleq u_7 + u_1 b - u_2 q \tag{31}$$

$$Z_{10} \triangleq Z_7 + u_4 \tag{32}$$

$$Z_{11} \triangleq u_2 Z_9 - u_3 Z_8 \tag{33}$$

$$Z_{12} \triangleq u_3 Z_7 - u_1 Z_9 + u_3 u_4 \underset{(32)}{=} u_3 Z_{10} - u_1 Z_9 \tag{34}$$

$$Z_{13} \triangleq u_1 Z_8 - u_2 Z_7 - u_2 u_4 \underset{(32)}{=} u_1 Z_8 - u_2 Z_{10} \tag{35}$$

and this enables us to express $^N\mathbf{a}^P$ as

$$^N\mathbf{a}^P \underset{(28\text{–}35)}{=} (\dot{u}_5 + \dot{u}_4 - \dot{u}_3 b + Z_{11})\hat{\mathbf{b}}_1 + (\dot{u}_6 + \dot{u}_3 q + Z_{12})\hat{\mathbf{b}}_2 + (\dot{u}_7 + \dot{u}_1 b - \dot{u}_2 q + Z_{13})\hat{\mathbf{b}}_3 \tag{36}$$

Equations (14) and Table 9.6.1 now permit us to write

$$F_1^\star = -\underset{(21)}{\dot{u}_1 B_1 + Z_1} - m_P b\,\underset{(36)}{(\dot{u}_7 + \dot{u}_1 b - \dot{u}_2 q + Z_{13})} \tag{37}$$

$$F_2^\star = -\underset{(21)}{\dot{u}_2 B_2 + Z_2} + m_P q\,\underset{(36)}{(\dot{u}_7 + \dot{u}_1 b - \dot{u}_2 q + Z_{13})} \tag{38}$$

$$\begin{aligned} F_3^\star = &-\underset{(21)}{\dot{u}_3 B_3 + Z_3} + m_P b\,\underset{(36)}{(\dot{u}_5 + \dot{u}_4 - \dot{u}_3 b + Z_{11})} \\ &- m_P q\,\underset{(36)}{(\dot{u}_6 + \dot{u}_3 q + Z_{12})} \end{aligned} \tag{39}$$

$$F_4^\star = -m_P\,\underset{(36)}{(\dot{u}_5 + \dot{u}_4 - \dot{u}_3 b + Z_{11})} \tag{40}$$

$$F_5^\star = -m_B\,\underset{(27)}{(\dot{u}_5 + Z_4)} - m_P\,\underset{(36)}{(\dot{u}_5 + \dot{u}_4 - \dot{u}_3 b + Z_{11})} \tag{41}$$

$$F_6^\star = -m_B\,\underset{(27)}{(\dot{u}_6 + Z_5)} - m_P\,\underset{(36)}{(\dot{u}_6 + \dot{u}_3 q + Z_{12})} \tag{42}$$

$$F_7^\star = -m_B\,\underset{(27)}{(\dot{u}_7 + Z_6)} - m_P\,\underset{(36)}{(\dot{u}_7 + \dot{u}_1 b - \dot{u}_2 q + Z_{13})} \tag{43}$$

and detailed expressions for $F_1, \ldots, F_7$ can be formed by substituting from Table 9.6.1 and Eq. (3) into Eqs. (15), with $\dot{q}$ replaced by u_4 [see Eq. (7)] in Eq. (3). This leads to

$$F_1 = F_2 = F_3 = F_5 = F_6 = F_7 = 0 \tag{44}$$

$$F_4 = -(\sigma q + \delta u_4) \tag{45}$$

Before using Eqs. (37)–(45) in conjunction with Eqs. (8.1.1) to write dynamical differential equations, we note that certain combinations of constants and functions appear repeatedly in Eqs. (37)–(43). Accordingly, we define $k_4, \ldots, k_8$ and Z_{14}, Z_{15}, Z_{16} as

$$k_4 \triangleq m_P b \tag{46}$$

$$k_5 \triangleq k_4 b \tag{47}$$

$$k_6 \triangleq B_1 + k_5 \tag{48}$$

$$k_7 \triangleq B_3 + k_5 \tag{49}$$

$$k_8 \triangleq m_B + m_P \tag{50}$$

and

$$Z_{14} \triangleq m_P q \tag{51}$$

$$Z_{15} \triangleq Z_{14} q \tag{52}$$

$$Z_{16} \triangleq k_4 q \tag{53}$$

whereupon we find that Eqs. (8.1.1) and (37)–(53) yield

$$-k_6\dot{u}_1 + Z_{16}\dot{u}_2 - k_4\dot{u}_7 = -Z_1 + k_4Z_{13} \tag{54}$$
$$Z_{16}\dot{u}_1 - (B_2 + Z_{15})\dot{u}_2 + Z_{14}\dot{u}_7 = -Z_2 - Z_{14}Z_{13} \tag{55}$$
$$-(k_7 + Z_{15})\dot{u}_3 + k_4\dot{u}_4 + k_4\dot{u}_5 - Z_{14}\dot{u}_6 = -Z_3 - k_4Z_{11} + Z_{14}Z_{12} \tag{56}$$
$$k_4\dot{u}_3 - m_P\dot{u}_4 - m_P\dot{u}_5 = m_PZ_{11} + \sigma q + \delta u_4 \tag{57}$$
$$k_4\dot{u}_3 - m_P\dot{u}_4 - k_8\dot{u}_5 = m_BZ_4 + m_PZ_{11} \tag{58}$$
$$-Z_{14}\dot{u}_3 - k_8\dot{u}_6 = m_BZ_5 + m_PZ_{12} \tag{59}$$
$$-k_4\dot{u}_1 + Z_{14}\dot{u}_2 - k_8\dot{u}_7 = m_BZ_6 + m_PZ_{13} \tag{60}$$

and these equations have precisely the form of Eqs. (2) if the elements X_{ij} and Y_i $(i,j = 1,\ldots,7)$ of X and Y are taken to be (only nonzero elements are displayed)

$$X_{11} \underset{(54)}{=} -k_6 \qquad X_{12} = X_{21} \underset{(54,\ 55)}{=} Z_{16} \qquad X_{17} = X_{71} \underset{(54,\ 60)}{=} -k_4 \tag{61}$$

$$X_{22} \underset{(55)}{=} -(B_2 + Z_{15}) \qquad X_{27} = X_{72} \underset{(55,\ 60)}{=} Z_{14} \tag{62}$$

$$X_{33} \underset{(56)}{=} -(k_7 + Z_{15}) \qquad X_{34} \underset{(56,\ 57)}{=} X_{43} = k_4$$

$$X_{35} \underset{(56,\ 58)}{=} X_{53} = k_4 \qquad X_{36} \underset{(56,\ 59)}{=} X_{63} = -Z_{14} \tag{63}$$

$$X_{44} \underset{(57)}{=} -m_P \qquad X_{45} \underset{(57,\ 58)}{=} X_{54} = -m_P \tag{64}$$

$$X_{55} \underset{(58)}{=} -k_8 \qquad X_{66} \underset{(59)}{=} -k_8 \qquad X_{77} \underset{(60)}{=} -k_8 \tag{65}$$

and

$$Y_1 \underset{(54)}{=} -Z_1 + k_4Z_{13} \tag{66}$$
$$Y_2 \underset{(55)}{=} -Z_2 - Z_{14}Z_{13} \tag{67}$$
$$Y_3 \underset{(56)}{=} -Z_3 - k_4Z_{11} + Z_{14}Z_{12} \tag{68}$$
$$Y_4 \underset{(57)}{=} m_PZ_{11} + \sigma q + \delta u_4 \tag{69}$$
$$Y_5 \underset{(58)}{=} m_BZ_4 + m_PZ_{11} \tag{70}$$
$$Y_6 \underset{(59)}{=} m_BZ_5 + m_PZ_{12} \tag{71}$$
$$Y_7 \underset{(60)}{=} m_BZ_6 + m_PZ_{13} \tag{72}$$

Since the matrix X is relatively sparse, one can uncouple Eqs. (54)–(60) by hand, proceeding as follows: Solve Eqs. (54), (55), and (60) for $\dot{u}_1$, $\dot{u}_2$, and $\dot{u}_7$; solve Eq. (59) for $\dot{u}_6$; substitute into Eq. (56); solve Eqs. (56), (57), and (58) for $\dot{u}_3$, $\dot{u}_4$, and $\dot{u}_5$. The resulting equations together with Eq. (7) can be written in the form of Eqs. (1), with $\nu = 8$, $x_i = u_i$ $(i = 1,\ldots,7)$, and $x_8 = q$, and a numerical integration of the equations can be performed, once the initial values of $u_1, \ldots, u_7$, and q, as well as the values of the parameters m_B, B_1, B_2, B_3, b, m_P, σ, and δ have been specified.

Calculations leading to values of θ for various values of t can be performed simultaneously with the numerical integration of the differential equations of motion on the basis of the considerations that follow.

With B_1, B_2, and B_3 as defined, $\underline{\mathbf{I}} \cdot {}^N\boldsymbol{\omega}^B$ is given by

$$\underline{\mathbf{I}} \cdot {}^N\boldsymbol{\omega}^B \underset{(9)}{=} B_1 u_1 \hat{\mathbf{b}}_1 + B_2 u_2 \hat{\mathbf{b}}_2 + B_3 u_3 \hat{\mathbf{b}}_3 \tag{73}$$

and substitution from this equation and Eqs. (10)–(13) into Eq. (5) yields

$$\mathbf{H} = H_1 \hat{\mathbf{b}}_1 + H_2 \hat{\mathbf{b}}_2 + H_3 \hat{\mathbf{b}}_3 \tag{74}$$

where

$$H_1 \triangleq B_1 u_1 + bZ_{17} \tag{75}$$

$$H_2 \triangleq B_2 u_2 - qZ_{17} \tag{76}$$

$$H_3 \triangleq B_3 u_3 + k_9[u_3 q^2 - b(u_4 - u_3 b)] \tag{77}$$

with

$$k_9 \triangleq \frac{m_B m_P}{k_8} \qquad Z_{17} \triangleq k_9(Z_9 - u_7) \tag{78}$$

Consequently,

$$\theta \underset{(4)}{=} \cos^{-1} \frac{H_1}{(H_1{}^2 + H_2{}^2 + H_3{}^2)^{1/2}} \tag{79}$$

The quantity $(H_1{}^2 + H_2{}^2 + H_3{}^2)^{1/2}$ is of interest not only in this connection but also because it must remain constant throughout the motion of the spacecraft [set $\mathbf{M} = \mathbf{0}$ in Eq. (9.4.9) and note Eq. (74)], and its evaluation at various instants during a numerical integration can thus serve as a check on the validity of the integration results.

Turning to some concrete examples in order to illustrate the uses of numerical simulations of motions, we let B be a uniform rectangular parallelepiped having edges of lengths 1.2 m, 1.225 m, and 1.3 m, these edges being respectively parallel to L_1, L_2, L_3, and assume that B has a mass density of 2760 kg/m^3 (the mass density of aluminum), so that $m_B = 2760 \times 1.2 \times 1.225 \times 1.3 = 5274.4$ kg, while $B_1 = 5274.4 \times (1.225^2 + 1.3^2)/12 = 1402.4$ kg m^2, $B_2 = 5274.4 \times (1.3^2 + 1.2^2)/12 = 1375.7$ kg m^2, $B_3 = 5274.4 \times (1.2^2 + 1.225^2)/12 = 1292.5$ kg m^2. We take $b = 1.0$ m and, after defining μ as

$$\mu \triangleq \frac{m_P}{m_B} \tag{80}$$

set $m_P = 52.744$ kg, so that $\mu = 0.01$. As for σ and δ, we choose these such that the oscillator formed by P, S, and D has a circular natural frequency (see Sec. 9.9) of 1 rad/s and is critically damped (see Problem 15.14), that is, $\sigma = 52.744$ N/m, and $\delta = 105.49$ Ns/m.

To explore what happens if the spacecraft is subjected to a small disturbance while it is performing a simple spinning motion, we begin by noting that, for all t prior to the disturbance,

$$ {}^N\boldsymbol{\omega}^B = \Omega\hat{\mathbf{b}}_1 \tag{81} $$

where Ω is a constant, called the nominal spin rate. The velocity of C, the mass center of the spacecraft, is given by

$$ {}^N\mathbf{v}^C = \mathbf{0} \tag{82} $$

and P remains on line L_2, so that

$$ q = 0 \tag{83} $$

From Eqs. (9) and (81), it then follows that

$$ u_1 = \Omega \qquad u_2 = u_3 = 0 \tag{84} $$

and Eqs. (7) and (83) imply that

$$ u_4 = 0 \tag{85} $$

Furthermore,

$$ \begin{aligned} {}^N\mathbf{v}^{B^\star} &\underset{(2.7.1)}{=} {}^N\mathbf{v}^C + {}^N\boldsymbol{\omega}^B \times \mathbf{p}^{CB^\star} \\ &= \underset{(82)}{\mathbf{0}} + \underset{(81)}{\Omega\hat{\mathbf{b}}_1} \times \underset{(10,\,83)}{\left(-\frac{m_P}{m_B+m_P}b\hat{\mathbf{b}}_2\right)} \\ &\underset{(80)}{=} -\Omega\frac{b\mu}{1+\mu}\hat{\mathbf{b}}_3 = -\Omega\frac{b}{101}\hat{\mathbf{b}}_3 \end{aligned} \tag{86} $$

so that, considering Eq. (11), we conclude that, for all t prior to a disturbance,

$$ u_5 = u_6 = 0 \tag{87} $$

and

$$ u_7 = -\frac{\Omega b}{101} \tag{88} $$

Equations (84), (85), (87), (88), and (83) furnish a precise characterization of the undisturbed motion, and it may be verified with the aid of Eqs. (20), (24)–(26), (29)–(35), and (51)–(53) that the equations of motion, that is, Eqs. (7) and (54)–(60), are, in fact, satisfied when $u_1,\ldots,u_7$, and q are given by Eqs. (84), (85), (87), (88), and (83), which means that the postulated motion is physically possible. Equations (75)–(79) show that $\theta = 0$ during this motion.

To study motions that ensue subsequent to a slight disturbance of the motion under consideration, we perform numerical simulations after assigning the following initial values to $u_1,\ldots,u_7$, and q:

$$ u_1(0) = \Omega + \varepsilon_1 \tag{89} $$

$$ u_i(0) = \varepsilon_i \qquad (i = 2,\ldots,6) \tag{90} $$

$$u_7(0) = -\frac{\Omega b}{101} + \varepsilon_7 \tag{91}$$

and

$$q(0) = \eta \tag{92}$$

where $\varepsilon_1, \ldots, \varepsilon_7$, and η are constants called *initial perturbations*. With $\Omega = 1.0$ rad/s, $\varepsilon_1 = 0$, $\varepsilon_2 = 0.1$ rad/s, $\varepsilon_3 = \cdots = \varepsilon_7 = \eta = 0$, this leads to the plot of θ versus t displayed in Fig. 9.6.2, which shows that the associated disturbance causes θ to have a nonzero initial value and to perform oscillations with decaying amplitudes. When the value of ε_2 is reduced to 0.05 rad/s, one-half of its former value, one might expect the amplitudes of the oscillations to be smaller than they were previously, and indeed they are, as can be seen in Fig. 9.6.3.

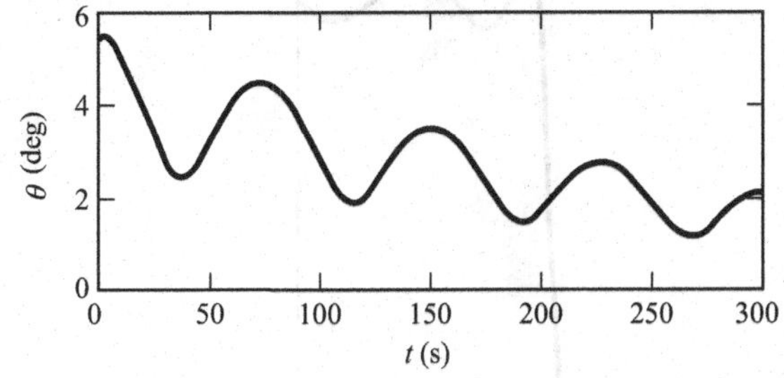

Figure 9.6.2

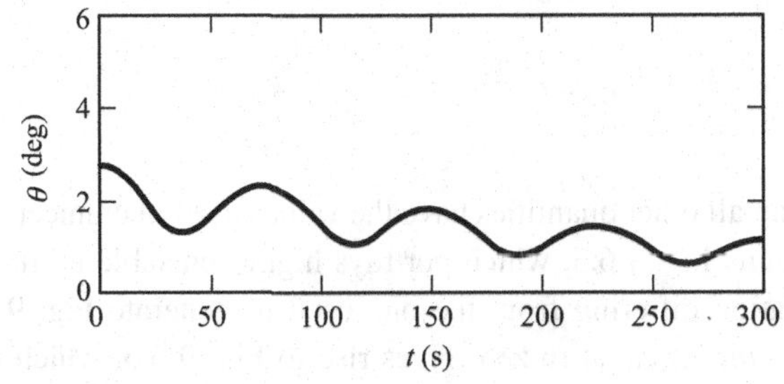

Figure 9.6.3

To see that P, S, and D do, in fact, serve as a nutation *damper*, one can perform a simulation differing in only one respect from the one that led to Fig. 9.6.2, namely, in that it applies when m_P is equal to zero. The resulting curve for θ versus t, shown in Fig. 9.6.4, reveals that oscillations in θ fail to decay when $m_P = 0$. In other words, there is no damping under these circumstances.

To see that P, S, and D can act as a nutation *generator*, rather than as a nutation damper, one can change the dimensions of the edges of B parallel to L_1, L_2, and L_3 to 0.5 m, 1.2 m, and 3.185 m, respectively, and set $\sigma = 0.52744$ N/m, so that the volume, and hence the mass, of B remains unchanged, but σ is reduced to $\frac{1}{100}$ of its

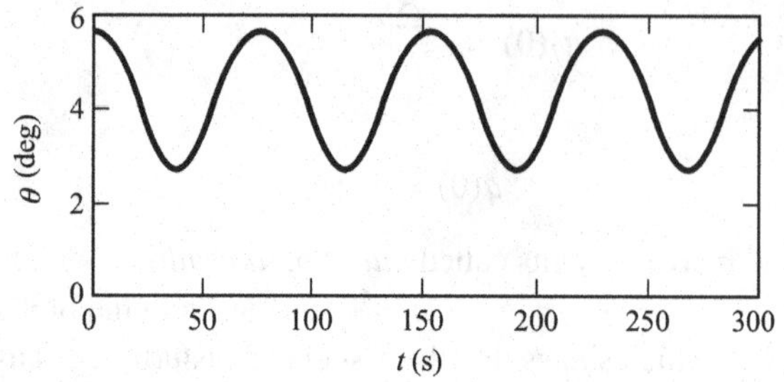

Figure 9.6.4

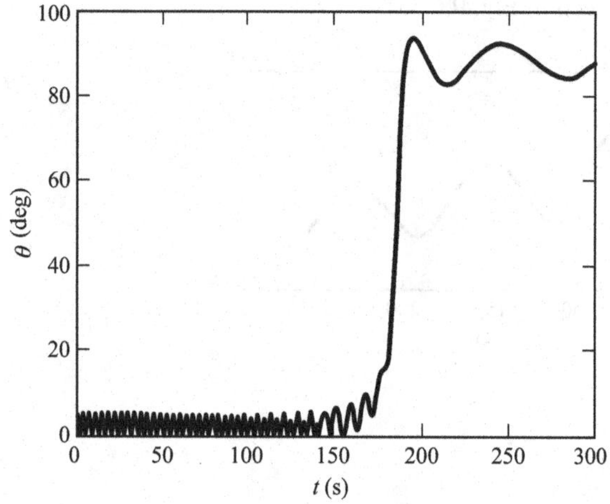

Figure 9.6.5

former value. Letting all other quantities have the values used in connection with Fig. 9.6.2, one then obtains Fig. 9.6.5, which portrays highly unstable attitude behavior. Moreover, a simulation differing from the one used to generate Fig. 9.6.5 only in that it applies when m_P is equal to zero gives rise to Fig. 9.6.6, which leads to the

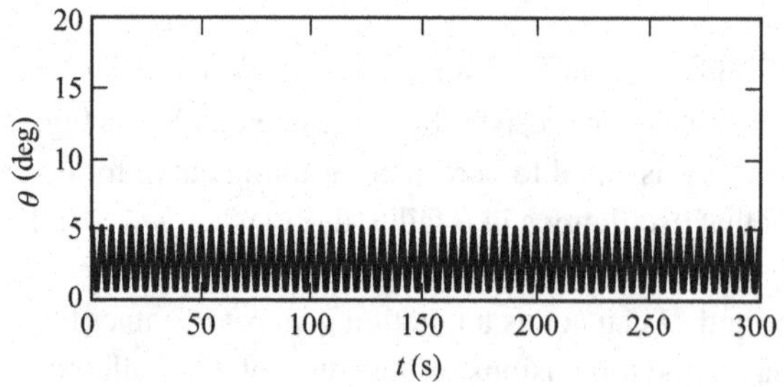

Figure 9.6.6

conclusion that P, S, and D are responsible for the unstable motion depicted in Fig. 9.6.5.

As a check on the validity of the computations underlying Figs. 9.6.2–9.6.6, the quantity $|\mathbf{H}| \triangleq (H_1{}^2 + H_2{}^2 + H_3{}^2)^{1/2}$ [see Eq. (74)] was evaluated throughout each of the simulations used to produce these figures. In every case, numerical integration error tolerances were chosen such that $|\mathbf{H}|$ remained constant to at least six significant figures.

The preceding results are intended to underscore the fact that simulations of motions can provide valuable physical insights. The extent to which they do so depends heavily on the analyst's ingenuity in choosing parameter values and initial conditions.

9.7 DETERMINATION OF CONSTRAINT FORCES AND CONSTRAINT TORQUES

Forces and torques that make no contributions to generalized active forces are called *constraint forces* and *constraint torques*, respectively. To determine these, one uses one or more dynamical equations in which constraint force and/or constraint torque measure numbers of interest come into evidence (see Sec. 8.3), substituting into such equations values of the generalized coordinates and motion variables obtained by solving the kinematical and dynamical differential equations governing the motion under consideration.

Example Figure 9.7.1 shows a linkage formed by uniform bars A, B, and C, each of length $2L$, having masses m_A, m_B, and m_C, respectively. The linkage is supported

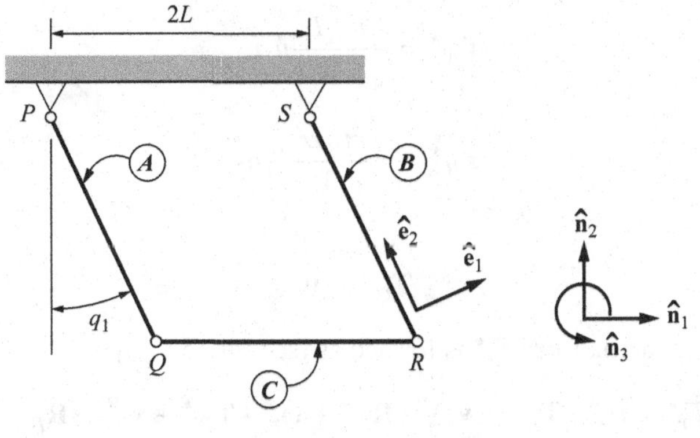

Figure 9.7.1

by horizontal pins at P and S, and is free to move in a vertical plane with one degree of freedom. The angle q_1, indicated in Fig. 9.7.1, serves as a generalized coordinate, and a generalized velocity u_1 may be introduced as

$$u_1 \triangleq \dot{q}_1 \tag{1}$$

The angular velocities of A, B, and C are given by

$$\boldsymbol{\omega}^A = u_1\hat{\mathbf{n}}_3 \qquad \boldsymbol{\omega}^B = u_1\hat{\mathbf{n}}_3 \qquad \boldsymbol{\omega}^C = \mathbf{0} \tag{2}$$

(see Fig. 9.7.1 for $\hat{\mathbf{n}}_3$) and the velocities of $A^\star$, $B^\star$, and $C^\star$, the mass centers of A, B, and C, respectively, are

$$\mathbf{v}^{A^\star} = \boldsymbol{\omega}^A \times (-L\hat{\mathbf{e}}_2) \underset{(2)}{=} Lu_1\hat{\mathbf{e}}_1 \tag{3}$$

$$\mathbf{v}^{B^\star} = \boldsymbol{\omega}^B \times (-L\hat{\mathbf{e}}_2) \underset{(2)}{=} Lu_1\hat{\mathbf{e}}_1 \tag{4}$$

$$\mathbf{v}^{C^\star} = \mathbf{v}^Q + \boldsymbol{\omega}^C \times (L\hat{\mathbf{n}}_1) = 2Lu_1\hat{\mathbf{e}}_1 + \underset{(2)}{\mathbf{0}} \times L\hat{\mathbf{n}}_1 = 2Lu_1\hat{\mathbf{e}}_1 \tag{5}$$

(see Fig. 9.7.1 for $\hat{\mathbf{e}}_1$ and $\hat{\mathbf{e}}_2$). Consequently, the partial angular velocities of A, B, and C and the partial velocities of $A^\star$, $B^\star$, and $C^\star$ associated with u_1 are found to be

$$\boldsymbol{\omega}_1^A = \hat{\mathbf{n}}_3 \qquad \boldsymbol{\omega}_1^B = \hat{\mathbf{n}}_3 \qquad \boldsymbol{\omega}_1^C = \mathbf{0} \tag{6}$$

$$\mathbf{v}_1^{A^\star} = L\hat{\mathbf{e}}_1 \qquad \mathbf{v}_1^{B^\star} = L\hat{\mathbf{e}}_1 \qquad \mathbf{v}_1^{C^\star} = 2L\hat{\mathbf{e}}_1 \tag{7}$$

The inertia forces for A, B, and C are [see Eq. (5.9.4)]

$$\mathbf{R}_A{}^\star = -m_A\mathbf{a}^{A^\star} = -m_AL(\dot{u}_1\hat{\mathbf{e}}_1 + u_1{}^2\hat{\mathbf{e}}_2) \tag{8}$$

$$\mathbf{R}_B{}^\star = -m_B\mathbf{a}^{B^\star} = -m_BL(\dot{u}_1\hat{\mathbf{e}}_1 + u_1{}^2\hat{\mathbf{e}}_2) \tag{9}$$

and

$$\mathbf{R}_C{}^\star = -m_C\mathbf{a}^{C^\star} = -2m_CL(\dot{u}_1\hat{\mathbf{e}}_1 + u_1{}^2\hat{\mathbf{e}}_2) \tag{10}$$

respectively, while the inertia torques for A, B, and C can be written [see Eq. (5.9.14)]

$$\mathbf{T}_A{}^\star = -\frac{m_AL^2}{3}\dot{u}_1\hat{\mathbf{n}}_3 \tag{11}$$

$$\mathbf{T}_B{}^\star = -\frac{m_BL^2}{3}\dot{u}_1\hat{\mathbf{n}}_3 \tag{12}$$

and

$$\mathbf{T}_C{}^\star = \mathbf{0} \tag{13}$$

The generalized inertia force $F_r{}^\star$ is formed as [see Eq. (5.9.7)]

$$F_r{}^\star = \boldsymbol{\omega}_r^A \cdot \mathbf{T}_A{}^\star + \mathbf{v}_r{}^{A^\star} \cdot \mathbf{R}_A{}^\star + \boldsymbol{\omega}_r^B \cdot \mathbf{T}_B{}^\star + \mathbf{v}_r{}^{B^\star} \cdot \mathbf{R}_B{}^\star + \boldsymbol{\omega}_r^C \cdot \mathbf{T}_C{}^\star + \mathbf{v}_r{}^{C^\star} \cdot \mathbf{R}_C{}^\star \tag{14}$$

Hence, $F_1{}^\star$ is given by

$$F_1{}^\star \underset{(14)}{=} \underset{(6,\,11)}{-\tfrac{1}{3}m_AL^2\dot{u}_1} \underset{(7,\,8)}{- m_AL^2\dot{u}_1} \underset{(6,\,12)}{- \tfrac{1}{3}m_BL^2\dot{u}_1} \underset{(7,\,9)}{- m_BL^2\dot{u}_1} + \underset{(6,\,13)}{0} \underset{(7,\,10)}{- 4m_CL^2\dot{u}_1}$$

$$= -\tfrac{4}{3}L^2(m_A + m_B + 3m_C)\dot{u}_1 \tag{15}$$

The generalized active force F_r is expressed as [see Eqs. (5.5.1) and (5.7.2)]

$$F_r = -g\hat{\mathbf{n}}_2 \cdot (m_A\mathbf{v}_r{}^{A^\star} + m_B\mathbf{v}_r{}^{B^\star} + m_C\mathbf{v}_r{}^{C^\star}) \tag{16}$$

and F_1 is found to be

$$F_1 \underset{(16,\,7)}{=} -gL(m_A + m_B + 2m_C)\mathrm{s}_1 \tag{17}$$

where $\mathrm{s}_1 \triangleq \sin q_1$.

Substitution from Eqs. (15) and (17) into

$$F_1 + F_1^\star \underset{(8.1.1)}{=} 0 \tag{18}$$

yields the differential equation

$$\dot{u}_1 = -p^2 \mathrm{s}_1 \tag{19}$$

where p is a constant defined as

$$p \triangleq \left[\frac{3g(m_A + m_B + 2m_C)}{4L(m_A + m_B + 3m_C)}\right]^{1/2} \tag{20}$$

Now, the set of forces exerted on A by the pin at P can be replaced with a couple whose torque is perpendicular to $\hat{\mathbf{n}}_3$, together with a force $R_1\hat{\mathbf{n}}_1 + R_2\hat{\mathbf{n}}_2 + R_3\hat{\mathbf{n}}_3$ applied to A at P. R_1 is to be determined for $t = 9$ s and $t = 10$ s, with $m_A = 1$ kg, $m_B = 2$ kg, $m_C = 3$ kg, $L = 1$ m, and $q_1(0) = 30°$, $u_1(0) = 0$.

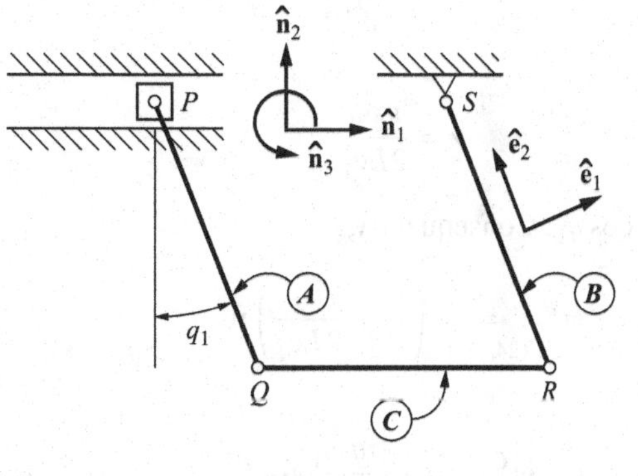

Figure 9.7.2

The constraint equation

$$\mathbf{v}^P \cdot \hat{\mathbf{n}}_1 = 0 \tag{21}$$

can be inspected [see Eqs. (6.5.1) and (6.5.2)] to identify a constraint force $R_1\hat{\mathbf{n}}_1$ applied to A at P in order to prevent P from moving in a direction parallel to $\hat{\mathbf{n}}_1$. Therefore, to bring R_1 (but not R_2, R_3, or the torque of the couple exerted on A by the pin) into evidence, one can imagine that P is attached to a slider that is free to move in a horizontal slot, as indicated in Fig. 9.7.2, and one can introduce a motion variable u_2 as

$$u_2 \triangleq \mathbf{v}^P \cdot \hat{\mathbf{n}}_1 = 0 \tag{22}$$

The angular velocities of A, B, and C then are given by

$$\boldsymbol{\omega}^A = u_1\hat{\mathbf{n}}_3 \tag{23}$$

and

$$\boldsymbol{\omega}^B = \beta\hat{\mathbf{n}}_3 \qquad \boldsymbol{\omega}^C = \gamma\hat{\mathbf{n}}_3 \tag{24}$$

where β and γ are functions of q_1, u_1, and u_2, which functions are found by setting $\mathbf{v}^S$, the velocity of point S, equal to zero after expressing $\mathbf{v}^S$ as

$$\mathbf{v}^S \underset{(2.7.1)}{=} \mathbf{v}^P + \boldsymbol{\omega}^A \times (-2L\hat{\mathbf{e}}_2) + \boldsymbol{\omega}^C \times (2L\hat{\mathbf{n}}_1) + \boldsymbol{\omega}^B \times (2L\hat{\mathbf{e}}_2) \tag{25}$$

In forming Eq. (25), the reason for treating link C as parallel to $\hat{\mathbf{n}}_1$, and for treating link A as parallel to B (that is, parallel to $\hat{\mathbf{e}}_2$), is that this reflects the configuration of the linkage during the motion of interest; as stated in Sec. 6.7, no additional generalized coordinates are to be introduced. Thus, one obtains

$$\mathbf{0} \underset{(25)}{=} \underset{(22)}{u_2\hat{\mathbf{n}}_1} + \underset{(23)}{2Lu_1\hat{\mathbf{e}}_1} + \underset{(24)}{2L\gamma\hat{\mathbf{n}}_2} - \underset{(24)}{2L\beta\hat{\mathbf{e}}_1} \tag{26}$$

and dot multiplications of Eq. (26) with $\hat{\mathbf{n}}_1$ and $\hat{\mathbf{e}}_2$ lead to

$$\beta = u_1 + \frac{u_2}{2L\mathrm{c}_1} \tag{27}$$

and

$$\gamma = \frac{u_2\mathrm{s}_1}{2L\mathrm{c}_1} \tag{28}$$

respectively, where $\mathrm{c}_1 \triangleq \cos q_1$. Consequently,

$$\boldsymbol{\omega}^B \underset{(24,\ 27)}{=} \left(u_1 + \frac{u_2}{2L\mathrm{c}_1}\right)\hat{\mathbf{n}}_3 \tag{29}$$

and

$$\boldsymbol{\omega}^C \underset{(24,\ 28)}{=} \frac{u_2\mathrm{s}_1}{2L\mathrm{c}_1}\hat{\mathbf{n}}_3 \tag{30}$$

Furthermore, the velocities of $A^\star$, $B^\star$, and $C^\star$, the mass centers of A, B, and C, respectively, are

$$\mathbf{v}^{A^\star} = \mathbf{v}^P + \boldsymbol{\omega}^A \times (-L\hat{\mathbf{e}}_2) = \underset{(22)}{u_2\hat{\mathbf{n}}_1} + \underset{(23)}{Lu_1\hat{\mathbf{e}}_1} \tag{31}$$

$$\mathbf{v}^{B^\star} = \boldsymbol{\omega}^B \times (-L\hat{\mathbf{e}}_2) \underset{(29)}{=} \left(Lu_1 + \frac{u_2}{2\mathrm{c}_1}\right)\hat{\mathbf{e}}_1 \tag{32}$$

$$\mathbf{v}^{C^\star} = \mathbf{v}^Q + \boldsymbol{\omega}^C \times (L\hat{\mathbf{n}}_1) \underset{(30)}{=} 2Lu_1\hat{\mathbf{e}}_1 + \left(\hat{\mathbf{n}}_1 + \frac{\mathrm{s}_1}{2\mathrm{c}_1}\hat{\mathbf{n}}_2\right)u_2 \tag{33}$$

The relationship given in Eq. (14) for generalized inertia forces remains applicable for $r = 1,2$, as do the expressions for inertia forces and inertia torques given in

Eqs. (8)–(13). On the other hand, the expression for generalized active forces appearing in Eq. (16) is replaced with

$$F_r = \mathbf{v}_r^P \cdot (R_1\hat{\mathbf{n}}_1 + R_2\hat{\mathbf{n}}_2 + R_3\hat{\mathbf{n}}_3) - g\hat{\mathbf{n}}_2 \cdot (m_A \mathbf{v}_r^{A\star} + m_B \mathbf{v}_r^{B\star} + m_C \mathbf{v}_r^{C\star}) \qquad (r = 1,2) \quad (34)$$

With the aid of Eqs. (23), (29), (30), and (31)–(33), one may verify that the partial angular velocities and partial velocities associated with u_1 remain unaltered [see Eqs. (6) and (7)]. Furthermore, in view of Eq. (22), one can write

$$\mathbf{v}_1^P = \mathbf{0} \quad (35)$$

Therefore, the expressions for $F_1^\star$ and F_1 given in Eqs. (15) and (17) remain unaltered subsequent to the introduction of u_2, as can be seen upon evaluating Eqs. (14) and (34) with $r = 1$. This outcome is to be expected because the single additional motion variable is in fact zero during the motion of interest. In contrast, $F_1^\star$ and F_1 are changed by introduction of the additional motion variable suggested in Problem 15.1.

Next, one can formulate the partial angular velocities of A, B, and C and the partial velocities of $A^\star$, $B^\star$, $C^\star$, and P associated with u_2 as

$$\boldsymbol{\omega}_2^A \underset{(23)}{=} \mathbf{0} \qquad \boldsymbol{\omega}_2^B \underset{(29)}{=} \frac{\hat{\mathbf{n}}_3}{2L\mathrm{c}_1} \qquad \boldsymbol{\omega}_2^C \underset{(30)}{=} \frac{\mathrm{s}_1\hat{\mathbf{n}}_3}{2L\mathrm{c}_1} \quad (36)$$

$$\mathbf{v}_2^{A\star} \underset{(31)}{=} \hat{\mathbf{n}}_1 \qquad \mathbf{v}_2^{B\star} \underset{(32)}{=} \frac{\hat{\mathbf{e}}_1}{2\mathrm{c}_1} \qquad \mathbf{v}_2^{C\star} \underset{(33)}{=} \hat{\mathbf{n}}_1 + \frac{\mathrm{s}_1\hat{\mathbf{n}}_2}{2\mathrm{c}_1} \quad (37)$$

and

$$\mathbf{v}_2^P \underset{(22)}{=} \hat{\mathbf{n}}_1 \quad (38)$$

Hence, the generalized inertia force $F_2^\star$ is given by

$$\begin{aligned} F_2^\star &\underset{(14)}{=} \underset{(36,\,11)}{0} \underset{(37,\,8)}{- m_A L(\dot{u}_1\mathrm{c}_1 - u_1{}^2\mathrm{s}_1)} \underset{(36,\,12)}{- \frac{m_B L}{6\mathrm{c}_1}\dot{u}_1} \underset{(37,\,9)}{- \frac{m_B L}{2\mathrm{c}_1}\dot{u}_1} \\ &\quad \underset{(36,\,13)}{+ \; 0} \underset{(37,\,10)}{- m_C L\left[\dot{u}_1\left(\mathrm{c}_1 + \frac{1}{\mathrm{c}_1}\right) - u_1{}^2\mathrm{s}_1\right]} \\ &= L(m_A + m_C)\mathrm{s}_1 u_1{}^2 - L\left\{m_A\mathrm{c}_1 + \frac{1}{\mathrm{c}_1}\left[\frac{2}{3}m_B + (1 + \mathrm{c}_1{}^2)m_C\right]\right\}\dot{u}_1 \end{aligned} \quad (39)$$

and the generalized active force F_2 is given by

$$F_2 \underset{(34)}{=} \underset{(38)}{R_1} \underset{(37)}{- g\,(m_B + m_C)\frac{\mathrm{s}_1}{2\mathrm{c}_1}} \quad (40)$$

Solved for R_1, the dynamical equation obtained by substituting from Eqs. (39) and (40) into

$$F_2 + F_2^\star \underset{(8.1.1)}{=} 0 \quad (41)$$

yields with the aid of Eq. (19)

$$R_1 = g(m_B + m_C)\frac{s_1}{2c_1} - L(m_A + m_C)s_1{u_1}^2 - L\left\{m_A c_1 + \frac{1}{c_1}\left[\frac{2}{3}m_B + (1 + {c_1}^2)m_C\right]\right\} p^2 s_1 \tag{42}$$

Table 9.7.1

t (s)	q_1 (deg)	u_1 (rad/s)
0	30.000	0.0000
1	−20.250	−0.8924
2	−2.852	1.2103
3	24.051	−0.7214
4	−29.470	−0.2250
5	15.716	1.0325
6	8.449	−1.1657
7	−26.989	0.5260
8	27.898	0.4427
9	−10.608	−1.1358
10	−13.736	1.0784

In Table 9.7.1, values are recorded for q_1 and u_1 as found by solving Eqs. (1) and (19) numerically with m_A, m_B, m_C, L, $q_1(0)$, and $u_1(0)$ as given. Using the tabulated values corresponding to $t = 9$ s, one finds that

$$s_1 = -0.1841 \qquad c_1 = 0.9829 \tag{43}$$

and

$$R_1 \underset{(42)}{=} 9.81(5)\frac{(-0.1841)}{2(0.9829)} - (4)(-0.1841)(-1.1358)^2 - \left\{0.9829 + \frac{1}{0.9829}\left[\frac{4}{3} + (1 + 0.9661)3\right]\right\} \underset{(20)}{(5.5181)} (-0.1841)$$
$$= 4.829 \text{ N} \tag{44}$$

Similarly, for $t = 10$ s, Eq. (42) and the last row of Table 9.7.1 yield $R_1 = 6.046$ N. Of course, the calculation of R_1 in accordance with Eq. (42) can be incorporated in the computer program used to solve Eqs. (1) and (19).

9.8 REAL SOLUTIONS OF A SET OF NONLINEAR, ALGEBRAIC EQUATIONS

The need to find real solutions of a set of nonlinear, algebraic equations can arise in connection with systems at rest in a Newtonian reference frame, steady motions, motions resembling states of rest, and various other problems of mechanics, as well as other areas of applied mathematics. For instance, referring to the example in Sec. 8.5, one may wish to find a value of the steering angle q_4 such that the system S remains at rest when M, R, a, b, L, T, and q_5 have preassigned values. To do so, one must solve Eq. (8.5.31), which one can do easily, despite the fact that the equation is nonlinear in q_4, for Eq. (8.5.31) leads directly to

$$q_4 = \tan^{-1}\left[\frac{T - MgR\sin\theta\sin q_5}{Tb/L - (MgRa/L)\sin\theta\cos q_5}\right] \tag{1}$$

Frequently, however, finding such solutions is not so straightforward. Consider, for example, the steady motion problem treated in the example in Sec. 8.6, and suppose that θ is to be determined for given values of $R\Omega^2/g$ and h/R. Since Eq. (8.6.17) cannot be solved readily for θ as a function of $R\Omega^2/g$ and h/R, one might resort to a numerical trial-and-error procedure consisting of evaluating the left-hand member of Eq. (8.6.17) for various values of θ until one has found a value that satisfies Eq. (8.6.17) to an acceptable degree of accuracy. But trial-and-error procedures become ineffective when employed for the solution of a *set* of nonlinear equations, as in the example in Sec. 8.7. Here, to determine q_1 and q_2 when $L\Omega^2/g$ is given, one must solve Eqs. (8.7.21) and (8.7.22) simultaneously, and one can do this by employing the procedures now to be described.

The most general set of n simultaneous equations in n unknowns $x_1,\dots,x_n$ can be written

$$f_i(x_1,\dots,x_n) = 0 \qquad (i = 1,\dots,n) \tag{2}$$

which alternatively may be written as

$$f(x) = 0 \tag{3}$$

where $f(x)$ denotes an $n \times 1$ column matrix whose i^{th} element is $f_i(x_1,\dots,x_n)$, x is an $n \times 1$ column matrix whose i^{th} element is x_i $(i = 1,\dots,n)$, and where the right hand member of Eq. (3) is an $n \times 1$ column matrix whose elements are all zero. This is done primarily so we can represent the results of an iterative process to approximate x numerically.

The most widely used approach for obtaining numerical solutions of nonlinear algebraic equations is the *Newton-Raphson method.* It involves an iterative procedure for finding x; in what follows, $\{x\}_r$ is used to denote the result obtained from the rth iteration. A solution for x, denoted by x^*, that satisfies $f(x^*) = 0$, is found by starting with an initial guess, $\{x\}_0$. A standard Newton-Raphson method uses the Taylor series approximation of $f(x)$, presuming that x is near x^*. Thus,

$$f(x^*) = f(\{x\}_0) + J(\{x\}_0)\{\Delta x\}_0 + \cdots \tag{4}$$

Terms of degree 2 and higher in $\{\Delta x\}_0$ are neglected, and $f(x^*)$ is set equal to zero, leading to the system of linear equations

$$J(\{x\}_0)\{\Delta x\}_0 = -f(\{x\}_0) \tag{5}$$

that is solved for $\{\Delta x\}_0$ to obtain the step taken to correct $\{x\}_0$ according to the relationship

$$\{x\}_1 = \{x\}_0 + \alpha\{\Delta x\}_0 \tag{6}$$

The *damping factor*, α, is a constant, the value of which is initially set equal to unity. The $n \times n$ matrix J, whose elements are given by

$$J_{ij} = \frac{\partial f_i}{\partial x_j} \qquad (i,j = 1,\ldots,n) \tag{7}$$

is called the *Jacobian*, and f is called the *error*, both of which depend on the current iterate of $\{x\}$.

The *Euclidean norm* of the error $f(x)$ is defined as

$$\|f(x)\| \triangleq \sqrt{\sum_{i=1}^{n} f_i{}^2(x)}$$

If the Euclidean norm of $f(\{x\}_1)$ is insufficiently close to zero, $\{x\}_0$ is replaced by $\{x\}_1$ in Eq. (5), so that

$$J\,(\{x\}_1)\,\{\Delta x\}_1 = -f\,(\{x\}_1) \tag{8}$$

and then

$$\{x\}_2 = \{x\}_1 + \alpha\{\Delta x\}_1 \tag{9}$$

This iterative process continues, with the rth step given by

$$J\,(\{x\}_r)\,\{\Delta x\}_r = -f\,(\{x\}_r) \tag{10}$$

and

$$\{x\}_{r+1} = \{x\}_r + \alpha\{\Delta x\}_r \tag{11}$$

until an estimate is found that is sufficiently close to the exact solution.

Unfortunately, the Newton-Raphson method is characterized by a strong sensitivity to the initial guess. Moreover, it may be that the rth step does not improve the solution (by, say, not lowering the error by a sufficiently large amount or not lowering it at all). To correct this the damping factor α may be decreased during the iterative process. A restricted-step Newton-Raphson iteration differs from the standard iteration in that if the full Newton step yields a value of $\|f\,(\{x\}_{r+1})\|$ that is an insufficient improvement over $\|f\,(\{x\}_r)\|$, perhaps even yielding a higher value, the damping factor α is halved, and $\{x\}_{r+1}$ is evaluated for the new value of α along with a recalculated value of $\|f\,(\{x\}_{r+1})\|$. This process is repeated until a sufficiently improved point is found. (When starting the solution of a new problem, α must be reset to unity.) The stopping criteria may include one or more of the following: (1) a value of the norm of the error less than some prescribed value, (2) a value of the norm of the step $\{\Delta x\}_r$ less than some prescribed value, (3) a value of α less than some prescribed value, and (4) a value of the number of iterations greater than some prescribed value.

The determination of $\{\Delta x\}_r$ at the rth step requires the solution of a linear system of algebraic equations of the form of Eq. (5). There are things that can go wrong at this step necessitating special measures. The Jacobian matrix may be ill-conditioned, meaning that the determinant is very small, which causes a loss in accuracy. It may be singular, meaning that the determinant is identically zero and that it is impossible to take a step. In either case, it is generally advisable to restart the procedure with a different initial guess.

The *continuation method* is another widely used approach for solving Eqs. (2). Rather than iteration, this method obtains a solution through numerical integration of a set of ordinary differential equations. Let $y_1(\tau),\ldots,y_n(\tau)$ be a set of n functions of a variable τ, with $0 \le \tau \le 1$; take

$$y_i(0) = k_i \qquad (i = 1,\ldots,n) \tag{12}$$

where k_i, a constant, is selected arbitrarily; and require that $y_1(\tau),\ldots,y_n(\tau)$ satisfy the equations

$$f_i(y_1,\ldots,y_n) = f_i(k_1,\ldots,k_n)(1-\tau) \qquad (i = 1,\ldots,n) \tag{13}$$

Then, as the right-hand sides of Eqs. (13) vanish at $\tau = 1$, the functions $y_1(\tau),\ldots,$ $y_n(\tau)$ satisfy, at $\tau = 1$, precisely the same equations as do $x_1,\ldots,x_n$ [see Eqs. (2)]. Now, $y_1(1),\ldots,y_n(1)$, and, hence, $x_1,\ldots,x_n$, may be found as follows: Differentiate Eqs. (13) with respect to τ, thus obtaining the set of first-order differential equations

$$\begin{aligned} \frac{\partial f_1}{\partial y_1}\frac{dy_1}{d\tau} + \cdots + \frac{\partial f_1}{\partial y_n}\frac{dy_n}{d\tau} &= -f_1(k_1,\ldots,k_n) \\ &\vdots \\ \frac{\partial f_n}{\partial y_1}\frac{dy_1}{d\tau} + \cdots + \frac{\partial f_n}{\partial y_n}\frac{dy_n}{d\tau} &= -f_n(k_1,\ldots,k_n) \end{aligned} \tag{14}$$

and perform a numerical integration of these equations (see Sec. 9.6), using Eqs. (12) as initial conditions and terminating the integration at $\tau = 1$. It is worth noting that the partial derivative $\partial f_i/\partial y_j$ in Eqs. (14) can be obtained from $\partial f_i/\partial x_j$ [see Eqs. (7)] simply by replacing x_j with y_j $(i,j = 1,\ldots,n)$. Moreover, Eq. (14) can be written in matrix form as

$$J\,(y(\tau))\;\dot{y} = -f(k) \tag{15}$$

where $f(k)$ denotes an $n \times 1$ column matrix whose i^{th} element is $f_i(k_1,\ldots,k_n)$, k is an $n \times 1$ column matrix whose i^{th} element is k_i, and where $\dot{y}$ is an $n \times 1$ column matrix whose i^{th} element is $dy_i/d\tau$ $(i = 1,\ldots,n)$.

As was stated previously, $k_1,\ldots,k_n$ may be assigned any values whatsoever. However, for certain choices of $k_1,\ldots,k_n$, it can occur that some of $y_1,\ldots,y_n$ do not possess real values for some values of τ in the interval $0 \le \tau \le 1$, in which event the numerical integration of the differential equations cannot be carried to completion. The method thus fails to converge in a manner similar to the Newton-Raphson method, in that the failure is typically manifested by the numerical integration procedure being unable to take an accurate step in pseudo-time τ. This would be either because the Jacobian is singular at that step, which causes it to be impossible to take a step at all, or because it is ill-conditioned, which leads to a loss of accuracy. When this happens, one simply changes one or more of $k_1,\ldots,k_n$. In general, results are obtained most expeditiously when $k_1,\ldots,k_n$ are good approximations to $x_1,\ldots,x_n$, respectively. Fortunately, in connection with physical problems, one often can make good guesses regarding $x_1,\ldots,x_n$ and hence assign suitable values to $k_1,\ldots,k_n$. Finally, it is worth noting that two or more distinct sets of values of $k_1,\ldots,k_n$ can lead to the same values of $x_1,\ldots,x_n$.

When there are serious difficulties in obtaining a converged solution, it may be helpful to generalize the continuation method. It may be generalized in a straightforward manner to a class of solution methods commonly known as *homotopy*. This generalization is not necessary unless one encounters serious convergence difficulties with the continuation method. It is often useful to generalize a continuation method into a homotopy method if (a) there are parameters that can take on values that are not of interest, for which the continuation method has no difficulties converging, and (b) those same parameters have actual values that are of interest for which convergence may not take place. These parameters are replaced by expressions that evaluate to their values at (a) when $\tau = 0$ and evaluate to their values at (b) when $\tau = 1$. This way, the parameters slowly change from values where the solution has no difficulties when τ is near zero to values where the solution is desired but the standard continuation method does not converge when τ is near unity.

Example If q_1, q_2, and q_3/L in Problem 13.6 are replaced with x_1, x_2, and x_3, respectively, then x_1, x_2, and x_3 are governed by equations having the form of Eqs. (2) with $n = 3$ and

$$f_1 \triangleq (1 - s_1)^2 + \left(\frac{3}{2} - s_2\right)^2 + (c_1 - c_2)^2 - \left(x_3 + \frac{1}{4}\right)^2 \tag{16}$$

$$f_2 \triangleq \left(x_3 + \frac{1}{4}\right) s_1 + 2x_3(c_2 s_1 - c_1) \tag{17}$$

$$f_3 \triangleq \left(x_3 + \frac{1}{4}\right) s_2 + x_3\,(2c_1 s_2 - 3c_2) \tag{18}$$

where s_i and c_i now stand for $\sin x_i$ and $\cos x_i$, respectively ($i = 1,2$).

Elements of the Jacobian matrix J, defined in Eq. (7), are given by

$$J_{11} = 2(s_1 c_2 - c_1) \tag{19}$$

$$J_{12} = 2c_1 s_2 - 3c_2 \tag{20}$$

$$J_{13} = -2\left(x_3 + \frac{1}{4}\right) \tag{21}$$

$$J_{21} = \left(x_3 + \frac{1}{4}\right) c_1 + 2x_3(c_1 c_2 + s_1) \tag{22}$$

$$J_{22} = -2x_3 s_1 s_2 \tag{23}$$

$$J_{23} = s_1 + 2(s_1 c_2 - c_1) \tag{24}$$

$$J_{31} = -2x_3 s_1 s_2 \tag{25}$$

$$J_{32} = \left(x_3 + \frac{1}{4}\right) c_2 + x_3\,(2c_1 c_2 + 3s_2) \tag{26}$$

$$J_{33} = s_2 + 2c_1 s_2 - 3c_2 \tag{27}$$

Table 9.8.1

r	q_1 (deg)	q_2 (deg)	q_3/L	$\|\Delta x\|$	$\|f\|$
0	0	0	0	–	3.1875
1	27.8294	41.7441	0.0607	0.8777	0.9174
2	54.7114	48.4336	0.5046	0.6564	0.5807
3	28.7884	42.0163	0.6343	0.4838	0.3127
4	35.4904	44.7307	0.6474	0.1269	0.0165
5	35.5489	44.9096	0.6538	0.0071	0.0001
6	35.5486	44.9081	0.6538	2.7244×10^{-5}	1.1540×10^{-9}
7	35.5486	44.9081	0.6538	4.8116×10^{-10}	2.2204×10^{-16}

A solution for x_1, x_2, and x_3 can be obtained with the Newton-Raphson method by carrying out the operations associated with Eqs. (10) and (11), starting with $r = 0$. One obtains the results in Tables 9.8.1, 9.8.2, and 9.8.3, where the 0th iteration is the initial guess. Note that very different initial guesses yield the same results in the first two cases, whereas a completely different result is obtained in the third case. Results in Tables 9.8.1 and 9.8.2 characterize an equilibrium configuration in which B_1 and B_2 (see Fig. P13.6) lie below the horizontal plane determined by the axes of the pins that support the bars, whereas results in Table 9.8.3 apply when the system is at rest

Table 9.8.2

r	q_1 (deg)	q_2 (deg)	q_3/L	$\|\Delta x\|$	$\|f\|$
0	9.9981	9.9981	0.1000	–	2.3306
1	31.0252	46.6682	0.1719	0.7413	0.7045
2	47.6889	41.1885	0.7263	0.6333	0.6775
3	36.8759	44.0996	0.6256	0.2198	0.0893
4	35.4975	44.9418	0.6536	0.0397	0.0032
5	35.5486	44.9081	0.6538	0.0011	2.2370×10^{-6}
6	35.5486	44.9081	0.6538	1.0826×10^{-6}	1.9169×10^{-12}
7	35.5486	44.9081	0.6538	6.5070×10^{-13}	5.6610×10^{-16}

Table 9.8.3

r	q_1 (deg)	q_2 (deg)	q_3/L	$\|\|\Delta x\|\|$	$\|\|f\|\|$
0	−99.9811	−99.9811	2.0000	–	5.1448
1	−124.8289	−90.1993	3.0890	1.1845	1.5125
2	−116.0555	−93.7939	2.9156	0.2396	0.0789
3	−114.7152	−94.7539	2.9103	0.0293	0.0012
4	−114.6908	−94.7724	2.9103	0.0005	3.6343×10^{-7}
5	−114.6908	−94.7724	2.9103	1.6092×10^{-7}	3.2406×10^{-14}
6	−114.6908	−94.7724	2.9103	1.4317×10^{-14}	9.9301×10^{-16}

with B_1 and B_2 above this plane. The value of α remains equal to unity for all iteration steps shown in Tables 9.8.1, 9.8.2, and 9.8.3.

One can conclude from these results that either or both norms, $\|\Delta x\|$ and $\|f\|$, can be used to stop the iteration. In these cases, both criteria would have led to approximately the same values of x_i. However, one should not expect this to be a general trend; one of the criteria may be very slow (or quick) to be satisfied, in which case a judgment must be made as to what to do about the solution. Recall that one may also stop iterating when the number of iterations exceeds some prescribed upper limit.

Alternatively, a solution can be obtained by employing the continuation method. The equations corresponding to Eqs. (14) are

$$2(\text{s}_1\text{c}_2 - \text{c}_1)\frac{dy_1}{d\tau} + (2\text{c}_1\text{s}_2 - 3\text{c}_2)\frac{dy_2}{d\tau} - 2\left(y_3 + \frac{1}{4}\right)\frac{dy_3}{d\tau}$$
$$= -\left[(1 - \sin k_1)^2 + \left(\frac{3}{2} - \sin k_2\right)^2 + (\cos k_1 - \cos k_2)^2 - \left(k_3 + \frac{1}{4}\right)^2\right] \tag{28}$$

$$\left[\left(y_3 + \frac{1}{4}\right)\text{c}_1 + 2y_3(\text{c}_1\text{c}_2 + \text{s}_1)\right]\frac{dy_1}{d\tau} - 2y_3\text{s}_1\text{s}_2\frac{dy_2}{d\tau} + [\text{s}_1 + 2(\text{s}_1\text{c}_2 - \text{c}_1)]\frac{dy_3}{d\tau}$$
$$= -\left[\left(k_3 + \frac{1}{4}\right)\sin k_1 + 2k_3(\cos k_2 \sin k_1 - \cos k_1)\right] \tag{29}$$

$$-2y_3\text{s}_1\text{s}_2\frac{dy_1}{d\tau} + \left[\left(y_3 + \frac{1}{4}\right)\text{c}_2 + y_3\,(2\text{c}_1\text{c}_2 + 3\text{s}_2)\right]\frac{dy_2}{d\tau}$$
$$+ [\text{s}_2 + (2\text{c}_1\text{s}_2 - 3\text{c}_2)]\,\frac{dy_3}{d\tau}$$
$$= -\left[\left(k_3 + \frac{1}{4}\right)\sin k_2 + k_3\,(2\cos k_1 \sin k_2 - 3\cos k_2)\right] \tag{30}$$

where s_i and c_i now denote $\sin y_i$ and $\cos y_i$, respectively ($i = 1,2$).

With $k_1 = k_2 = k_3 = 0$, numerical integration of Eqs. (28)–(30) in the interval $0 \leq \tau \leq 1$ leads to values of $y_1(1)$, $y_2(1)$, and $y_3(1)$, and thus to values of x_1, x_2, and x_3, such that

$$q_1 = 35.5486° \qquad q_2 = 44.9081° \qquad q_3 = 0.6538L \tag{31}$$

Identical results are obtained with $k_1 = k_2 = 0.1745$ (corresponding to $q_1 = q_2 = 10°$) and $k_3 = 0.1$; but taking $k_1 = k_2 = -1.745$ and $k_3 = 2$ produces

$$q_1 = -114.6908° \qquad q_2 = -94.7724° \qquad q_3 = 2.9103L \tag{32}$$

Equations (31) characterize an equilibrium configuration in which B_1 and B_2 (see Fig. P13.6) lie below the horizontal plane determined by the axes of the pins that support the bars, whereas Eqs. (32) apply when the system is at rest with B_1 and B_2 above this plane.

9.9 MOTIONS GOVERNED BY LINEAR DIFFERENTIAL EQUATIONS

Occasionally, one encounters a system whose motion is governed by one or more *linear* differential equations. More frequently, linear differential equations arise as a result of linearizations performed to take advantage of the fact that one can solve certain kinds of linear differential equations without resorting to numerical integration. The analyses dealing with the Foucault pendulum in Secs. 8.1 and 8.4 furnish cases in point. Because only one of Eqs. (8.1.30), (8.1.31), and (8.1.35) is a linear differential equation [the second of Eqs. (8.1.35)], this set of equations was integrated numerically to obtain the results reported in Table 8.1.1; but Eqs. (8.4.6)–(8.4.8), being linear differential equations, were solved in closed form.

One linear differential equation that arises frequently both in dynamics and in other areas of physics has the form

$$\ddot{x} + 2np\dot{x} + p^2 x = f(t) \tag{1}$$

where x is a function of time t, dots denote time differentiation, n and p are constants called, respectively, the *fraction of critical damping* and the *circular natural frequency*, and $f(t)$ is a specified function of t, called the *forcing function*. [See, for example, Eq. (9.5.21), where u_3 plays the role of x while $n = f(t) = 0$.] The general solution of Eq. (1) is

$$x = \frac{1}{a_1 - a_2}\Big\{[\dot{x}(0) - a_2 x(0)]e^{a_1 t} - [\dot{x}(0) - a_1 x(0)]e^{a_2 t} + \int_0^t f(\zeta)[e^{a_1(t-\zeta)} - e^{a_2(t-\zeta)}]d\zeta\Big\} \tag{2}$$

where $x(0)$ and $\dot{x}(0)$ denote, respectively, the values of x and $\dot{x}$ at $t = 0$, and a_1 and a_2 are defined as

$$a_1 \triangleq -p[n - (n^2 - 1)^{1/2}] \qquad a_2 \triangleq -p[n + (n^2 - 1)^{1/2}] \tag{3}$$

The set of n linear differential equations

$$\sum_{s=1}^{n}(M_{rs}\ddot{q}_s + K_{rs}q_s) = 0 \qquad (r = 1,\ldots,n) \tag{4}$$

governs motions of a holonomic system S under conditions stated in Problem 13.4. Given the constants M_{rs} and K_{rs} $(r, s = 1,\ldots,n)$, as well as $q_r(0)$ and $\dot{q}_r(0)$, the initial values of q_r and $\dot{q}_r$ $(r = 1,\ldots,n)$, respectively, one can find $q_1,\ldots,q_n$ for $t > 0$ by proceeding as follows.

Let M and K be $n \times n$ matrices respectively having M_{rs} and K_{rs} as the elements in row r, column s. M and K are called the *mass matrix* and the *stiffness matrix*, respectively.

Construct an upper-triangular $n \times n$ matrix V such that

$$M = V^T V \tag{5}$$

where V^T denotes the transpose of V. (When M is diagonal, then $V = M^{1/2}$, that is, V is the diagonal matrix whose elements are the square roots of corresponding elements of M; otherwise, use Cholesky decomposition.)

Find the eigenvalues $\lambda_1,\ldots,\lambda_n$ of the (necessarily symmetric) $n\times n$ matrix W defined as

$$W \triangleq (V^{-1})^T K V^{-1} \tag{6}$$

and determine the corresponding eigenvectors (each an $n \times 1$ matrix) $C^{(1)},\ldots,C^{(n)}$ of W. If not all eigenvalues of W are distinct, then, corresponding to an eigenvalue of multiplicity μ, construct μ eigenvectors that are orthogonal to each other.†

Define an $n \times 1$ matrix $B^{(i)}$, a scalar N_i, and an $n \times 1$ matrix $A^{(i)}$ $(i = 1,\ldots,n)$ as, respectively,

$$B^{(i)} \triangleq V^{-1}C^{(i)} \qquad (i = 1,\ldots,n) \tag{7}$$

$$N_i \triangleq \sqrt{[B^{(i)}]^T M B^{(i)}} \qquad (i = 1,\ldots,n) \tag{8}$$

and

$$A^{(i)} \triangleq \frac{B^{(i)}}{N_i} \qquad (i = 1,\ldots,n) \tag{9}$$

and let A be the $n \times n$ matrix having $A^{(1)},\ldots,A^{(n)}$ as columns; that is, define A as

$$A \triangleq [\ A^{(1)} \mid A^{(2)} \mid \cdots \mid A^{(n)}\ ] \tag{10}$$

The matrix A is called the *modal matrix normalized with respect to the mass matrix.*

Let $q(0)$ and $\dot{q}(0)$ be $n \times 1$ matrices respectively having $q_r(0)$ and $\dot{q}_r(0)$ as the elements in the rth row, and define $Q(0)$ and $\dot{Q}(0)$ as the $n \times 1$ matrices

$$Q(0) \triangleq A^T M q(0) \qquad \dot{Q}(0) \triangleq A^T M \dot{q}(0) \tag{11}$$

† The Gram-Schmidt orthogonalization procedure may be used to accomplish this task.

Introduce scalars $p_1,\ldots,p_n$ as

$$p_r \triangleq \lambda_r^{1/2} \qquad (r = 1,\ldots,n) \tag{12}$$

and define diagonal $n \times n$ matrices c, s, and p as

$$c \triangleq \begin{bmatrix} \cos p_1 t & 0 & \cdots & 0 \\ 0 & \cos p_2 t & \cdots & 0 \\ \cdots & \cdots & \cdots & \cdots \\ 0 & 0 & \cdots & \cos p_n t \end{bmatrix} \tag{13}$$

$$s \triangleq \begin{bmatrix} \sin p_1 t & 0 & \cdots & 0 \\ 0 & \sin p_2 t & \cdots & 0 \\ \cdots & \cdots & \cdots & \cdots \\ 0 & 0 & \cdots & \sin p_n t \end{bmatrix} \tag{14}$$

and

$$p \triangleq \begin{bmatrix} p_1 & 0 & \cdots & 0 \\ 0 & p_2 & \cdots & 0 \\ \cdots & \cdots & \cdots & \cdots \\ 0 & 0 & \cdots & p_n \end{bmatrix} \tag{15}$$

Finally, form the $n \times 1$ matrix Q defined as

$$Q \triangleq cQ(0) + p^{-1}s\dot{Q}(0) \tag{16}$$

and let q be the $n \times 1$ matrix whose elements are the generalized coordinates $q_1,\ldots,q_n$. Then q is given by

$$q = AQ \tag{17}$$

The elements $Q_1,\ldots,Q_n$ of Q are called *normal coordinates*. The system S can move in such a way that any one of the normal coordinates, say, the jth, varies with time while all of the rest of the normal coordinates vanish. This occurs when $Q_j(0)$ and/or $\dot{Q}_j(0)$ differ from zero, but $Q_k(0)$ and $\dot{Q}_k(0)$ vanish for $k \neq j$ $(k = 1,\ldots,n)$, for then Q_j is the only nonvanishing element of Q, as can be seen by reference to Eqs. (13)–(16). Under these circumstances, the expression for q produced by Eq. (17) involves only the jth column of the modal matrix A; that is,

$$q = A^{(j)}Q_j \tag{18}$$

and S is said to be *moving in the* j^{th} *normal mode*. To find initial conditions such that S moves in this manner, one needs only to assign arbitrary values to $Q_j(0)$ and $\dot{Q}_j(0)$, set $Q_k(0)$ and $\dot{Q}_k(0)$ equal to zero for $k \neq j$ $(k = 1,\ldots,n)$, and assign to $q(0)$ and $\dot{q}(0)$ the values $A^{(j)}Q_j(0)$ and $A^{(j)}\dot{Q}_j(0)$, respectively.

Equation (17) is equivalent to the set of equations

$$q_r = \sum_{s=1}^{n} A_r^{(s)} Q_s \qquad (r = 1,\ldots,n) \tag{19}$$

which show that q_r $(r = 1,\ldots,n)$ can be regarded as a sum, each of whose elements represents a motion of S in a normal mode [see Eq. (18)]. Moreover, Eqs. (19) furnish the basis for the constructing of *approximations* to $q_1, \ldots, q_n$ by *modal truncation*, that is, by letting ν be an integer smaller than n and writing

$$q_r \approx \sum_{s=1}^{\nu} A_r{}^{(s)} Q_s \qquad (r = 1, \ldots, n) \tag{20}$$

or, equivalently,

$$q \approx \widetilde{A}\,\widetilde{Q} \tag{21}$$

where $\widetilde{A}$ is the $n \times \nu$ matrix whose columns are the first ν columns of A, and $\widetilde{Q}$ is the $\nu \times 1$ matrix whose elements are the first ν elements of Q. Such an approximation can save a considerable amount of computational effort when ν is sufficiently small in comparison with n, for one needs only the ν eigenvalues $\lambda_1, \ldots, \lambda_\nu$ and the ν eigenvectors $C^{(1)}, \ldots, C^{(\nu)}$, rather than all n eigenvalues and eigenvectors, to evaluate $\widetilde{A}$ and $\widetilde{Q}$.

If $F_1(t), \ldots, F_n(t)$ are known functions of t, and $q_1, \ldots, q_n$ are governed by

$$\sum_{s=1}^{n} (M_{rs}\ddot{q}_s + K_{rs} q_s) = F_r(t) \qquad (r = 1, \ldots, n) \tag{22}$$

rather than by Eqs. (4), but M_{rs} and K_{rs} $(r, s = 1, \ldots, n)$ have the same meanings as heretofore, then q is again given by Eq. (17) if Q, rather than being defined as in Eq. (16), is taken to be

$$Q \triangleq cQ(0) + p^{-1}[s\dot{Q}(0) + \eta] \tag{23}$$

where η is an $n \times 1$ matrix whose r^{th} element is defined as

$$\eta_r \triangleq \sum_{j=1}^{n} A_j{}^{(r)} [\gamma_j{}^{(r)} \sin p_r t - \sigma_j{}^{(r)} \cos p_r t] \qquad (r = 1, \ldots, n) \tag{24}$$

with

$$\gamma_j{}^{(r)} \triangleq \int_0^t F_j(\zeta) \cos p_r \zeta \, d\zeta \qquad \sigma_j{}^{(r)} \triangleq \int_0^t F_j(\zeta) \sin p_r \zeta \, d\zeta \qquad (j, r = 1, \ldots, n) \tag{25}$$

Furthermore, approximate solutions of Eqs. (22) can be obtained by modal truncation, that is, by using Eq. (21) and taking for $\widetilde{Q}$ the $\nu \times 1$ matrix whose elements are the first ν elements of Q as given by Eq. (23).

When n is sufficiently small, say, less than 4, one can solve Eqs. (4) and (22) by methods simpler than the ones just set forth. Conversely, when n is relatively large, the use of normal modes is very effective, especially when good computer programs for performing matrix operations are readily available.

Derivations Differentiation of Eq. (2) with respect to t yields

$$\dot{x} = \frac{1}{a_1 - a_2} \Big\{ [\dot{x}(0) - a_2 x(0)] a_1 e^{a_1 t} - [\dot{x}(0) - a_1 x(0)] a_2 e^{a_2 t} + \int_0^t f(\zeta) [a_1 e^{a_1(t-\zeta)} - a_2 e^{a_2(t-\zeta)}] \, d\zeta \Big\} \tag{26}$$

and differentiating this equation one obtains

$$\ddot{x} = \frac{1}{a_1 - a_2}\left\{[\dot{x}(0) - a_2 x(0)]a_1{}^2 e^{a_1 t} - [\dot{x}(0) - a_1 x(0)]a_2{}^2 e^{a_2 t} + \int_0^t f(\zeta)[a_1{}^2 e^{a_1(t-\zeta)} - a_2{}^2 e^{a_2(t-\zeta)}]\,d\zeta\right\} + f(t) \tag{27}$$

Using Eqs. (2), (26), and (27), one thus finds that

$$\ddot{x} + 2np\dot{x} + p^2 x = \frac{1}{a_1 - a_2}\left\langle\left\{[\dot{x}(0) - a_2 x(0)]e^{a_1 t} + \int_0^t f(\zeta)e^{a_1(t-\zeta)}d\zeta\right\}(a_1{}^2 + 2npa_1 + p^2) - \left\{[\dot{x}(0) - a_1 x(0)]e^{a_2 t} + \int_0^t f(\zeta)e^{a_2(t-\zeta)}d\zeta\right\}(a_2{}^2 + 2npa_2 + p^2)\right\rangle + f(t) \tag{28}$$

Now, if a_1 and a_2 are given by Eqs. (3), then

$$a_1{}^2 + 2npa_1 + p^2 = a_2{}^2 + 2npa_2 + p^2 = 0 \tag{29}$$

Consequently,

$$\ddot{x} + 2np\dot{x} + p^2 x \underset{(28,\,29)}{=} f(t) \tag{30}$$

which is Eq. (1). Since x as given in Eq. (2) contains two arbitrary constants, it is thus seen to be the general solution of Eq. (1) when a_1 and a_2 are given by Eqs. (3).

To establish the validity of the procedure for finding functions $q_1, \ldots, q_n$ of t that satisfy Eqs. (4), we begin with some observations regarding the matrices M, K, $C^{(i)}$, $B^{(i)}$, and $A^{(i)}$ $(i = 1, \ldots, n)$, noting first that

$$[A^{(i)}]^T M A^{(i)} \underset{(9)}{=} \frac{[B^{(i)}]^T}{N_i} M \frac{B^{(i)}}{N_i} \underset{(8)}{=} 1 \qquad (i = 1, \ldots, n) \tag{31}$$

Second, if $B^{(i)}$ is the eigenvector of $M^{-1}K$ corresponding to the eigenvalue λ_i, that is, $M^{-1}KB^{(i)} = \lambda_i B^{(i)}$, and $C^{(i)}$ is the eigenvector of W corresponding to the eigenvalue $\overline{\lambda}_i$, that is, $WC^{(i)} = \overline{\lambda}_i C^{(i)}$, then it follows from Eq. (5) that $V^{-1}(V^{-1})^T KB^{(i)} = \lambda_i B^{(i)}$ or, when Eqs. (7) are taken into account, that $V^{-1}(V^{-1})^T KV^{-1}C^{(i)} = \lambda_i V^{-1}C^{(i)}$, while Eq. (6) enables one to write $(V^{-1})^T KV^{-1}C^{(i)} = \overline{\lambda}_i C^{(i)}$ or, after premultiplication with V^{-1}, $V^{-1}(V^{-1})^T KV^{-1}C^{(i)} = \overline{\lambda}_i V^{-1}C^{(i)}$ $(i = 1, \ldots, n)$. Consequently, $\lambda_i = \overline{\lambda}_i$ $(i = 1, \ldots, n)$, which is to say that $M^{-1}K$ and W have the same eigenvalues $\lambda_1, \ldots, \lambda_n$ and that $B^{(i)}$ is the eigenvector of $M^{-1}K$ corresponding to λ_i if $B^{(i)}$ is related to $C^{(i)}$, the eigenvector of W corresponding to λ_i $(i = 1, \ldots, n)$, as in Eqs. (7). Hence, we can write

$$M^{-1}KB^{(i)} = \lambda_i B^{(i)} \qquad (i = 1, \ldots, n) \tag{32}$$

or, after premultiplication with M,

$$KB^{(i)} \underset{(32)}{=} \lambda_i MB^{(i)} \qquad (i = 1,\ldots,n) \tag{33}$$

Consequently,

$$\begin{aligned}[B^{(i)}]^T KB^{(i)} &\underset{(33)}{=} \lambda_i [B^{(i)}]^T MB^{(i)} \\ &\underset{(8)}{=} \lambda_i N_i{}^2 \qquad (i = 1,\ldots,n)\end{aligned} \tag{34}$$

and

$$\frac{[B^{(i)}]^T}{N_i} K \frac{B^{(i)}}{N_i} \underset{(34)}{=} \lambda_i \tag{35}$$

or, in view of Eqs. (9),

$$[A^{(i)}]^T KA^{(i)} = \lambda_i \qquad (i = 1,\ldots,n) \tag{36}$$

Third, we establish the validity of the *orthogonality* relationships

$$[A^{(j)}]^T MA^{(i)} = 0 \qquad (i,j = 1,\ldots,n; i \neq j) \tag{37}$$

and

$$[A^{(j)}]^T KA^{(i)} = 0 \qquad (i,j = 1,\ldots,n; i \neq j) \tag{38}$$

as follows.

Because W is symmetric, the eigenvectors of W are orthogonal to each other if the eigenvalues of W are distinct, and infinitely many sets of orthogonal eigenvectors of W can be found if not all eigenvalues are distinct. Hence,

$$[C^{(j)}]^T C^{(i)} = 0 \qquad (i,j = 1,\ldots,n; i \neq j) \tag{39}$$

Now,

$$\begin{aligned}[B^{(j)}]^T MB^{(i)} &= [\underset{(7)}{V^{-1}C^{(j)}}]^T M [\underset{(7)}{V^{-1}C^{(i)}}] \\ &= [C^{(j)}]^T [V^T]^{-1} MV^{-1}C^{(i)} \\ &= [C^{(j)}]^T [V^T]^{-1} \underset{(5)}{V^T V} V^{-1}C^{(i)} \\ &= [C^{(j)}]^T C^{(i)} \underset{(39)}{=} 0 \qquad (i,j = 1,\ldots,n; i \neq j)\end{aligned} \tag{40}$$

and Eqs. (37) follow directly from these equations together with Eqs. (9). Furthermore,

$$[B^{(j)}]^T KB^{(i)} \underset{(33)}{=} \lambda_i [B^{(j)}]^T MB^{(i)} \underset{(40)}{=} 0 \qquad (i,j = 1,\ldots,n; i \neq j) \tag{41}$$

and the validity of Eqs. (38) thus is established when Eqs. (9) are taken into account.

As will be shown next, what is of interest in connection with Eqs. (4) is that Eqs. (31) and (37) imply that

$$A^T MA = U \tag{42}$$

while Eqs. (36) and (38) justify the conclusion that

$$A^T K A = \lambda \tag{43}$$

where A is the modal matrix defined in Eq. (10), U is the $n \times n$ unit matrix, and λ is the $n \times n$ diagonal matrix defined as

$$\lambda \triangleq \begin{bmatrix} \lambda_1 & & 0 \\ & \ddots & \\ 0 & & \lambda_n \end{bmatrix} \tag{44}$$

Specifically, Eqs. (4) are equivalent to the matrix differential equation

$$M\ddot{q} + Kq = 0 \tag{45}$$

so that, if q is set equal to AQ [see Eq. (17)], where Q is an, as yet, unknown $n \times 1$ matrix whose elements are functions of t, then

$$MA\ddot{Q} + KAQ \underset{(45)}{=} 0 \tag{46}$$

and, premultiplying this equation with A^T, we find in view of Eqs. (42) and (43) that

$$\ddot{Q} + \lambda Q = 0 \tag{47}$$

or, since λ is a diagonal matrix [see Eq. (44)], that

$$\ddot{Q}_r + \underset{(12)}{p_r{}^2} Q_r = 0 \qquad (r = 1, \ldots, n) \tag{48}$$

The general solution of these equations is

$$Q_r = Q_r(0)\cos p_r t + \frac{\dot{Q}_r(0)}{p_r}\sin p_r t \qquad (r = 1, \ldots, n) \tag{49}$$

where $Q_r(0)$ and $\dot{Q}_r(0)$ are, respectively, the initial values of Q_r and $\dot{Q}_r$ $(r = 1, \ldots, n)$. Moreover, if $Q(0)$ and $\dot{Q}(0)$ are the $n \times 1$ matrices whose elements are, respectively, $Q_1(0), \ldots, Q_n(0)$ and $\dot{Q}_1(0), \ldots, \dot{Q}_n(0)$, and c, s, and p are the matrices defined in Eqs. (13)–(15), then Eq. (16) is equivalent to Eqs. (49). What remains to be shown is that $Q(0)$ and $\dot{Q}(0)$ as defined in Eqs. (11) are, in fact, the initial values of Q and $\dot{Q}$, respectively. This is accomplished by referring to Eq. (17) to write

$$q(0) = AQ(0) \qquad \dot{q}(0) = A\dot{Q}(0) \tag{50}$$

and then premultiplying with $A^T M$, whereupon Eqs. (11) are obtained when Eq. (42) is taken into account.

Finally, we consider Eqs. (22), which are equivalent to the matrix equation

$$M\ddot{q} + Kq = F \tag{51}$$

where F is the $n \times 1$ matrix having $F_1(t), \ldots, F_n(t)$ as elements. Here, expressing q as in Eq. (17) and then premultiplying with A^T yields, with the aid of Eqs. (42) and (43),

$$\ddot{Q} + \lambda Q = A^T F \tag{52}$$

which is equivalent to

$$\ddot{Q}_r + p_r{}^2 Q_r \underset{(44,\ 12)\ (52)}{=} f_r(t) \qquad (r = 1, \ldots, n) \tag{53}$$

where $f_r(t)$ is the rth element of the $n \times 1$ matrix $A^T F$, that is

$$f_r(t) = \sum_{j=1}^{n} A_j{}^{(r)} F_j(t) \qquad (r = 1, \ldots, n) \tag{54}$$

Now, Eqs. (53) have the same form as Eq. (1) with $n = 0$. Referring to Eqs. (2) and (3), we can, therefore, write

$$\begin{aligned} Q_r &= \frac{1}{2ip_r}\left\{[\dot{Q}_r(0) + ip_r Q_r(0)]e^{ip_r t} - [\dot{Q}_r(0) - ip_r Q_r(0)]e^{-ip_r t}\right. \\ &\quad \left. + \int_0^t f_r(\zeta)[e^{ip_r(t-\zeta)} - e^{-ip_r(t-\zeta)}]d\zeta\right\} \\ &= Q_r(0)\cos p_r t + \frac{\dot{Q}_r(0)}{p_r}\sin p_r t \\ &\quad + \frac{1}{p_r}\int_0^t f_r(\zeta)\sin[p_r(t-\zeta)]d\zeta \qquad (r = 1, \ldots, n) \end{aligned} \tag{55}$$

or, after using Eqs. (54),

$$\begin{aligned} Q_r &= Q_r(0)\cos p_r t + \frac{\dot{Q}_r(0)}{p_r}\sin p_r t \\ &\quad + \frac{1}{p_r}\sum_{j=1}^{n} A_j{}^{(r)}\left[\sin p_r t \int_0^t F_j(\zeta)\cos p_r\zeta\, d\zeta\right. \\ &\quad \left. - \cos p_r t \int_0^t F_j(\zeta)\sin p_r\zeta\, d\zeta\right] \qquad (r = 1, \ldots, n) \end{aligned} \tag{56}$$

so that, with $\gamma_j{}^{(r)}$ and $\sigma_j{}^{(r)}$ $(j, r = 1, \ldots, n)$ as defined in Eqs. (25), and η_r $(r = 1, \ldots, n)$ as given by Eqs. (24), we have

$$Q_r = Q_r(0)\cos p_r t + \frac{1}{p_r}[\dot{Q}_r(0)\sin p_r t + \eta_r] \qquad (r = 1, \ldots, n) \tag{57}$$

which is equivalent to Eq. (23) if η is the $n \times 1$ matrix having $\eta_1, \ldots, \eta_n$ as elements.

Example Figure 9.9.1 shows a truss T consisting of eight members of length L and mass m and five members of length $\sqrt{2}L$ and mass $\sqrt{2}m$. All members have the same cross-sectional area Z and Young's modulus E, and spherical joints are used to connect members to each other and to a base fixed in a Newtonian reference frame N. Finally, $\hat{\mathbf{n}}_1$, $\hat{\mathbf{n}}_2$, and $\hat{\mathbf{n}}_3$ are mutually perpendicular unit vectors fixed in N.

When T is replaced with a *lumped-mass model* for purposes of vibrations analysis, that is, with a set S of eight particles $P_1, \ldots, P_8$ at the nodes of the truss, as indicated in Fig. 9.9.2, then the mass of each particle is taken to be equal to one-half of the sum of the masses of all members meeting at the node where the particle is placed. Thus, P_1 is assigned the mass $3m(1 + \sqrt{2})/2$, P_2 the mass $3m/2$, and so forth.

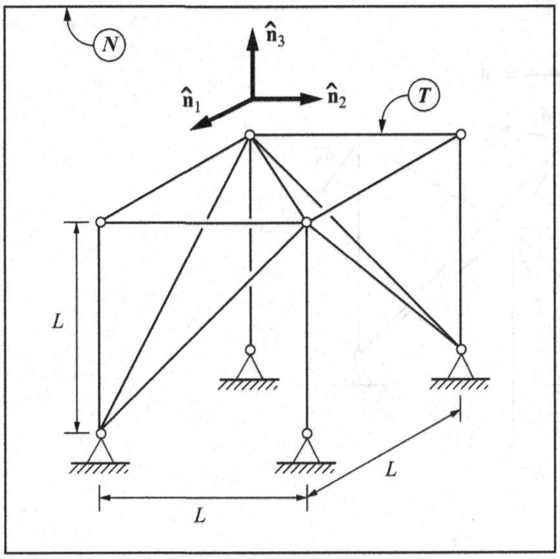

Figure 9.9.1

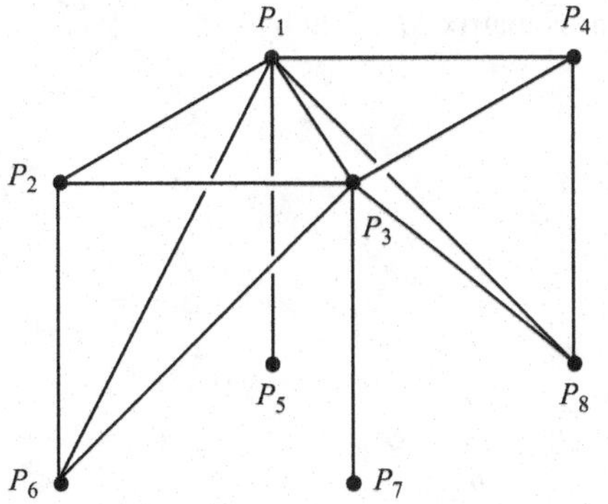

Figure 9.9.2

As for generalized coordinates, we note that S possess 12 degrees of freedom in N; let $\boldsymbol{\delta}_i$ be the position vector to P_i at time t from the point of N at which P_i $(i = 1,\ldots,4)$ is situated when T is undeformed, as shown in Fig. 9.9.3; and define $q_1,\ldots,q_{12}$ as

$$q_r = \left\{ \begin{array}{ll} \boldsymbol{\delta}_1 \cdot \hat{\mathbf{n}}_r & (r = 1,2,3) \\ \boldsymbol{\delta}_2 \cdot \hat{\mathbf{n}}_{r-3} & (r = 4,5,6) \\ \boldsymbol{\delta}_3 \cdot \hat{\mathbf{n}}_{r-6} & (r = 7,8,9) \\ \boldsymbol{\delta}_4 \cdot \hat{\mathbf{n}}_{r-9} & (r = 10,11,12) \end{array} \right\} \tag{58}$$

Linearized in $q_1,\ldots,q_{12}$, the dynamical equations governing all motions of S when gravitational forces are treated as negligible are precisely Eqs. (4) or, equivalently,

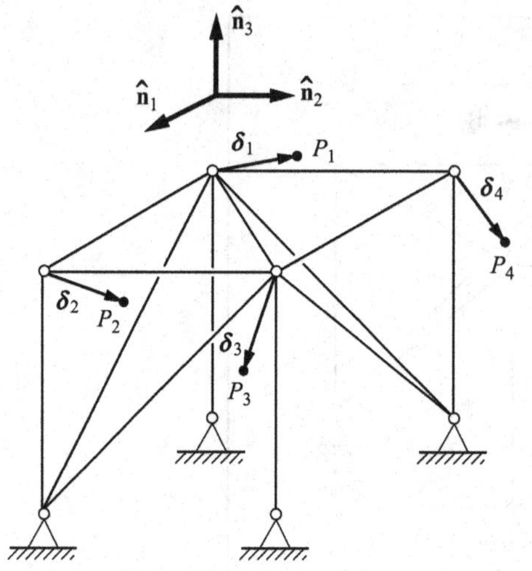

Figure 9.9.3

Eqs. (45), with $n = 12$ and the mass matrix M given by

$$M = m \begin{bmatrix} m_1 & & & & & & & & & & & \\ & m_1 & & & & & & & & & & \\ & & m_1 & & & & & & & & & \\ & & & m_2 & & & & & 0 & & & \\ & & & & m_2 & & & & & & & \\ & & & & & m_2 & & & & & & \\ & & & & & & m_1 & & & & & \\ & & & & & & & m_1 & & & & \\ & & & 0 & & & & & m_1 & & & \\ & & & & & & & & & m_2 & & \\ & & & & & & & & & & m_2 & \\ & & & & & & & & & & & m_2 \end{bmatrix} \tag{59}$$

where

$$m_1 \triangleq \frac{3(1+\sqrt{2})}{2} \qquad m_2 = \frac{3}{2} \tag{60}$$

while the stiffness matrix K assumes the form

$$K = k \begin{bmatrix} k_1 & k_2 & -k_2 & -1 & 0 & 0 & -k_2 & -k_2 & 0 & 0 & 0 & 0 \\ k_2 & k_1 & -k_2 & 0 & 0 & 0 & -k_2 & -k_2 & 0 & 0 & -1 & 0 \\ -k_2 & -k_2 & k_1 & 0 & 0 & 0 & 0 & 0 & 0 & 0 & 0 & 0 \\ -1 & 0 & 0 & 1 & 0 & 0 & 0 & 0 & 0 & 0 & 0 & 0 \\ 0 & 0 & 0 & 0 & 1 & 0 & 0 & -1 & 0 & 0 & 0 & 0 \\ 0 & 0 & 0 & 0 & 0 & 1 & 0 & 0 & 0 & 0 & 0 & 0 \\ -k_2 & -k_2 & 0 & 0 & 0 & 0 & k_1 & k_2 & k_2 & -1 & 0 & 0 \\ -k_2 & -k_2 & 0 & 0 & -1 & 0 & k_2 & k_1 & k_2 & 0 & 0 & 0 \\ 0 & 0 & 0 & 0 & 0 & 0 & k_2 & k_2 & k_1 & 0 & 0 & 0 \\ 0 & 0 & 0 & 0 & 0 & 0 & -1 & 0 & 0 & 1 & 0 & 0 \\ 0 & -1 & 0 & 0 & 0 & 0 & 0 & 0 & 0 & 0 & 1 & 0 \\ 0 & 0 & 0 & 0 & 0 & 0 & 0 & 0 & 0 & 0 & 0 & 1 \end{bmatrix} \tag{61}$$

with

$$k_1 \triangleq 1 + \frac{\sqrt{2}}{2} \qquad k_2 \triangleq \frac{\sqrt{2}}{4} \qquad k \triangleq \frac{EZ}{L} \tag{62}$$

Since M is diagonal [see Eq. (59)], the 12×12 upper triangular matrix V such that Eq. (5) is satisfied is the diagonal matrix

$$V = \sqrt{m} \begin{bmatrix} \sqrt{m_1} & & & & & & & & & & & \\ & \sqrt{m_1} & & & & & & & & & & \\ & & \sqrt{m_1} & & & & & & & & & \\ & & & \sqrt{m_2} & & & & & 0 & & & \\ & & & & \sqrt{m_2} & & & & & & & \\ & & & & & \sqrt{m_2} & & & & & & \\ & & & & & & \sqrt{m_1} & & & & & \\ & & & & & & & \sqrt{m_1} & & & & \\ & & & 0 & & & & & \sqrt{m_1} & & & \\ & & & & & & & & & \sqrt{m_2} & & \\ & & & & & & & & & & \sqrt{m_2} & \\ & & & & & & & & & & & \sqrt{m_2} \end{bmatrix} \tag{63}$$

and substitution from Eqs. (61) and (63) into Eq. (6) yields

$$W = \frac{k}{m}\begin{bmatrix} w_1 & w_2 & -w_2 & -w_3 & 0 & 0 & -w_2 & -w_2 & 0 & 0 & 0 & 0 \\ w_2 & w_1 & -w_2 & 0 & 0 & 0 & -w_2 & -w_2 & 0 & 0 & -w_3 & 0 \\ -w_2 & -w_2 & w_1 & 0 & 0 & 0 & 0 & 0 & 0 & 0 & 0 & 0 \\ -w_3 & 0 & 0 & w_4 & 0 & 0 & 0 & 0 & 0 & 0 & 0 & 0 \\ 0 & 0 & 0 & 0 & w_4 & 0 & 0 & -w_3 & 0 & 0 & 0 & 0 \\ 0 & 0 & 0 & 0 & 0 & w_4 & 0 & 0 & 0 & 0 & 0 & 0 \\ -w_2 & -w_2 & 0 & 0 & 0 & 0 & w_1 & w_2 & w_2 & -w_3 & 0 & 0 \\ -w_2 & -w_2 & 0 & 0 & -w_3 & 0 & w_2 & w_1 & w_2 & 0 & 0 & 0 \\ 0 & 0 & 0 & 0 & 0 & 0 & w_2 & w_2 & w_1 & 0 & 0 & 0 \\ 0 & 0 & 0 & 0 & 0 & 0 & -w_3 & 0 & 0 & w_4 & 0 & 0 \\ 0 & -w_3 & 0 & 0 & 0 & 0 & 0 & 0 & 0 & 0 & w_4 & 0 \\ 0 & 0 & 0 & 0 & 0 & 0 & 0 & 0 & 0 & 0 & 0 & w_4 \end{bmatrix} \tag{64}$$

where

$$w_1 \triangleq \frac{k_1}{m_1} \qquad w_2 \triangleq \frac{k_2}{m_1} \qquad w_3 \triangleq \frac{1}{\sqrt{m_1 m_2}} \qquad w_4 \triangleq \frac{1}{m_2} \tag{65}$$

The eigenvalues of W are

$$\lambda_1 = 0.0373322k/m \qquad \lambda_2 = 0.0668532k/m \qquad \lambda_3 = 0.0668532k/m \tag{66}$$

$$\lambda_4 = 0.244379k/m \qquad \lambda_5 = 0.488054k/m \qquad \lambda_6 = 0.495117k/m \tag{67}$$

$$\lambda_7 = 0.666667k/m \qquad \lambda_8 = 0.666667k/m \qquad \lambda_9 = 0.973587k/m \tag{68}$$

$$\lambda_{10} = 0.973587k/m \qquad \lambda_{11} = 0.986458k/m \qquad \lambda_{12} = 1.16287k/m \tag{69}$$

Note that $\lambda_2 = \lambda_3$, $\lambda_7 = \lambda_8$, and $\lambda_9 = \lambda_{10}$. Eigenvectors $C^{(1)}, C^{(2)}, \ldots, C^{(12)}$ of W, corresponding to $\lambda_1, \lambda_2, \ldots, \lambda_{12}$, respectively, are

$$C^{(1)} = \begin{bmatrix} 1 \\ 1 \\ 0.449838 \\ 0.681772 \\ 0.681772 \\ 0 \\ 1 \\ 1 \\ -0.449838 \\ 0.681772 \\ 0.681772 \\ 0 \end{bmatrix}, C^{(2)} = \begin{bmatrix} 0 \\ 0 \\ 0 \\ 0 \\ 0.715327 \\ 0 \\ -1 \\ 1 \\ 0 \\ -0.715327 \\ 0 \\ 0 \end{bmatrix}, \ldots, C^{(12)} = \begin{bmatrix} -1 \\ -1 \\ 0.282388 \\ 0.864687 \\ -0.864687 \\ 0 \\ 1 \\ 1 \\ 0.282388 \\ -0.864687 \\ 0.864687 \\ 0 \end{bmatrix} \tag{70}$$

so that Eqs. (7), (63), (70), and (60) lead to

$$B^{(1)} = \frac{1}{\sqrt{m}} \begin{bmatrix} 0.525493 \\ 0.525493 \\ 0.236386 \\ 0.556665 \\ 0.556665 \\ 0 \\ 0.525493 \\ 0.525493 \\ -0.236386 \\ 0.556665 \\ 0.556665 \\ 0 \end{bmatrix}, B^{(2)} = \frac{1}{\sqrt{m}} \begin{bmatrix} 0 \\ 0 \\ 0 \\ 0 \\ 0.584062 \\ 0 \\ -0.525493 \\ 0.525493 \\ 0 \\ -0.584062 \\ 0 \\ 0 \end{bmatrix}, \ldots, B^{(12)} = \frac{1}{\sqrt{m}} \begin{bmatrix} -0.525493 \\ -0.525493 \\ 0.148393 \\ 0.706014 \\ -0.706014 \\ 0 \\ 0.525493 \\ 0.525493 \\ 0.148393 \\ -0.706014 \\ 0.706014 \\ 0 \end{bmatrix} \tag{71}$$

which means that [see Eqs. (8), (71), and (59)]

$$N_1 = 2.50279, N_2 = 1.73879, \ldots, N_{12} = 2.67399 \tag{72}$$

and, from Eqs. (9), (71), and (72),

$$A^{(1)} = \frac{1}{\sqrt{m}} \begin{bmatrix} 0.20996 \\ 0.20996 \\ 0.094449 \\ 0.22242 \\ 0.22242 \\ 0 \\ 0.20996 \\ 0.20996 \\ -0.094449 \\ 0.22242 \\ 0.22242 \\ 0 \end{bmatrix}, A^{(2)} = \frac{1}{\sqrt{m}} \begin{bmatrix} 0 \\ 0 \\ 0 \\ 0 \\ 0.33590 \\ 0 \\ -0.30222 \\ 0.30222 \\ 0 \\ -0.33590 \\ 0 \\ 0 \end{bmatrix}, \ldots, A^{(12)} = \frac{1}{\sqrt{m}} \begin{bmatrix} -0.19652 \\ -0.19652 \\ 0.055495 \\ 0.26403 \\ -0.26403 \\ 0 \\ 0.19652 \\ 0.19652 \\ 0.055495 \\ -0.26403 \\ 0.26403 \\ 0 \end{bmatrix} \tag{73}$$

Referring to Eqs. (10) and (73), one thus finds that the modal matrix A is given by

$$A = \frac{1}{\sqrt{m}} \begin{bmatrix} \alpha_1 & 0 & -\alpha_5 & \alpha_6 & -\alpha_9 & -\alpha_{12} & 0 & 0 & -\alpha_{16} & 0 & -\alpha_{18} & -\alpha_{21} \\ \alpha_1 & 0 & \alpha_5 & \alpha_6 & -\alpha_9 & -\alpha_{12} & 0 & 0 & \alpha_{16} & 0 & -\alpha_{18} & -\alpha_{21} \\ \alpha_2 & 0 & 0 & \alpha_7 & \alpha_{10} & \alpha_{13} & 0 & 0 & 0 & 0 & \alpha_{19} & \alpha_{22} \\ \alpha_3 & 0 & -\alpha_4 & \alpha_8 & -\alpha_{11} & -\alpha_{14} & 0 & 0 & \alpha_{17} & 0 & \alpha_{20} & \alpha_{23} \\ \alpha_3 & \alpha_4 & 0 & -\alpha_8 & -\alpha_{11} & \alpha_{14} & 0 & 0 & 0 & -\alpha_{17} & \alpha_{20} & -\alpha_{23} \\ 0 & 0 & 0 & 0 & 0 & 0 & \alpha_{15} & 0 & 0 & 0 & 0 & 0 \\ \alpha_1 & -\alpha_5 & 0 & -\alpha_6 & -\alpha_9 & \alpha_{12} & 0 & 0 & 0 & -\alpha_{16} & -\alpha_{18} & \alpha_{21} \\ \alpha_1 & \alpha_5 & 0 & -\alpha_6 & -\alpha_9 & \alpha_{12} & 0 & 0 & 0 & \alpha_{16} & -\alpha_{18} & \alpha_{21} \\ -\alpha_2 & 0 & 0 & \alpha_7 & -\alpha_{10} & \alpha_{13} & 0 & 0 & 0 & 0 & -\alpha_{19} & \alpha_{22} \\ \alpha_3 & -\alpha_4 & 0 & -\alpha_8 & -\alpha_{11} & \alpha_{14} & 0 & 0 & 0 & \alpha_{17} & \alpha_{20} & -\alpha_{23} \\ \alpha_3 & 0 & \alpha_4 & \alpha_8 & -\alpha_{11} & -\alpha_{14} & 0 & 0 & -\alpha_{17} & 0 & \alpha_{20} & \alpha_{23} \\ 0 & 0 & 0 & 0 & 0 & 0 & 0 & \alpha_{15} & 0 & 0 & 0 & 0 \end{bmatrix} \tag{74}$$

where $\alpha_1, \ldots, \alpha_{23}$ have the values

$$\alpha_1 = 0.20996 \qquad \alpha_2 = 0.094449 \qquad \alpha_3 = 0.22242 \qquad \alpha_4 = 0.33590 \tag{75}$$

$$\alpha_5 = 0.30222 \qquad \alpha_6 = 0.16952 \qquad \alpha_7 = 0.14581 \qquad \alpha_8 = 0.26763 \tag{76}$$

$$\alpha_9 = 0.030231 \qquad \alpha_{10} = 0.35454 \qquad \alpha_{11} = 0.11284 \qquad \alpha_{12} = 0.040955 \tag{77}$$

$$\alpha_{13} = 0.33724 \qquad \alpha_{14} = 0.15916 \qquad \alpha_{15} = 0.81650 \qquad \alpha_{16} = 0.21618 \tag{78}$$

$$\alpha_{17} = 0.46958 \qquad \alpha_{18} = 0.15504 \qquad \alpha_{19} = 0.058777 \qquad \alpha_{20} = 0.32321 \tag{79}$$

$$\alpha_{21} = 0.19652 \qquad \alpha_{22} = 0.055495 \qquad \alpha_{23} = 0.26403 \tag{80}$$

Suppose that, at $t = 0$, $\boldsymbol{\delta}_1 = \boldsymbol{\delta}_2 = \boldsymbol{\delta}_3 = \boldsymbol{\delta}_4 = (L/10)\hat{\mathbf{n}}_1$, and $P_1, \ldots, P_4$ are at rest. Then, in accordance with Eqs. (58),

$$q(0) = \left(\frac{L}{10}\right) [\ 1\ \ 0\ \ 0\ \ 1\ \ 0\ \ 0\ \ 1\ \ 0\ \ 0\ \ 1\ \ 0\ \ 0\ ]^T \tag{81}$$

and

$$\dot{q}(0) = 0 \tag{82}$$

Consequently,

$$Q(0) \underset{(11,\ 74,\ 59,\ 81)}{=} \sqrt{m}L[0.2188\ \ -0.1598\ \ -0.1598\ \ 0\ \ -0.05575\ \ 0\ \ 0 \\ 0\ \ -0.007851\ \ -0.007851\ \ -0.01533\ \ 0]^T \tag{83}$$

$$\dot{Q}(0) \underset{(11,\ 82)}{=} 0 \tag{84}$$

and, after noting that [see Eqs. (12) and (66)–(69)]

$$p_1 = 0.1932\omega \qquad p_2 = 0.2586\omega \qquad p_3 = 0.2586\omega \tag{85}$$

$$p_4 = 0.4943\omega \qquad p_5 = 0.6986\omega \qquad p_6 = 0.7036\omega \tag{86}$$

$$p_7 = 0.8165\omega \qquad p_8 = 0.8165\omega \qquad p_9 = 0.9867\omega \tag{87}$$

$$p_{10} = 0.9867\omega \qquad p_{11} = 0.9932\omega \qquad p_{12} = 1.078\omega \tag{88}$$

where ω is defined as

$$\omega \triangleq \sqrt{\frac{k}{m}} \tag{89}$$

one has

$$Q \underset{(16,\ 13,\ 83\text{–}88)}{=} \sqrt{m}L \begin{bmatrix} 0.2188\cos(0.1932\omega t) \\ -0.1598\cos(0.2586\omega t) \\ -0.1598\cos(0.2586\omega t) \\ 0 \\ -0.05575\cos(0.6986\omega t) \\ 0 \\ 0 \\ 0 \\ -0.007851\cos(0.9867\omega t) \\ -0.007851\cos(0.9867\omega t) \\ -0.01533\cos(0.9932\omega t) \\ 0 \end{bmatrix} \tag{90}$$

and Eqs. (17), (73), and (90) thus lead, for example, to

$$\frac{q_1}{L} = \sum_{i=1}^{12} x_1{}^{(i)} \tag{91}$$

where

$$x_1{}^{(1)} = 0.04594\cos(0.1932\omega t) \qquad x_1{}^{(2)} = 0 \tag{92}$$
$$x_1{}^{(3)} = 0.04830\cos(0.2586\omega t) \qquad x_1{}^{(4)} = 0 \tag{93}$$
$$x_1{}^{(5)} = 0.001685\cos(0.6986\omega t) \qquad x_1{}^{(6)} = x_1{}^{(7)} = x_1{}^{(8)} = 0 \tag{94}$$
$$x_1{}^{(9)} = 0.001697\cos(0.9867\omega t) \qquad x_1{}^{(10)} = 0 \tag{95}$$
$$x_1{}^{(11)} = 0.002376\cos(0.9932\omega t) \qquad x_1{}^{(12)} = 0 \tag{96}$$

and

$$\frac{q_{10}}{L} = \sum_{i=1}^{12} x_{10}{}^{(i)} \tag{97}$$

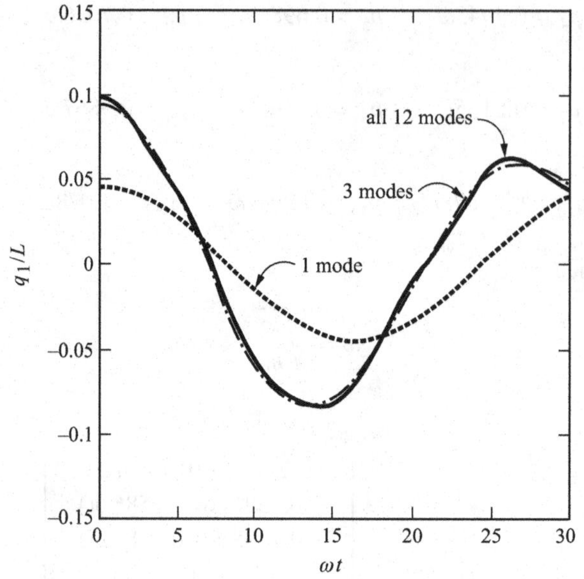

Figure 9.9.4

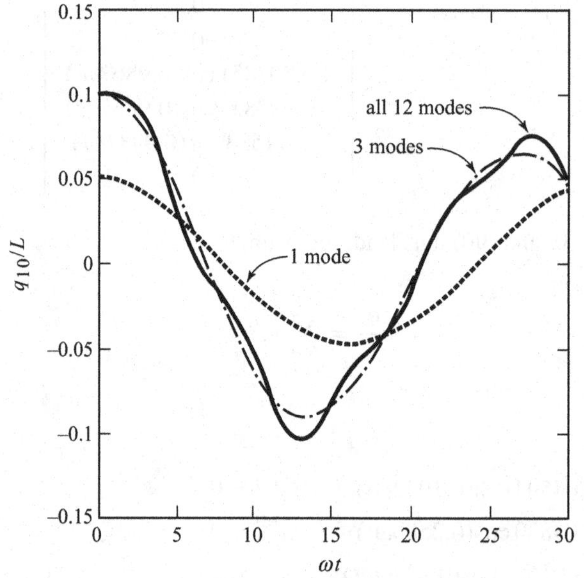

Figure 9.9.5

where

$$x_{10}{}^{(1)} = 0.04866\cos(0.1932\omega t) \qquad x_{10}{}^{(2)} = 0.05369\cos(0.2586\omega t) \tag{98}$$

$$x_{10}{}^{(3)} = x_{10}{}^{(4)} = 0 \qquad x_{10}{}^{(5)} = 0.006290\cos(0.6986\omega t) \tag{99}$$

$$x_{10}{}^{(6)} = x_{10}{}^{(7)} = 0 \tag{100}$$

$$x_{10}{}^{(8)} = x_{10}{}^{(9)} = 0 \qquad x_{10}{}^{(10)} = -0.003686\cos(0.9867\omega t) \tag{101}$$

$$x_{10}{}^{(11)} = -0.004954\cos(0.9932\omega t) \qquad x_{10}{}^{(12)} = 0 \tag{102}$$

Figure 9.9.4 contains three plots of q_1/L versus ωt. The curve labeled "all 12 modes" is based on Eq. (91), and the curves labeled "3 modes" and "1 mode" correspond, respectively, to using only the first three terms and the first term of Eq. (91); similarly for Fig. 9.9.5, q_{10}/L, and Eq. (97). As can be seen, use of only the first mode leads to rather poor approximations, whereas results obtained by truncation after the first three modes may be acceptable.

10 KINEMATICS OF ORIENTATION

Every dynamics problem involves considerations of kinematics, including, frequently, the orientation of a rigid body in a reference frame, which is the subject of the first three sections in Chapter 1. The topics dealt with there are given additional detailed attention in this chapter, in Secs. 10.1–10.3. Sections 10.2, 10.3, 10.4, and 10.5 deal with descriptions of orientation in terms of sets of nine direction cosines, three orientation angles, four Euler parameters, and three Wiener-Milenković parameters, respectively. The first five sections of this chapter deal with changes in the orientation of a rigid body in a reference frame, without regard to the time that necessarily elapses when a real body experiences a change in orientation. Time enters the discussion in the remaining four sections, in which the concept of angular velocity plays a central role. Kinematical differential equations governing direction cosines, orientation angles, Euler parameters, and Wiener-Milenković parameters are presented in Secs. 10.6, 10.7, 10.8, and 10.9, respectively.

10.1 EULER ROTATION

A motion of a rigid body or reference frame B relative to a rigid body or reference frame A is called a *simple rotation of B in A* if there exists a line L, called an *axis of rotation*, whose orientation relative to both A and B remains unaltered throughout the motion. This sort of motion is important because, as will be shown in Sec. 10.3, every change in the relative orientation of A and B can be produced by means of a simple rotation of B in A. In Sec. 1.1, simple rotation is illustrated by taking the axis of rotation to be parallel to unit vector $\hat{\mathbf{b}}_1$. The axis of rotation can in fact be oriented arbitrarily with respect to unit vectors $\hat{\mathbf{b}}_1$, $\hat{\mathbf{b}}_2$, $\hat{\mathbf{b}}_3$ fixed in B; this general case is sometimes called *Euler rotation*.

If $\mathbf{a}$ is any vector fixed in A (see Fig. 10.1.1), and $\mathbf{b}$ is a vector fixed in B and equal to $\mathbf{a}$ prior to the motion of B in A, then, when B has performed a simple rotation in A, $\mathbf{b}$ can be expressed in terms of the vector $\mathbf{a}$, a unit vector $\hat{\boldsymbol{\lambda}}$ parallel to L, and the radian measure θ of the angle between two lines, L_A and L_B, which are fixed in A and B, respectively, are perpendicular to L, and are parallel to each other initially. Specifically, if θ is regarded as positive when the angle between L_A and L_B is generated by a rotation of L_B relative to L_A about $\hat{\boldsymbol{\lambda}}$ during which a right-handed screw fixed in B with its axis parallel to $\hat{\boldsymbol{\lambda}}$ advances in the direction of $\hat{\boldsymbol{\lambda}}$ when B rotates relative to A, then

$$\mathbf{b} = \mathbf{a}\cos\theta - \mathbf{a}\times\hat{\boldsymbol{\lambda}}\sin\theta + \mathbf{a}\cdot\hat{\boldsymbol{\lambda}}\hat{\boldsymbol{\lambda}}(1-\cos\theta) \quad (1)$$

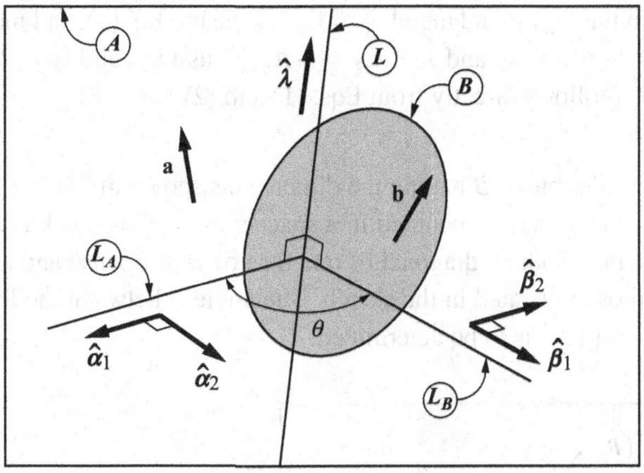

Figure 10.1.1

Equivalently, if a dyadic $\underline{\mathbf{C}}$ is defined as

$$\underline{\mathbf{C}} \triangleq \underline{\mathbf{U}}\cos\theta - \underline{\mathbf{U}} \times \hat{\boldsymbol{\lambda}}\sin\theta + \hat{\boldsymbol{\lambda}}\hat{\boldsymbol{\lambda}}(1-\cos\theta) \tag{2}$$

where $\underline{\mathbf{U}}$ is the unit (or identity) dyadic, then

$$\mathbf{b} = \mathbf{a} \cdot \underline{\mathbf{C}} \tag{3}$$

Derivations Let $\hat{\boldsymbol{\alpha}}_1$ and $\hat{\boldsymbol{\alpha}}_2$ be unit vectors fixed in A, with $\hat{\boldsymbol{\alpha}}_1$ parallel to L_A and $\hat{\boldsymbol{\alpha}}_2 = \hat{\boldsymbol{\lambda}} \times \hat{\boldsymbol{\alpha}}_1$; and let $\hat{\boldsymbol{\beta}}_1$ and $\hat{\boldsymbol{\beta}}_2$ be unit vectors fixed in B, with $\hat{\boldsymbol{\beta}}_1$ parallel to L_B and $\hat{\boldsymbol{\beta}}_2 = \hat{\boldsymbol{\lambda}} \times \hat{\boldsymbol{\beta}}_1$, as shown in Fig. 10.1.1. Then, if **a** and **b** are resolved into components parallel to $\hat{\boldsymbol{\alpha}}_1$, $\hat{\boldsymbol{\alpha}}_2$, $\hat{\boldsymbol{\lambda}}$ and $\hat{\boldsymbol{\beta}}_1$, $\hat{\boldsymbol{\beta}}_2$, $\hat{\boldsymbol{\lambda}}$, respectively, corresponding coefficients are equal to each other because $\hat{\boldsymbol{\alpha}}_1 = \hat{\boldsymbol{\beta}}_1$, $\hat{\boldsymbol{\alpha}}_2 = \hat{\boldsymbol{\beta}}_2$, and $\mathbf{a} = \mathbf{b}$ when $\theta = 0$. In other words, **a** and **b** can be expressed as

$$\mathbf{a} = p\hat{\boldsymbol{\alpha}}_1 + q\hat{\boldsymbol{\alpha}}_2 + r\hat{\boldsymbol{\lambda}} \tag{4}$$

and

$$\mathbf{b} = p\hat{\boldsymbol{\beta}}_1 + q\hat{\boldsymbol{\beta}}_2 + r\hat{\boldsymbol{\lambda}} \tag{5}$$

where p, q, and r are constants.

Expressed in terms of $\hat{\boldsymbol{\alpha}}_1$ and $\hat{\boldsymbol{\alpha}}_2$, the unit vectors $\hat{\boldsymbol{\beta}}_1$ and $\hat{\boldsymbol{\beta}}_2$ are given by

$$\hat{\boldsymbol{\beta}}_1 = \cos\theta\hat{\boldsymbol{\alpha}}_1 + \sin\theta\hat{\boldsymbol{\alpha}}_2 \tag{6}$$

and

$$\hat{\boldsymbol{\beta}}_2 = -\sin\theta\hat{\boldsymbol{\alpha}}_1 + \cos\theta\hat{\boldsymbol{\alpha}}_2 \tag{7}$$

so that, substituting into Eq. (5), one finds that

$$\mathbf{b} = (p\cos\theta - q\sin\theta)\hat{\boldsymbol{\alpha}}_1 + (p\sin\theta + q\cos\theta)\hat{\boldsymbol{\alpha}}_2 + r\hat{\boldsymbol{\lambda}} \tag{8}$$

The right-hand member of Eq. (8) is precisely what one obtains by carrying out the operations indicated in the right-hand member of Eq. (1), using Eq. (4), and making use of the relationships $\hat{\boldsymbol{\lambda}} \times \hat{\boldsymbol{\alpha}}_1 = \hat{\boldsymbol{\alpha}}_2$ and $\hat{\boldsymbol{\lambda}} \times \hat{\boldsymbol{\alpha}}_2 = -\hat{\boldsymbol{\alpha}}_1$. Thus the validity of Eq. (1) is established, and Eq. (3) follows directly from Eqs. (1) and (2).

Example A rectangular block B having the dimensions shown in Fig. 10.1.2 forms a portion of an antenna structure mounted in a spacecraft A. This block is subjected to a simple rotation in A about a diagonal of one face of B, the sense and amount of the rotation being those indicated in the sketch. The angle ϕ between the line OP in its initial and final positions is to be determined.

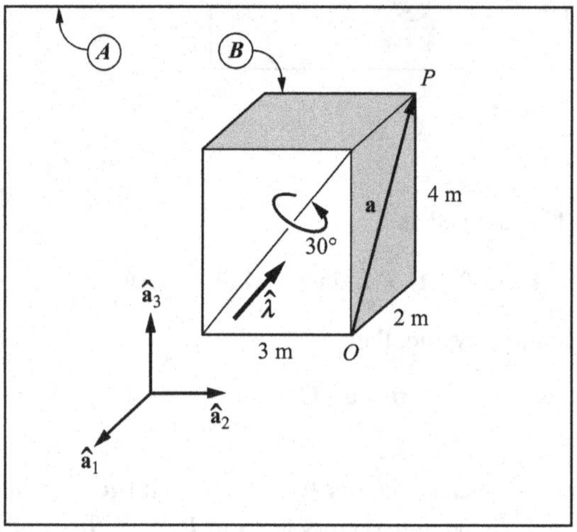

Figure 10.1.2

If $\hat{\mathbf{a}}_1$, $\hat{\mathbf{a}}_2$, and $\hat{\mathbf{a}}_3$ are unit vectors fixed in A and parallel to the edges of B prior to the rotation of B, then a unit vector $\hat{\boldsymbol{\lambda}}$ directed as shown in the sketch can be expressed as

$$\hat{\boldsymbol{\lambda}} = \frac{3\hat{\mathbf{a}}_2 + 4\hat{\mathbf{a}}_3}{5} \tag{9}$$

And if $\mathbf{a}$ denotes the position vector from O to P prior to the rotation of B, then

$$\mathbf{a} = -2\hat{\mathbf{a}}_1 + 4\hat{\mathbf{a}}_3 \tag{10}$$

$$\mathbf{a} \times \hat{\boldsymbol{\lambda}} = \frac{-12\hat{\mathbf{a}}_1 + 8\hat{\mathbf{a}}_2 - 6\hat{\mathbf{a}}_3}{5} \tag{11}$$

and

$$\mathbf{a} \cdot \hat{\boldsymbol{\lambda}}\hat{\boldsymbol{\lambda}} = \frac{48\hat{\mathbf{a}}_2 + 64\hat{\mathbf{a}}_3}{25} \tag{12}$$

Consequently, if $\mathbf{b}$ is the position vector from O to P subsequent to the rotation of B,

$$\begin{aligned}\underset{(1)}{\mathbf{b}} &= (-2\hat{\mathbf{a}}_1 + 4\hat{\mathbf{a}}_3)\cos\frac{\pi}{6} + \frac{12\hat{\mathbf{a}}_1 - 8\hat{\mathbf{a}}_2 + 6\hat{\mathbf{a}}_3}{5}\sin\frac{\pi}{6} \\ &\quad + \frac{48\hat{\mathbf{a}}_2 + 64\hat{\mathbf{a}}_3}{25}\left(1 - \cos\frac{\pi}{6}\right) \\ &= -0.532\hat{\mathbf{a}}_1 - 0.543\hat{\mathbf{a}}_2 + 4.407\hat{\mathbf{a}}_3 \end{aligned} \tag{13}$$

Since ϕ is the angle between $\mathbf{a}$ and $\mathbf{b}$,

$$\cos\phi = \frac{\mathbf{a}\cdot\mathbf{b}}{|\mathbf{a}||\mathbf{b}|} \tag{14}$$

where $|\mathbf{a}|$ and $|\mathbf{b}|$ denote the (equal) magnitudes of $\mathbf{a}$ and $\mathbf{b}$. Hence

$$\cos\phi = \frac{(-2)(-0.532) + 4(4.407)}{(4+16)^{1/2}(4+16)^{1/2}} = 0.935 \tag{15}$$

and $\phi = 20.77°$.

10.2 DIRECTION COSINES

If $\hat{\mathbf{a}}_1$, $\hat{\mathbf{a}}_2$, $\hat{\mathbf{a}}_3$, and $\hat{\mathbf{b}}_1$, $\hat{\mathbf{b}}_2$, $\hat{\mathbf{b}}_3$ are two dextral sets of orthogonal unit vectors, and nine quantities C_{ij} $(i,j = 1,2,3)$, called *direction cosines*, are defined as

$$C_{ij} \triangleq \hat{\mathbf{a}}_i \cdot \hat{\mathbf{b}}_j \qquad (i,j = 1,2,3) \tag{1}$$

then the two row matrices $\lfloor \hat{\mathbf{a}}_1 \quad \hat{\mathbf{a}}_2 \quad \hat{\mathbf{a}}_3 \rfloor$ and $\lfloor \hat{\mathbf{b}}_1 \quad \hat{\mathbf{b}}_2 \quad \hat{\mathbf{b}}_3 \rfloor$ are related to each other as follows:

$$\lfloor \hat{\mathbf{b}}_1 \quad \hat{\mathbf{b}}_2 \quad \hat{\mathbf{b}}_3 \rfloor = \lfloor \hat{\mathbf{a}}_1 \quad \hat{\mathbf{a}}_2 \quad \hat{\mathbf{a}}_3 \rfloor C \tag{2}$$

where C is a square matrix defined as

$$C \triangleq \begin{bmatrix} C_{11} & C_{12} & C_{13} \\ C_{21} & C_{22} & C_{23} \\ C_{31} & C_{32} & C_{33} \end{bmatrix} \tag{3}$$

If a superscript T is used to denote transposition, that is, if C^T is defined as

$$C^T \triangleq \begin{bmatrix} C_{11} & C_{21} & C_{31} \\ C_{12} & C_{22} & C_{32} \\ C_{13} & C_{23} & C_{33} \end{bmatrix} \tag{4}$$

then Eq. (2) can be replaced with the equivalent relationship

$$\lfloor \hat{\mathbf{a}}_1 \quad \hat{\mathbf{a}}_2 \quad \hat{\mathbf{a}}_3 \rfloor = \lfloor \hat{\mathbf{b}}_1 \quad \hat{\mathbf{b}}_2 \quad \hat{\mathbf{b}}_3 \rfloor C^T \tag{5}$$

The matrix C, called a *direction cosine matrix*, can be employed to describe the relative orientation of two reference frames or rigid bodies A and B. In that context, it can be advantageous to replace the symbol C with the more elaborate symbol ${}^AC^B$. In view of Eqs. (2) and (5), one must then regard the interchanging of superscripts as signifying transposition; that is,

$$^{B}C^{A} = (^{A}C^{B})^{T} \tag{6}$$

The direction cosine matrix $^{A}C^{B}$ plays a role in a number of useful relationships. For example, if $\mathbf{v}$ is any vector and $^{A}v_i$, and $^{B}v_i$ $(i = 1,2,3)$ are defined as

$$^{A}v_i \triangleq \mathbf{v} \cdot \hat{\mathbf{a}}_i \qquad (i = 1,2,3) \tag{7}$$

and

$$^{B}v_i \triangleq \mathbf{v} \cdot \hat{\mathbf{b}}_i \qquad (i = 1,2,3) \tag{8}$$

one can let $\lfloor ^{A}v \rfloor$ and $\lfloor ^{B}v \rfloor$ denote row matrices having the elements $^{A}v_1$, $^{A}v_2$, $^{A}v_3$ and $^{B}v_1$, $^{B}v_2$, $^{B}v_3$, respectively; furthermore, the transpose of each row matrix is a column matrix that can be represented by $\{^{A}v\}$ and $\{^{B}v\}$. $^{A}C^{B}$ can be used to "transform" the measure numbers of $\mathbf{v}$ in the following ways:

$$\lfloor ^{B}v \rfloor = \lfloor ^{A}v \rfloor \, ^{A}C^{B} \quad \lfloor ^{A}v \rfloor = \lfloor ^{B}v \rfloor \, ^{B}C^{A} \quad \{^{A}v\} = {}^{A}C^{B} \{^{B}v\} \quad \{^{B}v\} = {}^{B}C^{A} \{^{A}v\} \tag{9}$$

Similarly, if $\underline{\mathbf{D}}$ is any dyadic, and $^{A}D_{ij}$ and $^{B}D_{ij}$ $(i,j = 1,2,3)$ are defined as

$$^{A}D_{ij} \triangleq \hat{\mathbf{a}}_i \cdot \underline{\mathbf{D}} \cdot \hat{\mathbf{a}}_j \qquad (i,j = 1,2,3) \tag{10}$$

and

$$^{B}D_{ij} \triangleq \hat{\mathbf{b}}_i \cdot \underline{\mathbf{D}} \cdot \hat{\mathbf{b}}_j \qquad (i,j = 1,2,3) \tag{11}$$

while ^{A}D and ^{B}D denote square matrices having $^{A}D_{ij}$ and $^{B}D_{ij}$, respectively, as the elements in the ith rows and j^{th} columns, then

$$^{B}D = {}^{B}C^{A} \, ^{A}D \, ^{A}C^{B} \tag{12}$$

Use of the summation convention (a repeated subscript in a given term is understood to imply summation over the range of the repeated subscript) frequently makes it possible to formulate important relationships rather concisely. For example, if δ_{ij} is defined as

$$\delta_{ij} \triangleq 1 - \tfrac{1}{4}(i-j)^2[5 - (i-j)^2] \qquad (i,j = 1,2,3) \tag{13}$$

so that δ_{ij} is equal to unity when the subscripts have the same value, and equal to zero when the subscripts have different values, then use of the summation convention permits one to express a set of six relationships governing direction cosines as

$$C_{ik} \, C_{jk} = \delta_{ij} \qquad (i,j = 1,2,3) \tag{14}$$

or an equivalent set as

$$C_{ki} \, C_{kj} = \delta_{ij} \qquad (i,j = 1,2,3) \tag{15}$$

Alternatively, these relationships can be stated in matrix form as

$$C \, C^{T} = U \tag{16}$$

and

$$C^T C = U \tag{17}$$

where U denotes the unit (or identity) matrix, defined as

$$U \triangleq \begin{bmatrix} 1 & 0 & 0 \\ 0 & 1 & 0 \\ 0 & 0 & 1 \end{bmatrix} \tag{18}$$

Each element of the matrix C is equal to its cofactor in the determinant of C; and, if $|C|$ denotes the determinant, then

$$|C| = 1 \tag{19}$$

Consequently, C is an orthonormal matrix, that is, a matrix whose inverse and whose transpose are equal to each other. Moreover,

$$|C - U| = 0 \tag{20}$$

Hence, unity is an eigenvalue of every direction cosine matrix. In other words, for every direction cosine matrix C there exist row matrices $\lfloor \kappa_1 \quad \kappa_2 \quad \kappa_3 \rfloor$, called eigenvectors, which satisfy the equation

$$\lfloor \kappa_1 \quad \kappa_2 \quad \kappa_3 \rfloor C = \lfloor \kappa_1 \quad \kappa_2 \quad \kappa_3 \rfloor \tag{21}$$

Suppose now that $\hat{\mathbf{a}}_i$ and $\hat{\mathbf{b}}_i$ $(i = 1,2,3)$ are fixed in reference frames or rigid bodies A and B, respectively, and that B is subjected to a simple rotation in A (see Sec. 10.1); further, that $\hat{\mathbf{a}}_i = \hat{\mathbf{b}}_i$ $(i = 1,2,3)$ prior to the rotation, that $\hat{\boldsymbol{\lambda}}$ and θ are defined as in Sec. 10.1, and that λ_i is defined as

$$\lambda_i \triangleq \hat{\boldsymbol{\lambda}} \cdot \hat{\mathbf{a}}_i = \hat{\boldsymbol{\lambda}} \cdot \hat{\mathbf{b}}_i \qquad (i = 1,2,3) \tag{22}$$

Then the elements of C are given by

$$C_{11} = \cos\theta + {\lambda_1}^2(1 - \cos\theta) \tag{23}$$
$$C_{12} = -\lambda_3 \sin\theta + \lambda_1\lambda_2(1 - \cos\theta) \tag{24}$$
$$C_{13} = \lambda_2 \sin\theta + \lambda_3\lambda_1(1 - \cos\theta) \tag{25}$$
$$C_{21} = \lambda_3 \sin\theta + \lambda_1\lambda_2(1 - \cos\theta) \tag{26}$$
$$C_{22} = \cos\theta + {\lambda_2}^2(1 - \cos\theta) \tag{27}$$
$$C_{23} = -\lambda_1 \sin\theta + \lambda_2\lambda_3(1 - \cos\theta) \tag{28}$$
$$C_{31} = -\lambda_2 \sin\theta + \lambda_3\lambda_1(1 - \cos\theta) \tag{29}$$
$$C_{32} = \lambda_1 \sin\theta + \lambda_2\lambda_3(1 - \cos\theta) \tag{30}$$
$$C_{33} = \cos\theta + {\lambda_3}^2(1 - \cos\theta) \tag{31}$$

Equations (23)–(31) can be expressed more concisely after defining ϵ_{ijk} as

$$\epsilon_{ijk} \triangleq \tfrac{1}{2}(i-j)(j-k)(k-i) \qquad (i,j,k = 1,2,3) \tag{32}$$

(The quantity ϵ_{ijk} vanishes when two or three subscripts have the same value; it is equal to unity when the subscripts appear in cyclic order, that is, in the order 1, 2, 3, the order

2, 3, 1, or the order 3, 1, 2; and it is equal to negative unity in all other cases.) Using the summation convention, one then can replace Eqs. (23)–(31) with

$$C_{ij} = \delta_{ij} \cos\theta - \epsilon_{ijk}\lambda_k \sin\theta + \lambda_i\lambda_j(1 - \cos\theta) \qquad (i,j = 1,2,3) \tag{33}$$

Alternatively, C_{ij} can be expressed in terms of the dyadic $\underline{\mathbf{C}}$ defined in Eq. (10.1.2):

$$C_{ij} = \hat{\mathbf{a}}_i \cdot (\hat{\mathbf{a}}_j \cdot \underline{\mathbf{C}}) \qquad (i,j = 1,2,3) \tag{34}$$

All of these results simplify substantially when $\hat{\boldsymbol{\lambda}}$ is parallel to $\hat{\mathbf{a}}_i$, and hence to $\hat{\mathbf{b}}_i$ $(i = 1,2,3)$. If $C_i(\theta)$ denotes C for $\hat{\boldsymbol{\lambda}} = \hat{\mathbf{a}}_i = \hat{\mathbf{b}}_i$, then

$$C_1(\theta) = \begin{bmatrix} 1 & 0 & 0 \\ 0 & \cos\theta & -\sin\theta \\ 0 & \sin\theta & \cos\theta \end{bmatrix} \tag{35}$$

$$C_2(\theta) = \begin{bmatrix} \cos\theta & 0 & \sin\theta \\ 0 & 1 & 0 \\ -\sin\theta & 0 & \cos\theta \end{bmatrix} \tag{36}$$

$$C_3(\theta) = \begin{bmatrix} \cos\theta & -\sin\theta & 0 \\ \sin\theta & \cos\theta & 0 \\ 0 & 0 & 1 \end{bmatrix} \tag{37}$$

It was mentioned previously that unity is an eigenvalue of every direction cosine matrix. If the elements of a direction cosine matrix C are given by Eqs. (23)–(31), then the row matrix $\lfloor \lambda_1 \quad \lambda_2 \quad \lambda_3 \rfloor$ is one of the eigenvectors corresponding to the eigenvalue unity of C; that is,

$$\lfloor \lambda_1 \quad \lambda_2 \quad \lambda_3 \rfloor C = \lfloor \lambda_1 \quad \lambda_2 \quad \lambda_3 \rfloor \tag{38}$$

Equivalently, when $\underline{\mathbf{C}}$ is the dyadic defined in Eq. (10.1.2), then

$$\hat{\boldsymbol{\lambda}} \cdot \underline{\mathbf{C}} = \hat{\boldsymbol{\lambda}} \tag{39}$$

Derivations For any vector $\mathbf{v}$, the following is an identity:

$$\mathbf{v} = (\hat{\mathbf{a}}_1 \cdot \mathbf{v})\hat{\mathbf{a}}_1 + (\hat{\mathbf{a}}_2 \cdot \mathbf{v})\hat{\mathbf{a}}_2 + (\hat{\mathbf{a}}_3 \cdot \mathbf{v})\hat{\mathbf{a}}_3 \tag{40}$$

Hence, letting $\hat{\mathbf{b}}_1$ play the role of $\mathbf{v}$, one can express $\hat{\mathbf{b}}_1$ as

$$\hat{\mathbf{b}}_1 = (\hat{\mathbf{a}}_1 \cdot \hat{\mathbf{b}}_1)\hat{\mathbf{a}}_1 + (\hat{\mathbf{a}}_2 \cdot \hat{\mathbf{b}}_1)\hat{\mathbf{a}}_2 + (\hat{\mathbf{a}}_3 \cdot \hat{\mathbf{b}}_1)\hat{\mathbf{a}}_3 \tag{41}$$

or, by using Eq. (1), as

$$\hat{\mathbf{b}}_1 = C_{11}\hat{\mathbf{a}}_1 + C_{21}\hat{\mathbf{a}}_2 + C_{31}\hat{\mathbf{a}}_3 \tag{42}$$

Similarly,

$$\hat{\mathbf{b}}_2 = C_{12}\hat{\mathbf{a}}_1 + C_{22}\hat{\mathbf{a}}_2 + C_{32}\hat{\mathbf{a}}_3 \tag{43}$$

and

$$\hat{\mathbf{b}}_3 = C_{13}\hat{\mathbf{a}}_1 + C_{23}\hat{\mathbf{a}}_2 + C_{33}\hat{\mathbf{a}}_3 \tag{44}$$

These three equations are precisely what one obtains when forming expressions for $\hat{\mathbf{b}}_1$, $\hat{\mathbf{b}}_2$, $\hat{\mathbf{b}}_3$ in accordance with Eq. (2) and with the rules for matrix multiplication; and a similar line of reasoning leads to Eq. (5).

To see that the first of Eqs. (9) is valid one needs only to observe that

$$\begin{aligned} {}^B v_i &\underset{(8,\,2)}{=} \mathbf{v} \cdot (\hat{\mathbf{a}}_1 C_{1i} + \hat{\mathbf{a}}_2 C_{2i} + \hat{\mathbf{a}}_3 C_{3i}) \\ &= \mathbf{v} \cdot \hat{\mathbf{a}}_1 C_{1i} + \mathbf{v} \cdot \hat{\mathbf{a}}_2 C_{2i} + \mathbf{v} \cdot \hat{\mathbf{a}}_3 C_{3i} \\ &\underset{(7)}{=} {}^A v_1 C_{1i} + {}^A v_2 C_{2i} + {}^A v_3 C_{3i} \end{aligned} \tag{45}$$

and to recall the definitions of $\lfloor {}^A v \rfloor$ and $\lfloor {}^B v \rfloor$. The second of Eqs. (9) follows from the first after postmultiplying both sides with ${}^B C^A$ and making use of Eqs. (6) and (16); matrix transposition then leads to the third relationship, and premultiplication with ${}^B C^A$ finally yields the last of Eqs. (9). Similarly, Eq. (12) follows from

$$\begin{aligned} {}^B D_{ij} &\underset{(11,\,2)}{=} (\hat{\mathbf{a}}_1 C_{1i} + \hat{\mathbf{a}}_2 C_{2i} + \hat{\mathbf{a}}_3 C_{3i}) \cdot \underline{\mathbf{D}} \cdot (\hat{\mathbf{a}}_1 C_{1j} + \hat{\mathbf{a}}_2 C_{2j} + \hat{\mathbf{a}}_3 C_{3j}) \\ &\underset{(10)}{=} C_{1i}({}^A D_{11} C_{1j} + {}^A D_{12} C_{2j} + {}^A D_{13} C_{3j}) \\ &\quad + C_{2i}({}^A D_{21} C_{1j} + {}^A D_{22} C_{2j} + {}^A D_{23} C_{3j}) \\ &\quad + C_{3i}({}^A D_{31} C_{1j} + {}^A D_{32} C_{2j} + {}^A D_{33} C_{3j}) \end{aligned} \tag{46}$$

and from the definitions of ${}^A D$ and ${}^B D$, where C_{ij} $(i, j = 1,2,3)$ are elements of the matrix ${}^A C^B$.

As for Eqs. (14) and (15), these are consequences of (using the summation convention)

$$\hat{\mathbf{a}}_i \cdot \hat{\mathbf{a}}_j \underset{(1,\,5)}{=} C_{ik}\, C_{jk} \qquad (i, j = 1, 2, 3) \tag{47}$$

and

$$\hat{\mathbf{b}}_i \cdot \hat{\mathbf{b}}_j = C_{ki}\, C_{kj} \qquad (i, j = 1, 2, 3) \tag{48}$$

respectively, because $\hat{\mathbf{a}}_i \cdot \hat{\mathbf{a}}_j$ is equal to unity when $i = j$, and equal to zero when $i \neq j$, and similarly for $\hat{\mathbf{b}}_i \cdot \hat{\mathbf{b}}_j$; and Eqs. (16) and (17) can be seen to be equivalent to Eqs. (14) and (15), respectively, by referring to Eqs. (3), (4), and (18) when carrying out the indicated multiplications.

To verify that each element of C is equal to its cofactor in the determinant of C, note that

$$\hat{\mathbf{b}}_1 \underset{(2)}{=} C_{11}\hat{\mathbf{a}}_1 + C_{21}\hat{\mathbf{a}}_2 + C_{31}\hat{\mathbf{a}}_3 \tag{49}$$

and

$$\hat{\mathbf{b}}_2 \times \hat{\mathbf{b}}_3 \underset{(2)}{=} (C_{22}C_{33} - C_{32}C_{23})\hat{\mathbf{a}}_1 + (C_{32}C_{13} - C_{12}C_{33})\hat{\mathbf{a}}_2 + (C_{12}C_{23} - C_{22}C_{13})\hat{\mathbf{a}}_3 \tag{50}$$

so that, since $\hat{\mathbf{b}}_1 = \hat{\mathbf{b}}_2 \times \hat{\mathbf{b}}_3$ because $\hat{\mathbf{b}}_1$, $\hat{\mathbf{b}}_2$, $\hat{\mathbf{b}}_3$ form a dextral set of orthogonal unit vectors,

$$C_{11} = C_{22}C_{33} - C_{32}C_{23} \tag{51}$$

$$C_{21} = C_{32}C_{13} - C_{12}C_{33} \tag{52}$$

and

$$C_{31} = C_{12}C_{23} - C_{22}C_{13} \tag{53}$$

Thus each element in the first column of C [see Eq. (3)] is seen to be equal to its cofactor in C; and, using the relationships $\hat{\mathbf{b}}_2 = \hat{\mathbf{b}}_3 \times \hat{\mathbf{b}}_1$ and $\hat{\mathbf{b}}_3 = \hat{\mathbf{b}}_1 \times \hat{\mathbf{b}}_2$, one obtains corresponding results for the elements in the second and third columns of C. Furthermore, expanding C by cofactors of elements of the first row, and using Eq. (14) with $i = j = 1$, one arrives at Eq. (19).

The determinant of $C - U$ can be expressed as

$$\begin{aligned}|C - U| \underset{(3)}{=} & |C| + C_{11} + C_{22} + C_{33} + C_{12}C_{21} + C_{23}C_{32} + C_{31}C_{13} \\ & - (C_{11}C_{22} + C_{22}C_{33} + C_{33}C_{11}) - 1\end{aligned} \tag{54}$$

Hence, replacing C_{11}, C_{22}, and C_{33} with their respective cofactors in $|C|$, one finds that

$$|C - U| = |C| - 1 \underset{(19)}{=} 0 \tag{55}$$

in agreement with Eq. (20); and the existence of row matrices $\lfloor \kappa_1 \quad \kappa_2 \quad \kappa_3 \rfloor$ satisfying Eq. (21) is thus guaranteed.

The equality of $\hat{\boldsymbol{\lambda}} \cdot \hat{\mathbf{a}}_i$ and $\hat{\boldsymbol{\lambda}} \cdot \hat{\mathbf{b}}_i$ in Eq. (22) is a consequence of the fact that these two quantities are equal to each other prior to the rotation of B relative to A, that is, when $\hat{\mathbf{a}}_i = \hat{\mathbf{b}}_i$, and that neither $\hat{\boldsymbol{\lambda}} \cdot \hat{\mathbf{a}}_i$ nor $\hat{\boldsymbol{\lambda}} \cdot \hat{\mathbf{b}}_i$ changes during the rotation, since, by construction, $\hat{\boldsymbol{\lambda}}$ is parallel to a line whose orientation in both A and B remains unaltered during the rotation.

With **a** and **b** replaced by $\hat{\mathbf{a}}_j$ and $\hat{\mathbf{b}}_j$, respectively, Eq. (10.1.3) becomes

$$\hat{\mathbf{b}}_j = \hat{\mathbf{a}}_j \cdot \underline{\mathbf{C}} \tag{56}$$

Hence,

$$C_{ij} \underset{(1)}{=} \hat{\mathbf{a}}_i \cdot \hat{\mathbf{b}}_j \underset{(56)}{=} \hat{\mathbf{a}}_i \cdot (\hat{\mathbf{a}}_j \cdot \underline{\mathbf{C}}) \qquad (i, j = 1, 2, 3) \tag{57}$$

which is Eq. (34). Moreover, substituting for $\underline{\mathbf{C}}$ the expression given in Eq. (10.1.2), one finds that

$$C_{ij} = \hat{\mathbf{a}}_i \cdot \hat{\mathbf{a}}_j \cos\theta - \hat{\mathbf{a}}_i \cdot \hat{\mathbf{a}}_j \times \hat{\boldsymbol{\lambda}} \sin\theta + \hat{\mathbf{a}}_i \cdot \hat{\boldsymbol{\lambda}}\hat{\boldsymbol{\lambda}} \cdot \hat{\mathbf{a}}_j (1 - \cos\theta) \qquad (i, j = 1, 2, 3) \tag{58}$$

and this, together with Eq. (22) leads directly to Eqs. (23)–(31), or, in view of Eq. (32), to Eq. (33).

Equation (35) is obtained by setting $\lambda_1 = 1$ and $\lambda_2 = \lambda_3 = 0$ in Eqs. (23)–(31) and then using Eq. (3). Similarly, $\lambda_1 = 0$, $\lambda_2 = 1$, and $\lambda_3 = 0$ lead to Eq. (36), and $\lambda_1 = \lambda_2 = 0$, $\lambda_3 = 1$ yield Eq. (37). It is worth noting that these direction cosine matrices can also be obtained from Tables 1.3.1, 1.3.2, and 1.3.3, respectively, by setting q_1, q_2, and q_3 equal to θ.

Finally, Eq. (39) is derived from the observation that

$$\hat{\boldsymbol{\lambda}} \cdot \underline{\mathbf{C}} \underset{(10.1.2)}{=} \hat{\boldsymbol{\lambda}} \cdot \underline{\mathbf{U}} \cos\theta - \hat{\boldsymbol{\lambda}} \times \hat{\boldsymbol{\lambda}} \sin\theta + \hat{\boldsymbol{\lambda}}^2 \hat{\boldsymbol{\lambda}}(1 - \cos\theta)$$

$$= \hat{\boldsymbol{\lambda}} \cos\theta + \mathbf{0} + \hat{\boldsymbol{\lambda}}(1 - \cos\theta) = \hat{\boldsymbol{\lambda}} \tag{59}$$

since $\hat{\boldsymbol{\lambda}}$ is a unit vector, so that $\hat{\boldsymbol{\lambda}}^2 \triangleq \hat{\boldsymbol{\lambda}} \cdot \hat{\boldsymbol{\lambda}} = 1$; and the equivalence of Eqs. (38) and (39) follows from Eqs. (22) and (34).

Example In Fig. 10.2.1, B designates a uniform rectangular block that is part of a scanning platform mounted in a spacecraft A. Initially, the edges of B are parallel to unit vectors $\hat{\mathbf{a}}_1$, $\hat{\mathbf{a}}_2$, and $\hat{\mathbf{a}}_3$, which are fixed in A, and the platform is then subjected to a simple 90° rotation about a diagonal of B, as indicated in the sketch. If $\underline{\mathbf{I}}$ is the inertia dyadic of B for the mass center $B^\star$ of B (see Sec. 4.5), and ${}^A I_{ij}$ is defined as

$${}^A I_{ij} \triangleq \hat{\mathbf{a}}_i \cdot \underline{\mathbf{I}} \cdot \hat{\mathbf{a}}_j \qquad (i, j = 1, 2, 3) \tag{60}$$

what is the value of ${}^A I_{ij}$ $(i, j = 1, 2, 3)$ subsequent to the rotation?

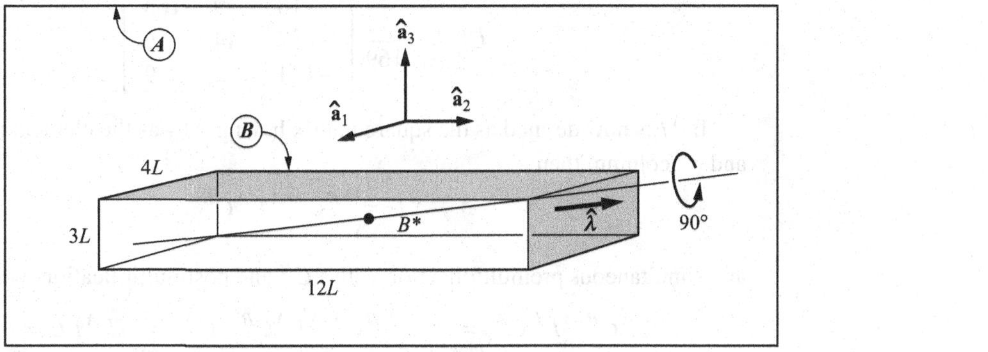

Figure 10.2.1

Let $\hat{\mathbf{b}}_i$ $(i = 1, 2, 3)$ be a unit vector fixed in B and equal to $\hat{\mathbf{a}}_i$ $(i = 1, 2, 3)$ prior to the rotation; and define ${}^B I_{ij}$ as

$${}^B I_{ij} \triangleq \hat{\mathbf{b}}_i \cdot \underline{\mathbf{I}} \cdot \hat{\mathbf{b}}_j \qquad (i, j = 1, 2, 3) \tag{61}$$

Then $\hat{\mathbf{b}}_1$, $\hat{\mathbf{b}}_2$, and $\hat{\mathbf{b}}_3$ are parallel to principal axes of inertia of B for $B^\star$, so that

$${}^B I_{12} = {}^B I_{21} = {}^B I_{23} = {}^B I_{32} = {}^B I_{31} = {}^B I_{13} = 0 \tag{62}$$

and, if m is the mass of B,

$${}^B I_{11} = \frac{m}{12}(12^2 + 3^2)L^2 = \frac{153}{12} mL^2 \tag{63}$$

$${}^B I_{22} = \frac{m}{12}(3^2 + 4^2)L^2 = \frac{25}{12} mL^2 \tag{64}$$

and

$$ {}^B I_{33} = \frac{m}{12}(4^2 + 12^2)L^2 = \frac{160}{12} mL^2 \tag{65} $$

Hence, if ${}^B I$ denotes the square matrix having ${}^B I_{ij}$ as the element in the ith row and j^{th} column, then

$$ {}^B I = \frac{mL^2}{12} \begin{bmatrix} 153 & 0 & 0 \\ 0 & 25 & 0 \\ 0 & 0 & 160 \end{bmatrix} \tag{66} $$

The unit vector $\hat{\boldsymbol{\lambda}}$ shown in Fig. 10.2.1 can be expressed as

$$ \hat{\boldsymbol{\lambda}} = \frac{4\hat{\mathbf{a}}_1 + 12\hat{\mathbf{a}}_2 + 3\hat{\mathbf{a}}_3}{(4^2 + 12^2 + 3^2)^{1/2}} = \frac{4}{13}\hat{\mathbf{a}}_1 + \frac{12}{13}\hat{\mathbf{a}}_2 + \frac{3}{13}\hat{\mathbf{a}}_3 \tag{67} $$

Consequently, λ_1, λ_2, and λ_3, if defined as in Eq. (22), are given by

$$ \lambda_1 = \tfrac{4}{13} \qquad \lambda_2 = \tfrac{12}{13} \qquad \lambda_3 = \tfrac{3}{13} \tag{68} $$

and, with $\theta = \pi/2$ rad, Eqs. (23)–(31) lead to the following expression for the direction cosine matrix ${}^A C^B$:

$$ {}^A C^B \underset{(3)}{=} \frac{1}{169} \begin{bmatrix} 16 & 9 & 168 \\ 87 & 144 & -16 \\ -144 & 88 & 9 \end{bmatrix} \tag{69} $$

If ${}^A I$ is now defined as the square matrix having ${}^A I_{ij}$ as the element in the ith row and j^{th} column, then

$$ {}^B I \underset{(12)}{=} {}^B C^A \, {}^A I \, {}^A C^B \tag{70} $$

and simultaneous premultiplication with ${}^A C^B$ and postmultiplication with ${}^B C^A$ gives

$$ {}^A C^B \, {}^B I \, {}^B C^A \underset{(70)}{=} {}^A C^B \, {}^B C^A \, {}^A I \, {}^A C^B \, {}^B C^A \underset{(16)}{=} U \, {}^A I \, U = {}^A I \tag{71} $$

Consequently,

$$ {}^A I \underset{(66,\,69,\,71)}{=} \frac{mL^2}{12 \times 169 \times 169} \times \tag{72} $$

$$ \begin{bmatrix} 16 & 9 & 168 \\ 87 & 144 & -16 \\ -144 & 88 & 9 \end{bmatrix} \begin{bmatrix} 153 & 0 & 0 \\ 0 & 25 & 0 \\ 0 & 0 & 160 \end{bmatrix} \begin{bmatrix} 16 & 87 & -144 \\ 9 & 144 & 88 \\ 168 & -16 & 9 \end{bmatrix} $$

$$ = \frac{mL^2}{12 \times 169 \times 169} \begin{bmatrix} 4{,}557{,}033 & -184{,}704 & -90{,}792 \\ -184{,}704 & 1{,}717{,}417 & -1{,}623{,}024 \\ -90{,}792 & -1{,}623{,}024 & 3{,}379{,}168 \end{bmatrix} $$

and

$$ {}^A I_{11} = \frac{4{,}557{,}033 \, mL^2}{12 \times 169 \times 169} \qquad {}^A I_{12} = -\frac{184{,}704 \, mL^2}{12 \times 169 \times 169} \tag{73} $$

and so forth.

10.3 ORIENTATION ANGLES

Both for physical and for analytical reasons it is sometimes desirable to describe the orientation of a rigid body B in a reference frame A in terms of three angles. For example, if B is the rotor of a gyroscope whose outer gimbal axis is fixed in a reference frame A, then the angles ϕ, θ, and ψ shown in Fig. 10.3.1 furnish a means for describing the orientation of B in A in a way that is particularly meaningful from a physical point of view.

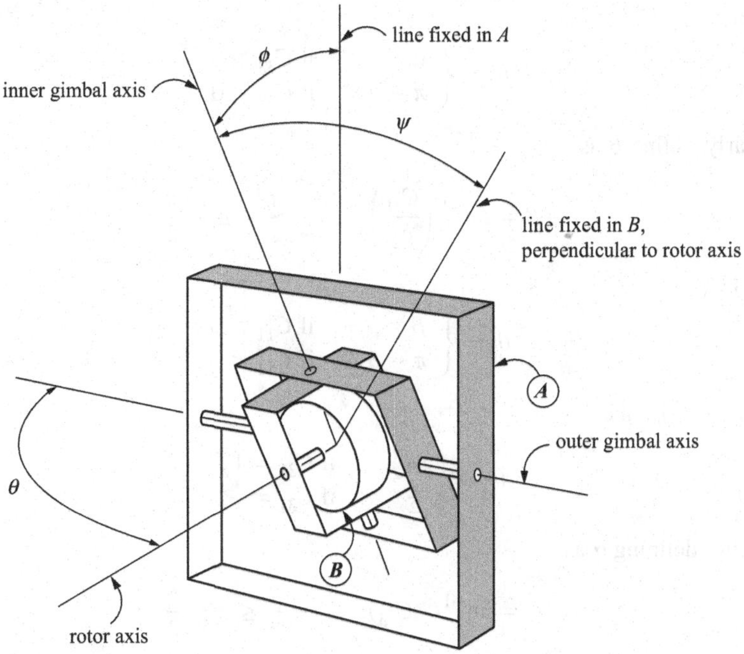

Figure 10.3.1

One scheme for bringing a rigid body B into a desired orientation in a reference frame A is to introduce $\hat{\mathbf{a}}_1$, $\hat{\mathbf{a}}_2$, $\hat{\mathbf{a}}_3$ and $\hat{\mathbf{b}}_1$, $\hat{\mathbf{b}}_2$, $\hat{\mathbf{b}}_3$ as dextral sets of orthogonal unit vectors fixed in A and B, respectively; align $\hat{\mathbf{b}}_i$ with $\hat{\mathbf{a}}_i$ $(i = 1,2,3)$; and subject B successively to a simple rotation about $\hat{\mathbf{a}}_1$ of amount θ_1, a simple rotation about $\hat{\mathbf{a}}_2$ of amount θ_2, and a simple rotation about $\hat{\mathbf{a}}_3$ of amount θ_3. (Recall that, for any unit vector $\hat{\boldsymbol{\lambda}}$, the phrase "simple rotation about $\hat{\boldsymbol{\lambda}}$" means a rotation of B relative to A during which a right-handed screw fixed in B with its axis parallel to $\hat{\boldsymbol{\lambda}}$ advances in the direction of $\hat{\boldsymbol{\lambda}}$.) Suitable values of θ_1, θ_2, and θ_3 can be found in terms of elements of the direction cosine matrix C (see Sec. 10.2), which, if s_i and c_i denote $\sin\theta_i$ and $\cos\theta_i$ $(i = 1,2,3)$ respectively, is given by

$$C = C_3(\theta_3)C_2(\theta_2)C_1(\theta_1) = \begin{bmatrix} c_2c_3 & s_1s_2c_3 - s_3c_1 & c_1s_2c_3 + s_3s_1 \\ c_2s_3 & s_1s_2s_3 + c_3c_1 & c_1s_2s_3 - c_3s_1 \\ -s_2 & s_1c_2 & c_1c_2 \end{bmatrix} \qquad (1)$$

where C_i are the direction cosine matrices indicated in Eqs. (10.2.35)–(10.2.37), respectively. Specifically, if $|C_{31}| \neq 1$, take

$$\theta_2 = \sin^{-1}(-C_{31}) \qquad -\frac{\pi}{2} < \theta_2 < \frac{\pi}{2} \tag{2}$$

Next, after evaluating c_2, define α as

$$\alpha \triangleq \sin^{-1}\left(\frac{C_{32}}{c_2}\right) \qquad -\frac{\pi}{2} \leq \alpha \leq \frac{\pi}{2} \tag{3}$$

and let

$$\theta_1 = \begin{cases} \alpha & \text{if } C_{33} \geq 0 \\ \pi - \alpha & \text{if } C_{33} < 0 \end{cases} \tag{4}$$

Similarly, define β as

$$\beta \triangleq \sin^{-1}\left(\frac{C_{21}}{c_2}\right) \qquad -\frac{\pi}{2} \leq \beta \leq \frac{\pi}{2} \tag{5}$$

and take

$$\theta_3 = \begin{cases} \beta & \text{if } C_{11} \geq 0 \\ \pi - \beta & \text{if } C_{11} < 0 \end{cases} \tag{6}$$

If $|C_{31}| = 1$, take

$$\theta_2 = \begin{cases} -\frac{\pi}{2} & \text{if } C_{31} = 1 \\ \frac{\pi}{2} & \text{if } C_{31} = -1 \end{cases} \tag{7}$$

and, after defining α as

$$\alpha \triangleq \sin^{-1}(-C_{23}) \qquad -\frac{\pi}{2} \leq \alpha \leq \frac{\pi}{2} \tag{8}$$

let

$$\theta_1 = \begin{cases} \alpha & \text{if } C_{22} \geq 0 \\ \pi - \alpha & \text{if } C_{22} < 0 \end{cases} \tag{9}$$

and

$$\theta_3 = 0 \tag{10}$$

In other words, two rotations suffice in this case.

A second method for accomplishing the same objective is to align $\hat{\mathbf{b}}_i$ with $\hat{\mathbf{a}}_i$ $(i = 1,2,3)$, and subject B successively to a simple rotation about $\hat{\mathbf{b}}_1$ of amount θ_1, a simple rotation about $\hat{\mathbf{b}}_2$ of amount θ_2, and a simple rotation about $\hat{\mathbf{b}}_3$ of amount θ_3. The matrix C relating $\hat{\mathbf{a}}_1, \hat{\mathbf{a}}_2, \hat{\mathbf{a}}_3$ to $\hat{\mathbf{b}}_1, \hat{\mathbf{b}}_2, \hat{\mathbf{b}}_3$ as in Eq. (10.2.2) subsequent to the last rotation is then given by

$$C = C_1(\theta_1)C_2(\theta_2)C_3(\theta_3) = \begin{bmatrix} c_2c_3 & -c_2s_3 & s_2 \\ s_1s_2c_3 + s_3c_1 & -s_1s_2s_3 + c_3c_1 & -s_1c_2 \\ -c_1s_2c_3 + s_3s_1 & c_1s_2s_3 + c_3s_1 & c_1c_2 \end{bmatrix} \tag{11}$$

where C_i $(i = 1,2,3)$ are the direction cosine matrices indicated in Eqs. (10.2.35)–(10.2.37), respectively. If $|C_{13}| \neq 1$, suitable values of θ_1, θ_2, and θ_3 are obtained by taking

$$\theta_2 = \sin^{-1}(C_{13}) \qquad -\frac{\pi}{2} < \theta_2 < \frac{\pi}{2} \tag{12}$$

$$\alpha \triangleq \sin^{-1}\left(\frac{-C_{23}}{c_2}\right) \qquad -\frac{\pi}{2} \leq \alpha \leq \frac{\pi}{2} \tag{13}$$

$$\theta_1 = \begin{cases} \alpha & \text{if } C_{33} \geq 0 \\ \pi - \alpha & \text{if } C_{33} < 0 \end{cases} \tag{14}$$

$$\beta \triangleq \sin^{-1}\left(\frac{-C_{12}}{c_2}\right) \qquad -\frac{\pi}{2} \leq \beta \leq \frac{\pi}{2} \tag{15}$$

$$\theta_3 = \begin{cases} \beta & \text{if } C_{11} \geq 0 \\ \pi - \beta & \text{if } C_{11} < 0 \end{cases} \tag{16}$$

whereas, if $|C_{13}| = 1$, one may take

$$\theta_2 = \begin{cases} \frac{\pi}{2} & \text{if } C_{13} = 1 \\ -\frac{\pi}{2} & \text{if } C_{13} = -1 \end{cases} \tag{17}$$

$$\alpha \triangleq \sin^{-1}(C_{32}) \qquad -\frac{\pi}{2} \leq \alpha \leq \frac{\pi}{2} \tag{18}$$

$$\theta_1 = \begin{cases} \alpha & \text{if } C_{22} \geq 0 \\ \pi - \alpha & \text{if } C_{22} < 0 \end{cases} \tag{19}$$

$$\theta_3 = 0 \tag{20}$$

so that once again, only two rotations are required.

The difference between these two procedures for bringing B into a desired orientation in A is that the first involves unit vectors fixed in the reference frame, whereas the second involves unit vectors fixed in the body. What the two methods have in common is that *three distinct* unit vectors are employed in both cases.

It is also possible to bring B into an arbitrary orientation relative to A by performing three successive rotations that involve only *two* distinct unit vectors, and these vectors may be fixed either in the reference frame or in the body. Specifically, if B is subjected successively to a simple rotation about $\hat{\mathbf{a}}_1$ of the amount θ_1, a simple rotation about $\hat{\mathbf{a}}_2$ of the amount θ_2, and again a simple rotation about $\hat{\mathbf{a}}_1$, but this time of the amount θ_3, then

$$C = C_1(\theta_3)C_2(\theta_2)C_1(\theta_1) = \begin{bmatrix} c_2 & s_1s_2 & c_1s_2 \\ s_2s_3 & -s_1c_2s_3 + c_3c_1 & -c_1c_2s_3 - c_3s_1 \\ -s_2c_3 & s_1c_2c_3 + s_3c_1 & c_1c_2c_3 - s_3s_1 \end{bmatrix} \tag{21}$$

and, if $|C_{11}| \neq 1$, one can take

$$\theta_2 = \cos^{-1}(C_{11}) \qquad 0 < \theta_2 < \pi \tag{22}$$

$$\alpha \triangleq \sin^{-1}\left(\frac{C_{12}}{s_2}\right) \qquad -\frac{\pi}{2} \le \alpha \le \frac{\pi}{2} \tag{23}$$

$$\theta_1 = \begin{cases} \alpha & \text{if } C_{13} \ge 0 \\ \pi - \alpha & \text{if } C_{13} < 0 \end{cases} \tag{24}$$

$$\beta \triangleq \sin^{-1}\left(\frac{C_{21}}{s_2}\right) \qquad -\frac{\pi}{2} \le \beta \le \frac{\pi}{2} \tag{25}$$

$$\theta_3 = \begin{cases} \beta & \text{if } C_{31} < 0 \\ \pi - \beta & \text{if } C_{31} \ge 0 \end{cases} \tag{26}$$

while, if $|C_{11}| = 1$, one may let

$$\theta_2 = \begin{cases} 0 & \text{if } C_{11} = 1 \\ \pi & \text{if } C_{11} = -1 \end{cases} \tag{27}$$

$$\alpha \triangleq \sin^{-1}(-C_{23}) \qquad -\frac{\pi}{2} \le \alpha \le \frac{\pi}{2} \tag{28}$$

$$\theta_1 = \begin{cases} \alpha & \text{if } C_{22} \ge 0 \\ \pi - \alpha & \text{if } C_{22} < 0 \end{cases} \tag{29}$$

$$\theta_3 = 0 \tag{30}$$

Finally, if B is subjected successively to a simple rotation about $\hat{\mathbf{b}}_1$ of the amount θ_1, a simple rotation about $\hat{\mathbf{b}}_2$ of the amount θ_2, and again a simple rotation about $\hat{\mathbf{b}}_1$, but this time of the amount θ_3, then

$$C = C_1(\theta_1)C_2(\theta_2)C_1(\theta_3) = \begin{bmatrix} c_2 & s_2 s_3 & s_2 c_3 \\ s_1 s_2 & -s_1 c_2 s_3 + c_3 c_1 & -s_1 c_2 c_3 - s_3 c_1 \\ -c_1 s_2 & c_1 c_2 s_3 + c_3 s_1 & c_1 c_2 c_3 - s_3 s_1 \end{bmatrix} \tag{31}$$

and, if $|C_{11}| \ne 1$, θ_1, θ_2, and θ_3 may be found by taking

$$\theta_2 = \cos^{-1}(C_{11}) \qquad 0 < \theta_2 < \pi \tag{32}$$

$$\alpha \triangleq \sin^{-1}\left(\frac{C_{21}}{s_2}\right) \qquad -\frac{\pi}{2} \le \alpha \le \frac{\pi}{2} \tag{33}$$

$$\theta_1 = \begin{cases} \alpha & \text{if } C_{31} < 0 \\ \pi - \alpha & \text{if } C_{31} \ge 0 \end{cases} \tag{34}$$

$$\beta \triangleq \sin^{-1}\left(\frac{C_{12}}{s_2}\right) \qquad -\frac{\pi}{2} \le \beta \le \frac{\pi}{2} \tag{35}$$

$$\theta_3 = \begin{cases} \beta & \text{if } C_{13} \ge 0 \\ \pi - \beta & \text{if } C_{13} < 0 \end{cases} \tag{36}$$

while, if $|C_{11}| = 1$, one can use

$$\theta_2 = \begin{cases} 0 & \text{if } C_{11} = 1 \\ \pi & \text{if } C_{11} = -1 \end{cases} \tag{37}$$

$$\alpha \triangleq \sin^{-1}(C_{32}) \qquad -\frac{\pi}{2} \le \alpha \le \frac{\pi}{2} \tag{38}$$

$$\theta_1 = \begin{cases} \alpha & \text{if } C_{22} \ge 0 \\ \pi - \alpha & \text{if } C_{22} < 0 \end{cases} \tag{39}$$

$$\theta_3 = 0 \tag{40}$$

The matrices in Eqs. (1) and (11) are intimately related to each other: either one may be obtained from the other by replacing θ_i with $-\theta_i$ $(i = 1,2,3)$ and transposing. The matrices in Eqs. (21) and (31) are related similarly. These facts have the following physical significance: If B is subjected to successive simple rotations of θ_1 about $\hat{\mathbf{a}}_1$, θ_2 about $\hat{\mathbf{a}}_2$, and θ_3 about $\hat{\mathbf{a}}_3$, then one can bring B back into its original orientation in A by next subjecting B to successive simple rotations of θ_1 about $-\hat{\mathbf{b}}_1$, θ_2 about $-\hat{\mathbf{b}}_2$, and θ_3 about $-\hat{\mathbf{b}}_3$, respectively. Similarly, employing only four unit vectors, one can subject B to successive rotations characterized by $\theta_1\hat{\mathbf{a}}_1, \theta_2\hat{\mathbf{a}}_2, \theta_3\hat{\mathbf{a}}_1, -\theta_1\hat{\mathbf{b}}_1, -\theta_2\hat{\mathbf{b}}_2, -\theta_3\hat{\mathbf{b}}_1$ without producing any ultimate change in the orientation of B in A. Furthermore, it does not matter whether the rotations involving unit vectors fixed in A are preceded or followed by those involving unit vectors fixed in B; that is, the sequences of successive rotations represented by $\theta_1\hat{\mathbf{b}}_1, \theta_2\hat{\mathbf{b}}_2, \theta_3\hat{\mathbf{b}}_3, -\theta_1\hat{\mathbf{a}}_1, -\theta_2\hat{\mathbf{a}}_2, -\theta_3\hat{\mathbf{a}}_3$ and by $\theta_1\hat{\mathbf{b}}_1, \theta_2\hat{\mathbf{b}}_2, \theta_3\hat{\mathbf{b}}_1, -\theta_1\hat{\mathbf{a}}_1, -\theta_2\hat{\mathbf{a}}_2, -\theta_3\hat{\mathbf{a}}_1$ also have no net effect on the orientation of B in A.

To indicate which set of three angles one is using, one can speak of *space-three angles* in connection with Eqs. (1)–(10), *body-three angles* for Eqs. (11)–(20), *space-two angles* for Eqs. (21)–(30), and *body-two angles* for Eqs. (31)–(40); and this terminology remains meaningful even when the angles and unit vectors employed are denoted by symbols other than those used in Eqs. (1)–(40). Moreover, once one has identified three angles in this way, one can always find appropriate replacements for Eqs. (1), (11), (21), or (31) by direct use of these equations. Suppose, for example, that $\hat{\mathbf{x}}$, $\hat{\mathbf{y}}$, $\hat{\mathbf{z}}$ and $\hat{\boldsymbol{\xi}}$, $\hat{\boldsymbol{\eta}}$, $\hat{\boldsymbol{\zeta}}$ are dextral sets of orthogonal unit vectors fixed in a reference frame A and in a rigid body B, respectively; that $\hat{\mathbf{x}} = \hat{\boldsymbol{\xi}}$, $\hat{\mathbf{y}} = \hat{\boldsymbol{\eta}}$, and $\hat{\mathbf{z}} = \hat{\boldsymbol{\zeta}}$, initially; that B is subjected to successive simple rotations about $\hat{\mathbf{z}}$ of amount γ, about $\hat{\mathbf{y}}$ of amount β, and about $\hat{\mathbf{x}}$ of amount α; and that it is required to find the elements L_{ij} $(i,j = 1,2,3)$ of the matrix L such that, subsequent to the last rotation,

$$[\hat{\boldsymbol{\xi}} \;\; \hat{\boldsymbol{\eta}} \;\; \hat{\boldsymbol{\zeta}}] = [\hat{\mathbf{x}} \;\; \hat{\mathbf{y}} \;\; \hat{\mathbf{z}}]\, L \tag{41}$$

Then, recognizing α, β, and γ as space-three angles, one can introduce $\hat{\mathbf{a}}_i$, $\hat{\mathbf{b}}_i$, and θ_i $(i = 1,2,3)$ as

$$\hat{\mathbf{a}}_1 \triangleq \hat{\mathbf{z}} \qquad \hat{\mathbf{a}}_2 \triangleq \hat{\mathbf{y}} \qquad \hat{\mathbf{a}}_3 \triangleq -\hat{\mathbf{x}} \tag{42}$$

$$\hat{\mathbf{b}}_1 \triangleq \hat{\boldsymbol{\zeta}} \qquad \hat{\mathbf{b}}_2 \triangleq \hat{\boldsymbol{\eta}} \qquad \hat{\mathbf{b}}_3 \triangleq -\hat{\boldsymbol{\xi}} \tag{43}$$

and

$$\theta_1 \triangleq \gamma \qquad \theta_2 \triangleq \beta \qquad \theta_3 \triangleq -\alpha \tag{44}$$

in which case the given sequence of rotations is represented by $\theta_1\hat{\mathbf{a}}_1$, $\theta_2\hat{\mathbf{a}}_2$, and $\theta_3\hat{\mathbf{a}}_3$; and L_{ij} then can be found by referring to Eq. (1) to express the scalar products associated with L_{ij} in terms of α, β and γ. For instance,

$$\begin{aligned} L_{21} &= \hat{\mathbf{y}} \cdot \hat{\boldsymbol{\xi}} = \hat{\mathbf{a}}_2 \cdot (-\hat{\mathbf{b}}_3) \\ &\underset{(10.2.1)}{=} -C_{23} \underset{(1)}{=} -\mathrm{c}_1\mathrm{s}_2\mathrm{s}_3 + \mathrm{c}_3\mathrm{s}_1 \\ &= \cos\gamma \sin\beta \sin\alpha + \cos\alpha \sin\gamma \end{aligned} \tag{45}$$

In the literature of dynamics one encounters a wide variety of direction cosine matrices. Twenty-four such matrices are tabulated in Appendix I. Each matrix therein is associated with a particular rotation sequence, the name of which indicates the ordering of the unit vectors involved. For example, the matrices in Eqs. (1), (11), (21), and (31) correspond, respectively, to rotation sequences referred to as space-three, *1-2-3*; body-three, *1-2-3*; space-two, *1-2-1*; and body-two, *1-2-1*.

Derivations Each orientation of B in A represented by a direction cosine matrix in Eqs. (1), (11), (21) and (31) is the result of three successive rotations. To establish the validity of Eq. (1), which is associated with a space-three, 1-2-3 rotation sequence, it is helpful to introduce a fictitious rigid body A' that remains fixed in A during the first rotation, but is then fixed in B while B performs the second rotation. For analytical purposes, the first rotation then can be regarded as a rotation of B relative to A', and the second rotation as one of A' relative to A.

If $\hat{\mathbf{a}}_i$, $\hat{\mathbf{b}}_i$, and $\hat{\mathbf{a}}'_i$ $(i = 1,2,3)$ are three dextral sets of orthogonal unit vectors fixed in A, B, and A', respectively, such that $\hat{\mathbf{a}}_i = \hat{\mathbf{b}}_i = \hat{\mathbf{a}}'_i$ $(i = 1,2,3)$ prior to the first rotation of B in A, and if ${}^{A'}C^B$, ${}^AC^{A'}$, and ${}^AC^B$ are the direction cosine matrices characterizing, respectively, the first, the second, and the single equivalent rotation, so that

$${}^{A'}C^B{}_{ij} = \hat{\mathbf{a}}'_i \cdot \hat{\mathbf{b}}_j \qquad {}^AC^{A'}{}_{ij} = \hat{\mathbf{a}}_i \cdot \hat{\mathbf{a}}'_j \qquad {}^AC^B{}_{ij} = \hat{\mathbf{a}}_i \cdot \hat{\mathbf{b}}_j \qquad (i,j = 1,2,3) \tag{46}$$

then ${}^AC^B$ is given by

$${}^AC^B = {}^AC^{A'}\,{}^{A'}C^B \tag{47}$$

The direction cosine matrix representing the first rotation, ${}^{A'}C^B$, is premultiplied by the direction cosine matrix characterizing the second rotation, ${}^AC^{A'}$. Repeated use of Eq. (47) permits one to construct formulas for a direction cosine matrix characterizing a single rotation that is equivalent to any number of successive rotations. In particular, for three successive rotations,

$${}^AC^B = {}^AC^{A''}\,{}^{A''}C^{A'}\,{}^{A'}C^B \tag{48}$$

where ${}^{A'}C^B$, ${}^{A''}C^{A'}$, and ${}^AC^{A''}$ are direction cosine matrices associated with the first, the second, and the third rotation, respectively. Equation (1) may be derived by using

Eq. (48) with

$$ {}^{A'}C^{B} = C_1(\theta_1) \underset{(10.2.35)}{=} \begin{bmatrix} 1 & 0 & 0 \\ 0 & c_1 & -s_1 \\ 0 & s_1 & c_1 \end{bmatrix} \tag{49} $$

$$ {}^{A''}C^{A'} = C_2(\theta_2) \underset{(10.2.36)}{=} \begin{bmatrix} c_2 & 0 & s_2 \\ 0 & 1 & 0 \\ -s_2 & 0 & c_2 \end{bmatrix} \tag{50} $$

and

$$ {}^{A}C^{A''} = C_3(\theta_3) \underset{(10.2.37)}{=} \begin{bmatrix} c_3 & -s_3 & 0 \\ s_3 & c_3 & 0 \\ 0 & 0 & 1 \end{bmatrix} \tag{51} $$

Equations (2)–(10) are immediate consequences of Eq. (1), and Eqs. (21)–(30) can be generated by procedures similar to those just employed.

To establish the validity of Eq. (11), which is associated with a body-three, 1-2-3 rotation sequence, one may introduce a fictitious rigid body B' that moves exactly like B during the first rotation, but remains fixed in A while B performs the second rotation. For analytical purposes, the first rotation then can be regarded as a rotation of B' relative to A, and the second rotation as one of B relative to B'.

If $\hat{\mathbf{a}}_i$, $\hat{\mathbf{b}}_i$, and $\hat{\mathbf{b}}_i'$ $(i = 1,2,3)$ are three dextral sets of orthogonal unit vectors fixed in A, B, and B', respectively, such that $\hat{\mathbf{a}}_i = \hat{\mathbf{b}}_i = \hat{\mathbf{b}}_i'$ $(i = 1,2,3)$ prior to the first rotation of B in A, and if ${}^{A}C^{B'}$, ${}^{B'}C^{B}$, and ${}^{A}C^{B}$ are the direction cosine matrices characterizing, respectively, the first, the second, and the single equivalent rotation, so that

$$ {}^{A}C^{B'}{}_{ij} = \hat{\mathbf{a}}_i \cdot \hat{\mathbf{b}}_j' \qquad {}^{B'}C^{B}{}_{ij} = \hat{\mathbf{b}}_i' \cdot \hat{\mathbf{b}}_j \qquad {}^{A}C^{B}{}_{ij} = \hat{\mathbf{a}}_i \cdot \hat{\mathbf{b}}_j \qquad (i,j = 1,2,3) \tag{52} $$

then ${}^{A}C^{B}$ is given by

$$ {}^{A}C^{B} = {}^{A}C^{B'}\,{}^{B'}C^{B} \tag{53} $$

The direction cosine matrix representing the first rotation, ${}^{A}C^{B'}$, is postmultiplied by the direction cosine matrix characterizing the second rotation, ${}^{B'}C^{B}$. Repeated appeals to Eq. (53) allow one to construct formulas for a direction cosine matrix characterizing a single rotation that is equivalent to any number of successive rotations. For three successive rotations,

$$ {}^{A}C^{B} = {}^{A}C^{B'}\,{}^{B'}C^{B''}\,{}^{B''}C^{B} \tag{54} $$

where ${}^{A}C^{B'}$, ${}^{B'}C^{B''}$, and ${}^{B''}C^{B}$ are direction cosine matrices associated with the first, the second, and the third rotation, respectively. Equation (11) can be obtained by using Eq. (54) with

$$ {}^{A}C^{B'} = C_1(\theta_1) \underset{(10.2.35)}{=} \begin{bmatrix} 1 & 0 & 0 \\ 0 & c_1 & -s_1 \\ 0 & s_1 & c_1 \end{bmatrix} \tag{55} $$

$$
{}^{B'}C^{B''} = C_2(\theta_2) \underset{(10.2.36)}{=} \begin{bmatrix} c_2 & 0 & s_2 \\ 0 & 1 & 0 \\ -s_2 & 0 & c_2 \end{bmatrix} \tag{56}
$$

and

$$
{}^{B''}C^{B} = C_3(\theta_3) \underset{(10.2.37)}{=} \begin{bmatrix} c_3 & -s_3 & 0 \\ s_3 & c_3 & 0 \\ 0 & 0 & 1 \end{bmatrix} \tag{57}
$$

Equations (12)–(20) follow directly from Eq. (11), and Eqs. (31)–(40) can be obtained by similar means.

Equations (1) and (11) can be used to show that B returns to its original orientation in A following successive simple rotations of θ_1 about $\hat{\mathbf{a}}_1$, θ_2 about $\hat{\mathbf{a}}_2$, θ_3 about $\hat{\mathbf{a}}_3$, θ_1 about $-\hat{\mathbf{b}}_1$, θ_2 about $-\hat{\mathbf{b}}_2$, and θ_3 about $-\hat{\mathbf{b}}_3$. The direction cosine matrix characterizing the first three rotations is, according to Eq. (1), $C_3(\theta_3)C_2(\theta_2)C_1(\theta_1)$, whereas the direction cosine matrix representing the final three rotations is, in view of Eq. (11), $C_1(-\theta_1)C_2(-\theta_2)C_3(-\theta_3)$. Thus, after all six rotations,

$$
\begin{aligned}
{}^{A}C^{B} &= C_3(\theta_3)C_2(\theta_2)C_1(\theta_1)C_1(-\theta_1)C_2(-\theta_2)C_3(-\theta_3) \\
&= C_3(\theta_3)C_2(\theta_2)C_1(\theta_1)\underset{(10.2.35)}{C_1{}^{T}(\theta_1)}\,\underset{(10.2.36)}{C_2{}^{T}(\theta_2)}\,\underset{(10.2.37)}{C_3{}^{T}(\theta_3)} \\
&\underset{(10.2.16)}{=} U
\end{aligned} \tag{58}
$$

In much the same way, other sequences of six successive rotations can be shown to have no net effect on the orientation of B in A.

As an alternative to employing Eqs. (48)–(51), one may establish the validity of the right-hand member of Eq. (1) through the use of Eq. (54), forming ${}^{A}C^{B'}$, ${}^{B'}C^{B''}$, and ${}^{B''}C^{B}$ with the aid of Eq. (10.2.35) and Eqs. (10.2.23)–(10.2.31). Specifically, to deal with the rotation of magnitude θ_1 about $\hat{\mathbf{a}}_1$, let

$$
{}^{A}C^{B'} \underset{(10.2.35)}{=} \begin{bmatrix} 1 & 0 & 0 \\ 0 & c_1 & -s_1 \\ 0 & s_1 & c_1 \end{bmatrix} \tag{59}
$$

Next, to construct a matrix ${}^{B'}C^{B''}$ that characterizes the rotation of magnitude θ_2 about $\hat{\mathbf{a}}_2$, let ${}^{A}\lambda$ and ${}^{B'}\lambda$ denote row matrices whose elements are $\hat{\mathbf{a}}_2 \cdot \hat{\mathbf{a}}_i$ and $\hat{\mathbf{a}}_2 \cdot \hat{\mathbf{b}}_i'$ $(i = 1,2,3)$, respectively. Then

$$
{}^{A}\lambda = \lfloor 0 \quad 1 \quad 0 \rfloor \tag{60}
$$

$$
{}^{B'}\lambda \underset{(10.2.9)}{=} {}^{A}\lambda\, {}^{A}C^{B'} \underset{(59,\ 60)}{=} \lfloor 0 \quad c_1 \quad -s_1 \rfloor \tag{61}
$$

and, from Eqs. (10.2.23)–(10.2.31), with $\lambda_1 = 0$, $\lambda_2 = c_1$, $\lambda_3 = -s_1$, and $\theta = \theta_2$,

$$
{}^{B'}C^{B''} = \begin{bmatrix} c_2 & s_1 s_2 & c_1 s_2 \\ -s_1 s_2 & 1 + s_1{}^2(c_2 - 1) & s_1 c_1(c_2 - 1) \\ -c_1 s_2 & s_1 c_1(c_2 - 1) & 1 + c_1{}^2(c_2 - 1) \end{bmatrix} \tag{62}
$$

A matrix ${}^{A}C^{B''}$ associated with a simple rotation that is equivalent to the first two rotations is now given by

$$
{}^{A}C^{B''} \underset{(53)}{=} {}^{A}C^{B'}\,{}^{B'}C^{B''} \underset{(59,\,62)}{=} \begin{bmatrix} c_2 & s_1 s_2 & c_1 s_2 \\ 0 & c_1 & -s_1 \\ -s_2 & s_1 c_2 & c_1 c_2 \end{bmatrix} \tag{63}
$$

and, to resolve $\hat{\mathbf{a}}_3$ into components required for the construction of a matrix ${}^{B''}C^{B}$, one may use the first of Eqs. (10.2.9) to obtain

$$
\lfloor 0 \quad 0 \quad 1 \rfloor\, {}^{A}C^{B''} \underset{(63)}{=} \lfloor -s_2 \quad s_1 c_2 \quad c_1 c_2 \rfloor \tag{64}
$$

after which Eqs. (10.2.23)–(10.2.31) yield

$$
{}^{B''}C^{B} = \tag{65}
$$

$$
\begin{bmatrix} s_2{}^2 + c_2{}^2 c_3 & -c_2[c_1 s_3 + s_1 s_2(1 - c_3)] & c_2[s_1 s_3 - c_1 s_2(1 - c_3)] \\ c_2[c_1 s_3 + s_1 s_2(1 - c_3)] & c_3(1 - s_1{}^2 c_2{}^2) + s_1{}^2 c_2{}^2 & s_2 s_3 + s_1 c_1 c_2{}^2(1 - c_3) \\ -c_2[s_1 s_3 + c_1 s_2(1 - c_3)] & -s_2 s_3 + s_1 c_1 c_2{}^2(1 - c_3) & 1 - (s_2{}^2 + s_1{}^2 c_2{}^2)(1 - c_3) \end{bmatrix}
$$

and substitution from Eqs. (59), (62), and (65) into Eq. (54) leads directly to the right-hand member of Eq. (1).

Example If unit vectors $\hat{\mathbf{a}}_1$, $\hat{\mathbf{a}}_2$, $\hat{\mathbf{a}}_3$ and $\hat{\mathbf{b}}_1$, $\hat{\mathbf{b}}_2$, $\hat{\mathbf{b}}_3$ are introduced as shown in Fig. 10.3.2, and the angles ϕ, θ, and ψ shown in Fig. 10.3.1 are renamed θ_1, θ_2, and θ_3, respectively, then θ_1, θ_2, and θ_3 are body-two, 1-2-1 angles such that Eqs. (31)–(40) can be used to discuss motions of B in A. However, as will be seen later in Sec. 10.7, it is undesirable to use these angles when dealing with motions during which the rotor axis becomes coincident, or even nearly coincident, with the outer gimbal axis. (Coincidence of these two axes is referred to as *gimbal lock*.) Therefore, it may be convenient to employ in the course of one analysis two modes of description of the orientation of B in A, switching from one to the other whenever θ_2 acquires a value lying in a previously designated range. The following sort of question then can arise: If ϕ_1, ϕ_2, and ϕ_3 are the space-three, 1-2-3 angles associated with $\hat{\mathbf{a}}_1$, $\hat{\mathbf{a}}_2$, $\hat{\mathbf{a}}_3$ and $\hat{\mathbf{b}}_1$, $\hat{\mathbf{b}}_2$, $\hat{\mathbf{b}}_3$, what are the values of these angles corresponding to $\theta_1 = 30°$, $\theta_2 = 45°$, $\theta_3 = 60°$?

Inspection of Eqs. (2)–(6) shows that the elements of C required for the evaluation of ϕ_1, ϕ_2, and ϕ_3 are C_{31}, C_{32}, C_{33}, C_{21}, and C_{11}. From Eq. (31),

$$
C_{31} = -\frac{\sqrt{3}}{2}\frac{\sqrt{2}}{2} = -0.612 \tag{66}
$$

$$
C_{32} = \frac{\sqrt{3}}{2}\frac{\sqrt{2}}{2}\frac{\sqrt{3}}{2} + \frac{1}{2}\frac{1}{2} = 0.780 \tag{67}
$$

$$
C_{33} = \frac{\sqrt{3}}{2}\frac{\sqrt{2}}{2}\frac{1}{2} - \frac{\sqrt{3}}{2}\frac{1}{2} = -0.127 \tag{68}
$$

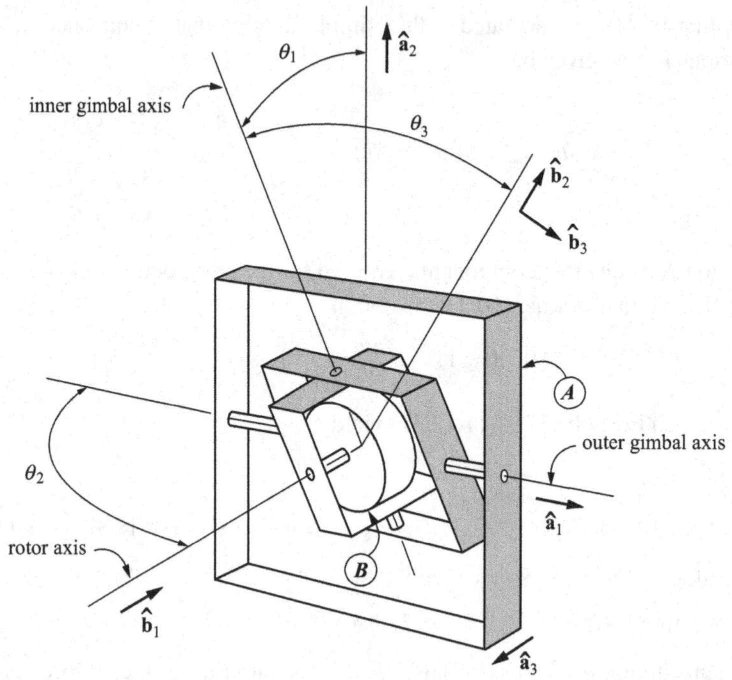

Figure 10.3.2

$$C_{21} = \frac{1}{2}\frac{\sqrt{2}}{2} = 0.354 \tag{69}$$

$$C_{11} = \frac{\sqrt{2}}{2} = 0.707 \tag{70}$$

Hence,

$$\phi_2 \underset{(2)}{=} \sin^{-1}(0.612) = 37.7° \tag{71}$$

$$\alpha \underset{(3)}{=} \sin^{-1}\left(\frac{0.780}{0.791}\right) = 80.4° \tag{72}$$

$$\phi_1 \underset{(4)}{=} 99.6° \qquad \beta \underset{(5)}{=} 26.6° \qquad \phi_3 \underset{(6)}{=} 26.6° \tag{73}$$

10.4 EULER PARAMETERS

The unit vector $\hat{\boldsymbol{\lambda}}$ and the angle θ introduced in Sec. 10.1 can be used to associate a vector $\boldsymbol{\epsilon}$, called the *Euler vector*, and four scalar quantities, $\epsilon_1, \ldots, \epsilon_4$, called *Euler parameters*, with a simple rotation of a rigid body B in a reference frame A by letting

$$\boldsymbol{\epsilon} \triangleq \hat{\boldsymbol{\lambda}} \sin\frac{\theta}{2} \tag{1}$$

$$\epsilon_i \triangleq \boldsymbol{\epsilon} \cdot \hat{\mathbf{a}}_i = \boldsymbol{\epsilon} \cdot \hat{\mathbf{b}}_i \qquad (i = 1,2,3) \tag{2}$$

and

$$\epsilon_4 \triangleq \cos\frac{\theta}{2} \tag{3}$$

where $\hat{\mathbf{a}}_1$, $\hat{\mathbf{a}}_2$, $\hat{\mathbf{a}}_3$ and $\hat{\mathbf{b}}_1$, $\hat{\mathbf{b}}_2$, $\hat{\mathbf{b}}_3$ are dextral sets of orthogonal unit vectors fixed in A and B respectively, with $\hat{\mathbf{a}}_i = \hat{\mathbf{b}}_i$ $(i = 1,2,3)$ prior to the rotation. (Where a discussion involves more than two bodies or reference frames, notations such as ${}^A\boldsymbol{\epsilon}^B$ and ${}^A\epsilon_i^B$ will be used.)

The Euler parameters are not independent of each other, for the sum of their squares is necessarily equal to unity:

$$\epsilon_1{}^2 + \epsilon_2{}^2 + \epsilon_3{}^2 + \epsilon_4{}^2 = \boldsymbol{\epsilon}^2 + \epsilon_4{}^2 = 1 \tag{4}$$

An indication of the utility of the Euler parameters may be gleaned from the fact that the elements of the direction cosine matrix C introduced in Sec. 10.2 assume a particularly simple and orderly form when expressed in terms of $\epsilon_1, \ldots, \epsilon_4$: If C_{ij} is defined as

$$C_{ij} \triangleq \hat{\mathbf{a}}_i \cdot \hat{\mathbf{b}}_j \qquad (i,j = 1,2,3) \tag{5}$$

then

$$C_{11} = \epsilon_1{}^2 - \epsilon_2{}^2 - \epsilon_3{}^2 + \epsilon_4{}^2 = 1 - 2\epsilon_2{}^2 - 2\epsilon_3{}^2 \tag{6}$$
$$C_{12} = 2(\epsilon_1\epsilon_2 - \epsilon_3\epsilon_4) \tag{7}$$
$$C_{13} = 2(\epsilon_3\epsilon_1 + \epsilon_2\epsilon_4) \tag{8}$$
$$C_{21} = 2(\epsilon_1\epsilon_2 + \epsilon_3\epsilon_4) \tag{9}$$
$$C_{22} = \epsilon_2{}^2 - \epsilon_3{}^2 - \epsilon_1{}^2 + \epsilon_4{}^2 = 1 - 2\epsilon_3{}^2 - 2\epsilon_1{}^2 \tag{10}$$
$$C_{23} = 2(\epsilon_2\epsilon_3 - \epsilon_1\epsilon_4) \tag{11}$$
$$C_{31} = 2(\epsilon_3\epsilon_1 - \epsilon_2\epsilon_4) \tag{12}$$
$$C_{32} = 2(\epsilon_2\epsilon_3 + \epsilon_1\epsilon_4) \tag{13}$$
$$C_{33} = \epsilon_3{}^2 - \epsilon_1{}^2 - \epsilon_2{}^2 + \epsilon_4{}^2 = 1 - 2\epsilon_1{}^2 - 2\epsilon_2{}^2 \tag{14}$$

Equations (6)–(14) may also be written in a matrix form. To do so, let

$$\epsilon \triangleq \lfloor \epsilon_1 \quad \epsilon_2 \quad \epsilon_3 \rfloor \tag{15}$$

and

$$\tilde{\epsilon} \triangleq \begin{bmatrix} 0 & -\epsilon_3 & \epsilon_2 \\ \epsilon_3 & 0 & -\epsilon_1 \\ -\epsilon_2 & \epsilon_1 & 0 \end{bmatrix} \tag{16}$$

Here we establish a notational convention that will be used again in subsequent sections, namely, placing a tilde over a symbol, say ϵ, denotes a skew-symmetric matrix whose off-diagonal elements are denoted by $\pm\epsilon_i$ $(i = 1,2,3)$, arranged according to Eq. (16). A skew-symmetric matrix $\tilde{\epsilon}$ is one that satisfies the relationship

$$\tilde{\epsilon}^T = -\tilde{\epsilon} \tag{17}$$

where a superscript T indicates matrix transposition. Then, it follows that

$$C = U(1 - 2\epsilon\epsilon^T) + 2\epsilon^T\epsilon + 2\epsilon_4\tilde{\epsilon} \tag{18}$$

where U denotes the identity matrix.

The Euler parameters can be expressed in terms of direction cosines in such a way that Eqs. (6)–(14) are satisfied identically. This is accomplished by taking

$$\epsilon_1 = \frac{C_{32} - C_{23}}{4\epsilon_4} \tag{19}$$

$$\epsilon_2 = \frac{C_{13} - C_{31}}{4\epsilon_4} \tag{20}$$

$$\epsilon_3 = \frac{C_{21} - C_{12}}{4\epsilon_4} \tag{21}$$

and

$$\epsilon_4 = \frac{1}{2}(1 + C_{11} + C_{22} + C_{33})^{1/2} \tag{22}$$

Since Eqs. (1) and (3) are satisfied if

$$\hat{\boldsymbol{\lambda}} = \frac{\epsilon_1\hat{\mathbf{a}}_1 + \epsilon_2\hat{\mathbf{a}}_2 + \epsilon_3\hat{\mathbf{a}}_3}{({\epsilon_1}^2 + {\epsilon_2}^2 + {\epsilon_3}^2)^{1/2}} \tag{23}$$

and

$$\theta = 2\cos^{-1}\epsilon_4 \qquad 0 \le \theta \le \pi \tag{24}$$

one can thus find a simple rotation such that the direction cosines associated with this rotation as in Eqs. (10.2.23)–(10.2.31) are equal to corresponding elements of any direction cosine matrix C that satisfies Eq. (10.2.2). In other words, every change in the relative orientation of two rigid bodies or reference frames A and B can be produced by means of a simple rotation of B in A. This proposition is known as *Euler's theorem on rotation.*

As an alternative to Eqs. (10.1.1) and (10.1.3), the relationship between a vector **a** fixed in a reference frame A and a vector **b** fixed in a rigid body B and equal to **a** prior to a simple rotation of B in A can be expressed in terms of $\boldsymbol{\epsilon}$ and ϵ_4 as

$$\mathbf{b} = \mathbf{a} + 2[\epsilon_4\boldsymbol{\epsilon} \times \mathbf{a} + \boldsymbol{\epsilon} \times (\boldsymbol{\epsilon} \times \mathbf{a})] \tag{25}$$

Derivations The equality of $\boldsymbol{\epsilon} \cdot \hat{\mathbf{a}}_i$ and $\boldsymbol{\epsilon} \cdot \hat{\mathbf{b}}_i$ [see Eq. (2)] follows from Eqs. (1) and (10.2.22); Eqs. (4) are consequences of Eqs. (1)–(3) and of the fact that $\hat{\boldsymbol{\lambda}}$ is a unit vector; and Eqs. (6)–(14) can be obtained from Eqs. (10.2.23)–(10.2.31) by replacing functions of θ with functions of $\theta/2$ and using Eq. (10.2.22) together with Eqs. (1)–(4). For example,

$$C_{11} \underset{(10.2.23)}{=} 2\cos^2\frac{\theta}{2} - 1 + 2{\lambda_1}^2\sin^2\frac{\theta}{2} \tag{26}$$

and

$$\lambda_1 \sin\frac{\theta}{2} \underset{(10.2.22)}{=} \hat{\boldsymbol{\lambda}} \cdot \hat{\mathbf{a}}_1 \sin\frac{\theta}{2} \underset{(1)}{=} \boldsymbol{\epsilon} \cdot \hat{\mathbf{a}}_1 \underset{(2)}{=} \epsilon_1 \tag{27}$$

while

$$\cos\frac{\theta}{2} \underset{(3)}{=} \epsilon_4 \tag{28}$$

Hence,

$$C_{11} = 2\epsilon_4{}^2 - 1 + 2\epsilon_1{}^2 \underset{(4)}{=} \epsilon_1{}^2 - \epsilon_2{}^2 - \epsilon_3{}^2 + \epsilon_4{}^2 \tag{29}$$

in agreement with Eq. (6). To see that the matrix form of Eqs. (18) is in fact a compact way of writing Eqs. (6)–(14), one need only substitute from Eqs. (10.2.3), (10.2.18), (15), and (16) into (18).

The validity of Eqs. (19)–(22) can be established by showing that the left-hand members of Eqs. (6)–(14) may be obtained by substituting from Eqs. (19)–(22) into the right-hand members. For example,

$$\begin{aligned} 1 - 2\epsilon_2{}^2 - 2\epsilon_3{}^2 \\ \underset{(20\text{–}22)}{=} \frac{2(1 + C_{11} + C_{22} + C_{33}) - C_{13}{}^2 + 2C_{13}C_{31} - C_{31}{}^2 - C_{21}{}^2 + 2C_{12}C_{21} - C_{12}{}^2}{2(1 + C_{11} + C_{22} + C_{33})} \\ \underset{(10.2.14)}{=} \frac{C_{11} + C_{22} + C_{33} + C_{13}C_{31} + C_{12}C_{21} + C_{11}{}^2}{1 + C_{11} + C_{22} + C_{33}} \end{aligned} \tag{30}$$

But, since each element of C is equal to its cofactor in $|C|$,

$$C_{13}C_{31} = C_{11}C_{33} - C_{22} \tag{31}$$

and

$$C_{12}C_{21} = C_{11}C_{22} - C_{33} \tag{32}$$

Consequently,

$$1 - 2\epsilon_2{}^2 - 2\epsilon_3{}^2 = \frac{C_{11} + C_{11}C_{33} + C_{11}C_{22} + C_{11}{}^2}{1 + C_{11} + C_{22} + C_{33}} = C_{11} \tag{33}$$

as required by Eq. (6).

To see that Eqs. (1) and (3) are satisfied if $\hat{\boldsymbol{\lambda}}$ and θ are given by Eqs. (23) and (24), note that

$$\cos\frac{\theta}{2} \underset{(24)}{=} \epsilon_4 \tag{34}$$

which is Eq. (3), and that

$$\sin\frac{\theta}{2} \underset{(24)}{=} (1 - \epsilon_4{}^2)^{1/2} \underset{(4)}{=} (\epsilon_1{}^2 + \epsilon_2{}^2 + \epsilon_3{}^2)^{1/2} \tag{35}$$

so that

$$\hat{\boldsymbol{\lambda}} \sin\frac{\theta}{2} \underset{(23)}{=} \epsilon_1\hat{\mathbf{a}}_1 + \epsilon_2\hat{\mathbf{a}}_2 + \epsilon_3\hat{\mathbf{a}}_3 \underset{(2)}{=} \boldsymbol{\epsilon} \tag{36}$$

as required by Eq. (1). Finally, Eq. (10.1.1) is equivalent to

$$\mathbf{b} = \mathbf{a} + \hat{\boldsymbol{\lambda}} \times \mathbf{a} \sin\theta + \hat{\boldsymbol{\lambda}} \times (\hat{\boldsymbol{\lambda}} \times \mathbf{a})(1 - \cos\theta) \tag{37}$$

$$\underset{(1)}{=} \mathbf{a} + 2\boldsymbol{\epsilon} \times \mathbf{a} \cos\frac{\theta}{2} + 2\boldsymbol{\epsilon} \times (\boldsymbol{\epsilon} \times \mathbf{a}) \underset{(3)}{=} \mathbf{a} + 2[\epsilon_4 \boldsymbol{\epsilon} \times \mathbf{a} + \boldsymbol{\epsilon} \times (\boldsymbol{\epsilon} \times \mathbf{a})]$$

in agreement with Eq. (25).

Example Triangle ABC in Fig. 10.4.1 can be brought into the position $A'B'C'$ by moving point A to A', without changing the orientation of the triangle, and then performing a simple rotation of the triangle while keeping A fixed at A'. To find $\hat{\boldsymbol{\lambda}}$, a unit vector parallel to the axis of rotation, and to determine θ, the associated angle

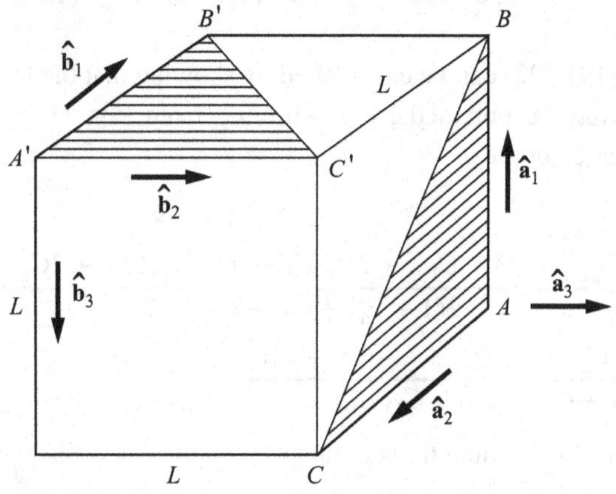

Figure 10.4.1

of rotation, let the unit vectors $\hat{\mathbf{a}}_i$ and $\hat{\mathbf{b}}_i$ $(i = 1,2,3)$ be directed as shown in Fig. 10.4.1, thus ensuring that $\hat{\mathbf{a}}_i = \hat{\mathbf{b}}_i$ $(i = 1,2,3)$ prior to the rotation; determine C_{ij} by evaluating $\hat{\mathbf{a}}_i \cdot \hat{\mathbf{b}}_j$; and use Eqs. (19)–(22) to form ϵ_i $(i = 1, \ldots, 4)$:

$$\epsilon_4 \underset{(22)}{=} \frac{1}{2}(1 + \hat{\mathbf{a}}_1 \cdot \hat{\mathbf{b}}_1 + \hat{\mathbf{a}}_2 \cdot \hat{\mathbf{b}}_2 + \hat{\mathbf{a}}_3 \cdot \hat{\mathbf{b}}_3)^{1/2}$$

$$= \frac{1}{2}(1 + 0 + 0 + 0)^{1/2} = \frac{1}{2} \tag{38}$$

$$\epsilon_1 \underset{(19)}{=} \frac{\hat{\mathbf{a}}_3 \cdot \hat{\mathbf{b}}_2 - \hat{\mathbf{a}}_2 \cdot \hat{\mathbf{b}}_3}{4\left(\frac{1}{2}\right)} = \frac{1 - 0}{2} = \frac{1}{2} \tag{39}$$

$$\epsilon_2 \underset{(20)}{=} \frac{\hat{\mathbf{a}}_1 \cdot \hat{\mathbf{b}}_3 - \hat{\mathbf{a}}_3 \cdot \hat{\mathbf{b}}_1}{4\left(\frac{1}{2}\right)} = \frac{-1 - 0}{2} = -\frac{1}{2} \tag{40}$$

$$\epsilon_3 \underset{(21)}{=} \frac{\hat{\mathbf{a}}_2 \cdot \hat{\mathbf{b}}_1 - \hat{\mathbf{a}}_1 \cdot \hat{\mathbf{b}}_2}{4\left(\frac{1}{2}\right)} = \frac{-1 - 0}{2} = -\frac{1}{2} \tag{41}$$

Then

$$\hat{\boldsymbol{\lambda}} \underset{(23)}{=} \frac{\frac{1}{2}\hat{\mathbf{a}}_1 - \frac{1}{2}\hat{\mathbf{a}}_2 - \frac{1}{2}\hat{\mathbf{a}}_3}{\left(\frac{3}{4}\right)^{1/2}} = \frac{\hat{\mathbf{a}}_1 - \hat{\mathbf{a}}_2 - \hat{\mathbf{a}}_3}{\sqrt{3}} \tag{42}$$

and

$$\theta \underset{(24)}{=} 2\cos^{-1}\frac{1}{2} = \frac{2\pi}{3}\text{rad} \tag{43}$$

10.5 WIENER-MILENKOVIĆ PARAMETERS

A vector $\boldsymbol{\mu}$, called the *Wiener-Milenković vector*, and three scalar quantities, μ_i ($i = 1,2,3$) called *Wiener-Milenković parameters*,† can be associated with a simple rotation of a rigid body B in a reference frame A (see Sec. 10.1) by letting

$$\boldsymbol{\mu} \triangleq 4\hat{\boldsymbol{\lambda}} \tan\frac{\theta}{4} \tag{1}$$

and

$$\mu_i \triangleq \boldsymbol{\mu} \cdot \hat{\mathbf{a}}_i = \boldsymbol{\mu} \cdot \hat{\mathbf{b}}_i \qquad (i = 1,2,3) \tag{2}$$

where $\hat{\boldsymbol{\lambda}}$ and θ have the same meaning as in Sec. 10.1, and $\hat{\mathbf{a}}_1$, $\hat{\mathbf{a}}_2$, $\hat{\mathbf{a}}_2$, and $\hat{\mathbf{b}}_1$, $\hat{\mathbf{b}}_2$, $\hat{\mathbf{b}}_3$ are dextral sets of orthogonal unit vectors fixed in A and B respectively, with $\hat{\mathbf{a}}_i = \hat{\mathbf{b}}_i$ ($i = 1,2,3$) prior to the rotation. (When a discussion involves more than two bodies or reference frames, notations such as ${}^A\boldsymbol{\mu}^B$ and ${}^A\mu_i^B$ will be used.)

The Wiener-Milenković parameters are intimately related to the Euler parameters (see Sec. 10.4):

$$\epsilon_i = \frac{\mu_i}{2\left[1 + \frac{1}{16}\left({\mu_1}^2 + {\mu_2}^2 + {\mu_3}^2\right)\right]} \qquad (i = 1,2,3)$$

$$\epsilon_4 = \frac{1 - \frac{1}{16}\left({\mu_1}^2 + {\mu_2}^2 + {\mu_3}^2\right)}{1 + \frac{1}{16}\left({\mu_1}^2 + {\mu_2}^2 + {\mu_3}^2\right)} \tag{3}$$

These relations may also be written in a matrix form. Introducing

$$\mu = \lfloor \mu_1 \quad \mu_2 \quad \mu_3 \rfloor \tag{4}$$

Thus, Eqs. (3) simplify to

$$\epsilon = \frac{\mu}{2(1 + \frac{1}{16}\mu\mu^T)}$$

$$\epsilon_4 = \frac{1 - \frac{1}{16}\mu\mu^T}{1 + \frac{1}{16}\mu\mu^T} \tag{5}$$

† O. A. Bauchau, *Flexible Multibody Dynamics* (Dordrecht, Netherlands: Springer, 2011), pp. 539–540.

Introducing

$$\widetilde{\mu} = \begin{bmatrix} 0 & -\mu_3 & \mu_2 \\ \mu_3 & 0 & -\mu_1 \\ -\mu_2 & \mu_1 & 0 \end{bmatrix} \tag{6}$$

and expressing the direction cosine matrix C in terms of Wiener-Milenković parameters (see Sec. 10.2), one may show that

$$C = FF \tag{7}$$

where

$$F = \frac{\left(1 - \frac{1}{16}\mu\mu^T\right)U + \frac{1}{8}\mu^T\mu + \frac{1}{2}\widetilde{\mu}}{1 + \frac{1}{16}\mu\mu^T} \tag{8}$$

Here F is an orthonormal matrix such that $FF^T = F^T F = U$.

The Wiener-Milenković parameters can be expressed in terms of direction cosines as follows.

$$\mu_1 = \frac{4(C_{32} - C_{23})}{1 + C_{11} + C_{22} + C_{33} + 2(1 + C_{11} + C_{22} + C_{33})^{1/2}} \tag{9}$$

$$\mu_2 = \frac{4(C_{13} - C_{31})}{1 + C_{11} + C_{22} + C_{33} + 2(1 + C_{11} + C_{22} + C_{33})^{1/2}} \tag{10}$$

$$\mu_3 = \frac{4(C_{21} - C_{12})}{1 + C_{11} + C_{22} + C_{33} + 2(1 + C_{11} + C_{22} + C_{33})^{1/2}} \tag{11}$$

Wiener-Milenković parameters have an advantage over Euler parameters in that there is one less parameter in the former set. Wiener-Milenković parameters, however, become singular at $\theta = 2\pi(1 + 2k)$ with k equal to an integer. Because the value of k is immaterial to the values of C_{ij}, multiples of 2π may be added to or subtracted from θ without consequence.

By imposing the restriction that $-\pi \le \theta \le \pi$, that is, when $\mu\mu^T \le 16$, one can easily *avoid* singularities. When μ_1, μ_2, and μ_3 are changing with time during a simulation, $\mu\mu^T$ may exceed 16. When this happens, the μ_i parameters can be simply rescaled to correspond to a rotation that differs by 2π. That is, when $\mu\mu^T > 16$, a scaled version of μ denoted by μ^* may be taken to be

$$\mu^* = -\frac{16\mu}{\mu\mu^T} \tag{12}$$

Thus, Wiener-Milenković parameters facilitate (a) a simple assessment of the parameters that determines whether they need to be rescaled, and (b) a simple rescaling law. $C(\mu^*)$ is thus identical to $C(\mu)$.

Derivations Given Eq. (10.4.1), one may express $\hat{\boldsymbol{\lambda}}$ as

$$\hat{\boldsymbol{\lambda}} = \frac{\boldsymbol{\epsilon}}{\sin\frac{\theta}{2}} \tag{13}$$

which, when substituted into Eq. (1), allows one to express

$$\mu = \frac{4\epsilon \tan\frac{\theta}{4}}{\sin\frac{\theta}{2}} = \frac{2\epsilon}{\cos^2\frac{\theta}{4}} \tag{14}$$

or equivalently

$$\mu = \frac{4\epsilon \tan\frac{\theta}{4}}{\sin\frac{\theta}{2}} = \frac{2\epsilon}{\cos^2\frac{\theta}{4}} \tag{15}$$

In view of Eq. (10.4.3), one may express

$$\cos^2\frac{\theta}{4} = \frac{1+\epsilon_4}{2} \tag{16}$$

Also, using Eq. (1), it may be seen that

$$\frac{\mu\mu^T}{16} = \lambda\lambda^T \tan^2\frac{\theta}{4} = \tan^2\frac{\theta}{4} \tag{17}$$

leading to

$$1 + \frac{\mu\mu^T}{16} = 1 + \tan^2\frac{\theta}{4} = \sec^2\frac{\theta}{4} = \frac{1}{\cos^2\frac{\theta}{4}} \tag{18}$$

and Eqs. (5). Substitution from Eqs. (5) into Eq. (10.4.18) yields Eq. (7) for C.

The validity of Eqs. (9)–(11) is established by first writing

$$\mu_i \underset{(15,\,16)}{=} \frac{4\epsilon_i}{1+\epsilon_4} \qquad (i = 1,2,3) \tag{19}$$

and then making appropriate substitutions from Eqs. (10.4.19)–(10.4.22). For example,

$$\begin{aligned} \mu_1 &\underset{(19)}{=} \frac{4\epsilon_1}{1+\epsilon_4} \underset{(10.4.19)}{=} \frac{C_{32}-C_{23}}{\epsilon_4(1+\epsilon_4)} \\ &\underset{(10.4.22)}{=} \frac{4(C_{32}-C_{23})}{1+C_{11}+C_{22}+C_{33}+2(1+C_{11}+C_{22}+C_{33})^{1/2}} \end{aligned} \tag{20}$$

in agreement with Eq. (9).

Finally, adding 2π to θ does not change C, but the values of μ_i do in fact change:

$$\mu^* = 4\lambda \tan\left(\frac{\theta+2\pi}{4}\right) = -\mu \cot^2\frac{\theta}{4} = -\frac{\mu}{\tan^2\frac{\theta}{4}} \tag{21}$$

which must lead to an identical value for all elements of C.

Example Given a set of Wiener-Milenković parameters $\mu_1 = 6$, $\mu_2 = 4$, and $\mu_3 = 5$, verify that θ is outside the range $-\pi \le \theta \le \pi$ and that when the scaling law Eq. (12) is applied, C does not change.

The matrix C for $\mu_1 = 6$, $\mu_2 = 4$, and $\mu_3 = 5$ is

$$C = \frac{1}{8649}\begin{bmatrix} 3401 & 7952 & -64 \\ -1808 & 841 & 8416 \\ 7744 & -3296 & 1993 \end{bmatrix} \tag{22}$$

However, because

$$\frac{\mu\mu^T}{16} = \frac{1}{16}\lfloor 6 \quad 4 \quad 5 \rfloor \begin{Bmatrix} 6 \\ 4 \\ 5 \end{Bmatrix} = \frac{77}{16} > 1 \tag{23}$$

it is thus established that $\theta > \pi$. Thus, the scaled values of the parameters become, by Eq. (12), $\mu_1^* = -96/77$, $\mu_2^* = -64/77$, $\mu_3^* = -80/77$, such that

$$\frac{\mu^* \mu^{*T}}{16} = \frac{16}{77} < 1 \tag{24}$$

but with C unchanged.

10.6 ANGULAR VELOCITY AND DIRECTION COSINES

If $\hat{\mathbf{a}}_1$, $\hat{\mathbf{a}}_2$, $\hat{\mathbf{a}}_3$, and $\hat{\mathbf{b}}_1$, $\hat{\mathbf{b}}_2$, $\hat{\mathbf{b}}_3$ are two dextral sets of orthogonal unit vectors fixed, respectively, in two reference frames or rigid bodies A and B that are moving relative to each other, then the direction cosine matrix C and its elements C_{ij} $(i,j = 1,2,3)$, defined in Sec. 10.2, are functions of time t. The time derivative of C, denoted by $\dot{C}$ and defined in terms of the time derivatives $\dot{C}_{ij}$ of C_{ij} $(i,j = 1,2,3)$ as

$$\dot{C} \triangleq \begin{bmatrix} \dot{C}_{11} & \dot{C}_{12} & \dot{C}_{13} \\ \dot{C}_{21} & \dot{C}_{22} & \dot{C}_{23} \\ \dot{C}_{31} & \dot{C}_{32} & \dot{C}_{33} \end{bmatrix} \tag{1}$$

can be expressed as the product of C and a skew-symmetric matrix $\widetilde{\omega}$ called an *angular velocity matrix* for B in A and defined as

$$\widetilde{\omega} \triangleq C^T \dot{C} \tag{2}$$

In other words, with $\widetilde{\omega}$ defined as in Eq. (2),

$$\dot{C} = C\,\widetilde{\omega} \tag{3}$$

The tilde is used in accordance with the notational convention established in Eq. (10.4.16). After defining ω_i as

$$\omega_i \triangleq {}^A\boldsymbol{\omega}^B \cdot \hat{\mathbf{b}}_i \qquad (i = 1,2,3) \tag{4}$$

where ${}^A\boldsymbol{\omega}^B$ is the angular velocity of B in A, $\widetilde{\omega}$ is expressed as

$$\widetilde{\omega} = \begin{bmatrix} 0 & -\omega_3 & \omega_2 \\ \omega_3 & 0 & -\omega_1 \\ -\omega_2 & \omega_1 & 0 \end{bmatrix} \tag{5}$$

The measure numbers ω_1, ω_2, and ω_3 are given by

$$\omega_1 = C_{13}\dot{C}_{12} + C_{23}\dot{C}_{22} + C_{33}\dot{C}_{32} \tag{6}$$
$$\omega_2 = C_{21}\dot{C}_{23} + C_{31}\dot{C}_{33} + C_{11}\dot{C}_{13} \tag{7}$$
$$\omega_3 = C_{32}\dot{C}_{31} + C_{12}\dot{C}_{11} + C_{22}\dot{C}_{21} \tag{8}$$

These equations can be expressed more concisely after defining η_{ijk} as

$$\eta_{ijk} \triangleq \tfrac{1}{2}\epsilon_{ijk}(\epsilon_{ijk} + 1) \qquad (i,j,k = 1,2,3) \tag{9}$$

where ϵ_{ijk} is given by Eq. (10.2.32). (The quantity η_{ijk} is equal to unity when the subscripts appear in cyclic order; otherwise it is equal to zero.) Using the summation convention for repeated subscripts, one then can replace Eqs. (6)–(8) with

$$\omega_i = \eta_{igh}\dot{C}_{jg}C_{jh} \qquad (i = 1,2,3) \tag{10}$$

Similarly, Eqs. (3) can be expressed as

$$\dot{C}_{ij} = C_{ig}\omega_h\epsilon_{ghj} \qquad (i,j = 1,2,3) \tag{11}$$

or, explicitly, as

$$\dot{C}_{11} = C_{12}\omega_3 - C_{13}\omega_2 \tag{12}$$
$$\dot{C}_{12} = C_{13}\omega_1 - C_{11}\omega_3 \tag{13}$$
$$\dot{C}_{13} = C_{11}\omega_2 - C_{12}\omega_1 \tag{14}$$
$$\dot{C}_{21} = C_{22}\omega_3 - C_{23}\omega_2 \tag{15}$$
$$\dot{C}_{22} = C_{23}\omega_1 - C_{21}\omega_3 \tag{16}$$
$$\dot{C}_{23} = C_{21}\omega_2 - C_{22}\omega_1 \tag{17}$$
$$\dot{C}_{31} = C_{32}\omega_3 - C_{33}\omega_2 \tag{18}$$
$$\dot{C}_{32} = C_{33}\omega_1 - C_{31}\omega_3 \tag{19}$$
$$\dot{C}_{33} = C_{31}\omega_2 - C_{32}\omega_1 \tag{20}$$

Equations (11) are known as *Poisson's kinematical equations.*

Derivations Premultiplication of $\widetilde{\omega}$ with C gives

$$C\,\widetilde{\omega} \underset{(2)}{=} CC^T\dot{C} \underset{(10.2.16)}{=} \dot{C} \tag{21}$$

in agreement with Eq. (3).

To see that $\widetilde{\omega}$ as defined in Eq. (2) is skew-symmetric, note that

$$\begin{aligned}\widetilde{\omega}^T + \widetilde{\omega} &\underset{(2)}{=} (C^T\dot{C})^T + C^T\dot{C} = \dot{C}^T C + C^T\dot{C} = \frac{d}{dt}(C^T C) \\ &\underset{(10.2.17)}{=} \frac{dU}{dt} = 0\end{aligned} \tag{22}$$

Hence,

$$\widetilde{\omega}^T = -\widetilde{\omega} \tag{23}$$

Equations (6)–(8) follow from Eqs. (2) and (5), that is, from

$$\begin{bmatrix} 0 & -\omega_3 & \omega_2 \\ \omega_3 & 0 & -\omega_1 \\ -\omega_2 & \omega_1 & 0 \end{bmatrix} = \begin{bmatrix} C_{11} & C_{21} & C_{31} \\ C_{12} & C_{22} & C_{32} \\ C_{13} & C_{23} & C_{33} \end{bmatrix} \begin{bmatrix} \dot{C}_{11} & \dot{C}_{12} & \dot{C}_{13} \\ \dot{C}_{21} & \dot{C}_{22} & \dot{C}_{23} \\ \dot{C}_{31} & \dot{C}_{32} & \dot{C}_{33} \end{bmatrix} \tag{24}$$

Example The quantities ω_1, ω_2, and ω_3 can be expressed in a simple and revealing form when a body B performs a motion of simple rotation (see Sec. 10.1) in a reference frame A. For, letting θ and λ_i $(i = 1,2,3)$ have the same meanings as in Secs. 10.1 and 10.2, and substituting from Eqs. (10.2.23)–(10.2.31) into Eq. (6), one obtains

$$\begin{aligned} \omega_1 &= [\lambda_2 \sin\theta + \lambda_3\lambda_1(1 - \cos\theta)](-\lambda_3 \cos\theta + \lambda_1\lambda_2 \sin\theta)\dot{\theta} \\ &\quad - [-\lambda_1 \sin\theta + \lambda_2\lambda_3(1 - \cos\theta)](\lambda_3{}^2 + \lambda_1{}^2) \sin\theta\, \dot{\theta} \\ &\quad + [1 - (\lambda_1{}^2 + \lambda_2{}^2)(1 - \cos\theta)](\lambda_1 \cos\theta + \lambda_2\lambda_3 \sin\theta)\dot{\theta} \\ &= [\lambda_1(\lambda_1{}^2 + \lambda_2{}^2 + \lambda_3{}^2) \\ &\quad + (1 - \lambda_1{}^2 - \lambda_2{}^2 - \lambda_3{}^2)(\lambda_1 \cos\theta + \lambda_2\lambda_3 \sin\theta - \lambda_2\lambda_3 \sin\theta \cos\theta)]\dot{\theta} \end{aligned} \tag{25}$$

which, since

$$\lambda_1{}^2 + \lambda_2{}^2 + \lambda_3{}^2 = 1 \tag{26}$$

reduces to

$$\omega_1 = \lambda_1 \dot{\theta} \tag{27}$$

Similarly,

$$\omega_2 = \lambda_2 \dot{\theta} \tag{28}$$

and

$$\omega_3 = \lambda_3 \dot{\theta} \tag{29}$$

10.7 ANGULAR VELOCITY AND ORIENTATION ANGLES

When the orientation of a rigid body B in a reference frame A is described by specifying the time dependence of orientation angles θ_1, θ_2, and θ_3 (see Sec. 10.3), the measure numbers ω_1, ω_2, and ω_3 [see Eqs. (10.6.4)] of the angular velocity of B in A (see Sec. 2.1) can be found by using the relationship

$$\lfloor \omega_1 \quad \omega_2 \quad \omega_3 \rfloor = \lfloor \dot{\theta}_1 \quad \dot{\theta}_2 \quad \dot{\theta}_3 \rfloor M \tag{1}$$

where M is a 3×3 matrix whose elements are functions of θ_1, θ_2, and θ_3. Conversely, if ω_1, ω_2, and ω_3 are known as functions of time, then θ_1, θ_2, and θ_3 can be evaluated by solving the differential equations

$$\lfloor \dot{\theta}_1 \quad \dot{\theta}_2 \quad \dot{\theta}_3 \rfloor = \lfloor \omega_1 \quad \omega_2 \quad \omega_3 \rfloor M^{-1} \tag{2}$$

For space-three, 1-2-3 angles, the matrices M and M^{-1} are

$$M = \begin{bmatrix} 1 & 0 & 0 \\ 0 & c_1 & -s_1 \\ -s_2 & s_1c_2 & c_1c_2 \end{bmatrix} \tag{3}$$

and

$$M^{-1} = \frac{1}{c_2}\begin{bmatrix} c_2 & 0 & 0 \\ s_1s_2 & c_1c_2 & s_1 \\ c_1s_2 & -s_1c_2 & c_1 \end{bmatrix} \tag{4}$$

For body-three, 1-2-3 angles,

$$M = \begin{bmatrix} c_2c_3 & -c_2s_3 & s_2 \\ s_3 & c_3 & 0 \\ 0 & 0 & 1 \end{bmatrix} \tag{5}$$

and

$$M^{-1} = \frac{1}{c_2}\begin{bmatrix} c_3 & c_2s_3 & -s_2c_3 \\ -s_3 & c_2c_3 & s_2s_3 \\ 0 & 0 & c_2 \end{bmatrix} \tag{6}$$

For space-two, 1-2-1 angles,

$$M = \begin{bmatrix} 1 & 0 & 0 \\ 0 & c_1 & -s_1 \\ c_2 & s_1s_2 & c_1s_2 \end{bmatrix} \tag{7}$$

and

$$M^{-1} = \frac{1}{s_2}\begin{bmatrix} s_2 & 0 & 0 \\ -s_1c_2 & c_1s_2 & s_1 \\ -c_1c_2 & -s_1s_2 & c_1 \end{bmatrix} \tag{8}$$

Finally, for body-two, 1-2-1 angles,

$$M = \begin{bmatrix} c_2 & s_2s_3 & s_2c_3 \\ 0 & c_3 & -s_3 \\ 1 & 0 & 0 \end{bmatrix} \tag{9}$$

and

$$M^{-1} = \frac{1}{s_2}\begin{bmatrix} 0 & 0 & s_2 \\ s_3 & s_2c_3 & -c_2s_3 \\ c_3 & -s_2s_3 & -c_2c_3 \end{bmatrix} \tag{10}$$

When c_2 vanishes, M as given by Eq. (3) or by Eq. (5) is a singular matrix, and M^{-1} is thus undefined. Hence, given ω_1, ω_2, and ω_3 one cannot use Eq. (2) to determine $\dot{\theta}_1$, $\dot{\theta}_2$, and $\dot{\theta}_3$ if θ_1, θ_2, and θ_3 are space-three or body-three angles and $c_2 = 0$. Similarly, if θ_1, θ_2, and θ_3 are space-two or body-two angles, Eq. (2) involves an undefined matrix when s_2 is equal to zero. These observations illustrate conditions under which use of a particular orientation angle sequence is undesirable, as is pointed out in the example in Sec. 10.3 with regards to the body-two, 1-2-1 sequence.

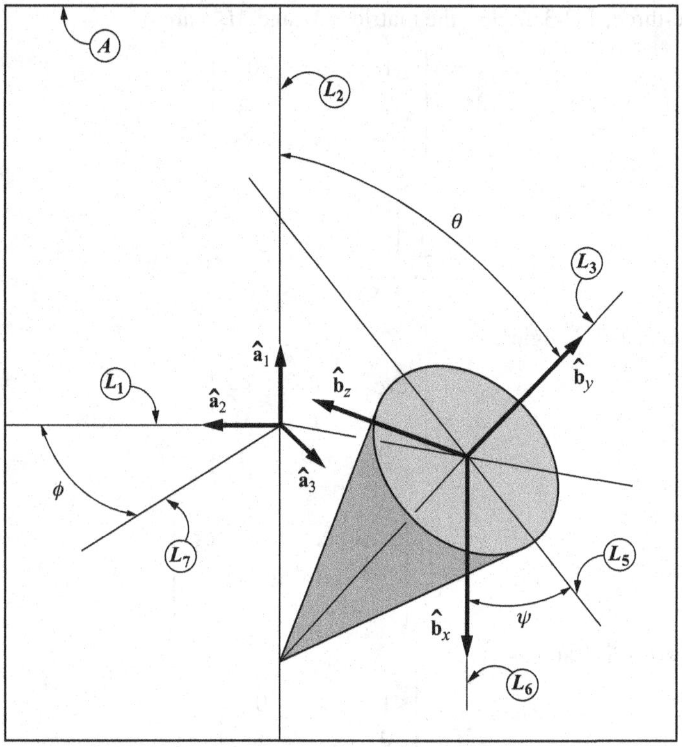

Figure 10.7.1

When the angles and unit vectors employed in an analysis are denoted by symbols other than those used in connection with Eqs. (1)–(10), appropriate replacements for these equations can be obtained directly from Eqs. (1)–(10) whenever the angles have been identified as regards type, that is, as being space-three angles, body-three angles, etc. Suppose, for example, that in the course of an analysis involving the cone shown in Fig. 2.4.1, and previously considered in the example in Sec. 2.4, unit vectors $\hat{\mathbf{b}}_x$, $\hat{\mathbf{b}}_y$, and $\hat{\mathbf{b}}_z$, fixed in B as shown in Fig. 10.7.1, have been introduced, and it is now desired to find ω_x, ω_y, and ω_z, defined as

$$\omega_x \triangleq \boldsymbol{\omega} \cdot \hat{\mathbf{b}}_x \qquad \omega_y \triangleq \boldsymbol{\omega} \cdot \hat{\mathbf{b}}_y \qquad \omega_z \triangleq \boldsymbol{\omega} \cdot \hat{\mathbf{b}}_z \tag{11}$$

where $\boldsymbol{\omega}$ denotes the angular velocity of B in A. This can be done easily by regarding ϕ, θ, and ψ as body-two, 1-2-1 angles, that is, by introducing unit vectors $\hat{\mathbf{a}}_1$, $\hat{\mathbf{a}}_2$, $\hat{\mathbf{a}}_3$ as shown in Fig. 10.7.1, defining $\hat{\mathbf{b}}_1$, $\hat{\mathbf{b}}_2$, and $\hat{\mathbf{b}}_3$ as

$$\hat{\mathbf{b}}_1 \triangleq \hat{\mathbf{b}}_y \qquad \hat{\mathbf{b}}_2 \triangleq \hat{\mathbf{b}}_z \qquad \hat{\mathbf{b}}_3 \triangleq \hat{\mathbf{b}}_x \tag{12}$$

and taking

$$\theta_1 = \phi \qquad \theta_2 = -\theta \qquad \theta_3 = -\psi \tag{13}$$

For it then follows immediately from Eqs. (1) and (9) that

$$\begin{bmatrix}\omega_y & \omega_z & \omega_x\end{bmatrix} = \begin{bmatrix}\dot{\phi} & -\dot{\theta} & -\dot{\psi}\end{bmatrix}\begin{bmatrix}\cos\theta & \sin\theta\sin\psi & -\sin\theta\cos\psi \\ 0 & \cos\psi & \sin\psi \\ 1 & 0 & 0\end{bmatrix} \tag{14}$$

so that

$$\omega_x = -\dot{\phi}\sin\theta\cos\psi - \dot{\theta}\sin\psi \tag{15}$$

$$\omega_y = \dot{\phi}\cos\theta - \dot{\psi} \tag{16}$$

$$\omega_z = \dot{\phi}\sin\theta\sin\psi - \dot{\theta}\cos\psi \tag{17}$$

In the dynamics literature, one encounters a wide variety of differential equations relating angular velocity measure numbers to orientation angles and their time derivatives. Twenty-four such sets of kinematical differential equations are tabulated in Appendix II.

Derivations From Eqs. (10.6.6) and (10.3.1),

$$\begin{aligned}\omega_1 &= (c_1 s_2 c_3 + s_3 s_1)\frac{d}{dt}(s_1 s_2 c_3 - s_3 c_1) \\ &\quad + (c_1 s_2 s_3 - c_3 s_1)\frac{d}{dt}(s_1 s_2 s_3 + c_3 c_1) + c_1 c_2 \frac{d}{dt}(s_1 c_2) \\ &= \dot{\theta}_1 - \dot{\theta}_3 s_2\end{aligned} \tag{18}$$

Similarly, from Eqs. (10.6.7) and (10.3.1),

$$\omega_2 = \dot{\theta}_2 c_1 + \dot{\theta}_3 s_1 c_2 \tag{19}$$

and from Eqs. (10.6.8) and (10.3.1)

$$\omega_3 = -\dot{\theta}_2 s_1 + \dot{\theta}_3 c_1 c_2 \tag{20}$$

These three equations are the three scalar equations corresponding to Eq. (1) when M is given by Eq. (3).

Equation (2) follows from Eq. (1) and from the definition of the inverse of a matrix; and the validity of Eq. (4) may be established by noting that the product of the right-hand members of Eqs. (3) and (4) is equal to U, the unit matrix. Proceeding similarly, but using Eq. (10.3.11), (10.3.21), or (10.3.31) in place of Eq. (10.3.1), and Eqs. (5) and (6), Eqs. (7) and (8), or Eqs. (9) and (10) in place of Eqs. (3) and (4), one can demonstrate the validity of Eqs. (5)–(10).

Example Figure 10.7.2 shows the gyroscopic system previously discussed in Sec. 10.3, where it was mentioned that one may wish to employ space-three, 1-2-3 angles ϕ_1, ϕ_2, and ϕ_3, as well as the body-two, 1-2-1 angles θ_1, θ_2, and θ_3 shown in Fig. 10.7.2, when analyzing motions during which θ_2 becomes small or equal to zero. Given θ_i and $\dot{\theta}_i$ $(i = 1,2,3)$, one must then be able to evaluate ϕ_i and $\dot{\phi}_i$ $(i = 1,2,3)$.

Suppose that, as in the example in Sec. 10.3, $\theta_1 = 30°$, $\theta_2 = 45°$, and $\theta_3 = 60°$ at a certain instant and that, furthermore, $\dot{\theta}_1 = 1.00$, $\dot{\theta}_2 = 2.00$, $\dot{\theta}_3 = 3.00$ rad/s. What are the values of $\dot{\phi}_1$, $\dot{\phi}_2$, and $\dot{\phi}_3$ at this instant?

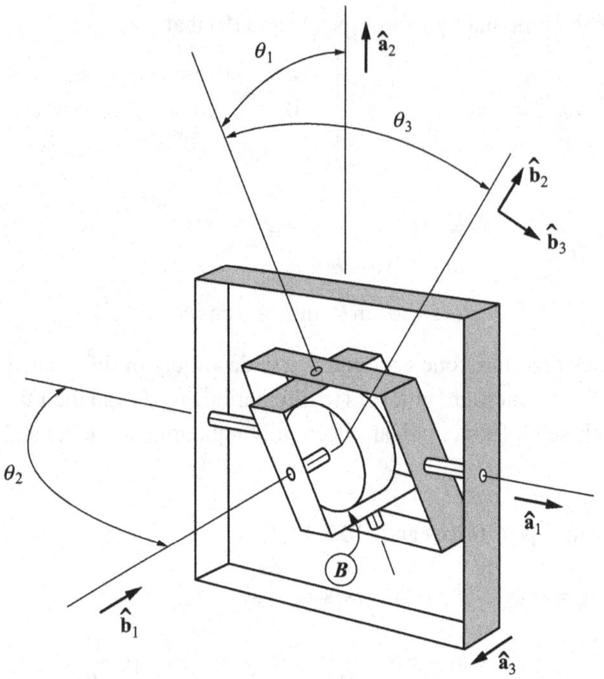

Figure 10.7.2

From Eqs. (2) and (4),

$$\lfloor \dot{\phi}_1 \quad \dot{\phi}_2 \quad \dot{\phi}_3 \rfloor = \frac{\lfloor \omega_1 \quad \omega_2 \quad \omega_3 \rfloor}{\cos\phi_2} \begin{bmatrix} \cos\phi_2 & 0 & 0 \\ \sin\phi_1 \sin\phi_2 & \cos\phi_1 \cos\phi_2 & \sin\phi_1 \\ \cos\phi_1 \sin\phi_2 & -\sin\phi_1 \cos\phi_2 & \cos\phi_1 \end{bmatrix} \tag{21}$$

or, using the values of ϕ_1 and ϕ_2 found previously,

$$\lfloor \dot{\phi}_1 \quad \dot{\phi}_2 \quad \dot{\phi}_3 \rfloor = \frac{\lfloor \omega_1 \quad \omega_2 \quad \omega_3 \rfloor}{0.791} \begin{bmatrix} 0.791 & 0 & 0 \\ 0.603 & -0.138 & 0.985 \\ -0.107 & -0.780 & -0.174 \end{bmatrix} \tag{22}$$

Now, from Eqs. (1) and (9),

$$\begin{aligned} \lfloor \omega_1 \quad \omega_2 \quad \omega_3 \rfloor &= \lfloor \dot{\theta}_1 \quad \dot{\theta}_2 \quad \dot{\theta}_3 \rfloor \begin{bmatrix} \cos\theta_2 & \sin\theta_2 \sin\theta_3 & \sin\theta_2 \cos\theta_3 \\ 0 & \cos\theta_3 & -\sin\theta_3 \\ 1 & 0 & 0 \end{bmatrix} \\ &= \lfloor 1.00 \quad 2.00 \quad 3.00 \rfloor \begin{bmatrix} 0.707 & 0.612 & 0.354 \\ 0 & 0.500 & -0.866 \\ 1 & 0 & 0 \end{bmatrix} \\ &= \lfloor 3.707 \quad 1.612 \quad -1.378 \rfloor \end{aligned} \tag{23}$$

Hence

$$\lfloor \dot{\phi}_1 \quad \dot{\phi}_2 \quad \dot{\phi}_3 \rfloor = \frac{\lfloor 3.707 \quad 1.612 \quad -1.378 \rfloor}{0.791} \begin{bmatrix} 0.791 & 0 & 0 \\ 0.603 & -0.138 & 0.985 \\ -0.107 & -0.780 & -0.174 \end{bmatrix}$$

$$= \lfloor 5.12 \quad 1.08 \quad 2.31 \rfloor \tag{24}$$

and

$$\dot{\phi}_1 = 5.12 \qquad \dot{\phi}_2 = 1.08 \qquad \dot{\phi}_3 = 2.31 \text{ rad/s} \tag{25}$$

The angular velocity ${}^A\boldsymbol{\omega}^B$ of B in A can be expressed in two ways, namely,

$$\begin{aligned} {}^A\boldsymbol{\omega}^B &= (\dot{\theta}_1 \cos\theta_2 + \dot{\theta}_3)\hat{\mathbf{b}}_1 + (\dot{\theta}_1 \sin\theta_2 \sin\theta_3 + \dot{\theta}_2 \cos\theta_3)\hat{\mathbf{b}}_2 \\ &\quad + (\dot{\theta}_1 \sin\theta_2 \cos\theta_3 - \dot{\theta}_2 \sin\theta_3)\hat{\mathbf{b}}_3 \\ &= (\dot{\phi}_1 - \dot{\phi}_3 \sin\phi_2)\hat{\mathbf{b}}_1 + (\dot{\phi}_2 \cos\phi_1 + \dot{\phi}_3 \sin\phi_1 \cos\phi_2)\hat{\mathbf{b}}_2 \\ &\quad + (-\dot{\phi}_2 \sin\phi_1 + \dot{\phi}_3 \cos\phi_1 \cos\phi_2)\hat{\mathbf{b}}_3 \end{aligned} \tag{26}$$

10.8 ANGULAR VELOCITY AND EULER PARAMETERS

If $\hat{\mathbf{a}}_1, \hat{\mathbf{a}}_2, \hat{\mathbf{a}}_3$, and $\hat{\mathbf{b}}_1, \hat{\mathbf{b}}_2, \hat{\mathbf{b}}_3$ are two dextral sets of orthogonal unit vectors fixed respectively in reference frames or rigid bodies A and B that are moving relative to each other, one can use Eqs. (10.4.19)–(10.4.22) to associate with each instant of time Euler parameters $\epsilon_1, \ldots, \epsilon_4$; and an Euler vector $\boldsymbol{\epsilon}$ then can be formed by reference to Eq. (10.4.2). In terms of $\boldsymbol{\epsilon}$ and ϵ_4, the angular velocity of B in A (see Sec. 2.1) can be expressed as

$${}^A\boldsymbol{\omega}^B = 2\left(\epsilon_4 \frac{{}^B d\boldsymbol{\epsilon}}{dt} - \dot{\epsilon}_4 \boldsymbol{\epsilon} - \boldsymbol{\epsilon} \times \frac{{}^B d\boldsymbol{\epsilon}}{dt}\right) \tag{1}$$

Conversely, if ${}^A\boldsymbol{\omega}^B$ is known as a function of time, the Euler parameters can be found by solving the differential equations

$$\frac{{}^B d\boldsymbol{\epsilon}}{dt} = \tfrac{1}{2}(\epsilon_4 {}^A\boldsymbol{\omega}^B + \boldsymbol{\epsilon} \times {}^A\boldsymbol{\omega}^B) \tag{2}$$

and

$$\dot{\epsilon}_4 = -\tfrac{1}{2} {}^A\boldsymbol{\omega}^B \cdot \boldsymbol{\epsilon} \tag{3}$$

Equations equivalent to Eqs. (1)–(3) can be formulated in terms of matrices ω, ϵ, and E defined as

$$\omega \triangleq \lfloor \omega_1 \quad \omega_2 \quad \omega_3 \quad 0 \rfloor \tag{4}$$

$$\epsilon \triangleq \lfloor \epsilon_1 \quad \epsilon_2 \quad \epsilon_3 \quad \epsilon_4 \rfloor \tag{5}$$

and

$$E = \begin{bmatrix} \epsilon_4 & -\epsilon_3 & \epsilon_2 & \epsilon_1 \\ \epsilon_3 & \epsilon_4 & -\epsilon_1 & \epsilon_2 \\ -\epsilon_2 & \epsilon_1 & \epsilon_4 & \epsilon_3 \\ -\epsilon_1 & -\epsilon_2 & -\epsilon_3 & \epsilon_4 \end{bmatrix} \tag{6}$$

These equations are

$$\omega = 2\dot{\epsilon}E \tag{7}$$

and

$$\dot{\epsilon} = \tfrac{1}{2}\omega E^T \tag{8}$$

Derivations Substitution from Eqs. (10.4.6)–(10.4.14) into Eqs. (10.6.6)–(10.6.8) gives

$$\begin{aligned}
\omega_1 &= 4(\epsilon_3\epsilon_1 + \epsilon_2\epsilon_4)(\dot{\epsilon}_1\epsilon_2 + \epsilon_1\dot{\epsilon}_2 - \dot{\epsilon}_3\epsilon_4 - \epsilon_3\dot{\epsilon}_4) \\
&\quad + 4(\epsilon_2\epsilon_3 - \epsilon_1\epsilon_4)(\epsilon_2\dot{\epsilon}_2 - \epsilon_3\dot{\epsilon}_3 - \epsilon_1\dot{\epsilon}_1 + \epsilon_4\dot{\epsilon}_4) \\
&\quad + 2(1 - 2\epsilon_1{}^2 - 2\epsilon_2{}^2)(\dot{\epsilon}_2\epsilon_3 + \epsilon_2\dot{\epsilon}_3 + \dot{\epsilon}_1\epsilon_4 + \epsilon_1\dot{\epsilon}_4) \\
&= 2(\dot{\epsilon}_1\epsilon_4 + \dot{\epsilon}_2\epsilon_3 - \dot{\epsilon}_3\epsilon_2 - \dot{\epsilon}_4\epsilon_1) && (9) \\
\omega_2 &= 2(\dot{\epsilon}_2\epsilon_4 + \dot{\epsilon}_3\epsilon_1 - \dot{\epsilon}_1\epsilon_3 - \dot{\epsilon}_4\epsilon_2) && (10) \\
\omega_3 &= 2(\dot{\epsilon}_3\epsilon_4 + \dot{\epsilon}_1\epsilon_2 - \dot{\epsilon}_2\epsilon_1 - \dot{\epsilon}_4\epsilon_3) && (11)
\end{aligned}$$

and these are three of the four scalar equations corresponding to Eq. (7). The fourth is

$$0 \underset{(7)}{=} 2(\dot{\epsilon}_1\epsilon_1 + \dot{\epsilon}_2\epsilon_2 + \dot{\epsilon}_3\epsilon_3 + \dot{\epsilon}_4\epsilon_4) \tag{12}$$

and this equation is satisfied because

$$\dot{\epsilon}_1\epsilon_1 + \dot{\epsilon}_2\epsilon_2 + \dot{\epsilon}_3\epsilon_3 + \dot{\epsilon}_4\epsilon_4 = \frac{1}{2}\frac{d}{dt}(\epsilon_1{}^2 + \epsilon_2{}^2 + \epsilon_3{}^2 + \epsilon_4{}^2) \underset{(10.4.4)}{=} 0 \tag{13}$$

Thus the validity of Eq. (7) is established; and Eq. (1) can be obtained by noting that

$$\begin{aligned}
{}^A\boldsymbol{\omega}^B &\underset{(2.3.6,\ 2.3.10)}{=} \omega_1\hat{\mathbf{b}}_1 + \omega_2\hat{\mathbf{b}}_2 + \omega_3\hat{\mathbf{b}}_3 \\
&\underset{(7)}{=} 2[(\dot{\epsilon}_1\epsilon_4 + \dot{\epsilon}_2\epsilon_3 - \dot{\epsilon}_3\epsilon_2 - \dot{\epsilon}_4\epsilon_1)\hat{\mathbf{b}}_1 \\
&\qquad + (\dot{\epsilon}_2\epsilon_4 + \dot{\epsilon}_3\epsilon_1 - \dot{\epsilon}_1\epsilon_3 - \dot{\epsilon}_4\epsilon_2)\hat{\mathbf{b}}_2 \\
&\qquad + (\dot{\epsilon}_3\epsilon_4 + \dot{\epsilon}_1\epsilon_2 - \dot{\epsilon}_2\epsilon_1 - \dot{\epsilon}_4\epsilon_3)\hat{\mathbf{b}}_3] \\
&= 2[\epsilon_4(\dot{\epsilon}_1\hat{\mathbf{b}}_1 + \dot{\epsilon}_2\hat{\mathbf{b}}_2 + \dot{\epsilon}_3\hat{\mathbf{b}}_3) - \dot{\epsilon}_4(\epsilon_1\hat{\mathbf{b}}_1 + \epsilon_2\hat{\mathbf{b}}_2 + \epsilon_3\hat{\mathbf{b}}_3) \\
&\qquad + (\dot{\epsilon}_2\epsilon_3 - \dot{\epsilon}_3\epsilon_2)\hat{\mathbf{b}}_1 + (\dot{\epsilon}_3\epsilon_1 - \dot{\epsilon}_1\epsilon_3)\hat{\mathbf{b}}_2 + (\dot{\epsilon}_1\epsilon_2 - \dot{\epsilon}_2\epsilon_1)\hat{\mathbf{b}}_3] \\
&\underset{(10.4.2)}{=} 2\left(\epsilon_4\frac{{}^B d\boldsymbol{\epsilon}}{dt} - \dot{\epsilon}_4\boldsymbol{\epsilon} - \boldsymbol{\epsilon}\times\frac{{}^B d\boldsymbol{\epsilon}}{dt}\right) && (14)
\end{aligned}$$

Postmultiplication of both sides of Eq. (7) with E^T gives

$$\omega E^T = 2\dot{\epsilon}EE^T \tag{15}$$

Now, using Eq. (10.4.4) and referring to Eq. (6), one finds that

$$EE^T = \begin{bmatrix} 1 & 0 & 0 & 0 \\ 0 & 1 & 0 & 0 \\ 0 & 0 & 1 & 0 \\ 0 & 0 & 0 & 1 \end{bmatrix} \tag{16}$$

Consequently,

$$\omega E^T = 2\dot{\epsilon} \tag{17}$$

in agreement with Eq. (8).

Finally,

$$\dot{\epsilon}_4 \underset{(8)}{=} -\tfrac{1}{2}(\omega_1\epsilon_1 + \omega_2\epsilon_2 + \omega_3\epsilon_3) \underset{(2.3.10,\ 10.4.2)}{=} -\tfrac{1}{2}{}^A\boldsymbol{\omega}^B \cdot \boldsymbol{\epsilon} \tag{18}$$

as in Eq. (3); and

$$\begin{aligned} \frac{{}^B d\boldsymbol{\epsilon}}{dt} &\underset{(10.4.2)}{=} \dot{\epsilon}_1\hat{\mathbf{b}}_1 + \dot{\epsilon}_2\hat{\mathbf{b}}_2 + \dot{\epsilon}_3\hat{\mathbf{b}}_3 \\ &\underset{(8)}{=} \tfrac{1}{2}[(\omega_1\epsilon_4 - \omega_2\epsilon_3 + \omega_3\epsilon_2)\hat{\mathbf{b}}_1 + (\omega_1\epsilon_3 + \omega_2\epsilon_4 - \omega_3\epsilon_1)\hat{\mathbf{b}}_2 \\ &\qquad - (\omega_1\epsilon_2 - \omega_2\epsilon_1 - \omega_3\epsilon_4)\hat{\mathbf{b}}_3] \end{aligned} \tag{19}$$

The right-hand member of this equation is equal to that of Eq. (2).

Example Suppose that B is an axisymmetric rigid body whose axis of symmetry is parallel to $\hat{\mathbf{b}}_3$. Then, if $\underline{\mathbf{I}}$ denotes the inertia dyadic of B for $B^\star$, the mass center of B, and if I and J are defined as

$$I \triangleq \hat{\mathbf{b}}_1 \cdot \underline{\mathbf{I}} \cdot \hat{\mathbf{b}}_1 = \hat{\mathbf{b}}_2 \cdot \underline{\mathbf{I}} \cdot \hat{\mathbf{b}}_2 \tag{20}$$

and

$$J \triangleq \hat{\mathbf{b}}_3 \cdot \underline{\mathbf{I}} \cdot \hat{\mathbf{b}}_3 \tag{21}$$

the angular momentum $\mathbf{H}$ of B in A with respect to $B^\star$ is given by

$$\mathbf{H} = I\omega_1\hat{\mathbf{b}}_1 + I\omega_2\hat{\mathbf{b}}_2 + J\omega_3\hat{\mathbf{b}}_3 \tag{22}$$

and the first time-derivative of $\mathbf{H}$ in A can be expressed as

$$\begin{aligned} \frac{{}^A d\mathbf{H}}{dt} &\underset{(2.3.1)}{=} \frac{{}^B d\mathbf{H}}{dt} + {}^A\boldsymbol{\omega}^B \times \mathbf{H} \\ &\underset{(22)}{=} [I\dot{\omega}_1 + (J - I)\omega_2\omega_3]\hat{\mathbf{b}}_1 + [I\dot{\omega}_2 - (J - I)\omega_3\omega_1]\hat{\mathbf{b}}_2 + J\dot{\omega}_3\hat{\mathbf{b}}_3 \end{aligned} \tag{23}$$

Hence, if B moves under the action of forces the sum of whose moments about $B^\star$ is equal to zero, and if A is an inertial reference frame, so that, in accordance with the angular momentum principle, ${}^A d\mathbf{H}/dt$ is equal to zero, then ω_1, ω_2, and ω_3 are governed by the differential equations

$$\dot{\omega}_1 - \frac{I - J}{I}\omega_2\omega_3 = 0 \tag{24}$$

$$\dot{\omega}_2 + \frac{I - J}{I}\omega_3\omega_1 = 0 \tag{25}$$

$$\dot{\omega}_3 = 0 \tag{26}$$

Letting $\overline{\omega}_i$ denote the value of ω_i $(i = 1,2,3)$ at $t = 0$, and defining a constant s as

$$s \triangleq \frac{I - J}{I}\overline{\omega}_3 \tag{27}$$

one can express the general solution of Eqs. (24)–(26) as

$$\omega_1 = \overline{\omega}_1 \cos st + \overline{\omega}_2 \sin st \tag{28}$$

$$\omega_2 = -\overline{\omega}_1 \sin st + \overline{\omega}_2 \cos st \tag{29}$$

$$\omega_3 = \overline{\omega}_3 \tag{30}$$

and, to determine the orientation of B in A, one then can seek the solution of the differential equations

$$\begin{aligned}\dot{\epsilon}_1 &\underset{(8)}{=} \tfrac{1}{2}(\omega_1\epsilon_4 - \omega_2\epsilon_3 + \omega_3\epsilon_2)\\ &\underset{(28\text{–}30)}{=} \tfrac{1}{2}[(\overline{\omega}_1 \cos st + \overline{\omega}_2 \sin st)\epsilon_4 + (\overline{\omega}_1 \sin st - \overline{\omega}_2 \cos st)\epsilon_3 + \overline{\omega}_3\epsilon_2]\end{aligned} \tag{31}$$

$$\begin{aligned}\dot{\epsilon}_2 &\underset{(8)}{=} \tfrac{1}{2}(\omega_1\epsilon_3 + \omega_2\epsilon_4 - \omega_3\epsilon_1)\\ &\underset{(28\text{–}30)}{=} \tfrac{1}{2}[(\overline{\omega}_1 \cos st + \overline{\omega}_2 \sin st)\epsilon_3 - (\overline{\omega}_1 \sin st - \overline{\omega}_2 \cos st)\epsilon_4 - \overline{\omega}_3\epsilon_1]\end{aligned} \tag{32}$$

$$\begin{aligned}\dot{\epsilon}_3 &\underset{(8)}{=} \tfrac{1}{2}(-\omega_1\epsilon_2 + \omega_2\epsilon_1 + \omega_3\epsilon_4)\\ &\underset{(28\text{–}30)}{=} \tfrac{1}{2}[-(\overline{\omega}_1 \cos st + \overline{\omega}_2 \sin st)\epsilon_2 + (-\overline{\omega}_1 \sin st + \overline{\omega}_2 \cos st)\epsilon_1 + \overline{\omega}_3\epsilon_4]\end{aligned} \tag{33}$$

$$\begin{aligned}\dot{\epsilon}_4 &\underset{(8)}{=} -\tfrac{1}{2}(\omega_1\epsilon_1 + \omega_2\epsilon_2 + \omega_3\epsilon_3)\\ &\underset{(28\text{–}30)}{=} -\tfrac{1}{2}[(\overline{\omega}_1 \cos st + \overline{\omega}_2 \sin st)\epsilon_1 + (-\overline{\omega}_1 \sin st + \overline{\omega}_2 \cos st)\epsilon_2 + \overline{\omega}_3\epsilon_3]\end{aligned} \tag{34}$$

using as initial conditions

$$\epsilon_1 = \epsilon_2 = \epsilon_3 = 0 \qquad \epsilon_4 = 1 \qquad \text{at } t = 0 \tag{35}$$

which means that the unit vectors $\hat{\mathbf{a}}_1$, $\hat{\mathbf{a}}_2$, and $\hat{\mathbf{a}}_3$ have been chosen such that $\hat{\mathbf{a}}_i = \hat{\mathbf{b}}_i$ $(i = 1,2,3)$ at $t = 0$.

Since Eqs. (31)–(34) have time-dependent coefficients, they cannot be solved for $\epsilon_1, \ldots, \epsilon_4$ by simple analytical procedures. However, attacking the physical problem at hand by a different method,† and defining a quantity p as

$$p \triangleq \left[\overline{\omega}_1{}^2 + \overline{\omega}_2{}^2 + \left(\overline{\omega}_3\frac{J}{I}\right)^2\right]^{1/2} \tag{36}$$

† T. R. Kane, P. W. Likins, and D. A. Levinson, *Spacecraft Dynamics* (New York: McGraw-Hill, 1983), Sec. 3.1.

one can show that ϵ_1, ϵ_2, ϵ_3, and ϵ_4 are given by

$$\epsilon_1 = \frac{\sin(pt/2)}{p}\left(\overline{\omega}_1 \cos\frac{st}{2} + \overline{\omega}_2 \sin\frac{st}{2}\right) \tag{37}$$

$$\epsilon_2 = \frac{\sin(pt/2)}{p}\left(-\overline{\omega}_1 \sin\frac{st}{2} + \overline{\omega}_2 \cos\frac{st}{2}\right) \tag{38}$$

$$\epsilon_3 = \overline{\omega}_3 \frac{J}{Ip} \sin\frac{pt}{2} \cos\frac{st}{2} + \cos\frac{pt}{2} \sin\frac{st}{2} \tag{39}$$

$$\epsilon_4 = -\overline{\omega}_3 \frac{J}{Ip} \sin\frac{pt}{2} \sin\frac{st}{2} + \cos\frac{pt}{2} \cos\frac{st}{2} \tag{40}$$

and it may be verified that these expressions do, indeed, satisfy Eqs. (31)–(35).

10.9 ANGULAR VELOCITY AND WIENER-MILENKOVIĆ PARAMETERS

If $\hat{\mathbf{a}}_1$, $\hat{\mathbf{a}}_2$, $\hat{\mathbf{a}}_3$ and $\hat{\mathbf{b}}_1$, $\hat{\mathbf{b}}_2$, $\hat{\mathbf{b}}_3$ are two dextral sets of orthogonal unit vectors fixed respectively in reference frames or rigid bodies A and B that are moving relative to each other, one can use Eqs. (10.5.9)–(10.5.11) to associate with each instant of time Wiener-Milenković parameters μ_1, μ_2, μ_3, and a vector $\boldsymbol{\mu}$ then can be formed by reference to Eq. (10.5.2). The angular velocity of B in A (see Sec. 2.1), expressed in terms of $\boldsymbol{\mu}$, is given by

$$^A\boldsymbol{\omega}^B = \frac{1}{\left(1+\frac{\mu^2}{16}\right)^2}\left[\left(1-\frac{\mu^2}{16}\right)\frac{^B d\boldsymbol{\mu}}{dt} + \frac{1}{8}\boldsymbol{\mu}\boldsymbol{\mu}\cdot\frac{^B d\boldsymbol{\mu}}{dt} - \frac{1}{2}\boldsymbol{\mu}\times\frac{^B d\boldsymbol{\mu}}{dt}\right] \tag{1}$$

Conversely, if $^A\boldsymbol{\omega}^B$ is known as a function of time, the vector $\boldsymbol{\mu}$ can be found by solving the differential equation

$$\frac{^B d\boldsymbol{\mu}}{dt} = \left(1-\frac{\mu^2}{16}\right){}^A\boldsymbol{\omega}^B + \frac{1}{8}\boldsymbol{\mu}\boldsymbol{\mu}\cdot{}^A\boldsymbol{\omega}^B - \frac{1}{2}{}^A\boldsymbol{\omega}^B\times\boldsymbol{\mu} \tag{2}$$

Equations equivalent to Eqs. (1) and (2) can be formulated in terms of the matrices μ and $\tilde{\mu}$ introduced in Eqs. (10.5.4) and (10.5.6), respectively, and the matrix ω defined as

$$\omega \triangleq \lfloor \omega_1 \quad \omega_2 \quad \omega_3 \rfloor \tag{3}$$

where ω_1, ω_2, and ω_3 have the same meanings as in Eqs. (10.6.4). The equations of interest are

$$\begin{aligned} \omega &\underset{(1)}{=} \frac{\dot{\mu}}{\left(1+\frac{\mu\mu^T}{16}\right)^2}\left[\left(1-\frac{\mu\mu^T}{16}\right)U + \frac{1}{8}\mu^T\mu + \frac{1}{2}\tilde{\mu}\right] \\ &\underset{(10.5.8)}{=} \frac{\dot{\mu}}{1+\frac{\mu\mu^T}{16}}F \end{aligned} \tag{4}$$

and

$$\dot{\mu} \underset{(2)}{=} \omega\left[\left(1-\frac{\mu\mu^T}{16}\right)U+\frac{1}{8}\mu^T\mu-\frac{1}{2}\tilde{\mu}\right]$$
$$\underset{(10.5.8)}{=} \omega F^T\left(1+\frac{\mu\mu^T}{16}\right) \tag{5}$$

Recall that $C = FF$ [see Eq. (10.5.7)], and $F^T F = FF^T = U$. Like its counterparts for the direction cosine matrix and for Euler parameters [see Eqs. (10.6.3) and (10.8.8)], Eq. (5) is, in general, an equation with variable coefficients. Since it is, moreover, nonlinear, one must usually resort to numerical methods to obtain solutions.

Derivations We begin with Eq. (10.8.1), that is,

$$^A\boldsymbol{\omega}^B = 2\left(\epsilon_4\frac{^Bd\boldsymbol{\epsilon}}{dt}-\dot{\epsilon}_4\boldsymbol{\epsilon}-\boldsymbol{\epsilon}\times\frac{^Bd\boldsymbol{\epsilon}}{dt}\right) \tag{6}$$

Solving Eq. (10.5.14) for $\boldsymbol{\epsilon}$ and then using μ^2 in place of $\mu\mu^T$ in Eq. (10.5.18) allows one to write

$$\boldsymbol{\epsilon} \underset{(10.5.14)}{=} \frac{1}{2}\boldsymbol{\mu}\cos^2\frac{\theta}{4} \underset{(10.5.18)}{=} \frac{\frac{1}{2}\boldsymbol{\mu}}{1+\frac{\mu^2}{16}} \tag{7}$$

Differentiating with respect to time in B, one obtains

$$\frac{^Bd\boldsymbol{\epsilon}}{dt} = \frac{\frac{1}{2}}{1+\frac{\mu^2}{16}}\frac{^Bd\boldsymbol{\mu}}{dt} - \frac{\frac{1}{32}\boldsymbol{\mu}}{\left(1+\frac{\mu^2}{16}\right)^2}2\boldsymbol{\mu}\cdot\frac{^Bd\boldsymbol{\mu}}{dt} = \frac{\frac{1}{2}}{1+\frac{\mu^2}{16}}\frac{^Bd\boldsymbol{\mu}}{dt} - \frac{\frac{1}{16}}{\left(1+\frac{\mu^2}{16}\right)^2}\boldsymbol{\mu}\boldsymbol{\mu}\cdot\frac{^Bd\boldsymbol{\mu}}{dt} \tag{8}$$

Now, Eq. (10.5.16) can be solved for ϵ_4, and μ^2 may be used in place of $\mu\mu^T$ once again in Eq. (10.5.18), to obtain

$$\epsilon_4 \underset{(10.5.16)}{=} 2\cos^2\frac{\theta}{4}-1 \underset{(10.5.18)}{=} \frac{2}{1+\frac{\mu^2}{16}}-1 = \frac{1-\frac{\mu^2}{16}}{1+\frac{\mu^2}{16}} \tag{9}$$

and the time derivative is given by

$$\dot{\epsilon}_4 = -\frac{\frac{1}{4}}{\left(1+\frac{\mu^2}{16}\right)^2}\boldsymbol{\mu}\cdot\frac{^Bd\boldsymbol{\mu}}{dt} \tag{10}$$

The three terms in Eq. (6) now can be assembled as follows.

$$\epsilon_4\frac{^Bd\boldsymbol{\epsilon}}{dt} = \underset{(9)}{\left(\frac{1-\frac{\mu^2}{16}}{1+\frac{\mu^2}{16}}\right)}\underset{(8)}{\left[\frac{\frac{1}{2}}{1+\frac{\mu^2}{16}}\frac{^Bd\boldsymbol{\mu}}{dt} - \frac{\frac{1}{16}}{\left(1+\frac{\mu^2}{16}\right)^2}\boldsymbol{\mu}\boldsymbol{\mu}\cdot\frac{^Bd\boldsymbol{\mu}}{dt}\right]} \tag{11}$$

$$\dot{\epsilon}_4\boldsymbol{\epsilon} = \underset{(10)}{-\frac{\frac{1}{4}}{\left(1+\frac{\mu^2}{16}\right)^2}\boldsymbol{\mu}\cdot\frac{^Bd\boldsymbol{\mu}}{dt}}\;\underset{(7)}{\frac{\frac{1}{2}\boldsymbol{\mu}}{1+\frac{\mu^2}{16}}} = -\frac{\frac{1}{8}}{\left(1+\frac{\mu^2}{16}\right)^3}\boldsymbol{\mu}\boldsymbol{\mu}\cdot\frac{^Bd\boldsymbol{\mu}}{dt} \tag{12}$$

$$\boldsymbol{\epsilon} \times \frac{{}^B d\boldsymbol{\epsilon}}{dt} = \underbrace{\frac{\frac{1}{2}\boldsymbol{\mu}}{1+\frac{\mu^2}{16}}}_{(7)} \times \left[\frac{\frac{1}{2}}{1+\frac{\mu^2}{16}} \frac{{}^B d\boldsymbol{\mu}}{dt} - \underbrace{\frac{\frac{1}{16}}{\left(1+\frac{\mu^2}{16}\right)^2}}_{(8)} \boldsymbol{\mu}\boldsymbol{\mu} \cdot \frac{{}^B d\boldsymbol{\mu}}{dt} \right]$$

$$= \frac{\frac{1}{4}}{\left(1+\frac{\mu^2}{16}\right)^2} \boldsymbol{\mu} \times \frac{{}^B d\boldsymbol{\mu}}{dt} - \mathbf{0} \tag{13}$$

Substitution from Eqs. (11)–(13) into (6) produces

$$\frac{1}{2} {}^A\boldsymbol{\omega}^B = \left(\frac{1-\frac{\mu^2}{16}}{1+\frac{\mu^2}{16}}\right) \left[\frac{\frac{1}{2}}{1+\frac{\mu^2}{16}} \frac{{}^B d\boldsymbol{\mu}}{dt} - \frac{\frac{1}{16}}{\left(1+\frac{\mu^2}{16}\right)^2} \boldsymbol{\mu}\boldsymbol{\mu} \cdot \frac{{}^B d\boldsymbol{\mu}}{dt} \right]$$

$$+ \frac{\frac{1}{8}}{\left(1+\frac{\mu^2}{16}\right)^3} \boldsymbol{\mu}\boldsymbol{\mu} \cdot \frac{{}^B d\boldsymbol{\mu}}{dt} - \frac{\frac{1}{4}}{\left(1+\frac{\mu^2}{16}\right)^2} \boldsymbol{\mu} \times \frac{{}^B d\boldsymbol{\mu}}{dt}$$

$$= \left(\frac{1-\frac{\mu^2}{16}}{1+\frac{\mu^2}{16}}\right) \frac{\frac{1}{2}}{1+\frac{\mu^2}{16}} \frac{{}^B d\boldsymbol{\mu}}{dt} + \frac{\frac{1}{16}}{\left(1+\frac{\mu^2}{16}\right)^2} \boldsymbol{\mu}\boldsymbol{\mu} \cdot \frac{{}^B d\boldsymbol{\mu}}{dt}$$

$$- \frac{\frac{1}{4}}{\left(1+\frac{\mu^2}{16}\right)^2} \boldsymbol{\mu} \times \frac{{}^B d\boldsymbol{\mu}}{dt}$$

$$= \frac{\frac{1}{2}}{\left(1+\frac{\mu^2}{16}\right)^2} \left[\left(1-\frac{\mu^2}{16}\right) \frac{{}^B d\boldsymbol{\mu}}{dt} + \frac{1}{8}\boldsymbol{\mu}\boldsymbol{\mu} \cdot \frac{{}^B d\boldsymbol{\mu}}{dt} - \frac{1}{2}\boldsymbol{\mu} \times \frac{{}^B d\boldsymbol{\mu}}{dt} \right] \tag{14}$$

which is equivalent to Eq. (1).

Our objective now is to eliminate the dot product and the cross product, each of which involve ${}^B d\boldsymbol{\mu}/dt$. To this end, we first solve Eq. (14) for ${}^B d\boldsymbol{\mu}/dt \times \boldsymbol{\mu}$,

$$\frac{{}^B d\boldsymbol{\mu}}{dt} \times \boldsymbol{\mu} = 2\left(1+\frac{\mu^2}{16}\right)^2 {}^A\boldsymbol{\omega}^B - 2\left(1-\frac{\mu^2}{16}\right) \frac{{}^B d\boldsymbol{\mu}}{dt} - \frac{1}{4}\boldsymbol{\mu}\boldsymbol{\mu} \cdot \frac{{}^B d\boldsymbol{\mu}}{dt} \tag{15}$$

Next, both sides of Eq. (14) are dot-multiplied by $\boldsymbol{\mu}$,

$$\frac{1}{2} {}^A\boldsymbol{\omega}^B \cdot \boldsymbol{\mu} = \frac{\frac{1}{2}}{\left(1+\frac{\mu^2}{16}\right)^2} \left[\left(1-\frac{\mu^2}{16}\right) \frac{{}^B d\boldsymbol{\mu}}{dt} \cdot \boldsymbol{\mu} + \frac{1}{8}\mu^2 \boldsymbol{\mu} \cdot \frac{{}^B d\boldsymbol{\mu}}{dt} - 0 \right]$$

$$= \frac{\frac{1}{2}}{\left(1+\frac{\mu^2}{16}\right)} \frac{{}^B d\boldsymbol{\mu}}{dt} \cdot \boldsymbol{\mu} \tag{16}$$

and the resulting expression is solved for ${}^B d\boldsymbol{\mu}/dt \cdot \boldsymbol{\mu}$,

$$\frac{{}^B d\boldsymbol{\mu}}{dt} \cdot \boldsymbol{\mu} = \left(1+\frac{\mu^2}{16}\right) {}^A\boldsymbol{\omega}^B \cdot \boldsymbol{\mu} \tag{17}$$

Then both sides of Eq. (14) are cross-multiplied by μ to obtain

$$\frac{1}{2}{}^{A}\boldsymbol{\omega}^{B}\times\mu = \frac{\frac{1}{2}}{\left(1+\frac{\mu^2}{16}\right)^2}\left[\left(1-\frac{\mu^2}{16}\right)\frac{{}^{B}d\mu}{dt}\times\mu + \mathbf{0} - \frac{1}{2}\left(\mu^2\frac{{}^{B}d\mu}{dt} - \mu\mu\cdot\frac{{}^{B}d\mu}{dt}\right)\right] \tag{18}$$

Substitution from Eq. (15) into (18) yields

$$\begin{aligned}{}^{A}\boldsymbol{\omega}^{B}\times\mu &= \frac{1}{\left(1+\frac{\mu^2}{16}\right)^2}\left\{2\left(1-\frac{\mu^2}{16}\right)\left[\left(1+\frac{\mu^2}{16}\right)^2{}^{A}\boldsymbol{\omega}^{B}\right.\right.\\ &\quad\left.\left.-\left(1-\frac{\mu^2}{16}\right)\frac{{}^{B}d\mu}{dt} - \frac{1}{8}\mu\mu\cdot\frac{{}^{B}d\mu}{dt}\right] - \frac{1}{2}\mu^2\frac{{}^{B}d\mu}{dt} + \frac{1}{2}\mu\mu\cdot\frac{{}^{B}d\mu}{dt}\right\}\\ &= 2\left(1-\frac{\mu^2}{16}\right){}^{A}\boldsymbol{\omega}^{B} + \frac{1}{\left(1+\frac{\mu^2}{16}\right)^2}\left\{\left[\frac{1}{2}-\frac{1}{4}\left(1-\frac{\mu^2}{16}\right)\right]\mu\mu\cdot\frac{{}^{B}d\mu}{dt}\right.\\ &\quad\left.-\left[\frac{1}{2}\mu^2+2\left(1-\frac{\mu^2}{16}\right)^2\right]\frac{{}^{B}d\mu}{dt}\right\}\\ &= 2\left(1-\frac{\mu^2}{16}\right){}^{A}\boldsymbol{\omega}^{B} + \frac{1}{1+\frac{\mu^2}{16}}\left(\frac{1}{4}\mu\mu\cdot\frac{{}^{B}d\mu}{dt}\right) - 2\frac{{}^{B}d\mu}{dt}\end{aligned} \tag{19}$$

Finally, substitution from Eq. (17) gives

$${}^{A}\boldsymbol{\omega}^{B}\times\mu = 2\left(1-\frac{\mu^2}{16}\right){}^{A}\boldsymbol{\omega}^{B} + \frac{1}{4}\mu\mu\cdot{}^{A}\boldsymbol{\omega}^{B} - 2\frac{{}^{B}d\mu}{dt} \tag{20}$$

or

$$\frac{{}^{B}d\mu}{dt} = \left(1-\frac{\mu^2}{16}\right){}^{A}\boldsymbol{\omega}^{B} + \frac{1}{8}\mu\mu\cdot{}^{A}\boldsymbol{\omega}^{B} - \frac{1}{2}{}^{A}\boldsymbol{\omega}^{B}\times\mu \tag{21}$$

which is Eq. (2).

Example The "spin-up" problem for an axisymmetric spacecraft B can be formulated most simply as follows: taking the axis of symmetry of B to be parallel to $\hat{\mathbf{b}}_3$, assuming that B is subjected to the action of a system of forces whose resultant moment about the mass center $B^\star$ of B is equal to $M\hat{\mathbf{b}}_3$, where M is a constant, and letting ω_1, ω_2, and ω_3 have the values

$$\omega_1 = \overline{\omega}_1 \qquad \omega_2 = \omega_3 = 0 \tag{22}$$

at time $t = 0$, determine the orientation of B in an inertial reference frame A for $t > 0$. (The reason for taking ω_2 equal to zero at $t = 0$ is that the unit vectors $\hat{\mathbf{b}}_1$ and $\hat{\mathbf{b}}_2$ can always be chosen such that $\hat{\mathbf{b}}_2$ is perpendicular to ${}^{A}\boldsymbol{\omega}^{B}$ at $t = 0$, in which case $\omega_2 = {}^{A}\boldsymbol{\omega}^{B}\cdot\hat{\mathbf{b}}_2 = 0$. As for ω_3, this is taken equal to zero at $t = 0$ because the satellite is presumed to have either no rotational motion or to be tumbling initially, tumbling here referring to a motion such that the angular velocity is perpendicular to the symmetry axis.)

Letting $\underline{\mathbf{I}}$ denote the inertia dyadic of B for $B^\star$, and defining I and J as

$$I \triangleq \hat{\mathbf{b}}_1 \cdot \underline{\mathbf{I}} \cdot \hat{\mathbf{b}}_1 = \hat{\mathbf{b}}_2 \cdot \underline{\mathbf{I}} \cdot \hat{\mathbf{b}}_2 \tag{23}$$

and

$$J \triangleq \hat{\mathbf{b}}_3 \cdot \underline{\mathbf{I}} \cdot \hat{\mathbf{b}}_3 \tag{24}$$

one can refer to Euler's dynamical equations stated in Problem 12.13 to obtain the following differential equations governing ω_1, ω_2, and ω_3:

$$\dot{\omega}_1 = \frac{I-J}{I}\omega_2\omega_3 \tag{25}$$

$$\dot{\omega}_2 = -\frac{I-J}{I}\omega_3\omega_1 \tag{26}$$

$$\dot{\omega}_3 = \frac{M}{J} \tag{27}$$

Since M and J are constants,

$$\omega_3 \underset{(27,\ 22)}{=} \frac{M}{J}t \tag{28}$$

and

$$\dot{\omega}_1 \underset{(25,\ 28)}{=} \frac{I-J}{I}\frac{M}{J}t\,\omega_2 \tag{29}$$

$$\dot{\omega}_2 \underset{(26,\ 28)}{=} -\frac{I-J}{I}\frac{M}{J}t\,\omega_1 \tag{30}$$

The solution of these equations is facilitated by introducing a function ϕ as

$$\phi \triangleq \frac{I-J}{I}\frac{M}{J}\frac{t^2}{2} \tag{31}$$

Then

$$\dot{\omega}_1 \underset{(29,\ 31)}{=} \dot{\phi}\,\omega_2 \tag{32}$$

$$\dot{\omega}_2 \underset{(30,\ 31)}{=} -\dot{\phi}\,\omega_1 \tag{33}$$

or

$$\frac{d\omega_1}{d\phi} \underset{(32)}{=} \omega_2 \tag{34}$$

$$\frac{d\omega_2}{d\phi} \underset{(33)}{=} -\omega_1 \tag{35}$$

so that

$$\frac{d^2\omega_1}{d\phi^2} + \omega_1 \underset{(34,\ 35)}{=} 0 \tag{36}$$

$$\omega_1 = C_1 \sin\phi + C_2 \cos\phi \tag{37}$$

and

$$\omega_2 \underset{(34,\ 37)}{=} C_1 \cos\phi - C_2 \sin\phi \tag{38}$$

where C_1 and C_2 are constants that can be evaluated by noting that ϕ [see Eq. (31)] vanishes at $t = 0$. That is,

$$\overline{\omega}_1 \underset{(22,\ 37)}{=} C_2 \tag{39}$$

and

$$0 \underset{(22,\ 38)}{=} C_1 \tag{40}$$

Consequently,

$$\omega_1 \underset{(37)}{=} \overline{\omega}_1 \cos\phi \tag{41}$$

and

$$\omega_2 \underset{(38)}{=} -\overline{\omega}_1 \sin\phi \tag{42}$$

Equations governing the Wiener-Milenković parameters μ_1, μ_2, and μ_3 now can be written by observing that

$$\omega = \begin{bmatrix} \overline{\omega}_1 \cos\phi & -\overline{\omega}_1 \sin\phi & \dfrac{M}{J}t \end{bmatrix} \tag{43}$$

and expanding Eq. (5) to obtain

$$\dot{\mu}_1 = \frac{1}{16}\left(\mu_1^2 - \mu_2^2 - \mu_3^2 + 16\right)\omega_1 + \frac{1}{8}\left(\mu_1\mu_2 - 4\mu_3\right)\omega_2 + \frac{1}{8}\left(4\mu_2 + \mu_1\mu_3\right)\omega_3 \tag{44}$$

$$\dot{\mu}_2 = \frac{1}{8}\left(\mu_1\mu_2 + 4\mu_3\right)\omega_1 + \frac{1}{16}\left(-\mu_1^2 + \mu_2^2 - \mu_3^2 + 16\right)\omega_2 + \frac{1}{8}\left(\mu_2\mu_3 - 4\mu_1\right)\omega_3 \tag{45}$$

$$\dot{\mu}_3 = \frac{1}{8}\left(\mu_1\mu_3 - 4\mu_2\right)\omega_1 + \frac{1}{8}\left(4\mu_1 + \mu_2\mu_3\right)\omega_2 + \frac{1}{16}\left(-\mu_1^2 - \mu_2^2 + \mu_3^2 + 16\right)\omega_3 \tag{46}$$

and, if $\hat{\mathbf{a}}_1$, $\hat{\mathbf{a}}_2$, and $\hat{\mathbf{a}}_3$ are chosen such that $\hat{\mathbf{a}}_i = \hat{\mathbf{b}}_i$ $(i = 1, 2, 3)$ at $t = 0$, then μ_1, μ_2, and μ_3 must satisfy the initial conditions

$$\mu_i(0) = 0 \qquad (i = 1, 2, 3) \tag{47}$$

Suppose now that one wished to study the behavior of the symmetry axis of B, say for $0 \le \overline{\omega}_1 t \le 10.0$, by plotting the angle θ between this axis and the line fixed in A with which the symmetry axis coincides initially. Once the dimensionless parameters J/I and $M/(J\overline{\omega}_1{}^2)$ have been specified, μ_1, μ_2, and μ_3 can be evaluated by integrating Eqs. (44)–(46) numerically, and θ is then given by

$$\theta = \cos^{-1}\left(\hat{\mathbf{a}}_3 \cdot \hat{\mathbf{b}}_3\right) = \cos^{-1} C_{33} \underset{(10.4.14,\ 10.5.3)}{=} \cos^{-1}\left[1 - \frac{\frac{1}{2}\left(\mu_1{}^2 + \mu_2{}^2\right)}{\left(1 + \frac{1}{16}\mu\mu^T\right)^2}\right] \tag{48}$$

Table 10.9.1 shows values of μ_i $(i = 1, 2, 3)$ and θ obtained in this way for $J/I = 0.5$

and $M/(J\overline{\omega}_1^{\,2}) = 0.1$; the numerical solution of Eqs. (44)–(46) is carried out with the scaling law derived in Section 10.5. A plot of θ versus $\overline{\omega}_1 t$ is shown in Fig. 10.9.1.

Table 10.9.1

$\overline{\omega}_1 t$	μ_1	μ_2	μ_3	θ (deg)
0.0	0.00	0.00	0.00	0
0.5	0.50	0.00	0.01	29
1.0	1.02	−0.02	0.05	57
1.5	1.57	−0.06	0.11	86
2.0	2.18	−0.15	0.18	115
2.5	2.88	−0.31	0.26	143
3.0	3.72	−0.59	0.31	169
3.5	−3.20	0.71	−0.19	157
4.0	−2.41	0.73	−0.01	129
4.5	−1.73	0.72	0.19	100
5.0	−1.12	0.66	0.40	71
5.5	−0.57	0.55	0.63	44
6.0	−0.04	0.37	0.87	20
6.5	0.48	0.10	1.12	26
7.0	0.98	−0.30	1.35	52
7.5	1.47	−0.87	1.53	80
8.0	1.91	−1.65	1.60	108
8.5	2.24	−2.69	1.42	136
9.0	−1.72	2.95	−0.59	155
9.5	−1.02	2.80	0.27	146
10.0	−0.39	2.56	1.09	120

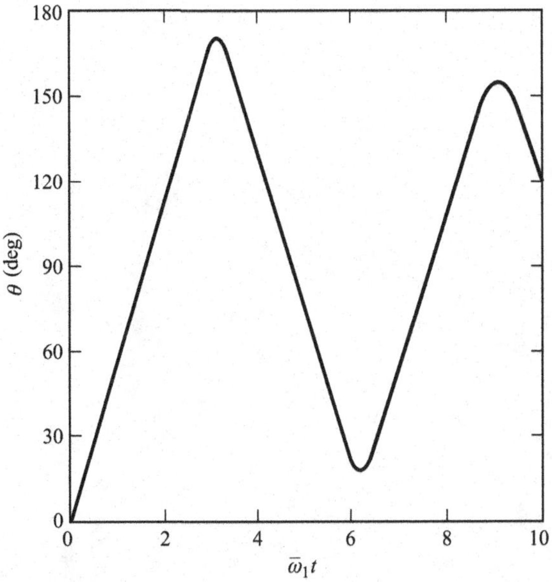

Figure 10.9.1

PROBLEM SETS

PROBLEM SET 1

(Secs. 1.1–1.13)

1.1 Referring to the example in Sec. 1.3 and to Fig. 1.3.1, define vectors **u**, **v**, and **w** as

$$\mathbf{u} = \hat{\mathbf{a}}_1 + 2\hat{\mathbf{a}}_2 + 3\hat{\mathbf{a}}_3$$
$$\mathbf{v} = \hat{\mathbf{b}}_1 + \hat{\mathbf{c}}_2 + \hat{\mathbf{d}}_3$$
$$\mathbf{w} = \hat{\mathbf{d}}_1 + 2\hat{\mathbf{d}}_2 + q_3\hat{\mathbf{d}}_3$$

For each of the vectors **u**, **v**, and **w**, for each of the variables q_1, q_2, and q_3, and for each of the reference frames $A, \ldots, D$, determine whether or not the vector is a function of the variable in the reference frame. Use the letters Y and N, standing for "yes" and "no," respectively, to indicate your results, as shown below.

Results

A	q_1	q_2	q_3
u	N	N	N
v	Y	Y	N
w	Y	Y	Y

B	q_1	q_2	q_3
u	Y	N	N
v	N	Y	N
w	N	Y	Y

C	q_1	q_2	q_3
u	Y	Y	N
v	N	Y	N
w	N	N	Y

D	q_1	q_2	q_3
u	Y	Y	Y
v	N	Y	Y
w	N	N	Y

1.2 Referring to Problem 1.1, and supposing that q_1 is a function of time t, whereas q_2 and q_3 are independent of t, determine in which, if any, of $A, \ldots, D$ the vector **v** is a function of t.

Result A

1.3 Referring to Problem 1.1, and supposing that q_2 is a function of time t, whereas q_1 and q_3 are independent of t, determine in which, if any, of $A, \ldots, D$ the vector **v** is a function of t.

Result A, B, C, D

1.4 Given any noncoplanar vectors $\mathbf{a}_1$, $\mathbf{a}_2$, $\mathbf{a}_3$ (not necessarily unit vectors), one can find (uniquely) three scalars, v_1, v_2, and v_3, such that an arbitrary vector **v** is given by

$\mathbf{v} = v_1\mathbf{a}_1 + v_2\mathbf{a}_2 + v_3\mathbf{a}_3$. [Note the difference between this expression and Eq. (1.6.1).] Show that

$$v_1 = \frac{\mathbf{v} \cdot \mathbf{a}_2 \times \mathbf{a}_3}{\mathbf{a}_1 \cdot \mathbf{a}_2 \times \mathbf{a}_3}$$

where, for any vectors $\boldsymbol{\alpha}$, $\boldsymbol{\beta}$, and $\boldsymbol{\gamma}$, the quantity $\boldsymbol{\alpha} \cdot \boldsymbol{\beta} \times \boldsymbol{\gamma}$ is called the *scalar triple product* of $\boldsymbol{\alpha}$, $\boldsymbol{\beta}$, and $\boldsymbol{\gamma}$. Verify that $\boldsymbol{\alpha} \times \boldsymbol{\beta} \cdot \boldsymbol{\gamma} = \boldsymbol{\alpha} \cdot \boldsymbol{\beta} \times \boldsymbol{\gamma}$ and that $\boldsymbol{\alpha} \cdot \boldsymbol{\beta} \times \boldsymbol{\gamma} = \boldsymbol{\gamma} \cdot \boldsymbol{\alpha} \times \boldsymbol{\beta} = \boldsymbol{\beta} \cdot \boldsymbol{\gamma} \times \boldsymbol{\alpha}$.

1.5 Referring to Problem 1.1, determine the magnitude of each of the following partial derivatives: ${}^A\partial\mathbf{v}/\partial q_1$, ${}^B\partial\mathbf{v}/\partial q_1$, ${}^C\partial\mathbf{v}/\partial q_2$, ${}^C\partial\mathbf{v}/\partial q_3$, ${}^D\partial\mathbf{v}/\partial q_2$, ${}^D\partial\mathbf{v}/\partial q_1$.

Results $(1 + \cos^2 q_2)^{1/2}$, 0, 1, 0, 1, 0

1.6 Referring to Problem 1.1, determine (*a*) the $\hat{\mathbf{a}}_1$, $\hat{\mathbf{a}}_2$, $\hat{\mathbf{a}}_3$ measure numbers of ${}^B\partial\mathbf{u}/\partial q_1$, (*b*) the $\hat{\mathbf{b}}_1$, $\hat{\mathbf{b}}_2$, $\hat{\mathbf{b}}_3$ measure numbers of ${}^B\partial\mathbf{u}/\partial q_1$, and (*c*) the $\hat{\mathbf{a}}_1$, $\hat{\mathbf{a}}_2$, $\hat{\mathbf{a}}_3$ measure numbers of ${}^A\partial\mathbf{u}/\partial q_1$.

Results (*a*) 0, 3, −2 (*b*) 0, $-2\sin q_1 + 3\cos q_1$, $-2\cos q_1 - 3\sin q_1$ (*c*) 0, 0, 0

1.7 The position vector $\mathbf{r}^{PQ}$ from a point P fixed in A to a point Q fixed in D, where A and D are two of the parallelepipeds introduced in the example in Sec. 1.3 and shown in Fig. 1.3.1, is given by

$$\mathbf{r}^{PQ} = \alpha\hat{\mathbf{a}}_1 + \beta\hat{\mathbf{b}}_2 + \gamma\hat{\mathbf{c}}_3$$

where α, β, and γ are the following functions of q_1, q_2, and q_3:

$$\alpha = q_1 + q_2 + q_3 \qquad \beta = {q_1}^2 + {q_2}^2 + {q_3}^2 \qquad \gamma = {q_1}^3 + {q_2}^3 + {q_3}^3$$

Determine the magnitude of ${}^D\partial\mathbf{r}^{PQ}/\partial q_3$ for $q_1 = \pi/2$ rad, $q_2 = q_3 = 0$.

Result $[1 + 3(\pi/2)^2 + (\pi/2)^4]^{1/2}$

***1.8** A vector $\mathbf{v}$ is a function of time t in a reference frame A. Show that the first time-derivative of $|\mathbf{v}|$, the magnitude of $\mathbf{v}$, satisfies the equation

$$|\mathbf{v}|\frac{d|\mathbf{v}|}{dt} = \mathbf{v} \cdot \frac{d\mathbf{v}}{dt}$$

and determine the first time-derivative in A of a unit vector $\hat{\mathbf{u}}$ that has the same direction as $\mathbf{v}$ for all t.

Result

$$\frac{{}^A d\hat{\mathbf{u}}}{dt} = \frac{1}{|\mathbf{v}|}\frac{{}^A d\mathbf{v}}{dt} - \frac{1}{|\mathbf{v}|^3}\mathbf{v}\mathbf{v} \cdot \frac{{}^A d\mathbf{v}}{dt} = \frac{1}{|\mathbf{v}|}(\underline{\mathbf{U}} - \hat{\mathbf{u}}\hat{\mathbf{u}}) \cdot \frac{{}^A d\mathbf{v}}{dt}$$

(The reader will become acquainted in Sec. 4.5 with dyadics, such as $\hat{\mathbf{u}}\hat{\mathbf{u}}$, and the unit dyadic $\underline{\mathbf{U}}$.)

1.9 In Fig. P1.9, $\hat{\mathbf{a}}_1$, $\hat{\mathbf{a}}_2$, $\hat{\mathbf{b}}_1$, and $\hat{\mathbf{b}}_2$ are coplanar unit vectors, with $\hat{\mathbf{a}}_1$ perpendicular to $\hat{\mathbf{a}}_2$, $\hat{\mathbf{b}}_1$ perpendicular to $\hat{\mathbf{b}}_2$, and θ the angle between $\hat{\mathbf{a}}_1$ and $\hat{\mathbf{b}}_1$. Letting A be a reference frame in which $\hat{\mathbf{a}}_1$ and $\hat{\mathbf{a}}_2$ are fixed, B a reference frame in which $\hat{\mathbf{b}}_1$ and $\hat{\mathbf{b}}_2$ are fixed, and $\mathbf{v}$ a vector given by

$$\mathbf{v} = f\hat{\mathbf{a}}_1 + g\hat{\mathbf{a}}_2$$

where f and g are functions of two scalar variables q_1 and q_2, show that, if θ also is a function of q_1 and q_2, then

$$\frac{{}^A\partial}{\partial q_1}\left(\frac{{}^B\partial \mathbf{v}}{\partial q_2}\right) - \frac{{}^B\partial}{\partial q_2}\left(\frac{{}^A\partial \mathbf{v}}{\partial q_1}\right) = (g\hat{\mathbf{a}}_1 - f\hat{\mathbf{a}}_2)\frac{\partial^2 \theta}{\partial q_1 \partial q_2}$$

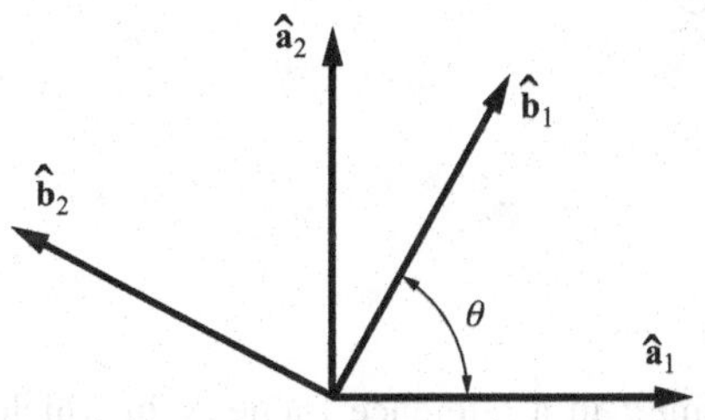

Figure P1.9

1.10 A circular disk C of radius r can rotate about an axis X fixed in a laboratory L, as shown in Fig. P1.10, and a rod R of length $3r$ is pinned to C, the axis Y of the pin passing through the center O of C. Letting $\mathbf{p}$ be the position vector from O to P, the endpoint of R, express the first time-derivative of $\mathbf{p}$ in L in terms of q_1, q_2, $\dot{q}_1$, $\dot{q}_2$, $\hat{\mathbf{c}}_1$, $\hat{\mathbf{c}}_2$, and $\hat{\mathbf{c}}_3$, where q_1 and q_2 are the radian measures of angles, as indicated in Fig. P1.10, $\dot{q}_1$ and $\dot{q}_2$ denote the time derivatives of q_1 and q_2, and $\hat{\mathbf{c}}_1$, $\hat{\mathbf{c}}_2$, $\hat{\mathbf{c}}_3$ are unit vectors fixed in C and directed as shown. Note that R performs a simple rotation relative to C, about $\hat{\mathbf{c}}_2$, through the angle q_2.

Suggestion: First verify that

$$\frac{{}^L\partial \hat{\mathbf{c}}_2}{\partial q_1} = \hat{\mathbf{c}}_3 \qquad \frac{{}^L\partial \hat{\mathbf{c}}_3}{\partial q_1} = -\hat{\mathbf{c}}_2$$

Result $r[\dot{q}_1(\hat{\mathbf{c}}_3 - 3\sin q_2 \hat{\mathbf{c}}_2) + 3\dot{q}_2(\cos q_2 \hat{\mathbf{c}}_3 + \sin q_2 \hat{\mathbf{c}}_1)]$

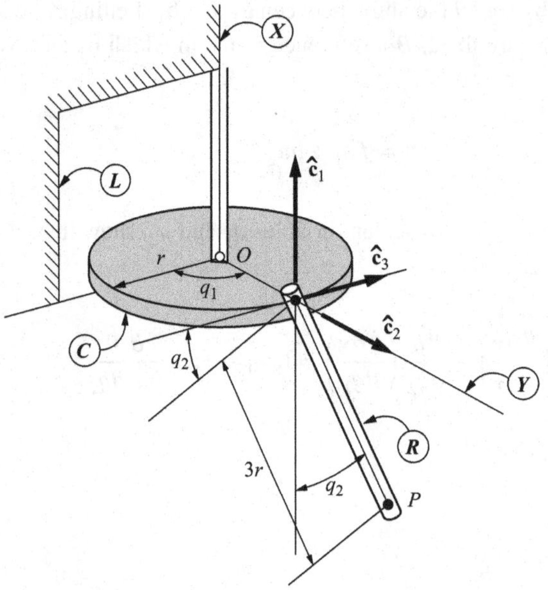

Figure P1.10

***1.11** A vector $\mathbf{v}$ is a function of time t in a reference frame N in which mutually perpendicular unit vectors $\hat{\mathbf{n}}_1$, $\hat{\mathbf{n}}_2$, $\hat{\mathbf{n}}_3$ are fixed. At time t^*,

$$\mathbf{v} = \hat{\mathbf{n}}_1 \qquad \frac{^N d\mathbf{v}}{dt} = \hat{\mathbf{n}}_2 \qquad \frac{^N d^2\mathbf{v}}{dt^2} = \hat{\mathbf{n}}_3$$

Letting $\hat{\mathbf{u}}$ be a unit vector that has the same direction as $\mathbf{v}$ at all times, determine the magnitude of $^N d^2\hat{\mathbf{u}}/dt^2$ at time t^*.

Result $\sqrt{2}$

1.12 Referring to the example in Sec. 1.12 and to Eq. (1.12.11), define s as

$$s \triangleq \mathbf{p} \cdot \hat{\mathbf{a}}_3 - R \sin t \sin 2t$$

and regard s as a scalar function of two vectors, $\mathbf{p}$ and $\hat{\mathbf{a}}_3$. Form ds/dt by differentiating both sides of this relationship, and evaluate ds/dt. Express $\partial s/\partial \mathbf{p}$, $\partial s/\partial \hat{\mathbf{a}}_3$, and $\partial s/\partial t$ in terms of $\mathbf{p}$, $\hat{\mathbf{a}}_3$, R, and t.

Results $ds/dt = \hat{\mathbf{a}}_3 \cdot {}^A d\mathbf{p}/dt - R(\cos t \sin 2t + 2 \sin t \cos 2t) = 0$, $\partial s/\partial \mathbf{p} = \hat{\mathbf{a}}_3$, $\partial s/\partial \hat{\mathbf{a}}_3 = \mathbf{p}$, $\partial s/\partial t = -R(\cos t \sin 2t + 2 \sin t \cos 2t)$

PROBLEM SET 2
(Secs. 2.1–2.5)

2.1 When a point P moves on a space curve C fixed in a reference frame A, a dextral set of orthogonal unit vectors $\hat{\mathbf{b}}_1$, $\hat{\mathbf{b}}_2$, $\hat{\mathbf{b}}_3$ can be generated by letting $\mathbf{p}$ be the position vector from a point O fixed on C to the point P and defining $\hat{\mathbf{b}}_1$, $\hat{\mathbf{b}}_2$, and $\hat{\mathbf{b}}_3$ as

$$\hat{\mathbf{b}}_1 \triangleq \mathbf{p}' \qquad \hat{\mathbf{b}}_2 \triangleq \frac{\mathbf{p}''}{|\mathbf{p}''|} \qquad \hat{\mathbf{b}}_3 \triangleq \hat{\mathbf{b}}_1 \times \hat{\mathbf{b}}_2$$

where primes denote differentiation in A with respect to the length s of the arc of C that connects O to P (see Fig. P2.1). The vector $\hat{\mathbf{b}}_1$ is called a *vector tangent*, $\hat{\mathbf{b}}_2$ the *vector principal normal*, and $\hat{\mathbf{b}}_3$ a *vector binormal* of C at P, and the derivatives in A of $\hat{\mathbf{b}}_1$, $\hat{\mathbf{b}}_2$, and $\hat{\mathbf{b}}_3$ with respect to s are given by the Serret-Frênet formulas

$$\hat{\mathbf{b}}_1' = \frac{\hat{\mathbf{b}}_2}{\rho} \qquad \hat{\mathbf{b}}_2' = -\frac{\hat{\mathbf{b}}_1}{\rho} + \lambda\hat{\mathbf{b}}_3 \qquad \hat{\mathbf{b}}_3' = -\lambda\hat{\mathbf{b}}_2$$

where ρ and λ, defined as

$$\rho \triangleq \frac{1}{|\mathbf{p}''|} \qquad \lambda \triangleq \rho^2\mathbf{p}' \cdot \mathbf{p}'' \times \mathbf{p}'''$$

are called the *principal radius of curvature* of C at P and the *torsion* of C at P, respectively.

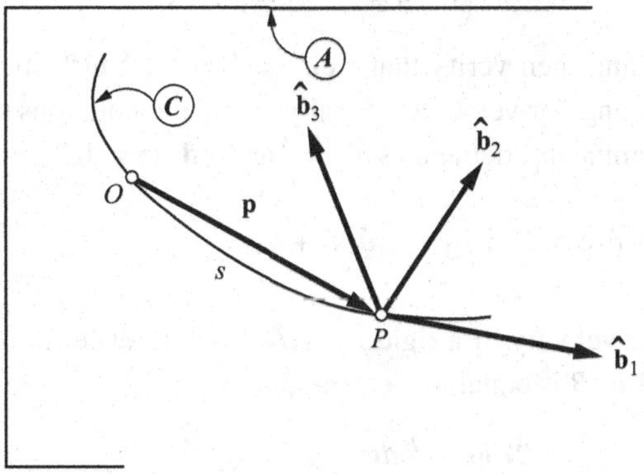

Figure P2.1

Letting B be a rigid body in which $\hat{\mathbf{b}}_1$, $\hat{\mathbf{b}}_2$, $\hat{\mathbf{b}}_3$ are fixed, express the angular velocity of B in A in terms of $\hat{\mathbf{b}}_1$, $\hat{\mathbf{b}}_2$, $\hat{\mathbf{b}}_3$, ρ, λ, and $\dot{s}$, the time derivative of s, taking advantage of the fact that

$$\frac{{}^A d\hat{\mathbf{b}}_i}{dt} = \hat{\mathbf{b}}_i' \, \dot{s} \qquad (i = 1,2,3)$$

Result $(\lambda\hat{\mathbf{b}}_1 + \hat{\mathbf{b}}_3/\rho)\dot{s}$

2.2 Table P2.2 shows the relationship between two dextral sets of orthogonal unit vectors, $\hat{\mathbf{a}}_1$, $\hat{\mathbf{a}}_2$, $\hat{\mathbf{a}}_3$ and $\hat{\mathbf{b}}_1$, $\hat{\mathbf{b}}_2$, $\hat{\mathbf{b}}_3$, fixed in rigid bodies A and B, respectively. The symbols s_i and c_i in the table stand for $\sin q_i$ and $\cos q_i$, respectively, where q_i $(i = 1,2,3)$ are the radian measures of certain angles.

Table P2.2

	$\hat{\mathbf{b}}_1$	$\hat{\mathbf{b}}_2$	$\hat{\mathbf{b}}_3$
$\hat{\mathbf{a}}_1$	$\mathrm{c}_2\mathrm{c}_3$	$\mathrm{s}_1\mathrm{s}_2\mathrm{c}_3 - \mathrm{s}_3\mathrm{c}_1$	$\mathrm{c}_1\mathrm{s}_2\mathrm{c}_3 + \mathrm{s}_3\mathrm{s}_1$
$\hat{\mathbf{a}}_2$	$\mathrm{c}_2\mathrm{s}_3$	$\mathrm{s}_1\mathrm{s}_2\mathrm{s}_3 + \mathrm{c}_3\mathrm{c}_1$	$\mathrm{c}_1\mathrm{s}_2\mathrm{s}_3 - \mathrm{c}_3\mathrm{s}_1$
$\hat{\mathbf{a}}_3$	$-\mathrm{s}_2$	$\mathrm{s}_1\mathrm{c}_2$	$\mathrm{c}_1\mathrm{c}_2$

Determine α_2 and β_2 such that $\boldsymbol{\omega}$, the angular velocity of B in A, is given by

$$\boldsymbol{\omega} = \alpha_1\hat{\mathbf{a}}_1 + \alpha_2\hat{\mathbf{a}}_2 + \alpha_3\hat{\mathbf{a}}_3 = \beta_1\hat{\mathbf{b}}_1 + \beta_2\hat{\mathbf{b}}_2 + \beta_3\hat{\mathbf{b}}_3$$

Results $\dot{q}_1\mathrm{c}_2\mathrm{s}_3 + \dot{q}_2\mathrm{c}_3$ $\qquad$ $\dot{q}_2\mathrm{c}_1 + \dot{q}_3\mathrm{s}_1\mathrm{c}_2$

2.3 Referring to Problem 1.1, use the definition of angular velocity in Sec. 2.1 to show that the angular velocity of B in D can be expressed as

$${}^{D}\boldsymbol{\omega}^{B} = -(\dot{q}_3 \sin q_2\hat{\mathbf{b}}_1 + \dot{q}_2\hat{\mathbf{b}}_2 + \dot{q}_3 \cos q_2\hat{\mathbf{b}}_3)$$

Next, form ${}^{D}\boldsymbol{\omega}^{C}$ and ${}^{C}\boldsymbol{\omega}^{B}$, and then verify that ${}^{D}\boldsymbol{\omega}^{B} = {}^{D}\boldsymbol{\omega}^{C} + {}^{C}\boldsymbol{\omega}^{B}$, in agreement with the addition theorem for angular velocities. Finally, using the notations $\mathrm{s}_i = \sin q_i$, $\mathrm{c}_i = \cos q_i$ $(i = 1,2,3)$, determine ω_i, defined as $\omega_i \triangleq {}^{A}\boldsymbol{\omega}^{D} \cdot \hat{\mathbf{d}}_i$ $(i = 1,2,3)$.

Results $\dot{q}_1\mathrm{c}_2\mathrm{c}_3 + \dot{q}_2\mathrm{s}_3$ $\qquad$ $-\dot{q}_1\mathrm{c}_2\mathrm{s}_3 + \dot{q}_2\mathrm{c}_3$ $\qquad$ $\dot{q}_1\mathrm{s}_2 + \dot{q}_3$

2.4 Letting $\boldsymbol{\omega}$ be the angular velocity of a rigid body B in a reference frame A, show that the angular velocity of A in B is equal to $-\boldsymbol{\omega}$, and that

$$\frac{{}^{A}d\boldsymbol{\omega}}{dt} = \frac{{}^{B}d\boldsymbol{\omega}}{dt}$$

***2.5** Letting $\boldsymbol{\beta}_1$ and $\boldsymbol{\beta}_2$ be nonparallel vectors fixed in a rigid body B, and using dots to denote differentiation with respect to time in a reference frame A, verify that the angular velocity $\boldsymbol{\omega}$ of B in A can be expressed as

$$\boldsymbol{\omega} = \frac{\dot{\boldsymbol{\beta}}_1 \times \dot{\boldsymbol{\beta}}_2}{\dot{\boldsymbol{\beta}}_1 \cdot \boldsymbol{\beta}_2} = \frac{1}{2}\left(\frac{\dot{\boldsymbol{\beta}}_1 \times \dot{\boldsymbol{\beta}}_2}{\dot{\boldsymbol{\beta}}_1 \cdot \boldsymbol{\beta}_2} + \frac{\dot{\boldsymbol{\beta}}_2 \times \dot{\boldsymbol{\beta}}_1}{\dot{\boldsymbol{\beta}}_2 \cdot \boldsymbol{\beta}_1}\right)$$

Suggestion: Make use of the fact that, for any vectors $\mathbf{a}$, $\mathbf{b}$, and $\mathbf{c}$, $\mathbf{a} \times (\mathbf{b} \times \mathbf{c}) = \mathbf{a} \cdot \mathbf{c}\mathbf{b} - \mathbf{a} \cdot \mathbf{b}\mathbf{c}$.

2.6 Referring to Problem 2.1, suppose that $\mathbf{q}$, the position vector from P to a point Q moving in A, is given by

$$\mathbf{q} = q_1\hat{\mathbf{b}}_1 + q_2\hat{\mathbf{b}}_2 + q_3\hat{\mathbf{b}}_3$$

where q_1, q_2, q_3 are functions of time t. Determine the quantities v_i $(i = 1,2,3)$ such that the first time-derivative in A of the position vector $\mathbf{r}$ from O to Q is equal to $v_1\hat{\mathbf{b}}_1 + v_2\hat{\mathbf{b}}_2 + v_3\hat{\mathbf{b}}_3$.

Results $\dot{q}_1 + \dot{s}(1 - q_2/\rho)$ $\quad \dot{q}_2 + \dot{s}[(q_1/\rho) - \lambda q_3]$ $\quad \dot{q}_3 + \dot{s}\lambda q_2$

2.7 Figure P2.7 shows a circular disk C of radius R in contact with a horizontal plane H that is fixed in a reference frame A rigidly attached to the Earth. Mutually perpendicular unit vectors $\hat{\mathbf{a}}_x$, $\hat{\mathbf{a}}_y$, and $\hat{\mathbf{a}}_z = \hat{\mathbf{a}}_x \times \hat{\mathbf{a}}_y$ are fixed in A, and $\hat{\mathbf{b}}_1$, $\hat{\mathbf{b}}_2$, $\hat{\mathbf{b}}_3$ form a dextral set of orthogonal unit vectors, with $\hat{\mathbf{b}}_1$ parallel to the tangent to the periphery of C at the point of contact between H and C, $\hat{\mathbf{b}}_2$ parallel to the line connecting this contact point to $C^\star$, the center of C, and $\hat{\mathbf{b}}_3$ normal to the plane of C.

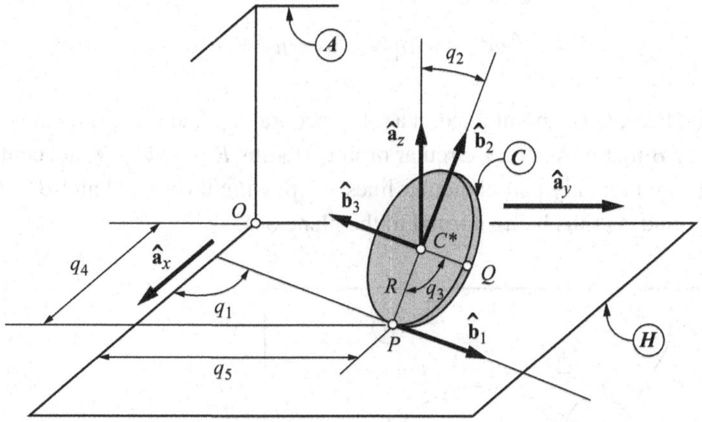

Figure P2.7

The orientation of C in A can be described in terms of the three angles q_1, q_2, q_3 indicated in Fig. P2.7, where Q is a point fixed on the periphery of C. The two quantities q_4 and q_5 characterize the position in A of the *path point* P.

The angular velocity of C in A can be expressed both as

$$^A\boldsymbol{\omega}^C = u_x\hat{\mathbf{a}}_x + u_y\hat{\mathbf{a}}_y + u_z\hat{\mathbf{a}}_z$$

and as

$$^A\boldsymbol{\omega}^C = u_1\hat{\mathbf{b}}_1 + u_2\hat{\mathbf{b}}_2 + u_3\hat{\mathbf{b}}_3$$

where u_x, u_y, u_z and u_1, u_2, u_3 are functions of q_i and $\dot{q}_i$ $(i = 1,2,3)$. Concomitantly, $\dot{q}_i$ $(i = 1,2,3)$ can be expressed as a function F_i of q_1, q_2, q_3, u_x, u_y, u_z or as a function G_i of q_1, q_2, q_3, u_1, u_2, u_3. Determine F_i and G_i $(i = 1,2,3)$. [The equations $\dot{q}_i = F_i$

and $\dot{q}_i = G_i$ $(i = 1,2,3)$ are called *kinematical differential* equations. When u_x, u_y, u_z or u_1, u_2, u_3 are known as functions of t, these equations can be solved for q_1, q_2, q_3 to obtain a description of the orientation of C in A.]

Results

$$
\begin{aligned}
F_1 &= (-u_x \mathrm{s}_1 + u_y \mathrm{c}_1) \tan q_2 + u_z \\
F_2 &= -u_x \mathrm{c}_1 - u_y \mathrm{s}_1 \\
F_3 &= (u_x \mathrm{s}_1 - u_y \mathrm{c}_1) \sec q_2 \\
G_1 &= u_2 \sec q_2 \\
G_2 &= -u_1 \\
G_3 &= -u_2 \tan q_2 + u_3
\end{aligned}
$$

[It is worth noting that G_i is simpler than F_i $(i = 1,2,3)$.]

2.8 Referring to Problem 2.7, and letting B be a reference frame in which $\hat{\mathbf{b}}_1$, $\hat{\mathbf{b}}_2$, $\hat{\mathbf{b}}_3$ are fixed, show that the angular velocity of B in A can be expressed as

$$
{}^A\boldsymbol{\omega}^B = u_1 \hat{\mathbf{b}}_1 + u_2 \hat{\mathbf{b}}_2 + u_2 \tan q_2 \hat{\mathbf{b}}_3
$$

2.9 In Fig. P2.9, O is a point fixed in a reference frame N, and $B^\star$ is the mass center of a rigid body B that moves on a circular orbit C (radius R) fixed in N and centered at O. A_1, A_2, A_3 are mutually perpendicular lines, A_1 passing through O and $B^\star$, A_2 tangent to C at $B^\star$, and A_3 thus being normal to the plane of C.

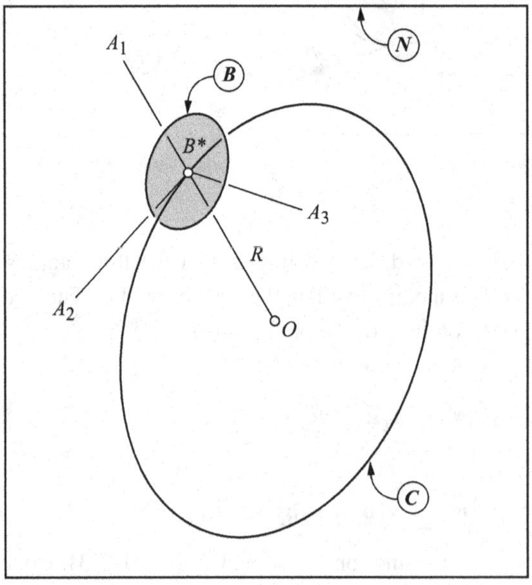

Figure P2.9

If $\hat{\mathbf{b}}_1$, $\hat{\mathbf{b}}_2$, $\hat{\mathbf{b}}_3$ form a dextral set of mutually perpendicular unit vectors fixed in B, the "attitude" of B in a reference frame A in which A_1, A_2, A_3 are fixed can be specified in terms of q_1, q_2, q_3, the radian measures of three angles generated as follows: Let $\hat{\mathbf{a}}_1$, $\hat{\mathbf{a}}_2$, $\hat{\mathbf{a}}_3$ be a dextral set of mutually perpendicular unit vectors fixed in A with $\hat{\mathbf{a}}_i$ parallel to A_i $(i = 1,2,3)$, align $\hat{\mathbf{b}}_i$ with $\hat{\mathbf{a}}_i$ $(i = 1,2,3)$, and then subject B to successive right-handed rotations relative to A, characterized by $q_1\hat{\mathbf{b}}_1$, $q_2\hat{\mathbf{b}}_2$, and $q_3\hat{\mathbf{b}}_1$ (note the last subscript). The quantities ω_i $(i = 1,2,3)$, defined as

$$\omega_i \triangleq \boldsymbol{\omega} \cdot \hat{\mathbf{b}}_i \qquad (i = 1,2,3)$$

where $\boldsymbol{\omega}$ is the angular velocity of B in N, then can be expressed as functions of q_i, $\dot{q}_i$ $(i = 1,2,3)$, and Ω, if the angular velocity of A in N is given by $\Omega\hat{\mathbf{a}}_3$.

Determine ω_i $(i = 1,2,3)$.

Results

$$\omega_1 = \dot{q}_1 c_2 + \dot{q}_3 - \Omega c_1 s_2$$
$$\omega_2 = \dot{q}_1 s_2 s_3 + \dot{q}_2 c_3 + \Omega(c_1 c_2 s_3 + c_3 s_1)$$
$$\omega_3 = \dot{q}_1 s_2 c_3 - \dot{q}_2 s_3 + \Omega(c_1 c_2 c_3 - s_3 s_1)$$

2.10 The angular acceleration ${}^A\boldsymbol{\alpha}^C$, where C is the circular disk in Problem 2.7, can be expressed both as

$${}^A\boldsymbol{\alpha}^C = \alpha_x\hat{\mathbf{a}}_x + \alpha_y\hat{\mathbf{a}}_y + \alpha_z\hat{\mathbf{a}}_z$$

and as

$${}^A\boldsymbol{\alpha}^C = \alpha_1\hat{\mathbf{b}}_1 + \alpha_2\hat{\mathbf{b}}_2 + \alpha_3\hat{\mathbf{b}}_3$$

Express α_x, α_y, α_z in terms of q_1, q_2, q_3, u_x, u_y, u_z, $\dot{u}_x$, $\dot{u}_y$, $\dot{u}_z$; and α_1, α_2, α_3 in terms of q_i, u_i, $\dot{u}_i$ $(i = 1,2,3)$.

Results

$$\alpha_x = \dot{u}_x \qquad \alpha_y = \dot{u}_y \qquad \alpha_z = \dot{u}_z$$
$$\alpha_1 = \dot{u}_1 + u_2(u_3 - u_2 \tan q_2)$$
$$\alpha_2 = \dot{u}_2 - u_1(u_3 - u_2 \tan q_2)$$
$$\alpha_3 = \dot{u}_3$$

2.11 Referring to Problem 1.10, determine the magnitudes of the angular acceleration of C in L, R in C, and R in L.

Results $|\ddot{q}_1|$, $|\ddot{q}_2|$, $[\ddot{q}_1^2 + \ddot{q}_2^2 + (\dot{q}_1\dot{q}_2)^2]^{1/2}$

***2.12** Equations (2.1.2), (2.2.1), (2.5.3), and (2.5.4) underlie one method for analyzing motions of planar linkages. The method consists of expressing the position vector from a hinge point of the linkage to the same hinge point as the sum of position vectors from one hinge point to an adjacent one, setting this sum equal to zero, differentiating the resulting equation repeatedly with respect to time t, and making use of the relationships

$$ {}^{A}\boldsymbol{\omega}^{B_i} \underset{(2.2.1)}{=} \omega_i \hat{\mathbf{k}} \qquad {}^{A}\boldsymbol{\alpha}^{B_i} \underset{(2.5.3)}{=} \alpha_i \hat{\mathbf{k}} $$

$$ \frac{{}^{A}d\hat{\boldsymbol{\beta}}_i}{dt} \underset{(2.1.2)}{=} \omega_i \hat{\mathbf{k}} \times \hat{\boldsymbol{\beta}}_i = \omega_i \hat{\boldsymbol{\beta}}_i{}' $$

$$ \frac{{}^{A}d\hat{\boldsymbol{\beta}}_i'}{dt} \underset{(2.1.2)}{=} \omega_i \hat{\mathbf{k}} \times \hat{\boldsymbol{\beta}}_i{}' = -\omega_i \hat{\boldsymbol{\beta}}_i $$

and

$$ \alpha_i \underset{(2.5.4)}{=} \frac{d\omega_i}{dt} $$

where A is a reference frame in which the plane of the linkage is fixed, B_i is a typical member of the linkage, $\hat{\mathbf{k}}$ is a unit vector normal to the plane of the linkage, $\hat{\boldsymbol{\beta}}_i$ is a unit vector parallel to the line connecting the hinge points of B_i, $\hat{\boldsymbol{\beta}}_i{}'$ is the unit vector $\hat{\mathbf{k}} \times \hat{\boldsymbol{\beta}}_i$, and ω_i and α_i are an angular speed of B_i (see Sec. 2.2) and a scalar angular acceleration of B_i (see Sec. 2.5), respectively.

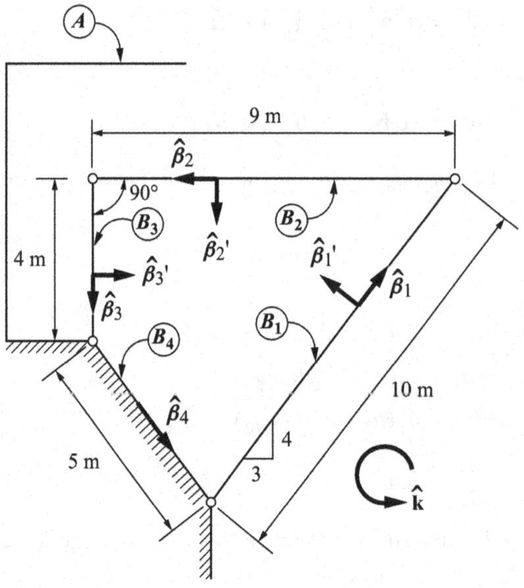

Figure P2.12

Figure P2.12 shows the configuration at time t^* of a four-bar linkage, one bar of which, B_4, is fixed in a reference frame A. At time t^*, the angular velocity of B_3 in A and the angular acceleration of B_1 in A are equal to $-6\hat{\mathbf{k}}$ rad/s and $5\hat{\mathbf{k}}$ rad/s^2, respectively.

Determine the angular velocities of B_1 and B_2 in A, and the angular accelerations of B_2 and B_3 in A, at time t^*, by making use of the equation

$$10\boldsymbol{\beta}_1 + 9\boldsymbol{\beta}_2 + 4\boldsymbol{\beta}_3 + 5\boldsymbol{\beta}_4 = \mathbf{0}$$

which is valid for all t.

Results

$$^A\boldsymbol{\omega}^{B_1} = -3\hat{\mathbf{k}} \text{ rad/s} \qquad ^A\boldsymbol{\omega}^{B_2} = -2\hat{\mathbf{k}} \text{ rad/s}$$

$$^A\boldsymbol{\alpha}^{B_2} = (34/3)\hat{\mathbf{k}} \text{ rad/s}^2 \qquad ^A\boldsymbol{\alpha}^{B_3} = (29/2)\hat{\mathbf{k}} \text{ rad/s}^2$$

PROBLEM SET 3
(Secs. 2.6–2.8)

3.1 Apply the definitions given in Eqs. (2.6.1) and (2.6.2) to formulate expressions for the velocity and the acceleration of the endpoint P of the rod R of Problem 1.10 in (*a*) reference frame L and (*b*) the circular disk C.

Results

$$\begin{aligned}
^L\mathbf{v}^P &= r[3\sin q_2(\dot{q}_2\hat{\mathbf{c}}_1 - \dot{q}_1\hat{\mathbf{c}}_2) + (\dot{q}_1 + 3\dot{q}_2\cos q_2)\hat{\mathbf{c}}_3] \\
^L\mathbf{a}^P &= r\{3(\ddot{q}_2\sin q_2 + \dot{q}_2^{\,2}\cos q_2)\hat{\mathbf{c}}_1 \\
&\quad -(3\ddot{q}_1\sin q_2 + 6\dot{q}_1\dot{q}_2\cos q_2 + \dot{q}_1^{\,2})\hat{\mathbf{c}}_2 \\
&\quad +[\ddot{q}_1 + 3\ddot{q}_2\cos q_2 - 3(\dot{q}_1^{\,2} + \dot{q}_2^{\,2})\sin q_2]\hat{\mathbf{c}}_3\} \\
^C\mathbf{v}^P &= 3r\dot{q}_2(\sin q_2\hat{\mathbf{c}}_1 + \cos q_2\hat{\mathbf{c}}_3) \\
^C\mathbf{a}^P &= 3r[(\ddot{q}_2\sin q_2 + \dot{q}_2^{\,2}\cos q_2)\hat{\mathbf{c}}_1 + (\ddot{q}_2\cos q_2 - \dot{q}_2^{\,2}\sin q_2)\hat{\mathbf{c}}_3]
\end{aligned}$$

3.2 A point P moves on a space curve C fixed in a reference frame A. Show that $\mathbf{v}$, the velocity of P in A, can be expressed as

$$\mathbf{v} = v\hat{\mathbf{b}}_1$$

where $\hat{\mathbf{b}}_1$ is a vector tangent of C at P (see Problem 2.1), and verify that $\mathbf{a}$, the acceleration of P in A, is given by

$$\mathbf{a} = \dot{v}\hat{\mathbf{b}}_1 + \frac{v^2}{\rho}\hat{\mathbf{b}}_2$$

where $\hat{\mathbf{b}}_2$ and ρ are, respectively, the vector principal normal and the principal radius of curvature of C at P.

3.3 Letting $u_4 \triangleq \dot{q}_4$ and $u_5 \triangleq \dot{q}_5$, with q_4 and q_5 defined as in Problem 2.7, determine v_i $(i = 1,2,3)$ such that the velocity of $C^\star$ in A is given by

$$^A\mathbf{v}^{C^\star} = v_1\hat{\mathbf{b}}_1 + v_2\hat{\mathbf{b}}_2 + v_3\hat{\mathbf{b}}_3$$

Express v_i $(i = 1,2,3)$ as functions of q_j and u_j $(j = 1,\ldots,5)$.

Results

$$v_1 = -Ru_2 \tan q_2 + u_4 c_1 + u_5 s_1$$
$$v_2 = (-u_4 s_1 + u_5 c_1) s_2$$
$$v_3 = Ru_1 + (u_4 s_1 - u_5 c_1) c_2$$

3.4 If a_i is defined as $a_i \triangleq {}^A\mathbf{a}^{C^\star} \cdot \hat{\mathbf{b}}_i$ $(i = 1,2,3)$, where A and $C^\star$ have the same meaning as in Problem 2.7 and ${}^A\mathbf{a}^{C^\star}$ denotes the acceleration of $C^\star$ in A, then a_1, a_2, a_3 can be expressed as functions of q_j, u_j, and $\dot{u}_j$ $(j = 1,\ldots,5)$, with u_4 and u_5 defined as in Problem 3.3. Determine these functions.

Results

$$a_1 = -\dot{u}_2 R \tan q_2 + \dot{u}_4 c_1 + \dot{u}_5 s_1 + u_1 u_2 R(1 + \sec^2 q_2)$$
$$a_2 = (-\dot{u}_4 s_1 + \dot{u}_5 c_1) s_2 - R{u_1}^2 - R{u_2}^2 \tan^2 q_2$$
$$a_3 = \dot{u}_1 R + (\dot{u}_4 s_1 - \dot{u}_5 c_1) c_2 + {u_2}^2 R \tan q_2$$

***3.5** Determine the velocity $\mathbf{v}$ and the acceleration $\mathbf{a}$ of the midpoint of bar B_2 of the linkage described in Problem 2.12 for time t^*.

Results $\mathbf{v} = -24\hat{\boldsymbol{\beta}}_2 + 9\hat{\boldsymbol{\beta}}_3$ m/s $\qquad \mathbf{a} = 76\hat{\boldsymbol{\beta}}_2 + 93\hat{\boldsymbol{\beta}}_3$ m/s^2

3.6 At time t, there exists precisely one point of the disk C of Problem 2.7 (see also Problems 2.8, 2.10, 3.3, and 3.4) that is in contact with the plane H. Calling this point $\hat{C}$, one can express the velocity and the acceleration of $\hat{C}$ in A at time t as

$${}^A\mathbf{v}^{\hat{C}} = \hat{v}_x \hat{\mathbf{a}}_x + \hat{v}_y \hat{\mathbf{a}}_y$$

and

$${}^A\mathbf{a}^{\hat{C}} = \hat{a}_1 \hat{\mathbf{b}}_1 + \hat{a}_2 \hat{\mathbf{b}}_2 + \hat{a}_3 \hat{\mathbf{b}}_3$$

respectively, where $\hat{v}_x$ and $\hat{v}_y$ are functions of q_j and u_j $(j = 1,\ldots,5)$, and $\hat{a}_i$ $(i = 1,2,3)$ are functions of q_j, u_j, and $\dot{u}_j$ $(j = 1,\ldots,5)$. Determine $\hat{v}_x$, $\hat{v}_y$, and $\hat{a}_i$ $(i = 1,2,3)$.

Suggestion: Use Eqs. (2.7.1) and (2.7.2), replacing the symbols P, Q, and B with $\hat{C}$, $C^\star$, and C, respectively.

Results

$$\hat{v}_x = (-u_2 \tan q_2 + u_3)Rc_1 + u_4$$
$$\hat{v}_y = (-u_2 \tan q_2 + u_3)Rs_1 + u_5$$
$$\hat{a}_1 = -\dot{u}_2 R \tan q_2 + \dot{u}_3 R + \dot{u}_4 c_1 + \dot{u}_5 s_1 + u_1 u_2 R \sec^2 q_2$$
$$\hat{a}_2 = (-\dot{u}_4 s_1 + \dot{u}_5 c_1)s_2 + R(u_3{}^2 - u_2{}^2 \tan^2 q_2)$$
$$\hat{a}_3 = (\dot{u}_4 s_1 - \dot{u}_5 c_1)c_2 + 2Ru_2(u_2 \tan q_2 - u_3)$$

3.7 Suppose that the quantities $\hat{v}_x$ and $\hat{v}_y$ of Problem 3.6 are differentiated with respect to t, the results are used to form a vector **a** defined as

$$\mathbf{a} \triangleq \frac{d\hat{v}_x}{dt}\hat{\mathbf{a}}_x + \frac{d\hat{v}_y}{dt}\hat{\mathbf{a}}_y$$

and quantities $\overline{a}_i$ $(i = 1,2,3)$ then are formed as

$$\overline{a}_i \triangleq \mathbf{a} \cdot \hat{\mathbf{b}}_i \qquad (i = 1,2,3)$$

Do you expect $\overline{a}_i$ to be equal to the quantity $\hat{a}_i$ of Problem 3.6? First explain your answer, then establish its validity by determining $\overline{a}_i$ $(i = 1,2,3)$.

3.8 If $\hat{A}$ is a point of a rigid body A, $\hat{B}$ a point of a rigid body B, $\hat{A}$ and $\hat{B}$ are in contact with each other at time t, and the velocities of $\hat{A}$ and $\hat{B}$ in any reference frame are equal to each other at time t, then A and B are said to be *rolling* on each other at time t. Alternatively, one can say that *no slipping* is taking place at the contact between A and B at time t. (If A and B are in contact with each other at more than one point, these contacts must be considered separately. The bodies A and B can be rolling on each other at some points while slipping is taking place at other points.)

When the circular disk C of Problem 2.7 rolls on plane H, the quantities u_4 and u_5 of Problem 3.3 can be expressed as functions of q_j $(j = 1,\ldots,5)$ and u_k $(k = 1,2,3)$. Determine these functions.

Results $u_4 = (u_2 \tan q_2 - u_3)Rc_1 \qquad u_5 = (u_2 \tan q_2 - u_3)Rs_1$

3.9 When the circular disk C of Problem 2.7 rolls on plane H, there exists at each instant one point $\hat{C}$ of C that is in contact with H. The acceleration of $\hat{C}$ in A is not, in general, equal to zero. Express the magnitude of this acceleration in terms of R, u_2, u_3, and $\dot{q}_3$.

Result $R|\dot{q}_3|(u_2{}^2 + u_3{}^2)^{1/2}$

3.10 When two rigid bodies A and B are rolling on each other (see Problem 3.8), the angular velocity ${}^A\boldsymbol{\omega}^B$ of B in A generally is not parallel to the plane P that is tangent to the surfaces of A and B at their points of contact with each other. When ${}^A\boldsymbol{\omega}^B$ is parallel to P, one speaks of *pure* rolling of A and B on each other.

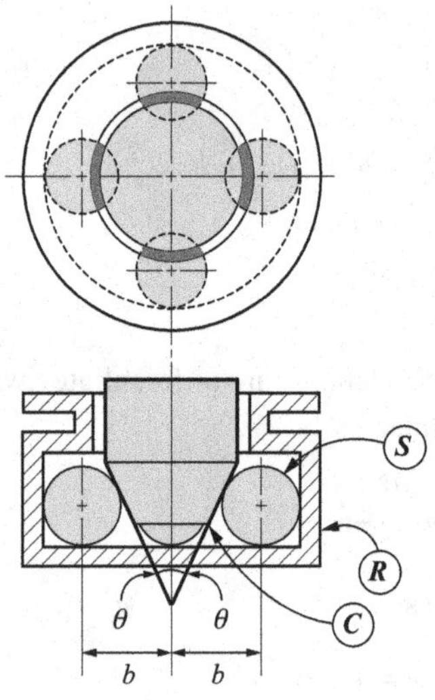

Figure P3.10

Figure P3.10 shows a shaft terminating in a truncated cone C of semivertex angle θ ($0 < \theta < \pi/4$ rad), this shaft being supported by a thrust bearing consisting of a fixed race R and four identical spheres S of radius r. When the shaft rotates, rolling takes place at the two contacts between R and S, as well as at the contact between S and C. Moreover, S and C perform a pure rolling motion on each other (which is desirable because it minimizes wear) if the dimension b is a suitable function of r and θ. Find this function.

Result $r(1 + \sin\theta)/(\cos\theta - \sin\theta)$

3.11 The concept of rolling (see Problem 3.8) comes into play in connection with gearing, where it can be invoked in conjunction with Eqs. (2.7.1) and (2.4.1) to discover relationships between angular speeds (see Sec. 2.2) of gears.

Figure P3.11 shows schematically how the drive shaft D of an automobile can be connected to the two halves, A and A', of an axle in such a way as to permit wheels attached rigidly to A and A' to rotate at different rates relative to the frame F that supports D, A, and A'. This is accomplished as follows: Bevel gears B and B', keyed to A and A', respectively, engage bevel gears b and b', respectively; b and b' are free to rotate on pins fixed in a casing C that can rotate about the (common) axis of A and A', and a bevel gear E, fastened rigidly to C, is driven by a bevel gear G that is keyed to D.

Letting P, P', and Q be points selected arbitrarily on the lines of contact between B and b, B' and b, and G and E, respectively, and assuming that rolling is taking place at these points, one can discover the relationship between angular speeds ${}^F\omega^A$, ${}^F\omega^{A'}$, and

${}^F\omega^D$, defined as

$$ {}^F\omega^A \triangleq {}^F\boldsymbol{\omega}^A \cdot \hat{\mathbf{N}} \qquad {}^F\omega^{A'} \triangleq {}^F\boldsymbol{\omega}^{A'} \cdot \hat{\mathbf{N}} \qquad {}^F\omega^D \triangleq {}^F\boldsymbol{\omega}^D \cdot \hat{\mathbf{n}} $$

where $\hat{\mathbf{N}}$ and $\hat{\mathbf{n}}$ are unit vectors directed as shown in Fig. P3.11, by reasoning as follows.

B and b have simple angular velocities in C, and, after expressing these as

$$ {}^C\boldsymbol{\omega}^B = {}^C\omega^B \hat{\mathbf{N}} \qquad {}^C\boldsymbol{\omega}^b = {}^C\omega^b \hat{\mathbf{n}} $$

one can let $\hat{B}$ and $\hat{b}$ be the points of B and b, respectively, that come into contact at P; introduce distances R and r as shown in Fig. P3.11; write for the velocities of $\hat{B}$ and $\hat{b}$ in C

$$ {}^C\mathbf{v}^{\hat{B}} \underset{(2.7.1)}{=} {}^C\boldsymbol{\omega}^B \times (-R\hat{\mathbf{n}}) = -R\,{}^C\omega^B\, \hat{\mathbf{N}} \times \hat{\mathbf{n}} $$

and

$$ {}^C\mathbf{v}^{\hat{b}} \underset{(2.7.1)}{=} {}^C\boldsymbol{\omega}^b \times (r\hat{\mathbf{N}}) = r\,{}^C\omega^b\, \hat{\mathbf{n}} \times \hat{\mathbf{N}} $$

and then ensure rolling at P by requiring that

$$ -R\,{}^C\omega^B\, \hat{\mathbf{N}} \times \hat{\mathbf{n}} = r\,{}^C\omega^b\, \hat{\mathbf{n}} \times \hat{\mathbf{N}} $$

which is guaranteed when

$$ R\,{}^C\omega^B = r\,{}^C\omega^b $$

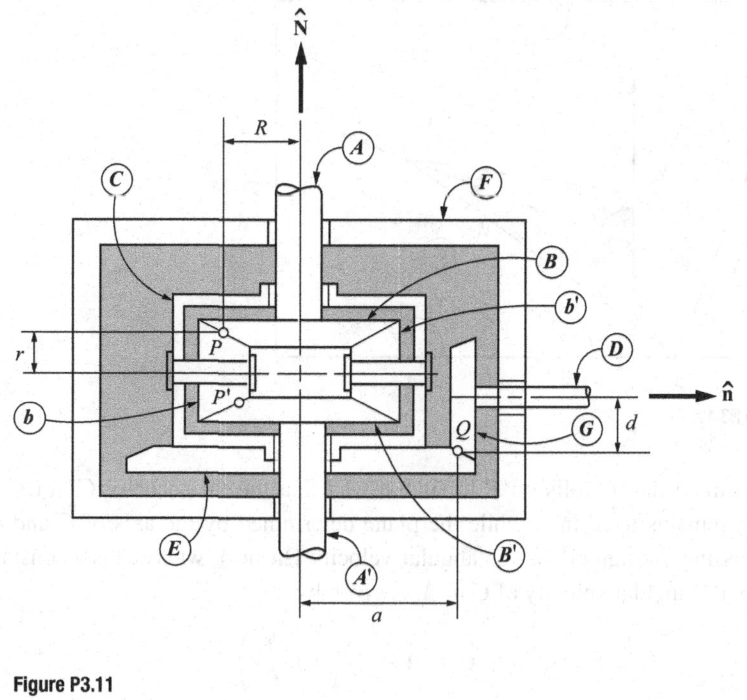

Figure P3.11

Similarly, by taking rolling at P' and at Q into account, one arrives at relationships between ${}^{C}\omega^{B'}$, and ${}^{C}\omega^{b}$, in one case, and ${}^{F}\omega^{E}$ and ${}^{F}\omega^{G}$, in the other case, where ${}^{C}\omega^{B'}$, ${}^{F}\omega^{E}$, and ${}^{F}\omega^{G}$ are angular speeds defined in terms of associated angular velocities as

$$ {}^{C}\omega^{B'} \triangleq {}^{C}\boldsymbol{\omega}^{B'} \cdot \hat{\mathbf{N}} \qquad {}^{F}\omega^{E} \triangleq {}^{F}\boldsymbol{\omega}^{E} \cdot \hat{\mathbf{N}} \qquad {}^{F}\omega^{G} \triangleq {}^{F}\boldsymbol{\omega}^{G} \cdot \hat{\mathbf{n}} $$

Next, one has

$$ {}^{F}\boldsymbol{\omega}^{B} \underset{(2.4.1)}{=} {}^{F}\boldsymbol{\omega}^{C} + {}^{C}\boldsymbol{\omega}^{B} $$

which, since B is attached rigidly to A, and all three vectors are parallel to $\hat{\mathbf{N}}$, implies that

$$ {}^{F}\omega^{A} = {}^{F}\omega^{C} + {}^{C}\omega^{B} $$

where ${}^{F}\omega^{C} \triangleq {}^{F}\boldsymbol{\omega}^{C} \cdot \hat{\mathbf{N}}$. Using Eq. (2.4.1) once more, this time in connection with B', C, and F, then brings one into position to find the relationship between ${}^{F}\omega^{A}$, ${}^{F}\omega^{A'}$, and ${}^{F}\omega^{D}$ by purely algebraic means.

Show that

$$ {}^{F}\omega^{D} = \frac{a}{2d}({}^{F}\omega^{A} + {}^{F}\omega^{A'}) $$

3.12 Figure P3.12 shows a right-circular, uniform, solid cone C in contact with a horizontal plane P that is fixed in a reference frame A. The base of C has a radius R, and C has a height $4h$.

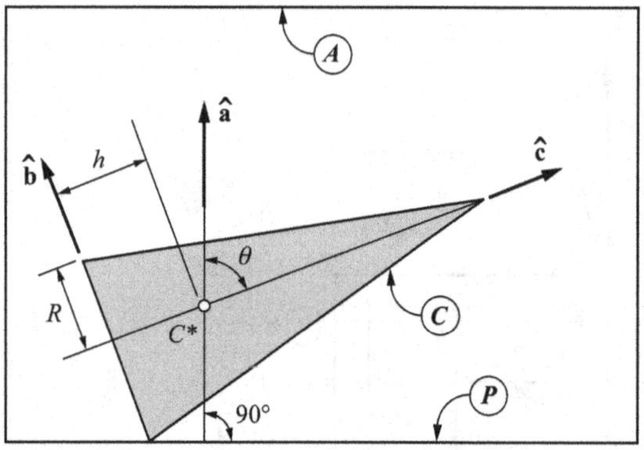

Figure P3.12

Assuming that C rolls on P in such a way that the mass center $C^\star$ of C (see Fig. P3.12) remains fixed in A while the plane determined by the axis of C and a vertical line passing through $C^\star$ has an angular velocity $\Omega\hat{\mathbf{a}}$ in A, where Ω is a constant, verify that $\boldsymbol{\omega}$, the angular velocity of C in A, is given by

$$ \boldsymbol{\omega} = \Omega \sin\theta \left(\hat{\mathbf{b}} + \frac{h}{R}\hat{\mathbf{c}} \right) $$

where θ is the inclination angle of the axis of C while $\hat{\mathbf{a}}$, $\hat{\mathbf{b}}$, and $\hat{\mathbf{c}}$ are unit vectors directed as shown in Fig. P3.12.

3.13 Referring to Problem 2.9, let P be a particle that moves on body B in such a way that the position vector $\mathbf{r}$ from $B^\star$ to P is given by

$$\mathbf{r} = c(\Omega^2 t^2 - 1)\hat{\mathbf{b}}_3$$

where c is a constant. Assuming that Ω is a constant, and letting $q_1 = q_2 = q_3 = \pi/2$ rad at time $t = 1/\Omega$, determine the acceleration of P in N for this instant.

Result $4c\Omega\dot{q}_1\hat{\mathbf{b}}_1 - \Omega(R\Omega + 4c\dot{q}_3)\hat{\mathbf{b}}_2 + 2c\Omega^2\hat{\mathbf{b}}_3$

3.14 The path point P mentioned in Problem 2.7 has a velocity both in A and in C, and these velocities can be expressed as

$${}^A\mathbf{v}^P = \widetilde{v}_x\,\hat{\mathbf{a}}_x + \widetilde{v}_y\,\hat{\mathbf{a}}_y + \widetilde{v}_z\,\hat{\mathbf{a}}_z$$

and

$${}^C\mathbf{v}^P = \widetilde{v}_1\,\hat{\mathbf{b}}_1 + \widetilde{v}_2\,\hat{\mathbf{b}}_2 + \widetilde{v}_3\,\hat{\mathbf{b}}_3$$

respectively, where $\widetilde{v}_x$, $\widetilde{v}_y$, $\widetilde{v}_z$, and $\widetilde{v}_i$ $(i = 1,2,3)$ are functions of q_j and u_j $(j = 1,\ldots,5)$, with u_4 and u_5 defined as in Problem 3.3. Determine $\widetilde{v}_x$, $\widetilde{v}_y$, $\widetilde{v}_z$, $\widetilde{v}_i$, and $\widetilde{a}_i$ $(i = 1,2,3)$, with $\widetilde{a}_i$ defined as

$$\widetilde{a}_i \triangleq {}^C\mathbf{a}^P \cdot \hat{\mathbf{b}}_i \qquad (i = 1,2,3)$$

where ${}^C\mathbf{a}^P$ is the acceleration of P in C.

Suggestion: To find ${}^C\mathbf{v}^P$ and ${}^C\mathbf{a}^P$, use Eqs. (2.8.1) and (2.8.2), replacing B with C, and $\overline{B}$ with $\hat{C}$ as defined in Problem 3.6.

Results

$$\widetilde{v}_x = u_4 \qquad \widetilde{v}_y = u_5 \qquad \widetilde{v}_z = 0$$
$$\widetilde{v}_1 = R(u_2 \tan q_2 - u_3) \qquad \widetilde{v}_2 = \widetilde{v}_3 = 0$$
$$\widetilde{a}_1 = R(\dot{u}_2 \tan q_2 - \dot{u}_3 - u_1 u_2 \sec^2 q_2)$$
$$\widetilde{a}_2 = R(u_2 \tan q_2 - u_3)^2$$
$$\widetilde{a}_3 = 0$$

***3.15** Figure P3.15 is a schematic representation of a robot arm consisting of three elements A, B, and C, the last of which holds a rigid body D rigidly. One end of A is a hub that is made to rotate about a vertical axis fixed in the Earth E. At a point P, B is connected to A by means of a motor (all parts of which are rigidly attached either to A or to B) that causes B to rotate relative to A about a horizontal axis fixed in A, passing through P, and perpendicular to the axis of A. Finally, C is connected to B by means of a rack-and-pinion drive that can make C slide relative to B.

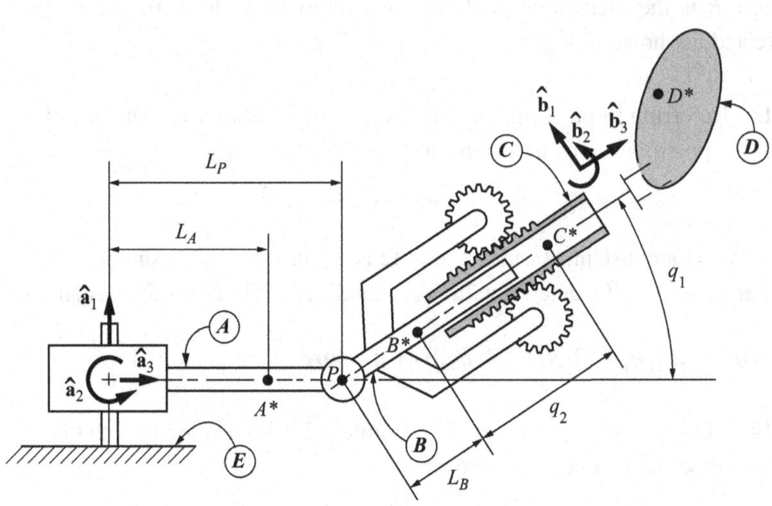

Figure P3.15

Letting L_A, L_B, and L_P denote distances as indicated in Fig. P3.15, where $A^\star$, $B^\star$, and $C^\star$ are points fixed on A, B, and C, respectively, introduce unit vectors $\hat{\mathbf{a}}_i$ and $\hat{\mathbf{b}}_i$ ($i = 1,2,3$) as shown in Fig. P3.15, let $p_i \triangleq \mathbf{p}^{C^\star D^\star} \cdot \hat{\mathbf{b}}_i$, where $\mathbf{p}^{C^\star D^\star}$ is the position vector from $C^\star$ to $D^\star$, the mass center of D, and define u_1, u_2, u_3 as

$$u_1 \triangleq {}^E\boldsymbol{\omega}^A \cdot \hat{\mathbf{a}}_1 \qquad u_2 \triangleq {}^A\boldsymbol{\omega}^B \cdot \hat{\mathbf{b}}_2 \qquad u_3 \triangleq {}^B\mathbf{v}^{C^\star} \cdot \hat{\mathbf{b}}_3$$

where ${}^E\boldsymbol{\omega}^A$ is the angular velocity of A in E, ${}^A\boldsymbol{\omega}^B$ is the angular velocity of B in A, and ${}^B\mathbf{v}^{C^\star}$ is the velocity of $C^\star$ in B. Denoting the radian measure of the angle between the axes of A and B by q_1, letting $\mathrm{s}_1 \triangleq \sin q_1$, $\mathrm{c}_1 \triangleq \cos q_1$, and designating as q_2 the distance from $B^\star$ to $C^\star$, determine ${}^E\boldsymbol{\omega}^A$, ${}^E\boldsymbol{\omega}^B$, ${}^E\boldsymbol{\omega}^C$, and ${}^E\boldsymbol{\omega}^D$, the angular velocities of A, B, C, and D in E; ${}^E\boldsymbol{\alpha}^A$, ${}^E\boldsymbol{\alpha}^B$, ${}^E\boldsymbol{\alpha}^C$, and ${}^E\boldsymbol{\alpha}^D$, the angular accelerations of A, B, C, and D in E; ${}^E\mathbf{v}^{A^\star}$, ${}^E\mathbf{v}^{B^\star}$, ${}^E\mathbf{v}^{C^\star}$, and ${}^E\mathbf{v}^{D^\star}$, the velocities of $A^\star$, $B^\star$, $C^\star$, and $D^\star$ in E; and ${}^E\mathbf{a}^{A^\star}$, ${}^E\mathbf{a}^{B^\star}$, ${}^E\mathbf{a}^{C^\star}$, and ${}^E\mathbf{a}^{D^\star}$, the accelerations of $A^\star$, $B^\star$, $C^\star$, and $D^\star$ in E. (Note that $\dot{q}_1 = u_2$ and $\dot{q}_2 = u_3$.)

Suggestion: To facilitate the writing of results, introduce $Z_1, \ldots, Z_{34}$ as

$$Z_1 \triangleq u_1\mathrm{c}_1 \qquad Z_2 \triangleq u_1\mathrm{s}_1 \qquad Z_3 \triangleq -Z_2u_2 \qquad Z_4 \triangleq Z_1u_2 \qquad Z_5 \triangleq -L_Au_1$$

$$Z_6 \triangleq -(L_P + L_B\mathrm{c}_1) \qquad Z_7 \triangleq u_2L_B \qquad Z_8 \triangleq Z_6u_1 \qquad Z_9 \triangleq L_B + q_2$$

$$Z_{10} \triangleq Z_6 - q_2\mathrm{c}_1 \qquad Z_{11} \triangleq u_2Z_9 \qquad Z_{12} \triangleq Z_{10}u_1 \qquad Z_{13} \triangleq -\mathrm{s}_1p_2$$

$$Z_{14} \triangleq Z_9 + p_3 \qquad Z_{15} \triangleq Z_{10} + \mathrm{s}_1p_1 - \mathrm{c}_1p_3 \qquad Z_{16} \triangleq \mathrm{c}_1p_2$$

$$Z_{17} \triangleq Z_{13}u_1 + Z_{14}u_2 \qquad Z_{18} \triangleq Z_{15}u_1 \qquad Z_{19} \triangleq Z_{16}u_1 - u_2p_1 + u_3$$

$$Z_{20} \triangleq u_1Z_5 \qquad Z_{21} \triangleq L_B\mathrm{s}_1u_2 \qquad Z_{22} \triangleq -Z_2Z_8 \qquad Z_{23} \triangleq Z_{21}u_1 + Z_2Z_7$$

$$Z_{24} \triangleq Z_1Z_8 - u_2Z_7 \qquad Z_{25} \triangleq Z_{21} - u_3\mathrm{c}_1 + q_2\mathrm{s}_1u_2$$

$$Z_{26} \triangleq 2u_2u_3 - Z_2Z_{12} \qquad Z_{27} \triangleq Z_{25}u_1 + Z_2Z_{11} - Z_1u_3$$

$$Z_{28} \triangleq Z_1 Z_{12} - u_2 Z_{11} \qquad Z_{29} \triangleq -Z_{16} u_2 \qquad Z_{30} \triangleq Z_{25} + u_2(\mathrm{c}_1 p_1 + \mathrm{s}_1 p_3)$$
$$Z_{31} \triangleq Z_{13} u_2 \qquad Z_{32} \triangleq Z_{29} u_1 + u_2(u_3 + Z_{19}) - Z_2 Z_{18}$$
$$Z_{33} \triangleq Z_{30} u_1 + Z_2 Z_{17} - Z_1 Z_{19} \qquad Z_{34} \triangleq Z_{31} u_1 + Z_1 Z_{18} - u_2 Z_{17}$$

Results

$$^E\boldsymbol{\omega}^A = u_1 \hat{\mathbf{a}}_1$$
$$^E\boldsymbol{\omega}^B = {}^E\boldsymbol{\omega}^C = {}^E\boldsymbol{\omega}^D = u_1 \mathrm{c}_1 \hat{\mathbf{b}}_1 + u_2 \hat{\mathbf{b}}_2 + u_1 \mathrm{s}_1 \hat{\mathbf{b}}_3 = Z_1 \hat{\mathbf{b}}_1 + u_2 \hat{\mathbf{b}}_2 + Z_2 \hat{\mathbf{b}}_3$$
$$^E\boldsymbol{\alpha}^A = \dot{u}_1 \hat{\mathbf{a}}_1 \qquad ^E\boldsymbol{\alpha}^B = {}^E\boldsymbol{\alpha}^C = {}^E\boldsymbol{\alpha}^D = (\dot{u}_1 \mathrm{c}_1 + Z_3)\hat{\mathbf{b}}_1 + \dot{u}_2 \hat{\mathbf{b}}_2 + (\dot{u}_1 \mathrm{s}_1 + Z_4)\hat{\mathbf{b}}_3$$
$$^E\mathbf{v}^{A^\star} = -L_A u_1 \hat{\mathbf{a}}_2 = Z_5 \hat{\mathbf{a}}_2 \qquad ^E\mathbf{v}^{B^\star} = u_2 L_B \hat{\mathbf{b}}_1 + Z_6 u_1 \hat{\mathbf{b}}_2 = Z_7 \hat{\mathbf{b}}_1 + Z_8 \hat{\mathbf{b}}_2$$
$$^E\mathbf{v}^{C^\star} = u_2 Z_9 \hat{\mathbf{b}}_1 + Z_{10} u_1 \hat{\mathbf{b}}_2 + u_3 \hat{\mathbf{b}}_3 = Z_{11} \hat{\mathbf{b}}_1 + Z_{12} \hat{\mathbf{b}}_2 + u_3 \hat{\mathbf{b}}_3$$
$$^E\mathbf{v}^{D^\star} = (Z_{13} u_1 + Z_{14} u_2)\hat{\mathbf{b}}_1 + Z_{15} u_1 \hat{\mathbf{b}}_2 + (Z_{16} u_1 - u_2 p_1 + u_3)\hat{\mathbf{b}}_3$$
$$= Z_{17} \hat{\mathbf{b}}_1 + Z_{18} \hat{\mathbf{b}}_2 + Z_{19} \hat{\mathbf{b}}_3$$
$$^E\mathbf{a}^{A^\star} = -L_A \dot{u}_1 \hat{\mathbf{a}}_2 + Z_{20} \hat{\mathbf{a}}_3$$
$$^E\mathbf{a}^{B^\star} = (\dot{u}_2 L_B + Z_{22})\hat{\mathbf{b}}_1 + (Z_6 \dot{u}_1 + Z_{23})\hat{\mathbf{b}}_2 + Z_{24} \hat{\mathbf{b}}_3$$
$$^E\mathbf{a}^{C^\star} = (\dot{u}_2 Z_9 + Z_{26})\hat{\mathbf{b}}_1 + (Z_{10} \dot{u}_1 + Z_{27})\hat{\mathbf{b}}_2 + (\dot{u}_3 + Z_{28})\hat{\mathbf{b}}_3$$
$$^E\mathbf{a}^{D^\star} = (Z_{13} \dot{u}_1 + Z_{14} \dot{u}_2 + Z_{32})\hat{\mathbf{b}}_1 + (Z_{15} \dot{u}_1 + Z_{33})\hat{\mathbf{b}}_2$$
$$+ (Z_{16} \dot{u}_1 - p_1 \dot{u}_2 + \dot{u}_3 + Z_{34})\hat{\mathbf{b}}_3$$

Note that two expressions are given for the angular velocities of B, C, and D in E, as well as for the velocities of $A^\star$, $B^\star$, $C^\star$, and $D^\star$ in E. In each case, the quantities u_1, u_2, u_3 appear explicitly in the first expression but are absent from the second, except in the case of $^E\mathbf{v}^{C^\star}$, where no simplification would result from replacing u_3 with another symbol. There are two reasons for writing each angular velocity and each velocity in these two ways. The first will become apparent in connection with Problem 4.19; the second is that both versions come into play when one seeks to write expressions for angular accelerations of rigid bodies and accelerations of mass centers in such a way that $\dot{u}_1$, $\dot{u}_2$, and $\dot{u}_3$ appear explicitly.

PROBLEM SET 4
(Secs. 3.1–3.9)

4.1 Mutually perpendicular unit vectors $\hat{\mathbf{a}}_1$, $\hat{\mathbf{a}}_2$, and $\hat{\mathbf{a}}_3 \triangleq \hat{\mathbf{a}}_1 \times \hat{\mathbf{a}}_2$ are fixed in a reference frame A, and a unit vector $\hat{\mathbf{b}}$ is fixed in a rigid body B, one of whose points, O, is fixed in A. B is brought into a general orientation in A, after $\hat{\mathbf{b}}$ has been aligned with $\hat{\mathbf{a}}_1$, by being subjected to successive rotations characterized by the vectors $\theta_1 \hat{\mathbf{a}}_1$, $\theta_2 \hat{\mathbf{a}}_2$, and $\theta_3 \hat{\mathbf{a}}_3$, where θ_1, θ_2, and θ_3 are the radian measures of angles. A particle P is free to move on a line L that is fixed in B, passes through O, and is parallel to $\hat{\mathbf{b}}$.

Letting $\mathbf{p}$ be the position vector from O to P, and defining x_i as

$$x_i \triangleq \mathbf{p} \cdot \hat{\mathbf{a}}_i \qquad (i = 1,2,3)$$

show that P is guaranteed to remain on L when x_1, x_2, and x_3 satisfy the holonomic constraint equations

$$x_1 s_3 - x_2 c_3 = 0 \qquad x_3 c_2 c_3 + x_1 s_2 = 0$$

where $s_i \triangleq \sin\theta_i$, $c_i \triangleq \cos\theta_i$ $(i = 2,3)$.

Letting q be an arbitrary quantity, verify that the constraint equations are satisfied identically if $x_1 = qc_2c_3$, $x_2 = qc_2s_3$, $x_3 = -qs_2$.

Determine x_1, x_2, and x_3 such that $\mathbf{p} = q\hat{\mathbf{b}}$.

4.2 Figure P4.2 shows a double pendulum consisting of two particles, P_1 and P_2, supported by rods of lengths L_1 and L_2 in such a way that the position vectors, $\mathbf{p}_1$ and $\mathbf{p}_2$, of P_1 and P_2 relative to a point O fixed in a reference frame A are at all times perpendicular to $\hat{\mathbf{a}}_z$, one of three mutually perpendicular unit vectors, $\hat{\mathbf{a}}_x$, $\hat{\mathbf{a}}_y$, $\hat{\mathbf{a}}_z$, fixed in A. Letting

$$x_i \triangleq \mathbf{p}_i \cdot \hat{\mathbf{a}}_x \qquad y_i \triangleq \mathbf{p}_i \cdot \hat{\mathbf{a}}_y \qquad z_i \triangleq \mathbf{p}_i \cdot \hat{\mathbf{a}}_z \qquad (i = 1,2)$$

construct functions $f_j(x_1, y_1, z_1, x_2, y_2, z_2)$ $(j = 1,\ldots,4)$ such that four holonomic constraint equations governing motions of P_1 and P_2 in A can be expressed as $f_j = 0$ $(j = 1,\ldots,4)$.

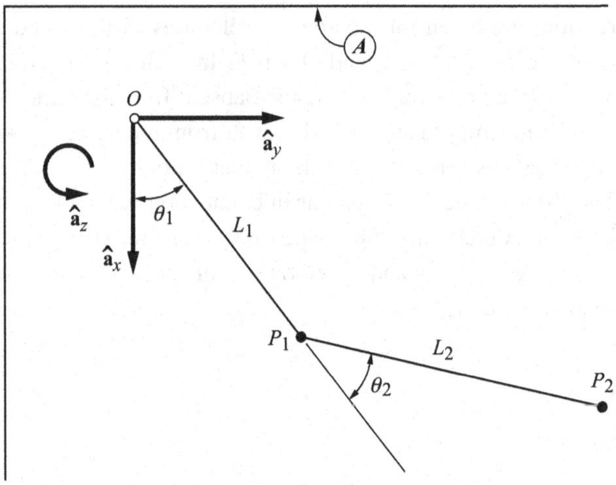

Figure P4.2

Results

$$f_1 = x_1{}^2 + y_1{}^2 - L_1{}^2$$
$$f_2 = (x_2 - x_1)^2 + (y_2 - y_1)^2 - L_2{}^2$$
$$f_3 = z_1 \qquad f_4 = z_2$$

4.3 Figure P4.3 shows two particles, P_1 and P_2, supported by a linkage in such a way that the position vectors, $\mathbf{p}_1$ and $\mathbf{p}_2$, of P_1 and P_2 relative to a point O fixed in a reference frame A are at all times perpendicular to $\hat{\mathbf{a}}_z$, one of three mutually perpendicular unit vectors $\hat{\mathbf{a}}_x$, $\hat{\mathbf{a}}_y$, $\hat{\mathbf{a}}_z$, fixed in A. Determine the number of holonomic constraint equations governing motions of P_1 and P_2 in A.

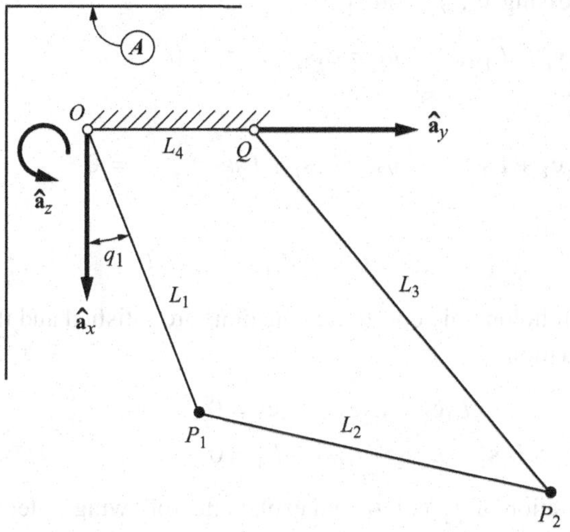

Figure P4.3

Result 5

4.4 Referring to Problem 4.1, and assuming that θ_1, θ_2, and θ_3 are prescribed functions of time, show that only one generalized coordinate is required to specify the configuration of P in A, and verify that q is such a generalized coordinate.

4.5 Referring to Problem 4.2, suppose that q_1 and q_2 are introduced by expressing x_i, y_i, and z_i $(i = 1,2)$ as

$$x_1 = q_1 \qquad y_1 = (L_1{}^2 - q_1{}^2)^{1/2} \qquad z_1 = 0$$

$$x_2 = q_2 \qquad y_2 = (L_1{}^2 - q_1{}^2)^{1/2} + [L_2{}^2 - (q_2 - q_1)^2]^{1/2} \qquad z_2 = 0$$

Verify that the four holonomic constraint equations considered in Problem 4.2 then are satisfied. Explain the following statement: "For motions during which y_1 and y_2 acquire negative values, q_1 and q_2 are not generalized coordinates." Finally, letting θ_1 and θ_2 be the radian measures of two angles as indicated in Fig. P4.2, show that θ_1 and θ_2 are generalized coordinates for P_1 and P_2 in A.

4.6 Referring to Problem 4.3, let

$$x_i \triangleq \mathbf{p}_i \cdot \hat{\mathbf{a}}_x \qquad y_i \triangleq \mathbf{p}_i \cdot \hat{\mathbf{a}}_y \qquad z_i \triangleq \mathbf{p}_i \cdot \hat{\mathbf{a}}_z \qquad (i = 1,2)$$

and, after noting that a single generalized coordinate suffices to characterize the configuration of P_1 and P_2 in A, attempt to express x_i, y_i, and z_i $(i = 1,2)$ as functions of the angle q_1 shown in Fig. P4.3 in such a way that all holonomic constraint equations are satisfied. Next, after expressing x_1, y_1, and z_1 as

$$x_1 = L_1 c_1 \qquad y_1 = L_1 s_1 \qquad z_1 = 0$$

and x_2, y_2, and z_2 as

$$x_2 = L_1 c_1 + L_2 c_2 \qquad y_2 = L_1 s_1 + L_2 s_2 \qquad z_2 = 0$$

and as

$$x_2 = L_3 c_3 \qquad y_2 = L_3 s_3 + L_4 \qquad z_2 = 0$$

show that in both cases all holonomic constraint equations are satisfied and that q_1, q_2, and q_3 satisfy the two equations

$$L_1 c_1 + L_2 c_2 - L_3 c_3 = 0$$
$$L_1 s_1 + L_2 s_2 - L_3 s_3 - L_4 = 0$$

Give a geometric interpretation of q_2 and q_3, and explain the following statement: "Any one of q_1, q_2, and q_3, but no more than one at a time, can be a generalized coordinate for P_1 and P_2 in A."

4.7 Determine the number of generalized coordinates of each of the following systems in a reference frame A: (*a*) Two rigid bodies attached to each other by means of a ball-and-socket joint, but otherwise free to move in A. (*b*) A rigid body B carrying a rotor that is free to rotate relative to B about an axis fixed in B while B is free to move in A. (*c*) A rigid body B carrying a rotor that is made to rotate relative to B at a prescribed, time-dependent rate while B is free to move in A. (*d*) The system of two particles in Problem 4.2. (*e*) The system of two particles in Problem 4.3.

Results (*a*) 9; (*b*) 7; (*c*) 6; (*d*) 2; (*e*) 1

4.8 When the disk C of Problem 2.7 moves subject to the configuration constraint that C must remain in contact with H, then $q_1, \ldots, q_5$ are generalized coordinates for C in A. Show that $u_1, \ldots, u_5$, defined as

$$u_i \triangleq {}^A\boldsymbol{\omega}^C \cdot \hat{\mathbf{b}}_i \qquad (i = 1,2,3) \qquad u_4 \triangleq {}^A\mathbf{v}^P \cdot \hat{\mathbf{a}}_x \qquad u_5 \triangleq {}^A\mathbf{v}^P \cdot \hat{\mathbf{a}}_y$$

are motion variables for C in A, provided that $|q_2| \neq \pi/2$ rad.

4.9 Referring to Problem 2.9, and defining motion variables u_1, u_2, u_3 as (*a*) $u_i \triangleq {}^N\boldsymbol{\omega}^B \cdot \hat{\mathbf{b}}_i$ $(i = 1,2,3)$ and (*b*) $u_i \triangleq {}^A\boldsymbol{\omega}^B \cdot \hat{\mathbf{b}}_i$ $(i = 1,2,3)$, determine Z_1, Z_2, Z_3 such that Eqs. (3.4.1) are satisfied.

Results

$$(a)\quad Z_1 = -\Omega c_1 s_2$$
$$Z_2 = \Omega(c_1 c_2 s_3 + c_3 s_1)$$
$$Z_3 = \Omega(c_1 c_2 c_3 - s_3 s_1)$$
$$(b)\quad Z_1 = Z_2 = Z_3 = 0$$

4.10 Referring to Problem 2.7, and considering only motions of rolling of C on H, let $u_1, \ldots, u_5$ be motion variables defined as in Problem 4.8. Determine A_{rs} and B_r ($r = 4,5;\ s = 1,2,3$) such that Eqs. (3.5.2) are satisfied.

Results

$$A_{41} = 0 \qquad A_{42} = R\cos q_1 \tan q_2 \qquad A_{43} = -R\cos q_1$$
$$A_{51} = 0 \qquad A_{52} = R\sin q_1 \tan q_2 \qquad A_{53} = -R\sin q_1$$
$$B_4 = B_5 = 0$$

4.11 Figure P4.11 shows two sharp-edged circular disks, C_1 and C_2, each of radius R, mounted at the extremities of a cylindrical shaft S of length $2L$, the axis of S coinciding with those of C_1 and C_2. The disks are supported by a plane P that is fixed in a reference frame A, and they can rotate freely relative to S.

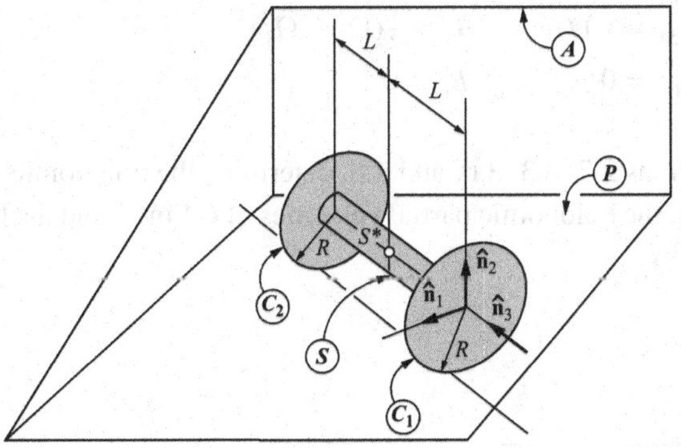

Figure P4.11

Six motion variables suffice to characterize all motions of C_1, C_2, and S in A. Defining these as

$$u_1 \triangleq {}^A\mathbf{v}^{S^\star} \cdot \hat{\mathbf{n}}_1 \qquad u_2 \triangleq {}^A\boldsymbol{\omega}^S \cdot \hat{\mathbf{n}}_2 \qquad u_3 \triangleq {}^A\boldsymbol{\omega}^S \cdot \hat{\mathbf{n}}_3$$
$$u_4 \triangleq {}^A\boldsymbol{\omega}^{C_1} \cdot \hat{\mathbf{n}}_3 \qquad u_5 \triangleq {}^A\boldsymbol{\omega}^{C_2} \cdot \hat{\mathbf{n}}_3 \qquad u_6 \triangleq {}^A\mathbf{v}^{S^\star} \cdot \hat{\mathbf{n}}_3$$

where $\hat{\mathbf{n}}_1$, $\hat{\mathbf{n}}_2$, and $\hat{\mathbf{n}}_3$ are mutually perpendicular unit vectors, with $\hat{\mathbf{n}}_2$ normal to P, as shown in Fig. P4.11, and $S^\star$ is the mass center of S, show that, when C_1 and C_2 roll on

P, then u_4, u_5, and u_6 each can be expressed in terms of u_1, u_2, and u_3 by means of an equation having the form of Eq. (3.5.2). Determine A_{rs} and B_r ($r = 4,5,6$; $s = 1,2,3$).

Results

$$A_{41} = -1/R \qquad A_{42} = L/R \qquad A_{43} = 0$$
$$A_{51} = -1/R \qquad A_{52} = -L/R \qquad A_{53} = 0$$
$$A_{61} = A_{62} = A_{63} = 0$$
$$B_4 = B_5 = B_6 = 0$$

4.12 Referring to Problem 4.11, suppose that motors connecting S to C_1 and C_2 are used to cause C_1 and C_2 to rotate in such a way that

$$^S\boldsymbol{\omega}^{C_1} = \Omega_1 \hat{\mathbf{n}}_3 \qquad ^S\boldsymbol{\omega}^{C_2} = \Omega_2 \hat{\mathbf{n}}_3$$

where Ω_1 and Ω_2 are prescribed functions of the time t. Show that the system formed by S, C_1, and C_2 possesses one degree of freedom in A if C_1 and C_2 roll on P, and determine A_{r1} and B_r ($r = 2,\ldots,6$) such that $u_2,\ldots,u_6$ satisfy Eqs. (3.5.2).

Results

$$A_{21} = 0 \qquad B_2 = \tfrac{1}{2}(R/L)(\Omega_1 - \Omega_2)$$
$$A_{31} = -1/R \qquad B_3 = -\tfrac{1}{2}(\Omega_1 + \Omega_2)$$
$$A_{41} = -1/R \qquad B_4 = \tfrac{1}{2}(\Omega_1 - \Omega_2)$$
$$A_{51} = -1/R \qquad B_5 = \tfrac{1}{2}(\Omega_2 - \Omega_1)$$
$$A_{61} = 0 \qquad B_6 = 0$$

4.13 Referring to Problems 2.7, 3.3, 3.6, and 4.8, determine the holonomic partial angular velocities of C in A, the holonomic partial velocities of $C^\star$ in A, and the holonomic partial velocities of $\hat{C}$ in A.

Results Table P4.13

Table P4.13

r	$^A\boldsymbol{\omega}_r^C$	$^A\mathbf{v}_r^{C^\star}$	$^A\mathbf{v}_r^{\hat{C}}$
1	$\hat{\mathbf{b}}_1$	$R\hat{\mathbf{b}}_3$	$\mathbf{0}$
2	$\hat{\mathbf{b}}_2$	$-R\tan q_2 \hat{\mathbf{b}}_1$	$-R\tan q_2 \hat{\mathbf{b}}_1$
3	$\hat{\mathbf{b}}_3$	$\mathbf{0}$	$R\hat{\mathbf{b}}_1$
4	$\mathbf{0}$	$\hat{\mathbf{a}}_x$	$\hat{\mathbf{a}}_x$
5	$\mathbf{0}$	$\hat{\mathbf{a}}_y$	$\hat{\mathbf{a}}_y$

4.14 With $u_1,\ldots,u_5$ defined as in Problem 4.8, and considering only motions of rolling of C on H, determine the nonholonomic partial angular velocities of C in A, the nonholonomic partial velocities of $C^\star$ in A, and the nonholonomic partial velocities of $\hat{C}$ in A. Do this by inspecting expressions for ${}^A\boldsymbol{\omega}^C$, ${}^A\mathbf{v}^{C^\star}$, and ${}^A\mathbf{v}^{\hat{C}}$; then check the results by using Eqs. (3.6.15) and (3.6.17) in conjunction with information available in Problems 4.10 and 4.13.

Results Table P4.14

Table P4.14

r	${}^A\tilde{\boldsymbol{\omega}}_r^C$	${}^A\tilde{\mathbf{v}}_r^{C^\star}$	${}^A\tilde{\mathbf{v}}_r^{\hat{C}}$
1	$\hat{\mathbf{b}}_1$	$R\hat{\mathbf{b}}_3$	$\mathbf{0}$
2	$\hat{\mathbf{b}}_2$	$\mathbf{0}$	$\mathbf{0}$
3	$\hat{\mathbf{b}}_3$	$-R\hat{\mathbf{b}}_1$	$\mathbf{0}$

4.15 Let $\hat{\mathbf{n}}_3$ be a unit vector fixed in a reference frame N and locally directed vertically upward. A particle P is required to move in such a way† as to maintain a constant angle γ between $\hat{\mathbf{n}}_3$ and the velocity ${}^N\mathbf{v}^P$ of P in N. (It may be possible to move a particle in this manner by fixing it in a small robotic flying vehicle.) Express the motion constraint in terms of ${}^N\mathbf{v}^P$, and determine whether the constraint equation is linear in ${}^N\mathbf{v}^P$. Let unit vectors $\hat{\mathbf{n}}_1$ and $\hat{\mathbf{n}}_2$ be locally horizontal, fixed in N, orthogonal to each other, and orthogonal to $\hat{\mathbf{n}}_3$, and choose motion variables u_1, u_2, and u_3 to characterize the unconstrained velocity of P in N as

$$ {}^N\mathbf{v}^P = u_1\hat{\mathbf{n}}_1 + u_2\hat{\mathbf{n}}_2 + u_3\hat{\mathbf{n}}_3 $$

Express the motion constraint in terms of these motion variables, and determine whether the resulting relationship is linear in the motion variables. Finally, after defining a constant a as

$$ a \triangleq \cos\gamma / \sin\gamma $$

show that the motion constraint can be expressed with the relationship

$$ u_3 = \pm|a|\sqrt{{u_1}^2 + {u_2}^2} $$

and relate each of the two possible signs to ranges of the angle γ.

Results $({}^N\mathbf{v}^P/\sqrt{{}^N\mathbf{v}^P \cdot {}^N\mathbf{v}^P}) \cdot \hat{\mathbf{n}}_3 - \cos\gamma = 0$, No; $u_3/({u_1}^2 + {u_2}^2 + {u_3}^2)^{1/2} - \cos\gamma = 0$, No; $u_3 \geq 0$ when $0 \leq \gamma \leq \pi/2$, $u_3 \leq 0$ when $\pi/2 \leq \gamma \leq \pi$

† Paul Appell, "Exemple de mouvement d'un point assujetti à liaison exprimée par une relation non linéaire entre les composantes de la vitesse," *Rendiconti del Circolo Matematico di Palermo* 32 (1911), pp. 48–50.

***4.16** The configuration of a system S in a reference frame A is characterized by generalized coordinates $q_1, \ldots, q_n$. Taking $u_r \triangleq \dot{q}_r$ $(r = 1, \ldots, n)$, and letting $\boldsymbol{\beta}$ be a vector fixed in a rigid body B belonging to S, show that

$$\frac{{}^A\partial \boldsymbol{\beta}}{\partial q_r} = \boldsymbol{\omega}_r \times \boldsymbol{\beta} \qquad (r = 1, \ldots, n)$$

where $\boldsymbol{\omega}_r$ is the rth partial angular velocity of B in A.

4.17 Referring to Problem 2.9, and defining motion variables u_1, u_2, u_3 as (*a*) $u_i \triangleq {}^N\boldsymbol{\omega}^B \cdot \hat{\mathbf{b}}_i$ $(i = 1,2,3)$, (*b*) $u_i \triangleq {}^A\boldsymbol{\omega}^B \cdot \hat{\mathbf{b}}_i$ $(i = 1,2,3)$, and (*c*) $u_i \triangleq \dot{q}_i$ $(i = 1,2,3)$, determine the partial angular velocities ${}^N\boldsymbol{\omega}_r^B$ $(r = 1,2,3)$ in each case.

If α_i is defined as ${}^N\boldsymbol{\alpha}^B \cdot \hat{\mathbf{b}}_i$ $(i = 1,2,3)$, where ${}^N\boldsymbol{\alpha}^B$ denotes the angular acceleration of B in N, then α_i can be expressed as a function of q_1, q_2, q_3, u_1, u_2, u_3, and $\dot{u}_1$, $\dot{u}_2$, $\dot{u}_3$. Which of the three definitions of u_i $(i = 1,2,3)$ given here leads to the simplest expressions for α_i $(i = 1,2,3)$?

Results

$$\begin{aligned}
(a)\quad & {}^N\boldsymbol{\omega}_r^B = \hat{\mathbf{b}}_r \qquad (r = 1,2,3)\\
(b)\quad & {}^N\boldsymbol{\omega}_r^B = \hat{\mathbf{b}}_r \qquad (r = 1,2,3)\\
(c)\quad & {}^N\boldsymbol{\omega}_1^B = c_2\hat{\mathbf{b}}_1 + s_2s_3\hat{\mathbf{b}}_2 + s_2c_3\hat{\mathbf{b}}_3\\
& {}^N\boldsymbol{\omega}_2^B = c_3\hat{\mathbf{b}}_2 - s_3\hat{\mathbf{b}}_3\\
& {}^N\boldsymbol{\omega}_3^B = \hat{\mathbf{b}}_1
\end{aligned}$$

Definition (*a*) leads to the simplest expression for α_i $(i = 1,2,3)$.

4.18 Show that the result developed in Problem 4.15 can be written as

$$({}^N\mathbf{v}^P \cdot \hat{\mathbf{n}}_3)^2 - a^2\left[({}^N\mathbf{v}^P \cdot \hat{\mathbf{n}}_1)^2 + ({}^N\mathbf{v}^P \cdot \hat{\mathbf{n}}_2)^2\right] = 0$$

Differentiate this relationship with respect to time to obtain one involving ${}^N\mathbf{a}^P$, the acceleration of P in N. Let $\dot{u}_r$ $(r = 1,2,3)$ be the time derivatives of the motion variables introduced in Problem 4.15. Express $\dot{u}_3$ in terms of $\dot{u}_1$ and $\dot{u}_2$, and, referring to Eqs. (3.7.1), determine $\widetilde{A}_{31}$, $\widetilde{A}_{32}$, and $\widetilde{B}_3$.

Results ${}^N\mathbf{a}^P \cdot \left[\hat{\mathbf{n}}_3 - a^2(u_1\hat{\mathbf{n}}_1 + u_2\hat{\mathbf{n}}_2)/u_3\right] = 0$, $\widetilde{A}_{31} = a^2u_1/u_3$, $\widetilde{A}_{32} = a^2u_2/u_3$, $\widetilde{B}_3 = 0$

***4.19** Referring to Problem 3.15, determine the partial angular velocities ${}^E\boldsymbol{\omega}_r^A$, ${}^E\boldsymbol{\omega}_r^B$, ${}^E\boldsymbol{\omega}_r^C$, and ${}^E\boldsymbol{\omega}_r^D$ $(r = 1,2,3)$ and the partial velocities ${}^E\mathbf{v}_r^{A\star}$, ${}^E\mathbf{v}_r^{B\star}$, ${}^E\mathbf{v}_r^{C\star}$, and ${}^E\mathbf{v}_r^{D\star}$ $(r = 1,2,3)$.

Results Tables P4.19(*a*), P4.19(*b*)

Table P4.19(a)

r	${}^E\boldsymbol{\omega}_r^A$	${}^E\boldsymbol{\omega}_r^B$	${}^E\boldsymbol{\omega}_r^C$	${}^E\boldsymbol{\omega}_r^D$
1	$\hat{\mathbf{a}}_1$	$c_1\hat{\mathbf{b}}_1 + s_1\hat{\mathbf{b}}_3$	$c_1\hat{\mathbf{b}}_1 + s_1\hat{\mathbf{b}}_3$	$c_1\hat{\mathbf{b}}_1 + s_1\hat{\mathbf{b}}_3$
2	$\mathbf{0}$	$\hat{\mathbf{b}}_2$	$\hat{\mathbf{b}}_2$	$\hat{\mathbf{b}}_2$
3	$\mathbf{0}$	$\mathbf{0}$	$\mathbf{0}$	$\mathbf{0}$

Table P4.19(b)

r	${}^E\mathbf{v}_r^{A\star}$	${}^E\mathbf{v}_r^{B\star}$	${}^E\mathbf{v}_r^{C\star}$	${}^E\mathbf{v}_r^{D\star}$
1	$-L_A\hat{\mathbf{a}}_2$	$Z_6\hat{\mathbf{b}}_2$	$Z_{10}\hat{\mathbf{b}}_2$	$Z_{13}\hat{\mathbf{b}}_1 + Z_{15}\hat{\mathbf{b}}_2 + Z_{16}\hat{\mathbf{b}}_3$
2	$\mathbf{0}$	$L_B\hat{\mathbf{b}}_1$	$Z_9\hat{\mathbf{b}}_1$	$Z_{14}\hat{\mathbf{b}}_1 - p_1\hat{\mathbf{b}}_3$
3	$\mathbf{0}$	$\mathbf{0}$	$\hat{\mathbf{b}}_3$	$\hat{\mathbf{b}}_3$

4.20 Referring to Problems 4.15 and 4.18, determine all nonholonomic partial accelerations of P in N, as well as the vector ${}^N\tilde{\mathbf{a}}_t^P$.

Results

$$ {}^N\tilde{\mathbf{a}}_1^P = \hat{\mathbf{n}}_1 + a^2u_1\hat{\mathbf{n}}_3/u_3 \qquad {}^N\tilde{\mathbf{a}}_2^P = \hat{\mathbf{n}}_2 + a^2u_2\hat{\mathbf{n}}_3/u_3 \qquad {}^N\tilde{\mathbf{a}}_t^P = \mathbf{0} $$

4.21 In Problem 4.8, five motion variables are defined for C in A. An alternative set of motion variables can be introduced by defining $u_1,\ldots,u_5$ as

$$ u_r \triangleq \dot{q}_r \qquad (r = 1,\ldots,5) $$

When C rolls on H, these motion variables satisfy constraint equations of the form of Eqs. (3.5.2), with $A_{43} = -R\cos q_1$, $A_{53} = -R\sin q_1$, and $A_{rs} = B_r = 0$ for $r = 4,5$ and $s = 1,2$. Furthermore, the velocity of $C^\star$ in A is given by

$$ {}^A\mathbf{v}^{C^\star} = -R[(u_1\sin q_2 + u_3)\hat{\mathbf{b}}_1 + u_2\hat{\mathbf{b}}_3] $$

To verify that it can be far more laborious to work with the right-hand members of Eqs. (3.9.7) than with the left-hand members, determine ${}^A\tilde{\mathbf{v}}_r^{C^\star} \cdot {}^A\mathbf{a}^{C^\star}$ $(r = 1,2,3)$, where ${}^A\mathbf{a}^{C^\star}$ is the acceleration of $C^\star$ in A, by (a) forming ${}^A\tilde{\mathbf{v}}_r^{C^\star}$ by inspection of the expression given here for ${}^A\mathbf{v}^{C^\star}$, forming ${}^A\mathbf{a}^{C^\star}$ by differentiating ${}^A\mathbf{v}^{C^\star}$ with respect to time t in A, and then dot-multiplying ${}^A\tilde{\mathbf{v}}_r^{C^\star}$ with ${}^A\mathbf{a}^{C^\star}$ $(r = 1,2,3)$, and (b) using the right-hand members of Eqs. (3.9.7).

Results

$$
\begin{aligned}
{}^{A}\tilde{\mathbf{v}}_1^{C\star} \cdot {}^{A}\mathbf{a}^{C\star} &= R^2(\dot{u}_1 \sin q_2 + 2u_1u_2 \cos q_2 + \dot{u}_3) \sin q_2 \\
{}^{A}\tilde{\mathbf{v}}_2^{C\star} \cdot {}^{A}\mathbf{a}^{C\star} &= R^2[\dot{u}_2 - u_1 \cos q_2(u_1 \sin q_2 + u_3)] \\
{}^{A}\tilde{\mathbf{v}}_3^{C\star} \cdot {}^{A}\mathbf{a}^{C\star} &= R^2(\dot{u}_1 \sin q_2 + 2u_1u_2 \cos q_2 + \dot{u}_3)
\end{aligned}
$$

PROBLEM SET 5
(Secs. 4.1–4.5)

5.1 Regarding Fig. P5.1 as showing two views of a body B formed by matter distributed uniformly (*a*) over a surface having no planar portions and (*b*) throughout a solid, determine (by integration) the coordinates $x^\star$, $y^\star$, $z^\star$ of the mass center of B.

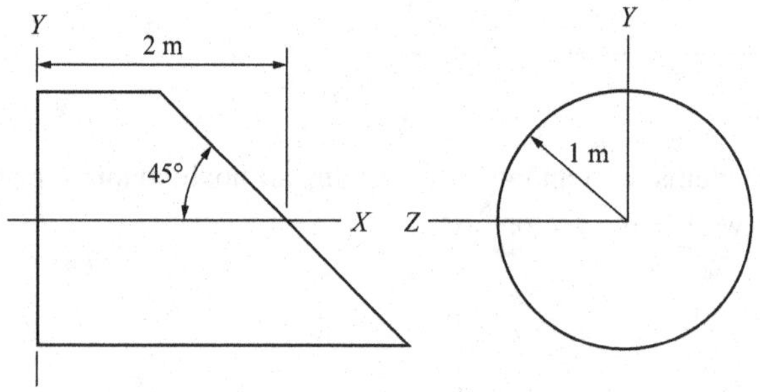

Figure P5.1

Results Table P5.1

Table P5.1

	$x^\star$ (m)	$y^\star$ (m)	$z^\star$ (m)
(*a*)	$\frac{9}{8}$	$-\frac{1}{4}$	0
(*b*)	$\frac{17}{16}$	$-\frac{1}{8}$	0

5.2 Regarding Fig. P5.2 as showing two views of a body B formed by matter distributed uniformly (*a*) over a surface having no planar portions and (*b*) throughout a solid, determine (without integration) the X-coordinate of the mass center of B.

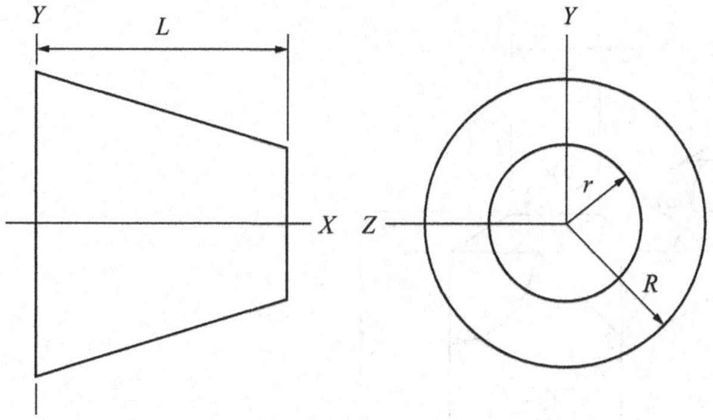

Figure P5.2

Results (*a*) $\dfrac{L}{3}\dfrac{R+2r}{R+r}$ (*b*) $\dfrac{L}{4}\dfrac{R^2+2rR+3r^2}{R^2+rR+r^2}$

***5.3** Prove the following theorems (known as "theorems of Pappus" or "Guldin's rules"):

When a plane curve C of length L is revolved about a line lying in the plane of C and not intersecting C, the area of the surface of revolution thus generated is equal to the product of L and the circumference of the circle described by the centroid of C.

When a plane region R of area A is revolved about a line lying in the plane of R and not passing through R, the volume of the solid of revolution thus generated is equal to the product of A and the circumference of the circle described by the centroid of R.

Use these theorems to locate the centroids of a semicircular curve and a semicircular sector, keeping in mind that the surface area and the volume of a sphere of radius R are equal to $4\pi R^2$ and $4\pi R^3/3$, respectively.

5.4 Parts A, B, C, D of the assembly shown in Fig. P5.4 are made of steel (7800 kg/m^3), sheet metal (17.00 kg/m^2), aluminum (2700 kg/m^3), and brass (8400 kg/m^3), respectively. (*a*) For $a = b = 0.3$ m, determine the coordinates $x^\star$, $y^\star$, $z^\star$ of the mass center of the assembly. (*b*) For $a = 0.3$ m, determine to three significant figures the range of values of b such that $x^\star = 0.400$ m.

Results (*a*) 0.434 m, 0.135 m, −0.300 m (*b*) 0.136 m $\le b \le$ 0.138 m

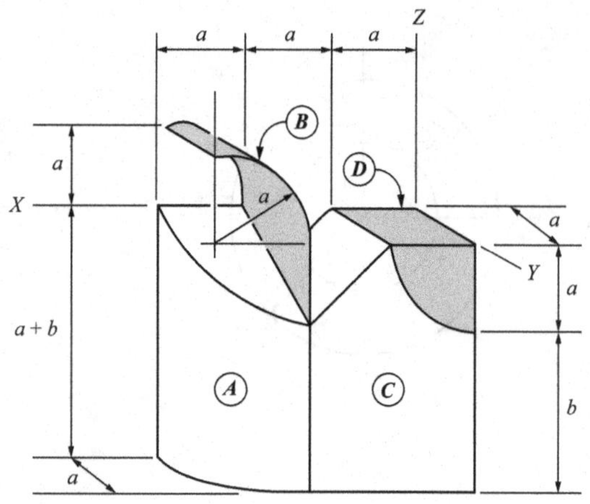

Figure P5.4

5.5 Show by means of an example that $\mathbf{I}_a$ as defined in Eq. (4.3.1) can be, but need not be, parallel to $\hat{\mathbf{n}}_a$.

5.6 Unit vectors $\hat{\mathbf{n}}_x$, $\hat{\mathbf{n}}_y$, and $\hat{\mathbf{n}}_z$ are respectively parallel to the axes OX, OY, OZ of a rectangular Cartesian coordinate system, and each unit vector points in the positive direction of the axis to which it is parallel. Letting S be a set of ν particles, m_i the mass of particle P_i, and x_i, y_i, and z_i the coordinates of P_i $(i = 1, \ldots, \nu)$, express I_x, the moment of inertia of S about the X-axis, and I_{yz}, the product of inertia of S relative to O for $\hat{\mathbf{n}}_y$ and $\hat{\mathbf{n}}_z$, in terms of the masses and coordinates of $P_1, \ldots, P_\nu$. Then answer the following questions: (*a*) Does it matter whether $\hat{\mathbf{n}}_x$, $\hat{\mathbf{n}}_y$, and $\hat{\mathbf{n}}_z$ form a right-handed or a left-handed set of unit vectors? (*b*) Would the results be altered if OX, OY, and OZ were not mutually perpendicular?

Results $I_x = \sum_{i=1}^{\nu} m_i (y_i^2 + z_i^2)$ $\quad I_{yz} = -\sum_{i=1}^{\nu} m_i y_i z_i$ $\quad$ (*a*) No $\quad$ (*b*) Yes

5.7 Show by means of examples that products of inertia can be positive, negative, or zero, and that radii of gyration can be equal to zero.

***5.8** A body B of mass m is modeled as matter distributed uniformly along a helix H constructed by drawing a straight line L on a rectangular sheet of paper having the dimensions shown in Fig. P5.8 and then bending the paper to form a right-circular cylinder. Letting $\hat{\mathbf{n}}_1$ and $\hat{\mathbf{n}}_2$ be unit vectors directed as shown, determine $\mathbf{I}_1$ and $\mathbf{I}_2$, the inertia vectors of B relative to the mass center of B for $\hat{\mathbf{n}}_1$ and $\hat{\mathbf{n}}_2$, respectively.

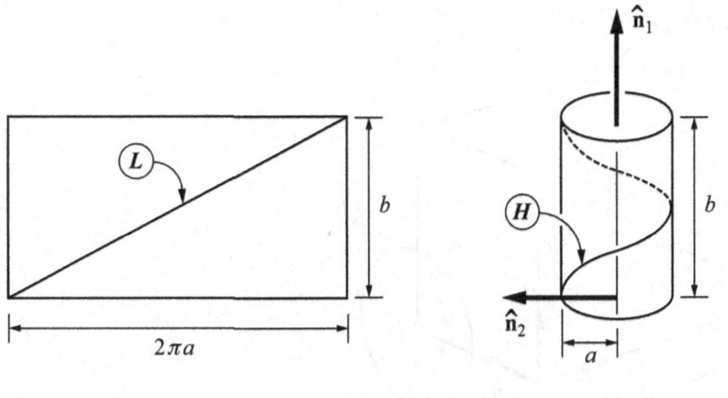

Figure P5.8

Results $\mathbf{I}_1 = ma^2\hat{\mathbf{n}}_1 + [mab/(2\pi)]\hat{\mathbf{n}}_1 \times \hat{\mathbf{n}}_2 \qquad \mathbf{I}_2 = (m/2)(a^2 + b^2/6)\hat{\mathbf{n}}_2$

5.9 A line L_a passes through a point O and is perpendicular to the unit vector $\hat{\mathbf{n}}_3$ of a set of mutually perpendicular unit vectors $\hat{\mathbf{n}}_1$, $\hat{\mathbf{n}}_2$, and $\hat{\mathbf{n}}_3$. Under these circumstances, I_a, the moment of inertia of a body B about line L_a, depends on the orientation of L_a in the plane that passes through O and is perpendicular to $\hat{\mathbf{n}}_3$.

Express the maximum and minimum values of I_a in terms of I_1, I_2, and I_{12}, the inertia scalars of B relative to O for $\hat{\mathbf{n}}_1$ and $\hat{\mathbf{n}}_2$, and, letting $\cos\theta\hat{\mathbf{n}}_1 + \sin\theta\hat{\mathbf{n}}_2$ be a unit vector parallel to L_a, show that

$$\tan 2\theta = \frac{2I_{12}}{I_1 - I_2}$$

when I_a has a maximum or minimum value.

Result

$$\frac{I_1 + I_2}{2} \pm \left[\left(\frac{I_1 - I_2}{2}\right)^2 + I_{12}{}^2\right]^{1/2}$$

5.10 In Fig. P5.10, $\hat{\mathbf{n}}_1$, $\hat{\mathbf{n}}_2$, and $\hat{\mathbf{n}}_3$ are mutually perpendicular unit vectors, and $B^\star$ designates the mass center of a body B. The inertia scalars of B relative to point O for $\hat{\mathbf{n}}_1$, $\hat{\mathbf{n}}_2$, and $\hat{\mathbf{n}}_3$ are shown in units of kg m^2 in Table P5.10.

Table P5.10

I_{jk}	1	2	3
1	260	72	−144
2	72	325	96
3	−144	96	169

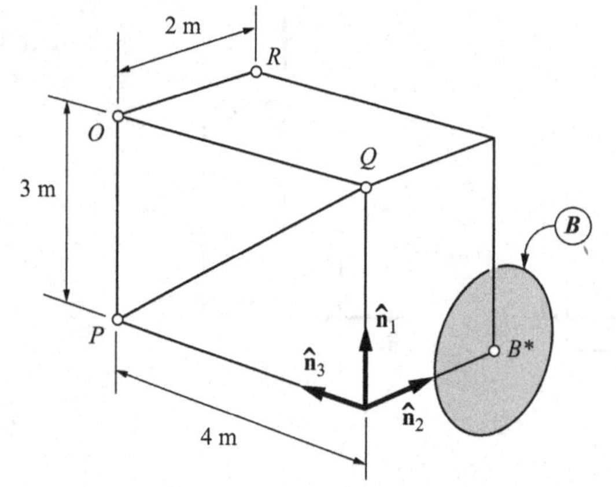

Figure P5.10

Determine the moment of inertia of B with respect to a line that is parallel to line PQ and passes through point O.

Result 340 kg m^2

5.11 Referring to Problem 5.10, and letting $\hat{\mathbf{e}}_1$, $\hat{\mathbf{e}}_2$, and $\hat{\mathbf{e}}_3$ be unit vectors defined as

$$\hat{\mathbf{e}}_1 \triangleq -\hat{\mathbf{n}}_3 \qquad \hat{\mathbf{e}}_2 \triangleq \hat{\mathbf{n}}_1 \qquad \hat{\mathbf{e}}_3 \triangleq \hat{\mathbf{n}}_2$$

form the inertia matrix of B relative to O for $\hat{\mathbf{e}}_1$, $\hat{\mathbf{e}}_2$, and $\hat{\mathbf{e}}_3$.

Result

$$\begin{bmatrix} 169 & 144 & -96 \\ 144 & 260 & 72 \\ -96 & 72 & 325 \end{bmatrix}$$

***5.12** Solve Problem 5.10 by performing multiplications of a row matrix with the matrix constructed in Problem 5.11.

5.13 If $\hat{\mathbf{a}}_1, \hat{\mathbf{a}}_2, \hat{\mathbf{a}}_3$ and $\hat{\mathbf{b}}_1, \hat{\mathbf{b}}_2, \hat{\mathbf{b}}_3$ are two sets of unit vectors, the unit vectors of each set are mutually perpendicular, and ${}^AC^B{}_{ij}$ is defined as

$${}^AC^B{}_{ij} \triangleq \hat{\mathbf{a}}_i \cdot \hat{\mathbf{b}}_j \qquad (i, j = 1, 2, 3)$$

then the 3×3 matrix ${}^AC^B$ having ${}^AC^B{}_{ij}$ as the jth element of the ith row is called a direction cosine matrix for the two sets of unit vectors. Letting $({}^AC^B)^T$ denote the transpose of ${}^AC^B$, show that ${}^AI$ and ${}^BI$, the inertia matrices of a set S of particles for a point O for $\hat{\mathbf{a}}_1, \hat{\mathbf{a}}_2, \hat{\mathbf{a}}_3$ and $\hat{\mathbf{b}}_1, \hat{\mathbf{b}}_2, \hat{\mathbf{b}}_3$, respectively, are related to each other as follows:

$${}^BI = ({}^AC^B)^T\, {}^AI\, {}^AC^B = {}^BC^A\, {}^AI\, {}^AC^B$$

***5.14** The time derivative of a dyadic $\underline{\mathbf{D}}$ in a reference frame A is defined as

$$\frac{{}^A d\underline{\mathbf{D}}}{dt} \triangleq \sum_{i=1}^{3}\sum_{j=1}^{3} \hat{\mathbf{a}}_i\hat{\mathbf{a}}_j \frac{d}{dt}(\hat{\mathbf{a}}_i \cdot \underline{\mathbf{D}} \cdot \hat{\mathbf{a}}_j)$$

where $\hat{\mathbf{a}}_1$, $\hat{\mathbf{a}}_2$, $\hat{\mathbf{a}}_3$ are mutually perpendicular unit vectors fixed in A. Show that the time derivatives of $\underline{\mathbf{D}}$ in two reference frames A and B are related by

$$\frac{{}^A d\underline{\mathbf{D}}}{dt} = \frac{{}^B d\underline{\mathbf{D}}}{dt} + {}^A\boldsymbol{\omega}^B \times \underline{\mathbf{D}} - \underline{\mathbf{D}} \times {}^A\boldsymbol{\omega}^B$$

where ${}^A\boldsymbol{\omega}^B$ is the angular velocity of B in A.

PROBLEM SET 6
(Secs. 4.6–4.9)

6.1 A point O of a rigid body B is fixed in a reference frame A. Show that $\mathbf{H}$, the angular momentum of B relative to O in A, is given by Eq. (4.5.28) if $\underline{\mathbf{I}}$ denotes the inertia dyadic of B relative to O and $\boldsymbol{\omega}$ is the angular velocity of B in A.

***6.2** A rigid body B moves in a reference frame A with an angular velocity $\boldsymbol{\omega}$. Show that the central angular momentum of B in A is parallel to $\boldsymbol{\omega}$ if and only if $\boldsymbol{\omega}$ is parallel to a central principal axis of B.

6.3 Letting $\mathbf{H}^{S/O}$, $\mathbf{H}^{S/S^\star}$, and $\mathbf{H}^{S^\star/O}$ denote, respectively, the angular momentum of a set S of particles relative to a point O fixed in a reference frame A, the central angular momentum of S in A, and the angular momentum in A, relative to O, of a particle whose motion is identical to that of the mass center $S^\star$ of S and whose mass is equal to the total mass of S, show that

$$\mathbf{H}^{S/O} = \mathbf{H}^{S/S^\star} + \mathbf{H}^{S^\star/O}$$

***6.4** Letting ${}^E\mathbf{H}^{C/C^\star}$, ${}^E\mathbf{H}^{D/C^\star}$, and ${}^A\mathbf{H}^{B/C^\star}$ denote angular momenta, show that

$${}^E\mathbf{H}^{C/C^\star} = {}^E\mathbf{H}^{D/C^\star} + {}^A\mathbf{H}^{B/C^\star}$$

if A, B, C, D, E, and $C^\star$ are defined as follows: A, a rigid body; B, a set of particles; C, the system formed by A and B; D, a rigid body that has the same motion as A and the same mass distribution as C; E, a reference frame; $C^\star$, the mass center of C.

***6.5** The mass center of a rigid body B is fixed in a rigid body A, but B is otherwise free to move relative to A. Letting C be the system formed by A and B, show that $\mathbf{H}$, the central angular momentum of C in a reference frame E, is given by

$$\mathbf{H} = \underline{\mathbf{I}}_B \cdot {}^A\boldsymbol{\omega}^B + \underline{\mathbf{I}}_C \cdot {}^E\boldsymbol{\omega}^A$$

where $\underline{\mathbf{I}}_B$ and $\underline{\mathbf{I}}_C$ are the central inertia dyadics of B and C, respectively, while ${}^A\boldsymbol{\omega}^B$ and ${}^E\boldsymbol{\omega}^A$ are the angular velocities of B in A, and A in E, respectively.

6.6 The body B of Problem 5.10 has a mass of 12 kg. Determine the moment of inertia of B about line PQ, and find the product of inertia of B relative to $B^\star$ for $\hat{\mathbf{n}}_1$ and $\hat{\mathbf{n}}_2$.

Results 3316/25 kg m^2; 0

6.7 Three identical uniform, square plates, each of mass m, are attached to each other as shown in Fig. P6.7. Determine the value of θ for which the radius of gyration of this assembly with respect to line L has a minimum value, and find this value.

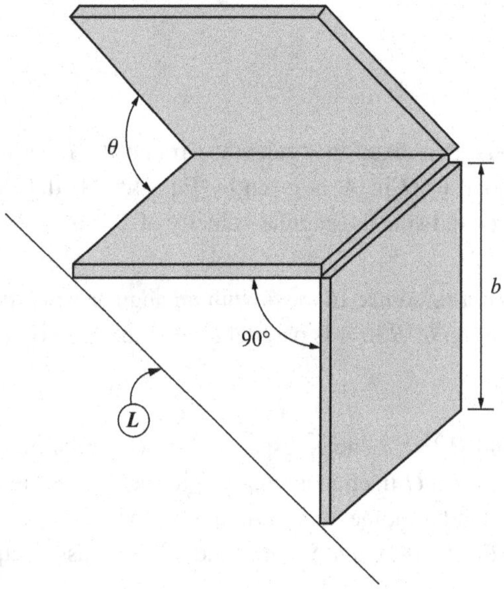

Figure P6.7

Results 340°; 0.696b

6.8 A thin-walled, right-circular, cylindrical shell has a radius R and height H. Determine the radius of gyration of the shell with respect to a line that passes through the mass center and is perpendicular to the axis of the shell.

Result $(R^2/2 + H^2/12)^{1/2}$

6.9 Verify each of the following statements and provide an illustrative example: (*a*) A central principal axis of a body is a principal axis for each point of the axis. (*b*) If a principal axis for a point other than the mass center passes through the mass center, it is a central principal axis. (*c*) A line that is a principal axis for two of its points is a central principal axis. (*d*) The three principal axes for a point on a central principal axis are parallel to central principal axes. (*e*) If two principal moments of inertia for a given point are equal to each other, then the moments of inertia with respect to all lines passing through this point and lying in the plane determined by the associated principal axes are equal to each other. (*f*) If the particles of a set S lie in a plane P, then the line L normal to P and intersecting P at a point O is a principal axis of S for O, and the moment of inertia of S about L is equal to the sum of the moments of inertia of S about any two orthogonal lines that lie in P and intersect at O.

6.10 Determine the smallest angle between line AB and any principal axis for point A of the thin, uniform, rectangular plate represented by the shaded portion of Fig. P6.10.

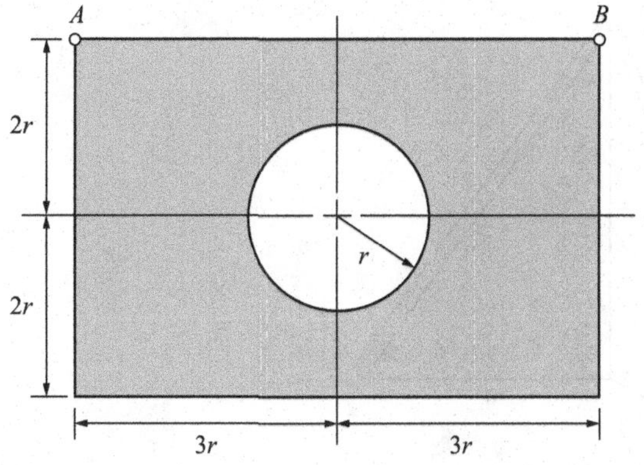

Figure P6.10

Result 30.02°

6.11 For I_{jk} $(j,k = 1,2,3)$ as in Eq. (4.5.22), show that no one of I_{jj} $(j = 1,2,3)$ can exceed the sum of the other two, and that $-I_{11}/2 \le I_{23} \le I_{11}/2, -I_{22}/2 \le I_{31} \le I_{22}/2$, and $-I_{33}/2 \le I_{12} \le I_{33}/2$.

***6.12** Defining C_1, C_2, and C_3 as

$$C_1 \triangleq I_1 + I_2 + I_3$$
$$C_2 \triangleq I_1 I_2 + I_2 I_3 + I_3 I_1 - I_{12}{}^2 - I_{23}{}^2 - I_{31}{}^2$$
$$C_3 \triangleq I_1 I_2 I_3 + 2I_{12} I_{23} I_{31} - I_1 I_{23}{}^2 - I_2 I_{31}{}^2 - I_3 I_{12}{}^2$$

where $I_j \triangleq I_{jj}$ and I_{jk} is the inertia scalar of a set of particles relative to a point for unit vectors $\hat{\mathbf{n}}_j$ and $\hat{\mathbf{n}}_k$ $(j,k = 1,2,3)$, show that the values of C_1, C_2, and C_3 are independent of the way $\hat{\mathbf{n}}_1$, $\hat{\mathbf{n}}_2$, and $\hat{\mathbf{n}}_3$ are chosen, so long as these vectors are mutually perpendicular.

Suggestion: Verify that Eq. (4.8.7) can be written

$$I_z^3 - C_1 I_z^2 + C_2 I_z - C_3 = 0$$

6.13 Four identical particles are placed at the points O, P, Q, R of Fig. P5.10. Determine the minimum radius of gyration of this set of particles, and find the smallest angle between the associated principal axis and line OP.

Results 1.436 m, 67.64°

6.14 Two identical, thin, uniform, right-triangular plates are attached to each other as shown in Fig. P6.14. When a/b is given, k, the minimum radius of gyration of this assembly, can be expressed as $k = nb$. Determine n for $a/b = 2$ and $a/b = 0.5$.

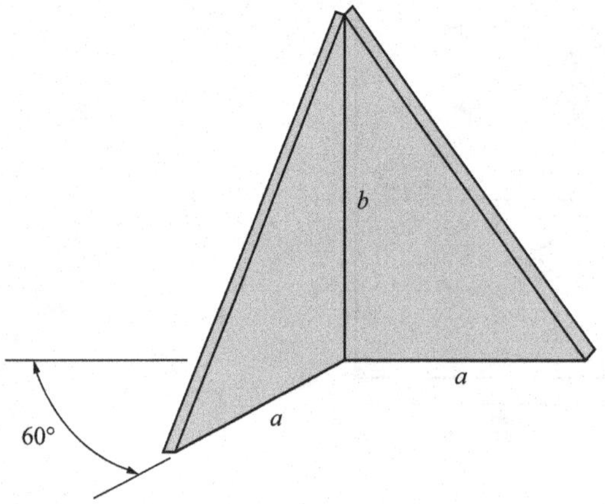

Figure P6.14

Results $\frac{1}{3}$, $\frac{1}{24}[35 - (241)^{1/2}]^{1/2}$

***6.15** Letting E be an inertia ellipsoid of a set S of particles for a point O, and letting R be the distance from O to a point P on the surface of E, show that $\mathbf{I}_a$, the inertia vector of S relative to O for $\hat{\mathbf{n}}_a$, a unit vector directed from O to P, is perpendicular to E at P.

Note: If $\hat{\mathbf{n}}_x$, $\hat{\mathbf{n}}_y$, $\hat{\mathbf{n}}_z$ are mutually perpendicular unit vectors, the position vector $\mathbf{r}^{OP}$ from a point O to a point P of a surface σ is expressed as

$$\mathbf{r}^{OP} = x\hat{\mathbf{n}}_x + y\hat{\mathbf{n}}_y + z\hat{\mathbf{n}}_z$$

and the equation of σ is written $f(x, y, z) = 0$, then ∇f, a vector defined as

$$\nabla f \triangleq \frac{\partial f}{\partial x}\hat{\mathbf{n}}_x + \frac{\partial f}{\partial y}\hat{\mathbf{n}}_y + \frac{\partial f}{\partial z}\hat{\mathbf{n}}_z$$

and called the *gradient* of f, is perpendicular to σ at P.

***6.16** For a set S of particles $P_1, \ldots, P_\nu$ moving in a reference frame A, a quantity G, known as a *Gibbs function* for S in A, is defined as

$$G \triangleq \frac{1}{2} \sum_{i=1}^{\nu} m_i \mathbf{a}_i{}^2$$

where m_i and $\mathbf{a}_i$ are the mass of P_i and the acceleration of P_i in A, respectively.

Letting B be a rigid body, express the Gibbs function for B in A in terms of the acceleration $\mathbf{a}$ of the mass center of B in A, the angular velocity $\boldsymbol{\omega}$ of B in A, the angular acceleration $\boldsymbol{\alpha}$ of B in A, the mass m of B, and the central inertia dyadic $\underline{\mathbf{I}}$ of B.

Result $G = \frac{1}{2}(m\mathbf{a}^2 + \boldsymbol{\alpha} \cdot \underline{\mathbf{I}} \cdot \boldsymbol{\alpha} + 2\boldsymbol{\alpha} \cdot \boldsymbol{\omega} \times \underline{\mathbf{I}} \cdot \boldsymbol{\omega} + \boldsymbol{\omega}^2 \boldsymbol{\omega} \cdot \underline{\mathbf{I}} \cdot \boldsymbol{\omega})$

PROBLEM SET 7
(Secs. 5.1–5.3)

7.1 Two forces, **A** and **B**, of equal magnitude act along the lines PQ and RS in Fig. P7.1 and are directed as there indicated. Determine the angle θ $(0 \leq \theta \leq 180°)$ between the resultant of this force system and the moment of the force system about point O.

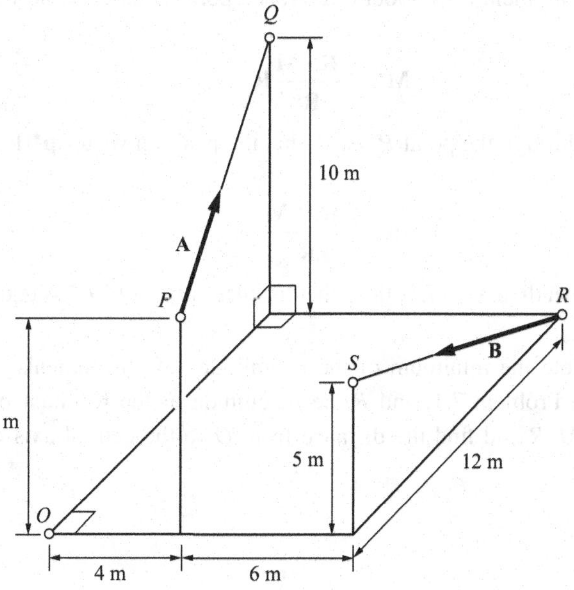

Figure P7.1

Result 117.12°

7.2 Referring to Problem 7.1, let **A** and **B** each have a magnitude of 10 N, and let **C** be a force applied at point O in such a way that **A**, **B**, and **C** form a couple. Determine the magnitude of the torque of this couple.

Result 90.56 N m

7.3 Show by means of an example that there exist sets of bound vectors such that the moment of the set of vectors about a point differs from the moment of the resultant about that point, no matter where the resultant is applied.

7.4 Show that the moments of a set S of bound vectors about all points of any line parallel to the resultant of S are equal to each other.

7.5 Letting **M** be the moment about a point O of a set S of bound vectors, and resolving **M** into two components, one parallel to **R**, the resultant of S, the other perpendicular to **R**, show that the magnitude of the first of these is independent of the location of O.

7.6 If a set S of bound vectors is not a couple, the points about which S has a moment $\mathbf{M}^\star$ of minimum magnitude lie on a line $L^\star$ that is parallel to **R**, the resultant of S. $L^\star$ is called the *central axis* of S.

Letting **M** be the moment of S about a reference point O selected arbitrarily, show that

$$\mathbf{M}^\star = \frac{\mathbf{R} \cdot \mathbf{M}}{\mathbf{R}^2}\mathbf{R}$$

and that $L^\star$ passes through the point $P^\star$ such that the position vector $\mathbf{p}^\star$ from O to $P^\star$ is given by

$$\mathbf{p}^\star = \frac{\mathbf{R} \times \mathbf{M}}{\mathbf{R}^2}$$

(Note that $\mathbf{p}^\star$ is perpendicular to $L^\star$; thus, the distance from O to $L^\star$ is equal to $|\mathbf{p}^\star|$.)

7.7 Letting M denote the minimum of the magnitudes of all moments of the force system described in Problem 7.1, and R the magnitude of the resultant of this force system, determine M/R and find the distance from O to the central axis of the force system.

Result 6 m; 11.72 m

7.8 A set of bound vectors consisting of a couple together with a single bound vector is called a *wrench* if the single bound vector is parallel to the torque of the couple. When the set of two forces of Problem 7.1 is replaced with a wrench consisting of a couple C together with a force **F**, what is the magnitude of the torque of C, and what is the distance from point O to the line of action of **F** if **A** and **B** each have a magnitude of 10 N?

Suggestion: Show that if the resultant of a set S of bound vectors is not equal to zero, then S can be replaced with a wrench consisting of the resultant of S, placed on the central axis of S, and a couple whose torque is equal to the moment of minimum magnitude for S.

Results 41.28 N m; 11.72 m

7.9 Show that a set of bound vectors whose lines of action intersect at a point O can be replaced with the resultant of the set, applied at O.

7.10 Show by means of an example that there exist sets of bound vectors, other than couples, that cannot be replaced with their resultants.

7.11 Letting S be a set of bound vectors whose lines of action are coplanar and whose resultant is not equal to zero, show that S can be replaced with its resultant, placed on the central axis of S.

7.12 Letting each of the forces in Problem 7.1 have a magnitude of 10 N, replace this set of forces with two forces $\mathbf{A}'$ and $\mathbf{B}'$, with $\mathbf{A}'$ acting along line OP. Determine the magnitudes of $\mathbf{A}'$ and $\mathbf{B}'$, and find the distance d from O to the line of action of $\mathbf{B}'$.

Results $|\mathbf{A}'| = 4.961$ N; $|\mathbf{B}'| = 5.838$ N; $d = 15.51$ m

***7.13** Let S be a set of coplanar forces $\mathbf{F}_i$ $(i = 1,\ldots,n)$ applied at points P_i $(i = 1,\ldots,n)$, respectively, and let $\mathbf{p}_i$ be the position vector from a point O lying in the plane of the forces to P_i. Let S' be the set of forces $\mathbf{F}_i{}'$ $(i = 1,\ldots,n)$ such that $\mathbf{F}_i{}'$ is obtained by rotating $\mathbf{F}_i$ about P_i through an angle θ (in the plane of the forces). Then S can be replaced with a single force $\mathbf{R}$, and S' with a single force $\mathbf{R}'$, and the lines of action of $\mathbf{R}$ and $\mathbf{R}'$ intersect at a point Z called the *astatic center* of S.

Determine $\mathbf{N}$ and ν such that $\mathbf{z}$, the position vector from O to Z, can be expressed as

$$\mathbf{z} = \mathbf{R} \times \mathbf{N} + \mathbf{R}\nu$$

Express the results in terms of $\mathbf{p}_i$ and $\mathbf{F}_i$ $(i = 1,\ldots,n)$, and note that $\mathbf{z}$ is independent of θ.

Suggestion: Take advantage of the fact that if $\mathbf{a}$ and $\mathbf{b}$ are the position vectors from a point O to points A and B, respectively, and lines L_A and L_B lie in the plane P determined by O, A, and B, with L_A perpendicular to $\mathbf{a}$ at A, and L_B perpendicular to $\mathbf{b}$ at B, then $\mathbf{c}$, the position vector from O to C, the intersection of L_A and L_B, is given by

$$\mathbf{c} = \mathbf{a} + \frac{(\mathbf{b}^2 - \mathbf{a} \cdot \mathbf{b})\mathbf{a} \times \hat{\mathbf{k}}}{\mathbf{a} \times \hat{\mathbf{k}} \cdot \mathbf{b}}$$

where $\hat{\mathbf{k}}$ is a unit vector normal to P.

Results $\mathbf{N} = \sum_{i=1}^{n} \mathbf{p}_i \times \mathbf{F}_i \Big/ \left(\sum_{i=1}^{n} \mathbf{F}_i\right)^2 \qquad \nu = \sum_{i=1}^{n} \mathbf{p}_i \cdot \mathbf{F}_i \Big/ \left(\sum_{i=1}^{n} \mathbf{F}_i\right)^2$

PROBLEM SET 8
(Secs. 5.4–5.9)

8.1 Referring to the example in Sec. 3.1, let line Y be vertical, and let P_1 and P_2 have masses m_1 and m_2, respectively. Define three sets of generalized velocities u_1, u_2, u_3 as

$$u_1 \triangleq {}^A\mathbf{v}^{P_1} \cdot \hat{\mathbf{a}}_x \qquad u_2 \triangleq {}^A\mathbf{v}^{P_1} \cdot \hat{\mathbf{a}}_y \qquad u_3 \triangleq \dot{q}_3$$
$$u_1 \triangleq {}^A\mathbf{v}^{P_1} \cdot \hat{\mathbf{e}}_x \qquad u_2 \triangleq {}^A\mathbf{v}^{P_1} \cdot \hat{\mathbf{e}}_y \qquad u_3 \triangleq \dot{q}_3$$
$$u_1 \triangleq \dot{q}_1 \qquad u_2 \triangleq \dot{q}_2 \qquad u_3 \triangleq \dot{q}_3$$

where $\hat{\mathbf{a}}_x$ and $\hat{\mathbf{a}}_y$ are unit vectors directed as shown in Fig. 3.1.2, $\hat{\mathbf{e}}_x$ and $\hat{\mathbf{e}}_y$ are unit vectors directed as shown in Fig. 2.6.1, and q_1, q_2, and q_3 are two distances and an angle, as indicated in Fig. 2.6.1. Assume that the panes of glass forming B are perfectly smooth, so that any forces they exert on P_1 and P_2 are parallel to $\hat{\mathbf{b}}_z$ in Fig. 2.6.1; and, regarding the mass of R as negligible, let the contact forces exerted by R on P_1 and P_2 be equal to $C\hat{\mathbf{e}}_x$ and $-C\hat{\mathbf{e}}_x$, respectively, where C is a function of $q_1, q_2, q_3, u_1, u_2, u_3$.

Letting S be the set of two particles P_1 and P_2, form expressions for the generalized active forces F_1, F_2, F_3 for S in reference frame A, doing so for each of the three sets of generalized velocities defined previously.

Results

$$F_1 = 0 \qquad F_2 = -(m_1 + m_2)g \qquad F_3 = -Lm_2 g c_3$$
$$F_1 = -(m_1 + m_2)g s_3 \qquad F_2 = -(m_1 + m_2)g c_3 \qquad F_3 = -Lm_2 g c_3$$
$$F_1 = 0 \qquad F_2 = -(m_1 + m_2)g \qquad F_3 = -Lm_2 g c_3$$

8.2 Referring to Problem 8.1, suppose that P_2 is replaced with a sharp-edged circular disk D (of mass m_2) whose axis is normal to the rod R and parallel to the plane in which R moves, as indicated in Fig. 3.5.1; further, that D comes into contact with the two panes of glass at the points D_1 and D_2. (The same assumptions were made in the example in Sec. 3.5.) Assume that D can be regarded as a particle on which the panes of glass exert contact forces equivalent to a force $Y\hat{\mathbf{e}}_y + Z\hat{\mathbf{e}}_z$ applied at $D^\star$. With u_1, u_2, u_3 defined as

$$u_1 \triangleq {}^A\mathbf{v}^{P_1} \cdot \hat{\mathbf{e}}_x \qquad u_2 \triangleq {}^A\mathbf{v}^{P_1} \cdot \hat{\mathbf{e}}_y \qquad u_3 \triangleq \dot{q}_3$$

and letting S be the set of two particles P_1 and D, determine (a) the generalized active forces F_1, F_2, and F_3 for S in A and (b) the generalized active forces $\widetilde{F}_1$ and $\widetilde{F}_2$ for S

in A. To find the results for (*b*), use Eqs. (5.4.2), then check the results by using Eqs. (5.4.4) together with the results from (*a*).

Results

$$
\begin{aligned}
(a)\quad F_1 &= -(m_1 + m_2)g s_3 \\
F_2 &= -(m_1 + m_2)g c_3 + Y \\
F_3 &= -L(m_2 g c_3 - Y) \\
(b)\quad \tilde{F}_1 &= -(m_1 + m_2)g s_3 \\
\tilde{F}_2 &= -m_1 g c_3
\end{aligned}
$$

8.3 Referring to Problem 2.7, suppose that C is of uniform density, so that $C^\star$ is the mass center of C, and let m be the mass of C. Let $P_x\hat{\mathbf{a}}_x + P_y\hat{\mathbf{a}}_y + P_z\hat{\mathbf{a}}_z$ be the contact force exerted by H on C at point P, and introduce motion variables $u_1, \ldots, u_5$ as

$$u_i \triangleq {}^A\boldsymbol{\omega}^C \cdot \hat{\mathbf{b}}_i \qquad (i = 1,2,3)$$

$$u_4 \triangleq \dot{q}_4 \qquad u_5 \triangleq \dot{q}_5$$

Determine (*a*) the generalized active forces $F_1, \ldots, F_5$ for C in A, assuming that slipping is taking place at P, and (*b*) the generalized active forces $\tilde{F}_1$, $\tilde{F}_2$, $\tilde{F}_3$ for C in A, assuming that C is rolling on H.

Results

$$
\begin{aligned}
(a)\quad F_1 &= -Rmg s_2 \\
F_2 &= -R\tan q_2(c_1 P_x + s_1 P_y) \\
F_3 &= R(c_1 P_x + s_1 P_y) \\
F_4 &= P_x \\
F_5 &= P_y \\
(b)\quad \tilde{F}_1 &= -Rmg s_2 \\
\tilde{F}_2 &= 0 \\
\tilde{F}_3 &= 0
\end{aligned}
$$

8.4 Referring to Problem 4.11, suppose that P is horizontal, C_1 and C_2 roll on P, and motors connecting S to C_1 and C_2 cause C_1 and C_2 to rotate relative to S. To account for the actions of the motors, let S exert contact forces on C_1 and C_2, and replace the set of all such forces acting on C_i with a couple of torque $\mathbf{M}_i$ together with a force $\mathbf{K}_i$ applied at the center of C_i, with $\mathbf{M}_i$ and $\mathbf{K}_i$ $(i = 1,2)$ given by

$$
\begin{aligned}
\mathbf{M}_1 &= \alpha_1\hat{\mathbf{n}}_1 + \alpha_2\hat{\mathbf{n}}_2 + \alpha_3\hat{\mathbf{n}}_3 \\
\mathbf{M}_2 &= \beta_1\hat{\mathbf{n}}_1 + \beta_2\hat{\mathbf{n}}_2 + \beta_3\hat{\mathbf{n}}_3 \\
\mathbf{K}_1 &= \gamma_1\hat{\mathbf{n}}_1 + \gamma_2\hat{\mathbf{n}}_2 + \gamma_3\hat{\mathbf{n}}_3 \\
\mathbf{K}_2 &= \delta_1\hat{\mathbf{n}}_1 + \delta_2\hat{\mathbf{n}}_2 + \delta_3\hat{\mathbf{n}}_3
\end{aligned}
$$

Determine the generalized active forces $\tilde{F}_1$, $\tilde{F}_2$, and $\tilde{F}_3$ in A for the system formed by S, C_1, and C_2. (Keep the law of action and reaction in mind.)

Results $\tilde{F}_1 = -(\alpha_3 + \beta_3)/R \qquad \tilde{F}_2 = L(\alpha_3 - \beta_3)/R \qquad \tilde{F}_3 = -(\alpha_3 + \beta_3)$

8.5 When the motors considered in Problem 8.4 are used to cause C_1 and C_2 to rotate in such a way that

$$^S\boldsymbol{\omega}^{C_1} = \Omega_1 \hat{\mathbf{n}}_3 \qquad ^S\boldsymbol{\omega}^{C_2} = \Omega_2 \hat{\mathbf{n}}_3$$

where Ω_1 and Ω_2 are prescribed functions of the time t, then the system formed by S, C_1, and C_2 possesses one degree of freedom in A if C_1 and C_2 roll on P, as was pointed out in Problem 4.12. Show that the associated nonholonomic generalized active force is equal to zero if P is horizontal.

8.6 The particle P introduced in Problem 4.15 operates in a terrestrial gravitational field. Letting m be the mass of P, and referring to Problem 4.20, determine the contributions $(\tilde{\tilde{F}}_r)_\gamma$ of all gravitational forces exerted on P by the Earth to the nonholonomic generalized active forces $\tilde{\tilde{F}}_r$ for P in N.

Results

$$(\tilde{\tilde{F}}_1)_\gamma = -mga^2 u_1/u_3 \qquad (\tilde{\tilde{F}}_2)_\gamma = -mga^2 u_2/u_3$$

8.7 In Fig. P8.7, which deals with the system previously considered in Problems 4.3 and 4.6, $\hat{\mathbf{a}}_x$ points vertically downward, while $\hat{\mathbf{a}}_y$ and $\hat{\mathbf{a}}_z$ are horizontal. Define a motion

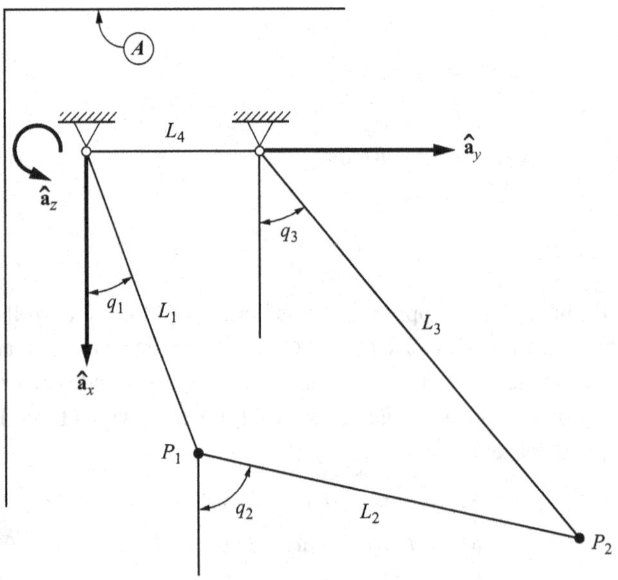

Figure P8.7

variable u_1 as $u_1 \triangleq \dot{q}_1$, let m_1 and m_2 be the masses of P_1 and P_2, respectively, and assume that the links supporting P_1 and P_2 are so light in comparison with P_1 and P_2 that the masses of the links can be neglected. Determine $(\tilde{F}_1)_\gamma$, the contribution of gravitational forces to the generalized active force $\tilde{F}_1$ for the particles P_1 and P_2 in A, on the basis of the following considerations.

The system S formed by P_1 and P_2 possesses only one degree of freedom in A. Hence, a single generalized coordinate, such as q_1, suffices to characterize the configuration of S in A. However, it is convenient to introduce the *pseudo-generalized coordinates* q_2 and q_3 (see Fig. P8.7), which must satisfy the configuration constraint equations

$$L_1 c_1 + L_2 c_2 - L_3 c_3 = 0$$

$$L_1 s_1 + L_2 s_2 - L_3 s_3 - L_4 = 0$$

and to let u_2 and u_3 be *pseudo-motion variables* defined as $u_2 \triangleq \dot{q}_2$ and $u_3 \triangleq \dot{q}_3$. Differentiation of the constraint equations with respect to time then permits one to express u_2 and u_3 as in Eqs. (3.5.2), and S then can be treated as if it were a simple nonholonomic system.

Result

$$(\tilde{F}_1)_\gamma = -gL_1 \left[m_1 s_1 + m_2 s_3 \frac{\sin(q_2 - q_1)}{\sin(q_2 - q_3)} \right]$$

8.8 Figure P8.8 shows 33 pin-connected rods, each of mass m and length L, suspended from a horizontal support. Contact forces of magnitudes Q, R, and S are applied to this system, as indicated in Fig. P8.8.

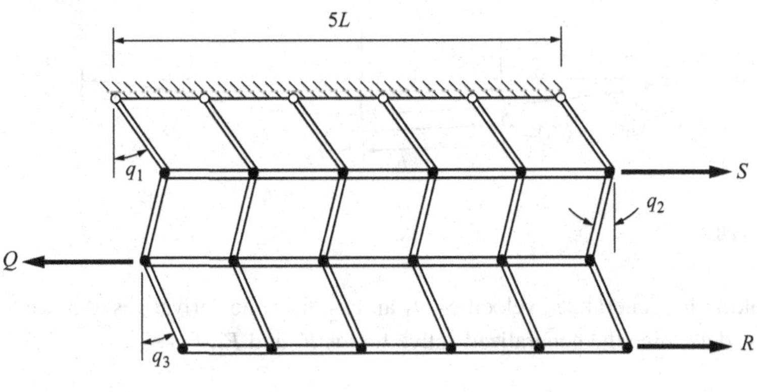

Figure P8.8

Using as generalized velocities the quantities $L\dot{q}_1$, $L\dot{q}_2$, and $L\dot{q}_3$, where q_1, q_2, and q_3 are the angles shown in Fig. P8.8, determine the generalized active forces F_1, F_2, and F_3.

Results

$$F_1 = (-Q + R + S)c_1 - 30mgs_1$$
$$F_2 = (Q - R)c_2 - 19mgs_2$$
$$F_3 = Rc_3 - 8mgs_3$$

8.9 A rigid block B of length $2b$ and mass m is supported by two elastic beams, each of length L and flexural rigidity EI. The beams are "built in" both at their supports and at the points P_1 and P_2 (Fig. P8.9).

Confining attention to planar motions of B during which q_1 and q_2, the vertical beam displacements of P_1 and P_2, respectively, remain small, one can treat B as a holonomic system possessing two degrees of freedom, and one can replace the set of contact forces exerted on B by the beam at the left with a couple of torque $M_1\hat{\beta}$, together with a force $V_1\hat{\alpha}$ applied at P_1, where $\hat{\alpha}$ and $\hat{\beta}$ are unit vectors directed as shown, and M_1 and V_1 are given by

$$M_1 = \frac{12EI}{L^2}\left(\frac{L}{3}\frac{q_2 - q_1}{2b} - \frac{q_1}{2}\right)$$
$$V_1 = \frac{12EI}{L^3}\left(q_1 - \frac{L}{2}\frac{q_2 - q_1}{2b}\right)$$

The set of contact forces exerted on B by the beam at the right can be replaced similarly.

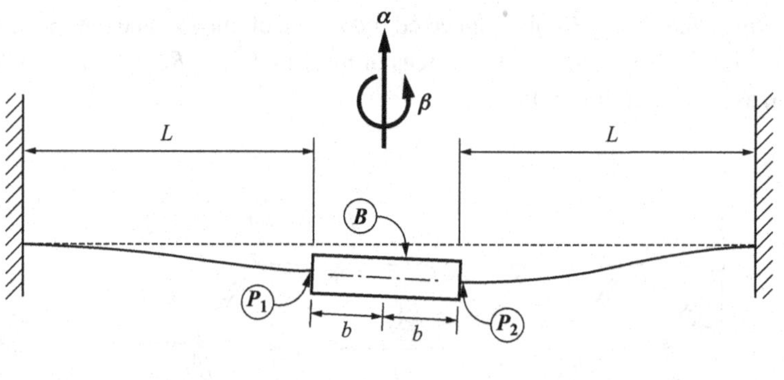

Figure P8.9

Taking for generalized velocities, u_1 and u_2, the time derivatives of q_1 and q_2, respectively, determine the generalized active forces F_1 and F_2.

Results

$$F_1 = \frac{12EI}{L^3}\left[-\left(1 + \frac{L}{2b} + \frac{L^2}{6b^2}\right)q_1 + \frac{L}{2b}\left(1 + \frac{L}{3b}\right)q_2\right] + \frac{mg}{2}$$
$$F_2 = \frac{12EI}{L^3}\left[\frac{L}{2b}\left(1 + \frac{L}{3b}\right)q_1 - \left(1 + \frac{L}{2b} + \frac{L^2}{6b^2}\right)q_2\right] + \frac{mg}{2}$$

8.10 Figure P8.10 is a schematic representation of a reduction gear consisting of a fixed bevel gear A, moving bevel gears B, C, C', D, D', and E, and an arm F. C and D are rigidly connected to each other, as are C' and D'. The number of teeth of each gear is shown in Table P8.10.

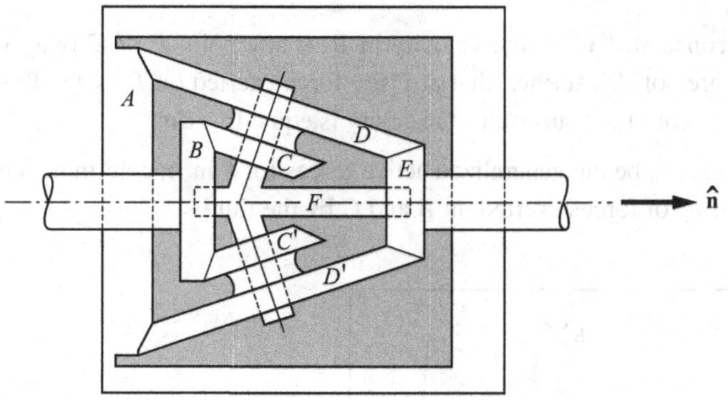

Figure P8.10

Table P8.10

A	B	C	C'	D	D'	E
60	30	30	30	61	61	20

Couples of torques $T_B\hat{\mathbf{n}}$ and $T_E\hat{\mathbf{n}}$, where $\hat{\mathbf{n}}$ is a unit vector directed as shown in Fig. P8.10, are applied to the shafts carrying B and E, respectively. Letting u_1 be a generalized velocity defined as $u_1 \triangleq {}^A\boldsymbol{\omega}^B \cdot \hat{\mathbf{n}}$, and assuming that rolling takes place at all points of contact between gears, determine the generalized active force F_1.

Result $F_1 = T_B + 244T_E$

8.11 A uniform sphere C of radius R is placed in a spherical cavity of a rigid body B, and B is free to move in a reference frame A, as indicated in Fig. P8.11. The radius of the cavity exceeds R only slightly, and the space between B and C is filled with a viscous fluid.

Regarding the system S formed by B and C as possessing nine degrees of freedom in A, let $u_1, \ldots, u_9$ be generalized velocities defined as

$$u_i \triangleq \begin{cases} {}^A\boldsymbol{\omega}^B \cdot \hat{\mathbf{b}}_i & (i = 1,2,3) \\ {}^A\boldsymbol{\omega}^C \cdot \hat{\mathbf{b}}_{i-3} & (i = 4,5,6) \\ \mathbf{v} \cdot \hat{\mathbf{b}}_{i-6} & (i = 7,8,9) \end{cases}$$

where ${}^A\boldsymbol{\omega}^B$ and ${}^A\boldsymbol{\omega}^C$ are, respectively, the angular velocities of B and C in A, $\mathbf{v}$ is the velocity of the center of C in A, and $\hat{\mathbf{b}}_1$, $\hat{\mathbf{b}}_2$, $\hat{\mathbf{b}}_3$ are mutually perpendicular unit vectors fixed in B. Assume that the force $d\sigma$ exerted on C by the fluid across a differential element $d\Sigma$ of the surface C is given by

$$d\sigma = -c\,{}^B\mathbf{v}^P\,da$$

where c is a constant, ${}^B\mathbf{v}^P$ is the velocity in B of any point P of C lying within $d\Sigma$, and da is the area of $d\Sigma$; further, that at P the force exerted on B by the fluid across a differential element of the surface of the cavity is equal to $-d\sigma$.

Letting $F_1, \ldots, F_9$ be the generalized active forces for S in A, determine the contributions to F_2 and F_4 of forces exerted on B and C by the fluid.

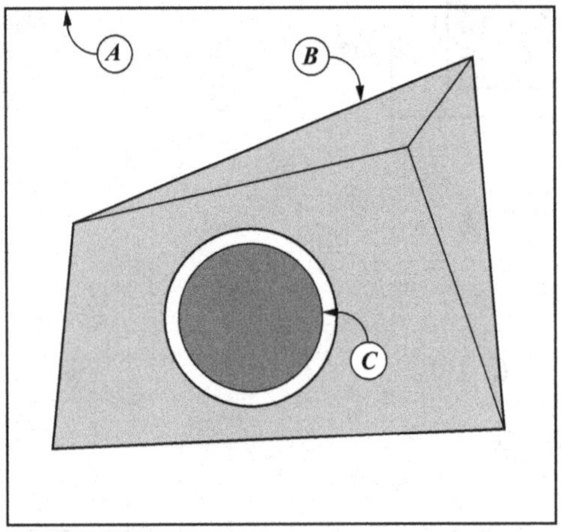

Figure P8.11

Results $\frac{8}{3}\pi c R^4(u_5 - u_2)$; $\quad \frac{8}{3}\pi c R^4(u_1 - u_4)$

***8.12** Referring to Problem 3.15 (see also Problem 4.19), let $A^\star$, $B^\star$, $C^\star$, $D^\star$ be the mass centers of A, B, C, D, respectively, and designate as m_A, m_B, m_C, m_D the masses of A, B, C, D, respectively. The pinion gears have negligible masses. To drive the robot arm, forces are transmitted from E to A via a motor (not shown in Fig. P3.15). The set of all such forces is equivalent to a couple of torque $\mathbf{T}^{E/A}$ together with a force $\mathbf{K}^{E/A}$ applied to A at a point on the axis of the hub. The set of forces exerted by A on B by means of the motor connecting A to B at P is equivalent to a couple of torque $\mathbf{T}^{A/B}$ together with a force $\mathbf{K}^{A/B}$ applied to B at P. Finally, the set of forces exerted by B on C through the rack-and-pinion drive is equivalent to a couple of torque $\mathbf{T}^{B/C}$ together with a force $\mathbf{K}^{B/C}$ applied to C at $C^\star$.

Expressing $\mathbf{T}^{E/A}$, $\mathbf{K}^{E/A}$, $\mathbf{T}^{A/B}$, $\mathbf{K}^{A/B}$, $\mathbf{T}^{B/C}$, and $\mathbf{K}^{B/C}$ as, respectively,

$$\mathbf{T}^{E/A} = T_1{}^{E/A}\hat{\mathbf{a}}_1 + T_2{}^{E/A}\hat{\mathbf{a}}_2 + T_3{}^{E/A}\hat{\mathbf{a}}_3$$
$$\mathbf{K}^{E/A} = K_1{}^{E/A}\hat{\mathbf{a}}_1 + K_2{}^{E/A}\hat{\mathbf{a}}_2 + K_3{}^{E/A}\hat{\mathbf{a}}_3$$
$$\mathbf{T}^{A/B} = T_1{}^{A/B}\hat{\mathbf{b}}_1 + T_2{}^{A/B}\hat{\mathbf{b}}_2 + T_3{}^{A/B}\hat{\mathbf{b}}_3$$
$$\mathbf{K}^{A/B} = K_1{}^{A/B}\hat{\mathbf{b}}_1 + K_2{}^{A/B}\hat{\mathbf{b}}_2 + K_3{}^{A/B}\hat{\mathbf{b}}_3$$
$$\mathbf{T}^{B/C} = T_1{}^{B/C}\hat{\mathbf{b}}_1 + T_2{}^{B/C}\hat{\mathbf{b}}_2 + T_3{}^{B/C}\hat{\mathbf{b}}_3$$
$$\mathbf{K}^{B/C} = K_1{}^{B/C}\hat{\mathbf{b}}_1 + K_2{}^{B/C}\hat{\mathbf{b}}_2 + K_3{}^{B/C}\hat{\mathbf{b}}_3$$

and letting g be the local gravitational force per unit mass, determine the generalized active forces F_1, F_2, and F_3 for the robot arm in E, where F_i is associated with the generalized velocity u_i $(i = 1,2,3)$ defined in Problem 3.15.

Results

$$F_1 = T_1{}^{E/A}$$
$$F_2 = T_2{}^{A/B} - g[(m_B L_B + m_C Z_9 + m_D Z_{14})c_1 - m_D p_1 s_1]$$
$$F_3 = K_3{}^{B/C} - g(m_C + m_D)s_1$$

8.13 Letting S be the set of two particles P_1 and P_2 considered in Problem 8.1, form expressions for the generalized inertia forces $F_1^\star$, $F_2^\star$, $F_3^\star$ for S in A, doing so for each of the three sets of generalized velocities defined in Problem 8.1. Comment briefly on the relative merits of the three sets of generalized velocities.

Results

$$F_1^\star = \{-(m_1 + m_2)(\dot{u}_1 \sec\omega t + 2\omega\dot{q}_1 \tan\omega t) + m_2 L[(\omega^2 + u_3{}^2)c_3 + \dot{u}_3 s_3]\}\sec\omega t$$
$$F_2^\star = -(m_1 + m_2)\dot{u}_2 + Lm_2(u_3{}^2 s_3 - \dot{u}_3 c_3)$$
$$F_3^\star = -m_2 L[L(\dot{u}_3 + \omega^2 s_3 c_3) - (\dot{u}_1 \sec\omega t + 2\omega\dot{q}_1 \tan\omega t)s_3 + \dot{u}_2 c_3]$$

$$F_1^\star = -(m_1 + m_2)(\dot{u}_1 - \omega^2 q_1 c_3 - u_2 u_3) + Lm_2(\omega^2 c_3{}^2 + u_3{}^2)$$
$$F_2^\star = -(m_1 + m_2)(\dot{u}_2 + u_3 u_1 + \omega^2 q_1 s_3) - Lm_2(\dot{u}_3 + \omega^2 s_3 c_3)$$
$$F_3^\star = -m_2 L[\dot{u}_2 + u_3 u_1 + \omega^2 q_1 s_3 + L(\dot{u}_3 + \omega^2 s_3 c_3)]$$

$$F_1^\star = -(m_1 + m_2)(\dot{u}_1 - \omega^2 q_1) + Lm_2(\omega^2 c_3 + \dot{u}_3 s_3 + u_3{}^2 c_3)$$
$$F_2^\star = -(m_1 + m_2)\dot{u}_2 + Lm_2(u_3{}^2 s_3 - \dot{u}_3 c_3)$$
$$F_3^\star = -m_2 L[L(\dot{u}_3 + \omega^2 s_3 c_3) - (\dot{u}_1 - \omega^2 q_1)s_3 + \dot{u}_2 c_3]$$

8.14 Making the same assumptions as in Problem 8.2, use Eqs. (5.9.2) to determine the generalized inertia forces $\tilde{F}_1^\star$ and $\tilde{F}_2^\star$ for S in A. Check the results by using Eqs. (5.9.5) in conjunction with results from Problem 8.13, and comment briefly on the relative merits of Eqs. (5.9.2) and Eqs. (5.9.5).

Results

$$\widetilde{F}_1^\star = (m_1 + m_2)(\omega^2 q_1 c_3 - \dot{u}_1) - m_1 u_2{}^2/L + m_2 L\omega^2 c_3{}^2$$
$$\widetilde{F}_2^\star = -m_1(\dot{u}_2 + \omega^2 q_1 s_3 - u_1 u_2/L)$$

8.15 Referring to Problems 4.15, 4.18, and 4.20, form all nonholonomic generalized inertia forces $\widetilde{\widetilde{F}}_r^\star$ for P in N.

Results

$$\widetilde{\widetilde{F}}_1^\star = -m\left[\left(1 + \frac{a^4 u_1{}^2}{u_3{}^2}\right)\dot{u}_1 + \frac{a^4 u_1 u_2}{u_3{}^2}\dot{u}_2\right]$$

$$\widetilde{\widetilde{F}}_2^\star = -m\left[\frac{a^4 u_1 u_2}{u_3{}^2}\dot{u}_1 + \left(1 + \frac{a^4 u_2{}^2}{u_3{}^2}\right)\dot{u}_2\right]$$

***8.16** The equations that follow furnish useful expressions for $\mathbf{T}^\star$, the inertia torque for a rigid body B in a reference frame A. Establish the validity of each of these equations.

(*a*) When $\boldsymbol{\omega}$, the angular velocity of B in A, can be expressed as $\boldsymbol{\omega} = \omega\hat{\mathbf{n}}_a$, where $\hat{\mathbf{n}}_a$ is a unit vector fixed in A, then

$$\mathbf{T}^\star = -\dot{\omega} I_a \hat{\mathbf{n}}_a + (\omega^2 I_{ca} - \dot{\omega} I_{ab})\hat{\mathbf{n}}_b - (\omega^2 I_{ab} + \dot{\omega} I_{ca})\hat{\mathbf{n}}_c$$

where $\hat{\mathbf{n}}_b$ and $\hat{\mathbf{n}}_c$ are unit vectors perpendicular to $\hat{\mathbf{n}}_a$ and to each other, and are oriented such that $\hat{\mathbf{n}}_a = \hat{\mathbf{n}}_b \times \hat{\mathbf{n}}_c$; I_a is the central moment of inertia of B with respect to a line parallel to $\hat{\mathbf{n}}_a$; I_{ab} is the central product of inertia of B for $\hat{\mathbf{n}}_a$ and $\hat{\mathbf{n}}_b$; and I_{ca} is the central product of inertia of B for $\hat{\mathbf{n}}_c$ and $\hat{\mathbf{n}}_a$.

(*b*) If $\hat{\mathbf{b}}_1$, $\hat{\mathbf{b}}_2$, $\hat{\mathbf{b}}_3$ form a right-handed set of mutually perpendicular unit vectors fixed in B, and $\omega_i \triangleq \boldsymbol{\omega} \cdot \hat{\mathbf{b}}_i$ $(i = 1,2,3)$, where $\boldsymbol{\omega}$ is the angular velocity of B in A, then

$$\begin{aligned}\mathbf{T}^\star = &-[I_{11}\dot{\omega}_1 + I_{12}\dot{\omega}_2 + I_{13}\dot{\omega}_3 + (I_{31}\omega_2 - I_{12}\omega_3)\omega_1 \\ &+ I_{23}(\omega_2{}^2 - \omega_3{}^2) + (I_{33} - I_{22})\omega_2\omega_3]\hat{\mathbf{b}}_1 \\ &- [I_{22}\dot{\omega}_2 + I_{23}\dot{\omega}_3 + I_{21}\dot{\omega}_1 + (I_{12}\omega_3 - I_{23}\omega_1)\omega_2 \\ &+ I_{31}(\omega_3{}^2 - \omega_1{}^2) + (I_{11} - I_{33})\omega_3\omega_1]\hat{\mathbf{b}}_2 \\ &- [I_{33}\dot{\omega}_3 + I_{31}\dot{\omega}_1 + I_{32}\dot{\omega}_2 + (I_{23}\omega_1 - I_{31}\omega_2)\omega_3 \\ &+ I_{12}(\omega_1{}^2 - \omega_2{}^2) + (I_{22} - I_{11})\omega_1\omega_2]\hat{\mathbf{b}}_3\end{aligned}$$

where I_{jk} is the central inertia scalar for $\hat{\mathbf{b}}_j$ and $\hat{\mathbf{b}}_k$ $(j,k = 1,2,3)$.

(*c*) If $\mathbf{H}$ is the central angular momentum of B in A, then

$$\mathbf{T}^\star = -\frac{{}^A d\mathbf{H}}{dt}$$

***8.17** $(\tilde{F}_r^\star)_B$, the contribution of a rigid body B to the generalized inertia force $\tilde{F}_r^\star$ $(r = 1,\ldots,p)$ in a reference frame A, can be expressed as

$$(\tilde{F}_r^\star)_B = {}^A\tilde{\boldsymbol{\omega}}_r^B \cdot \mathbf{T}^Q + {}^A\tilde{\mathbf{v}}_r^Q \cdot \mathbf{R}^\star \qquad (r = 1,\ldots,p) \tag{a}$$

where ${}^A\tilde{\boldsymbol{\omega}}_r^B$ and $\mathbf{R}^\star$ are defined as in connection with Eqs. (5.9.8), but $\mathbf{T}^Q$ and ${}^A\tilde{\mathbf{v}}_r^Q$ differ from $\mathbf{T}^\star$ and ${}^A\tilde{\mathbf{v}}_r^{B\star}$, respectively, being defined as follows: Let Q be a point of B, $\mathbf{r}^{QP_i}$ the position vector from Q to a generic particle P_i of B, m_i the mass of P_i, and ${}^A\mathbf{a}^{P_i}$ the acceleration of P_i $(i = 1,\ldots,\beta)$ in reference frame A. Then

$$\mathbf{T}^Q \triangleq -\sum_{i=1}^{\beta} m_i \mathbf{r}^{QP_i} \times {}^A\mathbf{a}^{P_i} \tag{b}$$

while ${}^A\tilde{\mathbf{v}}_r^Q$ is the rth nonholonomic partial velocity of Q in A. Furthermore, $\mathbf{T}^Q$ [compare with $\mathbf{T}^\star$ as given in Eq. (5.9.12)] can be expressed as

$$\mathbf{T}^Q = -m\mathbf{r}^{QB\star} \times {}^A\mathbf{a}^Q - {}^A\boldsymbol{\alpha}^B \cdot \underline{\mathbf{I}}^Q - {}^A\boldsymbol{\omega}^B \times \underline{\mathbf{I}}^Q \times {}^A\boldsymbol{\omega}^B \tag{c}$$

where m is the mass of B, $\mathbf{r}^{QB\star}$ is the position vector from Q to $B^\star$, ${}^A\mathbf{a}^Q$ is the acceleration of Q in A, ${}^A\boldsymbol{\alpha}^B$ and ${}^A\boldsymbol{\omega}^B$ are, respectively, the angular acceleration of B in A and the angular velocity of B in A, and $\underline{\mathbf{I}}^Q$ is the inertia dyadic of B relative to Q.

Establish the validity of Eqs. (a) and (c), and devise an example to illustrate the utility of these relationships.

***8.18** Referring to Problem 3.15 (see also Problems 4.19 and 8.12), let A, B, C, and D have the following mass distribution properties. The central principal axes of A are parallel to $\hat{\mathbf{a}}_1$, $\hat{\mathbf{a}}_2$, $\hat{\mathbf{a}}_3$ (see Fig. P3.15), and the associated moments of inertia have the values A_1, A_2, A_3, respectively. The central principal axes of B and C are parallel to $\hat{\mathbf{b}}_1$, $\hat{\mathbf{b}}_2$, $\hat{\mathbf{b}}_3$ (see Fig. P3.15), and the associated moments of inertia have the values B_1, B_2, B_3 and C_1, C_2, C_3, respectively. Central inertia scalars D_{ij} $(i,j = 1,2,3)$ for D are defined as

$$D_{ij} \triangleq \hat{\mathbf{b}}_i \cdot \underline{\mathbf{I}}^D \cdot \hat{\mathbf{b}}_j \qquad (i,j = 1,2,3)$$

where $\underline{\mathbf{I}}^D$ is the central inertia dyadic of D.

Letting $\mathbf{T}_A{}^\star$, $\mathbf{T}_B{}^\star$, $\mathbf{T}_C{}^\star$, and $\mathbf{T}_D{}^\star$ denote the inertia torques for A, B, C, and D in E, express $\mathbf{T}_A{}^\star$ in terms of $\hat{\mathbf{a}}_1$, $\hat{\mathbf{a}}_2$, $\hat{\mathbf{a}}_3$, and $\mathbf{T}_B{}^\star$, $\mathbf{T}_C{}^\star$, $\mathbf{T}_D{}^\star$ in terms of $\hat{\mathbf{b}}_1$, $\hat{\mathbf{b}}_2$, $\hat{\mathbf{b}}_3$. To facilitate the writing of results, introduce $k_1,\ldots,k_{16}$ and $Z_{35},\ldots,Z_{54}$ as

$$k_1 \triangleq B_2 - B_3 \qquad k_2 \triangleq B_3 - B_1 \qquad k_3 \triangleq B_1 - B_2 \qquad k_4 \triangleq C_2 - C_3$$
$$k_5 \triangleq C_3 - C_1 \qquad k_6 \triangleq C_1 - C_2 \qquad k_7 \triangleq D_{33} - D_{22} \qquad k_8 \triangleq D_{11} - D_{33}$$
$$k_9 \triangleq D_{22} - D_{11} \qquad k_{10} \triangleq B_1 + k_1 \qquad k_{11} \triangleq B_3 - k_3 \qquad k_{12} \triangleq C_1 + k_4$$
$$k_{13} \triangleq C_3 - k_6 \qquad k_{14} \triangleq D_{11} - k_7 \qquad k_{15} \triangleq D_{31} + D_{13} \qquad k_{16} \triangleq D_{33} + k_9$$

$$Z_{35} \triangleq \mathrm{c}_1 B_1 \qquad Z_{36} \triangleq Z_3 k_{10} \qquad Z_{37} \triangleq Z_2 Z_1 \qquad Z_{38} \triangleq -Z_{37} k_2$$
$$Z_{39} \triangleq \mathrm{s}_1 B_3 \qquad Z_{40} \triangleq Z_4 k_{11} \qquad Z_{41} \triangleq \mathrm{c}_1 C_1 \qquad Z_{42} \triangleq Z_3 k_{12}$$
$$Z_{43} \triangleq -Z_{37} k_5 \qquad Z_{44} \triangleq \mathrm{s}_1 C_3 \qquad Z_{45} \triangleq Z_4 k_{13} \qquad Z_{46} \triangleq {Z_1}^2$$
$$Z_{47} \triangleq {u_2}^2 \qquad Z_{48} \triangleq {Z_2}^2 \qquad Z_{49} \triangleq D_{11}\mathrm{c}_1 + D_{13}\mathrm{s}_1$$
$$Z_{50} \triangleq k_{14} Z_3 + k_{15} Z_4 - D_{12} Z_{37} + D_{23}(Z_{47} - Z_{48}) \qquad Z_{51} \triangleq D_{23}\mathrm{s}_1 + D_{21}\mathrm{c}_1$$
$$Z_{52} \triangleq D_{31}(Z_{48} - Z_{46}) + k_8 Z_{37} \qquad Z_{53} \triangleq D_{33}\mathrm{s}_1 + D_{31}\mathrm{c}_1$$
$$Z_{54} \triangleq k_{15} Z_3 + k_{16} Z_4 + D_{23} Z_{37} + D_{12}(Z_{46} - Z_{47})$$

Results

$$\begin{aligned}
\mathbf{T}_A{}^\star &= -\dot{u}_1 A_1 \hat{\mathbf{a}}_1 \\
\mathbf{T}_B{}^\star &= -(\dot{u}_1 Z_{35} + Z_{36})\hat{\mathbf{b}}_1 - (\dot{u}_2 B_2 + Z_{38})\hat{\mathbf{b}}_2 - (\dot{u}_1 Z_{39} + Z_{40})\hat{\mathbf{b}}_3 \\
\mathbf{T}_C{}^\star &= -(\dot{u}_1 Z_{41} + Z_{42})\hat{\mathbf{b}}_1 - (\dot{u}_2 C_2 + Z_{43})\hat{\mathbf{b}}_2 - (\dot{u}_1 Z_{44} + Z_{45})\hat{\mathbf{b}}_3 \\
\mathbf{T}_D{}^\star &= -(\dot{u}_1 Z_{49} + \dot{u}_2 D_{12} + Z_{50})\hat{\mathbf{b}}_1 - (\dot{u}_1 Z_{51} + \dot{u}_2 D_{22} + Z_{52})\hat{\mathbf{b}}_2 \\
&\quad - (\dot{u}_1 Z_{53} + \dot{u}_2 D_{32} + Z_{54})\hat{\mathbf{b}}_3
\end{aligned}$$

***8.19** Referring to Problem 3.15 (see also Problems 4.19, 8.12, and 8.18), determine the generalized inertia forces $F_1^\star$, $F_2^\star$, and $F_3^\star$ for the robot arm in E, where $F_i^\star$ is associated with the generalized velocity u_i $(i = 1, 2, 3)$.

Results

$$\begin{aligned}
F_1^\star &= -\dot{u}_1[A_1 + \mathrm{c}_1(Z_{35} + Z_{41} + Z_{49}) + \mathrm{s}_1(Z_{39} + Z_{44} + Z_{53}) \\
&\quad + m_A {L_A}^2 + m_B {Z_6}^2 + m_C {Z_{10}}^2 + m_D({Z_{13}}^2 + {Z_{15}}^2 + {Z_{16}}^2)] \\
&\quad - \dot{u}_2[\mathrm{c}_1 D_{12} + \mathrm{s}_1 D_{32} + m_D(Z_{13} Z_{14} - Z_{16} p_1)] - \dot{u}_3 m_D Z_{16} \\
&\quad - [\mathrm{c}_1(Z_{36} + Z_{42} + Z_{50}) + \mathrm{s}_1(Z_{40} + Z_{45} + Z_{54}) + m_B Z_6 Z_{23} \\
&\quad + m_C Z_{10} Z_{27} + m_D(Z_{13} Z_{32} + Z_{15} Z_{33} + Z_{16} Z_{34})] \\
F_2^\star &= -\dot{u}_1[Z_{51} + m_D(Z_{14} Z_{13} - p_1 Z_{16})] - \dot{u}_2[B_2 + C_2 + D_{22} + m_B {L_B}^2 \\
&\quad + m_C {Z_9}^2 + m_D({Z_{14}}^2 + {p_1}^2)] + \dot{u}_3 m_D p_1 - [Z_{38} + Z_{43} + Z_{52} \\
&\quad + m_B L_B Z_{22} + m_C Z_9 Z_{26} + m_D(Z_{14} Z_{32} - p_1 Z_{34})] \\
F_3^\star &= -\dot{u}_1 m_D Z_{16} + \dot{u}_2 m_D p_1 - \dot{u}_3(m_C + m_D) - (m_C Z_{28} + m_D Z_{34})
\end{aligned}$$

***8.20** Letting S be a simple nonholonomic system possessing generalized coordinates $q_1, \ldots, q_n$ and generalized velocities $u_1, \ldots, u_p$ in a reference frame A, and letting G be the Gibbs function for S in A (see Problem 6.16), show that the generalized inertia forces $\tilde{F}_1^\star, \ldots, \tilde{F}_p^\star$ for S in A can be expressed as

$$\tilde{F}_r^\star = -\frac{\partial G}{\partial \dot{u}_r} \qquad (r = 1, \ldots, p)$$

if G is regarded as a function of $q_1, \ldots, q_n$, $u_1, \ldots, u_p$, and $\dot{u}_1, \ldots, \dot{u}_p$. [The formula developed in Problem 6.16 can be used in conjunction with the present result to find the contribution of a rigid body B to $\tilde{F}_r^\star$ $(r = 1, \ldots, p)$. When using the formula for this purpose, one can omit the term $\boldsymbol{\omega}^2\boldsymbol{\omega} \cdot \underline{\mathbf{I}} \cdot \boldsymbol{\omega}$, for this term does not contain any of $\dot{u}_1, \ldots, \dot{u}_p$, and hence cannot contribute to $\partial G / \partial \dot{u}_r$ $(r = 1, \ldots, p)$.]

PROBLEM SET 9
(Secs. 6.1–6.7)

9.1 When two particles P_1 and P_2 orbit one another in accordance with *Kepler's laws*, the motion takes place in a plane that is fixed in a Newtonian reference frame N, as shown in Fig. P9.1. Let A be a reference frame having a simple angular velocity in N, ${}^N\boldsymbol{\omega}^A = \dot{\theta}\hat{\mathbf{a}}_3$, where unit vector $\hat{\mathbf{a}}_3$ is fixed in A and in N such that it is perpendicular to the plane in which P_1 and P_2 move. A unit vector $\hat{\mathbf{a}}_1$ is fixed in A such that it is perpendicular to $\hat{\mathbf{a}}_3$ and has the same direction as the position vector from P_1 to P_2. θ is the angle between $\hat{\mathbf{a}}_1$ and a line fixed in N in the plane of motion.

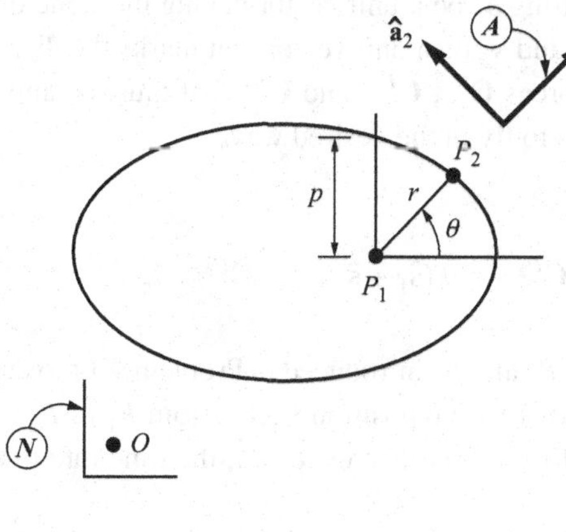

Figure P9.1

The position vectors $\mathbf{p}_1$ and $\mathbf{p}_2$ from a point O fixed in N to P_1 and P_2, respectively, are related to r, the distance between the two particles, as follows,

$$\mathbf{p}_2 - \mathbf{p}_1 = r\hat{\mathbf{a}}_1$$

and differentiation of this relationship with respect to the time t in N yields

$$^N\mathbf{v}^{P_2} - {}^N\mathbf{v}^{P_1} = \dot{r}\hat{\mathbf{a}}_1 + r\dot{\theta}\hat{\mathbf{a}}_2$$

where $^N\mathbf{v}^{P_i}$ is the velocity in N of P_i $(i = 1,2)$, and where $\hat{\mathbf{a}}_2 = \hat{\mathbf{a}}_3 \times \hat{\mathbf{a}}_1$.

Kepler's first law states that the orbit of P_2, a planet, is an ellipse with P_1, the Sun, at a focus. An ellipse (in fact, any conic section) can be described by the relationship

$$\frac{1}{r} = \frac{1 + e\cos\theta}{p} \triangleq B_1 + B_2\cos\theta$$

where the angle θ is zero at periapsis (the point where r is minimum) and is referred to as the true anomaly. The constants e and p are known respectively as the eccentricity of the conic section, and the semilatus rectum (or parameter). Constants B_1 and B_2 are defined as $B_1 \triangleq 1/p$ and $B_2 \triangleq e/p$. Regarding Kepler's first law as a configuration constraint, obtain a constraint equation involving the difference $^N\mathbf{v}^{P_2} - {}^N\mathbf{v}^{P_1}$, and inspect the constraint equation to identify the constraint forces $\mathbf{C}_1^{P_2}$ and $\mathbf{C}_1^{P_1}$ that must be applied to P_2 and P_1, respectively, in order to ensure satisfaction of Kepler's first law.

Results $\mathbf{C}_1^{P_2} = \lambda_1(\hat{\mathbf{a}}_1 - B_2\sin\theta\, r\hat{\mathbf{a}}_2)$ $\quad$ $\mathbf{C}_1^{P_1} = -\mathbf{C}_1^{P_2}$

9.2 Inside a laboratory module of the International Space Station, experiments are conducted in orbit with small free-flying vehicles. Each device is roughly spherical in shape, and can be propelled in any direction by means of cold compressed gas expelled through one or more of several rocket nozzles. Suppose that three such devices are involved in an experiment to study spacecraft formation flying. Regard the vehicles as particles, P_1, P_2, and P_3. The distance S_1 between P_1 and P_2 is required to be the same as the distance S_2 between P_2 and P_3. S_1, and therefore S_2, need not remain constant. The three particles need not be collinear. Letting $\hat{\mathbf{s}}_1$ be a unit vector having the same direction as the position vector from P_1 to P_2, and $\hat{\mathbf{s}}_2$ be a unit vector that marks the direction from P_2 to P_3, identify the constraint forces $\mathbf{C}^{P_1}$, $\mathbf{C}^{P_2}$, and $\mathbf{C}^{P_3}$ that must be applied to P_1, P_2, and P_3 in order for the vehicles to fly in the desired way.

Results

$$\mathbf{C}^{P_1} = \lambda\hat{\mathbf{s}}_1 \qquad \mathbf{C}^{P_2} = -\lambda(\hat{\mathbf{s}}_1 + \hat{\mathbf{s}}_2) \qquad \mathbf{C}^{P_3} = \lambda\hat{\mathbf{s}}_2$$

9.3 Suppose that the two particles P_1 and P_2 introduced in Problem 9.1 move according to *Kepler's second law*, which holds that the position vector from P_1 to P_2 sweeps out equal areas in equal increments of time. In other words, $\dot{A}$, the time rate of change of area swept out by $r\hat{\mathbf{a}}_1$, given by

$$\dot{A} = \tfrac{1}{2}r^2\dot{\theta}$$

is a constant. Treating Kepler's second law as a motion constraint, form a nonholonomic constraint equation that is linear in $^N\mathbf{v}^{P_2}$ and $^N\mathbf{v}^{P_1}$, and inspect the constraint equation to identify the constraint forces $\mathbf{C}_2^{P_2}$ and $\mathbf{C}_2^{P_1}$ that must be applied to P_2 and P_1, respectively, in order to ensure satisfaction of Kepler's second law.

Results $\mathbf{C}_2^{P_2} = \lambda_2 \hat{\mathbf{a}}_2 \qquad \mathbf{C}_2^{P_1} = -\mathbf{C}_2^{P_2}$

9.4 Referring to the example in Sec. 5.3, consider the constraint imposed by the rigid cross C on the orientation of the yoke Y' relative to yoke Y. Unit vector $\hat{\nu}'$ fixed in Y' must remain perpendicular to unit vector $\hat{\nu}$ fixed in Y. Express this restriction in terms of the two unit vectors and, by referring to Eqs. (6.5.1) and (6.5.3), identify the constraint torques that must be exerted on Y and Y' in order to enforce the restriction. Comment briefly on a comparison of your results with Eqs. (5.3.11) and (5.3.8).

Results $\mathbf{T}^Y = \mu\hat{\nu} \times \hat{\nu}' \qquad \mathbf{T}^{Y'} = -\mu\hat{\nu} \times \hat{\nu}'$

9.5 Two rigid rods A and B move in a plane fixed in a Newtonian reference frame N, as shown in Fig. P9.5. Rod A is pinned at point O to a support fixed in N, and A and B are fastened to each other by a pin at point P. Let $\hat{\mathbf{n}}_1$ and $\hat{\mathbf{n}}_2$ be two perpendicular unit vectors fixed in N, with $\hat{\mathbf{n}}_2$ parallel to a slot guiding a slider pinned to the endpoint Q of B. The slider is to be made to move in such a way that the velocity of Q in N is given by ${}^N\mathbf{v}^Q = v(t)\hat{\mathbf{n}}_2$, where $v(t)$ is a prescribed function of the time t.

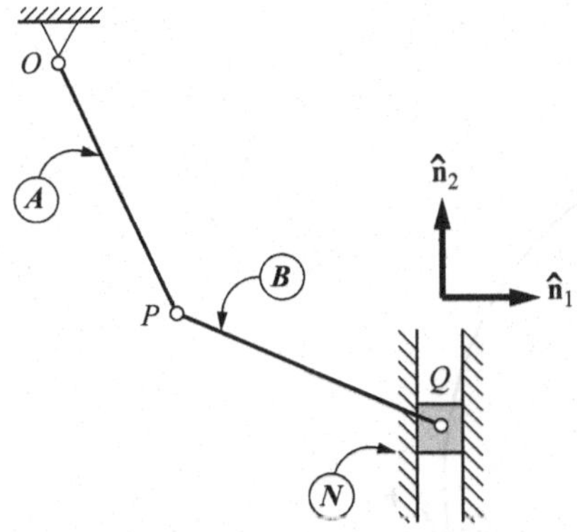

Figure P9.5

Considering the two constraints imposed on the motion of Q, identify a set of constraint forces by inspecting constraint equations (a) written in terms of ${}^N\mathbf{v}^Q$, and (b) expressed in terms of ${}^N\mathbf{v}^P$, the velocity of P in N, ${}^N\boldsymbol{\omega}^B$, the angular velocity of B in N, and $\mathbf{r}^{PQ}$, the position vector from P to Q. (c) Verify that the set of forces found in part (a) is equivalent to the set obtained in part (b).

Results (a) A constraint force $\mathbf{C}^Q = \lambda_1\hat{\mathbf{n}}_1 + \lambda_2\hat{\mathbf{n}}_2$ applied at Q. (b) A constraint force $\mathbf{C}^P = \lambda_1\hat{\mathbf{n}}_1 + \lambda_2\hat{\mathbf{n}}_2$ applied at P, together with a couple of torque $\mathbf{T}^B = \mathbf{r}^{PQ} \times (\lambda_1\hat{\mathbf{n}}_1 + \lambda_2\hat{\mathbf{n}}_2)$ acting on B.

9.6 Referring to Problems 4.15 and 4.18, regard N as a Newtonian reference frame and identify the constraint force **C** that must be applied to P in order to satisfy the motion constraint. Show that this force does not contribute to the nonholonomic generalized active forces $\tilde{\tilde{F}}_1$ and $\tilde{\tilde{F}}_2$ for P in N.

Result

$$\mathbf{C} = \lambda \left[\hat{\mathbf{n}}_3 - a^2(u_1\hat{\mathbf{n}}_1 + u_2\hat{\mathbf{n}}_2)/u_3\right]$$

9.7 The ends A and B of a uniform rod R of mass m are supported by a circular wire W that lies in a fixed vertical plane and has a radius r. At the center of W, A and B subtend an angle 2θ (Fig. P9.7).

(*a*) Letting u_1 be a generalized velocity defined as $u_1 \triangleq \dot{q}_1$, where q_1 is the angle shown in Fig. P9.7, and assuming that friction forces exerted on R by W at A and B are negligible, determine the generalized active force F_1.

(*b*) Letting $-sT_A\hat{\mathbf{a}}_2$ and $-sT_B\hat{\mathbf{b}}_2$ be friction forces exerted on R by W at A and B, respectively, where T_A and T_B are nonnegative quantities and $s \triangleq u_1/|u_1|$, while $\hat{\mathbf{a}}_2$ and $\hat{\mathbf{b}}_2$ are unit vectors directed as shown in Fig. P9.7, determine the generalized active force F_1.

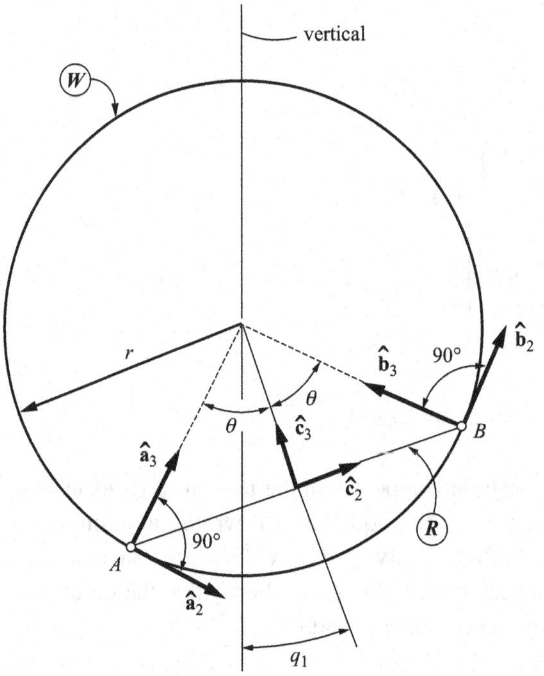

Figure P9.7

(*c*) The forces exerted on R by W at A and B in the radial directions at these points can be expressed as $N_A\hat{\mathbf{a}}_3$ and $N_B\hat{\mathbf{b}}_3$ respectively. To bring these into evidence in expressions for generalized active forces, let C be a reference frame in which the unit vectors $\hat{\mathbf{c}}_2$ and $\hat{\mathbf{c}}_3$ shown in Fig. P9.7 are fixed, introduce motion variables u_2 and u_3 such that the velocity of the center of R in C is given by $u_2\hat{\mathbf{c}}_2 + u_3\hat{\mathbf{c}}_3$, and determine the generalized active forces F_1, F_2, and F_3.

Results

$$(a) \quad F_1 = -mgr\cos\theta \sin q_1$$

$$(b) \quad F_1 = -r[s(T_A + T_B) + mg\cos\theta\sin q_1]$$

$$(c) \quad F_1 = -r[s(T_A + T_B) + mg\cos\theta\sin q_1]$$

$$F_2 = -s(T_A + T_B)\cos\theta + (N_A - N_B)\sin\theta - mg\sin q_1$$

$$F_3 = s(T_A - T_B)\sin\theta + (N_A + N_B)\cos\theta - mg\cos q_1$$

9.8 Referring to the example in Sec. 5.7, suppose that Q slides on P subject to the laws of Coulomb friction, and let μ' be the coefficient of kinetic friction for A and P. Under these circumstances, the quantities Q_1, Q_2, and Q_3 appearing in Eq. (5.7.14) can be expressed as functions of μ', R, M, g, e, f, θ, u_1, u_2, and a nonholonomic generalized active force $\widetilde{F}_3$ associated with a motion variable u_3 introduced for the purpose of bringing Q_1 into evidence. Taking $u_3 \triangleq {}^F\boldsymbol{\omega}^A \cdot \hat{\mathbf{a}}_3$, determine these functions.

Results

$$Q_1 = (eMg\cos\theta - \widetilde{F}_3)/[f - \mu' R u_2/({u_2}^2 + f^2{u_1}^2)^{1/2}]$$

$$Q_2 = -\mu' u_2 Q_1/({u_2}^2 + f^2{u_1}^2)^{1/2}$$

$$Q_3 = -\mu' u_1 f Q_1/({u_2}^2 + f^2{u_1}^2)^{1/2}$$

9.9 Figure P9.9 shows two views of a piston P of mass M in a cylinder C. R_1, R_2, and R_3 are piston rings of thickness t, and t is so small in comparison with R, the internal radius of C, that the contacts between R_1, R_2, R_3 and C can be regarded as taking place along circles of radius R. The axis of C is horizontal and parallel to a unit vector $\hat{\mathbf{i}}$, and P is subjected to the action of a set of contact forces equivalent to a force $E\hat{\mathbf{i}}$ applied to P at $P^\star$, the mass center of P.

Two motion variables, u_1 and u_2, defined as

$$u_1 \triangleq \boldsymbol{\omega} \cdot \hat{\mathbf{i}} \qquad u_2 \triangleq \mathbf{v}^\star \cdot \hat{\mathbf{i}}$$

where $\boldsymbol{\omega}$ is the angular velocity of P in C and $\mathbf{v}^\star$ is the velocity of $P^\star$ in C, characterize motions of P in C if P is free to rotate and translate. A third motion variable, u_3, defined as

$$u_3 \triangleq \mathbf{v}^\star \cdot \hat{\mathbf{j}}$$

where $\hat{\mathbf{j}}$ is a unit vector directed vertically downward, can be introduced to bring into evidence radial contact forces exerted on R_1, R_2, and R_3 by C.

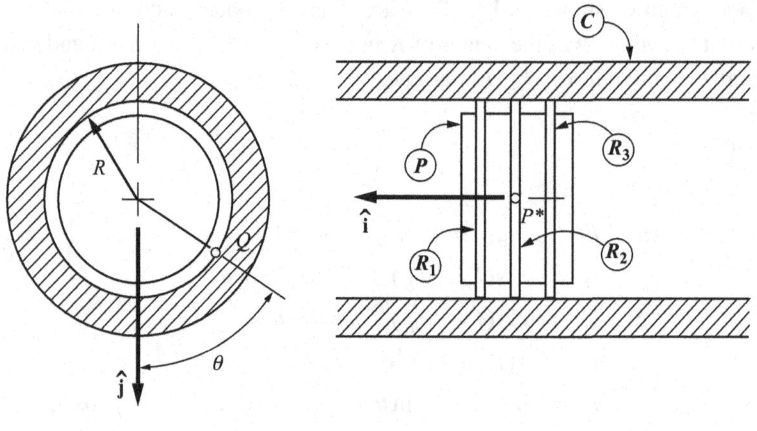

Figure P9.9

Assuming that n, the contact pressure (force per unit of length) at a point Q situated on any one of R_1, R_2, or R_3 as shown in Fig. P9.9, is given by

$$n = \alpha + \beta \cos\left(\frac{\theta}{2}\right) \qquad (-\pi \le \theta \le \pi)$$

where α and β are constants, and letting μ' be the coefficient of kinetic friction for C and R_1, R_2, R_3, determine the generalized active forces F_1, F_2, and F_3 for motions during which $u_3 = 0$.

Results

$$F_1 = -6\mu' R^3(\pi\alpha + 2\beta)u_1[(Ru_1)^2 + u_2{}^2]^{-1/2}$$
$$F_2 = -6\mu' R(\pi\alpha + 2\beta)u_2[(Ru_1)^2 + u_2{}^2]^{-1/2} + E$$
$$F_3 = -4R\beta + Mg$$

9.10 Figure P9.10 shows a rigid spacecraft B moving in a Newtonian reference frame N. A particle P that belongs to B is part of a science experiment.

The velocity ${}^N\mathbf{v}^{B^\star}$ in N of $B^\star$, the mass center of B, and the angular velocity ${}^N\boldsymbol{\omega}^B$ of B in N, can be written in terms of six generalized velocities,

$${}^N\mathbf{v}^{B^\star} = u_1\hat{\mathbf{b}}_1 + u_2\hat{\mathbf{b}}_2 + u_3\hat{\mathbf{b}}_3 \qquad {}^N\boldsymbol{\omega}^B = u_4\hat{\mathbf{b}}_1 + u_5\hat{\mathbf{b}}_2 + u_6\hat{\mathbf{b}}_3$$

where $\hat{\mathbf{b}}_1$, $\hat{\mathbf{b}}_2$, $\hat{\mathbf{b}}_3$ are dextral, mutually perpendicular unit vectors fixed in B.

P is coincident with a point $\overline{B}$ of B. Coincidence is maintained by a constraint force $\mathbf{C}$ applied to P by B; necessarily, therefore, P applies a force $-\mathbf{C}$ to B at $\overline{B}$. Frequently, a science experiment is conducted in orbit because the magnitude of $\mathbf{C}$ divided by the mass of P can be made six orders of magnitude smaller than what can be obtained in a terrestrial laboratory. As demonstrated in Sec. 6.6, $\mathbf{C}$ does not contribute to any of the generalized active forces $F_1, \ldots, F_6$ corresponding to $u_1, \ldots, u_6$. After expressing the position vector from $B^\star$ to $\overline{B}$ as

$$\mathbf{r} = r_1 \hat{\mathbf{b}}_1 + r_2 \hat{\mathbf{b}}_2 + r_3 \hat{\mathbf{b}}_3$$

and expressing $\mathbf{C}$ as

$$\mathbf{C} = \lambda_1 \hat{\mathbf{b}}_1 + \lambda_2 \hat{\mathbf{b}}_2 + \lambda_3 \hat{\mathbf{b}}_3$$

bring $\mathbf{C}$ into evidence by introducing additional motion variables u_7, u_8, and u_9. Determine (*a*) the contributions f_r of $\mathbf{C}$ to F_r by letting ${}^N\mathbf{v}^P = u_7 \hat{\mathbf{b}}_1 + u_8 \hat{\mathbf{b}}_2 + u_9 \hat{\mathbf{b}}_3$, and (*b*) the contributions g_r of $\mathbf{C}$ to F_r by letting ${}^B\mathbf{v}^P = u_7 \hat{\mathbf{b}}_1 + u_8 \hat{\mathbf{b}}_2 + u_9 \hat{\mathbf{b}}_3$, for $r = 1, \ldots, 9$. If one forms generalized active forces $F_1, \ldots, F_6$ prior to introducing u_7, u_8, and u_9, in which case(s) are $F_1, \ldots, F_6$ subsequently unchanged?

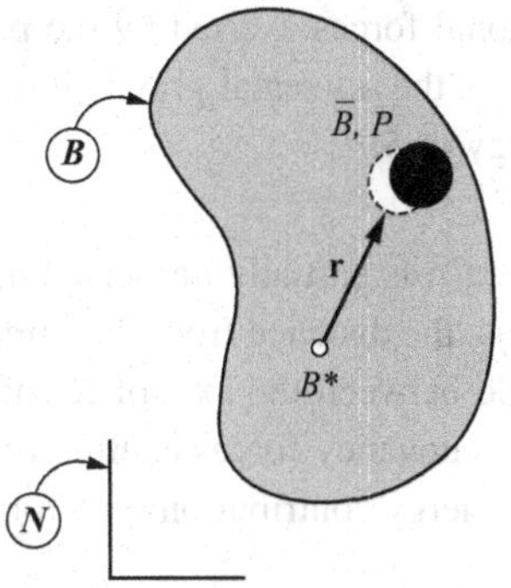

Figure P9.10

Results

(*a*) $f_1 = -\lambda_1$ $\quad f_2 = -\lambda_2$ $\quad f_3 = -\lambda_3$

$f_4 = r_3\lambda_2 - r_2\lambda_3$ $\quad f_5 = r_1\lambda_3 - r_3\lambda_1$ $\quad f_6 = r_2\lambda_1 - r_1\lambda_2$

$f_7 = \lambda_1$ $\quad f_8 = \lambda_2$ $\quad f_9 = \lambda_3$

(*b*) $g_1 = 0$ $\quad g_2 = 0$ $\quad g_3 = 0$

$g_4 = 0$ $\quad g_5 = 0$ $\quad g_6 = 0$

$g_7 = \lambda_1$ $\quad g_8 = \lambda_2$ $\quad g_9 = \lambda_3$

$F_1, \ldots, F_6$ are unchanged in case (*b*).

9.11 Spheres A and B in the examples in Secs. 6.5 and 6.6 are subject to a motion constraint requiring the velocities in E of their mass centers, $A^\star$ and $B^\star$, respectively, to be perpendicular to each other. According to Eqs. (6.5.28) the constraint is enforced by applying a force $\mathbf{C}_7^{A^\star} = \mu_7 {}^E\mathbf{v}^{B^\star}$ to $A^\star$, and a force $\mathbf{C}_7^{B^\star} = \mu_7 {}^E\mathbf{v}^{A^\star}$ to $B^\star$. This set of forces is not in evidence in $\tilde{\tilde{F}}_r$ $(r = 1, \ldots, 5)$, as may be verified by examining

Eqs. (6.6.61)–(6.6.64). Determine the contributions f_r to $\widetilde{F}_r$ $(r = 1,\ldots,6)$ of the pair of constraint forces, thereby bringing them into evidence.

Results

$$f_1 = \mu_7 u_4 \qquad f_2 = \mu_7 u_6 \qquad f_3 = 0 \qquad f_4 = \mu_7 u_1 \qquad f_5 = 0 \qquad f_6 = \mu_7 u_2$$

PROBLEM SET 10
(Secs. 7.1–7.3)

10.1 A system S of two particles P_1 and P_2 of masses m_1 and m_2 moves in such a way that the distance r between P_1 and P_2 is free to vary with time. Assuming that no forces act on P_1 and P_2 other than the gravitational forces exerted by the particles on each other, show that $-Gm_1m_2/r + C$, where G is the universal gravitational constant and C is any function of time, is a potential energy of S.

10.2 A uniform thin rod R of cross-sectional area A is partially immersed in a fluid of mass density ρ. R can move in such a way that q_1, the distance from the immersed end of R to the surface of the fluid, and q_2, the angle between the axis of R and the local vertical, are free to vary. Letting β be the set of buoyancy forces exerted on R by the fluid, show that $g\rho A({q_1}^2/2)\sec q_2$ is a potential energy contribution of β for R.

***10.3** Referring to Problem 2.7, suppose that C is of uniform density and has a mass m, and let γ be the set of gravitational forces acting on C. Verify that $mgR\cos q_2$ is a potential energy contribution of γ for C, and show that this function is a potential energy of C in A when C is rolling on H, but not when slipping is taking place at P, unless the contact between H and P is frictionless. (Expressions for generalized active forces are available in Problem 8.3.)

10.4 Show that $-g(m_1L_1c_1 + m_2L_3c_3)$ is a potential energy of the system of two particles considered in Problem 8.7. Verify that the expression for $(\widetilde{F}_1)_\gamma$ found in Problem 8.7 can be obtained by replacing $\widetilde{F}_r$ with $(\widetilde{F}_r)_\gamma$, and V with $-g(m_1L_1c_1 + m_2L_3c_3)$ in Eqs. (7.1.14).

10.5 Referring to Problem 8.8, determine with the aid of Eq. (7.2.2) a potential energy contribution V_γ of the gravitational forces acting on the 33 bars; use Eqs. (7.1.9) with V replaced with V_γ and F_r replaced with $(F_r)_\gamma$ to find the contributions $(F_r)_\gamma$ of the gravitational forces to the generalized active forces F_r $(r = 1,2,3)$. Comment briefly on the relative merits of the method used to obtain these contributions in Problem 8.8, on the one hand, and the method employed in the present problem, on the other hand.

10.6 Construct a potential energy V of the system described in Problem 8.9.

Result

$$V = \frac{6EI}{L^3}\left[\left(1 + \frac{L}{2b} + \frac{L^2}{6b^2}\right)({q_1}^2 + {q_2}^2) - \frac{L}{b}\left(1 + \frac{L}{3b}\right) q_1 q_2\right] - \frac{mg}{2}(q_1 + q_2)$$

10.7 Referring to Problem 9.7, suppose that the plane containing W is made to rotate with a prescribed angular speed about a vertical line passing through the center of W. Show that $-mgr\cos\theta\cos q_1$ is a potential energy of R so long as the contacts between R and W at A and B are perfectly smooth.

***10.8** Referring to the example in Sec. 5.7 (see Fig. 5.7.1), let q_1 be the angle between $\hat{\mathbf{n}}_2$ and $\hat{\mathbf{a}}_2$ (see Fig. 5.7.2) as before; define q_2 and q_3 as

$$q_2 \triangleq \mathbf{p}\cdot\hat{\mathbf{n}}_2 \qquad q_3 \triangleq \mathbf{p}\cdot\hat{\mathbf{n}}_3$$

where $\mathbf{p}$ is the position vector from a point fixed in F to point D; and introduce wheel rotation angles q_4 and q_5 as shown in Fig. P10.8, where L_B and L_C are lines fixed in B and C, respectively. Assume that no friction forces come into play at point Q in Fig. 5.7.1, and that B and C are driven by motors attached to A in such a way that the sets of contact forces exerted by A on B and by A on C are equivalent to couples of torques $\mathbf{T}_B$ and $\mathbf{T}_C$, respectively, with

$$\mathbf{T}_B = T_B\hat{\mathbf{a}}_3 \qquad \mathbf{T}_C = T_C\hat{\mathbf{a}}_3$$

where T_B and T_C are functions of $q_1,\ldots,q_5$.

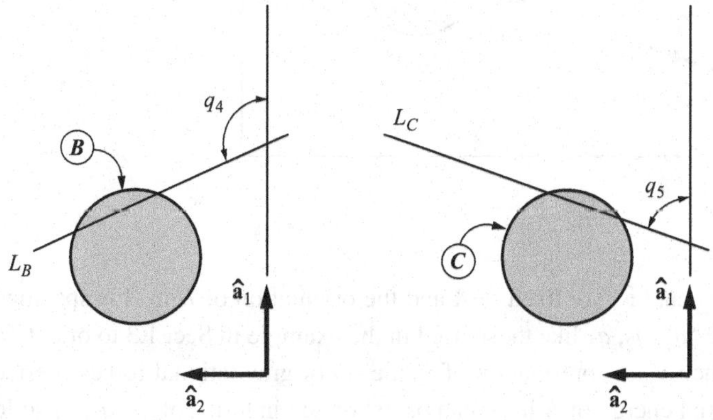

Figure P10.8

Show that $V(q_1,\ldots,q_5,t)$ is a potential energy of S in F if

$$\frac{\partial V}{\partial q_1} + \frac{b}{R}\left(\frac{\partial V}{\partial q_4} - \frac{\partial V}{\partial q_5}\right) = \frac{b}{R}(T_C - T_B) - Mge\sin\theta\,\mathrm{c}_1$$

$$\frac{\partial V}{\partial q_2}\mathrm{c}_1 + \frac{\partial V}{\partial q_3}\mathrm{s}_1 + \frac{1}{R}\left(\frac{\partial V}{\partial q_4} + \frac{\partial V}{\partial q_5}\right) = -\frac{1}{R}(T_C + T_B) - Mg\sin\theta\,\mathrm{s}_1$$

10.9 In Fig. P10.9, P is a particle fixed in a reference frame A, and $B^\star$ is the mass center of a rigid body B; $\hat{\mathbf{a}}_1$, $\hat{\mathbf{a}}_2$, $\hat{\mathbf{a}}_3$ form a right-handed set of mutually perpendicular unit vectors such that $\hat{\mathbf{a}}_1$ points from P to $B^\star$; $\hat{\mathbf{b}}_1$, $\hat{\mathbf{b}}_2$, $\hat{\mathbf{b}}_3$ form a similar set of unit vectors parallel to central principal axes of B.

Assume throughout what follows that the set γ of all gravitational forces exerted by P on B can be replaced with a couple of torque $\mathbf{T}$ given by

$$\mathbf{T} = \frac{3Gm}{R^3}\hat{\mathbf{a}}_1 \times \underline{\mathbf{I}} \cdot \hat{\mathbf{a}}_1$$

together with a force $\mathbf{F}$ applied at $B^\star$ and given by

$$\mathbf{F} = -\frac{GmM}{R^2}\hat{\mathbf{a}}_1$$

where G is the universal gravitational constant, m is the mass of P, R is the distance from P to $B^\star$, $\underline{\mathbf{I}}$ is the central inertia dyadic of B, and M is the mass of B.

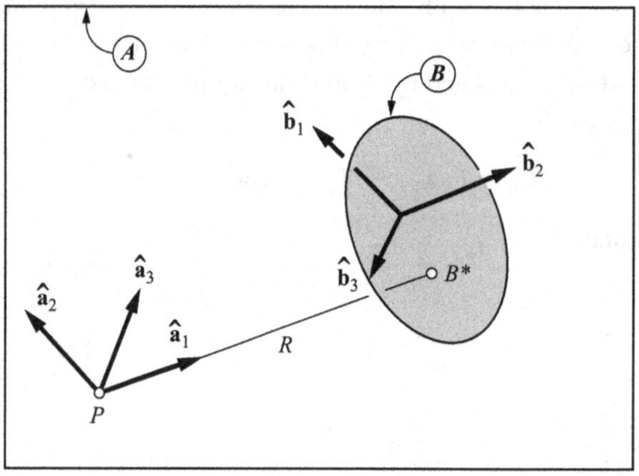

Figure P10.9

(a) If $\hat{\mathbf{a}}_1$, $\hat{\mathbf{a}}_2$, and $\hat{\mathbf{a}}_3$ are fixed in A and the orientation of B in A is specified in terms of three angles q_1, q_2, q_3 like those used in the example in Sec. 1.3 to orient D in A (see Fig. 1.3.1), then V_γ, a contribution of γ, the set of gravitational forces exerted by P on B, to a potential energy of B in A, can be expressed in terms of q_1, q_2, q_3 so long as $B^\star$ is kept fixed in A. Defining motion variables u_1, u_2, u_3 for B in A as

$$u_i \triangleq {}^A\boldsymbol{\omega}^B \cdot \hat{\mathbf{b}}_i \qquad (i = 1, 2, 3)$$

and letting $I_j \triangleq \hat{\mathbf{b}}_j \cdot \underline{\mathbf{I}} \cdot \hat{\mathbf{b}}_j$ $(j = 1, 2, 3)$, determine V_γ.

(b) Use the result of part (a) to determine $(F_1)_\gamma$, $(F_2)_\gamma$, $(F_3)_\gamma$, the contributions of γ to the generalized active forces for B in A corresponding to motion variables u_1, u_2, u_3 defined as

$$u_i \triangleq \dot{q}_i \qquad (i = 1, 2, 3)$$

(*c*) Assuming that $\hat{\mathbf{a}}_1$, $\hat{\mathbf{a}}_2$, and $\hat{\mathbf{a}}_3$ are fixed in A, but that $B^\star$ can move in A along line $PB^\star$, which means that B has four degrees of freedom in A, show that V_γ does not exist.

Results

$$(a) \qquad V_\gamma = -\frac{3Gm}{2R^3}[(I_1 - I_3)s_2{}^2 + (I_1 - I_2)c_2{}^2 s_3{}^2]$$

$$(b) \qquad (F_1)_\gamma = 0$$

$$(F_2)_\gamma = \frac{3Gm}{R^3} s_2 c_2 (I_1 c_3{}^2 + I_2 s_3{}^2 - I_3)$$

$$(F_3)_\gamma = \frac{3Gm}{R^3}(I_1 - I_2)c_2{}^2 s_3 c_3$$

***10.10** Referring to Problem 10.9, and assuming that, as in part (*c*), $B^\star$ is free to move in A along line $PB^\star$, suppose that γ is replaced with a couple of torque $\mathbf{T}$ given once again by

$$\mathbf{T} = \frac{3Gm}{R^3}\hat{\mathbf{a}}_1 \times \underline{\mathbf{I}} \cdot \hat{\mathbf{a}}_1$$

together with a force $\mathbf{F}$ applied at $B^\star$, but that $\mathbf{F}$ is given by

$$\mathbf{F} = -\frac{GmM}{R^2}(\hat{\mathbf{a}}_1 + \mathbf{f})$$

where

$$\mathbf{f} \triangleq \frac{3}{MR^2}\left\{\frac{1}{2}[I_1(1 - 3C_{11}{}^2) + I_2(1 - 3C_{12}{}^2) + I_3(1 - 3C_{13}{}^2)]\hat{\mathbf{a}}_1 \right.$$
$$+ (I_1 C_{21} C_{11} + I_2 C_{22} C_{12} + I_3 C_{23} C_{13})\hat{\mathbf{a}}_2$$
$$\left. + (I_1 C_{31} C_{11} + I_2 C_{32} C_{12} + I_3 C_{33} C_{13})\hat{\mathbf{a}}_3 \right\}$$

with $C_{ij} \triangleq \hat{\mathbf{a}}_i \cdot \hat{\mathbf{b}}_j$ $(i, j = 1, 2, 3)$. Find an expression for a contribution of γ to a potential energy of B in A.

Result

$$V_\gamma - \frac{GmM}{R} + \frac{Gm}{2R^3}(2I_1 - I_2 - I_3)$$

where V_γ is the potential energy contribution found in part (*a*) of Problem 10.9

10.11 A simple pendulum of mass m and length L is attached to a linear spring of natural length L' and spring constant $k = 5mg/L$, as shown in Fig. P10.11. A potential energy V of this system can be expressed as

$$V = mgL \sum_{i=1}^{\infty} a_i q^i$$

where $a_1, a_2, \ldots$ are constants. Determine $a_1, \ldots, a_4$.

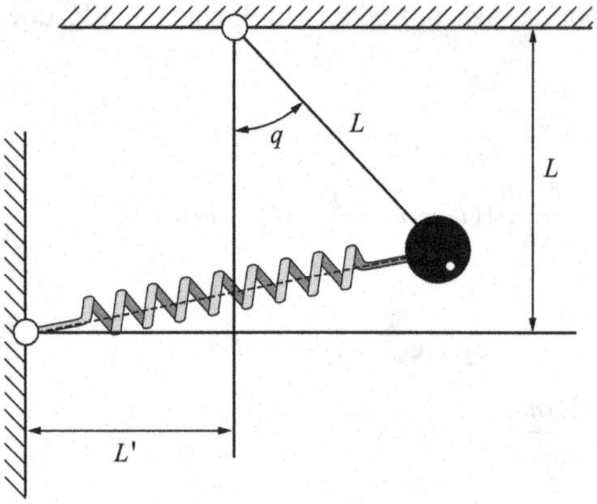

Figure P10.11

The generalized active force F_1 corresponding to $u_1 \triangleq \dot{q}$ can be expressed as

$$F_1 = mgL \sum_{i=1}^{\infty} b_i \, q^i$$

where $b_1, b_2, \ldots$ are constants. Determine b_1, b_2, and b_3, both by using the potential energy V and without reference to potential energy, and comment briefly on the relative merits of the two methods.

Results $a_1 = 0, a_2 = 3, a_3 = 0, a_4 = -\frac{7}{8}; b_1 = -6, b_2 = 0, b_3 = \frac{7}{2}$

10.12

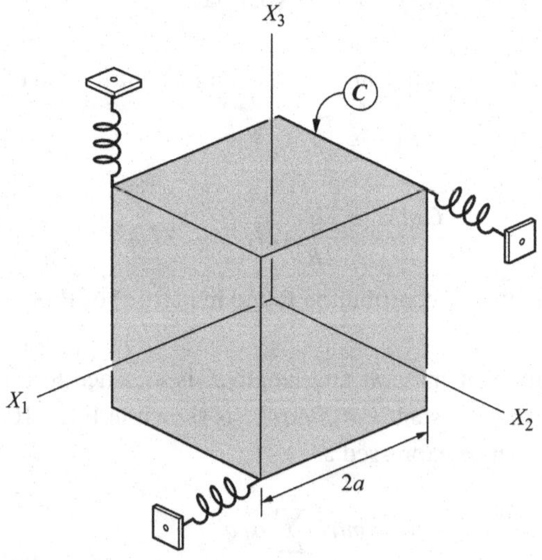

Figure P10.12

Three corners of a cube C are attached to fixed supports by means of identical, linear springs of spring constant k. When the springs are undeformed, their axes coincide with the edges of C, as indicated in Fig. P10.12.

To bring C into a general position, the center of C is displaced to a point whose coordinates relative to fixed axes X_1, X_2, X_3 (see Fig. P10.12) are aq_1, aq_2, aq_3, and three axes C_1, C_2, C_3, fixed in C and initially aligned with X_1, X_2, X_3, respectively, are brought into new orientations by means of successive right-handed rotations of C of amounts q_4 about C_1, q_5 about C_2, and q_6 about C_3.

Dropping all terms of third or higher degree in q_i $(i = 1,\ldots,6)$, find a potential energy contribution of the forces exerted on C by the springs.

Result $\frac{1}{2}ka^2[(q_1 - q_5 - q_6)^2 + (q_2 - q_6 - q_4)^2 + (q_3 - q_4 - q_5)^2]$

10.13 Two blocks are connected to each other and to a fixed support by means of springs and dashpots, as shown in Fig. P10.13. The springs have natural lengths L_1 and L_2, and the force transmitted by each dashpot is proportional to the speed of the piston relative to the cylinder, the constants of proportionality having the values α and β.

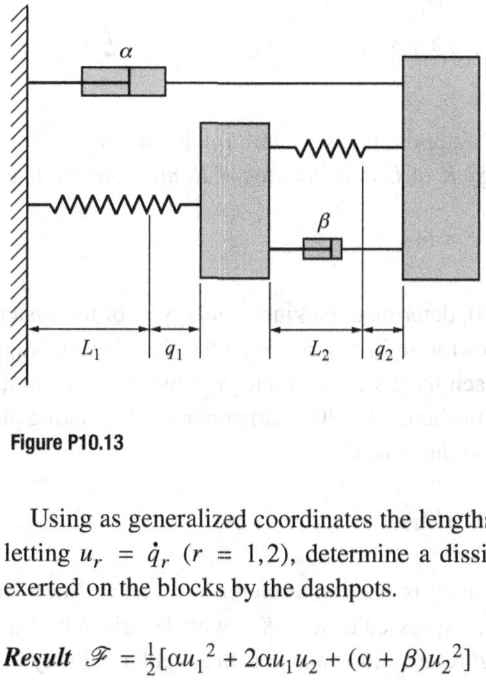

Figure P10.13

Using as generalized coordinates the lengths q_1 and q_2 indicated in Fig. P10.13, and letting $u_r = \dot{q}_r$ $(r = 1,2)$, determine a dissipation function $\mathscr{F}$ for the set of forces exerted on the blocks by the dashpots.

Result $\mathscr{F} = \frac{1}{2}[\alpha {u_1}^2 + 2\alpha u_1 u_2 + (\alpha + \beta){u_2}^2]$

***10.14** A rigid body B moves in a reference frame A while B is immersed in a fluid that exerts on B forces equivalent to a couple having a torque $-\alpha\boldsymbol{\omega}$, together with a force $-\beta\mathbf{v}$, applied at a point P of B, where α and β are constants, and $\boldsymbol{\omega}$ and $\mathbf{v}$ are the angular velocity of B in A and the velocity of P in A, respectively. Show that $\mathscr{F}$, a dissipation function for the forces exerted on B by the fluid, can be expressed as

$$\mathscr{F} = \tfrac{1}{2}(\alpha\boldsymbol{\omega}^2 + \beta\mathbf{v}^2)$$

Suggestion: Take advantage of the fact that here partial derivatives of $\boldsymbol{\omega}$ with respect to motion variables are partial angular velocities of B [see Eqs. (3.6.1)]; a similar statement applies to partial derivatives of $\mathbf{v}$ and partial velocities of P [see Eqs. (3.6.2)].

PROBLEM SET 11
(Secs. 7.4–7.6)

11.1 Referring to the example in Sec. 3.1, and letting P_1 and P_2 have masses m_1 and m_2, respectively, express the kinetic energies ${}^AK^S$ and ${}^BK^S$ of S in A and B, respectively, in terms of m_1, m_2, L, ω, q_1, q_2, q_3, and the first time-derivatives of q_1, q_2, q_3, the generalized coordinates shown in Fig. 3.2.1.

Results

$${}^AK^S = {}^BK^S + \frac{1}{2}\omega^2[m_1{q_1}^2 + m_2(q_1 + Lc_3)^2]$$

$${}^BK^S = \frac{1}{2}(m_1 + m_2)({\dot{q}_1}^2 + {\dot{q}_2}^2) - m_2L\left(\dot{q}_1 s_3 - \dot{q}_2 c_3 - \frac{L}{2}\dot{q}_3\right)\dot{q}_3$$

11.2 Referring to Problem 2.7, suppose that C is uniform, has a mass m, and is rolling on H. Express the kinetic energy K of C in A in terms of R, m, u_1, u_2, and u_3.

Result $K = (mR^2/8)(5{u_1}^2 + {u_2}^2 + 6{u_3}^2)$

11.3 Referring to Problem 3.10, determine the kinetic energy K of the system formed by the shaft and the four spheres for an instant at which the shaft has an angular speed ω, assuming that the spheres each have a mass m and are uniform and solid; the shaft has a moment of inertia J about its axis; $\theta = 30°$; and pure rolling is taking place at the contacts between the spheres and the cone C.

Result $K = \frac{1}{2}[J + 18mr^2(2 + \sqrt{3})/5]\omega^2$

***11.4** A point O of a rigid body B is fixed in a reference frame A. Show that K, the kinetic energy of B in A, can be expressed as $K = K_\omega$, with K_ω given by Eq. (7.4.3) if $\underline{\mathbf{I}}$ denotes the inertia dyadic of B relative to O and $\boldsymbol{\omega}$ is the angular velocity of B in A.

11.5 Figure P11.5 shows a uniform right-triangular plate of mass m and sides of lengths a and b, supported as follows: Vertex A is fixed and vertex B is attached to an inextensible string fastened at C, a point vertically above A, the length of the string being such that line AB is horizontal.

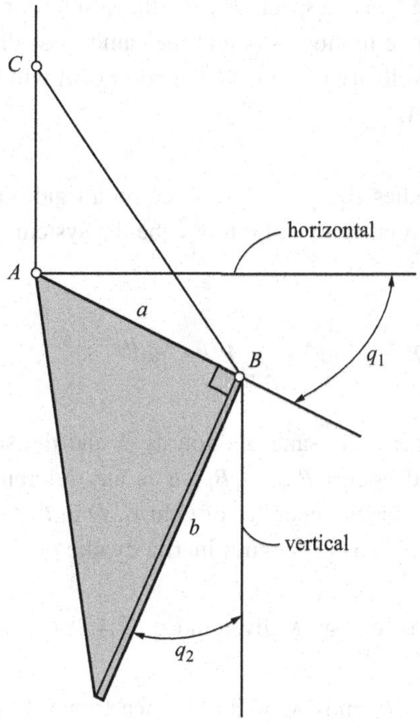

Figure P11.5

Letting q_1 and q_2 measure angles as indicated in Fig. P11.5, determine the kinetic energy K of the plate.

Result

$$K = \frac{m}{4}\left[\left(a^2 + \frac{b^2}{3}s_2{}^2\right)\dot{q}_1{}^2 + ab\,c_2\dot{q}_1\dot{q}_2 + \frac{b^2}{3}\dot{q}_2{}^2\right]$$

***11.6** At a certain instant, a rigid body B has a rotational kinetic energy K_ω and a central angular momentum $\mathbf{H}$. Show that

$$\frac{1}{2}\frac{\mathbf{H}^2}{I_{\max}} \le K_\omega \le \frac{1}{2}\frac{\mathbf{H}^2}{I_{\min}}$$

where $I_{\max}$ and $I_{\min}$ are, respectively, the maximum and the minimum central moments of inertia of B.

***11.7** Letting ${}^EK^C$, ${}^EK^D$, and ${}^AK^B$ denote kinetic energies, show that

$${}^EK^C = {}^EK^D + {}^AK^B + \sum_{i=1}^{\beta} m_i\, {}^E\mathbf{v}^{\bar{A}_i} \cdot {}^A\mathbf{v}^{P_i}$$

if A, B, C, D, E, ${}^E\mathbf{v}^{\overline{A}_i}$, and ${}^A\mathbf{v}^{P_i}$ are defined as follows: A, a rigid body; B, a set of β particles $P_1, \ldots, P_\beta$ of masses $m_1, \ldots, m_\beta$, respectively; C, the system formed by A and B; D, a rigid body that has the same motion as A and the same mass distribution as C; E, a reference frame; ${}^E\mathbf{v}^{\overline{A}_i}$, the velocity in E of $\overline{A}_i$, the point of A with which P_i coincides; ${}^A\mathbf{v}^{P_i}$, the velocity of P_i in A.

***11.8** The mass centers of k rigid bodies $B_1, \ldots, B_k$ are fixed on a rigid body A, but $B_1, \ldots, B_k$ are otherwise free to move relative to A. Letting C be the system formed by A and $B_1, \ldots, B_k$, show that

$$ {}^E K^C = {}^E K^D + {}^A K^B + {}^E\boldsymbol{\omega}^A \cdot \sum_{j=1}^{k} \underline{\mathbf{I}}^{B_j} \cdot {}^A\boldsymbol{\omega}^{B_j} $$

where D designates a rigid body that has the same motion as A and the same mass distribution as C; B is the set of rigid bodies $B_1, \ldots, B_k$; E is any reference frame; ${}^E K^C$, ${}^E K^D$, ${}^A K^B$ are, respectively, the kinetic energies of C in E, D in E, and B in A; ${}^E\boldsymbol{\omega}^A$ is the angular velocity of A in E; $\underline{\mathbf{I}}^{B_j}$ is the central inertia dyadic of B_j; ${}^A\boldsymbol{\omega}^{B_j}$ is the angular velocity of B_j in A.

Use this result to determine the kinetic energy K given in Eq. (7.4.28).

11.9 If S is a set of ν particles $P_1, \ldots, P_\nu$ moving in a reference frame A, and Q is a point moving in A, then the kinetic energy of S relative to Q in A, denoted by $K^{S/Q}$, is defined as

$$ K^{S/Q} \triangleq \frac{1}{2} \sum_{i=1}^{\nu} m_i (\mathbf{v}^{P_i/Q})^2 $$

where m_i is the mass of P_i, and $\mathbf{v}^{P_i/Q}$, called the velocity of P_i relative to Q in A, is the difference between $\mathbf{v}^{P_i}$, the velocity of P_i in A, and $\mathbf{v}^Q$, the velocity of Q in A; that is

$$ \mathbf{v}^{P_i/Q} \triangleq \mathbf{v}^{P_i} - \mathbf{v}^Q \qquad (i = 1, \ldots, \nu) $$

Letting $S^\star$ be the mass center of S, show that K^S, the kinetic energy of S in A, is given by

$$ K^S = K^{S/S^\star} + K^{S^\star} $$

where $K^{S^\star}$ is the kinetic energy in A of a particle whose mass is equal to the total mass of S and whose velocity in A is equal to the velocity of $S^\star$ in A.

Comment on the relationship between these facts and Eqs. (7.4.2)–(7.4.4).

***11.10** Referring to Problem 3.15 (see also Problems 4.19, 8.12, 8.18, and 8.19), determine K, the kinetic energy of the robot arm in E, and V_γ, a potential energy contribution of the set of gravitational forces acting on the robot arm.

Results

$$K = \tfrac{1}{2}[A_1u_1{}^2 + (B_1 + C_1)Z_1{}^2 + (B_2 + C_2)u_2{}^2 + (B_3 + C_3)Z_2{}^2$$
$$+ Z_1(D_{11}Z_1 + D_{12}u_2 + D_{13}Z_2) + u_2(D_{21}Z_1 + D_{22}u_2 + D_{23}Z_2)$$
$$+ Z_2(D_{31}Z_1 + D_{32}u_2 + D_{33}Z_2) + m_AZ_5{}^2 + m_B(Z_7{}^2 + Z_8{}^2)$$
$$+ m_C(Z_{11}{}^2 + Z_{12}{}^2 + u_3{}^2) + m_D(Z_{17}{}^2 + Z_{18}{}^2 + Z_{19}{}^2)]$$
$$V_\gamma = g[(m_BL_B + m_CZ_9 + m_DZ_{14})s_1 + m_Dp_1c_1]$$

11.11 Letting $(m_{rs})_B$ denote the contribution of a rigid body B to the inertia coefficient m_{rs}, show that

$$(m_{rs})_B = m\,\tilde{\mathbf{v}}_r \cdot \tilde{\mathbf{v}}_s + \tilde{\boldsymbol{\omega}}_r \cdot \underline{\mathbf{I}} \cdot \tilde{\boldsymbol{\omega}}_s \qquad (r,s = 1,\ldots,p)$$

where m is the mass of B, $\tilde{\mathbf{v}}_r$ is the rth nonholonomic partial velocity of the mass center of B in A, $\tilde{\boldsymbol{\omega}}_r$ is the rth nonholonomic partial angular velocity of B in A, and $\underline{\mathbf{I}}$ is the central inertia dyadic of B.

Suggestion: First show that

$$\tilde{\mathbf{v}}_r^{P_i} = \tilde{\mathbf{v}}_r + \tilde{\boldsymbol{\omega}}_r \times \mathbf{r}_i$$

where P_i is the ith particle of B and $\mathbf{r}_i$ is the position vector from the mass center of B to P_i $(r = 1,\ldots,p;\ i = 1,\ldots,\nu)$.

11.12 Determine the inertia coefficients for the system considered in the examples in Secs. 5.7 and 7.4.

Results

$$m_{11} = I_A + m_Aa^2 + m_B\left(\frac{R^2}{2} + 3b^2\right)$$
$$m_{22} = m_A + 3m_B$$
$$m_{12} = m_{21} = 0$$

11.13 When the inertia coefficient m_{rs} differs from zero, *dynamic coupling* is said to exist between the motion variables u_r and u_s.

Suppose that the mass center of a rigid body B is fixed in a reference frame A, and let $\hat{\mathbf{b}}_1$, $\hat{\mathbf{b}}_2$, $\hat{\mathbf{b}}_3$ be mutually perpendicular unit vectors parallel to central principal axes of B. Bring B into a general orientation in A by aligning $\hat{\mathbf{b}}_1$, $\hat{\mathbf{b}}_2$, $\hat{\mathbf{b}}_3$ with $\hat{\mathbf{a}}_1$, $\hat{\mathbf{a}}_2$, $\hat{\mathbf{a}}_3$, respectively, where $\hat{\mathbf{a}}_i$ $(i = 1,2,3)$ is a unit vector fixed in A, and then subjecting B to successive rotations characterized by the vectors $q_1\hat{\mathbf{b}}_1$, $q_2\hat{\mathbf{b}}_2$, $q_3\hat{\mathbf{b}}_3$. Define motion variables u_1, u_2, u_3 as (*a*) $u_r \triangleq \dot{q}_r$ $(r = 1,2,3)$ and (*b*) $u_r \triangleq \boldsymbol{\omega} \cdot \hat{\mathbf{b}}_r$ $(r = 1,2,3)$, where $\boldsymbol{\omega}$ is the angular velocity of B in A. Determine which of the motion variables are coupled dynamically.

Results (*a*) u_1 and u_2, u_1 and u_3; (*b*) None

11.14 Referring to Problem 8.14, determine $\tilde{F}_2^{\star}$ by using Eqs. (7.6.5). Comment briefly on the relative merits of using Eqs. (7.6.5), on the one hand, and Eqs. (5.9.2), on the other hand, to determine $\tilde{F}_2^{\star}$.

11.15 For the system considered in Problem 8.8, determine the generalized inertia force $F_1^{\star}$ by using (*a*) Eqs. (5.9.7), (*b*) Eqs. (7.6.7), and (*c*) the formula given in Problem 8.20. Comment briefly on the relative merits of the three methods.

Result

$$m[-29\dot{u}_1 + 19(c_1c_2 - s_1s_2)\dot{u}_2 - 8(s_3s_1 + c_3c_1)\dot{u}_3 - 19(s_2c_1 + c_2s_1)(u_2{}^2/L) - 8(s_1c_3 - c_1s_3)(u_3{}^2/L)]$$

PROBLEM SET 12
(Secs. 8.1–8.3)

12.1 Referring to the example in Sec. 3.1 (see also Problems 8.1 and 8.13), let line Y be vertical, let P_1 and P_2 have masses m_1 and m_2, respectively, and assume that the mass of R is negligible. Letting A be fixed relative to the Earth, and treating the Earth as a Newtonian reference frame, determine f_1, f_2, and f_3 such that the dynamical equations governing u_1, u_2, and u_3, defined as

$$u_1 \triangleq {}^A\mathbf{v}^{P_1} \cdot \hat{\mathbf{e}}_x \qquad u_2 \triangleq {}^A\mathbf{v}^{P_1} \cdot \hat{\mathbf{e}}_y \qquad u_3 \triangleq \dot{q}_3$$

where $\hat{\mathbf{e}}_x$ and $\hat{\mathbf{e}}_y$ are unit vectors directed as shown in Fig. 2.6.1, can be expressed as

$$\dot{u}_i = f_i \qquad (i = 1,2,3)$$

Results

$$f_1 = -gs_3 + \omega^2 q_1 c_3 + u_2 u_3 + (\omega^2 c_3{}^2 + u_3{}^2)Lm_2/(m_1 + m_2)$$
$$f_2 = -gc_3 - (\omega^2 q_1 s_3 + u_3 u_1)$$
$$f_3 = -\omega^2 s_3 c_3$$

12.2 For the system S formed by the particle P_1, disk D, and rod R considered in Problems 8.2 and 8.14, make the same assumptions as in Problem 12.1 regarding the line Y and the reference frame A. Determine f_1 and f_2 such that the dynamical equations governing u_1 and u_2 can be expressed as

$$\dot{u}_i = f_i \qquad (i = 1,2)$$

Results

$$f_1 = -gs_3 + \omega^2 q_1 c_3 + \frac{1}{m_1 + m_2}\left(m_2 L\omega^2 {c_3}^2 - \frac{m_1}{L}{u_2}^2\right)$$

$$f_2 = -gc_3 - \omega^2 q_1 s_3 + \frac{u_1 u_2}{L}$$

12.3 Regarding the Earth E, Moon M, and Sun S each as a particle, let Q be the mass center of E and S, and let F_E, F_M, F_S, and F_Q be reference frames in which E, M, S, and Q, respectively, are fixed and whose relative orientations do not vary with time. Furthermore, assume that F_S can be chosen in such a way that E moves in F_S on a circular path centered at S and traced out once per year.

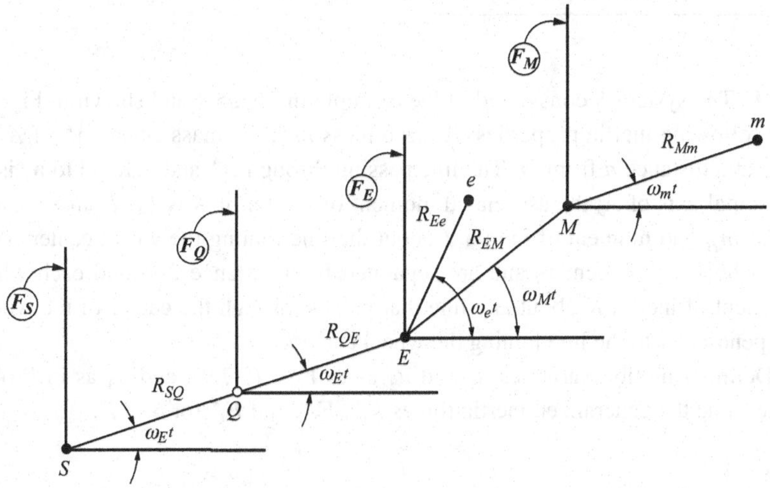

Figure P12.3

Assuming that F_Q is a Newtonian reference frame, assess the advisability of regarding F_E, F_M, and F_S as Newtonian reference frames for the purpose of analyzing motions of E, M, e, and m, where e and m designate a low-altitude satellite of E and a low-altitude satellite of M, respectively. In making these assessments, assume that S, E, M, e, and m are at all times coplanar, and let the distances and angular speeds indicated in Fig. P12.3 have the following values:

$$R_{SQ} = 4.5 \times 10^5 \text{ m} \qquad R_{QE} = 1.5 \times 10^{11} \text{ m}$$

$$R_{EM} = 4.0 \times 10^8 \text{ m} \qquad R_{Ee} = 7.0 \times 10^6 \text{ m}$$

$$R_{Mm} = 2.0 \times 10^6 \text{ m}$$

$$\omega_E = 2 \times 10^{-7} \text{ rad/s} \qquad \omega_M = 24 \times 10^{-7} \text{ rad/s}$$

$$\omega_e = 12 \times 10^{-4} \text{ rad/s} \qquad \omega_m = 10 \times 10^{-4} \text{ rad/s}$$

Results Table P12.3

Table P12.3

Moving object	Approximately Newtonian reference frame
E	F_S
M	F_S
e	F_S, F_E, F_M
m	F_S, F_E, F_M

12.4 The system S considered in the example in Sec. 5.7 and shown in Fig. 5.7.1 has the following inertia properties: A has a mass m_A; the mass center $A^\star$ of A is on line DE at a distance a from D. The line passing through $A^\star$ and parallel to $\hat{\mathbf{a}}_1$ is a central principal axis of A; the associated moment of inertia of A is I_A. B and C each have a mass m_B and moment of inertia J about the line joining the wheel centers (which are the wheels' mass centers and are separated by a distance $2b$), and each wheel has a moment of inertia K about any line that passes through the center of the wheel and is perpendicular to the line joining the wheel centers.

Defining motion variables u_1 and u_2 as in Eqs. (5.7.9), and u_3 as in Problem 9.8, determine the generalized inertia forces $\tilde{F}_1^\star$, $\tilde{F}_2^\star$, and $\tilde{F}_3^\star$ for S in F.

Results

$$\tilde{F}_1^\star = -(I_A + 2Jb^2/R^2 + 2K + m_A a^2 + 2m_B b^2)\dot{u}_1 - m_A a u_1 u_2$$
$$\tilde{F}_2^\star = -(m_A + 2m_B + 2J/R^2)\dot{u}_2 + m_A a {u_1}^2$$
$$\tilde{F}_3^\star = 0$$

12.5 The dynamical equations governing the generalized velocities u_1 and u_2 introduced in the example in Sec. 5.7 are to be formulated under the assumptions made in Problems 9.8 and 12.4. Determine f_1 and f_2 such that these equations can be expressed as

$$\dot{u}_i = f_i \qquad (i = 1,2)$$

Results

$$f_1 = \frac{fQ_3 + Mge\sin\theta\cos q_1 - m_A a u_1 u_2}{I_A + 2Jb^2/R^2 + 2K + m_A a^2 + 2m_B b^2}$$

$$f_2 = \frac{Q_2 + Mg\sin\theta\sin q_1 + m_A a {u_1}^2}{m_A + 2m_B + 2J/R^2}$$

where

$$Q_2 = \frac{-\mu' u_2 Q_1}{(u_2{}^2 + f^2 u_1{}^2)^{1/2}}$$

$$Q_3 = \frac{-\mu' u_1 f Q_1}{(u_2{}^2 + f^2 u_1{}^2)^{1/2}}$$

with

$$Q_1 = \frac{eMg\cos\theta}{f - \mu' R u_2/(u_2{}^2 + f^2 u_1{}^2)^{1/2}}$$

12.6 Referring to Problem 2.7 and letting C be a uniform disk *rolling* on H, determine f_1, f_2, and f_3 such that the dynamical equations governing u_1, u_2, and u_3 as defined in Problem 8.3 can be written

$$\dot{u}_i = f_i \qquad (i = 1, 2, 3)$$

Results

$$f_1 = \tfrac{1}{5}(u_2{}^2 \tan q_2 - 6u_2 u_3 - 4g s_2/R)$$
$$f_2 = 2u_3 u_1 - u_1 u_2 \tan q_2$$
$$f_3 = \tfrac{2}{3} u_1 u_2$$

12.7 A couple of torque $T\hat{\mathbf{a}}_z$ is applied to the link of length L_1 of Problem 4.3. Determine f_1 such that the dynamical equation governing u_1 as defined in Problem 8.7 can be written $\dot{u}_1 = f_1$.

Result

$$f_1 - \Bigg\{ [T + (\tilde{F}_1)_\gamma] \sin^2(q_2 - q_3) + m_2 L_1 L_3 \sin(q_1 - q_2)\left[\frac{L_1}{L_3} u_1(u_2 - u_1)\cos(q_1 - q_2) + u_3(u_3 - u_2)\cos(q_2 - q_3)\right]\Bigg\} \Bigg/ \Bigg\{ \left[m_1 \sin^2(q_3 - q_2) + m_2 \sin^2(q_1 - q_2)\right] L_1{}^2 \Bigg\}$$

where $(\tilde{F}_1)_\gamma$ is the generalized active force contribution found in Problem 8.7

12.8 The dynamical equations governing constrained motion of the particle P introduced in Problem 4.15 are to be formulated. Assume that the only forces exerted on P are terrestrial gravitational forces (see Problem 8.6) and the force needed to satisfy the motion constraint (see Problem 9.6). After referring to Problem 8.15, determine f_1 and

f_2 such that the equations of motion can be expressed as

$$\dot{u}_i = f_i \qquad (i = 1,2)$$

Results

$$f_1 = -g\,(\cos\gamma)^2\, u_1/u_3$$
$$f_2 = -g\,(\cos\gamma)^2\, u_2/u_3$$

***12.9** Referring to Problem 3.15 (see also Problems 4.19, 8.12, 8.18, 8.19, and 11.10), show that the dynamical equations of motion of the robot arm can be expressed as

$$\sum_{r=1}^{3} X_{sr}\dot{u}_r = Y_s \qquad (s = 1,2,3)$$

where $X_{rs} = X_{sr}$ $(r, s = 1,2,3;\ r \neq s)$. Determine X_{rs} and Y_s $(r, s = 1,2,3)$.

Results

$$X_{11} = -[A_1 + c_1(Z_{35} + Z_{41} + Z_{49}) + s_1(Z_{39} + Z_{44} + Z_{53}) + m_A L_A{}^2$$
$$+ m_B Z_6{}^2 + m_C Z_{10}{}^2 + m_D(Z_{13}{}^2 + Z_{15}{}^2 + Z_{16}{}^2)]$$
$$X_{12} = X_{21} = -[Z_{51} + m_D(Z_{13}Z_{14} - Z_{16}p_1)]$$
$$X_{13} = X_{31} = -m_D Z_{16}$$
$$X_{22} = -[B_2 + C_2 + D_{22} + m_B L_B{}^2 + m_C Z_9{}^2 + m_D(Z_{14}{}^2 + p_1{}^2)]$$
$$X_{23} = X_{32} = m_D p_1$$
$$X_{33} = -(m_C + m_D)$$
$$Y_1 = c_1(Z_{36} + Z_{42} + Z_{50}) + s_1(Z_{40} + Z_{45} + Z_{54}) + m_B Z_6 Z_{23}$$
$$+ m_C Z_{10} Z_{27} + m_D(Z_{13}Z_{32} + Z_{15}Z_{33} + Z_{16}Z_{34}) - T_1{}^{E/A}$$
$$Y_2 = Z_{38} + Z_{43} + Z_{52} + m_B L_B Z_{22} + m_C Z_9 Z_{26} + m_D(Z_{14}Z_{32} - p_1 Z_{34})$$
$$- T_2{}^{A/B} + g[(m_B L_B + m_C Z_9 + m_D Z_{14})c_1 - m_D p_1 s_1]$$
$$Y_3 = m_C Z_{28} + m_D Z_{34} - K_3{}^{B/C} + g(m_C + m_D)s_1$$

***12.10** Referring to Problem 3.15 (see also Problems 4.19, 8.12, 8.18, 8.19, 11.10, and 12.9), suppose that $D^\star$ lies on line $B^\star C^\star$ and that each central principal axis of D is parallel to one of $\hat{\mathbf{b}}_1$, $\hat{\mathbf{b}}_2$, $\hat{\mathbf{b}}_3$. Show that under these circumstances the dynamical equations of motion reduce to

$$\dot{u}_r = \frac{Y_r}{X_{rr}} \qquad (r = 1,2,3)$$

***12.11** Show that Eqs. (8.1.2) can be written

$$F_r + F_r^\star + \sum_{s=p+1}^{n} (F_s + F_s^\star)A_{sr} = 0 \qquad (r = 1,\ldots,p)$$

where $F_1, \ldots, F_n$ are *holonomic* generalized active forces for S in N, $F_1^\star, \ldots, F_n^\star$ are *holonomic* generalized inertia forces for S in N, and $A_{p+1,1}, \ldots, A_{n,p}$ have the same meaning as in Eqs. (3.5.2).

12.12 Figure P12.12 represents a one-cylinder reciprocating engine consisting of a counterweighted crank A, connecting rod B, piston C, and cylinder D. $A^\star$, $B^\star$, and $C^\star$ are the mass centers of A, B, and C, respectively, and A, B, and C have masses m_A, m_B, and m_C, respectively. The central radii of gyration of A and B with respect to axes normal to the middle plane of the mechanism have the values k_A and k_B, respectively.

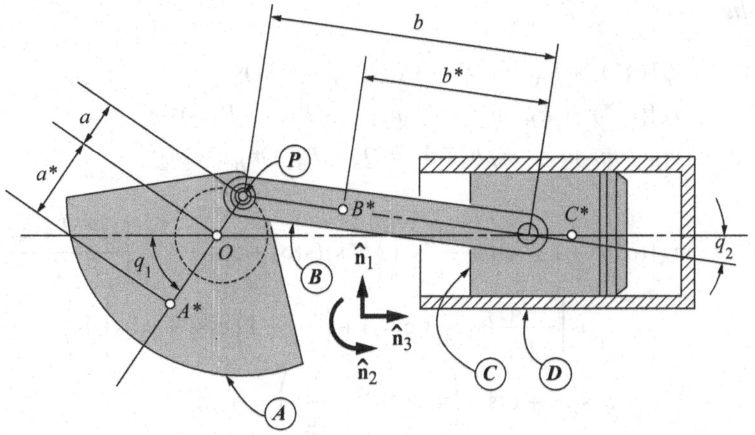

Figure P12.12

The system S formed by A, B, and C possesses one degree of freedom in a reference frame in which point O and the axis of the cylinder are fixed; the angle q_1 shown in Fig. P12.12, and u_1 defined as $u_1 \triangleq \dot{q}_1$, can serve, respectively, as a generalized coordinate and as a generalized velocity associated with this degree of freedom. However, it can be convenient, for example, for the purpose of bringing certain interaction forces into evidence (see Sec. 6.7), to introduce additional motion variables, u_2 and u_3, as

$$u_2 \triangleq \boldsymbol{\omega}^B \cdot \hat{\mathbf{n}}_2 \qquad u_3 \triangleq \mathbf{v}^{C^\star} \cdot \hat{\mathbf{n}}_3$$

Moreover, it is then helpful to define, in addition, an angle q_2 as indicated in Fig. P12.12.

Let the set of contact forces exerted on A by the crankshaft on which A is mounted be equivalent to a couple of torque $\alpha_1\hat{\mathbf{n}}_1 + \alpha_2\hat{\mathbf{n}}_2 + \alpha_3\hat{\mathbf{n}}_3$, together with a force $P_1\hat{\mathbf{n}}_1 + P_2\hat{\mathbf{n}}_2 + P_3\hat{\mathbf{n}}_3$ applied at O, and let the set of contact forces exerted on C by exploding gases and by the cylinder D be equivalent to a couple of torque $\beta_1\hat{\mathbf{n}}_1 + \beta_2\hat{\mathbf{n}}_2 + \beta_3\hat{\mathbf{n}}_3$, together with a force $Q_1\hat{\mathbf{n}}_1 + Q_2\hat{\mathbf{n}}_2 + Q_3\hat{\mathbf{n}}_3$ applied at $C^\star$. Finally, let the set of forces exerted by B on A be equivalent to a couple of torque $\gamma_1\hat{\mathbf{n}}_1 + \gamma_3\hat{\mathbf{n}}_3$, together with a force $R_1\hat{\mathbf{n}}_1 + R_2\hat{\mathbf{n}}_2 + R_3\hat{\mathbf{n}}_3$ applied at point P.

After verifying that u_2 and u_3 must satisfy the constraint equations

$$u_2 = -\frac{a\text{c}_1}{b\text{c}_2}u_1$$

$$u_3 = -\frac{a}{\text{c}_2}(\text{s}_1\text{c}_2 + \text{c}_1\text{s}_2)u_1$$

write dynamical equations governing all motions of S: (*a*) by using Eqs. (8.1.1) with $r = 1, 2, 3$ and (*b*) by using the equation in Problem 12.11 with $p = 1$. Neglect gravitational forces, and verify that the result of part (*b*) can be obtained from those of part (*a*) by using the first two equations of part (*a*) to eliminate R_3 from the third equation.

Results

(*a*)

$$m_A[(a^\star)^2 + k_A{}^2]\dot{u}_1 = \alpha_2 + a(R_1\text{c}_1 - R_3\text{s}_1)$$

$$m_B\{[(b^\star)^2 + k_B{}^2]\dot{u}_2 - b^\star\text{s}_2\dot{u}_3\} = b(R_1\text{c}_2 + R_3\text{s}_2)$$

$$(m_B + m_C)\dot{u}_3 - m_Bb^\star\text{s}_2\dot{u}_2 = Q_3 - R_3 - m_Bb^\star\text{c}_2u_2{}^2$$

(*b*)

$$m_A[(a^\star)^2 + k_A{}^2]\dot{u}_1 + \frac{m_Ba}{\text{c}_2}\left\langle\left\{b^\star\text{s}_2(\text{s}_1\text{c}_2 + \text{c}_1\text{s}_2) - \text{c}_1\left[\frac{(b^\star)^2 + k_B{}^2}{b}\right]\right\}\dot{u}_2\right.$$

$$\left. + \left[-\frac{m_C}{m_B}(\text{s}_1\text{c}_2 + \text{c}_1\text{s}_2) + \left(\frac{b^\star}{b} - 1\right)\text{c}_1\text{s}_2 - \text{s}_1\text{c}_2\right]\dot{u}_3\right\rangle$$

$$= \alpha_2 + a(\text{s}_1\text{c}_2 + \text{c}_1\text{s}_2)\left(m_Bb^\star u_2{}^2 - \frac{Q_3}{\text{c}_2}\right)$$

12.13 When a rigid body B moves in a Newtonian reference frame N under the action of a set of contact forces and distance forces equivalent to a couple of torque $\mathbf{T}$ together with a force $\mathbf{F}$ applied at the mass center $B^\star$ of B, the vectors $\mathbf{T}$ and $\mathbf{F}$ always can be expressed as

$$\mathbf{T} = \sum_{i=1}^{3} T_i\hat{\mathbf{b}}_i \qquad \mathbf{F} = \sum_{i=1}^{3} F_i\hat{\mathbf{n}}_i$$

where $\hat{\mathbf{b}}_1$, $\hat{\mathbf{b}}_2$, and $\hat{\mathbf{b}}_3$ form a dextral set of mutually perpendicular unit vectors parallel to central principal axes of B (but not necessarily fixed in B) and $\hat{\mathbf{n}}_1$, $\hat{\mathbf{n}}_2$, and $\hat{\mathbf{n}}_3$ are any three noncoplanar unit vectors fixed in N; the angular velocity $\boldsymbol{\omega}$ of B in N and the velocity $\mathbf{v}$ of $B^\star$ in N can be written

$$\boldsymbol{\omega} = \sum_{i=1}^{3} \omega_i\hat{\mathbf{b}}_i \qquad \mathbf{v} = \sum_{i=1}^{3} v_i\hat{\mathbf{n}}_i$$

Letting $v_1, v_2, v_3, \omega_1, \omega_2, \omega_3$ play the roles of generalized velocities, use Eqs. (8.1.1) to show that the associated dynamical equations are

$$F_1 = m\dot{v}_1 \qquad F_2 = m\dot{v}_2 \qquad F_3 = m\dot{v}_3$$

$$I_1\alpha_1 - (I_2 - I_3)\omega_2\omega_3 = T_1$$
$$I_2\alpha_2 - (I_3 - I_1)\omega_3\omega_1 = T_2$$
$$I_3\alpha_3 - (I_1 - I_2)\omega_1\omega_2 = T_3$$

where I_j is the moment of inertia of B about a line passing through $B^\star$ and parallel to $\hat{\mathbf{b}}_i$, and $\alpha_i \triangleq \boldsymbol{\alpha} \cdot \hat{\mathbf{b}}_i$, with $\boldsymbol{\alpha}$ the angular acceleration of B in N. The first three equations are an expression of *Euler's first law* for the mass center of a rigid body. The last three equations are known as *Euler's dynamical equations*, and represent a particular form of *Euler's second law.*

12.14 Referring to Sec. 7.6, show that one can use Eqs. (8.1.1) or Eqs. (8.1.2) to generate dynamical equations involving the kinetic energy K of a system S in a Newtonian reference frame, as follows.

For a holonomic system with $u_r \triangleq \dot{q}_r$ $(r = 1,\ldots,n)$,

$$\frac{d}{dt}\frac{\partial K}{\partial \dot{q}_r} - \frac{\partial K}{\partial q_r} = F_r \qquad (r = 1,\ldots,n)$$

These equations are known as *Lagrange's equations of the first kind.*

For a holonomic system with u_r defined as in Eqs. (3.4.1), so that Eqs. (3.6.5) apply,

$$\sum_{s=1}^{n}\left(\frac{d}{dt}\frac{\partial K}{\partial \dot{q}_s} - \frac{\partial K}{\partial q_s}\right) W_{sr} = F_r \qquad (r = 1,\ldots,n)$$

For a nonholonomic system with $u_r \triangleq \dot{q}_r$ $(r = 1,\ldots n)$,

$$\frac{d}{dt}\frac{\partial K}{\partial \dot{q}_r} - \frac{\partial K}{\partial q_r} + \sum_{s=p+1}^{n}\left(\frac{d}{dt}\frac{\partial K}{\partial \dot{q}_s} - \frac{\partial K}{\partial q_s}\right) C_{sr} = \tilde{F}_r \qquad (r = 1,\ldots,p)$$

where C_{sr} $(s = p+1,\ldots,n;\ r = 1,\ldots,p)$ has the same meaning as in Eqs. (7.1.13).

For a nonholonomic system with u_r defined as in Eqs. (3.4.1), so that Eqs. (3.6.5) apply,

$$\sum_{s=1}^{n}\left(\frac{d}{dt}\frac{\partial K}{\partial \dot{q}_s} - \frac{\partial K}{\partial q_s}\right)\left(W_{sr} + \sum_{k=p+1}^{n} W_{sk}A_{kr}\right) = \tilde{F}_r \qquad (r = 1,\ldots,p)$$

where A_{kr} $(k = p+1,\ldots,n;\ r = 1,\ldots,p)$ has the same meaning as in Eqs. (3.5.2). These last two sets of equations are called *Passerello-Huston equations.*

Considering the comment following Eq. (7.6.8), your own comments made in connection with Problems 11.14 and 11.15, and your solution of Problem 4.21, comment on the advisability of involving kinetic energy in the process of formulating dynamical equations.

12.15 When a system S possesses a potential energy V in a Newtonian reference frame N, a quantity $\mathscr{L}$, called the *Lagrangian* or the *kinetic potential* of S in N, is defined as

$$\mathscr{L} \triangleq K - V$$

where K is the kinetic energy of S in N.

Supposing that S is a holonomic system with generalized velocities defined as $u_r \triangleq \dot{q}_r$ $(r = 1,\ldots,n)$, where $q_1,\ldots,q_n$ are generalized coordinates for S in N, show that $\mathscr{L}$, regarded as a function of $q_1,\ldots,q_n$, $\dot{q}_1,\ldots,\dot{q}_n$, and the time t, satisfies the equations

$$\frac{d}{dt}\frac{\partial \mathscr{L}}{\partial \dot{q}_r} - \frac{\partial \mathscr{L}}{\partial q_r} = 0 \qquad (r = 1,\ldots,n)$$

These equations are called *Lagrange's equations of the second kind.*

***12.16** The choice of motion variables (see Sec. 3.4) can have a profound effect on the simplicity of dynamical equations of motion formulated according to Eqs. (8.1.1), (8.1.2), or (8.1.3). When a system contains a pair of rigid bodies A and B that are connected to each other and move in a Newtonian reference frame N, simplicity in the resulting equations is facilitated by selecting motion variables as indicated in the following three situations.†

When B is pin-connected to A and possesses one degree of freedom in A, the angular velocity ${}^A\boldsymbol{\omega}^B$ of B in A can be expressed as ${}^A\boldsymbol{\omega}^B = \sigma_1\hat{\mathbf{a}}$, where unit vector $\hat{\mathbf{a}}$ is fixed in A and in B, and is parallel to the axis of the pin. In this case it is advantageous to define a motion variable u_1 as

$$u_1 \triangleq {}^N\boldsymbol{\omega}^B \cdot \hat{\mathbf{a}} = {}^N\boldsymbol{\omega}^A \cdot \hat{\mathbf{a}} + \sigma_1$$

Moreover, the angular velocity ${}^N\boldsymbol{\omega}^B$ of B in N can be expressed as

$${}^N\boldsymbol{\omega}^B = {}^N\boldsymbol{\omega}^A + (u_1 - {}^N\boldsymbol{\omega}^A \cdot \hat{\mathbf{a}})\hat{\mathbf{a}} = {}^N\boldsymbol{\omega}^A \cdot (\underline{\mathbf{U}} - \hat{\mathbf{a}}\hat{\mathbf{a}}) + u_1\hat{\mathbf{a}}$$

where $\underline{\mathbf{U}}$ is the unit dyadic (see Sec. 4.5).

When B possesses two degrees of freedom in A and ${}^A\boldsymbol{\omega}^B$ can be expressed as ${}^A\boldsymbol{\omega}^B = \sigma_1\hat{\mathbf{a}} + \sigma_2\hat{\mathbf{b}}$, where $\hat{\mathbf{a}}$ and $\hat{\mathbf{b}}$ are (not necessarily perpendicular) unit vectors fixed in A and in B, respectively, it is beneficial to define two motion variables u_1 and u_2 as

$$u_1 \triangleq ({}^N\boldsymbol{\omega}^B - \sigma_2\hat{\mathbf{b}}) \cdot \hat{\mathbf{a}} = {}^N\boldsymbol{\omega}^A \cdot \hat{\mathbf{a}} + \sigma_1$$

$$u_2 \triangleq {}^N\boldsymbol{\omega}^B \cdot \hat{\mathbf{b}} = {}^N\boldsymbol{\omega}^A \cdot \hat{\mathbf{b}} + \sigma_1\hat{\mathbf{a}} \cdot \hat{\mathbf{b}} + \sigma_2$$

${}^N\boldsymbol{\omega}^B$ then can be expressed as

$$\begin{aligned}{}^N\boldsymbol{\omega}^B &= {}^N\boldsymbol{\omega}^A + (u_1 - {}^N\boldsymbol{\omega}^A \cdot \hat{\mathbf{a}})\hat{\mathbf{a}} + [u_2 - {}^N\boldsymbol{\omega}^A \cdot \hat{\mathbf{b}} - (u_1 - {}^N\boldsymbol{\omega}^A \cdot \hat{\mathbf{a}})\hat{\mathbf{a}} \cdot \hat{\mathbf{b}}]\hat{\mathbf{b}} \\ &= [{}^N\boldsymbol{\omega}^A \cdot (\underline{\mathbf{U}} - \hat{\mathbf{a}}\hat{\mathbf{a}}) + u_1\hat{\mathbf{a}}] \cdot (\underline{\mathbf{U}} - \hat{\mathbf{b}}\hat{\mathbf{b}}) + u_2\hat{\mathbf{b}}\end{aligned}$$

Hooke's joint (see Sec. 5.3) constitutes an example of this type of connection between two rigid bodies, in which case unit vectors $\hat{\mathbf{a}}$ and $\hat{\mathbf{b}}$ are perpendicular to each other as they are fixed in the rigid cross shown in Fig. 5.3.1.

When B possesses three degrees of freedom in A and ${}^A\boldsymbol{\omega}^B$ can be expressed as

† P. C. Mitiguy and T. R. Kane, *The International Journal of Robotics Research* 15, no. 5 (1996), pp. 522–532.

${}^A\boldsymbol{\omega}^B = \sigma_1\hat{\mathbf{i}} + \sigma_2\hat{\mathbf{j}} + \sigma_3\hat{\mathbf{k}}$, where $\hat{\mathbf{i}}, \hat{\mathbf{j}}$, and $\hat{\mathbf{k}}$ are nonparallel unit vectors, one can define three motion variables u_1, u_2, and u_3 as

$$u_1 \triangleq {}^N\boldsymbol{\omega}^B \cdot \hat{\mathbf{c}}_1 \qquad u_2 \triangleq {}^N\boldsymbol{\omega}^B \cdot \hat{\mathbf{c}}_2 \qquad u_3 \triangleq {}^N\boldsymbol{\omega}^B \cdot \hat{\mathbf{c}}_3$$

where $\hat{\mathbf{c}}_1$, $\hat{\mathbf{c}}_2$, and $\hat{\mathbf{c}}_3$ are nonparallel, noncoplanar unit vectors. If $\hat{\mathbf{c}}_i$ $(i = 1,2,3)$ are chosen to be mutually perpendicular, then ${}^N\boldsymbol{\omega}^B$ can be written

$${}^N\boldsymbol{\omega}^B = u_1\hat{\mathbf{c}}_1 + u_2\hat{\mathbf{c}}_2 + u_3\hat{\mathbf{c}}_3$$

The ball-and-socket joint in Problem 4.7 and the sphere contained in a spherical cavity in Problem 8.11 furnish two examples of this kind of connection between two rigid bodies.

Referring to Problem 8.11, let A play the part of a Newtonian reference frame, and suppose that the mass center $B^\star$ of rigid body B is coincident with the mass center $C^\star$ of uniform sphere C. Let $\hat{\mathbf{b}}_1$, $\hat{\mathbf{b}}_2$, and $\hat{\mathbf{b}}_3$ be mutually perpendicular unit vectors fixed in B, parallel to central principal axes of inertia of B, and denote corresponding central principal moments of inertia of B by I_1, I_2, and I_3. Represent the mass of B with m_B, the mass of C with m_C, and the central principal moment of inertia of C with J. Define three motion variables as $u_r \triangleq {}^A\boldsymbol{\omega}^B \cdot \hat{\mathbf{b}}_r$ $(r = 1,2,3)$, and three more as $u_r \triangleq {}^A\mathbf{v}^{B^\star} \cdot \hat{\mathbf{a}}_{r-6}$ $(r = 7,8,9)$, where mutually perpendicular unit vectors $\hat{\mathbf{a}}_1$, $\hat{\mathbf{a}}_2$, and $\hat{\mathbf{a}}_3$ are fixed in A. Assume that the set of forces exerted by the viscous fluid on B is equivalent to a couple of torque $K\,{}^B\boldsymbol{\omega}^C$, where K is a positive constant, and that B (but not C) is subject to an additional set of contact forces and distance forces equivalent to a force $R_1\hat{\mathbf{a}}_1 + R_2\hat{\mathbf{a}}_2 + R_3\hat{\mathbf{a}}_3$ applied at $B^\star$, together with a couple of torque $T_1\hat{\mathbf{b}}_1 + T_2\hat{\mathbf{b}}_2 + T_3\hat{\mathbf{b}}_3$. Obtain dynamical equations governing motions of B and C in A by defining motion variables u_4, u_5, and u_6 as (*a*) $u_r \triangleq {}^B\boldsymbol{\omega}^C \cdot \hat{\mathbf{b}}_{r-3}$ $(r = 4,5,6)$ and as (*b*) $u_r \triangleq {}^A\boldsymbol{\omega}^C \cdot \hat{\mathbf{b}}_{r-3}$ $(r = 4,5,6)$. Comment briefly on a comparison of the two sets of dynamical equations.

Results

$$\begin{aligned}
(a)\quad (I_1 + J)\dot{u}_1 + J\dot{u}_4 &= T_1 + (I_2 - I_3)u_2u_3 + J(u_3u_5 - u_2u_6)\\
(I_2 + J)\dot{u}_2 + J\dot{u}_5 &= T_2 + (I_3 - I_1)u_1u_3 + J(u_1u_6 - u_3u_4)\\
(I_3 + J)\dot{u}_3 + J\dot{u}_6 &= T_3 + (I_1 - I_2)u_1u_2 + J(u_2u_4 - u_1u_5)\\
J(\dot{u}_1 + \dot{u}_4) &= J(u_3u_5 - u_2u_6) - Ku_4\\
J(\dot{u}_2 + \dot{u}_5) &= J(u_1u_6 - u_3u_4) - Ku_5\\
J(\dot{u}_3 + \dot{u}_6) &= J(u_2u_4 - u_1u_5) - Ku_6\\
(m_B + m_C)\dot{u}_7 &= R_1\\
(m_B + m_C)\dot{u}_8 &= R_2\\
(m_B + m_C)\dot{u}_9 &= R_3
\end{aligned}$$

$$(b)\qquad \begin{aligned} I_1\dot{u}_1 &= T_1 + (I_2 - I_3)u_2u_3 + K(u_4 - u_1)\\ I_2\dot{u}_2 &= T_2 + (I_3 - I_1)u_1u_3 + K(u_5 - u_2)\\ I_3\dot{u}_3 &= T_3 + (I_1 - I_2)u_1u_2 + K(u_6 - u_3)\\ J\dot{u}_4 &= J(u_3u_5 - u_2u_6) - K(u_4 - u_1)\\ J\dot{u}_5 &= J(u_1u_6 - u_3u_4) - K(u_5 - u_2)\\ J\dot{u}_6 &= J(u_2u_4 - u_1u_5) - K(u_6 - u_3)\\ (m_B + m_C)\dot{u}_7 &= R_1\\ (m_B + m_C)\dot{u}_8 &= R_2\\ (m_B + m_C)\dot{u}_9 &= R_3 \end{aligned}$$

PROBLEM SET 13
(Secs. 8.4–8.9)

13.1 Referring to the discussion in Sec. 8.4 of motions of the bar B depicted in Fig. 8.4.3, suppose that $h = 3R/4$ and $\Omega^2 = 2g/R$. Find a value of $\overline{q}_1$ other than zero such that the equation $q_1 = \overline{q}_1$ describes a possible motion of B, and determine N such that the circular frequency of small-amplitude oscillations of B that ensue subsequent to a small disturbance of this motion is equal to $N(g/R)^{1/2}$.

Results $\overline{q}_1 = \cos^{-1}(9/10) \qquad N = (19/85)^{1/2}$

13.2 In Fig. P13.2, A, B, and C are the outer gimbal, the inner gimbal, and the rotor of a gyroscope, and P is a particle of mass m attached to the rotor axis at a distance h from the center of C. By means of an electric motor (not shown) that may be regarded as consisting of two parts, one rigidly attached to B, the other to C, C is driven relative to B in such a way that the angular velocity of C in B is given by

$$^{B}\boldsymbol{\omega}^{C} = s\hat{\mathbf{b}}_1$$

where s is a constant and $\hat{\mathbf{b}}_1$ is a unit vector directed as shown. The inertia properties of A, B, and C are characterized by the five quantities I, J_1, J_2, K_1, and K_2 defined in terms of the central inertia dyadics $\underline{\mathbf{I}}_A$, $\underline{\mathbf{I}}_B$, and $\underline{\mathbf{I}}_C$ of A, B, and C, respectively, as follows:

$$I \triangleq \hat{\mathbf{a}}_1 \cdot \underline{\mathbf{I}}_A \cdot \hat{\mathbf{a}}_1 \qquad J_r \triangleq \hat{\mathbf{b}}_r \cdot \underline{\mathbf{I}}_B \cdot \hat{\mathbf{b}}_r \qquad K_r \triangleq \hat{\mathbf{b}}_r \cdot \underline{\mathbf{I}}_C \cdot \hat{\mathbf{b}}_r \qquad (r = 1,2)$$

where $\hat{\mathbf{a}}_1$ and $\hat{\mathbf{b}}_2$ are unit vectors directed as shown, and $\hat{\mathbf{b}}_1$, $\hat{\mathbf{b}}_2$, and $\hat{\mathbf{b}}_3 \triangleq \hat{\mathbf{b}}_1 \times \hat{\mathbf{b}}_2$ are parallel to central principal axes of both B and C. Moreover, $\hat{\mathbf{b}}_3 \cdot \underline{\mathbf{I}}_B \cdot \hat{\mathbf{b}}_3 = J_1$ and $\hat{\mathbf{b}}_3 \cdot \underline{\mathbf{I}}_C \cdot \hat{\mathbf{b}}_3 = K_2$.

The system formed by A, B, C, and P can move in such a way that q_1, the angle between $\hat{\mathbf{a}}_1$ and $\hat{\mathbf{b}}_1$, remains equal to zero while the angular velocity of A is given by

$$\boldsymbol{\omega}^A = p\hat{\mathbf{a}}_1$$

where p is a constant. Letting $q_1{}^\star$ be a perturbation of q_1, determine ω^2 such that $q_1{}^\star$ is governed by the (linearized) differential equation

$$\ddot{q}_1{}^\star + \omega^2 q_1{}^\star \approx 0$$

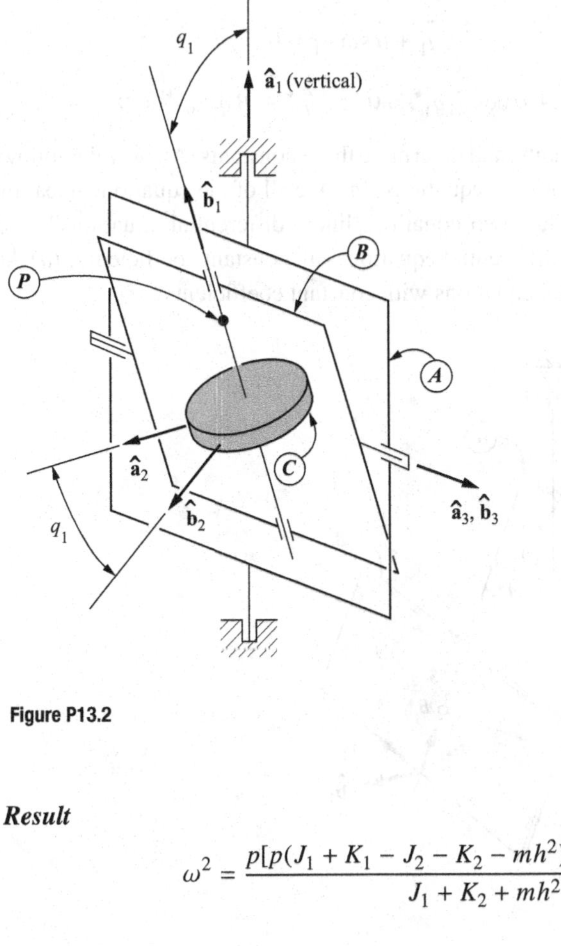

Figure P13.2

Result

$$\omega^2 = \frac{p[p(J_1 + K_1 - J_2 - K_2 - mh^2) + sK_1] - mgh}{J_1 + K_2 + mh^2}$$

13.3 The system S depicted in Fig. P13.3 consists of a rod A of mass m_A, one end of which is pinned to a fixed support at a point O while the other end supports a uniform plate B of mass m_B in such a way that B can rotate freely about the axis of A. The distances from O to $A^\star$ and $B^\star$, the mass centers of A and B, respectively, are L_A and L_B, respectively. Finally, A has a moment of inertia A_1 about any line normal to A and passing through $A^\star$, and the central principal moments of inertia of B have the values

B_1, B_2, and B_3, where the subscripts refer to the unit vectors $\hat{\mathbf{b}}_1$, $\hat{\mathbf{b}}_2$, and $\hat{\mathbf{b}}_3$ shown in Fig. P13.3.

S can move in such a way that the angles q_1 and q_2 (see Fig. P13.3) are given for all values of the time t by

$$q_1 = \widetilde{q}_1 \qquad q_2 = 0$$

where $\widetilde{q}_1$ is a function of t. Letting $q_1{}^\star$ and $q_2{}^\star$ be perturbations such that

$$q_1 = \widetilde{q}_1 + q_1{}^\star \qquad q_2 = q_2{}^\star$$

formulate differential equations linearized in $q_1{}^\star$, $q_2{}^\star$, $\dot{q}_1{}^\star$, and $\dot{q}_2{}^\star$. Use these equations to verify that $\widetilde{q}_1$, $q_1{}^\star$, and $q_2{}^\star$ satisfy the differential equations

$$\ddot{\widetilde{q}}_1 + \alpha \sin \widetilde{q}_1 \approx 0$$

$$\ddot{q}_1{}^\star + \alpha \cos \widetilde{q}_1 q_1{}^\star \approx 0 \qquad \ddot{q}_2{}^\star - \beta \dot{\widetilde{q}}_1{}^2 q_2{}^\star \approx 0$$

where α and β are constants, and determine these constants. Answer the following questions regarding the differential equations: (*a*) Are all of the equations linear differential equations? (*b*) Are the last two equations linear differential equations? (*c*) Is one or more of the equations a differential equation with constant coefficients? (*d*) Are the last two equations differential equations with constant coefficients?

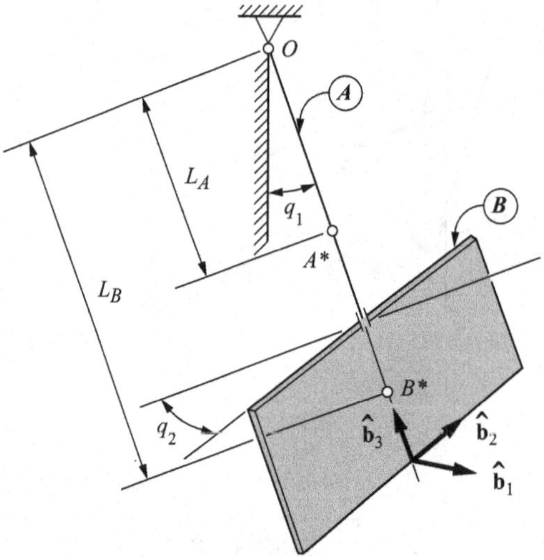

Figure P13.3

Results

$$\alpha = \frac{g(m_A L_A + m_B L_B)}{A_1 + B_1 + m_A {L_A}^2 + m_B {L_B}^2} \qquad \beta = \frac{B_2 - B_1}{B_3}$$

(*a*) No; (*b*) Yes; (*c*) Yes; (*d*) No

13.4 Consider a holonomic system S that consists of particles $P_1, \ldots, P_\nu$ and possesses generalized coordinates $q_1, \ldots, q_n$ in a Newtonian reference frame N. Define generalized velocities $u_1, \ldots, u_n$ for S in N as

$$u_r = \dot{q}_r \qquad (r = 1, \ldots, n)$$

and suppose that when $\mathbf{v}^{P_i}$, the velocity of P_i in N, is expressed as [see Eq. (3.6.2)]

$$\mathbf{v}^{P_i} = \sum_{r=1}^{n} \mathbf{v}_r^{P_i} u_r + \mathbf{v}_t^{P_i} \qquad (i = 1, \ldots, \nu)$$

then $\mathbf{v}_t^{P_i} = \mathbf{0}$ $(i = 1, \ldots, \nu)$ and $\mathbf{v}_r^{P_i}$ $(i = 1, \ldots, \nu;\ r = 1, \ldots, n)$ depend on the time t solely because $q_1, \ldots, q_n$ depend on t. Finally, assume that S possesses a potential energy V in N, that S can remain at rest in N, and that $q_1, \ldots, q_n$ have been chosen in such a way that, when S is at rest in N, the dynamical equations governing all motions of S in N are satisfied by $q_r \equiv 0$ $(r = 1, \ldots, n)$. Show that under these circumstances the dynamical equations of S in N, when linearized in q_r and $\dot{q}_r$ $(r = 1, \ldots, n)$, can be written

$$\sum_{s=1}^{n} (M_{rs}\ddot{q}_s + K_{rs}q_s) \approx 0 \qquad (r = 1, \ldots, n)$$

where M_{rs} $(r, s = 1, \ldots, n)$ are the values of the inertia coefficients of S in N (see Sec. 7.5) when $q_1 = \cdots = q_n = 0$, and K_{rs} denotes the value of $\partial^2 V/\partial q_r \partial q_s$ $(r, s = 1, \ldots, n)$ when $q_1 = \cdots = q_n = 0$.

Suggestion: Refer to Eq. (7.1.2) for F_r $(r = 1, \ldots, n)$, and make use of the facts that

$$\frac{\partial V}{\partial q_r} = \frac{\partial V(0)}{\partial q_r} + \sum_{s=1}^{n} \frac{\partial^2 V(0)}{\partial q_r \partial q_s} q_s + \cdots \qquad (r = 1, \ldots, n)$$

while

$$\mathbf{v}_r^{P_i} = \mathbf{v}_r^{P_i}(0) + \sum_{s=1}^{n} \frac{\partial \mathbf{v}_r^{P_i}(0)}{\partial q_s} q_s + \cdots \qquad (i = 1, \ldots, \nu;\ r = 1, \ldots .n)$$

from which it follows that the linearized velocity of P_i is given by

$$\mathbf{v}^{P_i} \approx \sum_{r=1}^{n} \mathbf{v}_r^{P_i}(0)\dot{q}_r \qquad (i = 1, \ldots, \nu)$$

***13.5** Referring to Problem 10.9, letting A be a Newtonian reference frame, and using as motion variables for B in A the quantities $\dot{q}_1$, $\dot{q}_2$, and $\dot{q}_3$, form linearized dynamical equations governing q_1, q_2, q_3. Do this by employing the result developed in Problem 13.4.

Result

$$\ddot{q}_1 \approx 0$$

$$\ddot{q}_2 + 3Gm\frac{I_3 - I_1}{I_2 R^3} q_2 \approx 0$$

$$\ddot{q}_3 - 3Gm\frac{I_1 - I_2}{I_3 R^3} q_3 \approx 0$$

13.6 Two uniform bars, B_1 and B_2, each of length L and mass m, are supported by pins, as indicated in Fig. P13.6, and are attached to each other by a light, linear spring of natural length $L/4$ and spring constant mg/L. Formulate equations that govern the angles q_1 and q_2 (see Fig. P13.6) when B_1 and B_2 are at rest, and verify that two sets of values satisfying these equations are $q_1 = 35.55°$, $q_2 = 44.91°$, and $q_1 = -114.69°$, $q_2 = -94.77°$.

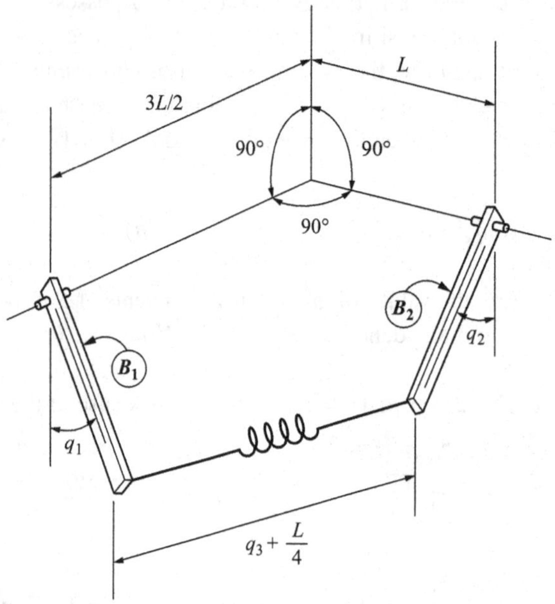

Figure P13.6

Suggestion: Introduce the stretch of the spring as a pseudo-generalized coordinate q_3 that must satisfy the configuration constraint equation

$$\left(q_3 + \frac{L}{4}\right)^2 = L^2 \left[(1 - s_1)^2 + \left(\frac{3}{2} - s_2\right)^2 + (c_1 - c_2)^2\right]$$

After defining motion variables u_1, u_2, and u_3 (u_3 is a pseudo-motion variable) as

$$u_r \triangleq \dot{q}_r \qquad (r = 1, 2, 3)$$

and differentiating the configuration constraint equation to obtain a constraint equation having the form of Eqs. (7.1.13), use Eqs. (8.5.4).

Results

$$\left(\frac{q_3}{L}+\frac{1}{4}\right)s_1+2\frac{q_3}{L}(c_2s_1-c_1)=0$$

$$\left(\frac{q_3}{L}+\frac{1}{4}\right)s_2+2\frac{q_3}{L}\left(c_1s_2-\frac{3}{2}c_2\right)=0$$

13.7 Two particles, P_1 and P_2, of masses m_1 and m_2, respectively, are supported by a light linkage, as indicated in Fig. P8.7, where $\hat{\mathbf{a}}_x$ and $\hat{\mathbf{a}}_y$ are unit vectors directed vertically downward and horizontally, respectively. Assuming that this system is at rest, verify that the parameters L_1, L_2, L_3, L_4, m_1, and m_2 are related to the variables q_1, q_2, and q_3 as follows:

$$L_1c_1+L_2c_2-L_3c_3=0$$
$$L_1s_1+L_2s_2-L_3s_3-L_4=0$$
$$m_1s_1+m_2s_3\frac{\sin(q_2-q_1)}{\sin(q_2-q_3)}=0$$

13.8 Referring to Problem 8.8, formulate three equations relating Q, R, S, q_1, q_2, q_3, and m when all rods are at rest.

Result

$$(-Q+R+S)c_1-30mgs_1=0$$
$$(Q-R)c_2-19mgs_2=0$$
$$Rc_3-8mgs_3=0$$

***13.9** Referring to Problem 8.10, formulate an equation relating T_B to T_E when B, C, C', D, D', E, and F are at rest.

Result $T_B+244T_E=0$

13.10 Referring to Problem 3.11, and letting $T\hat{\mathbf{N}}$, $T'\hat{\mathbf{N}}$, and $t\hat{\mathbf{n}}$ be the torques of couples applied to A, A', and D, repectively show that T, T', and t satisfy the equations

$$T=T'=-\frac{ta}{2d}$$

when the system is at rest.

13.11 Figure P13.11 shows four bars, each of length L, connected by hinges and linear springs, the springs having spring constants k_1 and k_2 and equal natural lengths $L/2$. Neglecting gravitational effects, show that when this system is at rest, then

$$\frac{k_1}{k_2}=\frac{1-\frac{1}{2}[2(1-\cos q)]^{-1/2}}{1-\frac{1}{2}[2(1+\cos q)]^{-1/2}}$$

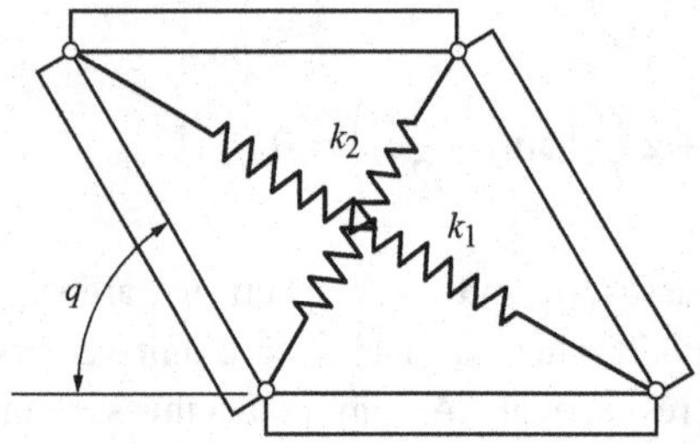

Figure P13.11

13.12 Referring to Problem 13.2, consider the steady motion characterized by the equations

$$q_1 = \overline{q}_1 \qquad \boldsymbol{\omega}^A = p\hat{\mathbf{a}}_1$$

where $\overline{q}_1$ and p are constants. Under these circumstances, p is called the rate of *precession* of the gyroscope, and the motion is termed a *steady precession*. Determine the relationship among m, h, J_1, J_2, K_1, K_2, $\overline{q}_1$, s, and p that prevails during steady precession.

Result $p^2(J_1 + K_1 - J_2 - K_2 - mh^2)\cos\overline{q}_1 + psK_1 - mgh = 0$

13.13 In Fig. P13.13, $A, \ldots, E$ are uniform, square plates, each having a mass m and sides of length L. These plates are attached to each other and to a uniform square plate F having a mass $5m$ and sides of length L, by means of smooth hinges.

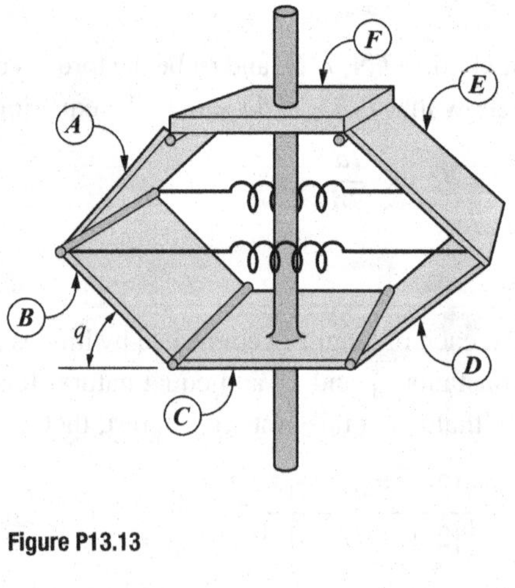

Figure P13.13

A vertical shaft, to which C is attached rigidly, passes through an opening in F, thus leaving F free to move up and down. Finally, two light, linear springs, each having a natural length L and spring constant k, connect the plates as shown.

One possible motion of this system is the following: The shaft is made to rotate with a constant angular speed Ω, and q (see Fig. P13.13) remains constant. Show that

$$\frac{L\Omega^2}{g} = \frac{6\cot q}{3+4\cos q}\left(4\frac{kL}{mg}\sin q - 7\right)$$

under these circumstances.

13.14 A rigid body B forms part of a simple nonholonomic system S possessing $n-m$ degrees of freedom in a Newtonian reference frame N. Letting $\mathbf{L}$ and $\mathbf{H}$ denote, respectively, the linear momentum of B in N and the central angular momentum of B in N, show that the contribution of B to the generalized momenta $p_1,\ldots,p_{n-m}$ can be expressed as

$$(p_r)_B = \tilde{\boldsymbol{\omega}}_r \cdot \mathbf{H} + \tilde{\mathbf{v}}_r^{\star} \cdot \mathbf{L} \qquad (r = 1,\ldots,n-m)$$

where $\tilde{\boldsymbol{\omega}}_r$ and $\tilde{\mathbf{v}}_r^{\star}$ are, respectively, the rth nonholonomic partial angular velocity of B in N and the rth nonholonomic partial velocity of the mass center of B in N.

13.15 A rigid body B is a part of a simple nonholonomic system S possessing $n-m$ degrees of freedom in a Newtonian reference frame N, and a subset σ of the contact forces acting on particles of S consists of forces applied to B and forming a couple of torque $\mathbf{T}$. Show that $(I_r)_\sigma$, the contribution of the forces of σ to the generalized impulse I_r, is given by

$$(I_r)_\sigma = \tilde{\boldsymbol{\omega}}_r^B(t_1) \cdot \int_{t_1}^{t_2} \mathbf{T}dt \qquad (r = 1,\ldots,n-m)$$

where $\tilde{\boldsymbol{\omega}}_r^B(t_1)$ is the value at time t_1 of the rth nonholonomic partial angular velocity of B in N, and t_2 differs so little from t_1 that the configuration of S in N does not change significantly during the time interval beginning at t_1 and ending at t_2.

13.16 Referring to Problem 4.11, let P be a horizontal plane, and let A be a Newtonian reference frame. Show that when C_1 and C_2 roll on P, $S^{\star}$ moves on a circle, and determine the radius of the circle when the system is set into motion by a blow applied to S along a line that intersects the axis of S at a distance s from $S^{\star}$. Assume that no slipping occurs, and express the result in terms of the radius R and length L shown in Fig. P4.11, as well as the mass m_S of S, the mass m_C of each of C_1 and C_2, the central transverse moment of inertia J of S, and the axial moment of inertia I of each of C_1 and C_2, letting C_1 and C_2 each have a transverse central moment of inertia $I/2$.

Result $\dfrac{I + J + 2(m_C + I/R^2)L^2}{s[m_S + 2(m_C + I/R^2)]}$

13.17 Four identical uniform rods, each of length $2L$, are connected by smooth pins, so as to form a square, and are resting on a smooth, horizontal surface when one of the rods is struck, the line of action of the blow passing through one corner of the square, as shown in Fig. P13.17. Determine the distance h such that the system moves as if it were a rigid body subsequent to being struck.

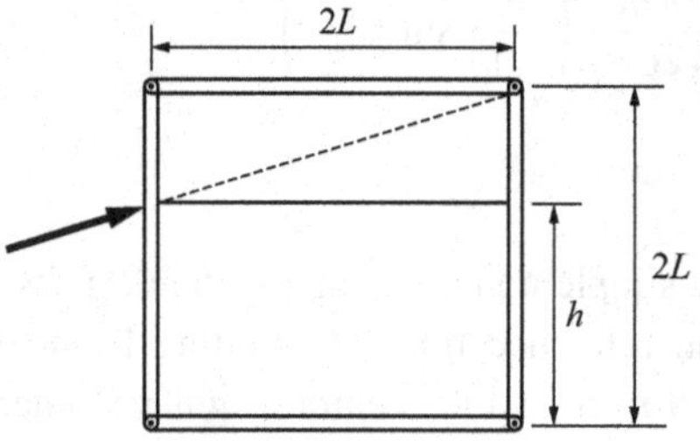

Figure P13.17

Result $h = 4L/3$

13.18 In Fig. P13.18, $\hat{\mathbf{n}}_1$, $\hat{\mathbf{n}}_2$, and $\hat{\mathbf{n}}_3$ are mutually perpendicular unit vectors fixed relative to a horizontal plane H in such a way that $\hat{\mathbf{n}}_2$ is perpendicular to H. S is a thin, uniform spherical shell of radius b that, at a certain instant, strikes H at a point A of H, the velocity $\mathbf{v}^\star$ of the center of S and the angular velocity $\boldsymbol{\omega}$ of S at this instant being given by

$$\mathbf{v}^\star = 10(-\hat{\mathbf{n}}_2 + \hat{\mathbf{n}}_3) \quad \text{m/s}$$

$$\boldsymbol{\omega} = 100(\hat{\mathbf{n}}_1 + 2\hat{\mathbf{n}}_2 + 5\hat{\mathbf{n}}_3) \quad \text{rad/s}$$

Thereafter, S bounces from A to B, from B to C, and so forth.

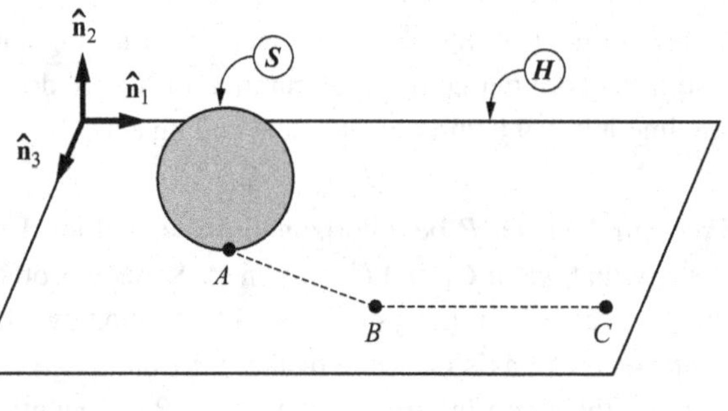

Figure P13.18

Letting the coefficient of restitution for S and H have the value 0.5, taking the coefficients of static friction and kinetic friction equal to 0.25 and 0.20, respectively, and setting $b = 0.03$ m, determine the distance from A to C.

Result 13.90 m

13.19 In Fig. P13.19(*a*), *A* and *B* are uniform rods, each having a length of 2 m and a mass of 3 kg. The two rods form a double pendulum moving in such a way that, at a certain instant, the free end of *B* strikes a horizontal surface *H* while the angular velocities of *A* and *B* have magnitudes of 0.1 rad/s and 0.2 rad/s, respectively, and are directed as shown.

(*a*) If e, the coefficient of restitution for *B* and *H*, has the value 0.5, what is the minimum value that μ, the coefficient of static friction for *B* and *H*, must have in order for *B* not to be sliding on *H* at the instant of separation?

(*b*) If $e = 0.5$, $\mu = 0.25$, and $\mu' = 0.20$, where μ' is the coefficient of kinetic friction for *B* and *H*, what are the angular velocities of *A* and *B* immediately after impact? Draw a sketch of the system, showing the angular velocities of *A* and *B*.

(*c*) For each of the sets of values of e, μ, and μ' in Table P13.19(*a*), determine whether or not slipping is taking place at the instant of separation, whether the kinetic energy increases or decreases during the collision, and the amount of kinetic energy change.

Results (*a*) 0.431 (*b*) Fig. P13.19(*b*) (*c*) Table P13.19(*b*)

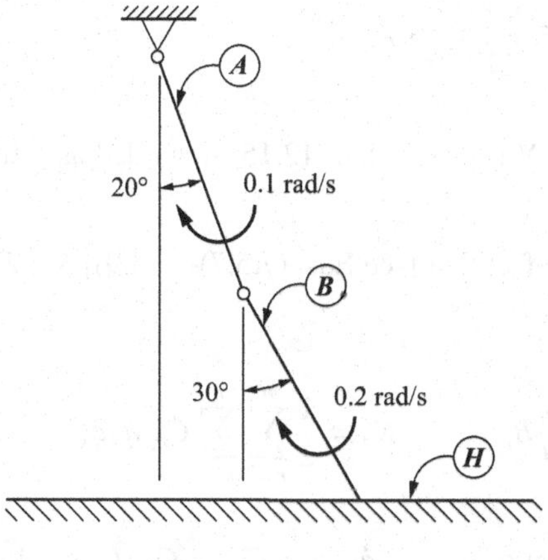

Figure P13.19(*a*)

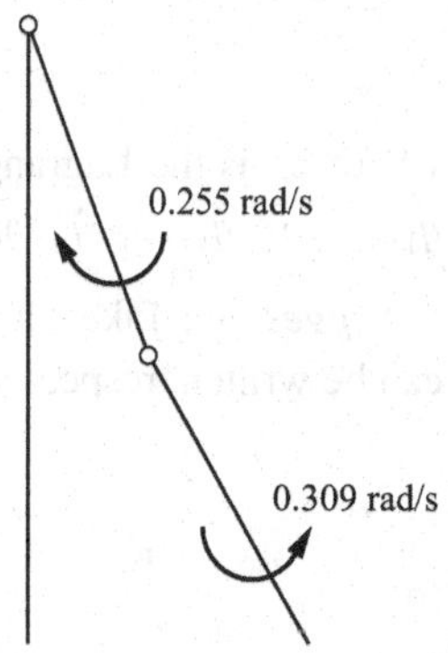

Figure P13.19(*b*)

Table P13.19(*a*)

e	μ	μ'
0.5	0.25	0.20
0.5	0.50	0.40
0.3	0.50	0.40
0.7	0.51	0.50

Table P13.19(*b*)

e	μ	μ'	Slipping	Kinetic energy change
0.5	0.25	0.20	Yes	0.03 N m decrease
0.5	0.50	0.40	No	0.16 N m increase
0.3	0.50	0.40	No	0.12 N m decrease
0.7	0.51	0.50	Yes	0.49 N m increase

PROBLEM SET 14
(Secs. 9.1–9.6)

14.1 Referring to Problem 8.7, let $L_1 = L_4 = 2$ m, $L_2 = L_3 = 3$ m, $m_1 = 4$ kg, $m_2 = 5$ kg, and suppose that the system is released from rest when $q_1 = 45°$. Determine the value of $\dot{q}_1$ for the first instant at which q_1 vanishes.

Note: It can be verified that q_2 and q_3 have the values 82.281° and 52.719°, respectively, when $q_1 = 45°$, and that $q_2 = 73.126°$, $q_3 = 16.874°$ when $q_1 = 0$.

Result -1.87 rad/s

14.2 Letting $q_1, \ldots, q_n$ be the generalized coordinates, and $u_r \triangleq \dot{q}_r$ $(r = 1, \ldots, n)$ the generalized velocities for a holonomic system in a reference frame N, show that H, the Hamiltonian of S in N (see Sec. 9.2), can be expressed as

$$H = \sum_{r=1}^{n} \frac{\partial \mathscr{L}}{\partial \dot{q}_r} \dot{q}_r - \mathscr{L}$$

where $\mathscr{L}$ is the Lagrangian of S in N (see Problem 12.15), regarded as a function of $q_1, \ldots, q_n, \dot{q}_1, \ldots, \dot{q}_n$, and t.

Suggestion: Take advantage of the fact that [see Eqs. (7.5.7)–(7.5.9)] K_0, K_1, and K_2 can be written, respectively, as

$$K_0 = A \qquad K_1 = \sum_{r=1}^{n} B_r \dot{q}_r \qquad K_2 = \frac{1}{2} \sum_{r=1}^{n} \sum_{s=1}^{n} C_{rs} \dot{q}_r \dot{q}_s$$

where A, B_r, and C_{rs} are independent of $\dot{q}_1, \ldots, \dot{q}_n$, and $C_{rs} = C_{sr}$ $(r, s = 1, \ldots, n)$.

14.3 The axis of a circular disk B of radius R (see Fig. P14.3) is fixed, and B is made to rotate about this axis with a constant angular speed ω. A vane V is fixed on B, the equation of the center line of V being $r = R \sin 2q$, where r is the distance from the axis of B to P, a generic point of the center line, and q is the angle between line OP and a line OQ that is fixed on B. Finally, a particle is free to move in V.

If the axis of B is vertical, and if the particle is inserted into V at O with a very small velocity (essentially zero), the particle moves toward the periphery of B and arrives there with a velocity having a magnitude v. Determine v.

Result $v = 2R\omega$

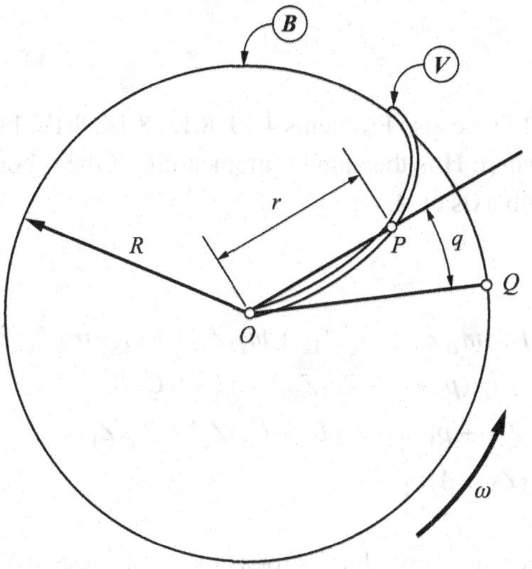

Figure P14.3

***14.4** Existence of an energy integral for a holonomic system S in a Newtonian reference frame N, when generalized velocities are defined as in Eqs. (3.4.1), requires satisfaction of Eqs. (7.1.10) and (7.6.3), as indicated in Table 9.2.1. When at least one of $Z_1, \ldots, Z_n$ in Eqs. (3.4.1) differs from zero, an alternative to Eq. (9.2.2) for forming a Hamiltonian is given by†

$$H = V + K_2 - K_0 - \sum_{r=1}^{n} p_r Z_r$$

where p_r is a generalized momentum of S in N (see Sec. 8.8). H is a constant when

$$\dot{H} = -\sum_{r=1}^{n} (F_r + F_r^\star)(u_r - Z_r) = 0$$

Determine the relationships that must be satisfied in order for $\dot{H}$ to vanish.

Results

$$\frac{\partial V}{\partial t} = 0 \qquad \sum_{i=1}^{\nu} m_i {}^N\mathbf{v}^{P_i} \cdot \frac{{}^N d}{dt}\left(\sum_{r=1}^{n} {}^N\mathbf{v}_r^{P_i} Z_r + {}^N\mathbf{v}_t^{P_i}\right) = 0$$

14.5 If the plane H in Problem 8.3 is perfectly smooth, the equations of motion of C possess an integral of the form

$$u_2 c_2 + \alpha u_3 s_2 = \beta$$

† D. L. Mingori, *Journal of Applied Mechanics* 62, no. 2 (1995), pp. 505–510.

where α is a definite constant and β is an arbitrary constant. Determine α.

Result $\alpha = 2$

***14.6** Referring to Problem 3.15 (see also Problems 4.19, 8.12, 8.18, 8.19, 11.10, 12.9, and 12.10), determine $\mathbf{H} \cdot \hat{\mathbf{a}}_1$, where $\mathbf{H}$ is the angular momentum of the robot arm in E with respect to a point on the hub axis of A.

Result

$$\begin{aligned}\mathbf{H} \cdot \hat{\mathbf{a}}_1 = &-m_A L_A Z_5 - L_P(m_B Z_8 + m_C Z_{12} + m_D Z_{18}) + c_1[-m_B L_B Z_8 \\ &- m_C Z_9 Z_{12} + m_D(p_2 Z_{19} - Z_{14} Z_{18}) + (B_1 + C_1)Z_1] \\ &+ s_1[m_D(-p_2 Z_{17} + p_1 Z_{18}) + (B_3 + C_3)Z_2] + Z_{49} Z_1 \\ &+ Z_{51} u_2 + Z_{53} Z_2 + A_1 u_1\end{aligned}$$

***14.7** When $q_1, \ldots, q_n$ are generalized coordinates for a holonomic system S in a Newtonian reference frame N, motion variables are defined as $u_r \triangleq \dot{q}_r$ $(r = 1, \ldots, n)$, and S possesses a kinetic potential $\mathscr{L}$ in N, $\mathscr{L}$ can be regarded as a function of $q_1, \ldots, q_n$, $\dot{q}_1, \ldots, \dot{q}_n$, and the time t (see Problem 12.15). Suppose that $q_{p+1}, \ldots, q_n$ are ignorable (see Sec. 9.4), and that α_s are the associated constant values of $\partial \mathscr{L} / \partial \dot{q}_s$ $(s = p + 1, \ldots, n)$. The momentum integrals can be used to construct relationships having the form of Eqs. (7.1.13), so that $\dot{q}_{p+1}, \ldots, \dot{q}_n$ are expressed as functions of $q_1, \ldots, q_p$, $\dot{q}_1, \ldots, \dot{q}_p$, $\alpha_{p+1}, \ldots, \alpha_n$, and t. Show that Eqs. (8.1.2) can be used to generate the dynamical equations

$$\tilde{F}_r + \tilde{F}_r^\star = \frac{d}{dt}\frac{\partial \mathscr{L}}{\partial \dot{q}_r} - \frac{\partial \mathscr{L}}{\partial q_r} = 0 \qquad (r = 1, \ldots, p)$$

where $\tilde{F}_r$ and $\tilde{F}_r^\star$ are associated with the motion variable $\dot{q}_r$ $(r = 1, \ldots, p)$.

A quantity $\mathscr{R}$, called a *Routhian* of S, can be defined as

$$\mathscr{R} \triangleq \mathscr{L} - \sum_{s=p+1}^{n} \alpha_s \dot{q}_s$$

After $\dot{q}_{p+1}, \ldots, \dot{q}_n$ have been eliminated from $\mathscr{R}$, it may be regarded as a function of $q_1, \ldots, q_p$, $\dot{q}_1, \ldots, \dot{q}_p$, $\alpha_{p+1}, \ldots, \alpha_n$, and t. Show that

$$\tilde{F}_r + \tilde{F}_r^\star = \frac{d}{dt}\frac{\partial \mathscr{R}}{\partial \dot{q}_r} - \frac{\partial \mathscr{R}}{\partial q_r} = 0 \qquad (r = 1, \ldots, p)$$

Suggestions: To obtain the dynamical equations involving $\mathscr{R}$, let $\tilde{\mathscr{L}}$ denote the kinetic potential after $\dot{q}_{p+1}, \ldots, \dot{q}_n$ have been eliminated, and regard $\tilde{\mathscr{L}}$ as a function of $q_1, \ldots, q_p$, $\dot{q}_1, \ldots, \dot{q}_p$, $\alpha_{p+1}, \ldots, \alpha_n$, and t. Verify that

$$\frac{\partial \tilde{\mathscr{L}}}{\partial \dot{q}_r} = \frac{\partial \mathscr{L}}{\partial \dot{q}_r} + \sum_{s=p+1}^{n} \frac{\partial \mathscr{L}}{\partial \dot{q}_s}\frac{\partial \dot{q}_s}{\partial \dot{q}_r} \qquad (r = 1, \ldots, p)$$

and

$$\frac{\partial \tilde{\mathscr{L}}}{\partial q_r} = \frac{\partial \mathscr{L}}{\partial q_r} + \sum_{s=p+1}^{n} \frac{\partial \mathscr{L}}{\partial \dot{q}_s} \frac{\partial \dot{q}_s}{\partial q_r} \qquad (r = 1, \ldots, p)$$

***14.8** Figure P14.8 shows a pendulum consisting of two particles, Q and P, connected by a light rod of length L. Q has mass M and can slide freely on a smooth horizontal rail. The mass of P is m. The angle between the rod and the vertical is denoted by q_1, and q_2 is the horizontal displacement of Q.

Construct an equation of motion by forming a Routhian and referring to Problem 14.7.

Construct an equation of motion by appealing to Eqs. (8.1.2), and verify that the result is identical to the relationship obtained with a Routhian.

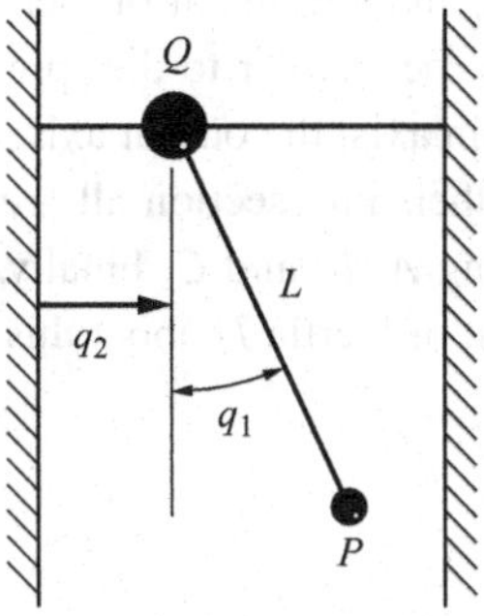

Figure P14.8

Result

$$\left(1 - \frac{m}{M+m} c_1{}^2\right) \ddot{q}_1 = -\frac{m}{M+m} s_1 c_1 \dot{q}_1{}^2 - \frac{g}{L} s_1$$

***14.9** The example in Sec. 8.1 involves a pendulum and generalized coordinates q_1 and q_2 that describe simple rotations about $\hat{\mathbf{e}}_1$ and $\hat{\mathbf{a}}_2$, respectively. As an alternative, let generalized coordinates θ_1 and θ_2 characterize, respectively, simple rotations about $\hat{\mathbf{e}}_3$ and $\hat{\mathbf{a}}_2$, define motion variables as $u_r \triangleq \dot{\theta}_r$ $(r = 1, 2)$, and regard the Earth, E, as a Newtonian reference frame. Construct a Routhian (see Problem 14.7) and use it to form an equation of motion. Verify that Eqs. (8.1.2) lead to the same result.

Result

$$\alpha_1 = mL^2 s_2{}^2 \dot{\theta}_1 \qquad \ddot{\theta}_2 = \left(\frac{\alpha_1}{mL^2}\right)^2 \frac{c_2}{s_2{}^3} - \frac{g}{L} s_2$$

14.10 Show without explicit use of differential equations of motion that the relationships

$$(3a^2 + b^2 s_2{}^2)\dot{q}_1{}^2 + b^2 \dot{q}_2{}^2 + 3abc_2\dot{q}_1\dot{q}_2 - 4gbc_2 = \alpha$$

and

$$2(3a^2 + b^2 s_2{}^2)\dot{q}_1 + 3abc_2\dot{q}_2 = \beta$$

where α and β are constants, are integrals of the equations of motion of the triangular plate in Problem 11.5.

14.11 Figure P14.11 shows a gyroscopic device consisting of a frame F, a torsion spring assembly S, a gimbal ring G, and a rotor R. These parts have the following inertia properties:

The point of intersection of the spin axis and the output axis is the mass center of R, and R has a moment of inertia J about the spin axis, and a moment of inertia I about any line passing through the mass center of R and perpendicular to the spin axis. The mass center of G coincides with that of R, and the spin axis, the output axis, and a line perpendicular to both of these and passing through their intersection all are principal axes of G, the corresponding moments of inertia being A, B, and C. Finally, the mass center of F lies on the input axis, and F has a moment of inertia D about this axis.

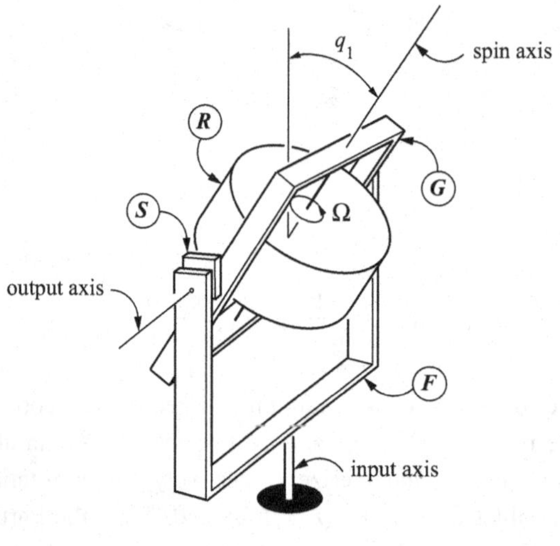

Figure P14.11

Consider the following class of motions of the device: F is free to rotate about the input axis, which is fixed; G can rotate about the output axis, but a resisting torque of magnitude kq_1 is associated with such rotations; and R is made to rotate with constant angular speed Ω in G (by means of a motor connecting R to G). Assuming that at time $t = 0$ the frame F is at rest, $q_1 = \pi/2$ rad, and $\dot{q}_1 = 0$, determine the value of the spring constant k such that $\dot{q}_1$ vanishes for $t > 0$ when $q_1 = \pi/4$ rad.

Result $\dfrac{(4J\Omega/\pi)^2}{3(A+C+I+J+2D)}$

14.12 Referring to the example in Sec. 9.5, suppose that the set of friction forces exerted on S by T is treated as equivalent to a couple of torque $-cu_1\hat{\mathbf{e}}_1$ (see Fig. 9.5.1), where c is a positive constant. Taking $u_1 = u_4 = 0$ and $u_5 = 4c(R-r)/(mr^2)$ at $t = 0$, show that u_4 approaches a limiting value $u_4{}^\star$ as t approaches infinity, and plot $u_4/u_4{}^\star$ versus $ct/(mr^2)$.

Result Figure P14.12

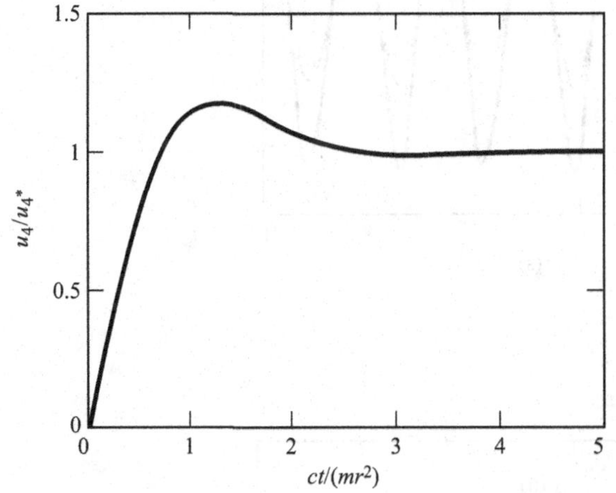

Figure P14.12

14.13 Considering the system S introduced in Problem 13.3, and letting $u_r \triangleq \dot{q}_r$ $(r = 1,2)$, (a) formulate exact dynamical equations of S. (b) Taking $m_A = 0.01$ kg, $m_B = 0.1$ kg, $L_A = 0.075$ m, $L_B = 0.2$ m, $A_1 = 5.0 \times 10^{-6}$ kg m^2, $B_1 = 2.5 \times 10^{-4}$ kg m^2, $B_2 = 5.0 \times 10^{-5}$ kg m^2, and $B_3 = 2.0 \times 10^{-4}$ kg m^2, perform a numerical simulation of the motion of S for $0 \le t \le 10$ s, using for initial conditions $u_1(0) = u_2(0) = 0$, $q_1(0) = 45°$, $q_2(0) = 1°$, and plot q_2 versus t. Leaving all other quantities unchanged, but taking $q_2(0) = 0.5°$, make another plot of q_2 versus t, and display the two curves on the same set of axes. (c) Repeat part (b) with $q_1(0) = 90°$.

Compare the set of two curves generated in part (b) with the set obtained in part (c), and briefly state your conclusions.

Results

$$(a)\quad \dot{u}_1 = \frac{2u_1u_2 s_2 c_2 (B_1 - B_2) - (m_A L_A + m_B L_B) g s_1}{A_1 + B_1 {c_2}^2 + B_2 {s_2}^2 + m_A {L_A}^2 + m_B {L_B}^2}$$

$$\dot{u}_2 = -{u_1}^2 s_2 c_2 (B_1 - B_2)/B_3$$

(*b*) Fig. P14.13(*a*)

(*c*) Fig. P14.13(*b*)

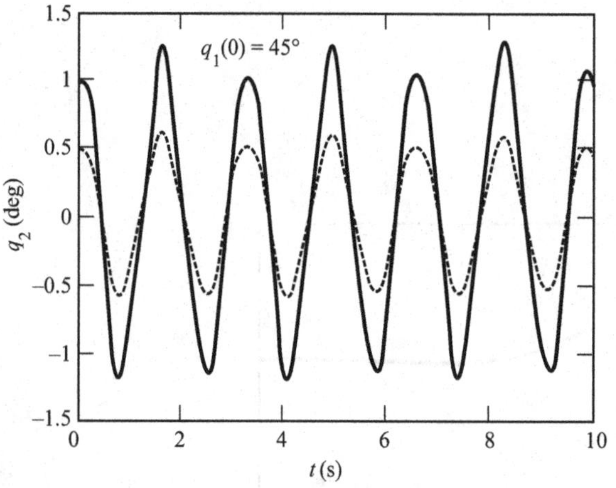

Figure P14.13(*a*)

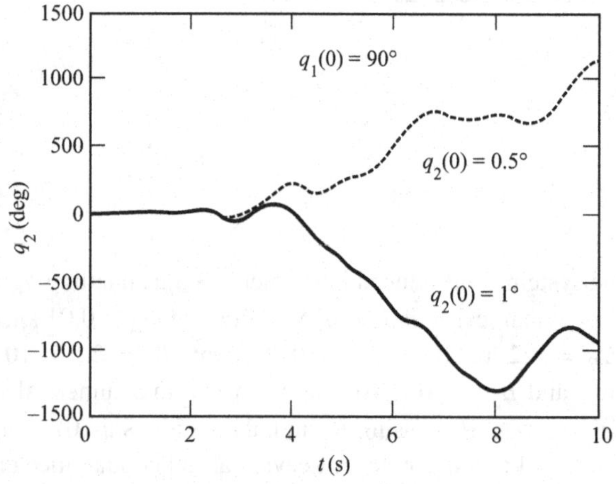

Figure P14.13(*b*)

***14.14** Referring to Problem 3.15 (see also Problems 4.19, 8.12, 8.18, 8.19, 11.10, 12.9, 12.10, and 14.6), take $L_A = 0.3$ m, $L_B = 0.5$ m, $L_P = 1.1$ m, $p_1 = 0.2$ m, $p_2 = 0.4$ m, $p_3 = 0.6$ m, $A_1 = 11$ kg m^2, $B_1 = 7$ kg m^2, $B_2 = 6$ kg m^2, $B_3 = 2$ kg m^2, $C_1 = 5$ kg m^2, $C_2 = 4$ kg m^2, $C_3 = 1$ kg m^2, $D_{11} = 2$ kg m^2, $D_{22} = 2.5$ kg m^2, $D_{33} = 1.3$ kg m^2, $D_{12} = 0.6$ kg m^2, $D_{23} = 0.75$ kg m^2, $D_{31} = -1.1$ kg m^2, $m_A = 87$ kg, $m_B = 63$ kg, $m_C = 42$ kg, $m_D = 50$ kg. (*a*) Letting $\dot{q}_3 = u_1$, and setting $u_1(0) = 0.1$ rad/s, $u_2(0) = 0.2$ rad/s, $u_3(0) = 0.3$ m/s, $q_1(0) = 30°$, $q_2(0) = 0.1$ m, $q_3(0) = 10°$, $T_1{}^{E/A} = T_2{}^{A/B} = K_3{}^{B/C} = 0$, determine the value of $K + V_\gamma$ (see Problem 11.10) at $t = 0$ and $t = 5$ s, and the value of $\mathbf{H} \cdot \hat{\mathbf{a}}_1$ (see Problem 14.6) at $t = 0$ and $t = 5$ s; (*b*) setting $T_2{}^{A/B} = 10$ N m and $K_3{}^{B/C} = 20$ N, but leaving all other quantities as in (*a*), determine the value of $K + V_\gamma$ at $t = 0$ and $t = 5$ s, and the value of $\mathbf{H} \cdot \hat{\mathbf{a}}_1$ at $t = 0$ and $t = 5$ s; (*c*) the robot arm can be brought from an initial state of rest in E to a final state of rest in E such that q_1, q_2, and q_3 have the specified values $q_1{}^\star$, $q_2{}^\star$, and $q_3{}^\star$, respectively, by using the following feedback control laws:

$$T_1{}^{E/A} = -\beta^{E/A}u_1 - \gamma^{E/A}(q_3 - q_3{}^\star)$$
$$T_2{}^{A/B} = -\beta^{A/B}u_2 - \gamma^{A/B}(q_1 - q_1{}^\star) + g[(m_B L_B + m_C Z_9 + m_D Z_{14})c_1 - m_D p_1 s_1]$$
$$K_3{}^{B/C} = -\beta^{B/C}u_3 - \gamma^{B/C}(q_2 - q_2{}^\star) + g(m_C + m_D)s_1$$

Here, $\beta^{E/A}$, $\gamma^{E/A}$, $\beta^{A/B}$, $\gamma^{A/B}$, $\beta^{B/C}$, and $\gamma^{B/C}$ are constant "gains," and the terms involving g in the expressions for $T_2{}^{A/B}$ and $K_3{}^{B/C}$ serve to counteract the effects of gravity. Using these control laws, the parameter values employed previously, and $u_1(0) = u_2(0) = u_3(0) = 0$, $q_1(0) = 30°$, $q_2(0) = 0.1$ m, $q_3(0) = 10°$, $q_1{}^\star = 60°$, $q_2{}^\star = 0.4$ m, $q_3{}^\star = 70°$, $\beta^{E/A} = 464$ N m s/rad, $\gamma^{E/A} = 306$ N m/rad, $\beta^{A/B} = 216$ N m s/rad, $\gamma^{A/B} = 285$ N m/rad, $\beta^{B/C} = 169$ N s/m, $\gamma^{B/C} = 56$ N/m, plot the values of q_1, q_2, and q_3 from $t = 0$ to $t = 15$ s.

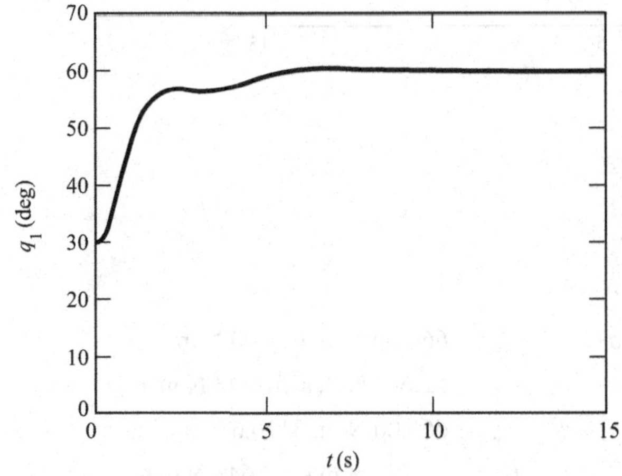

Figure P14.14(*a*)

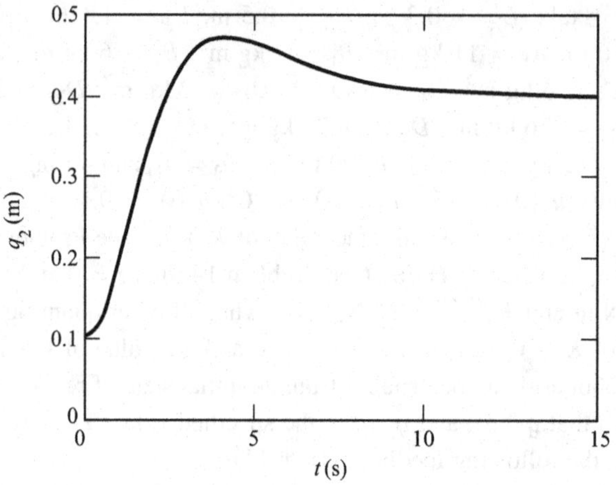

Figure P14.14(b)

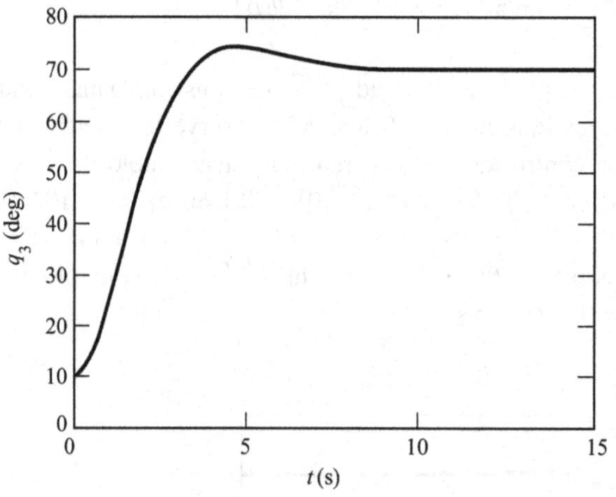

Figure P14.14(c)

Results

(a) $K + V_\gamma$: 666.00 N m, 666.00 N m

H_1 : 52.643 N m s, 52.643 N m s

(b) $K + V_\gamma$: 666.00 N m, 3116.6 N m

H_1 : 52.643 N m s, 52.643 N m s

(c) Figures P14.14(a), P14.14(b), P14.14(c)

PROBLEM SET 15
(Secs. 9.7–9.9)

15.1 Referring to the example in Sec. 9.7, replace the set of forces exerted on the left half of C by the right half with a couple together with a force $\sigma_1 \hat{\mathbf{n}}_1 + \sigma_2 \hat{\mathbf{n}}_2$, applying this force at the right end of the left half of C. For the values of m_A, m_B, m_C, L, $q_1(0)$, and $u_1(0)$ used in the example in Sec. 9.7, show that $\sigma_1 = -(g/16)\tan q_1$ and determine σ_1 for $t = 5$ s.

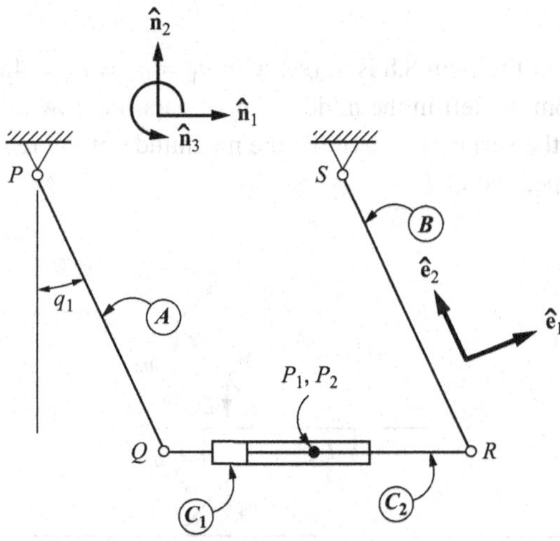

Figure P15.1

Suggestions: Regard C as composed of two identical bars, C_1 and C_2, each of length L and mass $m_C/2$, which can slide relative to each other as indicated in Fig. P15.1, and let P_1 and P_2 be the points of C_1 and C_2, respectively, corresponding to the midpoint of C. Note that during the motion of interest the velocity of P_2 in C_1 and the angular velocities of C_1 and C_2 can be expressed as

$$^{C_1}\mathbf{v}^{P_2} = v\hat{\mathbf{n}}_1 \qquad \boldsymbol{\omega}^{C_1} = \boldsymbol{\omega}^{C_2} = \gamma\hat{\mathbf{n}}_3$$

where v and γ are functions of q_1 and u_1. Introduce u_2 as $u_2 \triangleq \boldsymbol{\omega}^B \cdot \hat{\mathbf{n}}_3$, and verify that the velocities of P_1 and P_2 are given by

$$\mathbf{v}^{P_1} = 2Lu_1\hat{\mathbf{e}}_1 + L(u_2 - u_1)\mathrm{s}_1\hat{\mathbf{n}}_2$$

and

$$\mathbf{v}^{P_2} = 2Lu_2\hat{\mathbf{e}}_1 - L(u_2 - u_1)\mathrm{s}_1\hat{\mathbf{n}}_2$$

In addition, verify that F_r and $F_r^\star$ $(r = 1, 2)$ are given by

$$F_1 = 2L\mathrm{c}_1\sigma_1 - gL(m_A + m_C)\mathrm{s}_1 \qquad F_1^\star = -\tfrac{2}{3}L^2(2m_A + 3m_C)\dot{u}_1$$

$$F_2 = -2Lc_1\sigma_1 - gL(m_B + m_C)s_1 \qquad F_2^\star = -\tfrac{2}{3}L^2(2m_B + 3m_C)\dot{u}_1$$

Result $\sigma_1 = -0.173$ N

15.2 Consider expressions for F_1 and $F_1^\star$ associated with u_1 prior to introduction of u_2 in the example in Sec. 9.7 and subsequent to introduction of u_2 in Problem 15.1. Determine whether introduction of the additional motion variable in Problem 15.1 alters the following generalized forces: (*a*) F_1, (*b*) $F_1^\star$.

Results (*a*) Yes; (*b*) Yes

15.3 The system described in Problem 8.8 is at rest with $q_1 = q_2 = q_3 = 45°$. Letting B designate the third bar from the left in the middle row of bars (the row in which the bars make an angle q_2 with the vertical), determine the magnitude of the reaction of B on the pin supporting the upper end of B.

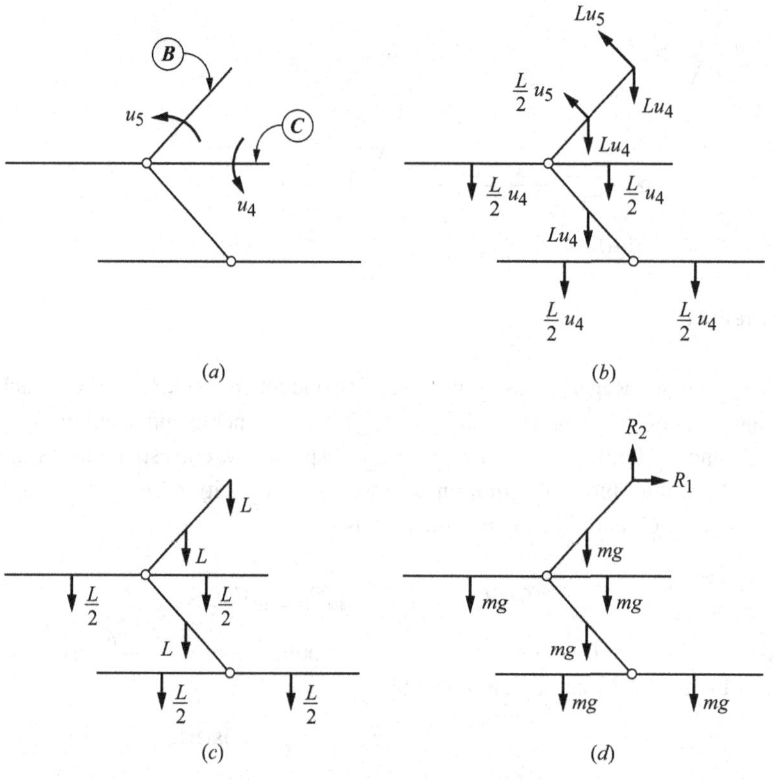

Figure P15.3

Suggestion: Regard the upper end of B as disconnected from the pin that supports this end, and note that the six bars shown in Fig. P15.3(*a*) are movable under these circumstances, whereas the rest of the bars are not movable so long as q_1, q_2, and q_3 have fixed values. Introduce motion variables u_4 and u_5 such that bars B and C have counterclockwise angular velocities, as indicated in Fig. P15.3(a). Record the velocities of the mass centers of the movable bars and the velocity of the upper end of B as in Fig. P15.3(b), and use this sketch to depict partial velocities associated with u_4 as in Fig. P15.3(c). Record the gravitational forces acting on the six bars, as well as the horizontal and vertical reaction force components at the upper end of B as in Fig. P15.3(d). Refer to Figs. P15.3(c) and P15.3(d) to perform by inspection the dot multiplications required to form the generalized active force F_4, thus verifying that $F_4 = L(-R_2 + 4mg)$. Make a sketch similar to Fig. P15.3(c) to depict the partial velocities associated with u_5, and use this sketch in conjunction with Fig. P15.3(d) to form an expression for the generalized active force F_5. Finally, appeal to Eqs. (8.5.1) to find R_1 and R_2.

Result $\sqrt{113}mg/2$

15.4 Relationships were introduced in Problem 12.13 to express Euler's first and second laws for a rigid body B. These laws can be restated in terms of vectors,

$$\mathbf{F} = m^N\mathbf{a}^{B^\star} \qquad \mathbf{T} = \frac{^N d}{dt}\,{}^N\mathbf{H}^{B/B^\star}$$

where m is the mass of B, ${}^N\mathbf{a}^{B^\star}$ is the acceleration in a Newtonian reference frame N of $B^\star$, the mass center of B, and where ${}^N\mathbf{H}^{B/B^\star}$ is the central angular momentum of B in N. The set of contact forces and distance forces acting on B is equivalent to a couple of torque $\mathbf{T}$ together with a force $\mathbf{F}$ applied at $B^\star$.

The foregoing equations underlie the *Newton-Euler method* for analyzing motions of a system S. One applies the equations in turn to each body B belonging to S, taking care when forming $\mathbf{F}$ and $\mathbf{T}$ to account for interaction forces exerted by objects in contact with B. Consequently, one frequently must expend effort to eliminate unknown interaction forces and interaction torques from the equations before arriving at dynamical equations governing motions of S in N.

To apply the Newton-Euler method to the linkage considered in the example in Sec. 9.7, one begins by writing the two foregoing vector equations for each of the three links A, B, C shown in Fig. 9.7.1. The resulting six vector equations give rise ultimately to 18 scalar equations, which would involve 18 motion-variable time derivatives $\dot{u}_1, \ldots, \dot{u}_{18}$ if the four pins were not present, because in that case each of the three links would move independently and possess six degrees of freedom in N. The equations can, however, be written in terms of only one motion-variable time derivative, $\dot{u}_1$, as may be verified by referring to Eqs. (9.7.8)–(9.7.13), and to the result obtained in part (c) of Problem 8.16, because the assembled linkage has in fact only one degree of freedom in N. (It may be concluded from this observation that 17 independent scalar constraint equations can be written to describe the constraints imposed by the four pins.)

The constraint forces and constraint torques exerted on A, B, and C by the horizontal pins at P, Q, R, and S are taken into account when $\mathbf{F}$ and $\mathbf{T}$ are constructed for each link. For example, the constraint force applied to A at P can be expressed as

$$\mathbf{C}^P = \lambda_1 \hat{\mathbf{e}}_1 + \lambda_2 \hat{\mathbf{e}}_2 + \lambda_3 \hat{\mathbf{e}}_3$$

and the torque of the couple exerted by this pin on A, which must be perpendicular to $\hat{\mathbf{e}}_3$, can be written as

$$\boldsymbol{\tau}^P = \lambda_4 \hat{\mathbf{e}}_1 + \lambda_5 \hat{\mathbf{e}}_2$$

Thus, five unknown scalars $\lambda_1, \ldots, \lambda_5$ are introduced in connection with the pin at P, and 15 more unknown scalars must enter the picture to characterize the interactions at the other three pins. The Newton-Euler method now appears to confront the analyst with the impossible task of using 18 equations to determine a total of 21 unknowns, including $\dot{u}_1$.

It turns out to be possible to construct three equations that involve only the three unknowns $\dot{u}_1$, λ_1, and λ_2. The constraint forces $\mathbf{C}^Q$, $\mathbf{C}^R$, and $\mathbf{C}^S$ applied, respectively, to C at Q, to B at R, and to B at S, can be expressed in terms of $\mathbf{C}^P$, ${}^N\mathbf{a}^{A^\star}$, ${}^N\mathbf{a}^{B^\star}$, and ${}^N\mathbf{a}^{C^\star}$ by appealing to the three vector relationships stating Euler's first law for the mass centers $A^\star$, $B^\star$, and $C^\star$. The resulting expressions for $\mathbf{C}^Q$, $\mathbf{C}^R$, and $\mathbf{C}^S$ are then used in forming moments about the mass centers as required in the three vector expressions of Euler's second law, and each of these equations is then dot-multiplied by $\hat{\mathbf{n}}_3$.

Record the three scalar equations thus obtained, and verify that they may be used to recover Eq. (9.7.19). Comment briefly on the relative merits of the method used to obtain Eq. (9.7.19) in the example of Sec. 9.7, on the one hand, and the method employed in the present problem, on the other hand.

Results

$$-\tfrac{2}{3} m_A L^2 \dot{u}_1 + 2L\lambda_1 = m_A g L s_1$$

$$-(m_A + m_C)L^2 s_1 \dot{u}_1 + L(s_1\lambda_1 + c_1\lambda_2) = (m_A + m_C)L^2 c_1 {u_1}^2 + \left(m_A + \tfrac{1}{2} m_C\right) gL$$

$$\left(m_A + \tfrac{2}{3} m_B + 2m_C\right) L^2 \dot{u}_1 - L\lambda_1 = -\left(m_A + \tfrac{1}{2} m_B + m_C\right) gL s_1$$

***15.5** Figure P15.5 shows a linkage containing five uniform rods A, B, C, D, and E, each of mass m and length $2L$. The linkage is suspended from a horizontal support by pins at P, S, and U; the rods are fastened together by pins at Q, R, and T. After defining $u_1 \triangleq \dot{q}_1$, where q_1 is the angle indicated in Fig. P15.5, employ the Newton-Euler method (see Problem 15.4) to construct dynamical equations of motion, and obtain an expression for $\dot{u}_1$ in which all measure numbers of constraint forces and constraint torques have been eliminated. Reproduce the expression for $\dot{u}_1$ by using Eqs. (8.1.1), and comment briefly on the relative merits of the two methods.

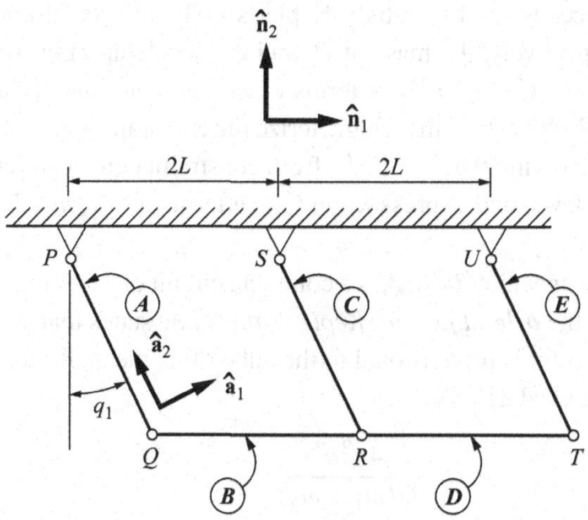

Figure P15.5

Suggestions: Express the constraint force applied to A at P as

$$\mathbf{C}^P = \mu_1 \hat{\mathbf{a}}_1 + \mu_2 \hat{\mathbf{a}}_2 + \mu_3 \hat{\mathbf{a}}_3$$

where unit vectors $\hat{\mathbf{a}}_1$, $\hat{\mathbf{a}}_2$, $\hat{\mathbf{a}}_3$ are fixed in A and directed as shown in Fig. P15.5. Likewise, let

$$\mathbf{C}^U = \mu_4 \hat{\mathbf{a}}_1 + \mu_5 \hat{\mathbf{a}}_2 + \mu_6 \hat{\mathbf{a}}_3$$

denote the constraint force applied to E at U, and similarly represent the constraint force applied to C at S, as well as the forces exerted by A and B on each other at Q, by C and B on each other at R, by C and D on each other at R, and by E and D on each other at T. After writing Euler's first law for each of the five rods, express in terms of $\mathbf{C}^P$ and $\mathbf{C}^U$ the remaining five constraint forces. Next, account for the constraint torques, all of which are necessarily perpendicular to $\hat{\mathbf{a}}_3$, acting at P, Q, R, S, T, and U, write Euler's second law for each of the five rods, and dot-multiply each of the five resulting vector equations with $\hat{\mathbf{a}}_3$ to obtain five scalar equations in terms of the five unknowns μ_1, μ_2, μ_4, μ_5, and $\dot{u}_1$. From these five equations, produce the desired relationship for $\dot{u}_1$.

Results

$$-\tfrac{2}{3} m L^2 \dot{u}_1 + 2L\mu_1 = mgL\mathrm{s}_1$$
$$4mL^2 \mathrm{s}_1 \dot{u}_1 - 2L(\mathrm{s}_1 \mu_1 + \mathrm{c}_1 \mu_2) = -4mL^2 \mathrm{c}_1 {u_1}^2 - 3mgL$$
$$\tfrac{40}{3} m L^2 \dot{u}_1 - 2L(\mu_1 + \mu_4) = -9mgL\mathrm{s}_1$$
$$4mL^2 \mathrm{s}_1 \dot{u}_1 - 2L(\mathrm{s}_1 \mu_4 + \mathrm{c}_1 \mu_5) = -4mL^2 \mathrm{c}_1 {u_1}^2 - 3mgL$$
$$-\tfrac{2}{3} m L^2 \dot{u}_1 + 2L\mu_4 = mgL\mathrm{s}_1$$
$$\dot{u}_1 = -\frac{7g}{12L}\mathrm{s}_1$$

15.6 Suppose that the two particles P_1 and P_2 introduced in Problem 9.1 are subject only to the constraint forces needed to satisfy Kepler's first and second laws. Letting m_i and ${}^N\mathbf{a}^{P_i}$ denote, respectively, the mass of P_i and the acceleration in N of P_i ($i = 1, 2$), express the quantity ${}^N\mathbf{a}^{P_2} - {}^N\mathbf{a}^{P_1}$ in terms of m_1, m_2, and the scalars λ_1 (see Problem 9.1) and λ_2 (see Problem 9.3) that characterize the constraint forces. Obtain two additional relationships involving ${}^N\mathbf{a}^{P_2} - {}^N\mathbf{a}^{P_1}$ from constraint equations representing Kepler's first and second laws, and express λ_1 and λ_2 in terms of m_1, m_2, B_1, B_2, r, θ, and $\dot{A}$.

The interval of time during which P_2 makes a complete circuit on an ellipse is denoted by T and is referred to as the *orbital period. Kepler's third law* states that the square of the period of an elliptical orbit is proportional to the cube of its mean distance from the Sun. This law can be expressed as

$$T^2 = \frac{4\pi^2 a^3}{G(m_1 + m_2)}$$

where a is called the semimajor axis of the ellipse, and G is the universal gravitational constant. After noting that the area of an ellipse is πab, and the parameter p (see Problem 9.1) can be written as $p = b^2/a$, where b is the semiminor axis of the ellipse, express the resultant constraint force acting on P_2, $\mathbf{R}_2 = \mathbf{C}_1^{P_2} + \mathbf{C}_2^{P_2}$, in terms of G.

Suggestions: Obtain a relationship for ${}^N\mathbf{a}^{P_2} - {}^N\mathbf{a}^{P_1}$ by making use of the kinematical equation for ${}^N\mathbf{v}^{P_2} - {}^N\mathbf{v}^{P_1}$ given in Problem 9.1, and by appealing to Newton's second law (see Eqs. 8.1.4).

Results

$${}^N\mathbf{a}^{P_2} - {}^N\mathbf{a}^{P_1} = \left(\frac{m_1 + m_2}{m_1 m_2}\right)[\lambda_1 \hat{\mathbf{a}}_1 + (\lambda_2 - \lambda_1 B_2 r \sin\theta)\hat{\mathbf{a}}_2]$$

$$\lambda_1 = -\frac{m_1 m_2}{(m_1 + m_2)} \frac{4B_1 \dot{A}^2}{r^2} \qquad \lambda_2 = \lambda_1 B_2 r \sin\theta$$

$$\mathbf{R}_2 = -\frac{G m_1 m_2}{r^2}\hat{\mathbf{a}}_1$$

15.7 The equations of motion obtained in Problem 12.8 contain no evidence of the constraint force $\mathbf{C}$ identified in Problem 9.6; in other words, the scalar λ is absent from these equations. Although this simplification can be a welcome result, it can happen that λ is of interest in its own right. Bring $\mathbf{C}$ into evidence by constructing relationships according to Eqs. (8.1.1) for $r = 1, 2, 3$, using the motion variables u_1, u_2, and u_3 introduced in Problem 4.15. After expressing $\dot{u}_3$ in terms of $\dot{u}_1$ and $\dot{u}_2$ such that the constraint introduced in Problem 4.15 is satisfied, determine λ.

Results

$$m\dot{u}_1 = -\lambda a^2 u_1/u_3 \qquad m\dot{u}_2 = -\lambda a^2 u_2/u_3 \qquad m\dot{u}_3 = \lambda - mg \qquad \lambda = mg(\sin\gamma)^2$$

***15.8** It is possible for one to make useful observations about constraint forces when their directions are known. Referring to Problems 4.15 and 9.6, show that **C** is perpendicular to ${}^N\mathbf{v}^P$. Referring also to Problems 8.6 and 15.7, show that the resultant force **R** defined as

$$\mathbf{R} \triangleq \mathbf{C} - mg\hat{\mathbf{n}}_3$$

is parallel to ${}^N\mathbf{v}^P$.

15.9 Referring to the particle P introduced in Problem 4.15, and to the results obtained in Problem 15.7, solve the differential equation governing u_3; express u_3 as a function of time t and $u_3(0)$, the initial value of u_3. Comment briefly on the vertical motion of P when $u_3(0) > 0$ and when $u_3(0) < 0$. Letting $u_1(0)$ and $u_2(0)$ denote initial values of u_1 and u_2 respectively, verify that the differential equations governing u_1 and u_2 are satisfied when u_1 and u_2 are given by

$$u_1(t) = \frac{u_1(0)}{u_3(0)} u_3(t) \qquad\qquad u_2(t) = \frac{u_2(0)}{u_3(0)} u_3(t)$$

Results $u_3(t) = -g(\cos\gamma)^2 t + u_3(0)$

***15.10** Referring to the example in Sec. 8.6, verify that there exist values of θ $(0 < \theta < \pi/2)$ such that the steady motion there considered cannot occur. For $0 \le h/R \le 2$, show that these values of θ correspond to the shaded region of Fig. P15.10.

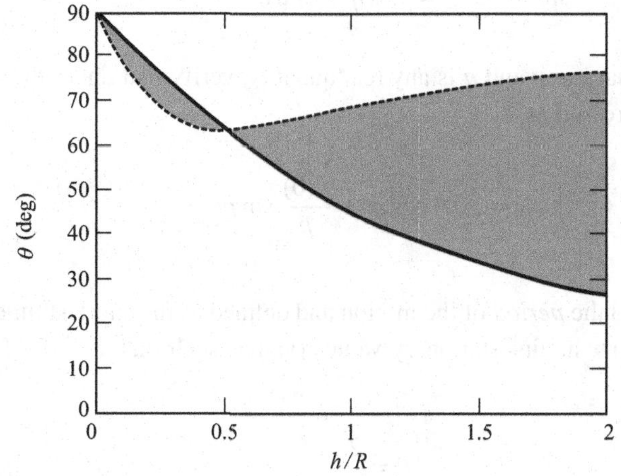

Figure P15.10

15.11 Taking $L\Omega^2/g = 3$, determine two pairs of constant values of q_1 and q_2 such that the equations of motion of the system considered in the example in Sec. 8.7 are satisfied, and sketch the system in the configuration associated with each pair of values.

Results $q_1 = 56.18°$, $q_2 = 226.30°$; $q_1 = 74.25°$, $q_2 = 78.34°$; Fig. P15.11

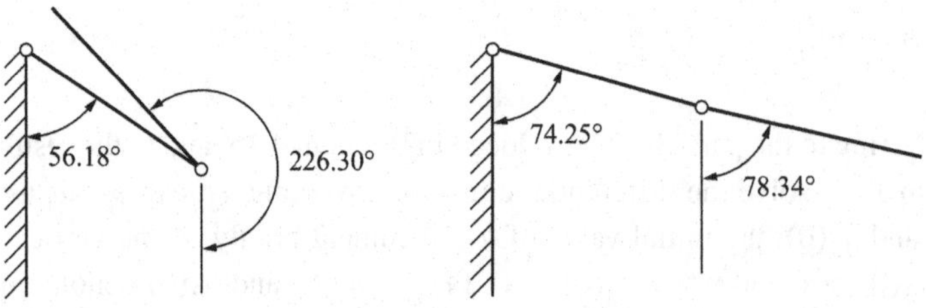

Figure P15.11

15.12 A system moving in accordance with Eq. (9.9.1) is said to be performing *undamped free vibrations* if

$$n = f(t) = 0$$

Making use of Euler's identities

$$e^{\pm iq} = \cos q \pm i \sin q$$

where i is the imaginary unit and q is any real quantity, verify that under these circumstances x can be expressed as

$$x = x(0) \cos pt + \frac{\dot{x}(0)}{p} \sin pt$$

Show that T, called the *period* of the motion and defined as the shortest time between two instants at which x attains stationary values (values such that $\dot{x} = 0$) of the same sign, is given by

$$T = \frac{2\pi}{p}$$

Taking $\dot{x}(0) = 0$, plot $x/x(0)$ versus pt for $0 \leq pt \leq 20$.

Result Figure P15.12

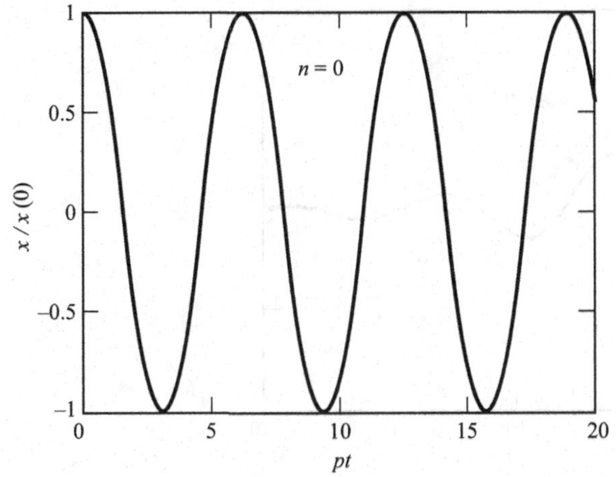

Figure P15.12

15.13 A system moving in accordance with Eq. (9.9.1) is said to be performing *underdamped free vibrations* if

$$0 < n < 1 \qquad f(t) = 0$$

Show that during such vibrations

$$x = \left\{ x(0)\cos[p(1-n^2)^{1/2}t] + \frac{\dot{x}(0) + npx(0)}{p(1-n^2)^{1/2}} \sin[p(1-n^2)^{1/2}t] \right\} e^{-npt}$$

and verify that the period T, defined as in Problem 15.12, is given by

$$T = \frac{2\pi}{p(1-n^2)^{1/2}}$$

Show that δ, the *logarithmic decrement*, defined as the natural logarithm of the ratio of $x(t)$ to $x(t+T)$, can be expressed as

$$\delta = npT$$

Taking $\dot{x}(0) = 0$ and $n = 0.2$, plot $x/x(0)$ versus pt for $0 \le pt \le 20$. Repeat with $n = 0.1$.

Results Figures P15.13(a), P15.13(b)

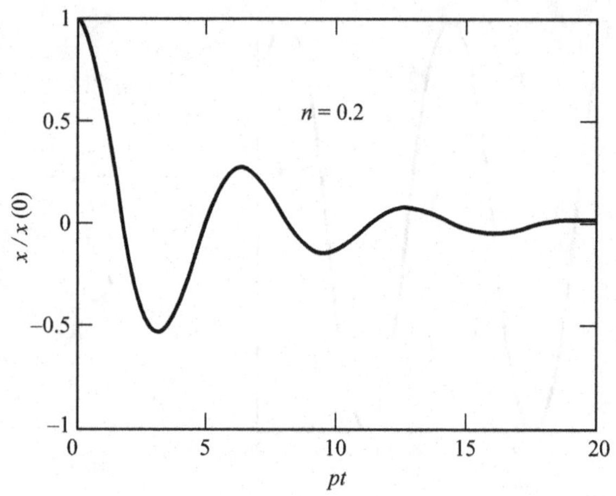

Figure P15.13(a)

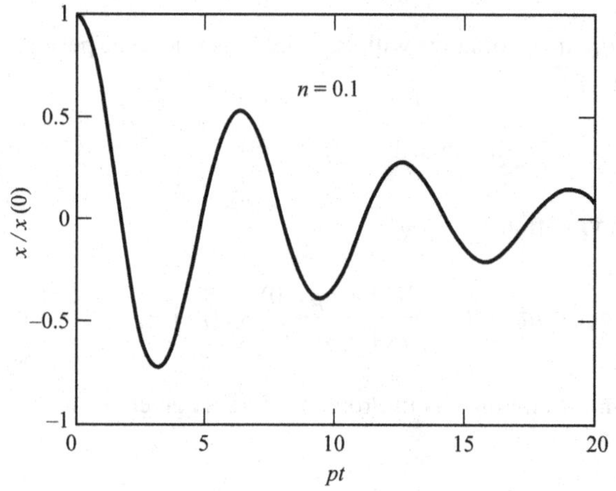

Figure P15.13(b)

15.14 A system moving in accordance with Eq. (9.9.1) is said to be performing *critically damped free vibrations* if

$$n = 1 \qquad f(t) = 0$$

Referring to Problem 15.13, use a limiting process to verify that

$$x = \{[\dot{x}(0) + px(0)]t + x(0)\}e^{-pt}$$

and show that there exists at most one value of t such that x has a stationary value.

Taking $\dot{x}(0) = 0$, plot $x/x(0)$ versus pt for $0 \le pt \le 20$.

Result Figure P15.14

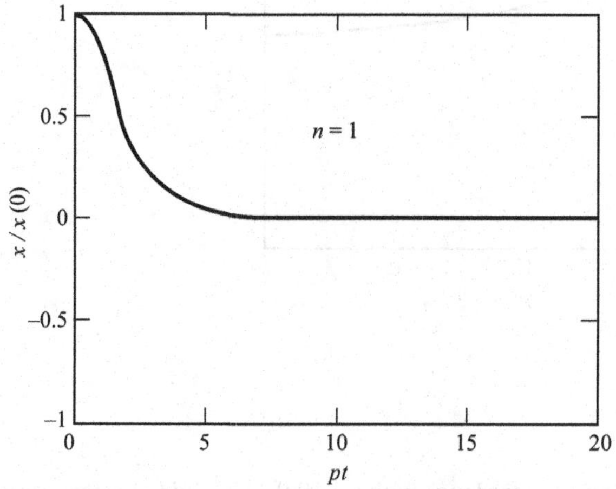

Figure P15.14

15.15 A system moving in accordance with Eq. (9.9.1) is said to be performing *over-damped free vibrations* if

$$n > 1 \qquad f(t) = 0$$

After verifying that x is given by

$$x = \frac{1}{a_1 - a_2}\{[\dot{x}(0) - a_2 x(0)]e^{a_1 t} - [\dot{x}(0) - a_1 x(0)]e^{a_2 t}\}$$

where a_1 and a_2 are *real* quantities given by Eqs. (9.9.3), show that, once again (see Problem 15.14), there exists at most one value of t such that x has a stationary value.

Taking $\dot{x}(0) = 0$ and $n = 3$, plot $x/x(0)$ versus pt for $0 \le pt \le 20$.

Result Figure P15.15

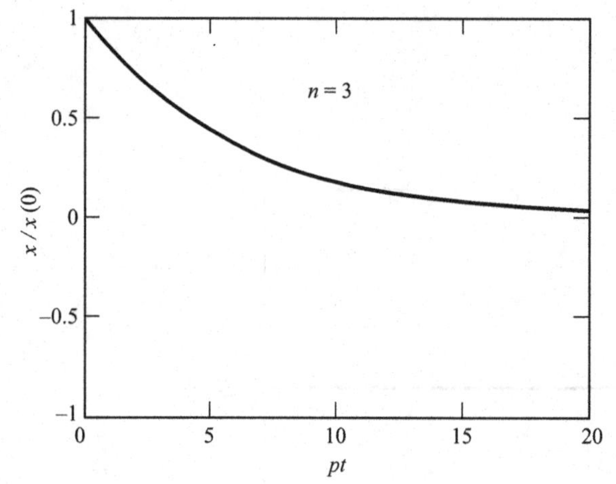

Figure P15.15

15.16 A system moving in accordance with Eq. (9.9.1) is said to be performing *damped harmonically forced vibrations* if

$$n > 0 \qquad f(t) = b\sin(\omega t + \phi)$$

where b, ω, and ϕ are constants called, respectively, the *amplitude*, the *circular frequency*, and the *phase angle* of the forcing function.

Show that x can be written as

$$x = x_S + x_T$$

where x_S and x_T, respectively called the *steady-state* response and the *transient* response, are given by

$$x_S = \frac{b(\beta_1 \sin\omega t + \beta_2 \cos\omega t)}{(p^2 - \omega^2)^2 + (2np\omega)^2}$$

and

$$x_T = \frac{\alpha_1 e^{a_1 t} - \alpha_2 e^{a_2 t}}{2p(n^2 - 1)^{1/2}}$$

with α_i and β_i $(i = 1, 2)$ defined as

$$\alpha_1 \triangleq \dot{x}(0) - a_2 x(0) + \frac{b}{a_1{}^2 + \omega^2}(a_1 \sin\phi + \omega\cos\phi)$$

$$\alpha_2 \triangleq \dot{x}(0) - a_1 x(0) + \frac{b}{a_2{}^2 + \omega^2}(a_2 \sin\phi + \omega\cos\phi)$$

$$\beta_1 \triangleq 2np\omega \sin\phi + (p^2 - \omega^2)\cos\phi$$

$$\beta_2 \triangleq (p^2 - \omega^2)\sin\phi - 2np\omega\cos\phi$$

and a_i $(i = 1,2)$ given by Eqs. (9.9.3). Verify that

$$\lim_{t\to\infty} x_T = 0$$

and that the absolute value of the maximum value of x_S, called the *steady-state amplitude*, is given by

$$|(x_S)_{\max}| = |b|[(p^2 - \omega^2)^2 + (2np\omega)^2]^{-1/2}$$

Taking $\dot{x}(0) = 0$, $n = 0.3$, $b = x(0)p^2$, $\phi = 0$, and $\omega = 2p$, plot $x/x(0)$ versus pt for $0 \le pt \le 20$. Repeat with $\omega = p$.

Results Figures P15.16(a), P15.16(b)

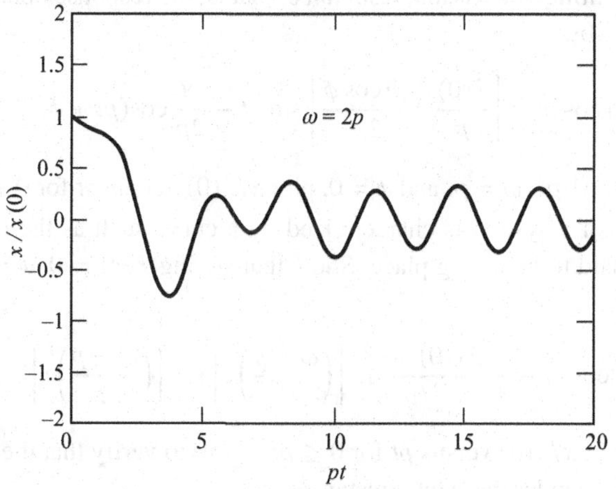

Figure P15.16(a)

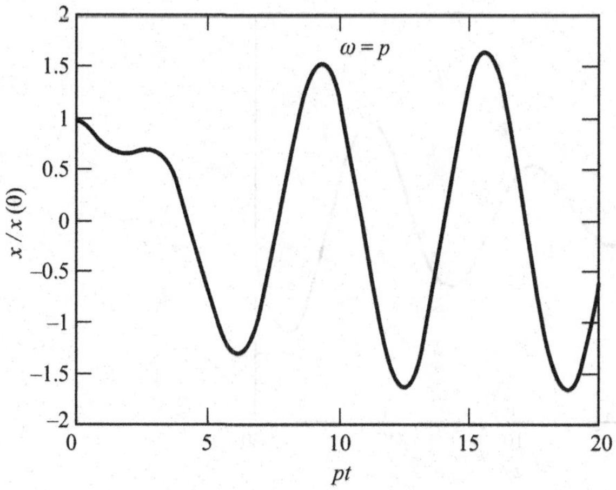

Figure P15.16(b)

15.17 A system moving in accordance with Eq. (9.9.1) is said to be performing *undamped harmonically forced vibrations* if

$$n = 0 \qquad f(t) = b \sin(\omega t + \phi)$$

where b, ω, and ϕ are constants (see Problem 15.16).

After verifying that, so long as ω differs from p, x is given by

$$x = x(0) \cos pt + \frac{\dot{x}(0)}{p} \sin pt$$
$$+ \frac{b}{p^2 - \omega^2} \left[\left(\sin \omega t - \frac{\omega}{p} \sin pt \right) \cos \phi + (\cos \omega t - \cos pt) \sin \phi \right]$$

show by means of a limiting process that *resonance*, that is, the response obtained when $\omega = p$, is characterized by

$$x = x(0) \cos pt + \left[\frac{\dot{x}(0)}{p} + \frac{b \cos \phi}{2p^2} \right] \sin pt - \frac{bt}{2p} \cos(pt + \phi)$$

Taking $\dot{x}(0) = 0$, $b = x(0)p^2$, $\omega = p$, and $\phi = 0$, plot $x/x(0)$ versus pt for $0 \le pt \le 20$.

When the behavior of a system is characterized by a curve such as the one in Fig. P15.17(*b*), *beats* are said to be taking place. Show that setting $\dot{x}(0) = 0$, $\phi = \pi/2$, and $b = x(0)p^2$ leads to

$$x = x(0) \cos pt - \frac{2x(0)}{1 - \omega^2/p^2} \sin \left[\left(\frac{\omega + p}{2} \right) t \right] \sin \left[\left(\frac{\omega - p}{2} \right) t \right]$$

Taking $\omega = 0.9p$, plot $x/x(0)$ versus pt for $0 \le pt \le 150$ to verify that the preceding equation describes beats under these circumstances.

Results Figures P15.17(*a*), P15.17(*b*)

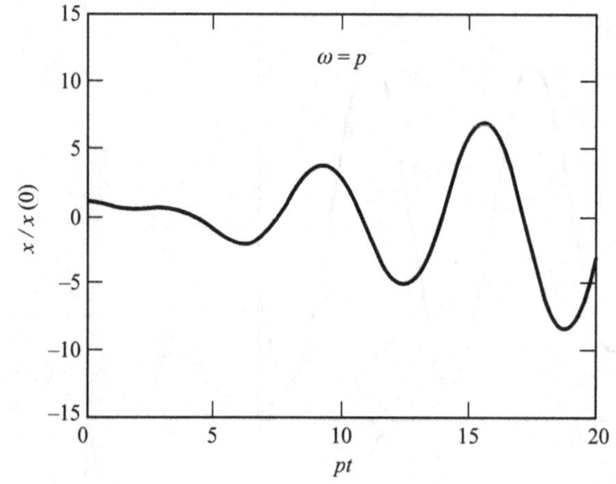

Figure P15.17(*a*)

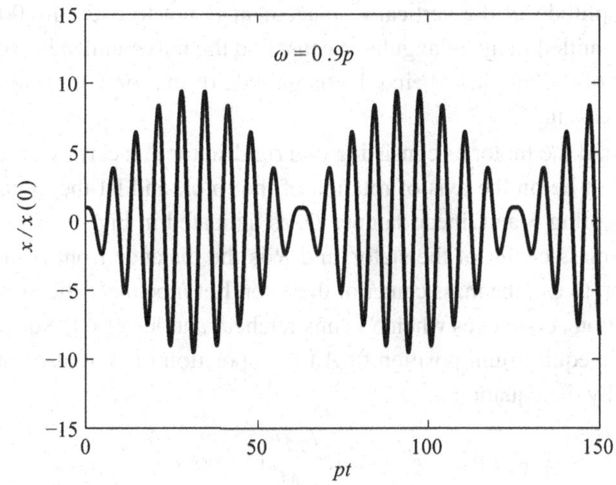

Figure P15.17(b)

15.18 A variable-speed motor is supported by a horizontal floor. When in operation, it performs vertical vibrations giving rise to the following observations.

When the rotor is turning at a constant angular speed, the amplitude of the vertical displacement of the motor casing from the casing's rest position is greatest if the angular speed is equal to 1000 rpm; the greatest amplitude has a value of 0.011 m. Once the motor has been turned off, so that the rotor remains at rest relative to the stator, vibrations with decaying amplitude take place with a frequency of 16 Hz.

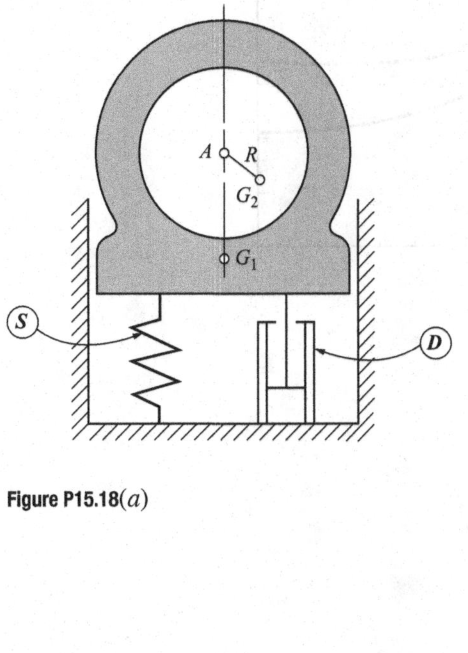

Figure P15.18(a)

Determine at what constant angular speeds below 3000 rpm the motor may not be operated if the amplitude of the vertical displacement is not to exceed 0.005 m. For operation in the permitted range of angular speeds, find the maximum value of the ratio of F to the weight of the motor, F being the magnitude of the force transmitted to the floor by the motor casing.

Suggestion: Regard the motor as consisting of a rigid stator that carries a rotor whose mass center does not lie on the axis of rotation of the rotor, and let the motor be supported by a linear spring S and linear damper D, as indicated in Fig. P15.18(a), where G_1 designates the mass center of the stator, and R is the distance from A, the axis of rotation of the rotor, to G_2, the mass center of the rotor. Let h be the static displacement of A from the position A occupies when S is unstretched, and let x be the displacement of A from the static equilibrium position of A to the position of A at any time. Verify that x is governed by the equation

$$\ddot{x} + 2np\dot{x} + p^2 x = -R\left(\frac{M'}{M}\right)\omega^2 \cos \omega t$$

where M and M' are the masses of the entire motor and of the rotor, respectively, ω is the (constant) angular speed of the rotor, $p^2 = g/h$, and $2np = C/M$, C being the damping constant associated with D. Assume that the transient response has vanished (see Problem 15.16), and plot the *displacement amplification factor* α, defined as $\alpha \triangleq |x_{\max}|/h$, and the *force amplification factor* β, defined as $\beta \triangleq F_{\max}/(Mg)$, as functions of ω/p for $0 \le \omega/p \le 3$ [see Fig. P15.18(b)].

Results $775 < \omega < 1731$ rpm; $[F/(Mg)]_{\max} = 7.1$

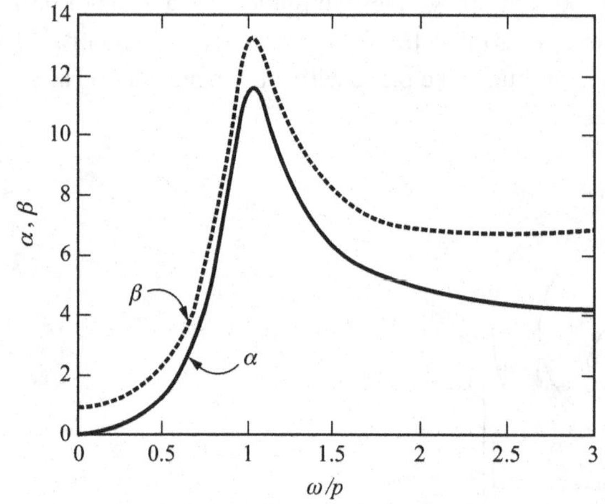

Figure P15.18(b)

***15.19** Referring to Problem 3.15 (see also Problems 4.19, 8.12, 8.18, 8.19, 11.10, 12.9, 12.10, 14.6, and 14.14), (*a*) let I denote the moment of inertia of A, B, C, and D with respect to the hub axis of A, let J be the moment of inertia of B, C, and D with respect to the joint axis at P, and take $M = m_C + m_D$. Show that (see Problem 12.9) $I = -X_{11}$, $J = -X_{22}$, and $M = -X_{33}$. (*b*) For the parameter values and values of $q_1(0)$, $q_2(0)$, and $q_3(0)$ used in Problem 14.14, determine I, J, and M. (*c*) Taking $T_1{}^{E/A}$ as in Problem 14.14 with $q_3{}^\star = 0$, and permitting only q_3 to vary, show that the equation of motion governing q_3, linearized in q_3, can be written

$$\ddot{x} + 2np\dot{x} + p^2 x = 0$$

where $x = q_3$, $2np = \beta^{E/A}/I$, and $p^2 = \gamma^{E/A}/I$; proceeding similarly in connection with q_1 and q_2, show that this equation applies also when $x = q_1$, $2np = \beta^{A/B}/J$, and $p^2 = \gamma^{A/B}/J$, in the case of q_1, and when $x = q_2$, $2np = \beta^{B/C}/M$, and $p^2 = \gamma^{B/C}/M$, in the case of q_2. (*d*) Referring to Problem 15.13, show that

$$n = \left[1 + \left(\frac{2\pi}{\delta}\right)^2\right]^{-1/2}$$

and determine δ and n such that $x(t+T)/x(t) = 1/100$. (*e*) Taking $T = 10$ s, determine $\beta^{E/A}$, $\gamma^{E/A}$, $\beta^{A/B}$, $\gamma^{A/B}$, $\beta^{B/C}$, and $\gamma^{B/C}$, each to the nearest whole number. (*f*) Using the parameters and initial conditions employed in part (*c*) of Problem 14.14, but assigning to $\beta^{E/A}$, $\gamma^{E/A}$, $\beta^{A/B}$, $\gamma^{A/B}$, $\beta^{B/C}$, and $\gamma^{B/C}$ the values found in part (*e*) of the present problem, plot the values of q_1, q_2, and q_3 from $t = 0$ to $t = 15$ s. (*g*) Recompute, to the nearest whole number, the values of $\beta^{A/B}$, $\gamma^{A/B}$, and $\beta^{B/C}$ with $T = 5$ s. Compare the values of $\beta^{E/A}$, $\gamma^{E/A}$, $\beta^{A/B}$, $\gamma^{A/B}$, $\beta^{B/C}$, and $\gamma^{B/C}$ now in hand with their counterparts in part (*c*) of Problem 14.14. Compare Figs. P14.14(*a*)–P14.14(*c*) with Figs. P15.19(*a*)–P15.19(*c*). What do you conclude?

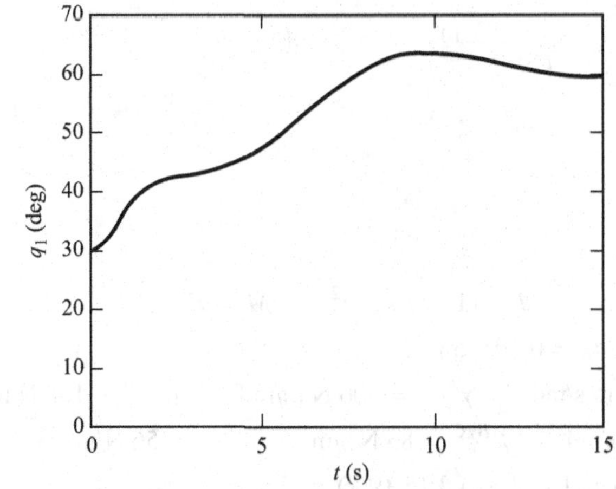

Figure P15.19(*a*)

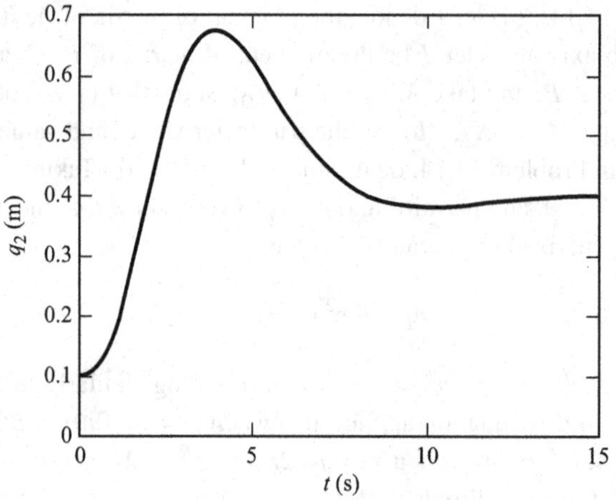

Figure P15.19(b)

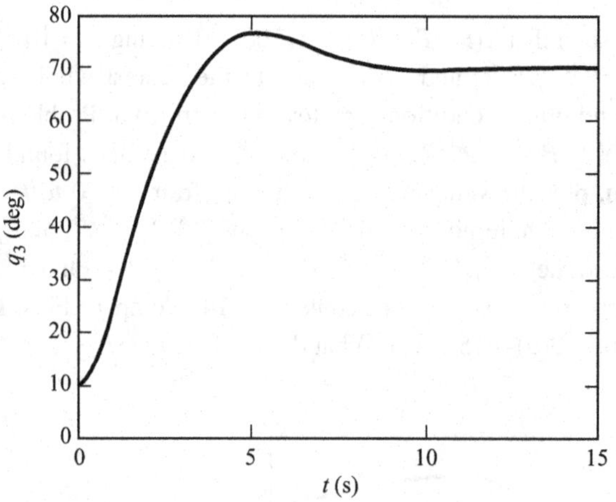

Figure P15.19(c)

Results

(b) $I = 503.61$ kg m^2 $J = 117.37$ kg m^2 $M = 92$ kg

(d) $\delta = 4.60517$ $n = 0.591155$

(e) $\beta^{E/A} = 464$ N m s/rad $\gamma^{E/A} = 306$ N m/rad $\beta^{A/B} = 108$ N m s/rad
$\gamma^{A/B} = 71$ N m/rad $\beta^{B/C} = 85$ N s/m $\gamma^{B/C} = 56$ N/m

(f) Figures P15.19(a), P15.19(b), P15.19(c)

(g) $\beta^{A/B} = 216$ N m s/rad $\gamma^{A/B} = 285$ N m/rad $\beta^{B/C} = 169$ N s/m

15.20 When a force $\mu kL\sin(\varepsilon\omega t)\hat{\mathbf{n}}_1$ is applied to the particle P_1 of the system considered in the example in Sec. 9.9, the differential equations governing the motion of S are Eqs. (9.9.51), with M and K as given in Eqs. (9.9.59) and (9.9.61), respectively, and

$$F = \mu kL\ [\sin(\varepsilon\omega t)\ \ 0\ \ 0\ \ 0\ \ 0\ \ 0\ \ 0\ \ 0\ \ 0\ \ 0\ \ 0\ \ 0]^T$$

Taking $q_r(0) = \dot{q}_r(0) = 0$ $(r = 1,\ldots,12)$, $\mu = 0.01$, and $\varepsilon = 0.1$, plot q_1/L versus ωt for $0 \le \omega t \le 60$, using 1, 3, and 12 modes. Next, using 12 modes, plot q_1/L versus ωt for $0 \le \omega t \le 60$ with $\mu = 0.01$ and $\varepsilon = 0.2586$. Comment briefly on the difference between the two curves for q_1/L versus ωt obtained with 12 modes for $\varepsilon = 0.1$ and $\varepsilon = 0.2586$.

Results Figures P15.20(a), P15.20(b)

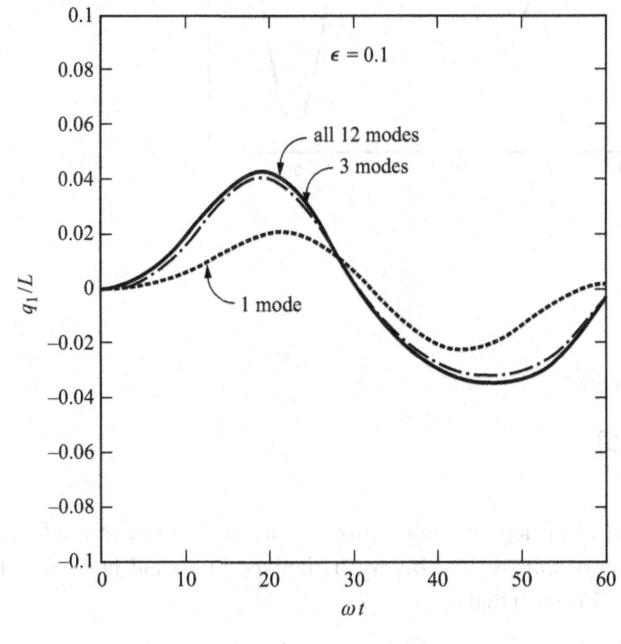

Figure P15.20(a)

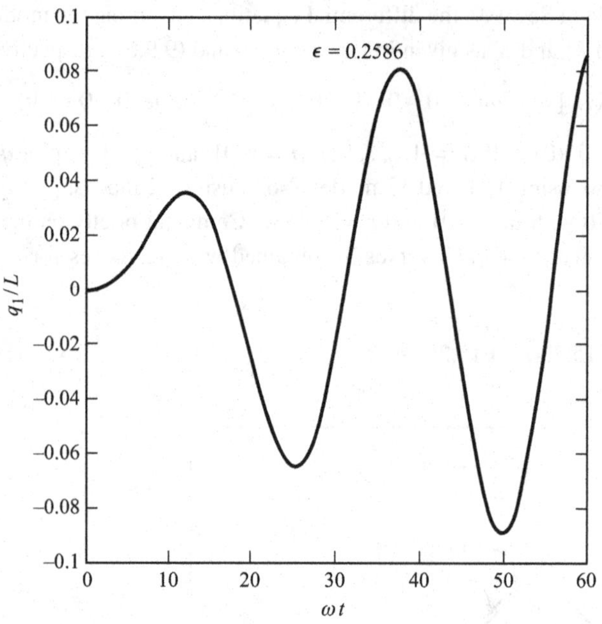

Figure P15.20(b)

PROBLEM SET 16
(Secs. 10.1–10.5)

16.1 A dextral set of orthogonal unit vectors $\hat{\mathbf{a}}_1$, $\hat{\mathbf{a}}_2$, $\hat{\mathbf{a}}_3$ is fixed in a reference frame A, and a similar set of unit vectors $\hat{\mathbf{b}}_1$, $\hat{\mathbf{b}}_2$, $\hat{\mathbf{b}}_3$ is fixed in a rigid body B. Initially the orientation of B in A is such that

$$\lfloor \hat{\mathbf{b}}_1 \quad \hat{\mathbf{b}}_2 \quad \hat{\mathbf{b}}_3 \rfloor = \lfloor \hat{\mathbf{a}}_1 \quad \hat{\mathbf{a}}_2 \quad \hat{\mathbf{a}}_3 \rfloor M$$

where

$$M = \begin{bmatrix} 0.9363 & -0.2896 & 0.1987 \\ 0.3130 & 0.9447 & -0.0981 \\ -0.1593 & 0.1540 & 0.9751 \end{bmatrix}$$

Body B is then subjected to a 30° rotation about $\hat{\boldsymbol{\lambda}}$, relative to A, where

$$\hat{\boldsymbol{\lambda}} = \frac{2\hat{\mathbf{a}}_1 + 3\hat{\mathbf{a}}_2 + 6\hat{\mathbf{a}}_3}{7}$$

Find the matrix N such that, subsequent to the rotation,

$$\lfloor \hat{\mathbf{b}}_1 \quad \hat{\mathbf{b}}_2 \quad \hat{\mathbf{b}}_3 \rfloor = \lfloor \hat{\mathbf{a}}_1 \quad \hat{\mathbf{a}}_2 \quad \hat{\mathbf{a}}_3 \rfloor N$$

Result

$$N = \begin{bmatrix} 0.6527 & -0.6053 & 0.4556 \\ 0.7103 & 0.6981 & -0.0903 \\ -0.2634 & 0.3825 & 0.8856 \end{bmatrix}$$

16.2 By definition, the eigenvalues of a direction cosine matrix C are values of a scalar quantity μ that satisfy the *characteristic equation*

$$|C - \mu U| = 0$$

Show that this equation can be expressed as

$$(1 - \mu)(1 + \mu B + \mu^2) = 0$$

where

$$B = 1 - (C_{11} + C_{22} + C_{33})$$

16.3 In Fig. P16.3, $\hat{\mathbf{a}}_1, \hat{\mathbf{a}}_2, \hat{\mathbf{a}}_3$ and $\hat{\mathbf{b}}_1, \hat{\mathbf{b}}_2, \hat{\mathbf{b}}_3$ are dextral sets of orthogonal unit vectors, with $\hat{\mathbf{a}}_1$ parallel to the line connecting a particle $\overline{P}$ and the mass center $B^\star$ of a rigid body B, and $\hat{\mathbf{b}}_1, \hat{\mathbf{b}}_2, \hat{\mathbf{b}}_3$ each parallel to a principal axis of inertia of B for $B^\star$.

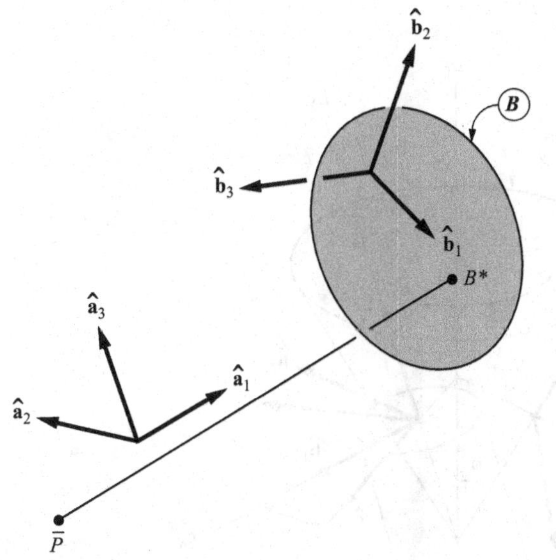

Figure P16.3

The system of gravitational forces exerted by $\overline{P}$ on B produces a moment $\mathbf{M}$ about $B^\star$. If the distance R between $\overline{P}$ and $B^\star$ exceeds the greatest distance from $B^\star$ to any point of B, $\mathbf{M}$ can be expressed approximately as

$$\mathbf{M} \approx \widetilde{\mathbf{M}} \triangleq 3G\overline{m}R^{-3}\, \hat{\mathbf{a}}_1 \times \underline{\mathbf{I}} \cdot \hat{\mathbf{a}}_1$$

where G is the universal gravitational constant, $\overline{m}$ is the mass of $\overline{P}$, and $\underline{\mathbf{I}}$ is the inertia dyadic of B for $B^\star$. $\tilde{\mathbf{M}}$ thus depends on the orientation of $\hat{\mathbf{a}}_1$ relative to $\hat{\mathbf{b}}_1$, $\hat{\mathbf{b}}_2$, $\hat{\mathbf{b}}_3$.

Letting ϵ_1, ϵ_2, ϵ_3, ϵ_4 be Euler parameters characterizing the relative orientation of $\hat{\mathbf{a}}_1$, $\hat{\mathbf{a}}_2$, $\hat{\mathbf{a}}_3$ and $\hat{\mathbf{b}}_1$, $\hat{\mathbf{b}}_2$, $\hat{\mathbf{b}}_3$, express $\tilde{\mathbf{M}}$ in terms of these parameters, the unit vectors $\hat{\mathbf{b}}_1$, $\hat{\mathbf{b}}_2$, $\hat{\mathbf{b}}_3$, and the principal moments of inertia I_1, I_2, I_3 of B for $B^\star$.

Result

$$\begin{aligned}\tilde{\mathbf{M}} = 6G\overline{m}R^{-3}[&2(\epsilon_1\epsilon_2 - \epsilon_3\epsilon_4)(\epsilon_3\epsilon_1 + \epsilon_2\epsilon_4)(I_3 - I_2)\hat{\mathbf{b}}_1 \\ &+ (\epsilon_3\epsilon_1 + \epsilon_2\epsilon_4)(1 - 2\epsilon_2{}^2 - 2\epsilon_3{}^2)(I_1 - I_3)\hat{\mathbf{b}}_2 \\ &+ (1 - 2\epsilon_2{}^2 - 2\epsilon_3{}^2)(\epsilon_1\epsilon_2 - \epsilon_3\epsilon_4)(I_2 - I_1)\hat{\mathbf{b}}_3]\end{aligned}$$

16.4 In Fig. P16.4, L_1 is the axis of a cone C of semivertex angle ϕ; L_2 is perpendicular to L_1 and fixed in C; L_3 is the axis of symmetry of a cylindrical body B that moves in such a way that L_3 always coincides with a generator of C; L_4 is perpendicular to L_1 and lies in the plane determined by L_1 and L_3; L_5 is perpendicular to L_3 and lies in the plane determined by L_1 and L_3; L_6 is perpendicular to L_3 and fixed in B; L_7 is a generator of C that intersects L_2; and L_8 is perpendicular to L_7 and lies in the plane determined by L_1 and L_7.

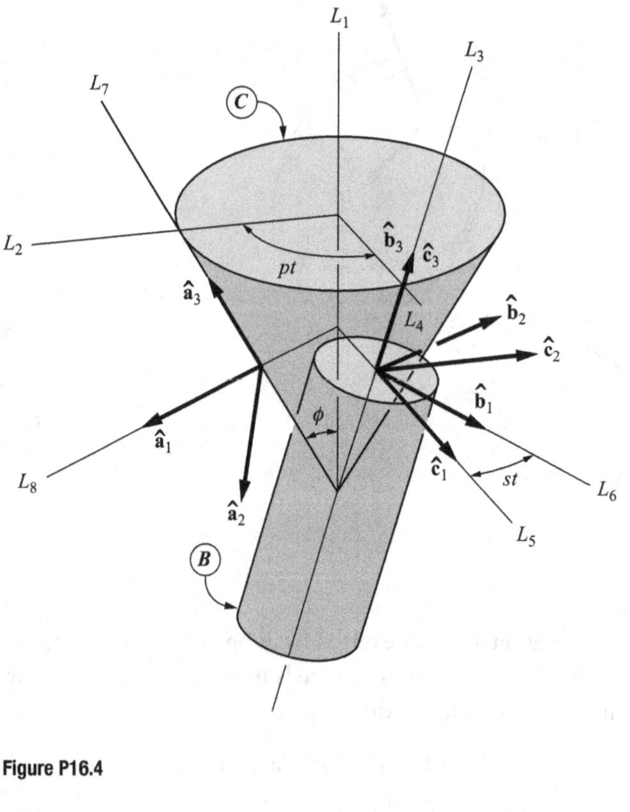

Figure P16.4

Assuming that B moves in such a way that the angle between L_2 and L_4 is equal to pt while the angle between L_5 and L_6 is equal to st, and letting $\hat{\mathbf{a}}_1, \hat{\mathbf{a}}_2, \hat{\mathbf{a}}_3$ and $\hat{\mathbf{b}}_1, \hat{\mathbf{b}}_2, \hat{\mathbf{b}}_3$ be dextral sets of orthogonal unit vectors directed as shown, express the Euler parameters $\epsilon_1, \ldots, \epsilon_4$ relating $\hat{\mathbf{a}}_1, \hat{\mathbf{a}}_2, \hat{\mathbf{a}}_3$ to $\hat{\mathbf{b}}_1, \hat{\mathbf{b}}_2, \hat{\mathbf{b}}_3$ as functions of ϕ, p, s, and t.

Suggestion: Introduce a set of unit vectors $\hat{\mathbf{c}}_1, \hat{\mathbf{c}}_2, \hat{\mathbf{c}}_3$ as shown, and use Eqs. (10.2.23)–(10.2.31) to generate a direction cosine matrix L such that

$$\lfloor \hat{\mathbf{c}}_1 \quad \hat{\mathbf{c}}_2 \quad \hat{\mathbf{c}}_3 \rfloor = \lfloor \hat{\mathbf{a}}_1 \quad \hat{\mathbf{a}}_2 \quad \hat{\mathbf{a}}_3 \rfloor L$$

Next, use Eq. (10.2.37) to find a matrix M such that

$$\lfloor \hat{\mathbf{b}}_1 \quad \hat{\mathbf{b}}_2 \quad \hat{\mathbf{b}}_3 \rfloor = \lfloor \hat{\mathbf{c}}_1 \quad \hat{\mathbf{c}}_2 \quad \hat{\mathbf{c}}_3 \rfloor M$$

Then

$$\lfloor \hat{\mathbf{b}}_1 \quad \hat{\mathbf{b}}_2 \quad \hat{\mathbf{b}}_3 \rfloor = \lfloor \hat{\mathbf{a}}_1 \quad \hat{\mathbf{a}}_2 \quad \hat{\mathbf{a}}_3 \rfloor N$$

where

$$N = LM$$

and the elements of N can be used in Eqs. (10.4.19)–(10.4.22) to find $\epsilon_1, \ldots, \epsilon_4$.

Result

$$\epsilon_1 = -\sin\phi \sin\frac{pt}{2}\cos\frac{st}{2}$$

$$\epsilon_2 = \sin\phi\sin\frac{pt}{2}\sin\frac{st}{2}$$

$$\epsilon_3 = \cos\phi\sin\frac{pt}{2}\cos\frac{st}{2} + \cos\frac{pt}{2}\sin\frac{st}{2}$$

$$\epsilon_4 = -\cos\phi\sin\frac{pt}{2}\sin\frac{st}{2} + \cos\frac{pt}{2}\cos\frac{st}{2}$$

16.5 Dextral sets of orthogonal unit vectors $\hat{\mathbf{a}}_1, \hat{\mathbf{a}}_2, \hat{\mathbf{a}}_3$ and $\hat{\mathbf{b}}_1, \hat{\mathbf{b}}_2, \hat{\mathbf{b}}_3$ are fixed in bodies A and B, respectively. At $t = 0$, $\hat{\mathbf{a}}_i = \hat{\mathbf{b}}_i$ $(i = 1, 2, 3)$. B then moves relative to A in such a way that the first time-derivatives of $\hat{\mathbf{b}}_1$ and $\hat{\mathbf{b}}_2$ in A are given by

$$\frac{{}^A d\hat{\mathbf{b}}_1}{dt} = -\cos pt\,\hat{\mathbf{b}}_3 \qquad \frac{{}^A d\hat{\mathbf{b}}_2}{dt} = \sin pt\,\hat{\mathbf{b}}_3$$

During this motion, the Euler parameters associated with any relative orientation of $\hat{\mathbf{a}}_1$, $\hat{\mathbf{a}}_2, \hat{\mathbf{a}}_3$ and $\hat{\mathbf{b}}_1, \hat{\mathbf{b}}_2, \hat{\mathbf{b}}_3$ must satisfy a set of differential equations given by Eqs. (10.8.8),

$$\lfloor \dot{\epsilon}_1 \quad \dot{\epsilon}_2 \quad \dot{\epsilon}_3 \quad \dot{\epsilon}_4 \rfloor = \tfrac{1}{2}\lfloor \omega_1 \quad \omega_2 \quad \omega_3 \quad 0 \rfloor E^T$$

where ω_1, ω_2, and ω_3 are in this case functions of pt, and where E^T is the transpose of a 4×4 matrix E whose elements depend solely on $\epsilon_1, \ldots, \epsilon_4$.

Find the set of differential equations governing the Euler parameters.

Suggestion: Use the given expressions for ${}^A d\hat{\mathbf{b}}_1/dt$ and ${}^A d\hat{\mathbf{b}}_2/dt$ together with Eqs. (10.6.4) and (2.1.1) to find $\omega_1, \omega_2, \omega_3$.

Result

$$\begin{aligned}
2\dot{\epsilon}_1 &= -\epsilon_3 \cos pt + \epsilon_4 \sin pt \\
2\dot{\epsilon}_2 &= \epsilon_3 \sin pt + \epsilon_4 \cos pt \\
2\dot{\epsilon}_3 &= \epsilon_1 \cos pt - \epsilon_2 \sin pt \\
2\dot{\epsilon}_4 &= -\epsilon_1 \sin pt - \epsilon_2 \cos pt
\end{aligned}$$

16.6 Letting $\theta_1, \theta_2, \theta_3$ be a set of body-two, 1-2-1 angles used to describe the orientation of a rigid body, express the associated Euler parameters in terms of these angles.

Result

$$\begin{aligned}
\epsilon_1 &= \cos(\theta_2/2)\sin[(\theta_1 + \theta_3)/2] \\
\epsilon_2 &= \sin(\theta_2/2)\cos[(\theta_1 - \theta_3)/2] \\
\epsilon_3 &= \sin(\theta_2/2)\sin[(\theta_1 - \theta_3)/2] \\
\epsilon_4 &= \cos(\theta_2/2)\cos[(\theta_1 + \theta_3)/2]
\end{aligned}$$

16.7 A dextral set of orthogonal unit vectors $\hat{\mathbf{a}}_1, \hat{\mathbf{a}}_2, \hat{\mathbf{a}}_3$ is fixed in a reference frame A; $\hat{\mathbf{b}}_1, \hat{\mathbf{b}}_2, \hat{\mathbf{b}}_3$ is a similar set fixed in a rigid body B; and $\hat{\mathbf{c}}_1, \hat{\mathbf{c}}_2, \hat{\mathbf{c}}_3$ is a third such set fixed in a rigid body C. Initially $\hat{\mathbf{a}}_i = \hat{\mathbf{b}}_i = \hat{\mathbf{c}}_i$ $(i = 1,2,3)$, and B is then subjected to successive rotations $\theta_1\hat{\mathbf{a}}_1, \theta_2\hat{\mathbf{a}}_2, \theta_3\hat{\mathbf{a}}_3$ while C undergoes successive rotations $\theta_3\hat{\mathbf{c}}_3, \theta_2\hat{\mathbf{c}}_2, \theta_1\hat{\mathbf{c}}_1$.

Show that $\hat{\mathbf{b}}_i = \hat{\mathbf{c}}_i$ $(i = 1,2,3)$ when all rotations have been completed.

16.8 At a certain instant the angles ϕ, θ, and ψ shown in Fig. P16.8 have the values $\phi = 63.03°$, $\theta = 20.56°$, and $\psi = -55.55°$. The angles are then changed in such a way that the rotor B experiences a 30° rotation about $\hat{\boldsymbol{\lambda}}$, relative to the frame A, where

$$\hat{\boldsymbol{\lambda}} = \frac{2\hat{\mathbf{a}}_1 + 3\hat{\mathbf{a}}_2 + 6\hat{\mathbf{a}}_3}{7}$$

Determine the changes $\Delta\phi$, $\Delta\theta$, and $\Delta\psi$ in the angles ϕ, θ, and ψ.

Suggestion: Use the results of Problem 16.1 after verifying that, when ϕ, θ, and ψ have the given values,

$$\lfloor \hat{\mathbf{b}}_1 \quad \hat{\mathbf{b}}_2 \quad \hat{\mathbf{b}}_3 \rfloor = \lfloor \hat{\mathbf{a}}_1 \quad \hat{\mathbf{a}}_2 \quad \hat{\mathbf{a}}_3 \rfloor M$$

with M as in Problem 16.1

Results $\Delta\phi = 6.62°$, $\Delta\theta = 28.69°$, $\Delta\psi = 2.52°$

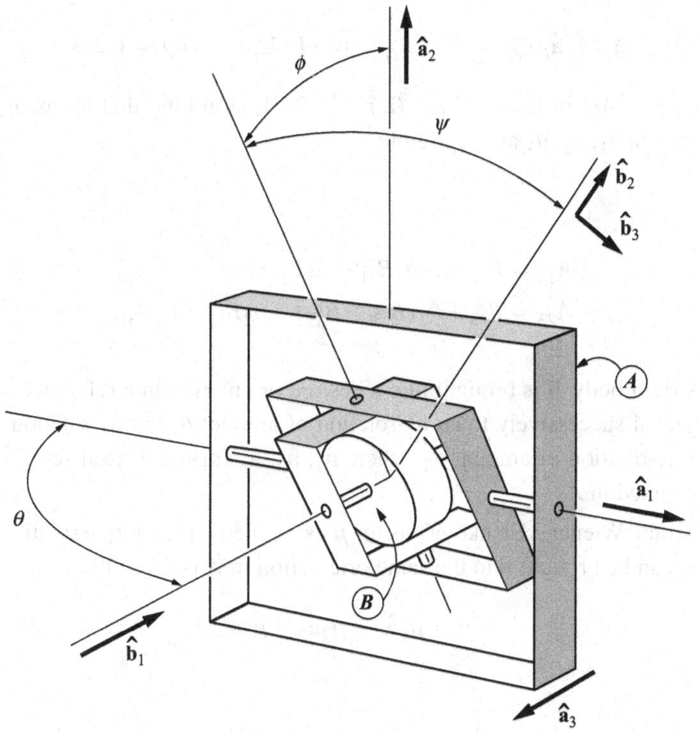

Figure P16.8

16.9 Dextral sets of orthogonal unit vectors $\hat{\mathbf{a}}_1$, $\hat{\mathbf{a}}_2$, $\hat{\mathbf{a}}_3$ and $\hat{\mathbf{b}}_1$, $\hat{\mathbf{b}}_2$, $\hat{\mathbf{b}}_3$ are fixed in a reference frame A and in a rigid body B, respectively. Initially $\hat{\mathbf{a}}_i = \hat{\mathbf{b}}_i$ $(i = 1, 2, 3)$, and B is then subjected to successive rotations $\theta_1 \hat{\mathbf{a}}_2$, $\theta_2 \hat{\mathbf{a}}_3$, and $\theta_3 \hat{\mathbf{a}}_2$.

Letting $\mathrm{s}_1 \triangleq \sin\theta_1$, $\mathrm{c}_1 \triangleq \cos\theta_1$, etc., determine M such that subsequent to the last rotation

$$\lfloor \hat{\mathbf{b}}_1 \quad \hat{\mathbf{b}}_2 \quad \hat{\mathbf{b}}_3 \rfloor = \lfloor \hat{\mathbf{a}}_1 \quad \hat{\mathbf{a}}_2 \quad \hat{\mathbf{a}}_3 \rfloor M$$

Result

$$M = \begin{bmatrix} \mathrm{c}_1\,\mathrm{c}_2\,\mathrm{c}_3 - \mathrm{s}_3\,\mathrm{s}_1 & -\mathrm{s}_2\,\mathrm{c}_3 & \mathrm{s}_1\,\mathrm{c}_2\,\mathrm{c}_3 + \mathrm{s}_3\,\mathrm{c}_1 \\ \mathrm{c}_1\,\mathrm{s}_2 & \mathrm{c}_2 & \mathrm{s}_1\,\mathrm{s}_2 \\ -\mathrm{c}_1\,\mathrm{c}_2\,\mathrm{s}_3 - \mathrm{c}_3\,\mathrm{s}_1 & \mathrm{s}_2\,\mathrm{s}_3 & -\mathrm{s}_1\,\mathrm{c}_2\,\mathrm{s}_3 + \mathrm{c}_3\,\mathrm{c}_1 \end{bmatrix}$$

16.10 Two rigid bodies, A and B, are attached to each other at a point P, and dextral sets of orthogonal unit vectors $\hat{\mathbf{a}}_1$, $\hat{\mathbf{a}}_2$, $\hat{\mathbf{a}}_3$ and $\hat{\mathbf{b}}_1$, $\hat{\mathbf{b}}_2$, $\hat{\mathbf{b}}_3$ are fixed in A and B, respectively. Initially $\hat{\mathbf{a}}_i = \hat{\mathbf{b}}_i$ $(i = 1, 2, 3)$, and B is then subjected to successive rotations $\theta_1 \hat{\mathbf{b}}_1$, $\theta_2 \hat{\mathbf{b}}_2$, $\theta_3 \hat{\mathbf{b}}_3$.

Letting $\underline{\mathbf{I}}$ denote the inertia dyadic of B for P, and defining A_{ij} and B_{ij} as

$$A_{ij} \triangleq \hat{\mathbf{a}}_i \cdot \underline{\mathbf{I}} \cdot \hat{\mathbf{a}}_j \qquad B_{ij} \triangleq \hat{\mathbf{b}}_i \cdot \underline{\mathbf{I}} \cdot \hat{\mathbf{b}}_j \qquad (i,j = 1,2,3)$$

express A_{11} and A_{23} in terms of B_{ij} $(i,j = 1,2,3)$, assuming that terms of second or higher degree in θ_1, θ_2, θ_3 are negligible.

Results

$$A_{11} = B_{11} + 2(\theta_2 B_{31} - \theta_3 B_{12})$$
$$A_{23} = B_{23} + \theta_1(B_{22} - B_{33}) - \theta_2 B_{12} + \theta_3 B_{31}$$

***16.11** A rigid body B is brought into a desired orientation in a reference frame A by being subjected successively to an $\hat{\mathbf{a}}_1$-rotation of amount θ_1, an $\hat{\mathbf{a}}_2$-rotation of amount θ_2, and an $\hat{\mathbf{a}}_3$-rotation of amount θ_3, where $\hat{\mathbf{a}}_1$, $\hat{\mathbf{a}}_2$, $\hat{\mathbf{a}}_3$ form a dextral set of orthogonal unit vectors fixed in A.

Show that the Wiener-Milenković vector $\boldsymbol{\mu}$ associated with a single rotation by means of which B can be brought into the same orientation in A is given by

$$\boldsymbol{\mu} = \mu_1 \hat{\mathbf{a}}_1 + \mu_2 \hat{\mathbf{a}}_2 + \mu_3 \hat{\mathbf{a}}_3$$

where

$$\mu_1 = \frac{8t_2t_3t_1^2 + \left(t_2^2 - 16\right)\left(t_3^2 - 16\right)t_1 - 128t_2t_3}{(t_2^2 + t_3^2)\,t_1^2 + 16t_2t_3t_1 + t_2^2t_3^2 + 256}$$

$$\mu_2 = \frac{-8t_1t_3t_2^2 + \left(t_1^2 - 16\right)\left(t_3^2 - 16\right)t_2 + 128t_1t_3}{(t_2^2 + t_3^2)\,t_1^2 + 16t_2t_3t_1 + t_2^2t_3^2 + 256}$$

$$\mu_3 = \frac{8t_1t_2t_3^2 + \left(t_1^2 - 16\right)\left(t_2^2 - 16\right)t_3 - 128t_1t_2}{(t_2^2 + t_3^2)\,t_1^2 + 16t_2t_3t_1 + t_2^2t_3^2 + 256}$$

and $t_i \triangleq 4\tan(\theta_i/4)$ $(i = 1,2,3)$.

16.12 The orientation of a rigid body B in a reference frame A is described by angles belonging to a body-three, 1-2-3 rotation sequence, given by, respectively, $\theta_1 = \Omega_1 t$, $\theta_2 = \Omega_2 t$, and $\theta_3 = \Omega_3 t$, where Ω_i $(i = 1,2,3)$ are constants. For $\Omega_i = i/10$ rad/s and $0 \le t \le 20$ s, (*a*) employ an expression involving orientation angles θ_i $(i = 1,2,3)$ to plot the direction cosine C_{31} versus t. (*b*) Plot time histories of the Wiener-Milenković parameters μ_i $(i = 1,2,3)$ that describe the orientation of B in A. (*c*) Use an expression involving Wiener-Milenković parameters to plot the time history of C_{31}, and verify that the results are the same as those obtained in part (*a*). (*d*) Plot $\mu\mu^T/16$ versus t, and discuss the behavior in the vicinity of the cusps near $t = 7.85$ s and $t = 19.26$ s.

Results Figures P16.12(*a*), P16.12(*b*)

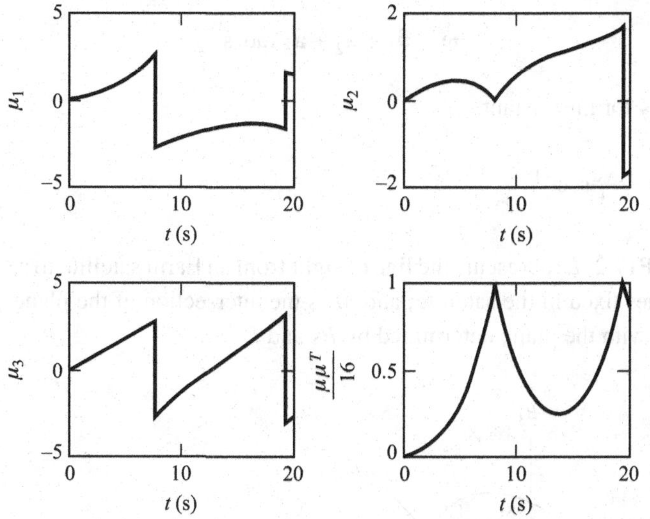

Figure P16.12(a)

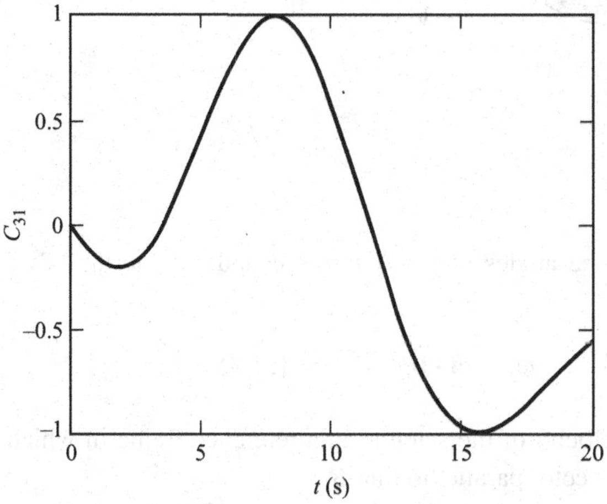

Figure P16.12(b)

PROBLEM SET 17
(Secs. 10.6–10.9)

17.1 Referring to Problem 16.8, and supposing that B moves in such a way that at a certain instant $\phi = \theta = \psi = \pi/4$ rad while the angular velocity $\boldsymbol{\omega}$ of B in A is given by

$$\boldsymbol{\omega} = \hat{\mathbf{a}}_1 + \hat{\mathbf{a}}_2 + \hat{\mathbf{a}}_3 \text{ rad/s}$$

determine $\dot{M}_{22}$ for this instant.

Result $\dot{M}_{22} = -\frac{\sqrt{2}}{4}\text{sec}^{-1}$

17.2 In Fig. P17.2, L represents the line of sight from an Earth satellite to a star; B_1, B_2, and B_3 are lines fixed in the satellite; and M is the intersection of the plane determined by B_1 and B_2 with the plane determined by B_3 and L.

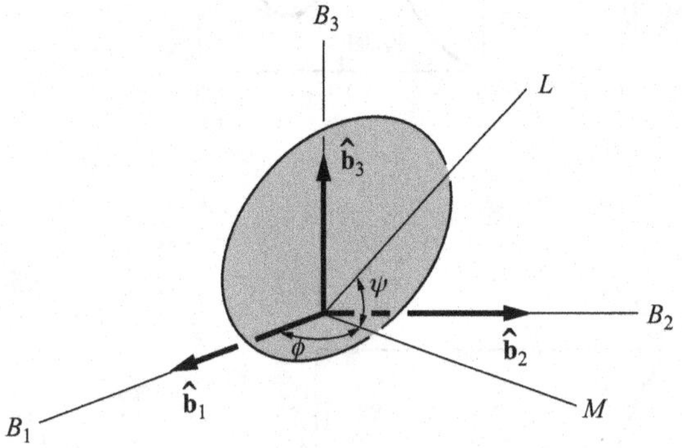

Figure P17.2

Letting ϕ and ψ be the angles shown, express $\dot{\phi}$ and $\dot{\psi}$ in terms of ϕ, ψ, and ω_i $(i = 1,2,3)$, where

$$\omega_i \triangleq \boldsymbol{\omega} \cdot \hat{\mathbf{b}}_i \qquad (i = 1,2,3)$$

and $\boldsymbol{\omega}$ is the angular velocity of the satellite in a reference frame in which L remains fixed, while $\hat{\mathbf{b}}_i$ is a unit vector parallel to line B_i.

Result

$$\dot{\phi} = (\omega_1 \cos\phi + \omega_2 \sin\phi)\tan\psi - \omega_3$$
$$\dot{\psi} = -\omega_1 \sin\phi + \omega_2 \cos\phi$$

17.3 Defining ω_i $(i = 1,2,3)$ and ϵ as in Eqs. (10.6.4) and (10.8.5), respectively, show that Eq. (10.8.8) is equivalent to

$$\dot{\epsilon} = \epsilon\Omega$$

where

$$\Omega = \frac{1}{2}\begin{bmatrix} 0 & -\omega_3 & \omega_2 & -\omega_1 \\ \omega_3 & 0 & -\omega_1 & -\omega_2 \\ -\omega_2 & \omega_1 & 0 & -\omega_3 \\ \omega_1 & \omega_2 & \omega_3 & 0 \end{bmatrix}$$

17.4 The angular velocity $\boldsymbol{\omega}$ of an axisymmetric rigid body B in an inertial reference frame A, when B is subjected to the action of a body-fixed, transverse torque of constant magnitude T, can be expressed as†

$$\boldsymbol{\omega} = \omega_1 \hat{\mathbf{b}}_1 + \omega_2 \hat{\mathbf{b}}_2 + \omega_3 \hat{\mathbf{b}}_3$$

and

$$\omega_1 = \overline{\omega}_1 \cos rt + (\overline{\omega}_2 + \mu) \sin rt$$

$$\omega_2 = -\mu - \overline{\omega}_1 \sin rt + (\overline{\omega}_2 + \mu) \cos rt$$

$$\omega_3 = \overline{\omega}_3$$

where $\hat{\mathbf{b}}_1$, $\hat{\mathbf{b}}_2$, $\hat{\mathbf{b}}_3$ are mutually perpendicular unit vectors fixed in B and parallel to central principal axes of inertia of B; $\hat{\mathbf{b}}_3$ is parallel to the symmetry axis of B; $\hat{\mathbf{b}}_1$ is parallel to the applied torque; $\overline{\omega}_1$, $\overline{\omega}_2$, $\overline{\omega}_3$ are the initial values of ω_1, ω_2, ω_3; r and μ are defined as

$$r \triangleq \overline{\omega}_3 \left(1 - \frac{J}{I}\right) \qquad \mu \triangleq \frac{T}{\overline{\omega}_3 (I - J)}$$

and I and J are, respectively, the transverse and the axial moment of inertia of B.

Letting ϕ be the angle between the symmetry axis of B and the line fixed in A with which the symmetry axis coincides initially, determine ϕ for $t = 1, 2, \ldots, 10$ s if $\overline{\omega}_1 = \overline{\omega}_2 = 1.0$ rad/s, $\overline{\omega}_3 = 1.5$ rad/s, $I = 60$ kg m^2, $J = 40$ kg m^2, and $T = 1.5$ N m.

Result Table P17.4

Table P17.4

t (s)	1	2	3	4	5	6	7	8	9	10
ϕ (deg)	77.5	108.0	46.7	37.3	105.2	82.9	3.6	76.1	108.3	51.2

17.5 The mass center $B^\star$ of a rigid body B moves in a circular orbit fixed in a reference frame C and centered at a point P, as shown in Fig. P17.5, where $\hat{\mathbf{a}}_1$, $\hat{\mathbf{a}}_2$, $\hat{\mathbf{a}}_3$ are mutually perpendicular unit vectors, $\hat{\mathbf{a}}_1$ being parallel to line $PB^\star$, $\hat{\mathbf{a}}_2$ pointing in the direction of motion of $B^\star$ in C, and $\hat{\mathbf{a}}_3$ thus being normal to the orbit plane. If θ_1, θ_2, θ_3 are space-three, 1-2-3 angles governing the orientation of B in a reference frame A in which $\hat{\mathbf{a}}_1$, $\hat{\mathbf{a}}_2$,

† Eugene Leimanis, *The General Problem of the Motion of Coupled Rigid Bodies about a Fixed Point* (New York: Springer-Verlag, 1965), p. 138.

$\hat{\mathbf{a}}_3$ are fixed, the angular velocity of B in C can be expressed as $\boldsymbol{\omega} = \omega_1\hat{\mathbf{a}}_1 + \omega_2\hat{\mathbf{a}}_2 + \omega_3\hat{\mathbf{a}}_3$; and $\dot{\theta}_i$ $(i = 1,2,3)$ can be expressed as a function f_i of θ_1, θ_2, θ_3, ω_1, ω_2, ω_3, and Ω, the angular speed of line $PB^\star$ in C.

Determine f_1, f_2, f_3, using the abbreviations $s_i \triangleq \sin\theta_i$, $c_i \triangleq \cos\theta_i$ $(i = 1,2,3)$.

Result

$$f_1 = (c_3\omega_1 + s_3\omega_2)/c_2$$
$$f_2 = -s_3\omega_1 + c_3\omega_2$$
$$f_3 = (c_3\omega_1 + s_3\omega_2)(s_2/c_2) + \omega_3 - \Omega$$

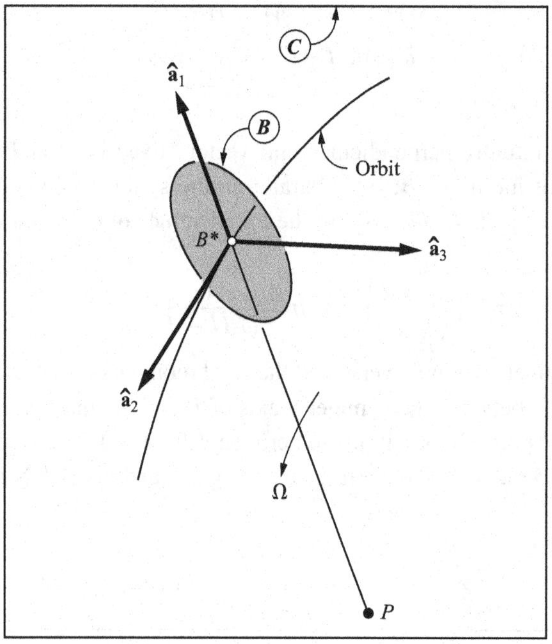

Figure P17.5

17.6 Derive Eq. (10.7.5) by using Eq. (2.4.1), selecting auxiliary reference frames such that each term in Eq. (2.4.1) represents the angular velocity of a body performing a motion of simple rotation and can, therefore, be expressed as in Eqs. (2.2.1) and (2.2.2). Noting that a similar derivation can be used to obtain Eq. (10.7.9), comment on the feasibility of using this approach to derive Eqs. (10.7.3) and (10.7.7).

17.7 The orientation of a rigid body B in a reference frame A is described in terms of the angles θ_1, θ_2, θ_3 of Problem 16.9. Letting $\omega_i \triangleq \boldsymbol{\omega} \cdot \hat{\mathbf{b}}_i$ $(i = 1,2,3)$, where $\boldsymbol{\omega}$ is the angular velocity of B in A, find the matrix L such that

$$\lfloor \omega_1 \quad \omega_2 \quad \omega_3 \rfloor = \lfloor \dot{\theta}_1 \quad \dot{\theta}_2 \quad \dot{\theta}_3 \rfloor L$$

Result

$$L = \begin{bmatrix} 0 & 1 & 0 \\ -s_1 & 0 & c_1 \\ c_1 s_2 & c_2 & s_1 s_2 \end{bmatrix}$$

17.8 Referring to Problem 16.12, obtain the time histories of $\mu_i(t)$ for $0 \le t \le 20$ s by numerical integration of the kinematical differential equations governing μ_i, using the initial values $\mu_i(0) = 0$ $(i = 1,2,3)$. Without scaling the Wiener-Milenković parameters, plot μ_1, μ_2, μ_3, and $\mu\mu^T/16$ versus t, and comment on differences between these results and those in Fig. P16.12(a).

Suggestion: Obtain the $\hat{\mathbf{b}}_1$, $\hat{\mathbf{b}}_2$, $\hat{\mathbf{b}}_3$ measure numbers of the angular velocity of B in A with the aid of Eqs. (10.7.1), after noting that $\dot{\theta}_i = \Omega_i$ $(i = 1,2,3)$.

Results Figure P17.8

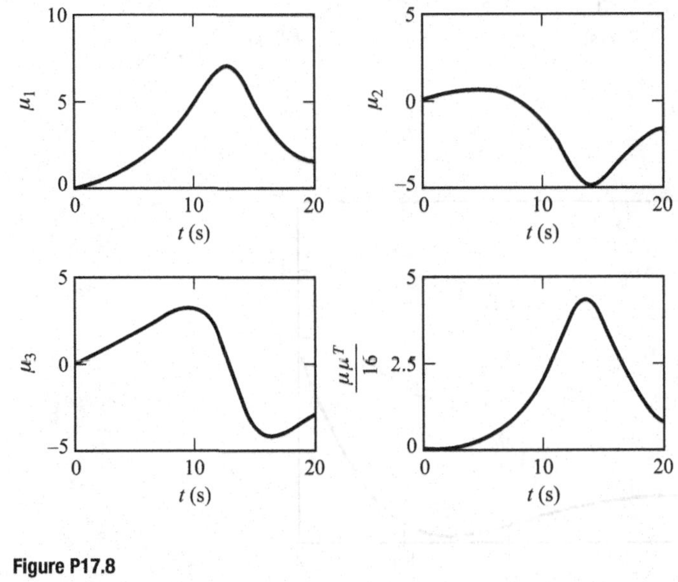

Figure P17.8

17.9 Consider a rigid body B rotating relative to a reference frame A, with angular velocity measure numbers $\omega_i = \boldsymbol{\omega} \cdot \hat{\mathbf{b}}_i = t^{i-1}/i$ rad/s $(i = 1,2,3)$. Let θ_i $(i = 1,2,3)$ be a set of body-three, 1-2-3 angles that describe the orientation of B in A. With initial values $\theta_1(0) = -93.2117°$, $\theta_2(0) = 25.2336°$, and $\theta_3(0) = 70.1541°$, obtain time histories of the angles by numerical integration of the governing ordinary differential equations. Plot θ_i $(i = 1,2,3)$, $\tan\theta_2(t)$, and $C_{31}(t)$ for $0 \le t \le 2$ s. Show that when

$\theta_2 = 90°$, the orientation is governed by only one angle, $\theta_1 + \theta_3$, and comment on the effects of this observation on the results.

Results Figures P17.9(a), P17.9(b). Note that $\tan\theta_2(t)$ becomes infinite near $t = 1.1$ s.

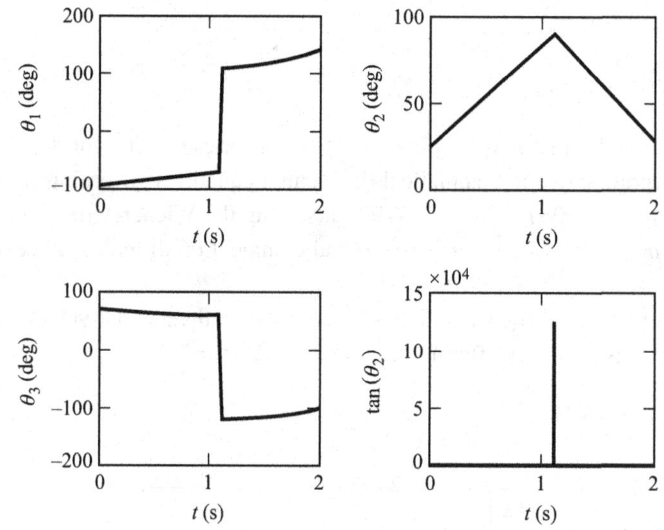

Figure P17.9(a)

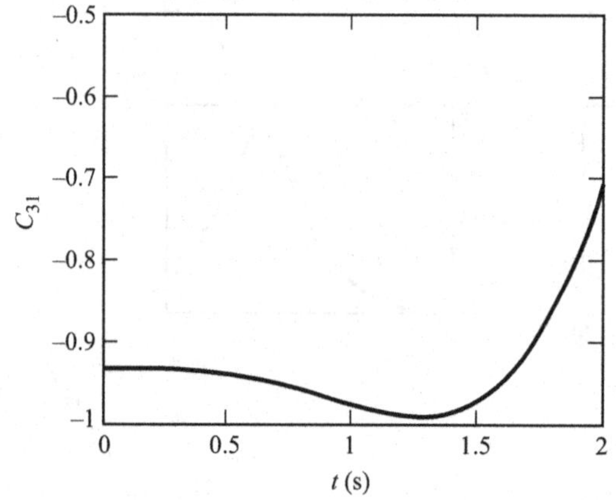

Figure P17.9(b)

17.10 Consider a rigid body B rotating relative to a reference frame A, with angular velocity measure numbers $\omega_i = \boldsymbol{\omega} \cdot \hat{\mathbf{b}}_i = t^{i-1}/i$ rad/s $(i = 1,2,3)$. Let μ_i $(i = 1,2,3)$ represent the Wiener-Milenković parameters that describe the orientation of B in A.

(*a*) Using the values of the angles at time $t = 0$ in Problem 17.9 together with Eqs. (10.5.9)–(10.5.11), determine the values of $\mu_i(0)$ $(i = 1,2,3)$. (*b*) Using the initial values of $\mu_i(0)$ $(i = 1,2,3)$ found in part (*a*), obtain time histories of μ_i through numerical solution of the kinematical differential equations for $0 \leq t \leq 5$ s. Plot the values of the Wiener-Milenković parameters without bothering to scale them, including $\mu\mu^T/16$. (*c*) Again obtain time histories of μ_i through numerical solution of the kinematical differential equations for $0 \leq t \leq 5$ s, this time using the scaling law [see Eq. (10.5.12)], and plot both scaled and unscaled values of each of the Wiener-Milenković parameters, including $\mu\mu^T/16$. Hint: You may have to write a special-purpose time-marching code in order to handle the scaling. (*d*) Verify that the result for $C_{31}(t)$ is the same as that obtained in Problem 17.9.

Results (*a*) $\mu_1(0) = -1.20515$, $\mu_2(0) = 1.29354$, $\mu_3(0) = 0.622935$ (*b*) Fig. P17.10, solid curves (*c*) Fig. P17.10, dots (*d*) Fig. P17.9(*b*)

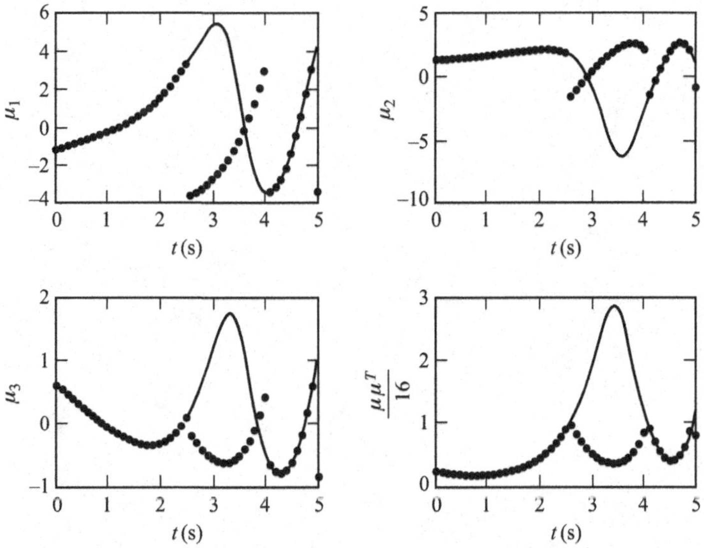

Figure P17.10

[illegible]

[illegible]

Appendix I DIRECTION COSINES AS FUNCTIONS OF ORIENTATION ANGLES

In this appendix, the direction cosines associated with each of 24 sets of angles describing the orientation of rigid body B in a reference frame A (see Sec. 10.3) are tabulated. To use the tables, proceed as follows: let $\hat{\mathbf{a}}_1$, $\hat{\mathbf{a}}_2$, $\hat{\mathbf{a}}_3$ be a dextral set of mutually perpendicular unit vectors fixed in the reference frame A, and let $\hat{\mathbf{b}}_1$, $\hat{\mathbf{b}}_2$, $\hat{\mathbf{b}}_3$ be a similar such set fixed in the body B. Regard $\hat{\mathbf{b}}_i$ as initially aligned with $\hat{\mathbf{a}}_i$ $(i = 1,2,3)$; select the type of rotation sequence of interest (that is, body-three, body-two, space-three, or space-two); letting θ_1, θ_2, θ_3 denote the amounts of the first, the second, and the third rotation, respectively, pick the rotation sequence of interest [for example, 3-1-2 (corresponding to a $\theta_1\hat{\mathbf{b}}_3$, $\theta_2\hat{\mathbf{b}}_1$, $\theta_3\hat{\mathbf{b}}_2$ body-three sequence or a $\theta_1\hat{\mathbf{a}}_3$, $\theta_2\hat{\mathbf{a}}_1$, $\theta_3\hat{\mathbf{a}}_2$ space-three sequence)]; finally, locate the table corresponding to the rotation sequence chosen. The nine entries in the table [with s_i and c_i standing, respectively, for $\sin\theta_i$ and $\cos\theta_i$ $(i = 1,2,3)$] are the elements C_{ij} of the associated direction cosine matrix, which elements are defined as $C_{ij} \triangleq \hat{\mathbf{a}}_i \cdot \hat{\mathbf{b}}_j$ $(i,j = 1,2,3)$. Moreover, by reading a row or column of the table, one can determine how to express any of $\hat{\mathbf{a}}_1$, $\hat{\mathbf{a}}_2$, $\hat{\mathbf{a}}_3$ in terms of $\hat{\mathbf{b}}_1$, $\hat{\mathbf{b}}_2$, $\hat{\mathbf{b}}_3$, or any of $\hat{\mathbf{b}}_1$, $\hat{\mathbf{b}}_2$, $\hat{\mathbf{b}}_3$ in terms of $\hat{\mathbf{a}}_1$, $\hat{\mathbf{a}}_2$, $\hat{\mathbf{a}}_3$. For example, in the table corresponding to the body-three, 3-1-2 sequence, the third row reveals that $\hat{\mathbf{a}}_3 = -\mathrm{c}_2\mathrm{s}_3\hat{\mathbf{b}}_1 + \mathrm{s}_2\hat{\mathbf{b}}_2 + \mathrm{c}_2\mathrm{c}_3\hat{\mathbf{b}}_3$, while the second column indicates that $\hat{\mathbf{b}}_2 = -\mathrm{s}_1\mathrm{c}_2\hat{\mathbf{a}}_1 + \mathrm{c}_1\mathrm{c}_2\hat{\mathbf{a}}_2 + \mathrm{s}_2\hat{\mathbf{a}}_3$.

Body-three, 1-2-3

	$\hat{\mathbf{b}}_1$	$\hat{\mathbf{b}}_2$	$\hat{\mathbf{b}}_3$
$\hat{\mathbf{a}}_1$	c_2c_3	$-c_2s_3$	s_2
$\hat{\mathbf{a}}_2$	$s_1s_2c_3 + s_3c_1$	$-s_1s_2s_3 + c_3c_1$	$-s_1c_2$
$\hat{\mathbf{a}}_3$	$-c_1s_2c_3 + s_3s_1$	$c_1s_2s_3 + c_3s_1$	c_1c_2

Body-three, 2-3-1

	$\hat{\mathbf{b}}_1$	$\hat{\mathbf{b}}_2$	$\hat{\mathbf{b}}_3$
$\hat{\mathbf{a}}_1$	c_1c_2	$-c_1s_2c_3 + s_3s_1$	$c_1s_2s_3 + c_3s_1$
$\hat{\mathbf{a}}_2$	s_2	c_2c_3	$-c_2s_3$
$\hat{\mathbf{a}}_3$	$-s_1c_2$	$s_1s_2c_3 + s_3c_1$	$-s_1s_2s_3 + c_3c_1$

Body-three, 3-1-2

	$\hat{\mathbf{b}}_1$	$\hat{\mathbf{b}}_2$	$\hat{\mathbf{b}}_3$
$\hat{\mathbf{a}}_1$	$-s_1s_2s_3 + c_3c_1$	$-s_1c_2$	$s_1s_2c_3 + s_3c_1$
$\hat{\mathbf{a}}_2$	$c_1s_2s_3 + c_3s_1$	c_1c_2	$-c_1s_2c_3 + s_3s_1$
$\hat{\mathbf{a}}_3$	$-c_2s_3$	s_2	c_2c_3

Body-three, 1-3-2

	$\hat{\mathbf{b}}_1$	$\hat{\mathbf{b}}_2$	$\hat{\mathbf{b}}_3$
$\hat{\mathbf{a}}_1$	c_2c_3	$-s_2$	c_2s_3
$\hat{\mathbf{a}}_2$	$c_1s_2c_3 + s_3s_1$	c_1c_2	$c_1s_2s_3 - c_3s_1$
$\hat{\mathbf{a}}_3$	$s_1s_2c_3 - s_3c_1$	s_1c_2	$s_1s_2s_3 + c_3c_1$

Body-three, 2-1-3

	$\hat{\mathbf{b}}_1$	$\hat{\mathbf{b}}_2$	$\hat{\mathbf{b}}_3$
$\hat{\mathbf{a}}_1$	$s_1s_2s_3 + c_3c_1$	$s_1s_2c_3 - s_3c_1$	s_1c_2
$\hat{\mathbf{a}}_2$	c_2s_3	c_2c_3	$-s_2$
$\hat{\mathbf{a}}_3$	$c_1s_2s_3 - c_3s_1$	$c_1s_2c_3 + s_3s_1$	c_1c_2

Body-three, 3-2-1

	$\hat{\mathbf{b}}_1$	$\hat{\mathbf{b}}_2$	$\hat{\mathbf{b}}_3$
$\hat{\mathbf{a}}_1$	c_1c_2	$c_1s_2s_3 - c_3s_1$	$c_1s_2c_3 + s_3s_1$
$\hat{\mathbf{a}}_2$	s_1c_2	$s_1s_2s_3 + c_3c_1$	$s_1s_2c_3 - s_3c_1$
$\hat{\mathbf{a}}_3$	$-s_2$	c_2s_3	c_2c_3

Body-two, 1-2-1

	$\hat{\mathbf{b}}_1$	$\hat{\mathbf{b}}_2$	$\hat{\mathbf{b}}_3$
$\hat{\mathbf{a}}_1$	c_2	s_2s_3	s_2c_3
$\hat{\mathbf{a}}_2$	s_1s_2	$-s_1c_2s_3 + c_3c_1$	$-s_1c_2c_3 - s_3c_1$
$\hat{\mathbf{a}}_3$	$-c_1s_2$	$c_1c_2s_3 + c_3s_1$	$c_1c_2c_3 - s_3s_1$

Body-two, 1-3-1

	$\hat{\mathbf{b}}_1$	$\hat{\mathbf{b}}_2$	$\hat{\mathbf{b}}_3$
$\hat{\mathbf{a}}_1$	c_2	$-s_2c_3$	s_2s_3
$\hat{\mathbf{a}}_2$	c_1s_2	$c_1c_2c_3 - s_3s_1$	$-c_1c_2s_3 - c_3s_1$
$\hat{\mathbf{a}}_3$	s_1s_2	$s_1c_2c_3 + s_3c_1$	$-s_1c_2s_3 + c_3c_1$

Body-two, 2-1-2

	$\hat{\mathbf{b}}_1$	$\hat{\mathbf{b}}_2$	$\hat{\mathbf{b}}_3$
$\hat{\mathbf{a}}_1$	$-s_1c_2s_3 + c_3c_1$	s_1s_2	$s_1c_2c_3 + s_3c_1$
$\hat{\mathbf{a}}_2$	s_2s_3	c_2	$-s_2c_3$
$\hat{\mathbf{a}}_3$	$-c_1c_2s_3 - c_3s_1$	c_1s_2	$c_1c_2c_3 - s_3s_1$

Body-two, 2-3-2

	$\hat{\mathbf{b}}_1$	$\hat{\mathbf{b}}_2$	$\hat{\mathbf{b}}_3$
$\hat{\mathbf{a}}_1$	$c_1c_2c_3 - s_3s_1$	$-c_1s_2$	$c_1c_2s_3 + c_3s_1$
$\hat{\mathbf{a}}_2$	s_2c_3	c_2	s_2s_3
$\hat{\mathbf{a}}_3$	$-s_1c_2c_3 - s_3c_1$	s_1s_2	$-s_1c_2s_3 + c_3c_1$

Body-two, 3-1-3

	$\hat{\mathbf{b}}_1$	$\hat{\mathbf{b}}_2$	$\hat{\mathbf{b}}_3$
$\hat{\mathbf{a}}_1$	$-s_1c_2s_3 + c_3c_1$	$-s_1c_2c_3 - s_3c_1$	s_1s_2
$\hat{\mathbf{a}}_2$	$c_1c_2s_3 + c_3s_1$	$c_1c_2c_3 - s_3s_1$	$-c_1s_2$
$\hat{\mathbf{a}}_3$	s_2s_3	s_2c_3	c_2

Body-two, 3-2-3

	$\hat{\mathbf{b}}_1$	$\hat{\mathbf{b}}_2$	$\hat{\mathbf{b}}_3$
$\hat{\mathbf{a}}_1$	$c_1c_2c_3 - s_3s_1$	$-c_1c_2s_3 - c_3s_1$	c_1s_2
$\hat{\mathbf{a}}_2$	$s_1c_2c_3 + s_3c_1$	$-s_1c_2s_3 + c_3c_1$	s_1s_2
$\hat{\mathbf{a}}_3$	$-s_2c_3$	s_2s_3	c_2

Space-three, 1-2-3

	$\hat{\mathbf{b}}_1$	$\hat{\mathbf{b}}_2$	$\hat{\mathbf{b}}_3$
$\hat{\mathbf{a}}_1$	c_2c_3	$s_1s_2c_3 - s_3c_1$	$c_1s_2c_3 + s_3s_1$
$\hat{\mathbf{a}}_2$	c_2s_3	$s_1s_2s_3 + c_3c_1$	$c_1s_2s_3 - c_3s_1$
$\hat{\mathbf{a}}_3$	$-s_2$	s_1c_2	c_1c_2

Space-three, 2-3-1

	$\hat{\mathbf{b}}_1$	$\hat{\mathbf{b}}_2$	$\hat{\mathbf{b}}_3$
$\hat{\mathbf{a}}_1$	c_1c_2	$-s_2$	s_1c_2
$\hat{\mathbf{a}}_2$	$c_1s_2c_3 + s_3s_1$	c_2c_3	$s_1s_2c_3 - s_3c_1$
$\hat{\mathbf{a}}_3$	$c_1s_2s_3 - c_3s_1$	c_2s_3	$s_1s_2s_3 + c_3c_1$

Space-three, 3-1-2

	$\hat{\mathbf{b}}_1$	$\hat{\mathbf{b}}_2$	$\hat{\mathbf{b}}_3$
$\hat{\mathbf{a}}_1$	$s_1s_2s_3 + c_3c_1$	$c_1s_2s_3 - c_3s_1$	c_2s_3
$\hat{\mathbf{a}}_2$	s_1c_2	c_1c_2	$-s_2$
$\hat{\mathbf{a}}_3$	$s_1s_2c_3 - s_3c_1$	$c_1s_2c_3 + s_3s_1$	c_2c_3

Space-three, 1-3-2

	$\hat{\mathbf{b}}_1$	$\hat{\mathbf{b}}_2$	$\hat{\mathbf{b}}_3$
$\hat{\mathbf{a}}_1$	c_2c_3	$-c_1s_2c_3 + s_3s_1$	$s_1s_2c_3 + s_3c_1$
$\hat{\mathbf{a}}_2$	s_2	c_1c_2	$-s_1c_2$
$\hat{\mathbf{a}}_3$	$-c_2s_3$	$c_1s_2s_3 + c_3s_1$	$-s_1s_2s_3 + c_3c_1$

Space-three, 2-1-3

	$\hat{\mathbf{b}}_1$	$\hat{\mathbf{b}}_2$	$\hat{\mathbf{b}}_3$
$\hat{\mathbf{a}}_1$	$-s_1s_2s_3 + c_3c_1$	$-c_2s_3$	$c_1s_2s_3 + c_3s_1$
$\hat{\mathbf{a}}_2$	$s_1s_2c_3 + s_3c_1$	c_2c_3	$-c_1s_2c_3 + s_3s_1$
$\hat{\mathbf{a}}_3$	$-s_1c_2$	s_2	c_1c_2

Space-three, 3-2-1

	$\hat{\mathbf{b}}_1$	$\hat{\mathbf{b}}_2$	$\hat{\mathbf{b}}_3$
$\hat{\mathbf{a}}_1$	c_1c_2	$-s_1c_2$	s_2
$\hat{\mathbf{a}}_2$	$c_1s_2s_3 + c_3s_1$	$-s_1s_2s_3 + c_3c_1$	$-c_2s_3$
$\hat{\mathbf{a}}_3$	$-c_1s_2c_3 + s_3s_1$	$s_1s_2c_3 + s_3c_1$	c_2c_3

Space-two, 1-2-1

	$\hat{\mathbf{b}}_1$	$\hat{\mathbf{b}}_2$	$\hat{\mathbf{b}}_3$
$\hat{\mathbf{a}}_1$	c_2	$s_1 s_2$	$c_1 s_2$
$\hat{\mathbf{a}}_2$	$s_2 s_3$	$-s_1 c_2 s_3 + c_3 c_1$	$-c_1 c_2 s_3 - c_3 s_1$
$\hat{\mathbf{a}}_3$	$-s_2 c_3$	$s_1 c_2 c_3 + s_3 c_1$	$c_1 c_2 c_3 - s_3 s_1$

Space-two, 1-3-1

	$\hat{\mathbf{b}}_1$	$\hat{\mathbf{b}}_2$	$\hat{\mathbf{b}}_3$
$\hat{\mathbf{a}}_1$	c_2	$-c_1 s_2$	$s_1 s_2$
$\hat{\mathbf{a}}_2$	$s_2 c_3$	$c_1 c_2 c_3 - s_3 s_1$	$-s_1 c_2 c_3 - s_3 c_1$
$\hat{\mathbf{a}}_3$	$s_2 s_3$	$c_1 c_2 s_3 + c_3 s_1$	$-s_1 c_2 s_3 + c_3 c_1$

Space-two, 2-1-2

	$\hat{\mathbf{b}}_1$	$\hat{\mathbf{b}}_2$	$\hat{\mathbf{b}}_3$
$\hat{\mathbf{a}}_1$	$-s_1 c_2 s_3 + c_3 c_1$	$s_2 s_3$	$c_1 c_2 s_3 + c_3 s_1$
$\hat{\mathbf{a}}_2$	$s_1 s_2$	c_2	$-c_1 s_2$
$\hat{\mathbf{a}}_3$	$-s_1 c_2 c_3 - s_3 c_1$	$s_2 c_3$	$c_1 c_2 c_3 - s_3 s_1$

Space-two, 2-3-2

	$\hat{\mathbf{b}}_1$	$\hat{\mathbf{b}}_2$	$\hat{\mathbf{b}}_3$
$\hat{\mathbf{a}}_1$	$c_1 c_2 c_3 - s_3 s_1$	$-s_2 c_3$	$s_1 c_2 c_3 + s_3 c_1$
$\hat{\mathbf{a}}_2$	$c_1 s_2$	c_2	$s_1 s_2$
$\hat{\mathbf{a}}_3$	$-c_1 c_2 s_3 - c_3 s_1$	$s_2 s_3$	$-s_1 c_2 s_3 + c_3 c_1$

Space-two, 3-1-3

	$\hat{\mathbf{b}}_1$	$\hat{\mathbf{b}}_2$	$\hat{\mathbf{b}}_3$
$\hat{\mathbf{a}}_1$	$-s_1 c_2 s_3 + c_3 c_1$	$-c_1 c_2 s_3 - c_3 s_1$	$s_2 s_3$
$\hat{\mathbf{a}}_2$	$s_1 c_2 c_3 + s_3 c_1$	$c_1 c_2 c_3 - s_3 s_1$	$-s_2 c_3$
$\hat{\mathbf{a}}_3$	$s_1 s_2$	$c_1 s_2$	c_2

Space-two, 3-2-3

	$\hat{\mathbf{b}}_1$	$\hat{\mathbf{b}}_2$	$\hat{\mathbf{b}}_3$
$\hat{\mathbf{a}}_1$	$c_1 c_2 c_3 - s_3 s_1$	$-s_1 c_2 c_3 - s_3 c_1$	$s_2 c_3$
$\hat{\mathbf{a}}_2$	$c_1 c_2 s_3 + c_3 s_1$	$-s_1 c_2 s_3 + c_3 c_1$	$s_2 s_3$
$\hat{\mathbf{a}}_3$	$-c_1 s_2$	$s_1 s_2$	c_2

Appendix II KINEMATICAL DIFFERENTIAL EQUATIONS IN TERMS OF ORIENTATION ANGLES

In this appendix, the kinematical differential equations associated with each of 24 sets of angles describing the orientation of a rigid body B in a reference frame A (see Sec. 10.7) are tabulated. To use the tables, proceed as follows: let $\hat{\mathbf{a}}_1$, $\hat{\mathbf{a}}_2$, $\hat{\mathbf{a}}_3$ be a dextral set of mutually perpendicular unit vectors fixed in the reference frame A, and let $\hat{\mathbf{b}}_1$, $\hat{\mathbf{b}}_2$, $\hat{\mathbf{b}}_3$ be a similar such set fixed in the body B. Regard $\hat{\mathbf{b}}_i$, as initially aligned with $\hat{\mathbf{a}}_i$ $(i = 1,2,3)$; select the type of rotation sequence of interest (that is, body-three, body-two, space-three, or space-two); letting θ_1, θ_2, θ_3 denote the amounts (in radians) of the first, the second, and the third rotation, respectively, pick the rotation sequence of interest [for example, 3-1-2 (corresponding to a $\theta_1\hat{\mathbf{b}}_3$, $\theta_2\hat{\mathbf{b}}_1$, $\theta_3\hat{\mathbf{b}}_2$ body-three sequence or a $\theta_1\hat{\mathbf{a}}_3$, $\theta_2\hat{\mathbf{a}}_1$, $\theta_3\hat{\mathbf{a}}_2$ space-three sequence)]; finally, locate the table corresponding to the rotation sequence chosen. The table contains the relationships between $\dot{\theta}_1$, $\dot{\theta}_2$, $\dot{\theta}_3$ and ω_1, ω_2, ω_3, where $\omega_i \triangleq {}^A\boldsymbol{\omega}^B \cdot \hat{\mathbf{b}}_i$ $(i = 1,2,3)$, ${}^A\boldsymbol{\omega}^B$ being the angular velocity of B in A.

Body-three, 1-2-3

$\omega_1 = \dot{\theta}_1 c_2 c_3 + \dot{\theta}_2 s_3$	$\dot{\theta}_1 = (\omega_1 c_3 - \omega_2 s_3)/c_2$
$\omega_2 = -\dot{\theta}_1 c_2 s_3 + \dot{\theta}_2 c_3$	$\dot{\theta}_2 = \omega_1 s_3 + \omega_2 c_3$
$\omega_3 = \dot{\theta}_1 s_2 + \dot{\theta}_3$	$\dot{\theta}_3 = (-\omega_1 c_3 + \omega_2 s_3) s_2/c_2 + \omega_3$

Body-three, 2-3-1

$\omega_1 = \dot{\theta}_1 s_2 + \dot{\theta}_3$	$\dot{\theta}_1 = (\omega_2 c_3 - \omega_3 s_3)/c_2$
$\omega_2 = \dot{\theta}_1 c_2 c_3 + \dot{\theta}_2 s_3$	$\dot{\theta}_2 = \omega_2 s_3 + \omega_3 c_3$
$\omega_3 = -\dot{\theta}_1 c_2 s_3 + \dot{\theta}_2 c_3$	$\dot{\theta}_3 = \omega_1 + (-\omega_2 c_3 + \omega_3 s_3) s_2/c_2$

Body-three, 3-1-2

$\omega_1 = -\dot{\theta}_1 c_2 s_3 + \dot{\theta}_2 c_3$	$\dot{\theta}_1 = (-\omega_1 s_3 + \omega_3 c_3)/c_2$
$\omega_2 = \dot{\theta}_1 s_2 + \dot{\theta}_3$	$\dot{\theta}_2 = \omega_1 c_3 + \omega_3 s_3$
$\omega_3 = \dot{\theta}_1 c_2 c_3 + \dot{\theta}_2 s_3$	$\dot{\theta}_3 = (\omega_1 s_3 - \omega_3 c_3) s_2/c_2 + \omega_2$

Body-three, 1-3-2

$\omega_1 = \dot{\theta}_1 c_2 c_3 - \dot{\theta}_2 s_3$	$\dot{\theta}_1 = (\omega_1 c_3 + \omega_3 s_3)/c_2$
$\omega_2 = -\dot{\theta}_1 s_2 + \dot{\theta}_3$	$\dot{\theta}_2 = -\omega_1 s_3 + \omega_3 c_3$
$\omega_3 = \dot{\theta}_1 c_2 s_3 + \dot{\theta}_2 c_3$	$\dot{\theta}_3 = (\omega_1 c_3 + \omega_3 s_3) s_2/c_2 + \omega_2$

Body-three, 2-1-3

$\omega_1 = \dot{\theta}_1 c_2 s_3 + \dot{\theta}_2 c_3$	$\dot{\theta}_1 = (\omega_1 s_3 + \omega_2 c_3)/c_2$
$\omega_2 = \dot{\theta}_1 c_2 c_3 - \dot{\theta}_2 s_3$	$\dot{\theta}_2 = \omega_1 c_3 - \omega_2 s_3$
$\omega_3 = -\dot{\theta}_1 s_2 + \dot{\theta}_3$	$\dot{\theta}_3 = (\omega_1 s_3 + \omega_2 c_3) s_2/c_2 + \omega_3$

Body-three, 3-2-1

$\omega_1 = -\dot{\theta}_1 s_2 + \dot{\theta}_3$	$\dot{\theta}_1 = (\omega_2 s_3 + \omega_3 c_3)/c_2$
$\omega_2 = \dot{\theta}_1 c_2 s_3 + \dot{\theta}_2 c_3$	$\dot{\theta}_2 = \omega_2 c_3 - \omega_3 s_3$
$\omega_3 = \dot{\theta}_1 c_2 c_3 - \dot{\theta}_2 s_3$	$\dot{\theta}_3 = \omega_1 + (\omega_2 s_3 + \omega_3 c_3) s_2/c_2$

Body-two, 1-2-1

$\omega_1 = \dot{\theta}_1 c_2 + \dot{\theta}_3$	$\dot{\theta}_1 = (\omega_2 s_3 + \omega_3 c_3)/s_2$
$\omega_2 = \dot{\theta}_1 s_2 s_3 + \dot{\theta}_2 c_3$	$\dot{\theta}_2 = \omega_2 c_3 - \omega_3 s_3$
$\omega_3 = \dot{\theta}_1 s_2 c_3 - \dot{\theta}_2 s_3$	$\dot{\theta}_3 = \omega_1 - (\omega_2 s_3 + \omega_3 c_3) c_2/s_2$

Body-two, 1-3-1

$\omega_1 = \dot{\theta}_1 c_2 + \dot{\theta}_3$	$\dot{\theta}_1 = (-\omega_2 c_3 + \omega_3 s_3)/s_2$
$\omega_2 = -\dot{\theta}_1 s_2 c_3 + \dot{\theta}_2 s_3$	$\dot{\theta}_2 = \omega_2 s_3 + \omega_3 c_3$
$\omega_3 = \dot{\theta}_1 s_2 s_3 + \dot{\theta}_2 c_3$	$\dot{\theta}_3 = \omega_1 + (\omega_2 c_3 - \omega_3 s_3) c_2/s_2$

Body-two, 2-1-2

$\omega_1 = \dot{\theta}_1 s_2 s_3 + \dot{\theta}_2 c_3$	$\dot{\theta}_1 = (\omega_1 s_3 - \omega_3 c_3)/s_2$
$\omega_2 = \dot{\theta}_1 c_2 + \dot{\theta}_3$	$\dot{\theta}_2 = \omega_1 c_3 + \omega_3 s_3$
$\omega_3 = -\dot{\theta}_1 s_2 c_3 + \dot{\theta}_2 s_3$	$\dot{\theta}_3 = (-\omega_1 s_3 + \omega_3 c_3) c_2/s_2 + \omega_2$

Body-two, 2-3-2

$\omega_1 = \dot{\theta}_1 s_2 c_3 - \dot{\theta}_2 s_3$	$\dot{\theta}_1 = (\omega_1 c_3 + \omega_3 s_3)/s_2$
$\omega_2 = \dot{\theta}_1 c_2 + \dot{\theta}_3$	$\dot{\theta}_2 = -\omega_1 s_3 + \omega_3 c_3$
$\omega_3 = \dot{\theta}_1 s_2 s_3 + \dot{\theta}_2 c_3$	$\dot{\theta}_3 = -(\omega_1 c_3 + \omega_3 s_3) c_2/s_2 + \omega_2$

Body-two, 3-1-3

$\omega_1 = \dot{\theta}_1 s_2 s_3 + \dot{\theta}_2 c_3$	$\dot{\theta}_1 = (\omega_1 s_3 + \omega_2 c_3)/s_2$
$\omega_2 = \dot{\theta}_1 s_2 c_3 - \dot{\theta}_2 s_3$	$\dot{\theta}_2 = \omega_1 c_3 - \omega_2 s_3$
$\omega_3 = \dot{\theta}_1 c_2 + \dot{\theta}_3$	$\dot{\theta}_3 = -(\omega_1 s_3 + \omega_2 c_3) c_2/s_2 + \omega_3$

Body-two, 3-2-3

$\omega_1 = -\dot{\theta}_1 s_2 c_3 + \dot{\theta}_2 s_3$	$\dot{\theta}_1 = (-\omega_1 c_3 + \omega_2 s_3)/s_2$
$\omega_2 = \dot{\theta}_1 s_2 s_3 + \dot{\theta}_2 c_3$	$\dot{\theta}_2 = \omega_1 s_3 + \omega_2 c_3$
$\omega_3 = \dot{\theta}_1 c_2 + \dot{\theta}_3$	$\dot{\theta}_3 = (\omega_1 c_3 - \omega_2 s_3) c_2/s_2 + \omega_3$

Space-three, 1-2-3

$\omega_1 = \dot{\theta}_1 - \dot{\theta}_3 s_2$	$\dot{\theta}_1 = \omega_1 + (\omega_2 s_1 + \omega_3 c_1) s_2/c_2$
$\omega_2 = \dot{\theta}_2 c_1 + \dot{\theta}_3 s_1 c_2$	$\dot{\theta}_2 = \omega_2 c_1 - \omega_3 s_1$
$\omega_3 = -\dot{\theta}_2 s_1 + \dot{\theta}_3 c_1 c_2$	$\dot{\theta}_3 = (\omega_2 s_1 + \omega_3 c_1)/c_2$

Space-three, 2-3-1

$\omega_1 = -\dot{\theta}_2 s_1 + \dot{\theta}_3 c_1 c_2$	$\dot{\theta}_1 = (\omega_1 c_1 + \omega_3 s_1) s_2/c_2 + \omega_2$
$\omega_2 = \dot{\theta}_1 - \dot{\theta}_3 s_2$	$\dot{\theta}_2 = -\omega_1 s_1 + \omega_3 c_1$
$\omega_3 = \dot{\theta}_2 c_1 + \dot{\theta}_3 s_1 c_2$	$\dot{\theta}_3 = (\omega_1 c_1 + \omega_3 s_1)/c_2$

Space-three, 3-1-2

$\omega_1 = \dot{\theta}_2 c_1 + \dot{\theta}_3 s_1 c_2$	$\dot{\theta}_1 = (\omega_1 s_1 + \omega_2 c_1) s_2/c_2 + \omega_3$
$\omega_2 = -\dot{\theta}_2 s_1 + \dot{\theta}_3 c_1 c_2$	$\dot{\theta}_2 = \omega_1 c_1 - \omega_2 s_1$
$\omega_3 = \dot{\theta}_1 - \dot{\theta}_3 s_2$	$\dot{\theta}_3 = (\omega_1 s_1 + \omega_2 c_1)/c_2$

Space-three, 1-3-2

$\omega_1 = \dot{\theta}_1 + \dot{\theta}_3 s_2$	$\dot{\theta}_1 = \omega_1 + (-\omega_2 c_1 + \omega_3 s_1) s_2/c_2$
$\omega_2 = \dot{\theta}_2 s_1 + \dot{\theta}_3 c_1 c_2$	$\dot{\theta}_2 = \omega_2 s_1 + \omega_3 c_1$
$\omega_3 = \dot{\theta}_2 c_1 - \dot{\theta}_3 s_1 c_2$	$\dot{\theta}_3 = (\omega_2 c_1 - \omega_3 s_1)/c_2$

Space-three, 2-1-3

$\omega_1 = \dot{\theta}_2 c_1 - \dot{\theta}_3 s_1 c_2$	$\dot{\theta}_1 = (\omega_1 s_1 - \omega_3 c_1) s_2/c_2 + \omega_2$
$\omega_2 = \dot{\theta}_1 + \dot{\theta}_3 s_2$	$\dot{\theta}_2 = \omega_1 c_1 + \omega_3 s_1$
$\omega_3 = \dot{\theta}_2 s_1 + \dot{\theta}_3 c_1 c_2$	$\dot{\theta}_3 = (-\omega_1 s_1 + \omega_3 c_1)/c_2$

Space-three, 3-2-1

$\omega_1 = \dot{\theta}_2 s_1 + \dot{\theta}_3 c_1 c_2$	$\dot{\theta}_1 = (-\omega_1 c_1 + \omega_2 s_1) s_2/c_2 + \omega_3$
$\omega_2 = \dot{\theta}_2 c_1 - \dot{\theta}_3 s_1 c_2$	$\dot{\theta}_2 = \omega_1 s_1 + \omega_2 c_1$
$\omega_3 = \dot{\theta}_1 + \dot{\theta}_3 s_2$	$\dot{\theta}_3 = (\omega_1 c_1 - \omega_2 s_1)/c_2$

Space-two, 1-2-1

$\omega_1 = \dot{\theta}_1 + \dot{\theta}_3 c_2$	$\dot{\theta}_1 = \omega_1 - (\omega_2 s_1 + \omega_3 c_1)c_2/s_2$
$\omega_2 = \dot{\theta}_2 c_1 + \dot{\theta}_3 s_1 s_2$	$\dot{\theta}_2 = \omega_2 c_1 - \omega_3 s_1$
$\omega_3 = -\dot{\theta}_2 s_1 + \dot{\theta}_3 c_1 s_2$	$\dot{\theta}_3 = (\omega_2 s_1 + \omega_3 c_1)/s_2$

Space-two, 1-3-1

$\omega_1 = \dot{\theta}_1 + \dot{\theta}_3 c_2$	$\dot{\theta}_1 = \omega_1 + (\omega_2 c_1 - \omega_3 s_1)c_2/s_2$
$\omega_2 = \dot{\theta}_2 s_1 - \dot{\theta}_3 c_1 s_2$	$\dot{\theta}_2 = \omega_2 s_1 + \omega_3 c_1$
$\omega_3 = \dot{\theta}_2 c_1 + \dot{\theta}_3 s_1 s_2$	$\dot{\theta}_3 = (-\omega_2 c_1 + \omega_3 s_1)/s_2$

Space-two, 2-1-2

$\omega_1 = \dot{\theta}_2 c_1 + \dot{\theta}_3 s_1 s_2$	$\dot{\theta}_1 = (-\omega_1 s_1 + \omega_3 c_1)c_2/s_2 + \omega_2$
$\omega_2 = \dot{\theta}_1 + \dot{\theta}_3 c_2$	$\dot{\theta}_2 = \omega_1 c_1 + \omega_3 s_1$
$\omega_3 = \dot{\theta}_2 s_1 - \dot{\theta}_3 c_1 s_2$	$\dot{\theta}_3 = (\omega_1 s_1 - \omega_3 c_1)/s_2$

Space-two, 2-3-2

$\omega_1 = -\dot{\theta}_2 s_1 + \dot{\theta}_3 c_1 s_2$	$\dot{\theta}_1 = -(\omega_1 c_1 + \omega_3 s_1)c_2/s_2 + \omega_2$
$\omega_2 = \dot{\theta}_1 + \dot{\theta}_3 c_2$	$\dot{\theta}_2 = -\omega_1 s_1 + \omega_3 c_1$
$\omega_3 = \dot{\theta}_2 c_1 + \dot{\theta}_3 s_1 s_2$	$\dot{\theta}_3 = (\omega_1 c_1 + \omega_3 s_1)/s_2$

Space-two, 3-1-3

$\omega_1 = \dot{\theta}_2 c_1 + \dot{\theta}_3 s_1 s_2$	$\dot{\theta}_1 = -(\omega_1 s_1 + \omega_2 c_1)c_2/s_2 + \omega_3$
$\omega_2 = -\dot{\theta}_2 s_1 + \dot{\theta}_3 c_1 s_2$	$\dot{\theta}_2 = \omega_1 c_1 - \omega_2 s_1$
$\omega_3 = \dot{\theta}_1 + \dot{\theta}_3 c_2$	$\dot{\theta}_3 = (\omega_1 s_1 + \omega_2 c_1)/s_2$

Space-two, 3-2-3

$\omega_1 = \dot{\theta}_2 s_1 - \dot{\theta}_3 c_1 s_2$	$\dot{\theta}_1 = (\omega_1 c_1 - \omega_2 s_1)c_2/s_2 + \omega_3$
$\omega_2 = \dot{\theta}_2 c_1 + \dot{\theta}_3 s_1 s_2$	$\dot{\theta}_2 = \omega_1 s_1 + \omega_2 c_1$
$\omega_3 = \dot{\theta}_1 + \dot{\theta}_3 c_2$	$\dot{\theta}_3 = (-\omega_1 c_1 + \omega_2 s_1)/s_2$

Appendix III INERTIA PROPERTIES OF UNIFORM BODIES

The information presented in Figs. A1–A28 applies variously to bodies modeled as matter distributed uniformly along a curve, over a surface, or throughout a solid. In each case, the mass center of the body under consideration is identified by the letter C, and the length of the curve (Figs. A1–A4), area of the surface (Figs. A5–A20), or volume of the solid (Figs. A21–A28) used to model the body is recorded. To distinguish figures dealing with surfaces from those involving curves or solids, shading is employed in Figs. A5–A20. For example, Figs. A18 and A24 apply to a spherical shell and a solid sphere, respectively.

The symbols I_1, I_2, and I_{12} in Figs. A1–A28 are defined as $I_j \triangleq \hat{\mathbf{n}}_j \cdot \underline{\mathbf{I}} \cdot \hat{\mathbf{n}}_j$ $(j = 1,2)$ and $I_{12} \triangleq \hat{\mathbf{n}}_1 \cdot \underline{\mathbf{I}} \cdot \hat{\mathbf{n}}_2$, where $\underline{\mathbf{I}}$ is the central inertia dyadic (see Sec. 4.5) of the body under consideration and $\hat{\mathbf{n}}_1$ and $\hat{\mathbf{n}}_2$ are orthogonal unit vectors shown in the figures. These unit vectors are parallel to central principal axes (see Sec. 4.8), except in the cases for which a product of inertia is reported (Figs. A5, A6, A12, A13, A17). A unit vector $\hat{\mathbf{n}}_3$, defined as $\hat{\mathbf{n}}_3 \triangleq \hat{\mathbf{n}}_1 \times \hat{\mathbf{n}}_2$, is parallel to a central principal axis in all cases; and I_3, the associated central principal moment of inertia, is given by $I_3 = I_1 + I_2$ for the bodies in Figs. A1–A17, but *not* for those in Figs. A18–A28. In connection with Figs. A18–A25, I_3 is equal to I_1; to obtain I_3 for the bodies in Figs. A26–A28, replace c with a in the expression for I_1. Finally, the symbol m denotes in every case the mass of the body under consideration.

Straight Line

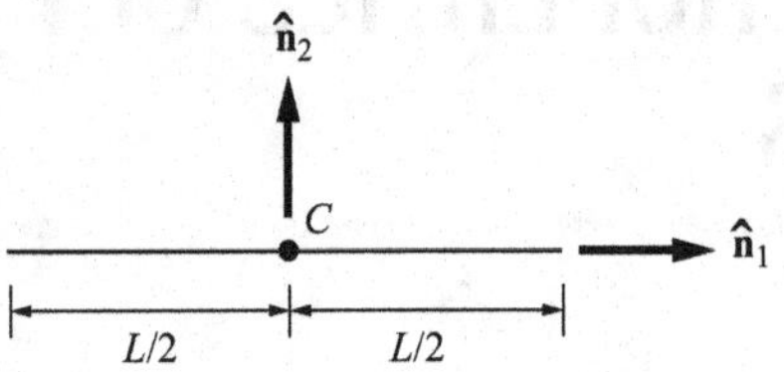

Figure A1

Length: L

$I_1 = 0$

$I_2 = \frac{mL^2}{12}$

Circular Arc

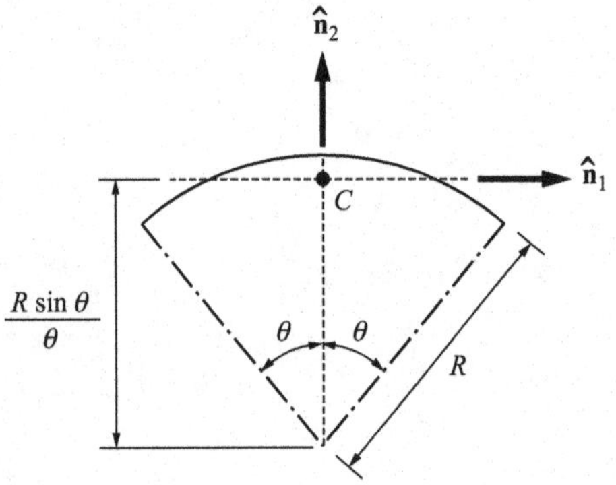

Figure A2

Length: $2R\theta$

$I_1 = \frac{mR^2}{2}\left(1 + \frac{\sin 2\theta}{2\theta} - \frac{2 \sin^2 \theta}{\theta^2}\right)$

$I_2 = \frac{mR^2}{2}\left(1 - \frac{\sin 2\theta}{2\theta}\right)$

Semicircle

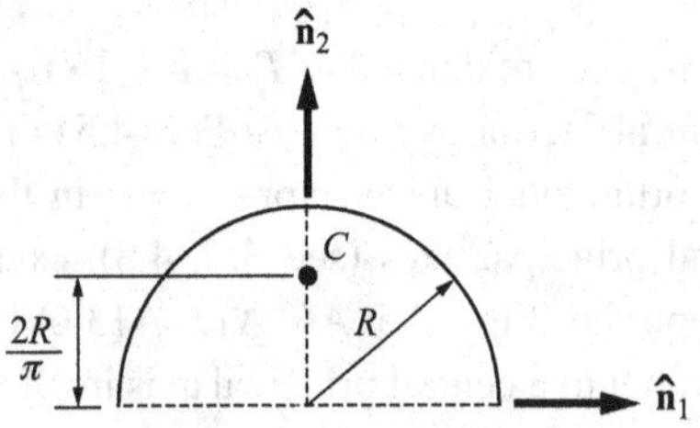

Figure A3

Length: πR

$I_1 = \frac{mR^2}{2}\left(1 - \frac{8}{\pi^2}\right)$

$I_2 = \frac{mR^2}{2}$

Circle

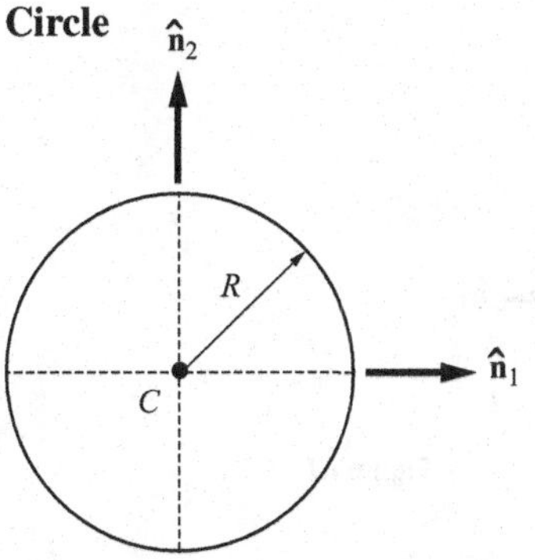

Figure A4

Length: $2\pi R$

$I_1 = I_2 = \frac{mR^2}{2}$

Trapezoid

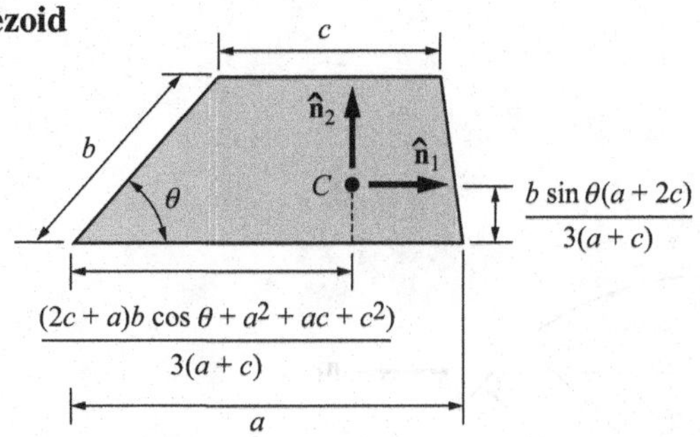

Figure A5

Area: $\frac{(a+c)b\sin\theta}{2}$

$I_1 = \frac{m}{18}\left(\frac{b\sin\theta}{a+c}\right)^2(a^2 + 4ac + c^2)$

$$I_2 = \frac{m}{18(a+c)^2}[b^2(a^2 + 4ac + c^2)\cos^2\theta - b(a^3 + 3a^2c - 3ac^2 - c^3)\cos\theta + a^4 + 2a^3c + 2ac^3 + c^4]$$

$I_{12} = -\frac{mb\sin\theta}{36(a+c)^2}(a^2 + 4ac + c^2)(2b\cos\theta + c - a)$

Parallelogram

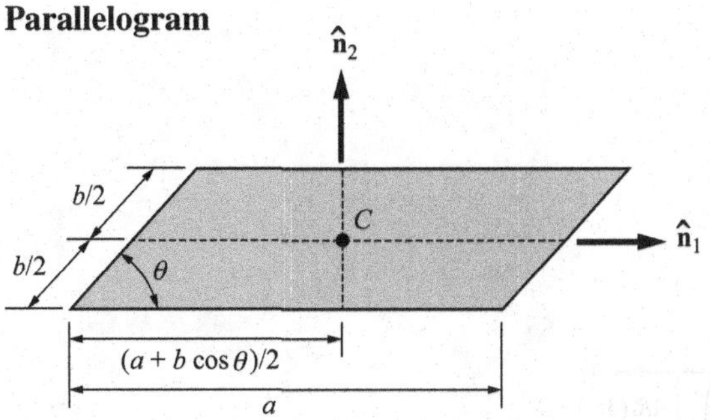

Figure A6

Area: $ab\sin\theta$

$I_1 = \frac{mb^2\sin^2\theta}{12}$

$I_2 = \frac{m(a^2+b^2\cos^2\theta)}{12}$

$I_{12} = -\frac{mb^2\sin\theta\cos\theta}{12}$

Rectangle

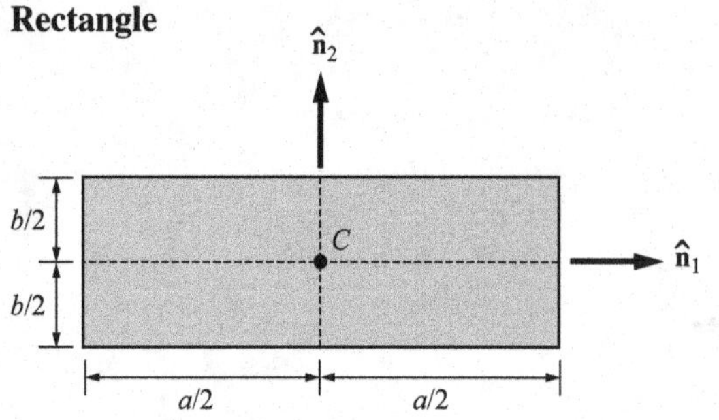

Figure A7

Area: ab

$I_1 = \frac{mb^2}{12}$

$I_2 = \frac{ma^2}{12}$

Circular Sector

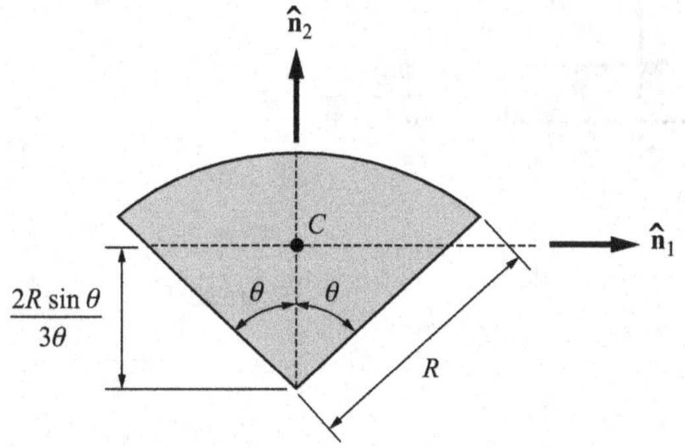

Figure A8

Area: θR^2

$I_1 = \frac{mR^2}{4}\left(1 + \frac{\sin 2\theta}{2\theta} - \frac{16\sin^2\theta}{9\theta^2}\right)$

$I_2 = \frac{mR^2}{4}\left(1 - \frac{\sin 2\theta}{2\theta}\right)$

Semicircle

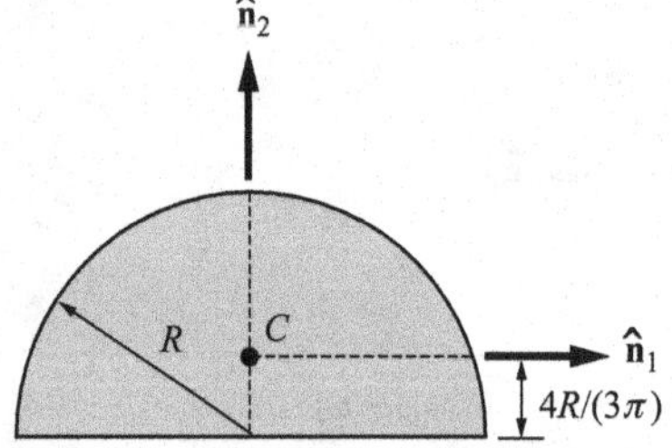

Figure A9

Area: $\frac{\pi R^2}{2}$

$I_1 = \frac{mR^2}{4}\left(1 - \frac{64}{9\pi^2}\right)$

$I_2 = \frac{mR^2}{4}$

Circle

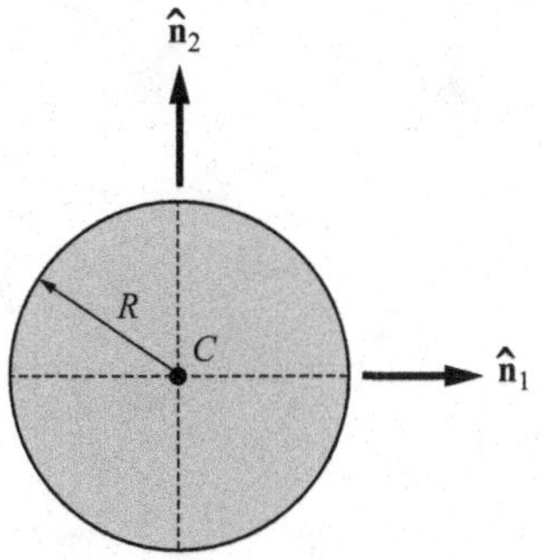

Figure A10

Area: πR^2

$I_1 = I_2 = \frac{mR^2}{4}$

Circular Segment

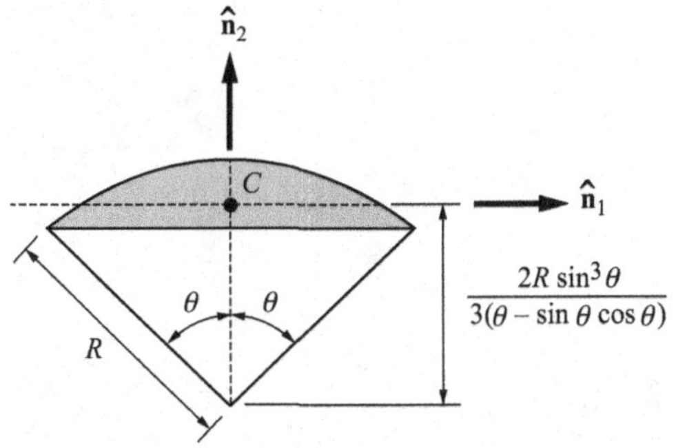

Figure A11

Area: $R^2\left(\theta - \frac{\sin 2\theta}{2}\right)$

$I_1 = \frac{mR^2}{4}\left[1 + \frac{2\sin^3\theta\cos\theta}{\theta - \sin\theta\cos\theta} - \frac{16\sin^6\theta}{9(\theta - \sin\theta\cos\theta)^2}\right]$

$I_2 = \frac{mR^2}{12}\left(3 - \frac{2\sin^3\theta\cos\theta}{\theta - \sin\theta\cos\theta}\right)$

Triangle

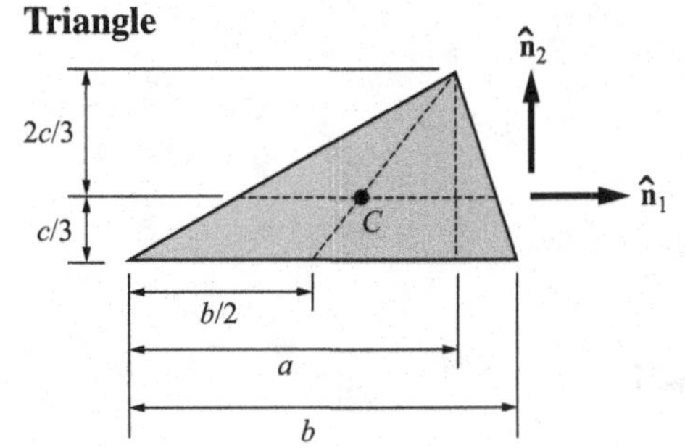

Figure A12

Area: $\frac{bc}{2}$

$I_1 = \frac{mc^2}{18}$

$I_2 = \frac{m(a^2 - ab + b^2)}{18}$

$I_{12} = \frac{mc(b - 2a)}{36}$

Right Triangle

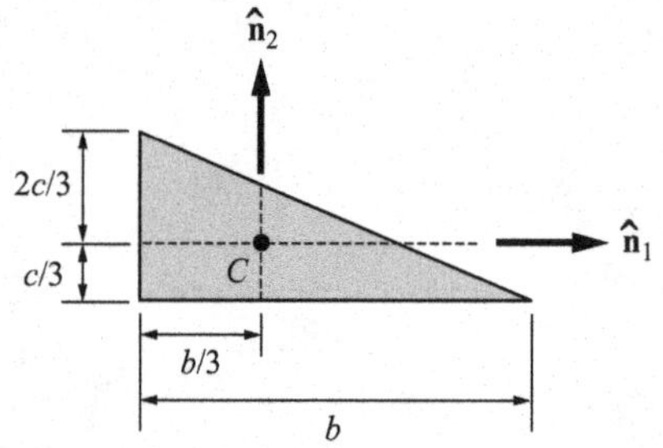

Figure A13

Area: $\frac{bc}{2}$

$I_1 = \frac{mc^2}{18}$

$I_2 = \frac{mb^2}{18}$

$I_{12} = \frac{mbc}{36}$

Isosceles Triangle

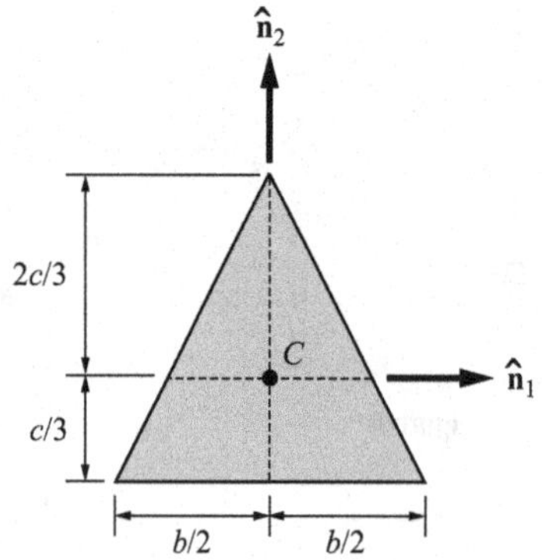

Figure A14

Area: $\frac{bc}{2}$

$I_1 = \frac{mc^2}{18}$

$I_2 = \frac{mb^2}{24}$

Ellipse

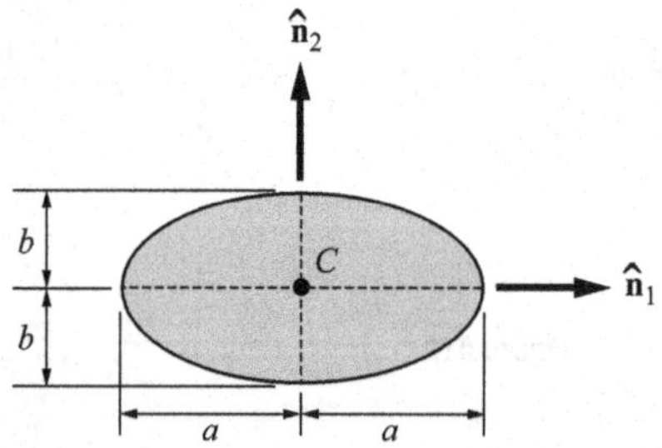

Figure A15

Area: πab

$I_1 = \frac{mb^2}{4}$

$I_2 = \frac{ma^2}{4}$

Semiellipse

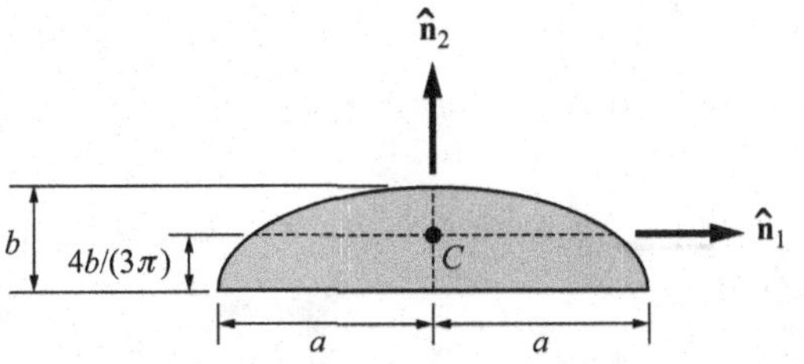

Figure A16

Area: $\frac{\pi ab}{2}$

$I_1 = \frac{mb^2}{4}\left(1 - \frac{64}{9\pi^2}\right)$

$I_2 = \frac{ma^2}{4}$

Semiparabola

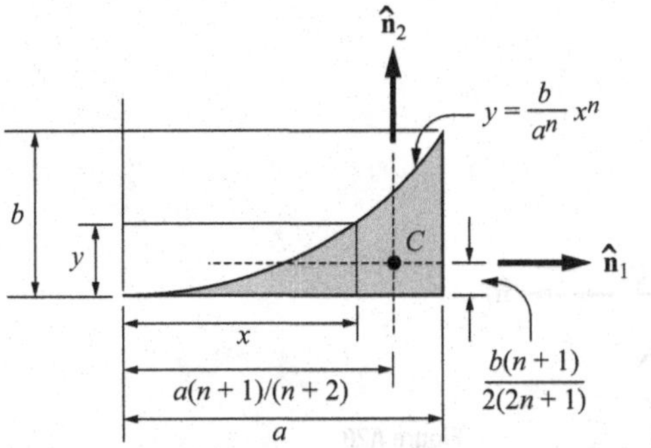

Figure A17

Area: $\frac{ab}{n+1}$

$I_1 = \frac{mb^2(n+1)(7n^2+4n+1)}{12(3n+1)(2n+1)^2}$

$I_2 = \frac{ma^2(n+1)}{(n+3)(n+2)^2}$

$I_{12} = -\frac{mabn}{4(n+2)(2n+1)}$

Sphere

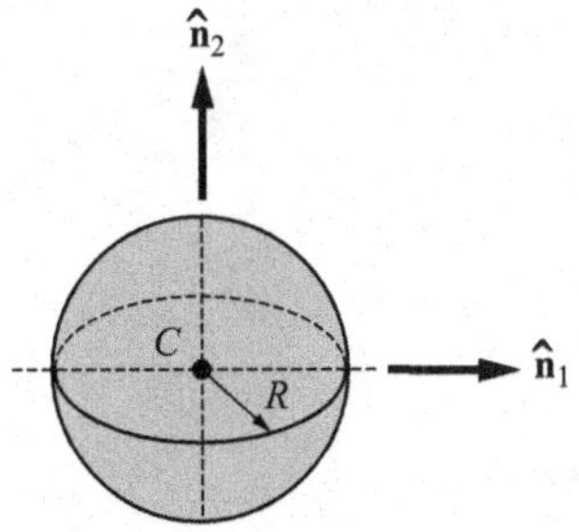

Figure A18

Area: $4\pi R^2$

$I_1 = I_2 = \frac{2mR^2}{3}$

Hemisphere

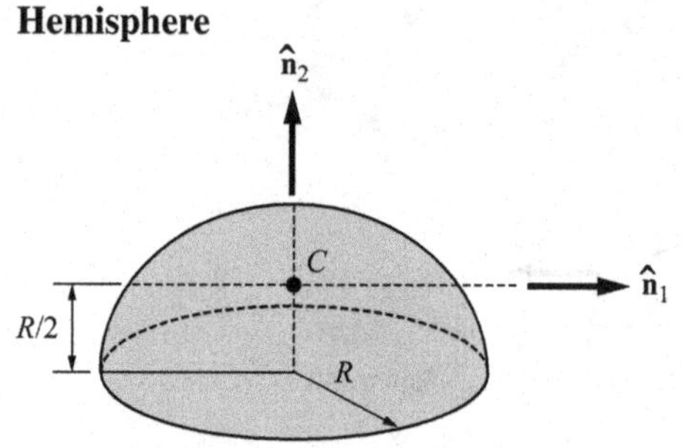

Figure A19

Area: $2\pi R^2$

$I_1 = \frac{5mR^2}{12}$

$I_2 = \frac{2mR^2}{3}$

Right Circular Cone

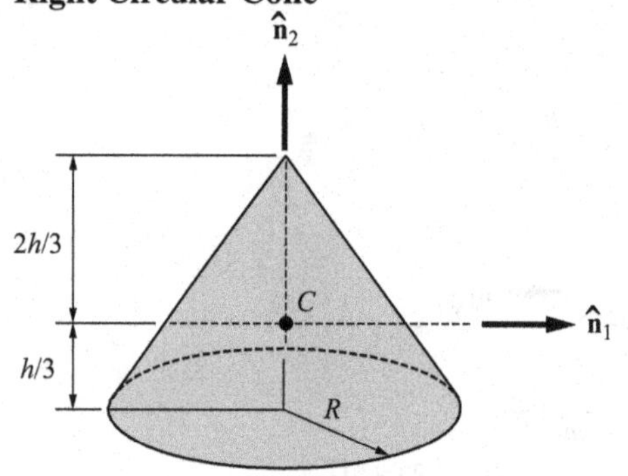

Figure A20

Area: $\pi R(h^2 + R^2)^{1/2}$

$I_1 = \frac{m}{2}\left(\frac{R^2}{2} + \frac{h^2}{9}\right)$

$I_2 = \frac{mR^2}{2}$

Right Circular Cylinder

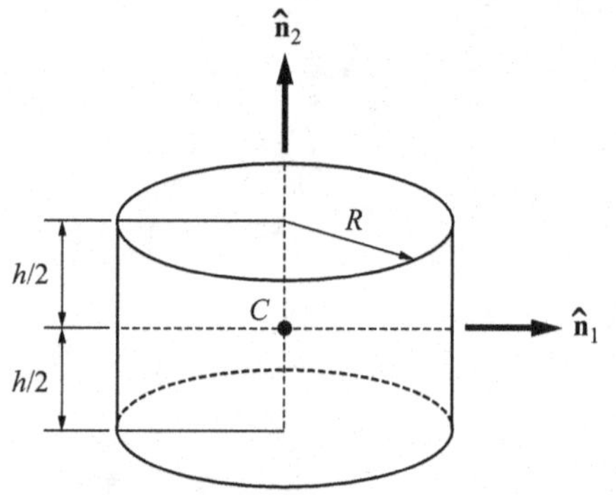

Figure A21

Volume: $\pi h R^2$

$I_1 = \frac{m(3R^2+h^2)}{12}$

$I_2 = \frac{mR^2}{2}$

Right Circular Cone

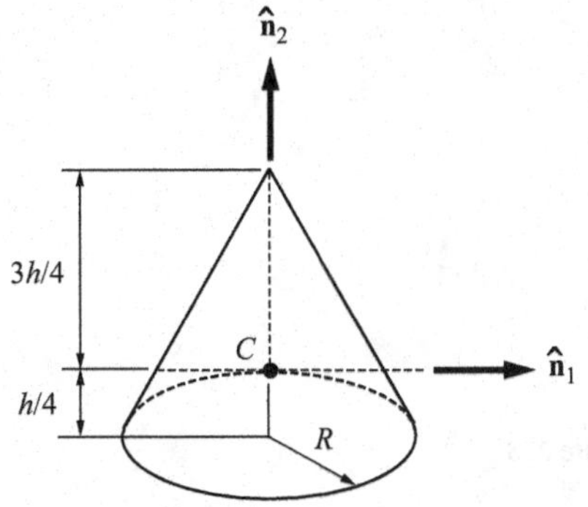

Figure A22

Volume: $\frac{\pi h R^2}{3}$

$I_1 = \frac{3m(4R^2+h^2)}{80}$

$I_2 = \frac{3mR^2}{10}$

Hemisphere

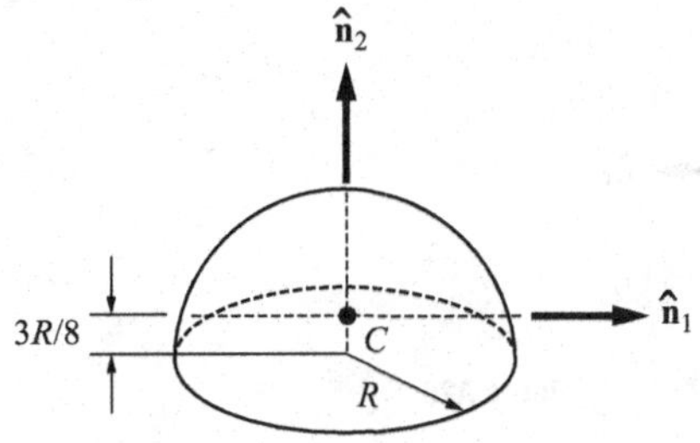

Figure A23

Volume: $\frac{2\pi R^3}{3}$

$I_1 = \frac{83mR^2}{320}$

$I_2 = \frac{2mR^2}{5}$

Sphere

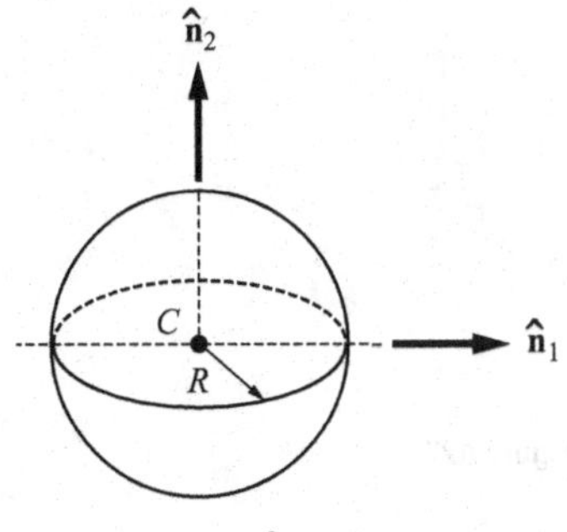

Figure A24

Volume: $\frac{4\pi R^3}{3}$

$I_1 = I_2 = \frac{2mR^2}{5}$

Paraboloid of Revolution

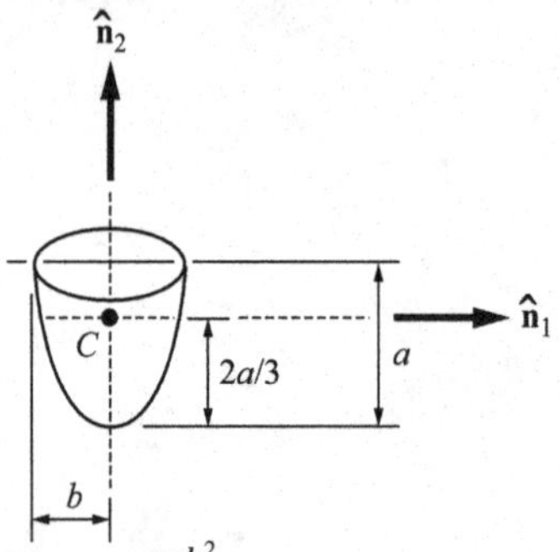

Figure A25

Volume: $\frac{\pi a b^2}{2}$

$I_1 = \frac{m(a^2+3b^2)}{18}$

$I_2 = \frac{mb^2}{3}$

Rectangular Parallelepiped

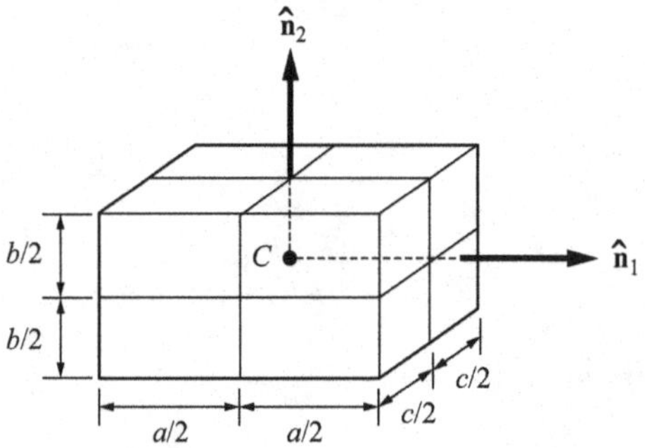

Figure A26

Volume: abc

$I_1 = \frac{m(b^2+c^2)}{12}$

$I_2 = \frac{m(c^2+a^2)}{12}$

Right Rectangular Pyramid

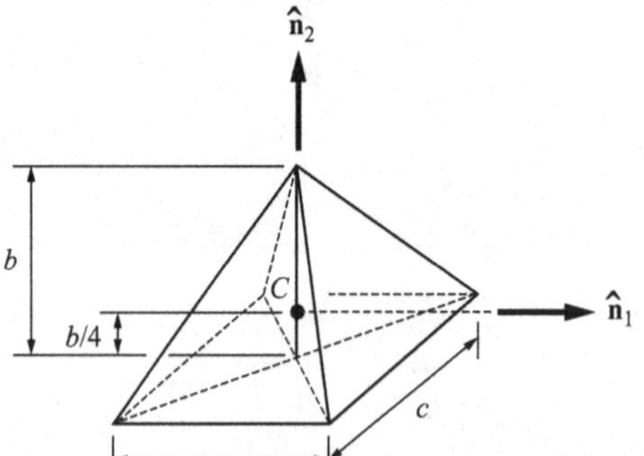

Figure A27

Volume: $\frac{abc}{3}$

$I_1 = \frac{m(3b^2+4c^2)}{80}$

$I_2 = \frac{m(c^2+a^2)}{20}$

Ellipsoid

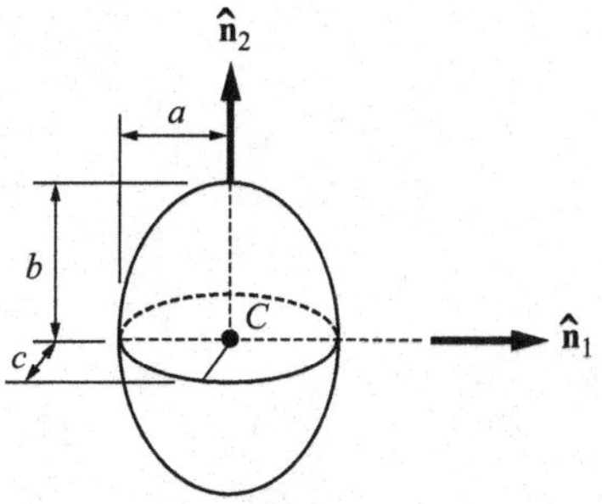

Figure A28

Volume: $\frac{4\pi abc}{3}$

$I_1 = \frac{m(b^2+c^2)}{5}$

$I_2 = \frac{m(c^2+a^2)}{5}$

Index

Printed in the United States
By Bookmasters